CAMBRIDGE
Brighter Thinking

A Level Mathematics for AQA

Student Book 2 (Year 2)

Stephen Ward, Paul Fannon, Vesna Kadelburg and Ben Woolley

CAMBRIDGE
UNIVERSITY PRESS

Shaftesbury Road, Cambridge CB2 8EA, United Kingdom

One Liberty Plaza, 20th Floor, New York, NY 10006, USA

477 Williamstown Road, Port Melbourne, VIC 3207, Australia

314–321, 3rd Floor, Plot 3, Splendor Forum, Jasola District Centre, New Delhi – 110025, India

103 Penang Road, #05-06/07, Visioncrest Commercial, Singapore 238467

Cambridge University Press is part of the University of Cambridge.

It furthers the University's mission by disseminating knowledge in the pursuit of education, learning and research at the highest international levels of excellence.

www.cambridge.org
Information on this title:
www.cambridge.org/9781316644256 (Paperback)
www.cambridge.org/9781316644690 (Paperback with Digital Access edition)

First published 2017

20 19 18 17 16 15 14 13 12

Printed in Great Britain by Ashford Colour Press Ltd.

A catalogue record for this publication is available from the British Library

ISBN 978-1-316-64425-6 Paperback
ISBN 978-1-316-64469-0 Paperback with Digital Access edition

Additional resources for this publication at www.cambridge.org/education

Message from AQA

This textbook has been approved by AQA for use with our qualification. This means that we have checked that it broadly covers the specification and we are satisfied with the overall quality. Full details of our approval process can be found on our website.

We approve textbooks because we know how important it is for teachers and students to have the right resources to support their teaching and learning. However, the publisher is ultimately responsible for the editorial control and quality of this book.

Please note that when teaching the A/AS Level Mathematics (7356, 7357) course, you must refer to AQA's specification as your definitive source of information. While this book has been written to match the specification, it cannot provide complete coverage of every aspect of the course.

A wide range of other useful resources can be found on the relevant subject pages of our website: www.aqa.org.uk

IMPORTANT NOTE
AQA has not approved any Cambridge with Digital Access content.

Contents

Introduction

You have probably been told that mathematics is very useful, yet it can often seem like a lot of techniques that just have to be learnt to answer examination questions. You are now getting to the point where you will start to see where some of these techniques can be applied in solving real problems. However as well as seeing how maths can be useful we hope that anyone working through this book will realise that it can also be incredibly frustrating, surprising and ultimately beautiful.

The book is woven around three key themes from the new curriculum:

Proof

Maths is valued because it trains you to think logically and communicate precisely. At a high level maths is far less concerned about answers and more about the clear communication of ideas. It is not about being neat – although that might help! It is about creating a coherent argument which other people can easily follow but find difficult to refute. Have you ever tried looking at your own work? If you cannot follow it yourself it is unlikely anybody else will be able to understand it. In maths we communicate using a variety of means – feel free to use combinations of diagrams, words and algebra to aid your argument. And once you have attempted a proof, try presenting it to your peers. Look critically (but positively) at some other people's attempts. It is only through having your own attempts evaluated and trying to find flaws in other proofs that you will develop sophisticated mathematical thinking. This is why we have included lots of common errors in our Work it out boxes – just in case your friends don't make any mistakes!

Problem solving

Maths is valued because it trains you to look at situations in unusual, creative ways, to persevere and to evaluate solutions along the way. We have been heavily influenced by a great mathematician and maths educator George Polya who believed that students were not just born with problem-solving skills – they were developed by seeing problems being solved and reflecting on their solutions before trying similar problems. You may not realise it but good mathematicians spend most of their time being stuck. You need to spend some time on problems you can't do, trying out different possibilities. If after a while you have not cracked it then look at the solution and try a similar problem. Don't be disheartened if you cannot get it immediately – in fact, the longer you spend puzzling over a problem the more you will learn from the solution. You may never need to integrate a rational function in the future, but we firmly believe that the problem solving skills you will develop by trying it can be applied to many other situations.

Modelling

Maths is valued because it helps us solve real-world problems. However, maths describes ideal situations and the real world is messy! Modelling is about deciding on the important features needed to describe the essence of a situation and turning that into a mathematical form, then using it to make predictions, compare to reality and possibly improve the model. In many situations the technical maths is actually the easy part – especially with modern technology. Deciding which features of reality to include or ignore and anticipating the consequences of these decisions is the hard part. Yet it is amazing how some fairly drastic assumptions – such as pretending a car is a single point or that people's votes are independent – can result in models which are surprisingly accurate.

More than anything else this book is about making links – links between the different chapters, the topics covered and the themes above, links to other subjects and links to the real world. We hope that you will grow to see maths as one great complex but beautiful web of interlinking ideas.

Maths is about so much more than examinations, but we hope that if you take on board these ideas (and do plenty of practice!) you will find maths examinations a much more approachable and possibly even enjoyable experience. However always remember that the results of what you write down in a few hours by yourself in silence under exam conditions is not the only measure you should consider when judging your mathematical ability – it is only one variable in a much more complicated mathematical model!

How to use this book

Throughout this book you will notice particular features that are designed to aid your learning. This section provides a brief overview of these features.

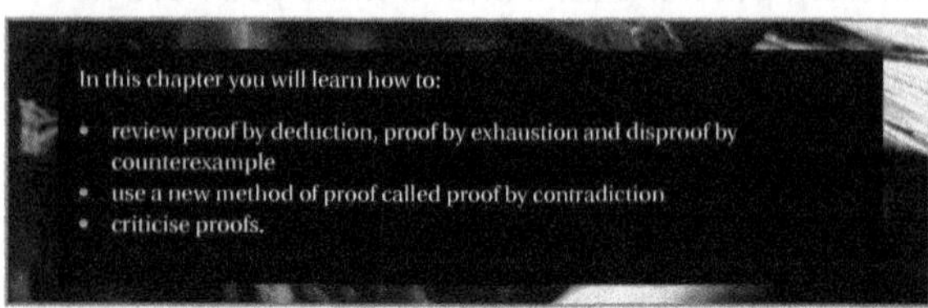

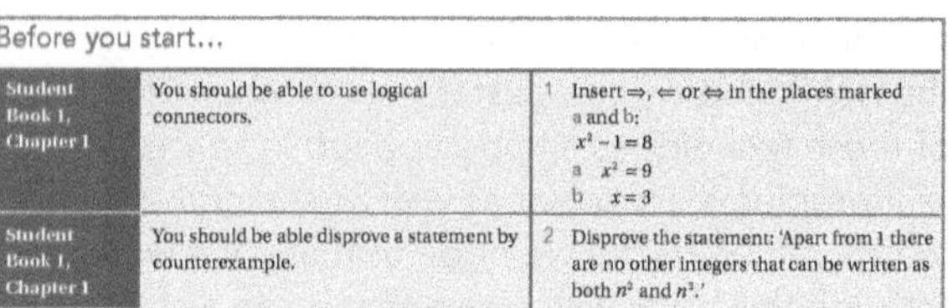

Before you start...

Student Book 1, Chapter 1	You should be able to use logical connectors.	1 Insert ⇒, ⇐ or ⇔ in the places marked a and b: $x^2-1=8$ a $x^2=9$ b $x=3$
Student Book 1, Chapter 1	You should be able disprove a statement by counterexample.	2 Disprove the statement: 'Apart from 1 there are no other integers that can be written as both n^2 and n^3.'

Learning objectives
A short summary of the content that you will learn in each chapter.

Before you start
Points you should know from your previous learning and questions to check that you're ready to start the chapter.

WORKED EXAMPLE

The left-hand side shows you how to set out your working. The right-hand side explains the more difficult steps and helps you understand why a particular method was chosen.

Key point

A summary of the most important methods, facts and formulae.

PROOF

Step-by-step walkthroughs of standard proofs and methods of proof.

Common error

Specific mistakes that are often made. These typically appear next to the point in the Worked example where the error could occur.

WORK IT OUT

Can you identify the correct solution and find the mistakes in the two incorrect solutions?

Tip

Useful guidance, including ways of calculating or checking answers and using technology.

Each chapter ends with a **Checklist of learning and understanding** and a **Mixed practice exercise**, which includes **past paper questions** marked with the icon.

In between chapters, you will find extra sections that bring together topics in a more synoptic way.

Focus on...

Unique sections relating to the preceding chapters that develop your skills in proof, problem-solving and modelling.

Cross-topic review exercise

Questions covering topics from across the preceding chapters, testing your ability to apply what you have learned.

You will find practice papers towards the end of the book, as well as a glossary of key terms (picked out in colour within the chapters), and answers to all questions. Full worked solutions can be found on the Cambridge Elevate digital platform, along with a digital version of this Student Book.

Maths is all about making links, which is why throughout this book you will find signposts emphasising connections between different topics, applications and suggestions for further research.

Rewind

Reminders of where to find useful information from earlier in your study.

Focus on...

Links to problem-solving, modelling or proof exercises that relate to the topic currently being studied.

Fast forward

Links to topics that you may cover in greater detail later in your study.

Did you know?

Interesting or historical information and links with other subjects to improve your awareness about how mathematics contributes to society.

Some of the links point to the material available only through the **Cambridge Elevate** digital platform.

Elevate

A support sheet for each chapter contains further worked examples and exercises on the most common question types. Extension sheets provide further challenge for the most ambitious.

Gateway to A Level

GCSE transition material which provides a summary of facts and methods you need to know before you start a new topic, with worked examples and practice questions.

Colour coding of exercises

The questions in the exercises are designed to provide careful progression, ranging from basic fluency to practice questions. They are uniquely colour-coded, as shown here.

1. A sequence is defined by $u_n = 2 \times 3^{n-1}$. Use the principle of mathematical induction to prove that $u_1 + u_2 + \ldots + u_n = 3^n - 1$.
2. Show that $1^2 + 2^2 + \ldots + n^2 = \frac{n(n+1)(2n+1)}{6}$
3. Show that $1^3 + 2^3 + \ldots + n^3 = \frac{n^2(n+1)^2}{4}$
4. Prove by induction that $\frac{1}{1\times 2} + \frac{1}{2\times 3} + \frac{1}{3\times 4} + \ldots + \frac{1}{n(n+1)} = \frac{n}{n+1}$
5. Prove by induction that $\frac{1}{1\times 3} + \frac{1}{3\times 5} + \frac{1}{5\times 7} + \ldots + \frac{1}{(2n-1)\times(2n+1)} = \frac{n}{2n+1}$
6. Prove that $1 \times 1! + 2 \times + 3 \times 3! \ldots + n \times n! = (n+1)! - 1$
7. Use the principle of mathematical induction to show that $1^2 - 2^2 + 3^2 - 4^2 + \ldots + (-1)^{n-1} n^2 = (-1)^{n-1}\frac{n(n+1)}{2}$.
8. Prove that $(n+1) + (n+2) + (n+3) + \ldots + (2n) = \frac{1}{2}n(3n+1)$
9. Prove that $\sum_{k=1}^{n} k\,2^k = (n-1)2^{n+1} + 2$

Black – practice questions which come in several parts, each with subparts **i** and **ii**. You only need attempt subpart **i** at first; subpart **ii** is essentially the same question, which you can use for further practice if you got part **i** wrong, for homework, or when you revisit the exercise during revision.

Green – practice questions at a basic level
Blue – practice questions at an intermediate level
Red – practice questions at an advanced level
Yellow – designed to encourage reflection and discussion

Working with the large data set

As part of your course you will work with a large data set. At the time of this Student Book's publication (2017), this data set is on the purchasing of different types of food in different parts of the country in different years – this context may change in future years, but the techniques for working with large data sets will stay the same. This is an opportunity to explore statistics in real life. As well as using the ideas from Chapters 20 and 22 you will use these data to look at four key themes.

Practical difficulties with data

The real world is messy. Often there are difficulties with being overwhelmed by too much data, or there are errors, missing items or ambiguous labels. For example, how do you deal with the fact that milk drinks and milk substitutes are combined together in some years if you want to compare regions over time? When grouping data for a histogram, how big a difference does where you choose to put the class boundaries make?

Using technology

Modern statistics is heavily based on using technology. You will use spreadsheets and graphing packages, looking at the common tools which help simplify calculations and present data effectively. One important technique you can employ with modern technology is simulation. You will look more closely at hypothesis testing by using the data set to simulate the effect of sampling on making inferences about the population.

Thinking critically about statistics

Why might you want to use a pie chart rather than a histogram? Whenever you calculate statistics or represent data sets graphically, some information is lost and some is highlighted. In modern statistics it is important to ask critical questions about how evidence provided by statistics is used to support arguments. One important part of this is the idea of validating statistics. For example, what does it mean when it says that there has been a 100% decrease in the amount of sterilised milk purchased? You will look at ways to interrogate the data to try to understand it more.

Statistical problem solving

Technology can often do calculations for you. The art of modern statistics is deciding what calculations to do on what data. You rarely have exactly the data you want, so you have to make indirect inferences from the data you do have. For example, you probably will not see a newspaper headline saying 'the correlation coefficient between amount of bread purchased and amount of confectionary purchased is –0.52', but you might see one saying 'Filling up on carbs reduces snacking!'. Deciding on an appropriate statistical technique to determine whether bread purchases influence confectionary purchases and then interpreting results is a skill which is hard to test but very valuable in real world statistics.

There are lots of decisions to make. Should you use the mass of butter bought? Or the mass compared to 2001? Or the mass as a percentage of all fats? The answer to your main question depends on such decisions.

You will explore all of these themes with examples and questions in the Elevate Section. You need to get used to working with the variables and contexts presented in the large data set.

1 Proof and mathematical communication

In this chapter you will learn how to:

- review proof by deduction, proof by exhaustion and disproof by counterexample
- use a new method of proof called proof by contradiction
- criticise proofs.

Before you start...

Student Book 1, Chapter 1	You should be able to use logical connectors.	1 Insert $\Rightarrow$, $\Leftarrow$ or $\Leftrightarrow$ in the places marked a and b: $x^2 - 1 = 8$ a $x^2 = 9$ b $x = 3$
Student Book 1, Chapter 1	You should be able disprove a statement by counterexample.	2 Disprove the statement: 'Apart from 1 there are no other integers that can be written as both n^2 and n^3.'
Student Book 1, Chapter 1	You should be able to prove a statement by deduction.	3 Prove that the sum of any two odd numbers is always even.
Student Book 1, Chapter 1	You should be able to prove a statement by exhaustion.	4 Use proof by exhaustion to prove that 17 is a prime number.

Developing proof

One of the purposes of this chapter is to provide revision of the material from Student Book 1. It draws on all chapters from that book but, in particular, it builds on the fundamental ideas of proof from Chapter 1. This chapter introduces a new and very powerful method of proof that mathematicians often rely on: proof by contradiction.

Section 1: A reminder of methods of proof

In Student Book 1, Chapter 1, you met proof by deduction, proof by exhaustion and disproof by counterexample. The questions in Exercise 1A show how these methods can be used in topics from throughout Student Book 1.

EXERCISE 1A

1 Use a counterexample to show that this statement is not true:

$\sin 2x = 1 \Rightarrow x = 45°$

2 The velocity of a particle after time t is given by $v = t^2 + 3$. Prove that the particle never returns to its original position.

3 **a** Prove from first principles that if $y = x^2$ then $\frac{dy}{dx} = 2x$.

b Use a counterexample to show that if $\frac{dy}{dx} = 2x$ then it is not necessarily true that $y = x^2$.

4 Prove that $\binom{n}{1} = n$.

5 Prove by exhaustion that any square number is either a multiple of 4 or one more than a multiple of 4.

6 A set of data has mean A, mode B and median C.

Consider this statement: $B < A \Rightarrow C < A$

Prove this statement or use a counterexample to disprove it.

7 The diagram shows a triangle OAB where A and B lie on the circle with centre O.

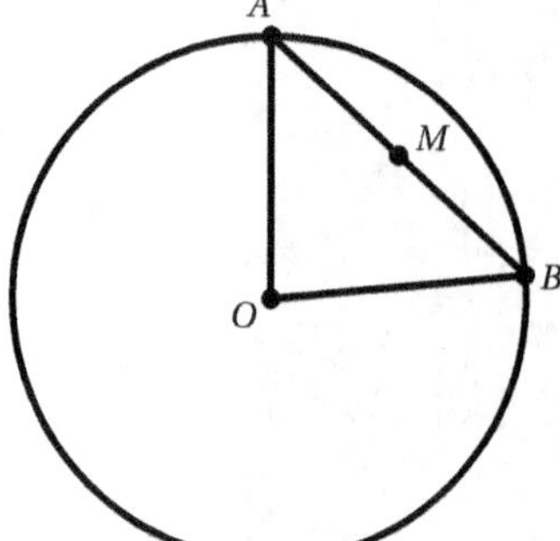

M is the midpoint of AB.

θ is the angle AOM.

a Use the cosine rule to prove that $AB = \sqrt{2r^2 - 2r^2 \cos(2\theta)}$.

b Show that $AM = r \sin\theta$.

c Hence prove that $\cos 2\theta = 1 - 2\sin^2\theta$.

8 In quadrilateral $OABC$, O is the origin and **a**, **b**, **c** are the position vectors of points A, B and C.

P is the midpoint of OA, Q is the midpoint of AB, R is the midpoint of BC and S is the midpoint of OC.

a Show that $\overline{PQ} = \frac{1}{2}\mathbf{b}$.

b Hence prove that $PQRS$ is a parallelogram.

c If $PQRS$ is a rectangle, what can be said about the quadrilateral $OABC$?

9 **a** Use a counterexample to disprove the statement $\ln(x+y) \equiv \ln x + \ln y$.

b Prove that if $\ln(x+y) \equiv \ln x + \ln y$, then $y = \frac{x}{x-1}$.

c If $y = \frac{x}{x-1}$ does it mean that $\ln(x+y) \equiv \ln x + \ln y$?

10 **a** Show that $a^3 + 1 \equiv (a+1)(a^2 + ka + 1)$ where k is a constant to be determined.

b Hence prove that $a^3 + 1$ is prime if and only if $a = 1$.

11 Prove algebraically that if $X \sim \mathrm{B}(n, p)$ then the sum of the probabilities of the different values X can take is one.

Section 2: Proof by contradiction

Proof by contradiction starts from the opposite of the statement you are trying to prove, and shows that this results in an impossible conclusion.

WORKED EXAMPLE 1.1

Use proof by contradiction to prove that there are an infinite number of prime numbers.

Assume that there is a largest prime number, P.

> Proof by contradiction always starts by assuming the opposite of what you want to prove.

Construct another number N that is the product of all the prime numbers up to and including P.

> Now set about trying to find a larger prime than P.

Consider $N+1$. This is one greater than a number divisible by all the primes up to and including P, so it cannot be divisible by any of the primes up to and including P.

Therefore $N+1$ is either itself prime, or is divisible by primes larger than P.

Either way, a prime larger than P has been discovered which contradicts the premise that there is a largest prime number.

> Here the contradiction to the original assumption (that there is a largest prime) occurs.

Therefore there are an infinite number of prime numbers.

Did you know?

See the Bold-Shaddow of Vrania's Glory,
Immortall in His Race, no lesse in Story:
An Artist without Error; from whose Lyne,
Both Earth and Heav'ns, in sweet Proportions twine:
Behold Great EUCLID. But, behold Him well!
For 'tis in Him Divinity doth dwell.
G. Wharton.

A variation on the proof in Worked example 1.1 can be found in Euclid's masterpiece *The Elements*, a textbook written in around 300 BCE but still in use in many schools in the first half of the 20th century!

WORKED EXAMPLE 1.2

Using the fact that if a^2 is even then so is a, prove that $\sqrt{2}$ is irrational.

Start by assuming the opposite of what you want to prove.

Assume that $\sqrt{2} = \frac{p}{q}$ where p and q are integers with no common factors.

A number is rational if it can be written as a fraction – and this can always be cancelled down until the numerator and denominator share no common factors.

Squaring both sides:

$$2 = \frac{p^2}{q^2}$$

$$\Leftrightarrow p^2 = 2q^2 \quad (1)$$

This means that p^2 is even so p must also be even.

Using the fact given.

Therefore $p = 2k$, so $p^2 = 4k^2$.

Substituting into (1):

$$4k^2 = 2q^2$$

$$\Leftrightarrow 2k^2 = q^2$$

This means that q^2 is even, so q must be even.

Using the given fact again.

This has shown that both p and q are even, so they share a factor of 2. This contradicts the original assertion, so it must be incorrect.

Here the contradiction (to the fact that p and q share no common factors) arises.

Therefore $\sqrt{2}$ cannot be written as $\frac{p}{q}$.

EXERCISE 1B

1. Prove that if n^2 is even then n is also even.
2. Prove that $\sqrt{3}$ is irrational.
3. Prove that there is an infinite number of even numbers.
4. Prove that the sum of a rational and irrational number is irrational.
5. Prove that if ab is even, with a, b integers, then at least one of them is even.
6. Prove that $\sqrt[3]{2}$ is irrational.
7. Prove that $\log_2 3$ is irrational.

Elevate

See Support Sheet 1 for an example of the same type as Q7 and for further practice questions on proof by contradiction.

8 Suppose that n is a composite integer. Prove that n has a prime factor less than or equal to $\sqrt{n}$.

9 Prove that if any 25 dates are chosen, some three must be within the same month.

10 Prove that $a^2 - 4b^2$ is never 2 if a and b are whole numbers.

11 **a** Show that if $x = \frac{p}{q}$ is a solution to the equation $x^3 + x + 1 = 0$ then $p^3 + pq^2 + q^3 = 0$.

b Explain why there is no solution to this equation when p is odd and q is odd.

c Prove that there are no rational solutions to $x^3 + x + 1 = 0$.

12 Prove that if a triangle has three sides, a, b and c, such that $a^2 + b^2 = c^2$, then it is a right-angled triangle.

Section 3: Criticising proofs

In Student Book 1, Chapter 1, you were introduced to the notation used in logic:

- $A \Leftrightarrow B$ means that statements A and B are equivalent.
- $A \Rightarrow B$ means if A is true then so is B.
- $A \Leftarrow B$ means if B is true then so is A.

When looking at proof (including solving equations, which is a type of proof!) you have probably looked out for errors in areas such as arithmetic or algebra. You now also need to look for errors in logic.

WORKED EXAMPLE 1.3

Yas was solving the equation $2\log_{10} x = 4$. Find the errors in her working.

Line 1: $2\log_{10} x = 4$

Line 2: $\Leftrightarrow \log_{10}(x^2) = 4$

Line 3: $\Rightarrow x^2 = 10^4 = 10\,000$

Line 4: $\Leftrightarrow x = \pm 1000$

In line 2 the symbol should be $\Rightarrow$: if x is negative line 2 is correct but line 1 is not possible. — This is an error in logic.

In line 3 the symbol should be $\Leftrightarrow$: if $x^2 = 10^4$ then $\log_{10}(x^2) = 4$. — This is an error in logic.

In line 4 the square root of 10 000 should be 100. — This is an arithmetic error.

Because some of the implications go only one way, the final solution might not work in the original equation. It should be checked. — This is an error in logic.

EXERCISE 1C

1 Lambert was asked to solve the equation $x = \sqrt{3x+4}$. Here is his working.

Line 1: $x = \sqrt{3x+4}$

Line 2: $\Leftrightarrow\ x^2 = 3x+4$

Line 3: $\Leftrightarrow\ x^2 - 3x - 4 = 0$

Line 4: $\Leftrightarrow\ (x-4)(x+1) = 0$

Line 5: $\Leftrightarrow\ x = 4$ or $x = -1$

a By checking his solutions, find the correct solution.

b In which line of working is his mistake?

2 Craig was asked to solve the equation $x^2 = 3x$. Here is his solution.

$$x^2 = 3x$$
$$\Leftrightarrow \quad x = 3$$

a Show that $x = 0$ is also a solution to the original equation.

b What logical symbol should Craig have used in the second line?

3 Freja was asked to solve the equation $x - \frac{1}{x-3} = 1 + \frac{5-2x}{x-3}$. Here is her working.

Line 1: $x - \frac{1}{x-3} = 1 + \frac{5-2x}{x-3}$

Line 2: $\Leftrightarrow\ x - 1 = \frac{6-2x}{x-3}$

Line 3: $\Leftrightarrow\ (x-1)(x-3) = 6-2x$

Line 4: $\Leftrightarrow\ x^2 - 4x + 3 = 6 - 2x$

Line 5: $\Leftrightarrow\ x^2 - 2x - 3 = 0$

Line 6: $\Leftrightarrow\ (x-3)(x+1) = 0$

Line 7: $\Leftrightarrow\ x = 3$ or $x = -1$

a By checking her solutions find the correct solution.

b In which line of working is her mistake?

4 Jamie was asked to solve $\log_2(-x) + \log_2(2-x) = 3$.

Here is her working.

Line 1: $\log_2(-x) + \log_2(2-x) = 3$

Line 2: $\Leftrightarrow\ \log_2(-x(2-x)) = 3$

Line 3: $\Leftrightarrow\ \log_2(x^2 - 2x) = 3$

Line 4: $\Leftrightarrow\ x^2 - 2x = 2^3$

Line 5: $\Leftrightarrow\ x^2 - 2x - 8 = 0$

Line 6: $\Leftrightarrow\ (x-4)(x+2) = 0$

Line 7: $\Leftrightarrow\ x = 4$ or $x = -2$

In which line of working did Jamie make a mistake?

5 Andrew was asked to prove the statement: 'The function $y = x^3 - 3x$ has a minimum at $x = 1$.'

Here is his working.

Line 1: $\frac{dy}{dx} = 3x^2 - 3 = 0$

Line 2: $x^2 = 1$

Line 3: $x = 1$

Line 4: $\frac{d^2y}{dx^2} = \frac{d}{dx}\left(\frac{dy}{dx}\right)$

Line 5: $= \frac{d}{dx}(0)$

Line 6: $= 0$

Line 7: So it is a minimum.

Describe the errors in this proof.

6 Ann was asked to answer the question: '$x + q$ is a factor of $x^3 + px + q$. Find the remaining factor.'

If $x + q$ is a factor of $x^3 + px + q$ then:

$$x^3 + px + q \equiv (x + q)(x^2 + bx + 1)$$
$$\equiv x^3 + x^2(b + q) + x(bq + 1) + q$$

Comparing coefficients of x^2: $0 = b + q \Rightarrow b = -q$

Therefore the remaining factor is $x^2 - qx + 1$.

Brian claims that Ann's solution isn't complete. Explain why he is correct, and give a full solution.

7 Find the error in this proof that $\sqrt{16}$ is irrational.

Line 1: Assume that $\sqrt{16} = \frac{p}{q}$ where p and q are integers with no common factors.

Line 2: Squaring both sides gives $16 = \frac{p^2}{q^2}$.

Line 3: So $p^2 = 16q^2$. (1)

Line 4: This means that p^2 is even so p must also be even.

Line 5: You can then write that $p = 2k$, so $p^2 = 4k^2$.

Line 6: Substituting this into equation (1) gives $4k^2 = 16q^2$.

Line 7: So $k^2 = 4q^2$.

Line 8: This means that q^2 is even, so q must be even.

Line 9: But you have shown that both p and q are even, so they share a factor of 2.

Line 10: This contradicts the original assertion, so $\sqrt{16}$ cannot be written as $\frac{p}{q}$.

Checklist of learning and understanding

- You should be able to apply counterexamples, proof by exhaustion and proof by deduction to material from Student Book 1.
- Proof by contradiction is a method of proof that works by showing that assuming the opposite of the required statement leads to an impossible situation.
- When criticising proofs, look out for flaws in logic as well as mistakes in algebra or arithmetic.

Mixed practice 1

1. Which symbol should be used to replace '?' in this working?

   $x^2 = 8x$

   ? $\quad x = 8$

   A $\Rightarrow$ **B** $\Leftrightarrow$ **C** $\Leftarrow$ **D** $\equiv$

2. Prove that $n^2 - n$ is always even.

3. Prove that $\sqrt{5}$ is irrational.

4. Prove that there is an infinite number of square numbers.

5. Find the error(s) in this working to solve $\tan x = 2\sin x$ for $0° \leqslant x < 360°$.

   Line 1: $\dfrac{\sin x}{\cos x} = 2\sin x$

   Line 2: $\Leftrightarrow \sin x = 2\sin x\cos x$

   Line 3: $\Leftrightarrow 1 = 2\cos x$

   Line 4: $\Leftrightarrow 0.5 = \cos x$

   Line 5: $\Leftrightarrow x = 60°$

6. Fermat says that if x is prime then $2x+1$ is prime.

   Which value of x provides a counterexample to this statement?

   A 9 **B** 11 **C** 13 **D** 15

7. Prove that $\log_2 5$ is irrational.

8. $OABC$ is a parallelogram with O at the origin and **a**, **b**, **c** are the position vectors of points A, B and C.

   P is the midpoint of BC and Q is the point on OB such that $OQ : QB$ is $2 : 1$.

   Prove that AQP is a straight line.

9. Prove that if a and b are whole numbers $a^2 - b^2$ is either odd or a multiple of 4.

10. **a** Prove that if $f(x)$ is a polynomial of finite order, with integer coefficients, and n is an integer, then $f(n)$ is an integer.

    b Use a counterexample to show that this statement is not always correct.

    If $f(x)$ is a polynomial, where $f(n)$ is an integer whenever n is an integer, then $f(x)$ must have integer coefficients.

11 Consider this working to solve $x+\frac{4x}{x-2}=\frac{8}{x-2}$.

Line 1: $\Leftrightarrow \ x(x-2)+4x=8$

Line 2: $\Leftrightarrow \ x^2-2x+4x=8$

Line 3: $\Leftrightarrow \ x^2+2x=8$

Line 4: $\Leftrightarrow \ (x-1)^2=9$

Line 5: $\Rightarrow \ x-1=3$

Line 6: $\Leftrightarrow \ x=4$

a In which lines are there mistakes?

b Rewrite the solution correctly, making appropriate use of logical connectors.

12 This proof is trying to demonstrate that there are an arbitrary number of consecutive composite (non-prime) numbers.

Line 1: Consider $n!$ for $n \geqslant r \geqslant 1$.

Line 2: $n!+r$ is divisible by r.

Line 3: So the numbers $n!+1, n!+2, \ldots, n!+n$ are not prime.

Line 4: Therefore, this is a list of n consecutive composite numbers.

Which is the first line to contain an error?

13 Prove that if a, b and c are integers such that $a^2+b^2=c^2$, then either a or b is even.

14 **a** By considering a right-angled triangle, prove that if A is an acute angle then

$$\tan(90°-A)=\frac{1}{\tan A}$$

b Hence prove that $\tan 10° \times \tan 20° \times \tan 30° \ldots \times \tan 80°$ is a rational number.

15 Prove that for values of x between 90° and 180° $\sin x - \cos x \geqslant 1$.

Elevate

See Extension Sheet 1 to complete the details of a couple of famous proofs.

2 Functions

In this chapter you will learn how to:

- distinguish between mappings and functions
- determine whether a function is one-to-one or many-to-one
- find the domain and range of a function
- find composite functions
- find the inverse of a function.

Before you start...

GCSE	You should be able to interpret function notation.	1 Given that $f(x) = 2 - x$, evaluate: a f(3) b f(−4).
Student Book 1, Chapter 1	You should be able to use interval notation to write inequalities.	2 Use interval notation to write these inequalities. a $x > 3$ and $x \leqslant 6$ b $x < 3$ or $x \geqslant 6$
Student Book 1, Chapter 3	You should be able to complete the square.	3 Express $f(x) = x^2 + 5x + 3$ in the form $(x + a)^2 + b$. Hence state the coordinates of the turning point of f(x).
Student Book 1, Chapter 3	You should be able to solve quadratic inequalities.	4 Solve the inequality $x^2 - 4x - 5 > 0$.
Student Book 1, Chapter 7	You should be able to rearrange exponential and log expressions.	5 Make x the subject of each equation. a $y = e^{2x-1}$ b $y = \ln(3x + 4)$
Student Book 1, Chapter 13	You should be able to establish where a function is increasing/decreasing.	6 Find the range of x-values for which $f(x) = x^{\frac{3}{2}} - 2x$ is an increasing function.

Why study functions?

Doubling, adding five, finding the largest prime factor – these are all instructions that can be applied to numbers to produce a numerical result. This idea comes up a lot in mathematics. The formal study of it leads to the concept of a function.

You will use functions whenever you need to express how one quantity changes with another, whether it is how the strength of the gravitational force varies with distance, or how the amount of paint needed depends on the area of the wall.

Section 1: Mappings and functions

A **mapping** is any rule that assigns to each input value (x) one or more output values (y). For example:

- $y = x^2 + 2$
- $x^2 + y^2 = 4$
- y = a factor of x.

In the first of these, each x-value maps to a single y-value. But in the second and the third this is not always the case. In the second example, when $x = 0$, y could be 2 or –2. In the third example, when $x = 6$, y could be 1, 2, 3 or 6.

Mappings such as $y = x^2 + 2$, where there is only one y-value for each x-value, are called **functions**.

The easiest way to identify if a mapping is a function is to look at its graph, and apply the **vertical line test**.

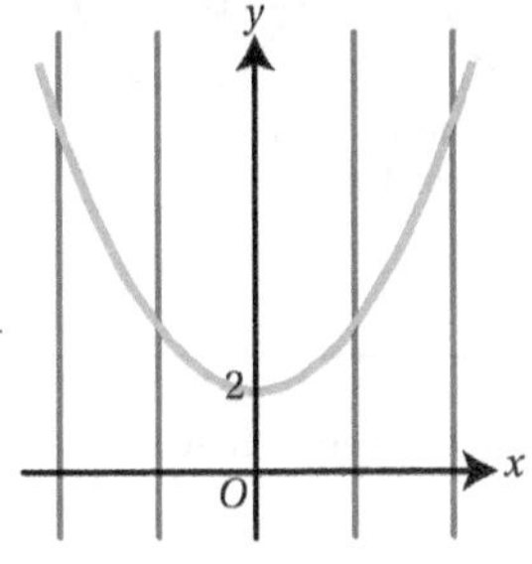

$y = x^2 + 2$: a function

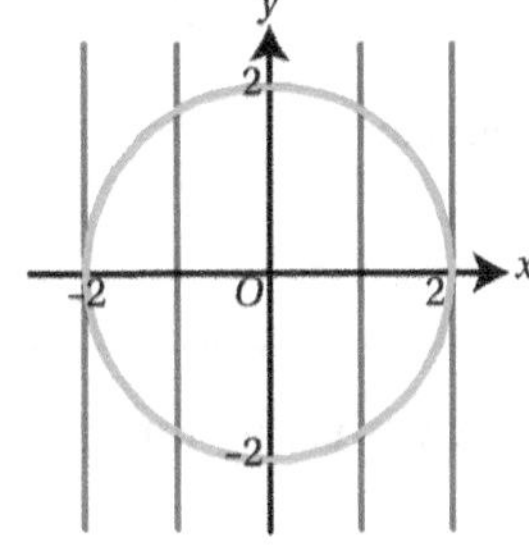

$x^2 + y^2 = 4$: not a function

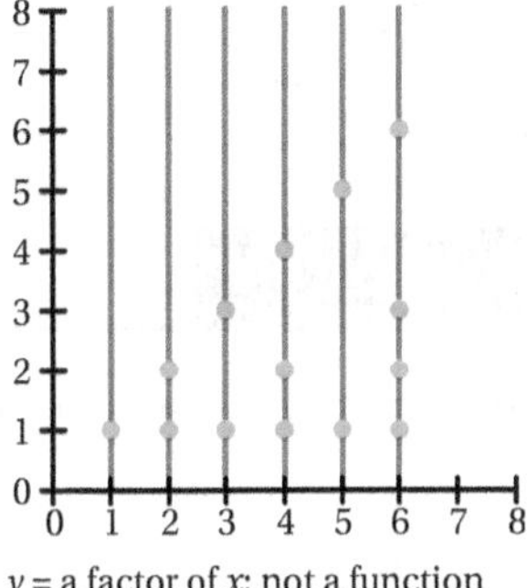

y = a factor of x: not a function

Key point 2.1

- A mapping is a function if every x-value maps to a single y-value.
- Vertical line test: if a mapping is a function, any vertical line will cross its graph at most once.

The most common way to describe a function is by using function notation, for example $f(x) = x^2 + 2$. An alternative way of writing this is $f : x \mapsto x^2 + 2$. For a specific value of the function, you would write $f(3) = 11$, or $f : 3 \mapsto 11$. 11 is the **image** of 3.

Once you have decided that a mapping is a function, you can ask whether each output comes from just one input. To check this, apply the **horizontal line test**.

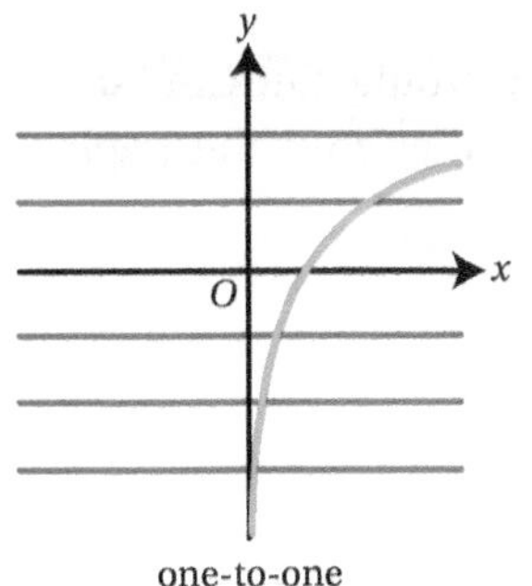

one-to-one

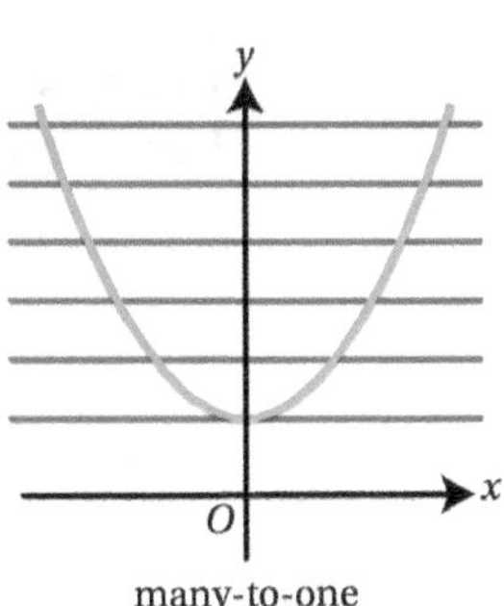

many-to-one

Gateway to A level

See Gateway to A Level Section F for a reminder about basic techniques with functions.

Key point 2.2

- A function is:
  - **one-to-one** if every y-value corresponds to only one x-value.
  - **many-to-one** if there are some y-values that come from more than one x-value.
- Horizontal line test: if a function is one-to-one, any horizontal line will cross the graph at most once.

A mapping in which a single input corresponds to more than one output (so is not a function) is called **one-to-many**.

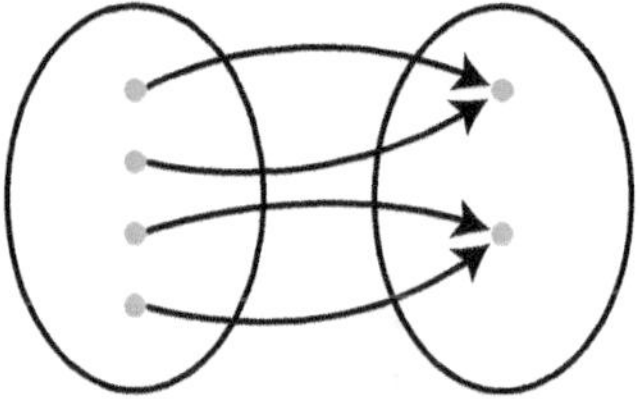
many-to-one function

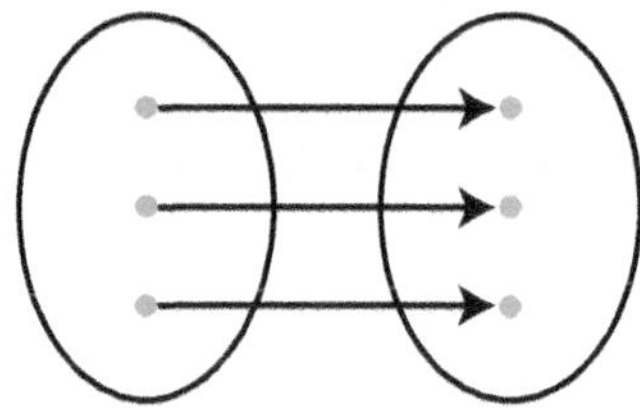
one-to-one function

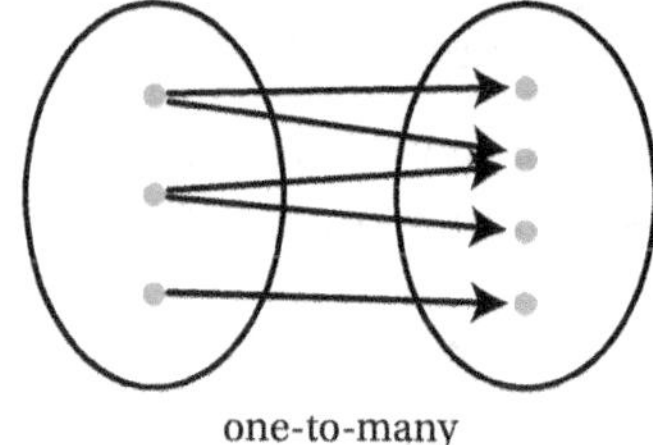
one-to-many (not a function)

WORKED EXAMPLE 2.1

Which of these graphs represent functions? Classify those that are functions as one-to-one or many-to-one.

a

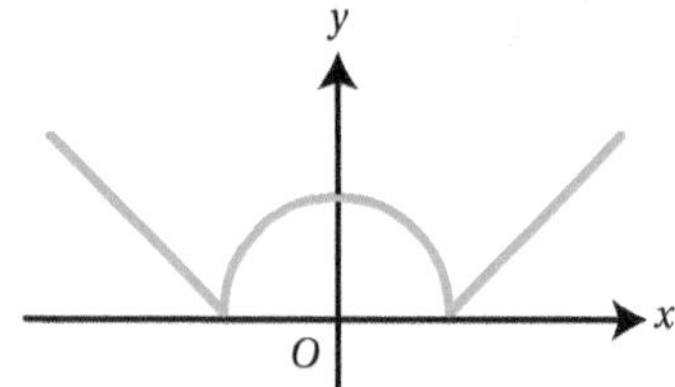

b

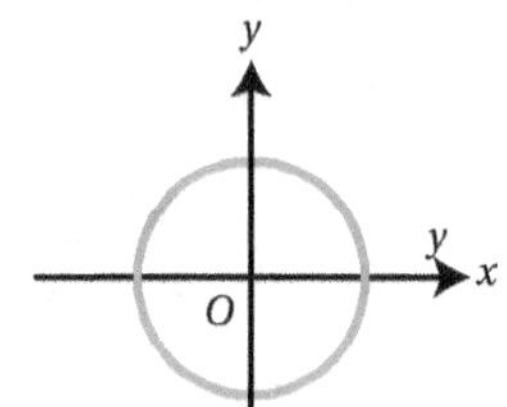

c

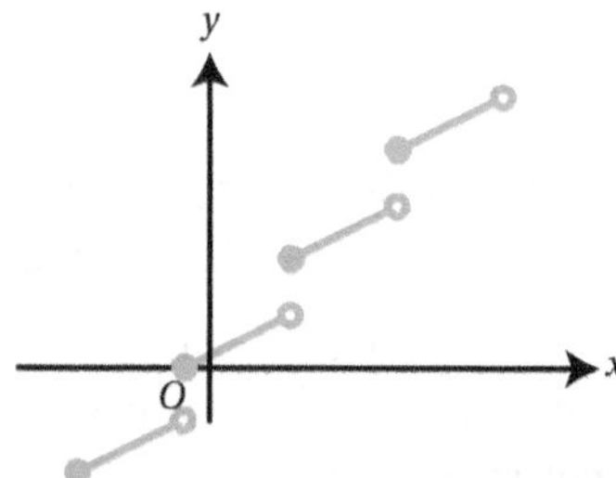

a

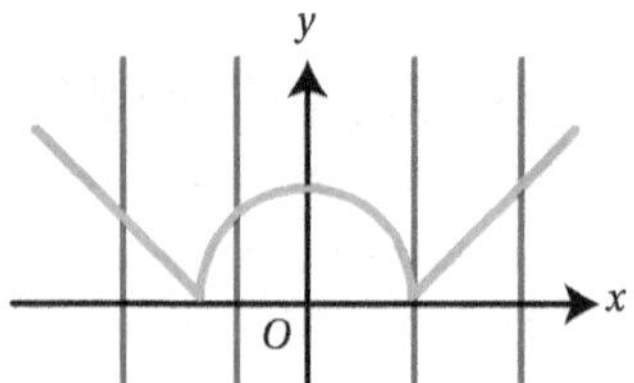

Draw several vertical lines and see how many times they cross the graph.

Any vertical line meets the graph at most once, therefore it is a function.

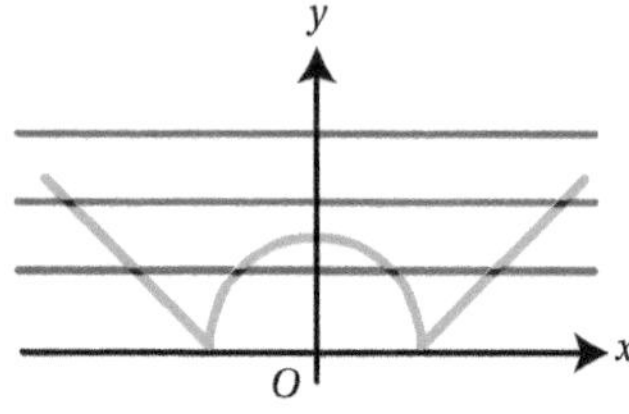

Draw several horizontal lines and see how many times they cross the graph.

Some horizontal lines meet the graph at more than one point, therefore it is a many-to-one function.

Continues on next page

b

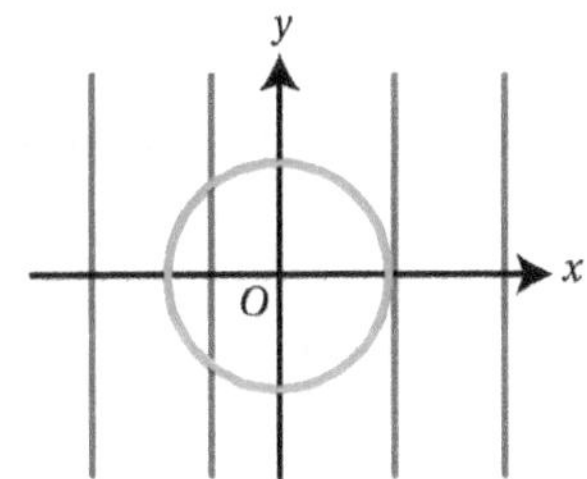

Draw several vertical lines and see how many times they cross the graph.

Some vertical lines meet the graph more than once, therefore it is not a function.

c

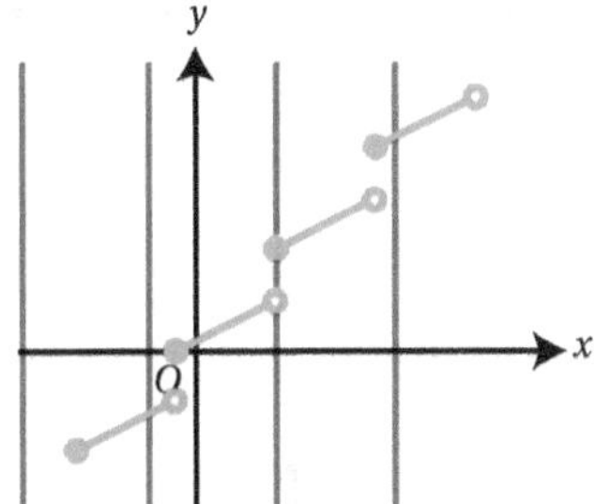

Draw several vertical lines and see how many times they cross the graph. Be careful: a vertical line through an open circle does not count as an intersection.

Any vertical line meets the graph at most once, therefore it is a function.

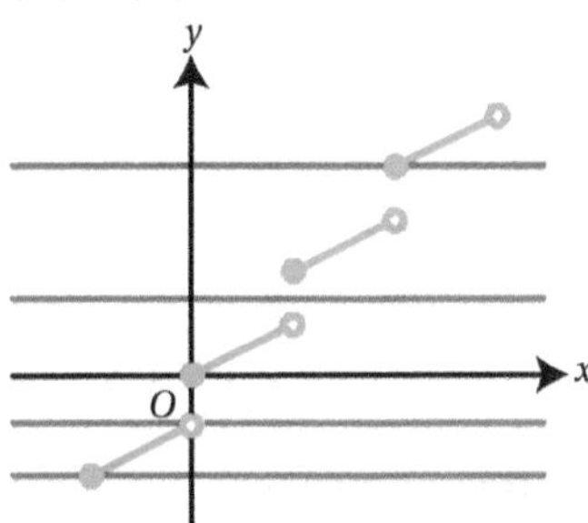

Draw several horizontal lines and see how many times they cross the graph.

Any horizontal line meets the graph at most once, therefore it is a one-to-one function.

Tip

Remember that an open circle on a graph means that point is not a part of the graph, and a solid circle means that it is.

When a mapping is given by an equation linking x and y you may not know how to draw its graph. In that case, use the equation and ask whether there are any values of x for which you can find more than one value of y.

WORKED EXAMPLE 2.2

Determine whether each mapping is a function.

a $x + y^2 = 5$ **b** $x + y^3 = 9$

a For $x = 1$, y could be 2 or -2. So this is not a function.

Is there more than one value of y for each value of x? Think of an example.

b $x + y^3 = 9$

$\Leftrightarrow y^3 = 9 - x$

$\Leftrightarrow y = \sqrt[3]{9 - x}$

There is only one value of y for each value of x.

So this is a one-to-one function.

Is there more than one value of y for each value of x?

Find the value of y for a given value of x. Finding the cube root does not introduce other possible values (unlike finding the square root, which does).

EXERCISE 2A

1 Classify each graph as a mapping or a function. If you decide a graph is a function, state if it is one-to-one or many-to-one.

a
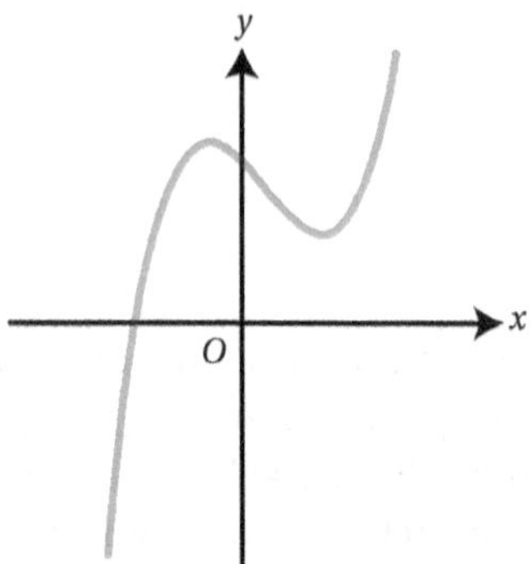

b
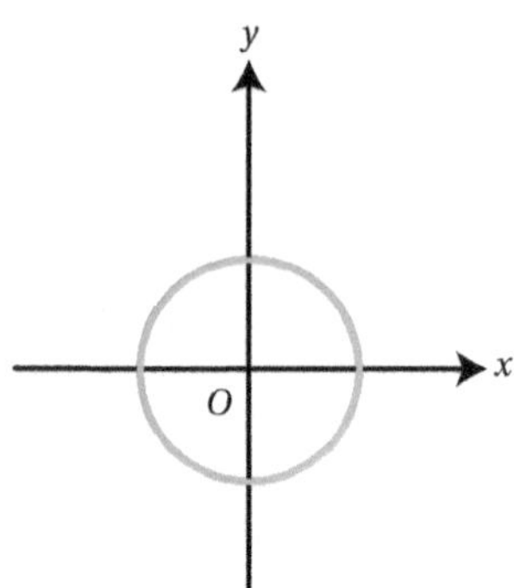

c
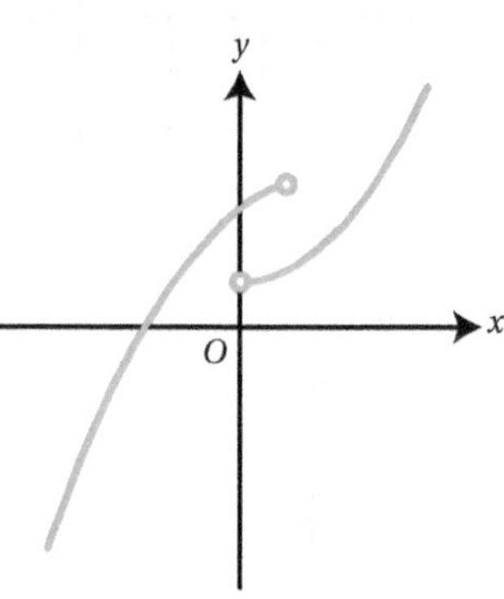

d
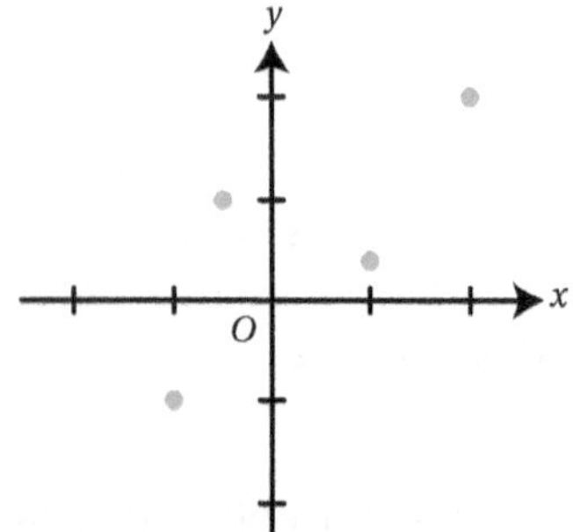

e
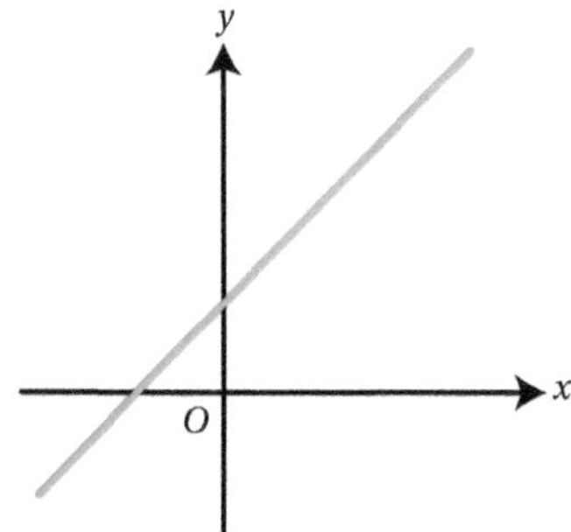

f
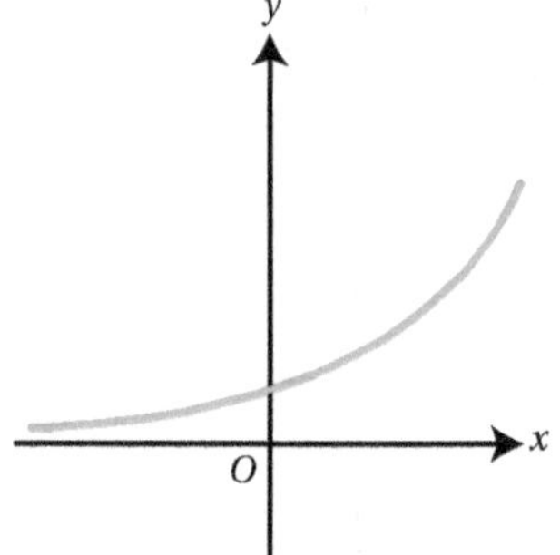

2 Determine whether each mapping is a function (x denotes the input and y the output).

a i $y=\sqrt{x}$ ii $y=\sqrt[3]{x}$ b i $f(x)=5x^2+3x$ ii $f(x)=9-x^2$

c i $y^2=x$ ii $x^2+y^2=1$

3 Classify each function as one-to-one or many-to-one.

a i $y=x^3$ ii $y=x^4$ b i $f: x \mapsto \sin x$ ii $f: x \mapsto \tan x$

c i $f(x)=\ln(x^2)$ ii $f(x)=e^{-3x}$

4 Determine whether each mapping is a function (x denotes the input and y the output).

a i $y^3=x$ ii $\sqrt{y}=x$ b i $\sin y=x$ ii $\tan y=x$

c i $y^2-3y=x$ ii $\dfrac{y-1}{y+2}=x$

5 Which of these statements is correct?

A The square root of 4 is 2.

B The square root of 4 is ± 2.

Would it be possible to define the 'square root' function so that the square root of 4 is -2?

Section 2: Domain and range

As well as telling you what to do with the input, for a function to be completely defined it needs to tell you what type of input is allowed to go into the function.

Key point 2.3

The set of allowed input values is called the **domain** of the function.

WORKED EXAMPLE 2.3

Sketch the graph of $f(x) = x + 1$ over each domain.

a $x \in \mathbb{R}, x > 2$ **b** $x \in \mathbb{Z}$

a

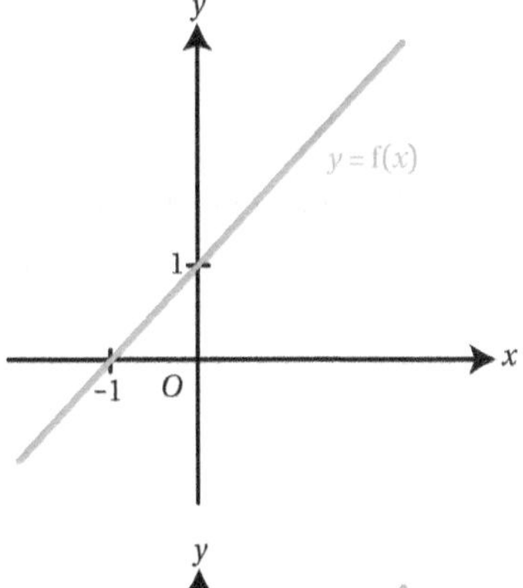

Sketch the graph where all real numbers are allowed for x.

Discard the part of the graph outside the required domain. Since the endpoint is not included, label it with an open circle.

b

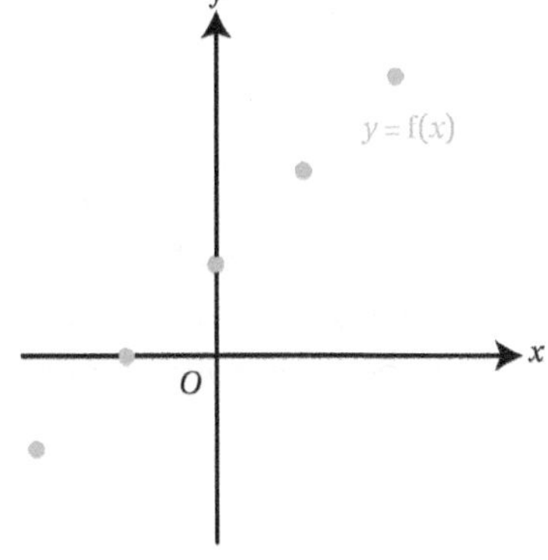

Use the same original graph as in **a**, but this time it only exists at whole numbers, so label them with solid circles.

The largest possible domain of a function will be all real numbers unless it contains a mathematical operation that can't accept certain types of number. The three most common restrictions on the domain are:

- you cannot divide by zero
- you cannot find the square root of a negative number
- you cannot take the logarithm of a non-positive number.

Fast forward

In Sections 3 and 4 of this chapter, you'll see that you don't always use the largest possible domain when forming composite or inverse functions.

WORKED EXAMPLE 2.4

Find the largest possible domain of $f(x) = \ln(4 - 2x)$.

Need $4 - 2x > 0$ — You can only take the log of a positive number.

So the domain is $x < 2$.

Sometimes you need to apply more than one restriction.

WORKED EXAMPLE 2.5

What is the largest possible domain of $h : x \mapsto \frac{1}{x-2} + \sqrt{x+3}$? Write your answer using interval notation.

Need $x - 2 \neq 0$ — Look for any of the three potential problems listed.

$\therefore x \neq 2$ — There will be division by zero when $x - 2 = 0$.

Need $x + 3 \geqslant 0$ — There will be the square root of a negative number when $x + 3 < 0$.

$\therefore x \geqslant -3$

Domain:

$x \geqslant -3$ and $x \neq 2$ — Both these restrictions are needed.

$x \in [-3, 2) \cup (2, \infty)$ — This describes two intervals: from −3 to 2 and from 2 to infinity, excluding 2.

Tip

Remember that $f : x \mapsto x + 3$, for example, is just an alternative notation for $f(x) = x + 3$.

Once you have fixed the domain of a function, you can ask what values can come out of the function.

Key point 2.4

The set of all possible outputs of a function is called the **range**.

Tip

Be aware that the range will depend upon the domain.

The easiest way of finding the range is to sketch the graph.

WORKED EXAMPLE 2.6

Find the range of $f(x) = x^2 + 3$ if the domain is:

a $x \in \mathbb{R}$ **b** $x > 2$.

a

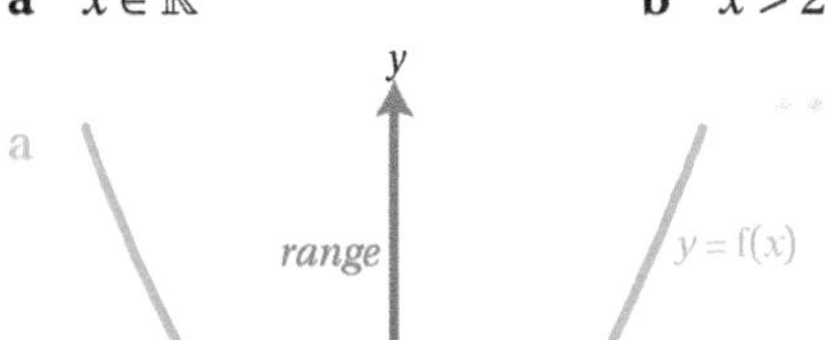

Sketch the graph $y = f(x)$ for the given domain.

Range: $f(x) \geqslant 3$

Use the graph to state which y-values can occur.

b

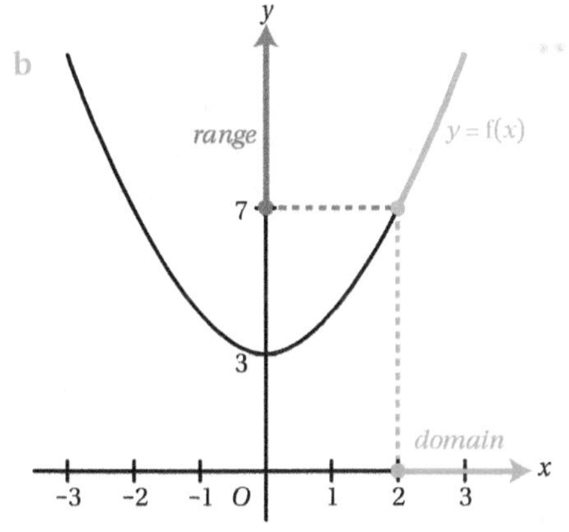

Sketch the graph $y = f(x)$ for the given domain.

Range: $f(x) > 7$

Use the graph to state which y-values can occur.

Tip

Graph-plotting software or graphical calculators can be useful when investigating the domain and range of functions.

Tip

Although it is the y values of the graph of $y = f(x)$ that you are considering when finding the range, always make sure you give the range in terms of $f(x)$ rather than y, for example, $f(x) > 7$, not $y > 7$.

EXERCISE 2B

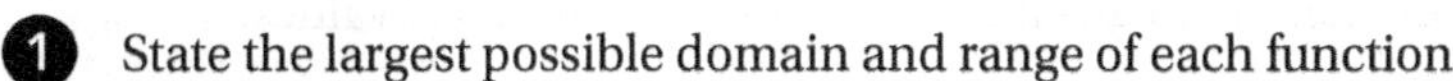

1 State the largest possible domain and range of each function.

a $f(x) = 2^x$ **b** $f(x) = a^x, a > 0$

c $f(x) = \log_{10} x$ **d** $f(x) = \log_b x, b > 0$

Rewind

Exponential functions were covered in Student Book 1, Chapter 8.

2 Find the largest possible domain of each function.

a **i** $f(x)=\frac{1}{x+2}$ **ii** $f(x)=\frac{5}{x-7}$

b **i** $f(x)=\frac{3}{(x-2)(x+4)}$ **ii** $g(x)=\frac{x}{x^2-9}$

c **i** $r(x)=\sqrt{x^3-1}$ **ii** $h(x)=\sqrt{x+3}$

d **i** $f(a)=\frac{1}{\sqrt{a-1}}$ **ii** $f(a)=\frac{5a}{\sqrt{2-5a}}$

e **i** $a(x)=\frac{1}{x}+\frac{2}{x+1}$ **ii** $f(x)=\sqrt{x+1}+\frac{1}{x+2}$

f **i** $f(x)=\sqrt{x}+\frac{1}{x+7}-x^3+5$ **ii** $f(x)=e^x+\sqrt{2x+3}-\frac{1}{x^2+4}-2$

3 Find the range of each function.

a **i** $f(x)=7-x^2, x\in\mathbb{R}$ **ii** $f(x)=x^2+3, x\in\mathbb{R}$

b **i** $g(x)=x^2+3, x\geqslant 3$ **ii** $g(x)=x^2-1, x<-3$

c **i** $h(x)=x-2, x<5, x\in\mathbb{Z}$ **ii** $h(x)=x+1, x>3, x\in\mathbb{Z}$

d **i** $q(x)=\frac{1}{x}, x\geqslant -1, x\neq 0$ **ii** $q(x)=3\sqrt{x}, x>0$

4 Find the largest possible domain and range of each function.

a **i** $f(x)=x^2-4x-1$ **ii** $f(x)=x^2+2x+5$

b **i** $g: x\mapsto 5-x^2$ **ii** $g: x\mapsto 3-2x^2$

c **i** $f(x)=\sqrt{x^2-5}$ **ii** $f(x)=\sqrt{9-x^2}$

d **i** $f: x\mapsto 2\sqrt{x^2-6x+8}$ **ii** $f: x\mapsto 4\sqrt{x^2+2x-3}$

Rewind

These questions require knowledge of completing the square and the solution of quadratic inequalities, covered in Student Book 1, Chapter 3.

5 **a** Write $2x^2+6x-3$ in the form $a(x+p)^2+q$.

b Hence state the range of the function $f: x\mapsto 2x^2+6x-3$.

6 Find the domain and range of the function $g(x)=\ln(6+4x)$.

7 The function f is given by $f(x)=\sqrt{\ln(x-4)}$. Find the domain of the function.

8 Find the largest possible domain of the function $f(x)=\frac{4^{\sqrt{x-1}}}{x+2}-\frac{1}{x^2-5x+6}+x^2+1$.

9 **a** Sketch the graph of $y=6-x-2x^2$.

b Hence find the largest possible domain of the function $f: x\mapsto\sqrt{6-x-2x^2}$.

10 Find the largest possible domain of the function $g(x)=\ln(x^2+3x+2)$.

11 Find the largest set of values of x such that the function f given by $f(x)=\sqrt{\frac{8x-4}{x-12}}$ takes real values.

12 **a** State the domain of the function $f(x)=\sqrt{x-a}+\ln(b-x)$ if:

i $a<b$ **ii** $a>b$.

b Write an expression for $f(a)$.

Section 3: Composite functions

After applying a function to a number, you can apply another function to the result. The resulting function is called a **composite function**.

Key point 2.5

If you apply the function g to x and then the function f to the result, you write:

$$f(g(x)) \text{ or } fg(x) \text{ or } f \circ g(x)$$

Common error

If you have the composite function $fg(x)$, make sure you apply g first and then f. Don't work left to right!

Be careful: $f(g(x))$ and $g(f(x))$ are not the same function.

It can be useful to refer to $g(x)$ as the inner function and $f(x)$ as the outer function.

WORKED EXAMPLE 2.7

If $f(x) = x^2$ and $g(x) = x - 3$ find:

a $f \circ g(1)$ **b** $fg(x)$ **c** $gf(x)$.

a $g(1) = 1 - 3 = -2$

$f(-2) = (-2)^2 = 4$

$\therefore f(g(1)) = 4$

Evaluate $g(1)$ and then apply f to the result.

Note that you do not need to work out the general expression for $f \circ g(x)$.

b $f(g(x)) = f(x - 3)$

$= (x - 3)^2$

Replace x in $f(x)$ with the expression for $g(x)$.

c $g(f(x)) = g(x^2)$

$= x^2 - 3$

Replace x in $g(x)$ with the expression for $f(x)$.

WORK IT OUT 2.1

Two functions are defined by $f(x) = 3x - 2$ and $g(x) = x^2 - 1$. Find $f \circ g(5)$.

Which is the correct solution? Identify the errors made in the incorrect solutions.

Solution 1	Solution 2	Solution 3
$f(5) = 15 - 2 = 13$ $g(5) = 25 - 1 = 24$ $\therefore f(g(x)) = 13 \times 24$ $= 312$	$g(5) = 25 - 1 = 24$ $fg(x) = f(24)$ $= 72 - 2$ $= 70$	$f(5) = 15 - 2 = 13$ $f \circ g(5) = 13^2 - 1$ $= 168$

You can also compose a function with itself.

Key point 2.6

$f \circ f(x)$ can also be written as $f^2(x)$.

WORKED EXAMPLE 2.8

Given that $f(x) = \frac{2}{x-3}$ find and simplify expression for $f^2(x)$.

$f^2(x) = f(f(x))$

$= \frac{2}{\frac{2}{x-3} - 3}$

$f^2(x)$ means do $f(x)$, then apply f to the result.

$= \frac{2(x-3)}{2-3(x-3)}$

$= \frac{2x-6}{11-3x}$

Multiply top and bottom of the fraction by $(x-3)$ to simplify the denominator.

For the composite function $fg(x)$ to exist, the range of $g(x)$ must lie entirely within the domain of $f(x)$, otherwise you would be trying to put values into $f(x)$ which it cannot take.

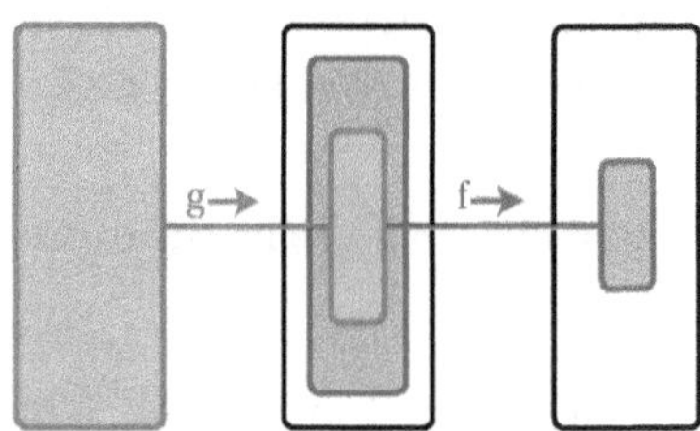

Tip

Whichever notation is being used, remember the correct order: the function nearest to x acts first!

WORKED EXAMPLE 2.9

The functions f and g are defined by $f : x \mapsto x^2 - 5, x \in \mathbb{R}$ and $g : x \mapsto \sqrt{x+3}, x \geqslant -3$.

a Explain why the composite function $g(f(x))$ is not defined.

b Find the largest possible domain for which $g(f(x))$ is defined. In this case, state the range of $g(f(x))$.

a Need $f(x) \geqslant -3$.

Check whether the output from f is within the domain of g.

But, for example, $f(0) = -5$, and this is not in the domain of g.

It is enough to find one counterexample to show that $f(x) \geqslant -3$ is not true for all $x \in \mathbb{R}$.

b For largest possible domain,

You need $f(x) \geqslant -3$.

$x^2 - 5 \geqslant -3$

$x^2 \geqslant 2$

$x \leqslant -\sqrt{2}$ or $x \geqslant \sqrt{2}$

This is a quadratic inequality; the solution consists of two separate intervals.

On this domain, the range is $g(f(x)) \geqslant 0$.

When $\sqrt{x+3}$ is defined, it can take any non-negative real value.

A more complex problem is to recover one of the original functions when you are given a composite function. The best way to do this is to use a substitution.

WORKED EXAMPLE 2.10

If $f(x+1) = 4x^2 + x$ find $f(x)$.

$y = x+1$ — Substitute $y =$ the inner function

$x = y-1$ — Rearrange to get $x = \ldots$

$f(y) = 4(y-1)^2 + (y-1)$ — Replace all xs.

$= 4y^2 - 8y + 4 + y - 1$

$= 4y^2 - 7y + 3$

$\therefore f(x) = 4x^2 - 7x + 3$ — Write the answer in terms of x.

EXERCISE 2C

1 Given that $f(x) = x^2 + 1$ and $g(x) = 3x + 2$ find:

a **i** $g(f(0))$ **ii** $fg(1)$ **b** **i** $f \circ g(-2)$ **ii** $g(f(3))$.

2 Given that $f(x) = x^2 + 1$ and $g(x) = 3x + 2$ find:

a **i** $g(f(x))$ **ii** $f^2(x)$ **b** **i** $gg(x)$ **ii** $f \circ g(x)$.

3 **a** Given that $f(x) = x^2 + 1$ and $g(x) = 3x + 2$ find:

i $g^2(\sqrt{a} + 1)$ **ii** $f^2(y-1)$.

b Given that $f(x) = 4 - x$ and $g(x) = x^2$, find:

i $fg(x-2)$ **ii** $gf(3-x)$.

4 Given that $f(x) = x^2 + 1$ and $g(x) = 3x + 2$ find:

a $ggf(y)$ **b** $gfg(z)$.

5 Find $f(x)$, given these conditions.

a **i** $f(2a) = 4a^2$ **ii** $f\left(\frac{b}{3}\right) = \frac{b^3}{27}$

b **i** $f(x+1) = 3x - 2$ **ii** $f(x-2) = x^2 + x$

c **i** $f(1-y) = 5 - y$ **ii** $f(y^3) = y^2$

d **i** $f(e^k) = \ln k$ **ii** $f(3n+2) = \ln(n+1)$

6 Given that $f(x) = x^2 + 1$ and $g(x) = 3x + 2$, solve the equation $fg(x) = gf(x)$.

7 Given that $f(x) = 3x + 1$ and $g(x) = \frac{x}{x^2 + 25}$, solve the equation $gf(x) = 0$.

8 The function f is defined by $f : x \mapsto x^3$. Find an expression for $g(x)$ in terms of x in each case.

a $(f \circ g)(x) = 2x + 3$ **b** $(g \circ f)(x) = 2x + 3$

9 Functions f and g are defined by $f(x) = \sqrt{x^2 - 2x}$ and $g(x) = 3x + 4$. The composite function $f \circ g$ is undefined for $x \in (a, b)$.

a Find the value of a and the value of b.

b Find the range of $f \circ g$.

10 $f(x) = x - 1, x > 3$ and $g(x) = x^2, x \in \mathbb{R}$

a Explain why $g \circ f$ exists but $f \circ g$ does not.

b Find the largest possible domain for g so that $f \circ g$ is defined.

11 Let f and g be two functions. Given that $(f \circ g)(x) = \frac{x+2}{3}$ and $g(x) = 2x + 5$, find $f(x - 1)$.

Section 4: Inverse functions

Functions transform an input into an output, but sometimes you can reverse this process to allow you to say which input produced a given output. When this is possible, you do it by finding the **inverse function**, usually labelled f^{-1}.

For example, if $f(x) = 3x$ then $f^{-1}(12)$ is a number that, when put into f, produces the output 12. In other words, you are looking for a number x such that $f(x) = 12$. Hence $f^{-1}(12) = 4$.

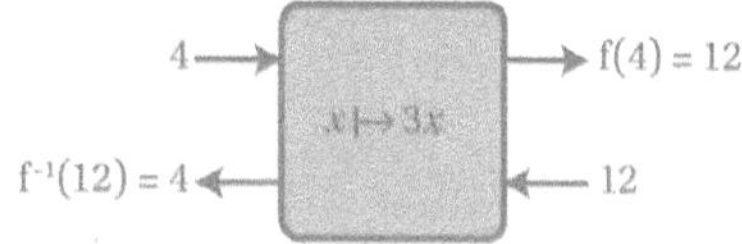

Common error

Make sure you don't get confused about this notation. With numbers, the superscript '–1' denotes reciprocal, for example, $x^{-1} = \frac{1}{x}$, $3^{-1} = \frac{1}{3}$.

With functions, f^{-1} denotes the inverse function of f.

Finding the inverse function

To find the inverse function you must rearrange the formula to find the input (x) in terms of the output (y).

Key point 2.7

To find the inverse function $f^{-1}(x)$ given an expression for $f(x)$:

1 start with $y = f(x)$
2 rearrange to get x (the input) in terms of y (the output)
3 give $f^{-1}(x)$ by replacing the ys with xs.

WORKED EXAMPLE 2.11

Find the inverse function of $f(x) = 3\ln(x+4)$.

$y = 3\ln(x+4)$	Start with $y = f(x)$.
$\frac{y}{3} = \ln(x+4)$	Rearrange to make x the subject.
$e^{\frac{y}{3}} = x+4$	Express both sides as powers of e to remove ln.
$x = e^{\frac{y}{3}} - 4$	
$\therefore f^{-1}(x) = e^{\frac{x}{3}} - 4$	Write the resulting function in terms of x.

WORK IT OUT 2.2

Find the inverse function of $f(x) = \frac{3x-1}{x+4}$.

Which is the correct solution? Identify the errors made in the incorrect solutions.

Solution 1	Solution 2	Solution 3
$y = \frac{3x-1}{x+4}$ $xy + 4y = 3x - 1$ $x(y-3) = -1 - 4y$ $x = \frac{4y+1}{3-y}$ $\therefore f^{-1}(x) = \frac{4x+1}{3-x}$	$f(x) = \frac{3x-1}{x+4}$ $\therefore f^{-1}(x) = \frac{x+4}{3x-1}$	$y = \frac{3x-1}{x+4}$ $xy + 4y = 3x - 1$ $xy = 3x - 1 - 4y$ $x = 3x - 1 - 4$ $\therefore f^{-1}(x) = 3x - 5$

The relationship between f and f^{-1}

Once you know how to find inverse functions there are a couple of very important facts you need to know about them.

When you are finding the inverse function you switch the inputs and the outputs, so on the graph you switch the x- and y-axes.

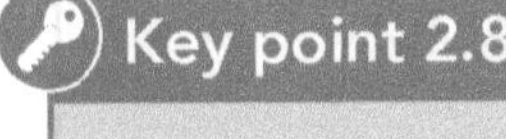

Key point 2.8

The graph of $y = f^{-1}(x)$ is a reflection of the graph of $y = f(x)$ in the line $y = x$.

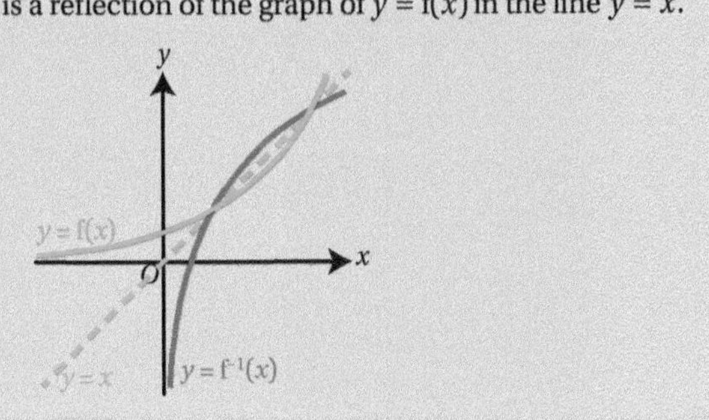

When you apply a function and then its inverse you get back to where you started.

Key point 2.9

$ff^{-1}(x) = f^{-1}f(x) = x$, for all x.

WORKED EXAMPLE 2.12

The graph of $y = h(x)$ is shown. Sketch the graphs of:

a $y = h^{-1}(x)$ **b** $y = h \circ h^{-1}(x)$.

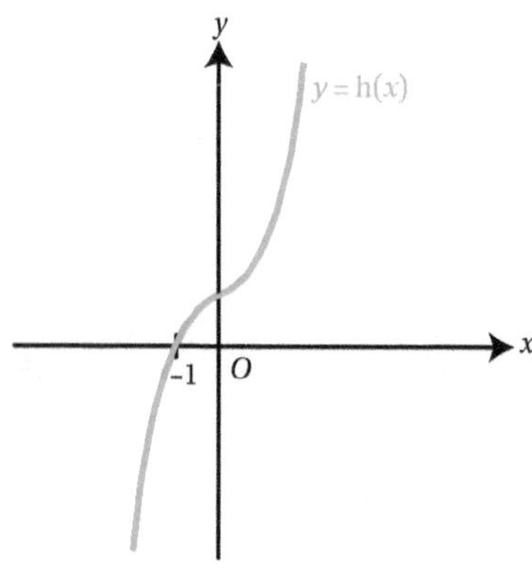

a

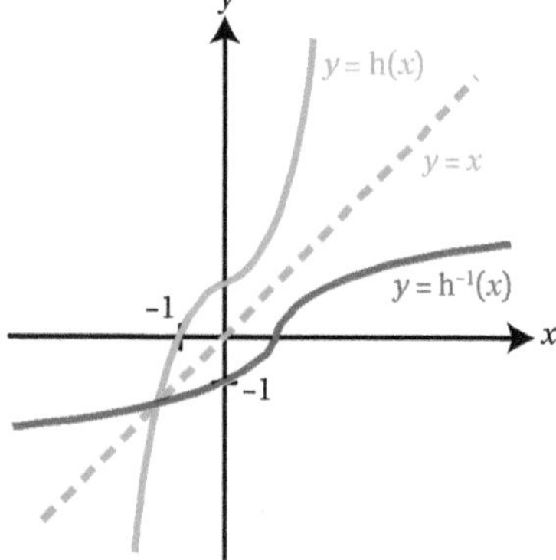

The graph of $y = h^{-1}(x)$ is a reflection in the line $y = x$ of $y = h(x)$.

b

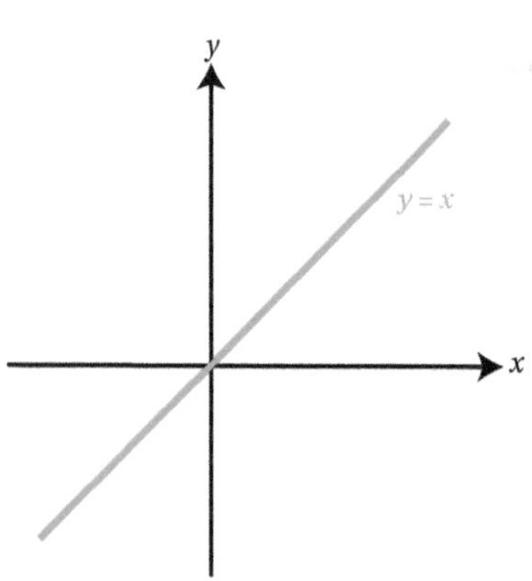

Simplify $y = h \circ h^{-1}(x)$ to $y = x$.

The fact that the graphs of f and f^{-1} are reflections of each other gives you a very useful technique to solve some equations that would otherwise involve complicated (or impossible) algebra.

WORKED EXAMPLE 2.13

The diagram shows the graph of the function $f(x) = \frac{1}{27}x^3 + x - 8$.

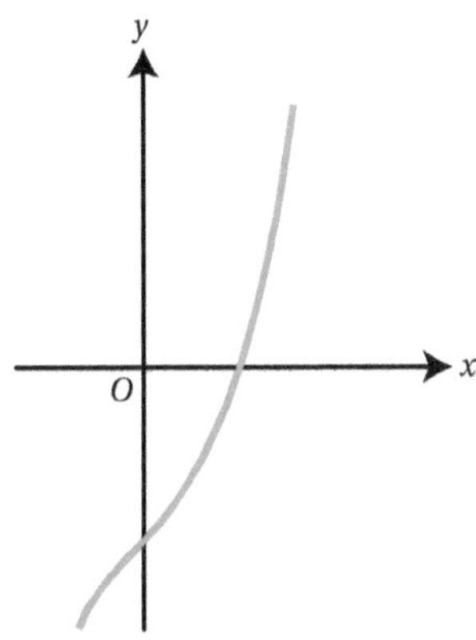

a On the same axes, sketch the graph of $y = f^{-1}(x)$.

b Solve the equation $f(x) = f^{-1}(x)$.

a

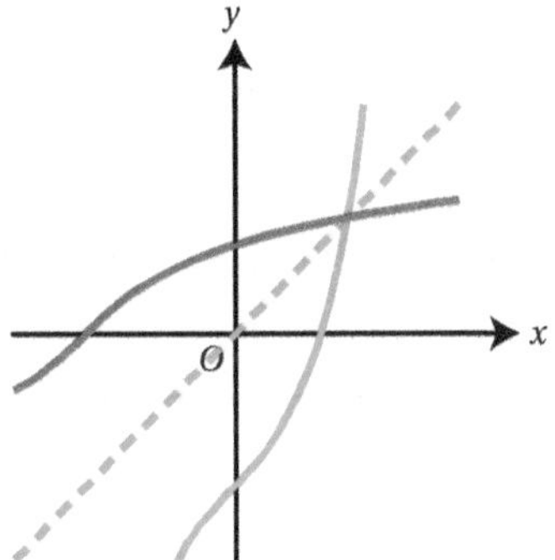

The graph of $f^{-1}(x)$ is the reflection of the graph of $y = f(x)$ in the line $y = x$.

b

$$f(x) = f^{-1}(x)$$
$$\Leftrightarrow f(x) = x$$
$$\therefore \frac{1}{27}x^3 + x - 8 = x$$
$$\frac{1}{27}x^3 = 8$$
$$x^3 = 8 \times 27$$
$$x = 2 \times 3 = 6$$

Finding the equation for the inverse function would involve solving a complicated cubic equation.

Luckily, you can see from the graph that the graphs of f and f^{-1} intersect on the line $y = x$. This means that solving the equation $f = f^{-1}$ is equivalent to solving the equation $f(x) = x$.

The reflection in the line $y = x$ swaps the domain and the range of a function, because it swaps x and y coordinates.

Key point 2.10

- The domain of $f^{-1}(x)$ is the same as the range of $f(x)$.
- The range of $f^{-1}(x)$ is the same as the domain of $f(x)$.

This means that, when asked to find the domain and range, you can work either with f or with f^{-1}, whichever one is easier.

WORKED EXAMPLE 2.14

The function f is defined by $f(x) = \dfrac{1+x}{3-x}$ for $x \neq 3$.

a Find an expression for $f^{-1}(x)$ and state its domain and range.
b Find the range of f.

a $y = \dfrac{1+x}{3-x}$ — Set $y = f(x)$.

$y(3-x) = 1+x$ — Make x the subject.
$3y - yx = 1+x$
$3y - 1 = x + xy$
$3y - 1 = x(1+y)$
$x = \dfrac{3y-1}{1+y}$

$\therefore f^{-1}(x) = \dfrac{3x-1}{1+x}$ — Replace y with x.

The domain of f^{-1} is $x \neq -1$. — In the expression for f^{-1}, the denominator cannot be zero.

The range of f^{-1} is $f^{-1}(x) \neq 3$. — The range of f^{-1} is the same as the domain of f, which was given in the question.

b The range of f is $f(x) \neq -1$. — The range of f is the same as the domain of f^{-1}, which you found in part **a**.

EXERCISE 2D

1 Find $f^{-1}(x)$ if:

a i $f(x) = 3x+1$ ii $f(x) = 7x-3$

b i $f(x) = \dfrac{2x}{3x-2}, x \neq \dfrac{2}{3}$ ii $f(x) = \dfrac{x}{2x+1}$

c i $f(x) = \dfrac{x-a}{x-b}, x \neq b$ ii $f(x) = \dfrac{ax-1}{bx-1}, x \neq \dfrac{1}{b}$

d i $f(a) = 1-a$ ii $f(y) = 3y+2$

e i $f(x) = \sqrt{3x-2}, x \geqslant \dfrac{2}{3}$ ii $f(x) = \sqrt{2-5x}, x \leqslant \dfrac{2}{5}$

f i $f(x) = \ln(1-5x), x < 0.2$ ii $f(x) = \ln(2x+2), x > -1$

g i $f(x) = 7e^{\frac{x}{2}}$ ii $f(x) = 9e^{10x}$

h i $f(x) = x^2 - 10x + 6, x < 5$ ii $f(x) = x^2 + 6x - 1, x > 0$

2 Sketch the inverse of each function.

a

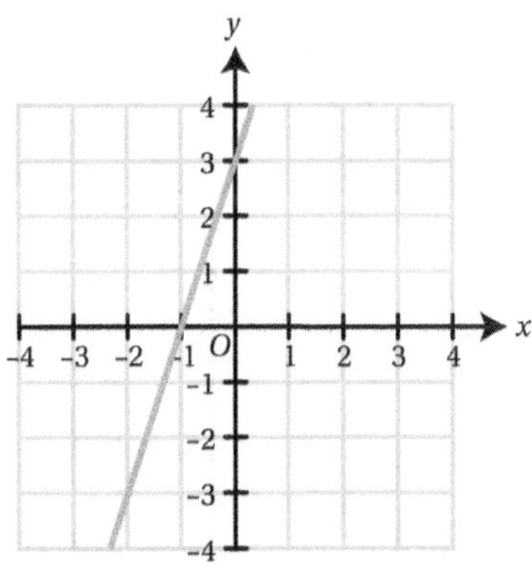

b

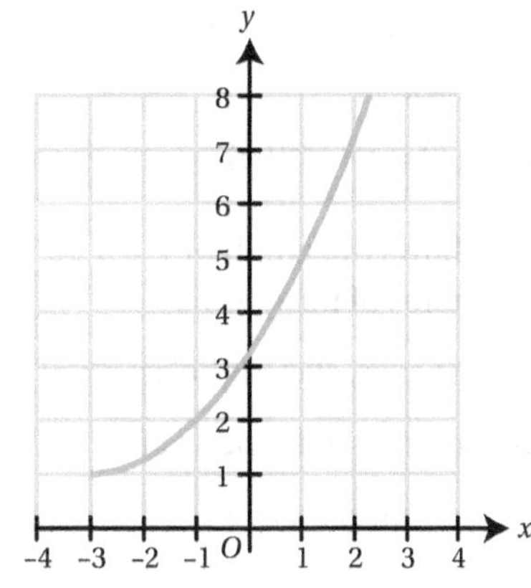

c

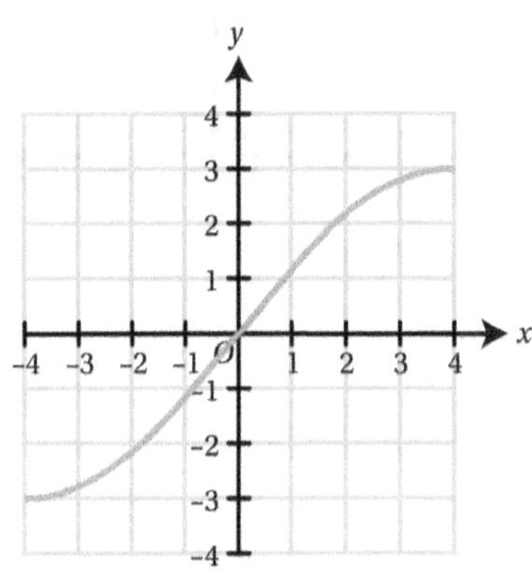

d

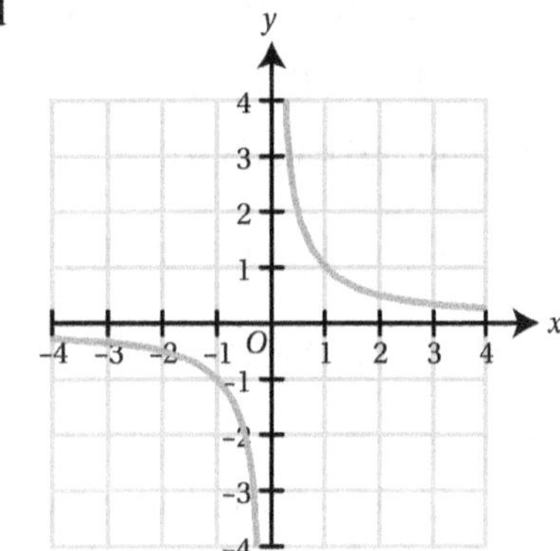

3 For each function, find the expression for f^{-1} and hence state the largest possible domain and range for both f and f^{-1}.

a **i** $f(x) = \dfrac{3x-1}{x+2}$ **ii** $f(x) = \dfrac{2x+3}{x-2}$

b **i** $f(x) = 2 - 3\sqrt{3x-1}$ **ii** $f(x) = \dfrac{1}{2}\sqrt{4-x} + 1$

c **i** $f(x) = 3 + \ln(4x-3)$ **ii** $f(x) = 2\ln(x+3) - 1$

d **i** $f(x) = 3 - 2e^{x-2}$ **ii** $f(x) = 3e^{5-2x} + 1$

4 This table gives selected values of the one-to-one function $f(x)$.

x	−1	0	1	2	3	4
$f(x)$	−4	−1	3	0	7	2

a Evaluate $ff(2)$

b Evaluate $f^{-1}(3)$

Elevate

See Support Sheet 2 for a further example of finding inverse functions and their domains, and for more practice questions.

5 The function f is defined by $f : x \mapsto \sqrt{3-2x}$ for $x \leqslant \dfrac{3}{2}$.

Evaluate $f^{-1}(7)$.

6 Given that $f(x) = 3e^{2x}$:

a find the inverse function $f^{-1}(x)$

b state the domain and range of f^{-1}.

7 Given functions $f: x \mapsto 2x+3$ and $g: x \mapsto x^3$, find the function $(f \circ g)^{-1}$.

8 Let f and g be to functions such that $f \circ g$ is defined, and suppose that both f^{-1} and g^{-1} both exist. Let $h(x) = f \circ g(x)$.

Prove that $h^{-1}(x) = g^{-1} \circ f^{-1}(x)$.

9 The diagram shows the graph of $y = f(x)$. The lines $y = -9$ and $y = 9$ are the asymptotes of the graph.

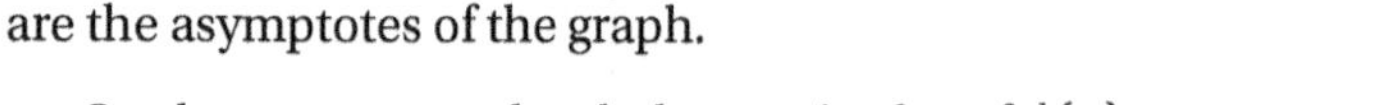

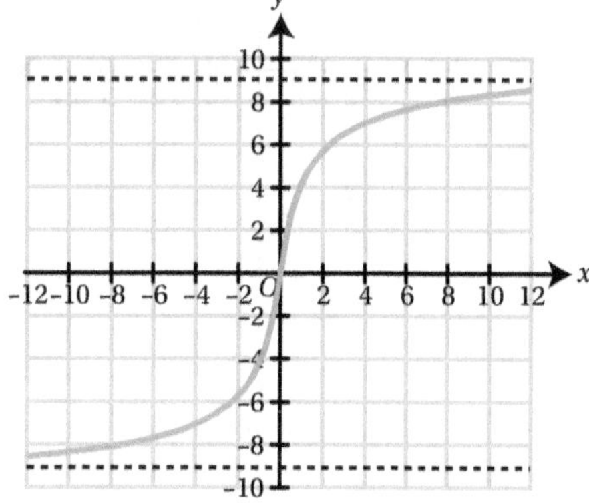

a On the same axes, sketch the graph of $y = f^{-1}(x)$.

b State the domain and range of f^{-1}.

c Solve the equation $f(x) = f^{-1}(x)$.

10 The functions f and g are defined by $f: x \mapsto e^{2x}$, $g: x \mapsto x+1$.

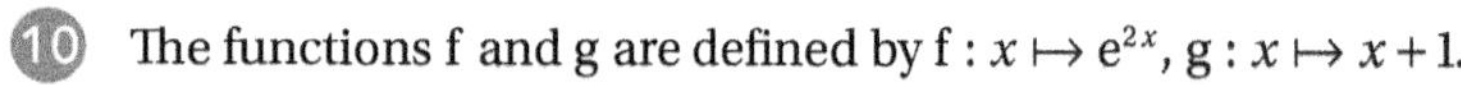

a Calculate $f^{-1}(3) \times g^{-1}(3)$.

b Show that $(f \circ g)^{-1}(3) = \ln\sqrt{3} - 1$.

11 The function f is defined for $x \leqslant 0$ by $f(x) = \dfrac{x^2 - 4}{x^2 + 9}$. Find an expression for $f^{-1}(x)$.

12 Let $f(x) = \ln(x-1) + \ln 3$, for $x > 1$.

a Find $f^{-1}(x)$.

b Let $g(x) = e^x$. Find $(g \circ f)(x)$, giving your answer in the form $ax+b$, where $a, b \in \mathbb{Z}$.

13 The function f is defined by $f(x) = \sqrt[3]{x^3 + 30x - 45}$ for $x \in \mathbb{R}$. The graph of $y = f(x)$ is shown in the diagram.

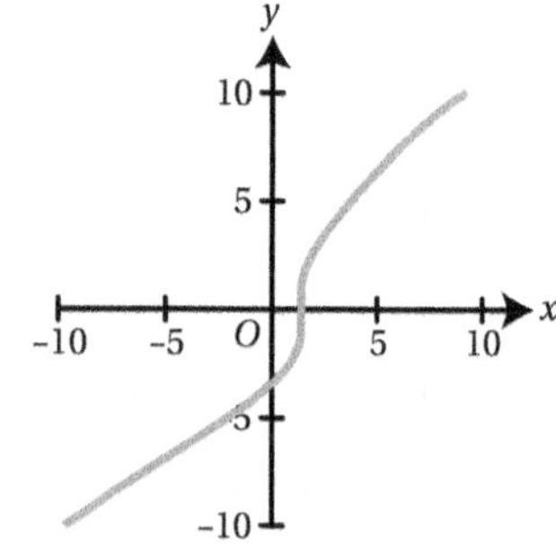

a On the same set of axes, sketch the graph of $y = f^{-1}(x)$.

b Solve the equation $f(x) = f^{-1}(x)$.

When does the inverse function exist?

All functions have inverse mappings, but these inverse mappings are not necessarily themselves functions. Since an inverse function is a reflection in the line $y = x$, for the result to pass the vertical line test, the original function must pass the horizontal line test. But, as you saw in Section 1, this means it must be a one-to-one function.

Key point 2.11

Only one-to-one functions have inverse functions.

Fast forward

This idea will be applied in Chapter 7 when inverses of trigonometric functions are defined.

This leads to one of the most important uses of domains. By restricting the domain you can turn any function into a one-to-one function, which allows you to find its inverse function.

WORKED EXAMPLE 2.15

a Find the largest value of k such that the function $f(x) = (x-3)^2$, $x \leqslant k$ is one-to-one.
b Find $f^{-1}(x)$ for this value of k and state its range.

a

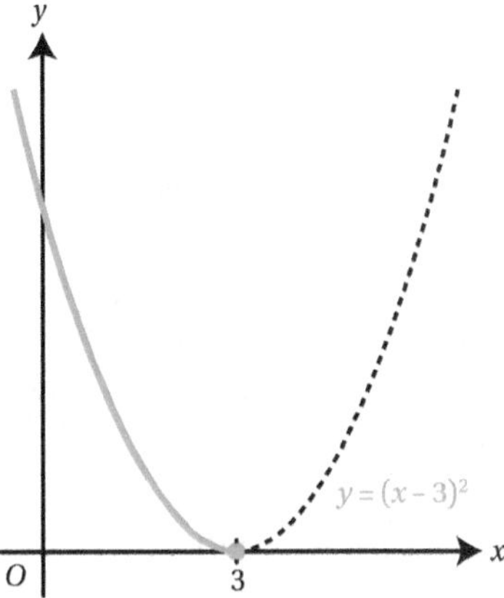

Sketch the graph of $y = (x-3)^2, x \in \mathbb{R}$... but eliminate the points towards the right of the graph that cause the horizontal line test to fail.

$x \leqslant 3$

Decide which section remains.

b

$$y = (x-3)^2$$
$$\pm\sqrt{y} = x-3$$
$$x = 3 \pm \sqrt{y}$$

Follow the standard procedure for finding inverse functions.

Since $x \leqslant 3$, $x = 3 - \sqrt{y}$

Take the negative root, as $x \leqslant 3$.

$\therefore f^{-1}(x) = 3 - \sqrt{x}$

Write $f^{-1}(x)$.

The range of f^{-1} is $f(x) \leqslant 3$.

The range of f^{-1} is the domain of f.

WORK IT OUT 2.3

Find the inverse function of $f(x) = x^2 - 3$, $x \in \mathbb{R}$.

Which is the correct solution? Identify the errors made in the incorrect solutions.

Solution 1	Solution 2	Solution 3
$y = x^2 - 3$ $x^2 = y + 3$ $x = \pm\sqrt{y+3}$ $\therefore f^{-1}(x) = \pm\sqrt{x+3}$	$y = x^2 - 3$ $x^2 = y + 3$ $x = \sqrt{y+3}$ $\therefore f^{-1}(x) = \sqrt{x+3}$	It doesn't exist.

It should be clear from the horizontal line test that a function f is one-to-one if f either increases or decreases throughout its domain. As soon as there is a turning point, the function is no longer one-to-one (and therefore has no inverse).

WORKED EXAMPLE 2.16

$f(x) = x^2 - 8x^{\frac{3}{2}} + 18x$, for all real x.

Prove that $f(x)$ has an inverse function.

$f'(x) = 2x - 12x^{\frac{1}{2}} + 18 = 2\left(x - 6x^{\frac{1}{2}} + 9\right)$ — f must either be increasing or decreasing so find f′ to determine which.

$= 2\left(x^{\frac{1}{2}} - 3\right)^2$

$\geqslant 0$ for all x — Complete the square. Note that this is a common way of showing that a function is non-negative. (In this case it actually factorises.)

$\therefore$ f is one-to-one for all x. — Although $f'(9) = 0$, the gradient is never negative and so f is one-to-one.

So f has an inverse function.

EXERCISE 2E

1 Find the value of k that gives the largest possible domain such that the inverse function exists. For this domain, find the inverse function.

a $y = x^2, x \leqslant k$

b $y = (x+1)^2 + 2, x > k$

c $y = 5 + 2x - x^2, x \leqslant k$

d $y = x^2 + 4x + 3, x > k$

2 For each function shown in the diagrams, determine the largest possible domain of the given form for which the inverse function exists. For that domain, sketch the inverse function.

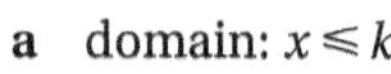

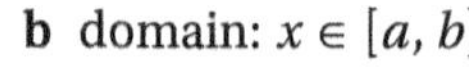

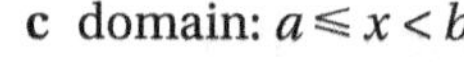

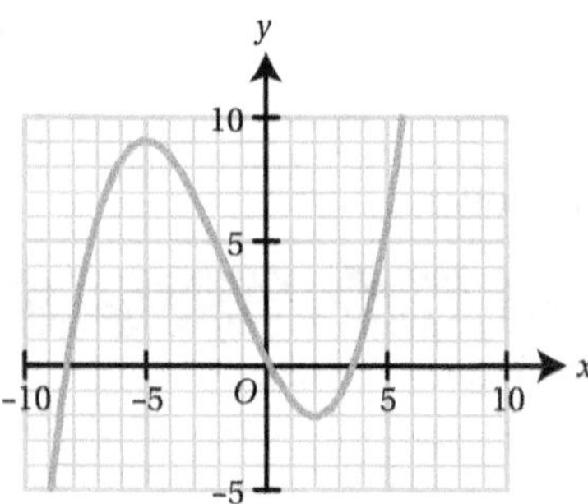

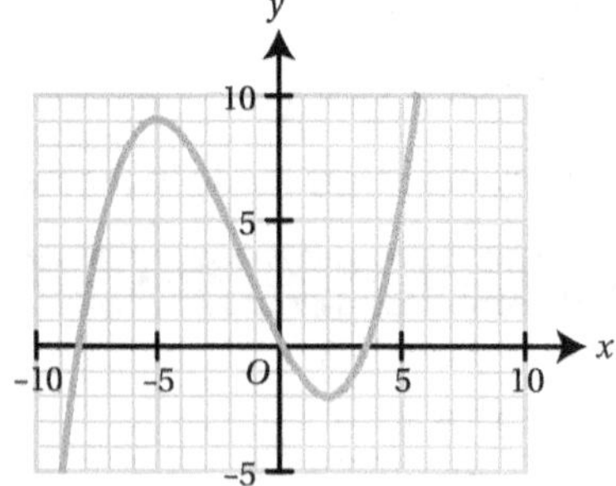

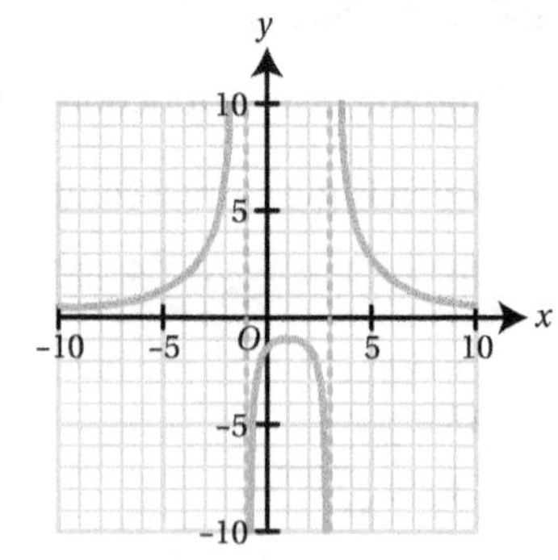

3 Given that $f(x) = e^x - 3x$ for $x \in \mathbb{R}$:

a find $f'(x)$

b explain why $f^{-1}(x)$ does not exist.

The domain of f is changed to $x \leqslant k$ so that f^{-1} now exists.

c Find the largest possible value of k.

4 A function is self-inverse if $f(x) = f^{-1}(x)$ for all x in the domain.

Find the value of the constant k so that $g(x) = \dfrac{3x-5}{x+k}$ is a self-inverse function.

5 A function is defined by $f(x) = x^3 + 3ax^2 + 3ax + 1$ where a is a constant.

Find the set of values of a for which $f^{-1}(x)$ exists for all x.

Checklist of learning and understanding

- A **mapping** is a **function** if every x-value maps to a single y-value. The output value corresponding to the input x is called the **image** of x.
- A function is:
  - **one-to-one** if every y-value corresponds to only one x-value
  - **many-to-one** if there are some y-values that come from more than one x-value.
- The vertical and horizontal line tests can be applied to graphs of mappings.
  - **Vertical line test**: if a mapping is a function, any vertical line will cross its graph at most once.
  - **Horizontal line test**: if a function is one-to-one, any horizontal line will cross the graph at most once.
- The set of allowed input values of a function is called the **domain**.
- The set of all possible outputs of a function is called the **range**.
- The composite function formed by applying g to x and then f to the result is written as $f(g(x))$ or $fg(x)$ or $f \circ g(x)$.
- The inverse, f^{-1}, of a function f is such that $ff^{-1}(x) = f^{-1}f(x) = x$, for all x.
- To find the inverse function $f^{-1}(x)$ given an expression for $f(x)$:
  1. start with $y = f(x)$
  2. rearrange to get x (the input) in terms of y (the output)
  3. give $f^{-1}(x)$ by replacing the ys with xs.
- Only one-to-one functions have inverse functions.
- The graph of $y = f^{-1}(x)$ is a reflection of the graph of $y = f(x)$ in the line $y = x$.
  - The domain of $f^{-1}(x)$ is the same as the range of $f(x)$.
  - The range of $f^{-1}(x)$ is the same as the domain of $f(x)$.

Mixed practice 2

1. $f(x) = \frac{1}{x}, x \neq 0$

   Which one of these statements about the inverse function, $f^{-1}(x)$ is true?

   A $f^{-1}(x) = \frac{1}{f(x)}$ **B** $f^{-1}(x) = f(x)$ **C** $f^{-1}(x) = x$ **D** $f^{-1}(x)$ doesn't exist

2. Find the inverse of each function.

   a $f(x) = \log_3(x+3), x > -3$ **b** $f(x) = 3e^{x^3-1}$

3. The diagram shows three graphs.

   A is part of the graph of $y = x$.

   B is part of the graph of $y = 2^x$.

   C is the reflection of graph B in line A.

   Write down:

   a the equation of C in the form $y = f(x)$

   b the coordinates of the point where C cuts the x-axis.

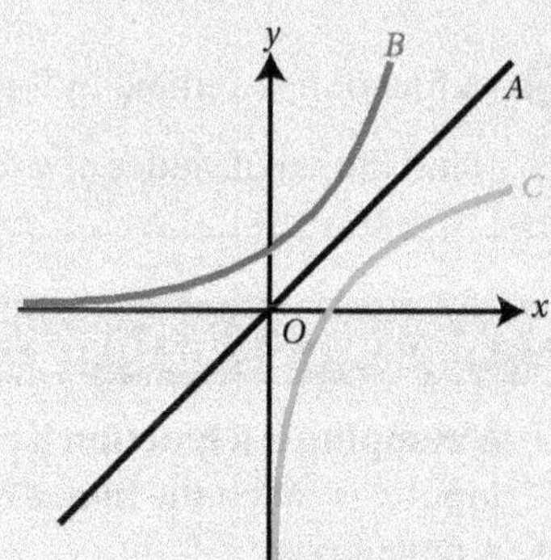

4. Let $f(x) = \sqrt{x}$, and $g(x) = 2x$. Solve the equation $(f^{-1} \circ g)(x) = \frac{1}{4}$.

5. If $f(x) = x^2 + 1, x \geqslant 3$ and $g(x) = 5 - x$:

   a evaluate f(3)

   b find an simplify an expression for $gf(x)$

   c state the geometric relationship between the graphs of $y = f(x)$ and $y = f^{-1}(x)$

   d **i** find an expression for $f^{-1}(x)$

   ii find the range of $f^{-1}(x)$

   iii find the domain of $f^{-1}(x)$

   e solve the equation $f(x) = g(3x)$.

6. The functions f and g are defined with their respective domains by:

   $f(x) = \sqrt{2x-5}$, for $x \geqslant 2.5$

   $g(x) = \frac{10}{x}$, for real values of x, $x \neq 0$

   a State the range of f.

   b **i** Find $fg(x)$.

   ii Solve the equation $fg(x) = 5$.

   c The inverse of f is f^{-1}.

   i Find $f^{-1}(x)$.

   ii Solve the equation $f^{-1}(x) = 7$.

7 The functions f and g are defined with their respective domains by $f(x) = x^2$ for all real values of x and $g(x) = \frac{1}{2x+1}$ for real values of x, $x \neq -0.5$.

a Explain why f does not have an inverse.

b The inverse of g is g^{-1}. Find $g^{-1}(x)$.

c State the range of g^{-1}.

d Solve the equation $fg(x) = g(x)$.

[© AQA 2011]

8 The function f is given by $f(x) = x^2 - 6x + 10$, for $x \geqslant 3$.

a Write $f(x)$ in the form $(x - p)^2 + q$.

b Find the inverse function $f^{-1}(x)$.

c State the domain of $f^{-1}(x)$.

9 $f(x) = 2x + 1$ and $g(x) = \frac{x+3}{x-1}$, $x \neq 1$

a Find and simplify:

i $f(7)$

ii the range of $f(x)$

iii $f(z)$

iv $fg(x)$

v $ff(x)$.

b Explain why $gf(x)$ does not exist.

c **i** Find the form of $g^{-1}(x)$.

ii State the geometric relationship between the graphs of $y = g(x)$ and $y = g^{-1}(x)$.

iii State the domain of $g^{-1}(x)$.

iv State the range of $g^{-1}(x)$.

10 The functions f and g are defined over the domain of all real numbers by $f(x) = x^2 + 4x + 9$ and $g(x) = e^x$.

a Write $f(x) = x^2 + 4x + 9$, $x \in \mathbb{R}$ in the form $f(x) = (x + p)^2 + q$.

b Hence sketch the graph of $y = x^2 + 4x + 9$, labelling all axis intercepts and the coordinates of the turning point.

c State the range of $f(x)$ and $g(x)$.

d Hence or otherwise find the range of $h(x) = e^{2x} + 4e^x + 9$.

11 The curve with equation $y = \mathrm{f}(x)$, where $\mathrm{f}(x) = \ln(2x - 3)$, $x > \frac{3}{2}$, is shown in the sketch.

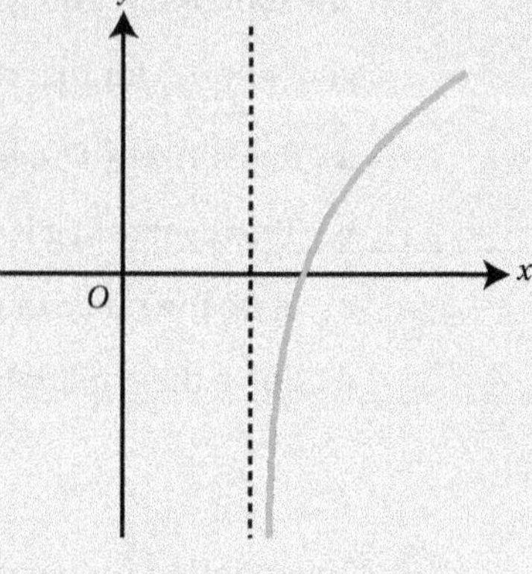

a The inverse of f is f^{-1}.

i Find $\mathrm{f}^{-1}(x)$.

ii State the range of f^{-1}.

iii Sketch the curve with equation $y = \mathrm{f}^{-1}(x)$, indicating the value of the y-coordinate of the point where the curve intersects the y-axis.

b The function g is defined by $\mathrm{g}(x) = \mathrm{e}^{2x} - 4$, for all real values of x.

i Find $\mathrm{gf}(x)$, giving your answer in the form $(ax - b)^2 - c$, where a, b and c are integers.

ii Write down an expression for $\mathrm{fg}(x)$, and hence find the exact solution of the equation $\mathrm{fg}(x) = \ln 5$.

[AQA 2013]

12 $\mathrm{h}(x) = x^2 - 6x + 2$

a Write $\mathrm{h}(x)$ in the form $(x - p)^2 + q$.

b Hence or otherwise find the range of $\mathrm{h}(x)$.

c By using the largest possible domain of the form $x \geqslant k$, find the inverse function $\mathrm{h}^{-1}(x)$.

13 a Show that if $\mathrm{g}(x) = \frac{1}{x}$ then $\mathrm{gg}(x) = x$.

b A function satisfies the identity $\mathrm{f}(x) + 2\mathrm{f}\left(\frac{1}{x}\right) = 2x + 1$. By replacing all instances of x with $\frac{1}{x}$ find another identity that $\mathrm{f}(x)$ satisfies.

c By solving these two identities simultaneously, express $\mathrm{f}(x)$ in terms of x.

14 The functions $\mathrm{f}(x)$ and $\mathrm{g}(x)$ are given by $\mathrm{f}(x) = \sqrt{x - 2}$ and $\mathrm{g}(x) = x^2 + x$. The function $(\mathrm{f} \circ \mathrm{g})(x)$ is defined for $x \in \mathbb{R}$ except for the interval $a < x < b$.

a Calculate the values of a and of b.

b Find the range of $\mathrm{f} \circ \mathrm{g}$.

15 An odd function is any function $\mathrm{f}(x)$ that satisfies $\mathrm{f}(x) = -\mathrm{f}(-x)$.

a Show that $\mathrm{f}(x) = x^3$ is an odd function.

b What type of symmetry must the graph of any odd function have?

c Given any function $\mathrm{g}(x)$, show that $\mathrm{g}(x) - \mathrm{g}(-x)$ is an odd function.

An even function is any function which satisfies $\mathrm{f}(x) = \mathrm{f}(-x)$.

d Show that $\mathrm{f}(x) = |x|$ is an even function.

e What type of symmetry must the graph of any even function have?

f Given any function $\mathrm{g}(x)$ show that $\mathrm{g}(x) + \mathrm{g}(-x)$ is an even function.

g Hence or otherwise show that any function can be written as the sum of an even function and an odd function.

Elevate

See Extension Sheet 2 for a selection of more challenging problems.

3 Further transformations of graphs

In this chapter you will learn how to:

- draw a graph after two (or more) transformations
- find the equation of a graph after a combination of transformations
- sketch graphs of functions involving the modulus (absolute value)
- use modulus graphs to solve equations and inequalities.

Before you start...

Student Book 1, Chapter 5	You should be able to recognise a graph transformation from the equation.	1 The graph of $y = f(x)$ is shown in the diagram. Sketch the graph of: a $y = f(x+2)$ b $y = -f(x)$.
Student Book 1, Chapter 5	You should be able to change the equation of a graph to achieve a given transformation.	2 A graph has equation $y = x^2 - 3x$. Find the equation of the graph after: a a translation of 5 units in the positive y-direction, followed by b a horizontal stretch with scale factor 2.
Student Book 1, Chapter 1	You should be able to use interval notation to express solutions of inequalities.	3 Solve each inequality and write the solution using interval notation. a $3x - 2 \geqslant 7$ b $2x + 5 < 1$ and $3 - 2x < 9$

Combining transformations

In this chapter the transformations you met in Student Book 1, Chapter 5, will be combined to produce a sequence of transformations of the original graph.

The modulus function will also be introduced; this is used in many contexts where a quantity needs to be positive, for example, the total distance travelled must be the sum of positive quantities, even though the particle may change direction. You have also seen this idea used to find the area between a curve and the x-axis when the enclosed region is below the axis.

Section 1: Combined transformations

In this section you will look at what happens when two transformations are applied to a graph. An important question to consider is, does the order in which the two transformations are done affect the outcome? To investigate this question, first consider transformations of a single point.

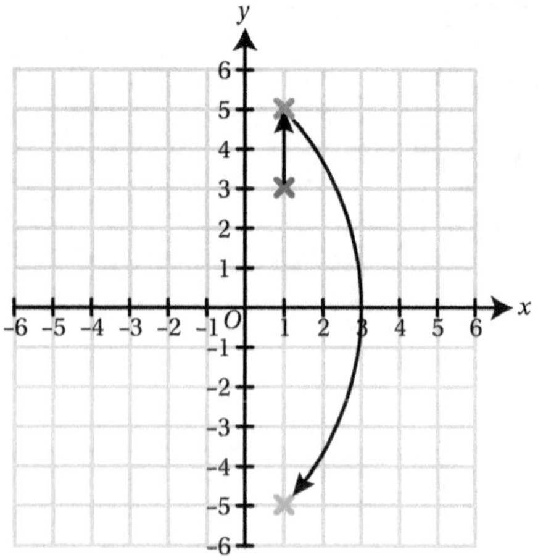

The point $(1, 3)$ is translated two units up and then reflected in the x-axis. The new point is $(1, -5)$.

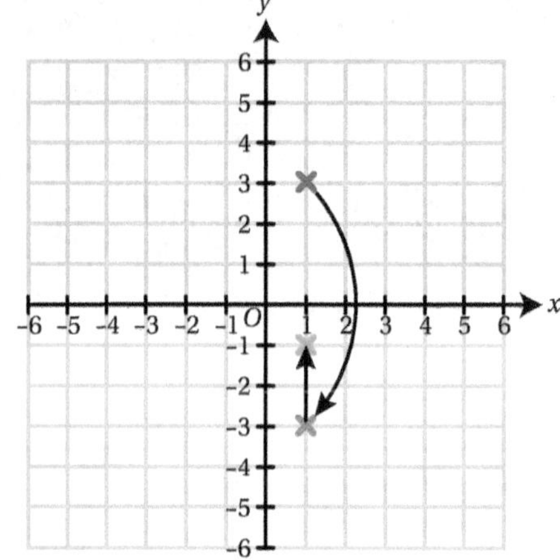

The point $(1, 3)$ is reflected in the x-axis first and then translated two units up. The new point is $(1, -1)$.

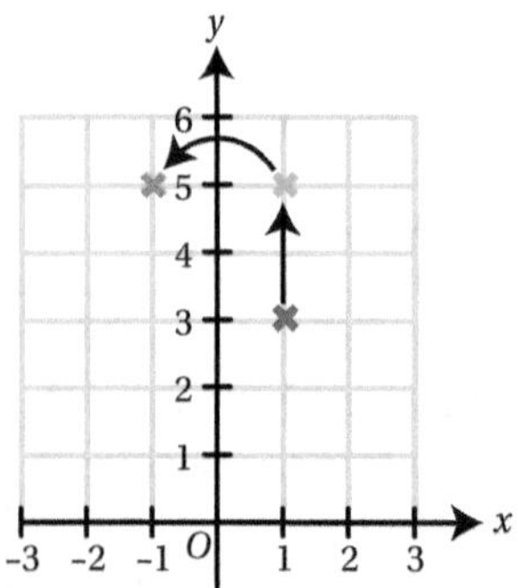

The point $(1, 3)$ is translated two units up and then reflected in the y-axis. The new point is $(-1, 5)$.

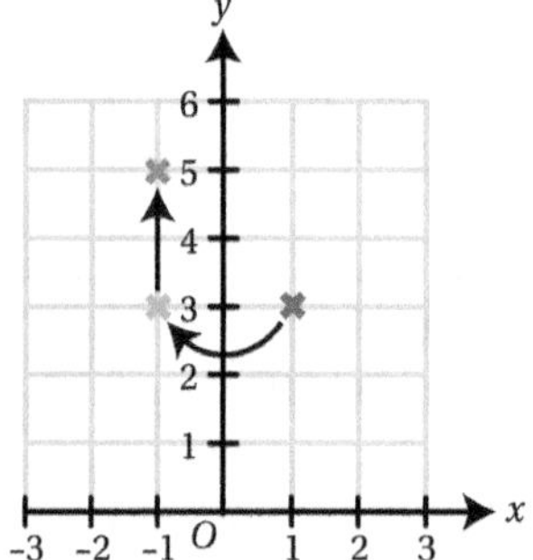

The point $(1, 3)$ is reflected in the y-axis first and then translated two units up. The new point is $(-1, 5)$.

These examples suggest some rules for combining transformations.

Key point 3.1

When two vertical transformations or two horizontal transformations are combined, changing the order affects the outcome.

When one vertical and one horizontal transformation are combined, the outcome does not depend on the order.

Combining one vertical and one horizontal transformation

WORKED EXAMPLE 3.1

The diagram shows the graph of the function $y = \mathrm{f}(x)$.

On separate diagrams, draw the graph of:

a $y = 2\mathrm{f}(x+3)$ **b** $y = -\mathrm{f}\left(\frac{x}{2}\right)$.

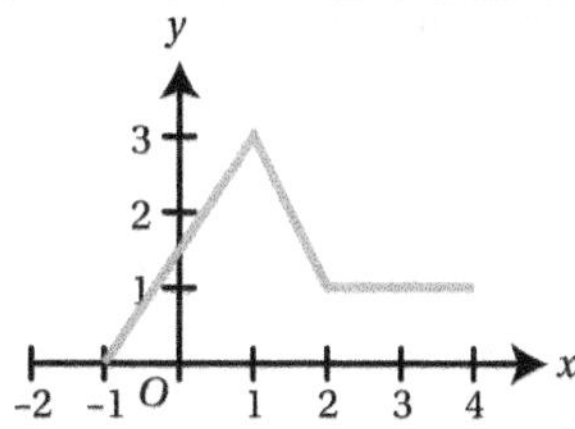

Both parts involve one horizontal and one vertical transformation. Since the order doesn't matter, deal with the transformation in brackets first.

a $y = \mathrm{f}(x+3)$: translate 3 units to the left

x is replaced by $x + 3$, so the graph is translated to the left.

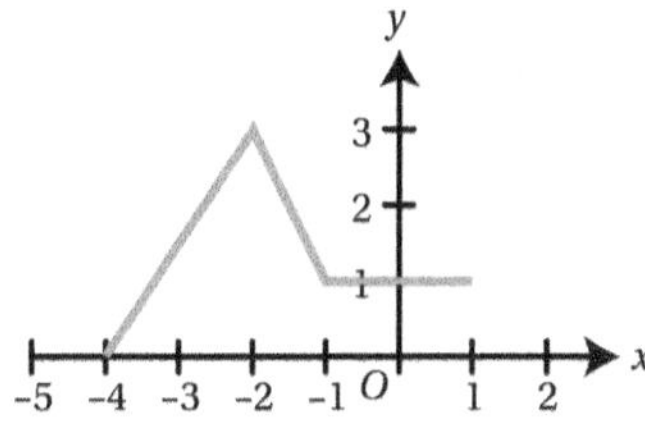

$y = 2\mathrm{f}(x+3)$: vertical stretch with scale factor 2

Multiplying the whole expression by 2 results in all y-values doubling.

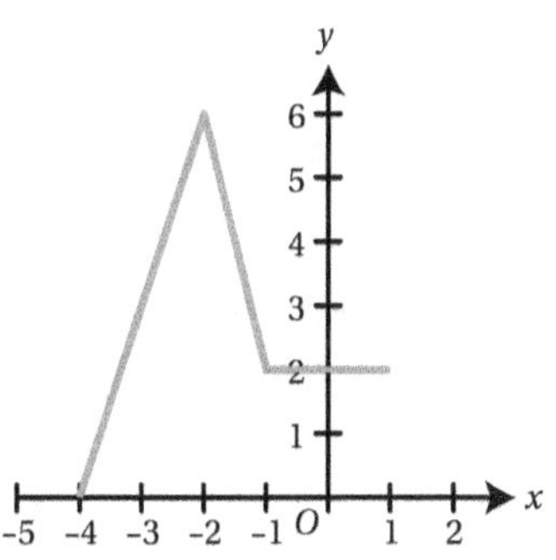

b $y = \mathrm{f}\left(\frac{x}{2}\right)$: horizontal stretch with scale factor 2

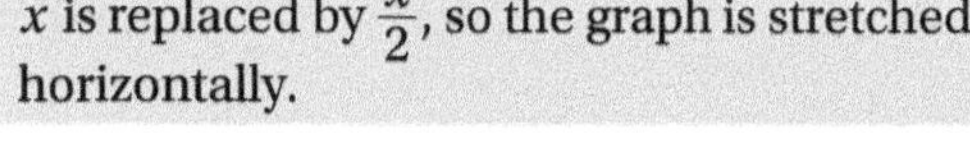

x is replaced by $\frac{x}{2}$, so the graph is stretched horizontally.

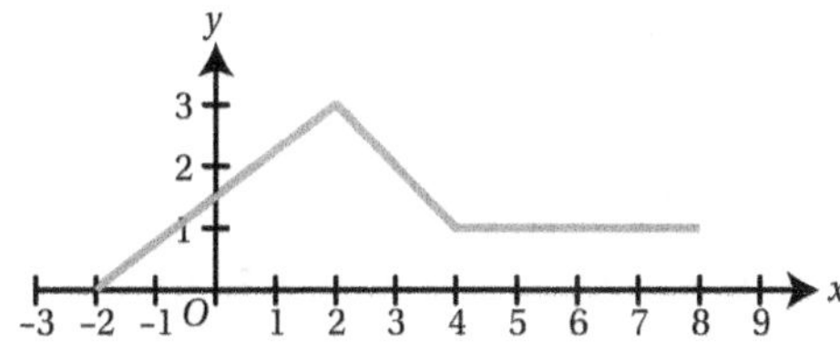

$y = -\mathrm{f}\left(\frac{x}{2}\right)$: reflection in the x-axis

Making the y-coordinate negative results in a reflection in the x-axis.

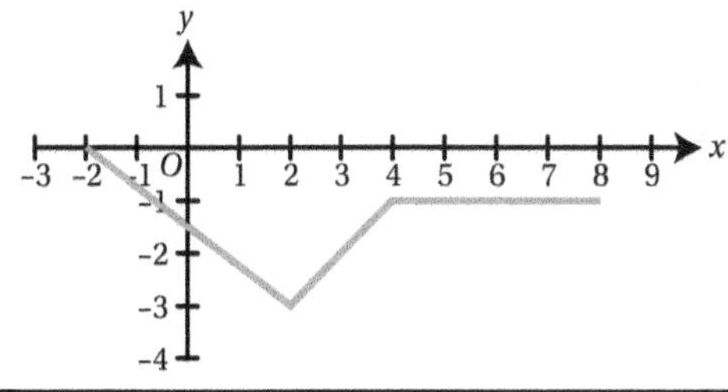

Combining two vertical transformations

To transform the graph of $y = \mathrm{f}(x)$ into the graph of $y = p\mathrm{f}(x) + c$ first multiply $\mathrm{f}(x)$ by p and then add on c. The flow chart shows the order of operations and the corresponding transformations.

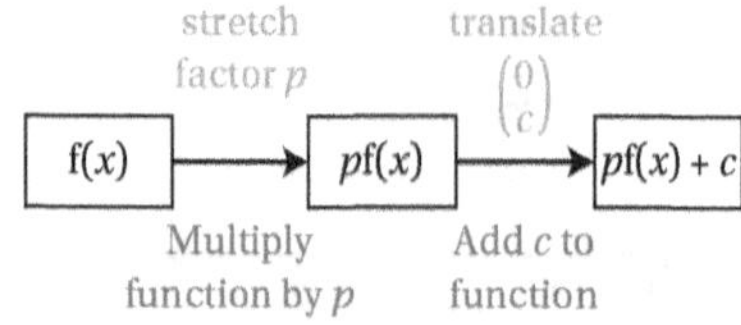

Notice that this follows the normal order of operations: multiplication is done before addition.

WORKED EXAMPLE 3.2

A graph has equation $y = x^2 - 3$. Find the equation of the graph after these transformations.

a A vertical stretch with scale factor 2 followed by a translation with vector $\begin{pmatrix} 0 \\ 5 \end{pmatrix}$.

b A translation with vector $\begin{pmatrix} 0 \\ 5 \end{pmatrix}$ followed by a vertical stretch with scale factor 2.

a Vertical stretch: $y = x^2 - 3$ becomes $y = 2x^2 - 6$.

Vertical stretch: multiply the function by 2.

Vertical translation: $y = 2x^2 - 6$ becomes $y = 2x^2 - 1$.

Vertical translation: add 5 to the whole expression.

The final equation is $y = 2x^2 - 1$.

You could perform the two transformations in one go: The new equation is $y = 2\mathrm{f}(x) + 5 = 2(x^2 - 3) + 5 = 2x^2 - 1$.

b Vertical translation: $y = x^2 - 3$ becomes $y = x^2 + 2$.

This time add 5 to the expression first.

Vertical stretch: $y = x^2 + 2$ becomes $y = 2x^2 + 4$.

Then multiply the whole expression by 2.

The final equation is $y = 2x^2 + 4$.

Again, you can write this in one go:

$y = (\mathrm{f}(x) + 5) \times 2 = 2(x^2 - 3 + 5) = 2x^2 + 4$.

Combining two horizontal transformations

To transform the graph of $y = \mathrm{f}(x)$ into the graph of $y = \mathrm{f}(qx + d)$ you first replace x with $x + d$ and then replace **all** occurrences of x by qx. The flow chart shows the order of operations and the corresponding transformations.

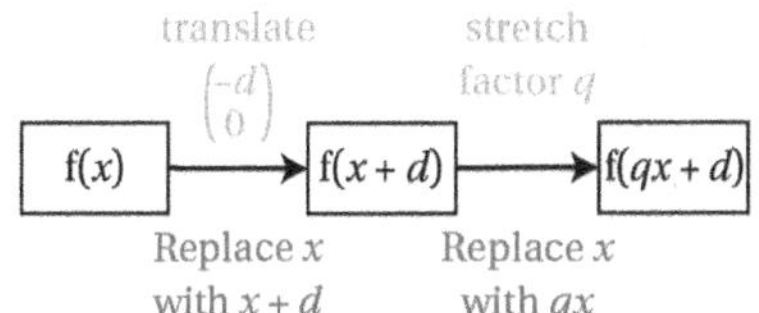

Notice that the transformations are in the 'wrong' order: the addition is done before the multiplication.

WORKED EXAMPLE 3.3

The diagram shows the graph of $y = \mathrm{f}(x)$. Sketch the graph of $y = \mathrm{f}(2x+1)$.

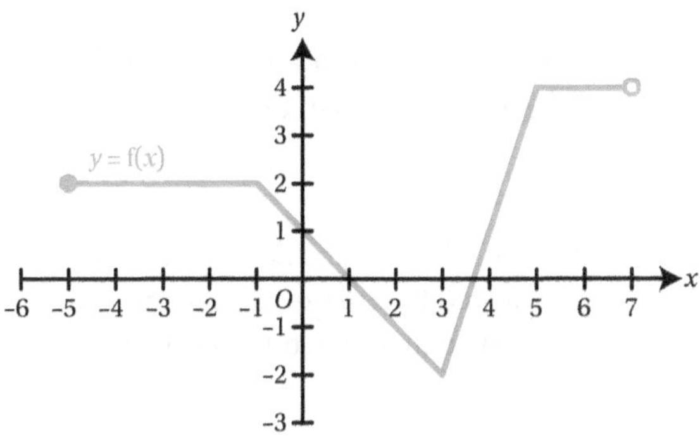

This is a combination of two horizontal transformations, so deal with the addition first.

x is replaced by $x+1 \rightarrow$ horizontal translation 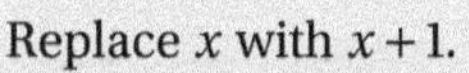 $\begin{pmatrix}-1\\0\end{pmatrix}$

Replace x with $x+1$.

Change $y = \mathrm{f}(x)$ to $y = \mathrm{f}(x+1)$.

$y = \mathrm{f}(x+1)$

x is replaced by $2x \rightarrow$ horizontal stretch, scale factor 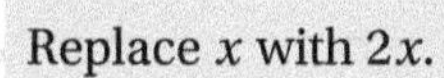 $\frac{1}{2}$

Replace x with $2x$.

Change $y = \mathrm{f}(x+1)$ to $y = \mathrm{f}(2x+1)$.

This graph shows the final answer.

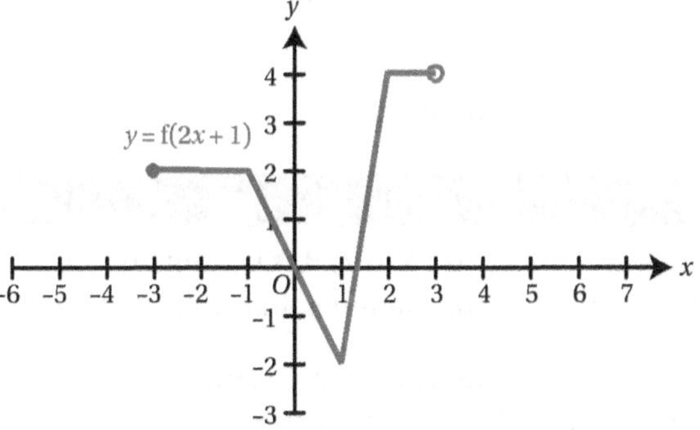

If you want to perform a horizontal stretch before a translation you need to use brackets correctly in the equation.

WORKED EXAMPLE 3.4

The graph of $y = \sin x$ is transformed by a horizontal stretch with scale factor 2 and then translated 3 units to the right. Find the equation of the resulting graph.

Replace x by $\frac{x}{2}$:

$y = \sin x$ is changed to $y = \sin\left(\frac{x}{2}\right)$.

A horizontal stretch with scale factor q is achieved by replacing x by $\frac{x}{q}$.

Replace x by $x - 3$:

$y = \sin\left(\frac{x}{2}\right)$ is changed to $y = \sin\left(\frac{x-3}{2}\right)$.

A horizontal translation by d units is achieved by replacing x by $x - d$.

The final equation is $y = \sin\left(\frac{x}{2} - \frac{3}{2}\right)$.

Note that if the translation had been performed before the stretch, the resulting equation would have been $y = \sin\left(\frac{x}{2} - 3\right)$.

Make sure that you know the correct order of resolving transformations.

Key point 3.2

- Vertical transformations follow the normal order of operations as applied to arithmetic.
- Horizontal transformations are done in the opposite order.

WORK IT OUT 3.1

Describe the sequence of two transformations that transform the graph of $y = f(x)$ to the graph of $y = f(2x + 4)$.

Which is the correct solution? Identify the errors made in the incorrect solutions.

Solution 1	Solution 2	Solution 3
Add 4 to x; this is a horizontal translation 4 units to the left. Replace x by $2x$; this is a horizontal stretch with scale factor $\frac{1}{2}$.	Replace x by $2x$; this is a horizontal stretch with scale factor $\frac{1}{2}$. Add 4 to x; this is a horizontal translation 4 units to the left.	Add 4 to x; this is a horizontal translation 4 units to the left. Replace x by $2x$; this is a horizontal stretch with scale factor 2.

EXERCISE 3A

1 Here are the graphs of $y = \text{f}(x)$ and $y = \text{g}(x)$.

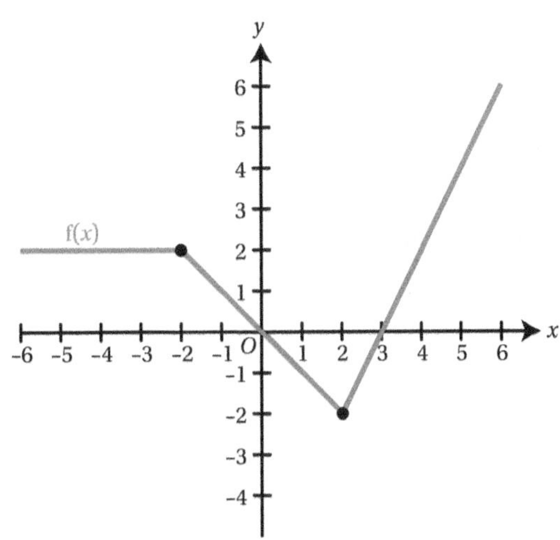

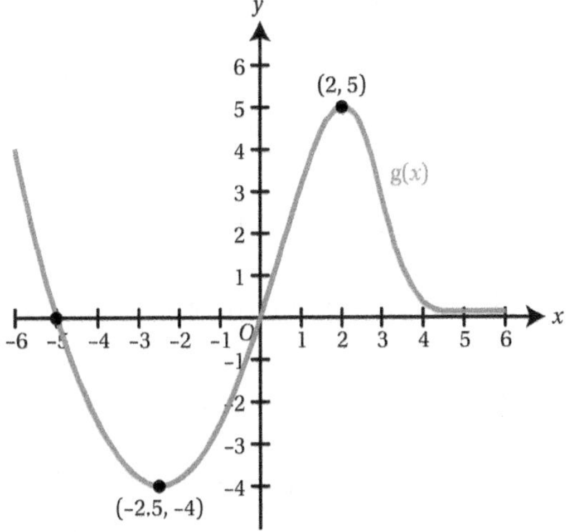

Sketch the graphs of:

a **i** $2\text{f}(x)-1$ **ii** $\frac{1}{2}\text{g}(x)+3$ **b** **i** $4-\text{f}(x)$ **ii** $2-2\text{g}(x)$

c **i** $3(\text{f}(x)-2)$ **ii** $\frac{1-\text{g}(x)}{2}$ **d** **i** $\text{f}(2x-1)$ **ii** $\text{g}(2x+3)$

e **i** $\text{f}\left(\frac{3-x}{2}\right)$ **ii** $\text{g}\left(\frac{x-3}{2}\right)$.

2 Given that $\text{f}(x) = x^2$, express each of these functions as $a\text{f}(x)+b$ and hence describe the sequence of transformations mapping $\text{f}(x)$ to the given function.

a **i** $\text{k}(x) = 2x^2 - 6$ **ii** $\text{k}(x) = 5x^2 + 4$ **b** **i** $\text{h}(x) = 5 - 3x^2$ **ii** $\text{h}(x) = 4 - 8x^2$

3 Given that $\text{f}(x) = 2x^2 - 4$, give the function $\text{g}(x)$ that represents the graph of $\text{f}(x)$ after:

a **i** translation $\begin{pmatrix} 0 \\ 2 \end{pmatrix}$, followed by a vertical stretch of scale factor 3

ii translation $\begin{pmatrix} 0 \\ 6 \end{pmatrix}$, followed by a vertical stretch of scale factor $\frac{1}{2}$

b **i** vertical stretch of scale factor $\frac{1}{2}$, followed by a translation $\begin{pmatrix} 0 \\ 6 \end{pmatrix}$

ii vertical stretch of scale factor $\frac{7}{2}$, followed by a translation $\begin{pmatrix} 0 \\ 10 \end{pmatrix}$

c **i** reflection through the horizontal axis, followed by a translation $\begin{pmatrix} 0 \\ -1 \end{pmatrix}$

ii reflection through the horizontal axis, followed by a translation $\begin{pmatrix} 0 \\ 2 \end{pmatrix}$

d **i** reflection through the horizontal axis, followed by a vertical stretch of scale factor $\frac{1}{2}$, followed by a translation $\begin{pmatrix} 0 \\ 3 \end{pmatrix}$

ii reflection through the horizontal axis followed by a translation $\begin{pmatrix} 0 \\ -6 \end{pmatrix}$, followed by a vertical stretch, scale factor $\frac{3}{2}$.

4 Given that $f(x) = x^2$, express each of these functions as $f(ax + b)$ and hence describe the transformation mapping $f(x)$ to the given function.

a **i** $g(x) = x^2 + 2x + 1$ **ii** $g(x) = x^2 - 6x + 9$

b **i** $k(x) = 4x^2 + 8x + 4$ **ii** $k(x) = 9x^2 - 6x + 1$

5 Given that $f(x) = 2x^2 - 4$, give the function $g(x)$ that represents the graph of $f(x)$ after:

a **i** translation $\begin{pmatrix} 1 \\ 0 \end{pmatrix}$, followed by a horizontal stretch of scale factor $\frac{1}{4}$

ii translation $\begin{pmatrix} -2 \\ 0 \end{pmatrix}$, followed by a horizontal stretch of scale factor $\frac{1}{2}$

b **i** horizontal stretch of scale factor $\frac{1}{2}$, followed by a translation $\begin{pmatrix} -4 \\ 0 \end{pmatrix}$

ii horizontal stretch of scale factor $\frac{2}{3}$, followed by a translation $\begin{pmatrix} 1 \\ 0 \end{pmatrix}$

c **i** translation $\begin{pmatrix} -3 \\ 0 \end{pmatrix}$, followed by a reflection in the y-axis

ii reflection through the vertical axis, followed by a translation $\begin{pmatrix} -3 \\ 0 \end{pmatrix}$.

6 Find the resulting equation after the graph of $y = \sin x$ is transformed through each sequence of transformations.

a A vertical translation c units up, then a vertical stretch with scale factor p relative to the x-axis.

b A vertical stretch with scale factor p, followed by a vertical translation c units up.

c A horizontal stretch with scale factor q relative to the y-axis, then horizontal translation d units to the left.

d A horizontal translation d units to the left, followed by a horizontal stretch with scale factor q.

7 **a** The graph of $y = x^2$ is transformed through a horizontal stretch with scale factor 2, followed by a vertical stretch with scale factor 4. Find the equation of the resulting graph. Can you explain why this is the case?

b The graph of $y = e^x$ is transformed through a horizontal translation 2 units to the left, followed by a vertical stretch with scale factor q. The equation of the resulting graph is again $y = e^x$. Find the value of q.

c The graph of $y = \ln x$ is translated 2 units up. What transformation (other than a translation 2 units down) will return the graph to its original position?

Use technology to sketch the graphs and see why this is the case.

8 The diagram shows the graph of $y = f(x)$.

On separate axes sketch the graphs of:

a $y = f(2x) - 3$ **b** $y = 1 - 3f(x)$.

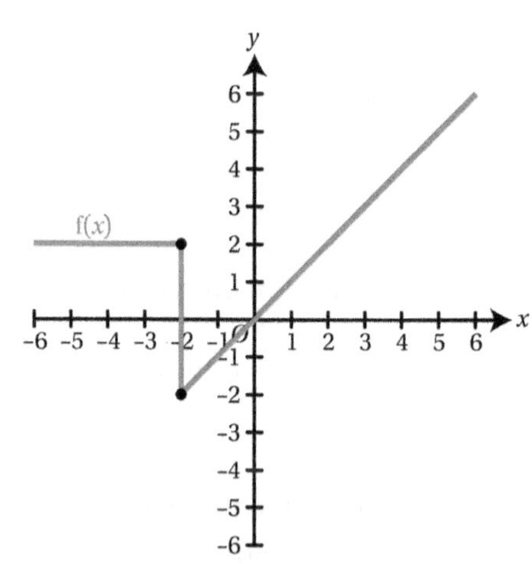

9 Sketch each graph.

a $y = \ln x$ **b** $y = 3\ln(x+2)$ **c** $y = \ln(2x-1)$

In each case, indicate clearly the positions of the vertical asymptote and the x-intercept.

10 The graph of $y = x^2 - 3x$ is translated 2 units to the right and then reflected in the x-axis. Find the equation of the resulting graph, in the form $y = ax^2 + bx + c$.

11 The graph of function $f(x) = ax + b$ is transformed by this sequence:

- translation by $\begin{pmatrix} 1 \\ 2 \end{pmatrix}$
- reflection in $y = 0$
- horizontal stretch with scale factor $\frac{1}{3}$.

The resultant function is $g(x) = 4 - 15x$.

Find the values of a and b.

12 The graph of function $f(x) = ax^2 + bx + c$ is transformed by this sequence:

- reflection through $x = 0$
- translation by $\begin{pmatrix} -1 \\ 3 \end{pmatrix}$
- horizontal stretch with scale factor 2.

The resultant function is $g(x) = 4x^2 + ax - 6$.

Find the values of a, b and c.

13 Given that $f(x) = 2^x + x$, give in simplest terms the formula for $h(x)$, which is obtained by transforming $f(x)$ by this sequence of transformations:

- vertical stretch, scale factor 8
- translation by $\begin{pmatrix} 1 \\ 4 \end{pmatrix}$
- horizontal stretch, scale factor $\frac{1}{2}$.

14 **a** Describe a sequence of two transformations that transforms the graph of $y = f(x)$ to the graph of $y = 3f\left(\frac{x}{2}\right)$.

b Sketch the graph of $y = 3\ln\left(\frac{x}{2}\right)$.

c Sketch the graph of $y = 3\ln\left(\frac{x}{2}+1\right)$ marking clearly the positions of any asymptotes and x-intercepts.

Section 2: The modulus function

The **modulus** (or **absolute value**) of a number is an operation that leaves positive numbers alone but makes negative numbers positive. You use $|x|$ to denote the modulus of number x; for example, $|5| = 5$ and $|-3| = 3$.

Key point 3.3

The modulus function can be defined as:

$$|x| = \begin{cases} x & \text{for } x \geqslant 0 \\ -x & \text{for } x < 0 \end{cases}$$

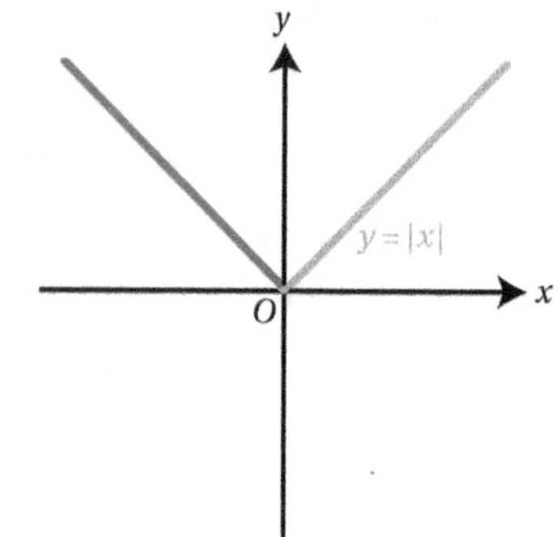

The graph of $y = |x|$ is shown. It is useful to call the red branch the **reflected branch** (as it is the reflection of the graph $y = x$ in the x-axis) and the blue branch the **unreflected branch.**

The domain of $|x|$ is all real numbers, whilst the range is all positive numbers and zero.

You can combine this with the rules for transforming graphs to sketch some other functions involving the modulus.

Did you know?

You can think of the modulus function as giving the distance of a number from the zero on the number line. A similar idea was applied to vectors in Student Book 1. But the distance between two objects is not always easy to define; for example, what length gives the 'distance' between two points on the surface of the Earth? This question leads to the idea of a metric. You might want to find out about the *Minkowski Metric* for finding the distance between two points in space–time in the Theory of Relativity.

WORKED EXAMPLE 3.5

Sketch the graphs of:

a $y=|x-3|$ **b** $y=|x+2|-5.$

In each case find the y-intercept of the graph.

a y-intercept: $|0-3|=|-3|=3$

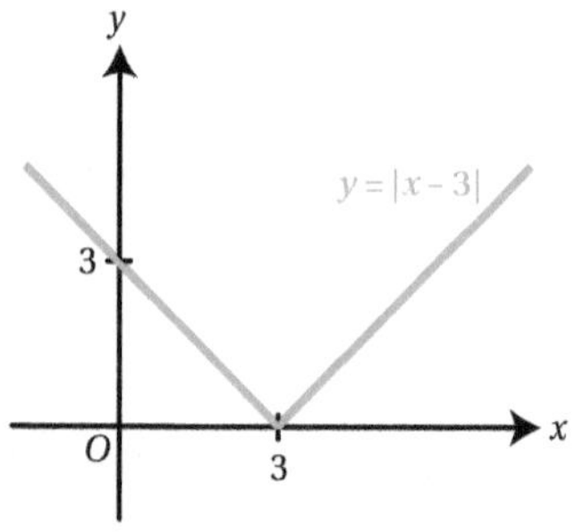

x has been replaced by $x-3$, so the graph of $y=|x|$ is shifted three units to the right.

b y-intercept: $|0+2|-5=2-5=-3$

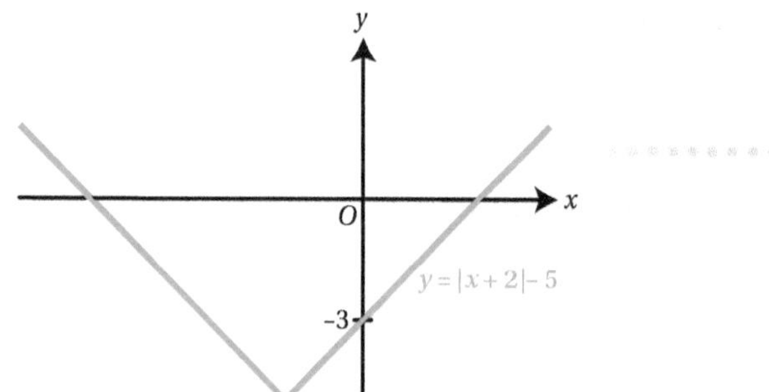

x has been replaced by $x+2$, so the graph of $y=|x|$ is shifted two units to the left.

Subtracting 5 then shifts the graph 5 units down.

You can also think about applying the modulus function to the graph of the function inside the modulus symbol. Any parts of the original graph that are below the x-axis will have their y-values changed from negative to positive so those parts of the graph will be reflected in the x-axis.

Tip

In practice it is usually easier to sketch modulus graphs by reflecting parts of the original graph that are below the x-axis rather than thinking of transforming $y = |x|$.

Key point 3.4

To sketch a graph of $y = |f(x)|$, start with the graph of $y = f(x)$ and reflect in the x-axis any parts that are below the x-axis.

WORKED EXAMPLE 3.6

Sketch the graph of $y = |5 - 2x|$, indicating the intercepts with the coordinate axes.

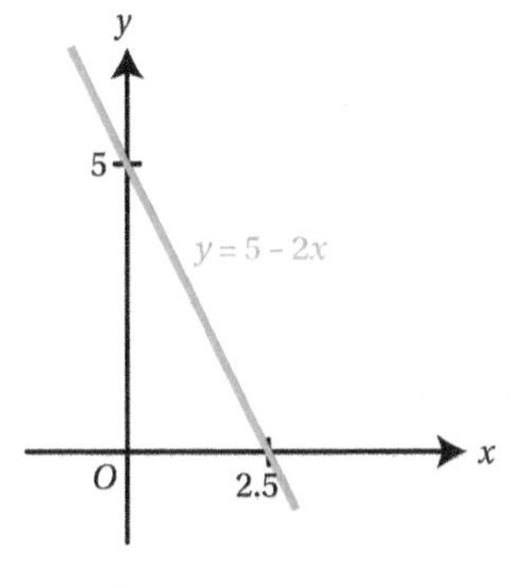

Start by drawing the graph of $y = 5 - 2x$.

This crosses the x-axis at $(2.5, 0)$ and the y-axis at $(0, 5)$.

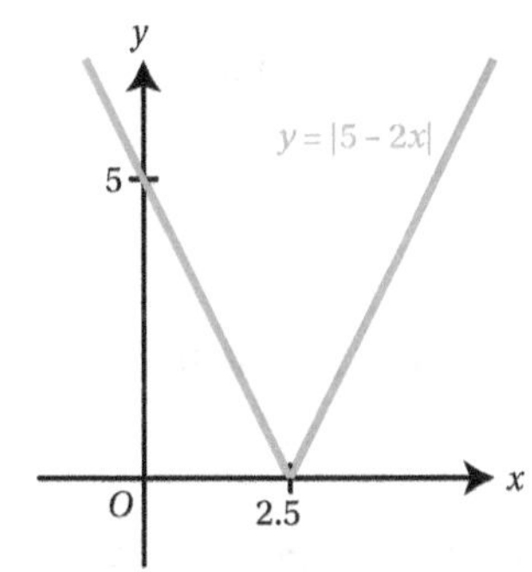

The part of the graph to the right of the x-intercept has negative y-values; taking the modulus will make those positive so this part of the graph needs to be reflected in the x-axis.

Using modulus notation in inequalities

The modulus function is often used as an alternative way of writing some inequalities; for example, you can write $-3 < x < 3$ concisely as $|x| < 3$.

You can extend this notation to write other inequalities that represent a single interval. For example, $|x - 5| < 3$ means $-3 < x - 5 < 3$, which can be rearranged into $2 < x < 8$.

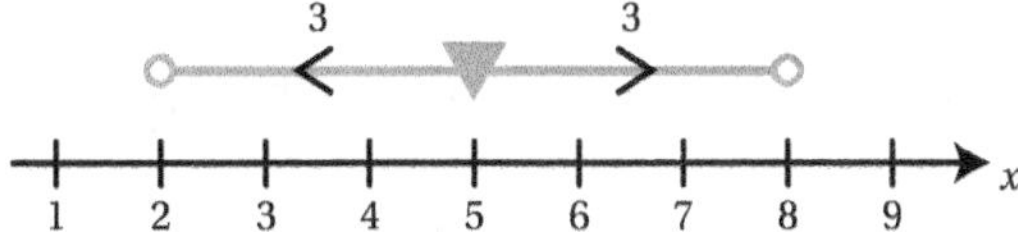

The number 5 is in the middle of the interval $(2, 8)$, 3 units away from each end. So the inequality $|x - 5| < 3$ can be read as: 'the distance between x and 5 is less than 3.'

Fast forward

You will use inequalities of this form with sequences (Chapter 4) and the binomial expansion (Chapter 6).

Rewind

You have already used the modulus of a vector to represent the distance between two points in Student Book 1, Chapter 15.

Similarly, the inequality $|x| > 3$ represents the region outside the interval $-3 < x < 3$, i.e. the region $x > 3$ or $x < -3$.

Key point 3.5

The modulus inequality:

- $|x-a| < b$ is equivalent to $-b < x-a < b$
- $|x-a| > b$ is equivalent to $x-a > b$ or $x-a < -b$

Fast forward

If you study Further Mathematics, you will extend the modulus notation to measure distances between points in the complex plane.

WORKED EXAMPLE 3.7

a Write the inequality:

i $|x+4| < 6$ in the form $p < x < q$

ii $|x-2| > 7$ in the form $x > p$ or $x < q$.

b Write the interval $[-3, 7]$ using an inequality of the form $|x-a| \leqslant b$.

a i $-6 < x+4 < 6$ — $x+4$ is between -6 and 6.

$-6-4 < x < 6-4$ — Rearrange to leave x on its own.

$-10 < x < 2$

ii $x-2 > 7$ or $x-2 < -7$ — $x-2$ is greater than 7 or less than -7.

$x > 7+2$ or $x < -7+2$ — Rearrange again.

$x > 9$ or $x < -5$

b $a = \frac{(-3)+7}{2} = 2$ — a is the number in the middle of the interval.

$b = 7-2 = 5$ — b is the distance from the middle of the interval to one end.

So $|x-2| \leqslant 5$. — The end-points of the interval are included.

EXERCISE 3B

1 Sketch each graph, showing the axes intercepts.

a **i** $y = |x-3|$ **ii** $y = |x+5|$ **b** **i** $y = |3x+5|$ **ii** $y = |2x-1|$

c **i** $y = |4-2x|$ **ii** $y = |-3x+2|$ **d** **i** $y = |x|+1$ **ii** $y = |x|-2$

2 Write the equation of each graph in the form $y = |ax + b|$.

a i

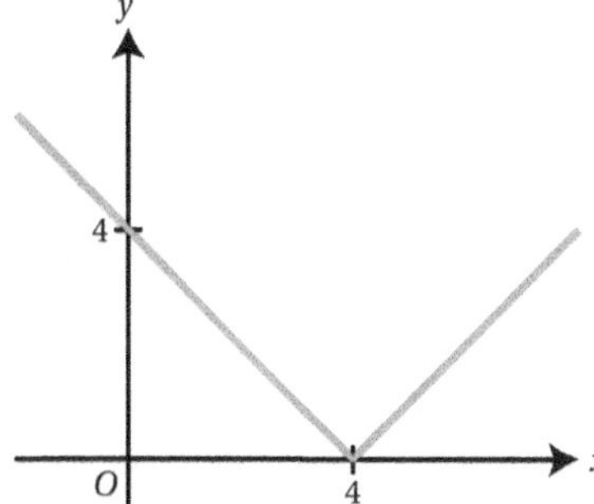

ii

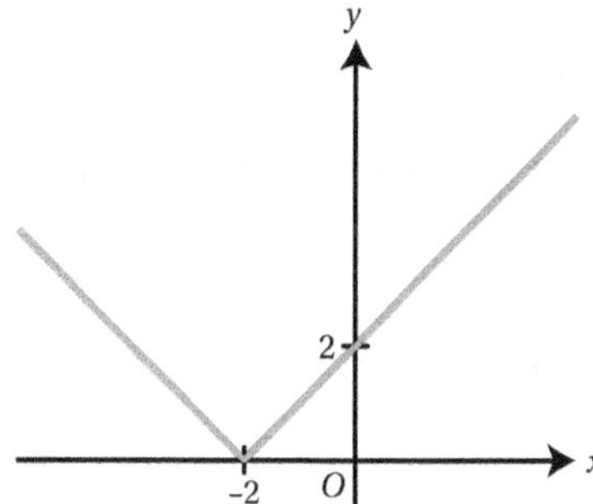

b i

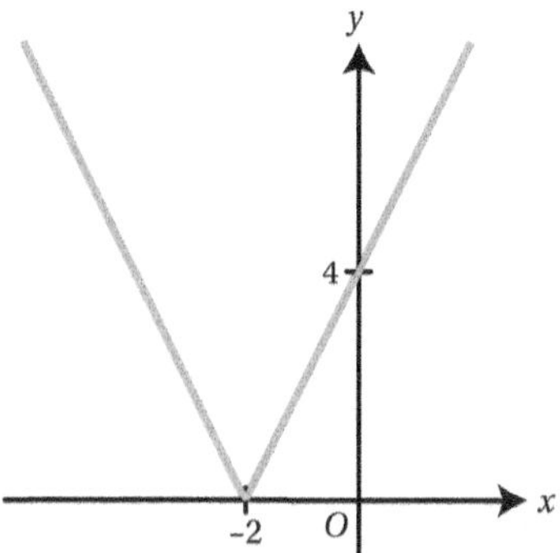

ii

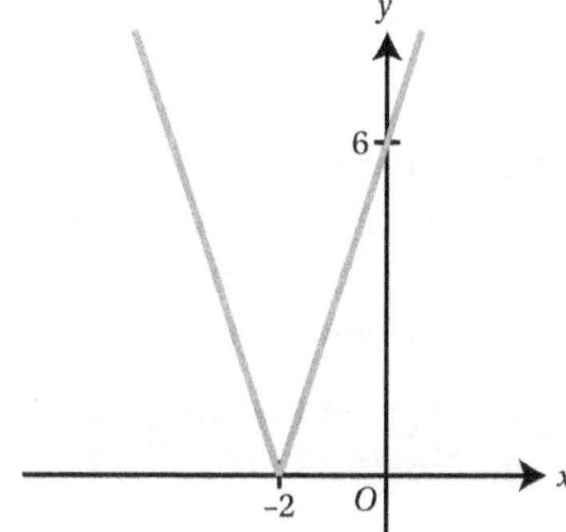

c i

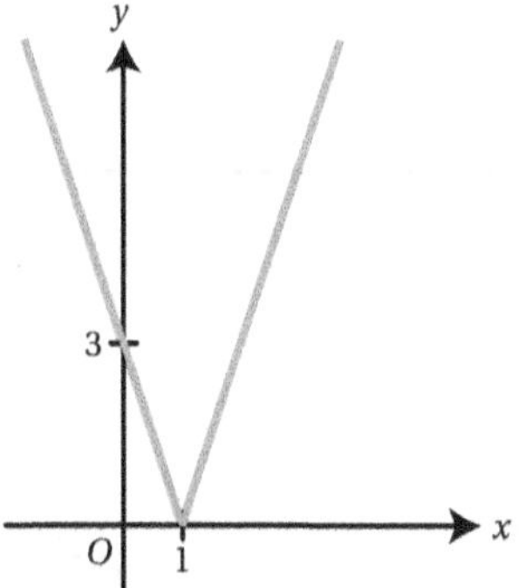

ii

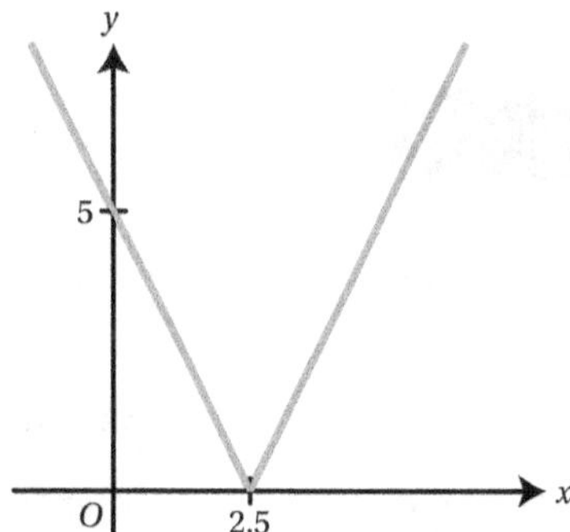

3 Write these inequalities in the form $a < x < b$.

a i $|x-5| < 8$ **ii** $|x-2| < 9$ **b i** $|x+1| < 5$ **ii** $|x+6| < 3$

4 Write these inequalities in the form $x > a$ or $x < b$.

a i $|x-2| > 4$ **ii** $|x-1| > 7$ **b i** $|x+5| > 11$ **ii** $|x+4| > 2$

5 Write these statements in the form $|x-a| < b$.

a i $x \in (5, 13)$ **ii** $x \in (11, 25)$ **b i** $x \in (-14, 10)$ **ii** $x \in (-16, 2)$

6 Write these statements in the form $|x-a| > b$.

a i $x \in (-\infty, 3) \cup (11, \infty)$ **ii** $x \in (-\infty, 5) \cup (19, \infty)$

b i $x \in (-\infty, -9) \cup (-3, \infty)$ **ii** $x \in (-\infty, -14) \cup (6, \infty)$

7 Sketch the graph of:

a $y=|x-2|$ b $y=|x-2|+3$

showing the coordinates of any axes intercepts.

8 Sketch the graph of $y=|2x+3|$, labelling any intercepts with the coordinate axes.

9 Sketch the graph of $y=5-|x-2|$. Give the coordinates of the points where the graph crosses the axes.

10 Sketch the graph of $y=x|x|$.

Section 3: Modulus equations and inequalities

You can use graphs to solve equations and inequalities involving the modulus function.

You need to use the graph to decide whether the intersection is on the reflected or the unreflected part of the graph. If it is on the unreflected part you can rewrite the equation without the modulus sign in. If it is on the reflected part you need to replace the modulus sign by a minus sign.

Key point 3.5

When solving an equation involving a modulus function, sketch the graph first.

WORKED EXAMPLE 3.8

Solve the equation $\frac{x}{2}=|x-1|$.

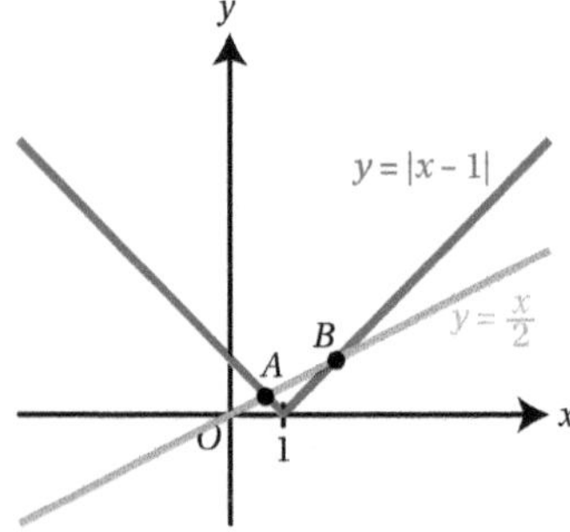

Sketch the graphs of $y=\frac{x}{2}$ and $y=|x-1|$.

There are two intersection points:

A: the blue graph intersects the reflected part of the red graph

B: the blue line intersects the unreflected part of the red graph.

Write a separate equation for each.

A: $\frac{x}{2}=-(x-1)$

For the reflected part, make the equation negative.

$\frac{3x}{2}=1$

$x=\frac{2}{3}$

B: $\frac{x}{2}=x-1$

For the unreflected part just remove the modulus sign.

$-\frac{x}{2}=-1$

$x=2$

So the solution is $x=\frac{2}{3}$ or 2

You can also solve the intersection of two modulus graphs.

WORKED EXAMPLE 3.9

Solve the equation $|x+1| = |2x-1|$.

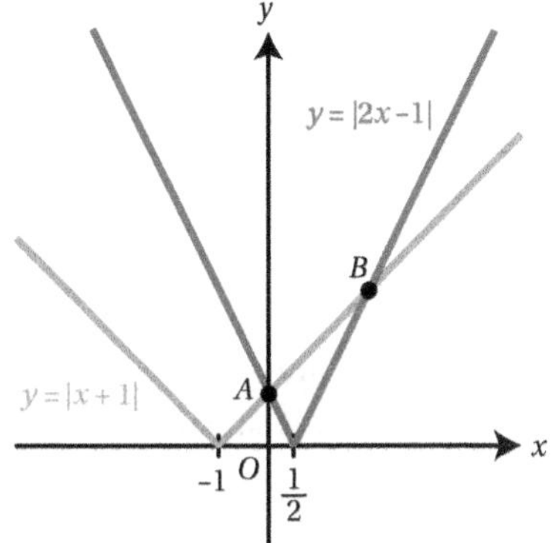

Sketch the graphs of $y=|x+1|$ and $y=|2x-1|$.

There are two intersections:

A: the unreflected blue line and the reflected red line

B: the unreflected blue line and the unreflected red line.

A: $x+1 = -(2x-1)$

$x+1 = -2x+1$

$3x = 0$

$x = 0$

You need the reflected part of the red graph, so make the equation negative.

B: $x+1 = 2x-1$

$x+1 = 2x-1$

$x = 2$

You need the unreflected parts for both graph, so remove the modulus signs.

The solution is $x = 0$ or 2.

WORK IT OUT 3.2

Solve the equation $|2x-1| = |3-x|$.

Which is the correct solution? Identify the errors made in the incorrect solutions.

Solution 1	Solution 2	Solution 3
$2x-1 = 3-x$ $3x = 4$ $x = \frac{4}{3}$ $2x-1 = -3+x$ $x = -2$ Solution: $x = \frac{4}{3}$ or -2	$2x-1 = 3-x$ $3x = 4$ $x = \frac{4}{3}$ $-(2x-1) = -(3-x)$ $-3x = -4$ $x = \frac{4}{3}$ Solution: $x = \frac{4}{3}$	$2x-1 = 3-x$ $x = -2$ $2x-1 = 3+x$ $x = 4$ Solution: $x = -2$ or 4

To solve inequalities, sketch the graphs and find their intersections. Then decide on which parts of the graph the inequality is satisfied.

Rewind

This is the same method that you use to solve quadratic inequalities – see Student Book 1, Chapter 3.

WORKED EXAMPLE 3.10

Solve $|2x-3| > x+4$

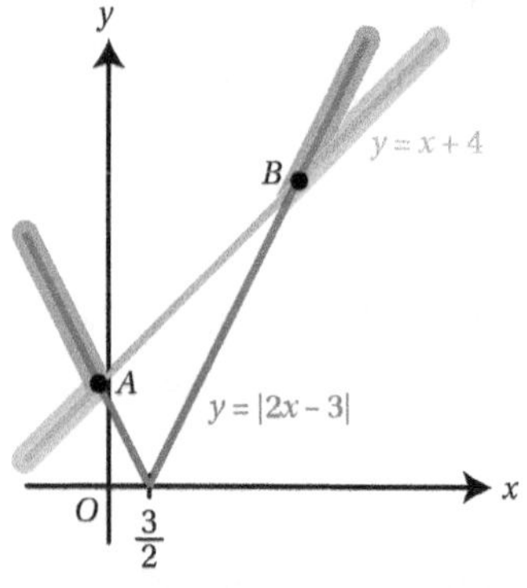

Sketch the graphs of $y = |2x-3|$ and $y = x+4$, highlighting where the inequality is satisfied.

A: $x+4 = -(2x-3)$

$x+4 = -2x+3$

$3x = -1$

$x = -\frac{1}{3}$

Find the intersection points.

A is on the reflected part of the red line.

B: $x+4 = 2x-3$

$x = 7$

B is on the unreflected part of the red line.

$x \in \left(-\infty, -\frac{1}{3}\right) \cup (7, \infty)$

Describe the highlighted region in terms of x. Remember that you can use either inequality notation or interval notation to do this.

EXERCISE 3C

1 Solve each equation.

a i $|x| = 4$ ii $|x| = 18$

b i $|2x-4| = 4$ ii $|3x+1| = 2$

c i $|2x+4| = 4-x$ ii $|5-2x| = x+3$

d i $|3x-4| = 8-x$ ii $|5+2x| = 3-2x$

e i $|3-2x| = |x+1|$ ii $|4+x| = |5-3x|$

f i $|6-x| = |5+x|$ ii $|4+3x| = |x|$

2 Solve each inequality. Write your answer using the stated notation.

a Use interval notation.

i $|x|>5$ ii $|x|>2$

b Use inequality notation.

i $|x|<3$ ii $|x|<10$

c Use interval notation.

i $|2x+1|\geqslant 4$ ii $|3x-2|\leqslant 3$

d Use inequality notation.

i $|2x-5|<x+1$ ii $|5-3x|>2x$

e Use interval notation.

i $|2x+1|>|x+4|$ ii $|3x-4|>|2x+1|$

f Use inequality notation.

i $|x+4|\geqslant|2x|$ ii $|1+3x|\leqslant|x+3|$

3 a Solve the equation $|4x+1|=x+3$.

b Solve the inequality $|4x+1|<x+3$.

4 a Solve the equation $|2x-5|=|x+2|$.

b Solve the inequality $|2x-5|\geqslant|x+2|$. Write your answer using interval notation.

Elevate

See Support Sheet 3 for a further example of modulus inequalities and for more practice questions.

5 a Sketch the graph of $y=|3x-7|$.

b Hence solve the inequality $|3x-7|<1-x$.

6 Solve the inequality $|3x+1|>2x$.

7 Given that $k>0$, find in terms of k the solution of the inequality $|x-k|\leqslant|2x-k|$.

8 Solve the equation $|x+k|=|x|+k$, where $k>0$.

Checklist of learning and understanding

- Two (or more) transformations of graphs can be combined: translations, stretches and reflections, both horizontal and vertical.
- The order in which transformations occur may affect the outcome.
  - One horizontal and one vertical transformation can be done in either order.
  - Changing the order of two horizontal or two vertical transformation affects the outcome.
  - For a function of the form $y=p\mathrm{f}(x)+c$ the stretch is performed before the translation.
  - For a function of the form $y=\mathrm{f}(qx+d)$ the translation is performed before the stretch.
- The modulus function can be used to reflect the part of the graph below the x-axis so that the whole graph is above it.
- To solve equations and inequalities involving the modulus function always use graphs. You need to decide whether the solutions are on the reflected or unreflected part of the graph.

Mixed practice 3

1 The graph of $y = f(x)$ is shown.

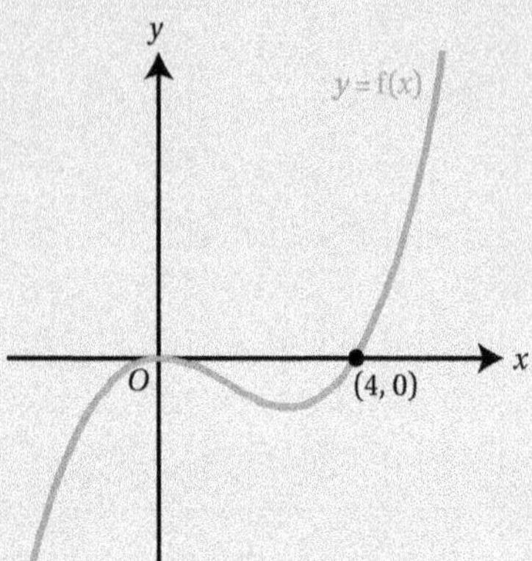

Sketch on separate diagrams the graphs of:

a $y = 3f(x-2)$ **b** $y = 3 - f(2x)$.

2 Solve the equation $|2x-3| = 5$.

Choose from these options.

A $x = 4$ or -4 **B** $x = 4$ or -1 **C** $x = 4$ **D** $x = -4$

3 The graph of $y = x^3 - 1$ is transformed by applying a translation with vector $\begin{pmatrix} 2 \\ 0 \end{pmatrix}$ followed by a vertical stretch with scale factor 2. Find the equation of the resulting graph in the form $y = ax^3 + bx^2 + cx + d$.

4 Find two transformations whose composition transforms the graph of $y = (x-1)^2$ to the graph of $y = 3(x+2)^2$.

5 **a** On the same set of axes sketch the graphs of $y = x$ and $y = |2x-1|$.

b Hence solve the inequality $|2x-1| < x$.

6 The diagram shows the graphs of $y = |2x-3|$ and $y = |x|$.

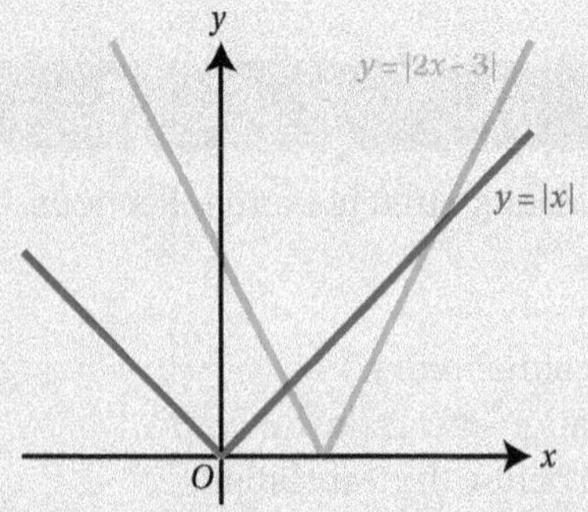

a Find the x-coordinates of the points of intersection of the graphs of $y = |2x-3|$ and $y = |x|$.

b Hence, or otherwise, solve the inequality $|2x-3| \geqslant |x|$.

[© AQA 2013]

7 Describe a sequence of two transformations that map the graph of $y = f(x)$ onto the graph of $y = f(3x - 2)$.

Choose from these options.

A Horizontal stretch, scale factor 3 followed by translation left by 2.

B Horizontal stretch, scale factor $\frac{1}{3}$ followed by translation right by 2.

C Translation left by 2 followed by horizontal stretch, scale factor 3.

D Translation right by 2 followed by horizontal stretch, scale factor $\frac{1}{3}$.

8 Solve the inequality $|2x - 1| > |x - 6|$.

9 **a** Describe two transformations that transform the graph of $y = x^2$ to the graph of $y = 3x^2 - 12x + 12$.

b Describe two transformations that transform the graph of $y = x^2 + 6x - 1$ to the graph of $y = x^2$.

c Hence describe a sequence of transformations that transforms the graph of $y = x^2 + 6x - 1$ to the graph of $y = 3x^2 - 12x + 12$.

10 **a** State the sequence of three transformations that transforms the graph of $y = |x|$ to the graph of $y = 5 - 3|x|$. Hence sketch the graph of $y = 5 - 3|x|$.

b Solve the equation $|2x - 1| = 5 - 3|x|$.

c Write down the solution of the inequality $|2x - 1| \leqslant 5 - 3|x|$.

11 **a** Describe a transformation that transforms the graph of $y = f(x)$ to the graph of $y = f(x + 2)$.

b Sketch on the same diagram the graphs of:

i $y = \ln(x + 2)$ **ii** $y = 3 - \ln(x + 2)$.

Mark clearly any asymptotes and x-intercepts on your sketches.

c The graph of the function $y = g(x)$ has been translated and then reflected in the x-axis to produce the graph of $y = h(x)$.

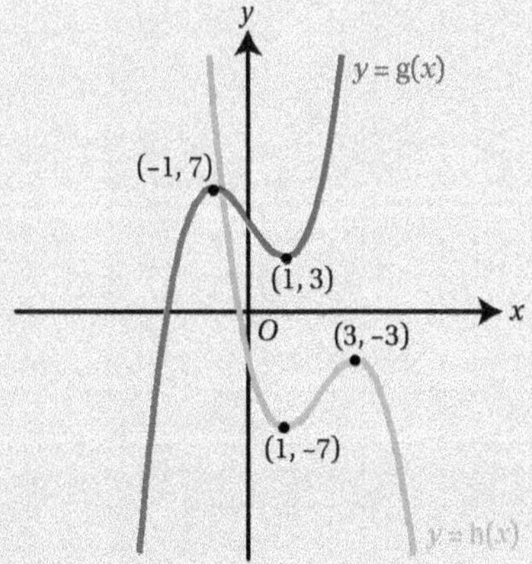

i State the translation vector.

ii If $g(x) = x^3 - 3x + 5$ find constants a, b, c, d such that $h(x) = ax^3 + bx^2 + cx + d$.

  **a** Describe a sequence of geometrical transformations that maps the graph of $y = \ln x$ onto the graph of $y = 4\ln(x+1) - 2$.

b Find the exact values of the coordinates of the points where the graph of $y = 4\ln(x+1) - 2$ crosses the coordinate axes.

[© AQA 2008]

13 Solve the equation $x|x| = x^2$.

14 Sketch the graph of $y = x + |x|$.

 **Elevate**

See Extension Sheet 3 for more challenging questions on modulus graphs and equations.

4 Sequences and series

In this chapter you will learn how to:

- determine the behaviour of some sequences
- use a sigma notation for series
- work with sequences with a constant difference between terms
- work with finite series with a constant difference between terms
- work with sequences with a constant ratio between terms
- work with finite and infinite series with a constant ratio between terms
- apply sequences to real-life problems.

Before you start...

GCSE	You should be able to find the formula for the *n*th term of a linear sequence.	1 Find the formula for the *n*th term of each sequence. a 2, 5, 8, 11, ... b 15, 11, 7, 3, ...
GCSE	You should be able to use term-to-term rules to generate sequences.	2 Find the second and third terms of the sequence defined by: $u_{n+1} = 3u_n - 2, u_1 = 4$
GCSE	You should be able to solve linear simultaneous equations.	3 Solve the simultaneous equations: $a + 4b = 8$ $3a + 5b = 3$
Student Book 1, Chapter 3	You should be able to solve quadratic equations and inequalities.	4 Find the smallest positive integer that satisfies the inequality $3x^2 + 7x > 163$.
Student Book 1, Chapter 7	You should be able to solve exponential equations and inequalities.	5 Find the smallest integer value of *n* such that $3.5 \times 1.2^n > 75$.
Chapter 3	You should be able to use modulus notation.	6 List all integers *r* that satisfy $\left\lvert \frac{3r}{5} \right\rvert < 2$.

Modelling with sequences

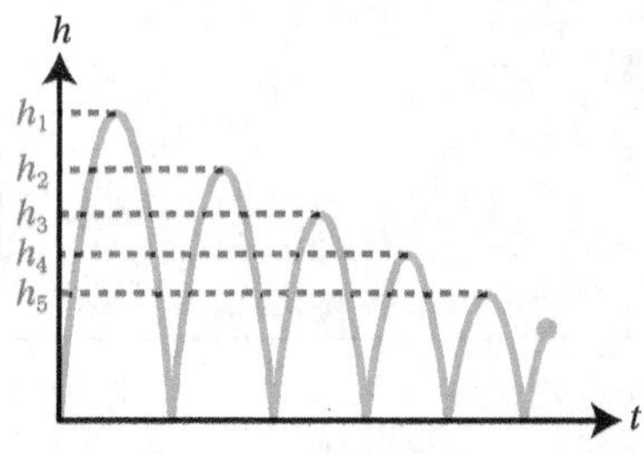

When you drop a ball, it will bounce a little lower each time it hits the ground. The heights the ball reaches after each bounce form a sequence. Although the idea of a sequence may just seem to be about abstract number patterns, it has a remarkable number of applications in the real world – from calculating mortgages to estimating harvests on farms.

Section 1: General sequences

You can describe sequence such as 2, 5, 8, 11, 14, ... in two different ways:

- by a **term-to-term rule:** $u_1 = 2, u_{n+1} = u_n + 3$
- by a **position-to-term rule** (also called the formula for the nth term): $u_n = 3n - 1$.

Gateway to A level

For a reminder of how these two types of rules work, see Gateway to A Level Section X. There is also further practice at the beginning of Exercise 4A.

Tip

u_n is the standard notation for the nth term of a sequence. So in the sequence 2, 5, 8, 11, 14, ..., $u_1 = 2$, $u_2 = 5$, $u_3 = 8$ etc.

Did you know?

Term-to-term rules can involve more than one previous term. You may already know the example of the famous Fibonacci sequence, $u_{n+2} = u_{n+1} + u_n$ with $u_1 = u_2 = 1$. This is based on a model Leonardo Fibonacci made for the breeding of rabbits, and has found many applications from describing the pattern made by the scales on pine cones to a proof of the infinity of prime numbers. There is also a beautiful link to the golden ratio: $\frac{1 \pm \sqrt{5}}{2}$.

Increasing, decreasing and periodic sequences

Sequences can behave in many different ways but you need to be aware of three possibilities.

Key point 4.1

- An **increasing sequence** is one where each term is larger than the previous one: $u_{n+1} > u_n$ for all n.
- A **decreasing sequence** is one where each term is smaller than the previous one: $u_{n+1} < u_n$ for all n.
- A **periodic sequence** in one where the terms start repeating after a while: $u_{n+k} = u_n$ for some number k (the period of the sequence).

WORKED EXAMPLE 4.1

Find the first five terms of each sequence, and describe the behaviour of the sequence.

a $x_{n+1} = 5x_n - 1, x_1 = 2$ **b** $u_n = \frac{16}{2^n}$ **c** $a_{n+1} = \frac{1}{a_n}, a_1 = 3$

a $x_1 = 2$ — Substitute each term into the formula to get the next term.

$x_2 = 10 - 1 = 9$

$x_3 = 45 - 1 = 44$

$x_4 = 220 - 1 = 219$

$x_5 = 1095 - 1 = 1094$

The sequence is increasing. — The terms are clearly getting larger.

b $u_1 = \frac{16}{2^1} = 8$ — Substitute $n = 1, 2, 3 \ldots$ to calculate the terms.

$u_2 = \frac{16}{2^2} = 4$

$u_3 = \frac{16}{2^3} = 2$

$u_4 = \frac{16}{2^4} = 1$

$u_5 = \frac{16}{2^5} = \frac{1}{2}$

The sequence is decreasing. — The terms are halving each time so are getting smaller.

c $a_1 = 3$ — Substitute each term into the formula to get the next term.

$a_2 = \frac{1}{3}$

$a_3 = \frac{1}{\frac{1}{3}} = 3$

$a_4 = \frac{1}{3}$

$a_5 = 3$

The sequence is periodic, with period 2. — The sequence starts to repeat after two terms: $u_3 = u_1$

In some cases, you might need to provide a proof that a sequence is increasing or decreasing, rather than just listing a few terms and making a conclusion.

Focus on...

Focus on ... Problem solving 1 looks further at the technique of using small cases to establish patterns in sequences.

Key point 4.2

To prove that a sequence is increasing or decreasing, find $u_{n+1} - u_n$.

- If $u_{n+1} - u_n > 0$ then $u_{n+1} > u_n$ and the sequence is increasing,
- If $u_{n+1} - u_n < 0$ then $u_{n+1} < u_n$ and the sequence is decreasing.

WORKED EXAMPLE 4.2

The nth term of a sequence is defined by $u_n = \frac{n+1}{2^n}$.

Prove that the sequence is decreasing.

$u_{n+1} - u_n = \frac{(n+1)+1}{2^{n+1}} - \frac{n+1}{2^n}$

You need to show that $u_{n+1} - u_n < 0$. Create a common denominator and combine the fractions.

$= \frac{n+2}{2^{n+1}} - \frac{2(n+1)}{2 \times 2^n}$

$= \frac{n+2}{2^{n+1}} - \frac{2n+2}{2^{n+1}}$

$= \frac{-n}{2^{n+1}} < 0$ for all $n \geq 1$

Since $2^{n+1} > 0$ and $-n < 0$, the fraction is negative.

$\therefore u_n$ is decreasing.

The limit of a sequence

In Worked example 4.1, the sequence in part **b** decreases but the terms remain positive although they get closer and closer to zero. The sequence **converges** to zero. By contrast, the sequence in part **a** **diverges** – the terms increase without a limit.

You can use your calculator to investigate the long-term behaviour of sequences, by generating a large number of terms until you can see what is going on. But you also need to be able to show that a convergent sequence has a particular limit, by solving an equation.

If a sequence is converging, then, in the limit as $n \to \infty$, consecutive terms u_n and u_{n+1} will be the same (and equal to the limit).

Fast forward

You will see in Chapter 14 how you can use convergent sequences to solve some equations.

Tip

Use the table function on your calculator if the sequence is given by the nth term formula, or the ANS button if you have a term-to-term rule.

Key point 4.3

To find the limit, L, of a convergent sequence defined by a term-to-term rule, set $u_{n+1} = u_n = L$ and solve for L.

WORKED EXAMPLE 4.3

A convergent sequence is defined by $u_{n+1} = \frac{2}{3}u_n + 5, u_1 = 6$.

By setting up and solving an equation, find the limit, L, as $n \to \infty$.

Let $u_{n+1} = u_n = L$

Replace u_{n+1} and u_n by L.

$L = \frac{2}{3}L + 5$

Solve the equation for L.

$\frac{1}{3}L = 5$

$L = 15$

The limit as $n \to \infty$ is 15.

EXERCISE 4A

1 Write out the first five terms of each sequence, defined by a term-to-term rule.

a i $u_{n+1} = u_n + 5, u_1 = 3.1$ ii $u_{n+1} = u_n - 3.8, u_1 = 10$

b i $u_{n+1} = 3u_n + 1, u_1 = 0$ ii $u_{n+1} = 9u_n - 10, u_1 = 1$

c i $u_{n+1} = \frac{u_n + 2}{u_n}, u_1 = 1$ ii $u_{n+1} = -\frac{3}{u_n}, u_1 = 3$

d i $u_{n+1} = u_n + 4, u_4 = 12$ ii $u_{n+1} = u_n - 2, u_6 = 3$

2 Write out the first five terms of each sequence, defined by an nth term formula.

a i $u_n = 3n + 2$ ii $u_n = 1.5n - 6$ b i $u_n = n^3 - 1$ ii $u_n = 5n^2$

c i $u_n = 3^n$ ii $u_n = 8 \times (0.5)^n$ d i $u_n = n^n$ ii $u_n = \sin(90n)$

3 Using your calculator, determine the behaviour of each sequence.

a i $x_{n+1} = 0.8x_n + 2, x_1 = 5$ ii $x_{n+1} = 1.6x_n - 4, x_1 = 5$

b i $u_{n+1} = -\frac{4}{u_n}, u_1 = 2$ ii $u_{n+1} = 10 - u_n, u_1 = 3$

4 A sequence is defined by $u_n = n2^n$.

a Write down u_1 and u_2.

b Show that $\frac{u_{n+1}}{u_n} = a + \frac{a}{n}$ for a constant a. Work out the value of a.

5 A sequence is defined by $x_{n+1} = \frac{x_n + 6}{x_n}$ with $x_1 = 2$.

a Find the fifth term of the sequence. b Describe the long-term behaviour of the sequence.

6 A sequence is defined by $u_{n+1} = ku_n + 9, u_1 = 5$, where k is a constant.

The limit of the sequence as $n \to \infty$ is 15.

Find the value of k.

7 a A sequence is defined by $x_1 = 1, x_{n+1} = ax_n + x_n^2$ where a is a constant.

Show that $x_3 = 2a^2 + 3a + 1$.

b Given that $x_3 = 3$ and that the sequence is increasing, find the value of a.

8 A sequence is defined by $u_{n+1} = \frac{6u_n + 2}{4 - 13u_n}, u_1 = 0$.

a Find u_2, u_3, u_4, u_5 and u_6. b State the value of u_{102}.

9 A sequence is defined by $u_{n+1} = \frac{u_n^2 + 1}{2}$.

a If u_1 is chosen so that the sequence converges, find the limit.

b Determine, explaining your reasoning fully, the range of value of u_1 for which the sequence converges.

Elevate

See Extension Sheet 4 for some more challenging questions on the long term behaviour of sequences and series.

10 The nth term of a sequence is defined by $u_n = \frac{n}{n+1}$.

Prove that the sequence is increasing.

Section 2: General series and sigma notation

If 5% interest is paid on money in a bank account each year, these interest payments form a sequence. While it may be useful to know how much interest is paid in each year, you might be even more interested to know how much will be accumulated altogether. This is one of many examples of a situation where you might want to sum a sequence. The sum of a sequence up to a certain point is called a **series**. The symbol S_n is often used to denote the sum of the first n terms of a sequence; that is:

$$S_n = u_1 + u_2 + u_3 + \ldots + u_n$$

Instead of seeing $S_n = u_1 + u_2 + u_3 + \ldots + u_n$, you will often see exactly the same expression in a shorter form, using **sigma notation**.

Key point 4.4

Sigma notation is a shorthand way to describe a series:

$$\sum_{r=1}^{r=n} \mathrm{f}(r) = \mathrm{f}(1) + \mathrm{f}(2) + \ldots + \mathrm{f}(n)$$

where the Greek capital sigma means 'add up'.

- r is a placeholder – it shows what changes with each new term.
- $r = 1$ is the first value taken by r – where counting starts.
- $r = n$ is the last value taken by r – where counting ends.

Tip

Don't be intimidated by this notation. If you are in doubt, try writing out the first few terms longhand.

There is nothing special about the letter r here; you could use any letter but r and k are the most usual. You may also see the r or k being missed out above and below the sigma.

WORKED EXAMPLE 4.4

$S_n = \sum_{2}^{n} r^2$ Find the value of S_4.

$S_4 = 2^2 + 3^2 + 4^2$ — Put the starting value, $r = 2$, into the expression to be summed, r^2. The end value has not been reached, so put in $r = 3$. The end value has still not been reached, so put in $r = 4$.

$= 4 + 9 + 16$ — The end value has been reached, so stop and evaluate.

$= 29$

WORKED EXAMPLE 4.5

Write the series $\frac{1}{2} + \frac{1}{3} + \frac{1}{4} + \frac{1}{5} + \frac{1}{6}$ in sigma notation.

General term $= \frac{1}{r}$ — Describe the general term of the series in the variable r.

Starts at $r = 2$ — Note the first value of r.

Ends at $r = 6$ — Note the final value of r.

$\frac{1}{2} + \frac{1}{3} + \frac{1}{4} + \frac{1}{5} + \frac{1}{6} = \sum_{2}^{6} \frac{1}{r}$ — Summarise in sigma notation.

EXERCISE 4B

1 Evaluate each expression.

a **i** $\sum_{2}^{4} 3r$ **ii** $\sum_{5}^{7} (2r+1)$ **b** **i** $\sum_{3}^{6} (2^r - 1)$ **ii** $\sum_{-1}^{4} 1.5^r$

c **i** $\sum_{a=1}^{a=4} b(a+1)$ **ii** $\sum_{q=-3}^{q=2} pq^2$

2 Write each expression in sigma notation. Be aware that there is more than one correct answer.

a **i** $2+3+4+\ldots+43$ **ii** $6+8+10+\ldots+60$

b **i** $\frac{1}{4}+\frac{1}{8}+\frac{1}{16}+\ldots+\frac{1}{128}$ **ii** $2+\frac{2}{3}+\frac{2}{9}+\ldots+\frac{2}{243}$

c **i** $14a+21a+28a+\ldots+70a$ **ii** $0+1+2^b+3^b+\ldots+19^b$

3 A sequence is defined by $u_1 = 1, u_{n+1} = u_n^2 - 2u_n + 3$.

Find $\sum_{r=1}^{5} u_r$.

4 A sequence is defined by $u_1 = k, u_{n+1} = 3u_n - 2$ where k is a constant.

a Show that $u_3 = 9k - 8$. **b** Given that $\sum_{r=1}^{4} u_r = 24$, find the value of k.

5 Find the exact value of $\sum_{k=1}^{5} \ln(3^k)$, giving your answer in the form $a \ln 3$.

6 **a** Find the exact value of $\sum_{k=1}^{6} \sin\left(\frac{k\pi}{3}\right)$. **b** Hence find the exact value of $\sum_{k=1}^{92} \sin\left(\frac{k\pi}{3}\right)$.

Section 3: Arithmetic sequences

Arithmetic sequences have a common difference between consecutive terms: to get from one term to the next you add the common difference (which may be negative). For example:

- $1, 5, 9, 13, 17, \ldots$ Add 4 to each term.
- $20, 17, 14, 11, \ldots$ Subtract 3 from each term.

In general, if the first term is a and the common difference is d, then

1st term	2nd term	3rd term	4th term	5th term
a	$a+d$	$a+2d$	$a+3d$	$a+4d$

This shows how to form the nth term of an arithmetic sequence.

Key point 4.5

The nth term of an arithmetic sequence with first term a and common difference d is:

$$u_n = a + (n-1)d$$

Common error

The formula for the nth term involves $(n-1)d$ and not nd. So, for example, $u_8 = a + 7d$ and not $u_8 = a + 8d$.

You should already be familiar with these sequences and know how to find a formula for the nth term, although you might have used a slightly different method in the past.

Gateway to A level

For practice at identifying arithmetic sequences and finding their nth term, see Gateway to A Level Section X.

WORKED EXAMPLE 4.6

An arithmetic progression has first term 5 and common difference 7.

Find the term with the value 355.

$u_n = a + (n-1)d$ — You need to find n when $u_n = 355$.

$355 = 5 + 7(n-1)$ — Use $u_n = a + (n-1)d$ with $a = 5$ and $d = 7$.

$350 = 7(n-1)$

$50 = n - 1$

$n = 51$ — Solve the equation.

So 355 is the 51st term.

Tip

Arithmetic progression is just another way of saying arithmetic sequence.

Elevate

See Support Sheet 4 for a further example of using simultaneous equations to find a and d.

You will often need to set up simultaneous equations to find a and d.

WORKED EXAMPLE 4.7

The fifth term of an arithmetic sequence is 7 and the eighth term is 16.

a Find the first term, a, and common difference, d.
b Hence find the 100th term.

a $u_5 = a(5-1)d$ — Use $u_n = a + (n-1)d$ with $n = 5$.

$7 = a + 4d \quad (1)$ — You know that $u_5 = 7$.

$u_8 = a + 7d$ — Repeat for the eighth term.

$16 = a + 7d \quad (2)$

$(2) - (1)$:

$9 = 3d$ — Solve simultaneously.

$d = 3$

$\therefore a = -5$

$u_n = a + (n-1)d$

$= -5 + (n-1) \times 3$ — Now form the nth term formula. Note that you could tidy up by expanding the brackets and simplifying: $u_n = 3n - 8$. However, there is no need as you are not asked to state the formula at all.

b $u_{100} = -5 + (100-1) \times 3$ — Substitute $n = 100$.

$= 292$

WORKED EXAMPLE 4.8

The first three terms of an arithmetic sequence are 12, x^2, $5x$.

Find the possible values of x.

$u_3 - u_2 = u_2 - u_1$ — The difference between terms must be constant ($= d$).

$5x - x^2 = x^2 - 12$

$2x^2 - 5x - 12 = 0$ — Rearrange and solve the quadratic equation.

$(2x+3)(x-4) = 0$

$x = -\frac{3}{2}$ or 4

EXERCISE 4C

1 Find the general formula for the arithmetic sequence, given these conditions.

a **i** first term 9, common difference 3 **ii** first term 57, common difference 0.2

b **i** first term 12, common difference −1 **ii** first term 18, common difference $-\frac{1}{2}$

c **i** first term 1, second term 4 **ii** first term 9, second term 19

d **i** first term 4, second term 0 **ii** first term 27, second term 20

e **i** third term 5, eighth term 60 **ii** fifth term 8, eighth term 38

2 How many terms are there in each sequence?

a **i** $1, 3, 5, \ldots, 65$ **ii** $18, 13, 8, \ldots, -122$

b **i** first term 8, common difference 9, last term 899
ii first term 0, ninth term 16, last term 450

3 An arithmetic sequence has 5 and 13 as its first two terms, respectively.

a Write down, in terms of n, an expression for the nth term, u_n.

b Find the number of terms of the sequence that are less than 400.

4 The 10th term of an arithmetic sequence is 61 and the 13th term is 79.

Find the value of the 20th term.

5 The eighth term of an arithmetic sequence is 74 and the 15th term is 137.

Which term has the value 227?

6 The heights of the rungs above ground in a ladder forms an arithmetic sequence. The third rung is 70 cm above the ground and the 10th rung is 210 cm above the ground. Given that the top rung is 350 cm above the ground, how many rungs does the ladder have?

7 The first four terms of an arithmetic sequence are $2, a-b, 2a+b+7$, and $a-3b$, where a and b are constants. Find a and b.

8 A book starts at page 1 and is numbered on every page.

a Show that the page numbers for the first 11 pages contain 13 digits.

b If the total number of digits used for page numbers is 1260, how many pages are there in the book?

Section 4: Arithmetic series

When you add up terms of an arithmetic sequence you get an **arithmetic series.** Just as it was useful to have a formula for the nth term, so it is useful to have a formula that will find the sum of the first n terms. There are two different versions of this.

Key point 4.6

For an arithmetic series with first term a and common difference d:

$$S_n = \frac{n}{2}[2a+(n-1)d]$$

or

$$S_n = \frac{n}{2}(a+l)$$

where l is the last term of the series.

These will be given in your formula book.

Focus on...

The first formula is proved in Focus on ... Proof 1.

The second formula follows immediately from the first, since the last term, l, is just the nth term of the arithmetic sequence:

$$\begin{aligned} S_n &= \frac{n}{2}[2a+(n-1)d] \\ &= \frac{n}{2}[a+a+(n-1)d] \\ &= \frac{n}{2}[a+l] \end{aligned}$$

You will need the first formula when you know or want d, and the second when you know or want l.

WORKED EXAMPLE 4.9

Find the sum of the first 30 terms of an arithmetic progression with first term 8 and common difference 0.5.

$S_{30} = \frac{30}{2}[2\times 8+(30-1)0.5]$ — Use $S_n = \frac{n}{2}[2a+(n-1)d]$ with $n=30, a=8, d=0.5$

$= 457.5$

You must be able to work backwards too and find how many terms there are in the series, given the sum of the series. Remember that the number of terms can only be a positive integer.

WORKED EXAMPLE 4.10

An arithmetic sequence has first term 5 and common difference 10. The sum of the first n terms is 720.

Find the value of n.

$S_n = \frac{n}{2}[2a + (n-1)d]$ — Use $S_n = \frac{n}{2}[2a+(n-1)d]$ with $S_n = 720, a = 5, d = 10$

$720 = \frac{n}{2}[2\times 5 + (n-1)\times 10]$

$720 = \frac{n}{2}(10 + 10n - 10)$ — Simplify and then solve the equation.

$720 = 5n^2$

$n^2 = 144$

$n = \pm 12$

$\therefore n = 12$ — n must be positive.

Sometimes you have to interpret the question carefully to see that it is about an arithmetic series.

WORKED EXAMPLE 4.11

Find the sum of all the multiples of 3 between 100 and 1000.

$S = 102 + 105 + 108 + \ldots + 999$ — Write out the first few terms to see what is happening.

This is an arithmetic series with $a = 102$ and $d = 3$.

$u_n = a + (n-1)d$

$999 = 102 + 3(n-1)$ — You need to know how many terms there are in this series, so set $u_n = 999$ in the formula $u_n = a + (n-1)d$ and solve for n.

$897 = 3(n-1)$

$n = 300$

$S_{300} = \frac{300}{2}(102 + 999)$ — Now use $S_n = \frac{n}{2}(a+l)$, with $n = 300, a = 102, l = 999$.

$= 165\,150$

EXERCISE 4D

1 Find the sum of each arithmetic sequence.

a **i** $12, 33, 54, \ldots$ (17 terms) **ii** $-100, -85, -70, \ldots$ (23 terms)

b **i** $3, 15, \ldots, 459$ **ii** $2, 11, \ldots, 650$

c **i** $28, 23, \ldots, -52$ **ii** $100, 97, \ldots, 40$

d **i** $15, 15.5, \ldots, 29.5$ **ii** $\frac{1}{12}, \frac{1}{6}, \ldots, 1.5$

2 An arithmetic sequence has first term 4 and common difference 8. How many terms are required to get a sum of:

a i 676 ii 4096 b i $x^2, x > 0$ ii $100x, x > 0$?

3 For the arithmetic series $2+5+8+\ldots$ find the value of n for which $S_n = 1365$.

4 The sum of the first n terms of a series is given by $S_n = 2n^2 - n$, where $n \in \mathbb{Z}^+$.

a Find the first three terms of the series.

b Find an expression for the nth term of the series.

5 The second term of an arithmetic sequence is 7. The sum of the first four terms is 12.

Find the first term, a, and the common difference, d, of the sequence.

6 The fourth term of an arithmetic sequence is 17. The sum of the first 20 terms is 990.

Find the first term, a, and the common difference, d, of the sequence.

7 The sum of the first three terms of an arithmetic sequence is −12, and the sum of the first 12 terms is 114.

Find the first term, a, and common difference, d.

8 Prove that the sum of the first n odd numbers is n^2.

9 Find the largest possible value of the sum of the arithmetic sequence 85, 78, 71, ...

10 Find the least number of terms for which the sum of the series $-6+1+8+15+\ldots$ is greater than 10 000.

11 The sum of the first n terms of an arithmetic sequence is $S_n = 3n^2 - 2n$. Find the nth term u_n.

12 A circular disc is cut into twelve sectors whose areas are in an arithmetic sequence. The angle of the largest sector is twice the angle of the smallest sector.

Find the size of the angle of the smallest sector.

13 a Find the sum of all multiples of 7 between 1 and 1000.

b Hence find the sum of all integers between 1 and 1000 that are not divisible by 7.

14 The ratio of the fifth term to the 12th term of a sequence in an arithmetic progression is $\frac{6}{13}$. If each term of this sequence is positive, and the product of the first term and the third term is 32, find the sum of the first 100 terms of this sequence.

15 a Find an expression for the sum of the first 20 terms of the series

$S_{20} = \ln x + \ln x^4 + \ln x^7 + \ln x^{10} + \ldots$

b Hence find the exact value of x for which $S_{20} = 2950$.

16 Find the sum of all three-digit numbers that are multiples of 14 but not 21.

Section 5: Geometric sequences

A **geometric sequence** has a common ratio between consecutive terms: to get from one term to the next you multiply by the common ratio. For example:

- $1, 2, 4, 8, 16, \ldots \times 2$
- $100, 50, 25, 12.5, 6.25, \ldots \times \frac{1}{2}$
- $1, -3, 27, -81, \ldots \times -3$

In general, if the first term is a and the common ratio is r, then:

1st term	2nd term	3rd term	4th term	5th term
a	ar	ar^2	ar^3	ar^4

From this you can see how to form the nth term of a geometric sequence.

Key point 4.7

The nth term of a geometric sequence with first term a and common ratio r is:

$$u_n = ar^{n-1}$$

Common error

The formula for the nth term involves r^{n-1} and not r^n.
So, for example, $u_8 = ar^7$ and not $u_8 = ar^8$.

WORKED EXAMPLE 4.12

For the geometric sequence $324, -108, 36, -12, \ldots$

a find a formula for the nth term

b hence find the 10th term of the sequence.

a $a = 324$

$r = \frac{-108}{324} = -\frac{1}{3}$

If the common ratio is not obvious, divide the second term by the first term (or u_3 by u_2, etc.) to find it.

$\therefore u_n = 324\left(-\frac{1}{3}\right)^{n-1}$

Now use $u_n = ar^{n-1}$.

b $u_{10} = 324\left(-\frac{1}{3}\right)^{10-1}$

Substitute in $n = 10$.

$= 324\left(-\frac{1}{3}\right)^{9}$

$= -\frac{4}{243}$

As with arithmetic sequences, you will often need to set up simultaneous equations.

Elevate

See Support Sheet 4 for a further example of using simultaneous equations to find a and r.

WORKED EXAMPLE 4.13

The seventh term of a geometric sequence is 13. The ninth term is 52.

Find the possible values of the common ratio.

$u_7 = ar^6$ — Use $u_n = ar^{n-1}$ with $n = 7$.

$13 = ar^6 \quad (1)$ — But you know that $u_7 = 13$.

$52 = ar^8 \quad (2)$ — Repeat for the ninth term.

$(2) \div (1)$: — Solve simultaneously. Note that dividing the equations is a quick way of eliminating a and solving for r.

$$\frac{52}{13} = \frac{ar^8}{ar^6}$$

$$4 = r^2$$

$$r = \pm 2$$

Tip

Notice that the question asked for values rather than value, so you should expect to get at least two answers.

WORKED EXAMPLE 4.14

Three consecutive terms of a geometric sequence with common ratio $r \neq 0$ are

$$x-2, x+2, 5x-2$$

a Find the possible values of x.

b For each value of x, find the common ratio of the sequence.

a $\frac{x+2}{x-2} = \frac{5x-2}{x+2}$ — The ratio between terms must be constant $(= r)$.

$(x+2)(x+2) = (5x-2)(x-2)$ — Rearrange and solve the quadratic.

$$x^2 + 4x + 4 = 5x^2 - 12x + 4$$

$$4x^2 - 16x = 0$$

$$x(x-4) = 0$$

$$x = 0 \text{ or } 4$$

b When $x = 0$,

$r = \frac{x+2}{x-2} = \frac{2}{-2} = -1$ — Find the ratio of the first two terms for each value of x in turn.

When $x = 4$,

$$r = \frac{4+2}{4-2} = \frac{6}{2} = 3$$

When you are asked which term satisfies a particular condition, you normally need to use logarithms.

Rewind

See Student Book 1, Chapter 7, if you need a reminder of how to solve exponential equations.

WORKED EXAMPLE 4.15

A geometric sequence has first term 10 and common ratio $\frac{1}{3}$.

What is the first term that is less than 10^{-6}?

$u_n < 10^{-6}$ — Express the condition for the nth term as an inequality.

$10\times\left(\frac{1}{3}\right)^{n-1} < 10^{-6}$ — Use $u_n = ar^{n-1}$ with $a = 10$ and $r = \frac{1}{3}$ The unknown is in the power so solve using logarithms.

$\left(\frac{1}{3}\right)^{n-1} < 10^{-7}$

$\ln\left(\frac{1}{3}\right)^{n-1} < \ln 10^{-7}$

$(n-1)\ln\left(\frac{1}{3}\right) < \ln 10^{-7}$

$(n-1) > \frac{\ln 10^{-7}}{\ln\left(\frac{1}{3}\right)}$ — $\ln\left(\frac{1}{3}\right)$ is negative so when you divide by it, you must remember to change the inequality sign.

$n-1 > 14.7$

$n > 15.7$ (3 s.f.)

$\therefore n = 16$ — n is an integer, so you need the first integer greater than 15.7.

EXERCISE 4E

1 Find an expression for the nth term of each geometric sequence.

a **i** $6, 12, 24, \ldots$ **ii** $12, 18, 27, \ldots$ **b** **i** $20, 5, 1.25, \ldots$ **ii** $1, \frac{1}{2}, \frac{1}{4}, \ldots$

c **i** $1, -2, 4, \ldots$ **ii** $5, -5, 5, \ldots$ **d** **i** $2, 2\sqrt{3}, 6, 6\sqrt{3}, \ldots$ **ii** $4, \frac{4}{\sqrt{2}}, 2, \frac{2}{\sqrt{2}}, \ldots$

e **i** $a, ax, ax^2, \ldots$ **ii** $3, 6x, 12x^2, \ldots$

2 How many terms are there in each geometric sequence?

a **i** $6, 12, 24, \ldots, 24\,576$ **ii** $20, 50, \ldots, 4882.8125$ **b** **i** $1, -3, \ldots, -19\,683$ **ii** $2, -4, 8, \ldots, -1024$

c **i** $\frac{1}{2}, \frac{1}{4}, \ldots, \frac{1}{1024}$ **ii** $3, 2, \frac{4}{3}, \ldots, \frac{128}{729}$

3 The second term of a geometric sequence is 6 and the fifth term is 162.

a Find a formula for the nth term of the sequence.

b Hence find the 10th term.

4 The third term of a geometric progression is 12 and the fifth term is 48.

a Find the possible values of the first term and the common ratio.

b Hence find the two possible values of the eighth term.

5 The first three terms of a geometric sequence are $a, a+14, 9a$.

a Find the possible values of a.

b In each case, find the tenth term of the sequence.

6 The third term of a geometric sequence is 112 and the sixth term is 7168.

Which term has the value 1 835 008?

7 A geometric sequence has first term 2 and common ratio −3. $u_n = -4374$.

a Show that $(n-1)\ln 3 = \ln 2187$.

b Hence find the value of n.

8 A sequence is defined by $u_{n+1} = \frac{3u_n}{2}, u_1 = 4$.

What is the first term of the sequence to exceed 1 million?

9 What is the first term of the sequence $\frac{2}{5}, \frac{4}{25}, \frac{8}{125}, \ldots$ to be less than 10^{-9}?

10 The difference between the fourth and the third terms of a geometric sequence equals one quarter of the difference between the second and first terms. Find the possible values of the common ratio.

11 The three terms $a, 1, b$ are in arithmetic progression. The three terms $1, a, b$ are in geometric progression. Find the values of a and b, given that $a \neq b$.

12 The sum of the first n terms of an arithmetic sequence is given by the formula $S_n = 4n^2 - 2n$. Three terms of this sequence, u_2, u_m and u_{32}, are consecutive terms in a geometric sequence. Find m.

Section 6: Geometric series

Just as for an arithmetic series, a **geometric series** is the sum of the terms in a geometric sequence.

Key point 4.8

The formula for the sum S_n of the first n terms in a geometric sequence is:

$$S_n = \frac{a(1-r^n)}{1-r}$$

or equivalently:

$$S_n = \frac{a(r^n-1)}{r-1}$$

The first formula will be given in your formula book.

Tip

In general, use the first of these formulae when the common ratio is less than one and the second when the common ratio is greater than 1. This avoids having to work with negative numbers.

Focus on...

This formula is proved in Focus on ... Proof 1.

WORKED EXAMPLE 4.16

Find the exact value of the sum of the first six terms of the geometric sequence with first term 8 and common difference $\frac{1}{2}$.

$$S_6 = \frac{8\left(1-\left(\frac{1}{2}\right)^6\right)}{1-\frac{1}{2}}$$

$a = 8, r = \frac{1}{2}$ and $n = 6$. Use the first sum formula as $r < 1$.

$$= \frac{8\left(1-\frac{1}{64}\right)}{\frac{1}{2}}$$

$$= 16\left(\frac{63}{64}\right)$$

$$= \frac{63}{4}$$

Again, you will need to use logarithms if you are asked to find the term number.

WORKED EXAMPLE 4.17

How many terms are needed for the sum of the geometric series $3+6+12+24+\ldots$ to exceed 100 000?

$a = 3$

$r = 2$

You need n, but you know a and r.

$$S_n > 100\,000$$

$$\frac{3(2^n-1)}{2-1} > 100\,000$$

Use the second sum formula ($r > 1$) and express the condition as an inequality.

$$3(2^n-1) > 100\,000$$

$$2^n > \frac{100\,003}{3}$$

$$\ln 2^n > \ln\left(\frac{100\,003}{3}\right)$$

$$n\ln 2 > \ln\left(\frac{100\,003}{3}\right)$$

$$n > \frac{\ln\left(\frac{100\,003}{3}\right)}{\ln 2}$$

$$n > 15.02$$

The unknown is in the power, so use logarithms to solve the inequality.

$\therefore n = 16$

But n is a whole number, so 16 terms are needed.

EXERCISE 4F

1 Find the sum of each geometric series. (There may be more than one possible answer!)

a i $7, 35, 175, \ldots$ (10 terms) ii $1152, 576, 288, \ldots$ (12 terms)

b i $16, 24, 36, \ldots, 182.25$ ii $1, 1.1, 1.21, \ldots, 1.771\,561$

c i first term 8, common ratio −3, last term 52 488

ii first term −6, common ratio −3, last term 13 122

d i third term 24, fifth term 6, 12 terms.

ii ninth term 50, thirteenth term 0.08, last term 0.0032.

2 Find the possible values of the common ratio if:

a i first term is 11, sum of the first two terms is 12.65 ii first term is 1, sum of the first two terms is 3.7

b i first term is 12, sum of the first three terms is 16.68 ii first term is 10, sum of the first three terms is 21.9.

3 The nth term, u_n, of a geometric sequence is given by $u_n = 3\times 5^{n+2}, n \in \mathbb{Z}^+$.

a Find the common ratio r.

b Hence, or otherwise, find an expression for S_n, the sum of the first n terms of this sequence.

4 A series is defined by $S = \sum_{k=1}^{10} 5\left(\frac{2}{3}\right)^k$.

a State the values of a and r. b Hence find the value of S to three significant figures.

5 The first term of a geometric sequence is 6 and the sum of the first three terms is 29. Find the common ratio.

6 The sum of the first three terms of a geometric sequence is $23\frac{3}{4}$ and the sum of the first four terms is $40\frac{5}{8}$. Find the first term and the common ratio.

7 The sum of the first four terms of a geometric sequence is 520, the sum of the first five terms is 844, and the sum of the first six terms is 1330.

a Find the common ratio. b Find the sum of the first two terms.

8 The fifth term of a geometric sequence is 128 and the sixth term is 512.

a Find the common ratio and first term.

b Find the smallest value of n so that the sum of the first n terms exceeds 500 000.

9 The sum of the first three terms of a geometric sequence with common ratio $r > 0$ is 147 and the sum of the first six terms is 9555.

a Show that $r^6 - 65r^3 + 64 = 0$. b Hence find the sum of the first 10 terms.

10 a Show that in a geometric sequence with common ratio r, the ratio of the sum of the first n terms to the sum of the next n is $1 : r^n$.

b The sum of the seventh term and four times the fifth term equals the eighth term.

Find the ratio of the sum of the first 10 terms to the sum of the next 10 terms.

Section 7: Infinite geometric series

If you keep adding together terms of any arithmetic sequence the sum increases (or decreases if d is negative) without limit. This can happen when you add the terms of a geometric sequence too, but there is also a possibility that the sum will converge.

Looking at the formula for the sum of a geometric sequence,

$$S_n = \frac{a(1-r^n)}{1-r}$$

you can see that the only part that is affected by making n bigger is r^n. When you raise most numbers to a large power the result gets bigger and bigger. The exception is when r is a number between -1 and 1; in this case, r^n gets smaller as n increases – in fact it tends to 0.

Key point 4.9

As n increases the sum of a geometric series converges to

$$S_\infty = \frac{a}{1-r} \text{ if } |r|<1$$

This is called the sum to infinity of the series.

This will be given in your formula book.

Tip

The condition that $|r|<1$ is just as important as the formula itself.

WORKED EXAMPLE 4.18

The sum to infinity of a geometric sequence is 5 and the second term is $-\frac{6}{5}$. Find the common ratio.

$5 = \frac{a}{1-r}$ (1) — Use $S_\infty = \frac{a}{1-r}$ with $S_\infty = 5$

$-\frac{6}{5} = ar$ (2) — Use $u_2 = ar$ with $u_2 = -\frac{6}{5}$

From (2): — Solve simultaneously.

$a = -\frac{6}{5r}$

Substituting into (1):

$$\frac{-6}{5r(1-r)} = 5$$

$$-6 = 25(r-r^2)$$

$$0 = 25r^2 - 25r - 6$$

$$0 = (5r-6)(5r+1)$$

$$r = \frac{6}{5} \text{ or } -\frac{1}{5}$$

But since the sum to infinity exists, $-1<r<1$ — Check that the series actually converges for the values of r found.

$\therefore r = -\frac{1}{5}$

Remember that some questions may focus on the condition for the sequence to converge as well as the value that it converges to.

WORKED EXAMPLE 4.19

The geometric series $(2-x)+(2-x)^2+(2-x)^3 \ldots$ converges. Find the range of possible values of x.

$r=(2-x)$ — Identify r.

Since the series converges:
$-1<2-x<1$ — Use the fact that the series converges.

$-3<-x<-1$
$1<x<3$ — Solve the inequality.

EXERCISE 4G

1. Find the value of each infinite geometric series, or state that it is divergent.

   a i $9+3+1+\frac{1}{3}\ldots$ ii $56+8+1\frac{1}{7}\ldots$

   b i $0.3+0.03+0.003\ldots$ ii $0.78+0.0078+0.000\,078\ldots$

   c i $0.01+0.02+0.04\ldots$ ii $\frac{19}{10\,000}+\frac{19}{1000}+\frac{19}{100}\ldots$

   d i $10-2+0.4\ldots$ ii $6-4+\frac{8}{3}\ldots$

   e i $10-40+160\ldots$ ii $4.2-3.36+2.688\ldots$

2. Find the values of x that allow each geometric series to converge.

   a i $9+9x+9x^2\ldots$ ii $-2-2x-2x^2$

   b i $1+3x+9x^2\ldots$ ii $1+10x+100x^2\ldots$

   c i $-2-10x-50x^2\ldots$ ii $8+24x+72x^2\ldots.$

   d i $40+10x+2.5x^2\ldots$ ii $144+12x+x^2\ldots$

   e i $243-81x+27x^2\ldots$ ii $1-\frac{5}{4}x+\frac{25}{16}x^2\ldots$

   f i $3-\frac{6}{x}+\frac{12}{x^2}\ldots$ ii $18-\frac{9}{x}+\frac{1}{x^2}\ldots$

   g i $5+5(3-2x)+5(3-2x)^2\ldots$ ii $7+\frac{7(2-x)}{2}+\frac{7(2-x)^2}{4}\ldots$

   h i $1+\left(3-\frac{2}{x}\right)+\left(3-\frac{2}{x}\right)^2\ldots$ ii $1+\frac{1+x}{x}+\frac{(1+x)^2}{x^2}\ldots$

   i i $7+7x^2+7x^4\ldots$ ii $12-48x^3+192x^6\ldots$

3. Find the sum to infinity of the geometric sequence $-18, 12, -8\ldots$

4. The first and fourth terms of a geometric series are 18 and $-\frac{2}{3}$ respectively.

   Find the sum to infinity of the series.

5 The fifth term of a geometric sequence is 12 and the seventh term is 3.

Find the two possible values of the sum to infinity of the series.

6 A geometric series has first term a and common ratio r, with $a > 0$ and $r > 0$. The second term is 4 and the eighth term is $\frac{1}{2}$.

a Find a and r.

b Find the exact value of the sum to infinity of the series, giving your answer in the form $m\left(\sqrt{n}+1\right)$, where m and n are constants to be stated.

7 A geometric sequence has all positive terms. The sum of the first two terms is 15 and the sum to infinity is 27. Find the value of:

a the common ratio

b the first term.

8 The sum to infinity of a geometric series is 32. The sum of the first four terms is 30 and all the terms are positive.

Find the difference between the sum to infinity and the sum of the first eight terms.

9 Consider the infinite geometric series $1+\left(\frac{2x}{3}\right)+\left(\frac{2x}{3}\right)^2 \ldots$.

a For what values of x does the series converge?

b Find the sum of the series if $x = 1.2$.

10 The sum of an infinite geometric sequence is 13.5, and the sum of the first three terms is 13. Find the first term.

11 An infinite geometric series is given by $\sum_{k=1}^{\infty} 2(4-3x)^k$.

a Find the values of x for which the series has a finite sum.

b When $x = 1.2$, find the minimum number of terms needed to give a sum that is greater than 1.328.

12 $f(x) = \sum_{r=0}^{\infty} 2^r x^r$

Evaluate:

a $f\left(\frac{1}{3}\right)$

b $f\left(\frac{2}{3}\right)$

13 The common ratio of the terms in a geometric series is 2^x.

a State the set of values of x for which the sum to infinity of the series exists.

b If the first term of the series is 35, find the value of x for which the sum to infinity is 40.

14 $\sum_{k=0}^{\infty} e^{-(2k+1)x} = \frac{3}{8}$

a Show that $3e^{2x} - 8e^x - 3 = 0$.

b Hence find the exact value of x.

15 Show that for a convergent geometric series with first term $a > 0$ and sum to infinity S_∞, $0 < a < 2S_\infty$.

Section 8: Mixed arithmetic and geometric questions

Tip

With questions like that in Worked example 4.20, think carefully about whether the amount you are calculating is for the beginning or the end of a year.

Be very careful when dealing with sequences and series questions. It is vital that you:

- identify whether it is a geometric or an arithmetic sequence
- identify whether it is asking for a term in the sequence or the sum of terms in the sequence
- translate the information given in the question into equations.

WORKED EXAMPLE 4.20

A savings account pays 2.4% annual compound interest, added at the end of each year. If £200 is paid into the account at the start of the first year, how much will there be in the account at the start of the seventh year?

$2.4\% = 1 + \frac{2.4}{100} = 1.024$

Each year the balance of the account is increased by the same percentage, so this gives a geometric sequence. It is a good idea to write out the first few terms in questions like this to make sure you know what is happening.

The balance of the account at the beginning of:

Year 1 : 200

Year 2 : 200×1.024

Year 3 : 200×1.024^2

This is a geometric sequence with $a = 200$ and $r = 1.024$.

$u_7 = ar^6$
$= 200 \times 1.024^6$
$= £230.58$

The balance at the start of the seventh year is u_7.

WORK IT OUT 4.1

A car loses value at the rate of 20% per year. It was bought for £15 000 at the beginning of 2017. In which year will its value fall below £1000?

Which is the correct solution? Identify the errors made in the incorrect solutions.

Solution A	Solution B	Solution C
$a = 15\,000,\ r = 0.8$	$a = 15\,000,\ r = 0.8$	$a = 15\,000 \times 0.8,\ r = 0.8$
$u_n < 1000$	$u_n < 1000$	$u_n < 1000$
$15\,000 \times 0.8^{n-1} < 1000$	$15\,000 \times 0.8^{n-1} < 1000$	$15\,000 \times 0.8^{n} < 1000$
$0.8^{n-1} < \frac{1}{15}$	$0.8^{n-1} < \frac{1}{15}$	$0.8^{n} < \frac{1}{15}$
$(n-1)\ln 0.8 < \ln \frac{1}{15}$	$(n-1)\ln 0.8 < \ln \frac{1}{15}$	$n \ln 0.8 < \ln \frac{1}{15}$
$n > \frac{\ln \frac{1}{15}}{\ln 0.8} + 1$	$n > \frac{\ln \frac{1}{15}}{\ln 0.8} + 1$	$n > \frac{\ln \frac{1}{15}}{\ln 0.8}$
$n > 13.1$	$n > 13.1$	$n > 12.1$
$\therefore n = 14$	$2017 + 13 = 2030$	$2017 + 12 = 2029$
$2017 + 14 = 2031$		

EXERCISE 4H

1. Philippa invests £1000 at 3% compound interest for 6 years.

   a How much interest does she get paid in the sixth year?

   b How much does she get back after 6 years?

2. Lars starts a job on an annual salary of £32 000 and is promised an annual increase of £1500.

   a How much will his 20th year's salary be?

   b After how many complete years will he have earned a total of £1 million?

3. A sum of £5000 is invested at a compound interest rate of 6.3% per annum.

   a Write down an expression for the value of the investment after n full years.

   b What will be the value of the investment at the end of 5 years?

   c The value of the investment will exceed £10 000 after n full years.

      i Write an inequality to represent this information.

      ii Calculate the minimum value of n.

4. Each row of seats in a theatre has 200 more seats than the row in front of it. There are 50 seats in the front row and the designer wants the capacity to be at least 8000.

   a How many rows are required?

   b Assuming the rows are equally spread, what percentage of people are seated in the front half of the theatre?

5. A sum of £100 is invested.

   a If the interest is compounded annually at a rate of 5% per year, find the total value V of the investment after 20 years.

   b If the interest is compounded monthly at a rate of $\frac{5}{12}$ % per month, find the minimum number of months for the value of the investment to exceed V.

6. A marathon is a 26 mile race. In a training regime for a marathon a runner runs 1 mile on his first day of training and each day increases his distance by $\frac{1}{4}$ of a mile.

   a After how many days has he run for a total of 26 miles?

   b On which day does he first run over 26 miles?

7. Aaron and Blake each open a savings account. Aaron deposits £100 in the first month and then increases his deposits by £10 each month.

   Blake deposits £50 in the first month and then increases his deposits by 5% each month.

   No interest is paid on balances.

   Show that the first time Blake will have more money in his account than Aaron is after 73 months.

8. A ball is dropped from 2 m in the air. Each time it bounces up to a height of 80% of its previous height.

   a How high does it bounce on the fourth bounce?

   b How far has it travelled when it hits the ground for the ninth time?

   c Give one reason why this model is unlikely to be accurate after 20 bounces.

9 Samantha puts £1000 into a bank account at the beginning of each year, starting in 2010. At the end of each year 4% interest is added to the account.

a Show that at the beginning of 2012 there is $1000+1000\times1.04+1000\times(1.04)^2$ in the account.

b Find an expression for the amount in the account at the beginning of year n.

c When Samantha has a total of at least £50 000 in her account at the beginning of a year she will start looking for a house. In which year will this happen?

Checklist of learning and understanding

- A sequence can be defined by a formula for the nth term or by a term-to-term rule.
- A sequence is:
  - **increasing** if each term is larger than the previous one: $u_{n+1} > u_n$
  - **decreasing** if each term is smaller than the previous one: $u_{n+1} < u_n$
  - **periodic** if the terms start repeating after a while: $u_{n+k} = u_n$ for some number k (the period of the sequence).
- A **series** is a sum of terms in a sequence and it can be described in a shorthand way using **sigma notation**.

$$\sum_{r=1}^{n} f(r) = f(1)+f(2)+\ldots+f(n)$$

- **Arithmetic sequences** have a constant difference, d, between consecutive terms.
  - If you know the first term, a, the nth term is:

    $$u_n = a+(n-1)d$$

  - If you know the first term and the common difference, the sum of all n terms in the sequence is:

    $$S_n = \frac{n}{2}[2a+(n-1)d]$$

  - If you know the first and last term (l), the sum of all n terms in the sequence is:

    $$S_n = \frac{n}{2}(a+l)$$

- **Geometric sequences** have a constant ratio, r, between consecutive terms.
  - If you know the first term, a, the nth term is:

    $$u_n = ar^{n-1}$$

  - The sum of the first n terms is:

    $$S_n = \frac{a(1-r^n)}{1-r}$$

    or equivalently:

    $$S_n = \frac{a(r^n-1)}{r-1}$$

  - If $|r|<1$, the series **converges** and the **sum to infinity** is given by:

    $$S_\infty = \frac{a}{1-r}$$

Mixed practice 4

1 The nth term of a geometric sequence is $u_n = r^n$.

The sum to infinity is 5.

Find the value of r.

Choose from these options.

A $-\frac{4}{5}$ B $\frac{5}{6}$ C $-\frac{5}{4}$ D $\frac{1}{2}$

2 The third term of an arithmetic sequence is 4 and the sum of the first 20 terms is –445. Find the first term and the common difference.

3 The fourth term of an arithmetic sequence is 9.6 and the ninth term is 15.6. Find the sum of the first nine terms.

4 The third term of a geometric sequence is 192 and the sixth term is 3.

a Find the first term and the common ratio. b Find the sum to infinity.

5 The fifth term of an arithmetic sequence is three times the second term. Find the ratio:

$$\frac{\text{common difference}}{\text{first term}}$$

6 The nth term of a geometric sequence is u_n, where

$u_n = 3 \times 4^n$

a Find the value of u_1 and show that $u_2 = 48$.

b Write down the common ratio of the geometric sequence.

c i Show that the sum of the first 12 terms of the geometric sequence is $4^k - 4$, where k is an integer.

ii Hence find the value of $\sum_{n=2}^{12} u_n$.

[© AQA 2007]

7 The sum of the first three terms of a geometric sequence is 19. The sum to infinity is 27.

Find the common ratio, r.

Choose from these options.

A $r = \frac{2}{3}$ B $r = \frac{19}{27}$ C $r = \sqrt[3]{\frac{19}{27}}$ D $r = \frac{3}{4}$

8 Find the exact value of $\sum_{r=0}^{\infty} \frac{\sqrt{2}}{4^r}$.

Choose from these options.

A $\frac{4\sqrt{2}}{3}$ B $\frac{\sqrt{2}}{3}$ C $\frac{\sqrt{2}}{4}$ D $\frac{4}{4-\sqrt{2}}$

9 Which is the first term of the sequence $\frac{1}{3}, \frac{1}{9}, \ldots, \frac{1}{3^n}$ less than 10^{-6}?

10 Ben builds a pyramid out of toy bricks. The top row contains one brick, the second row contains three bricks and each row after that contains two more bricks than the previous row.

a How many bricks are there in the nth row?

b If a total of 36 bricks are used how many rows are there?

c In Ben's largest ever pyramid he noticed that the total number of bricks was four more than four times the number of bricks in the last row. What is the total number of bricks?

11 Kenny is offered two investment plans, each requiring an initial investment of £10 000.

Plan A offers a fixed return of £800 per year.

Plan B offers a return of 5% each year, reinvested in the plan.

Over what period of time is plan A better than plan B?

12 The first three terms of a geometric sequence are $2x+4, x+5, x+1$.

a Find the two possible values of x.

b Given that it exists, find the sum to infinity of the series.

13 Evaluate $\sum_{i=0}^{\infty} \frac{2^i + 4^i}{6^i}$.

14 Find the sum of all the integers between 300 and 600 that are divisible by 7.

15 A sequence is defined by $u_{n+1} = ku_n - 3, u_1 = 2$.

a If $u_3 \leqslant -1$, find the range of possible values of the constant k.

b Given that $k = -\frac{1}{4}$, find the limit of the sequence as n tends to infinity.

16 The nth term of a sequence is u_n. The sequence is defined by

$u_{n+1} = pu_n + q$

where p and q are constants.

The first two terms of the sequence are given by $u_1 = 96$ and $u_2 = 72$.

The limit of u_n as n tends to infinity is 24.

a Show that $p = \frac{2}{3}$.

b Find the value of u_3.

[© AQA 2013]

17 An arithmetic series has first term a and common difference d.

The sum of the first five terms of the series is 575.

a Show that $a + 2d = 115$.

b Given also that the 10th term of the series is 87, find the value of d.

c The nth term of the series is u_n. Given that $u_k > 0$ and $u_{k+1} < 0$, find the value of $\sum_{n=1}^{k} u_n$.

[© AQA 2014]

18 Prove that the sequence defined by $u_n = \dfrac{n^2-2}{n+1}$ is increasing.

19 A geometric sequence and an arithmetic sequence both start with a first term of 1. The third term of the arithmetic sequence is the same as the second term of the geometric sequence. The fourth term of the arithmetic sequence is the same as the third term of the geometric sequence. Find the possible values of the common difference of the arithmetic sequence.

20 The first, second and fourth terms of a geometric sequence form consecutive terms of an arithmetic sequence.

Given that the sum to infinity of the geometric exists, find the exact value of the common ratio.

21 Find an expression for the sum of the first 23 terms of the series

$\ln\dfrac{a^3}{\sqrt{b}} + \ln\dfrac{a^3}{b} + \ln\dfrac{a^3}{b\sqrt{b}} + \ln\dfrac{a^3}{b^2}\dots$

giving your answer in the form $\ln\dfrac{a^m}{b^n}$, where $m, n \in \mathbb{Z}$.

22 A student writes '1' on the first line of a page, then the next two integers '2, 3' on the second line of the page then the next three integers '4, 5, 6' on the third line. He continues this pattern.

a How many integers are there on the nth line?

b What is the last integer on the nth line?

c What is the first integer on the nth line?

d Show that the sum of all the integers on the nth line is $\dfrac{n}{2}(n^2+1)$.

e The sum of all the integers on the last line of the page is 16 400. How many lines are on the page?

23 Selma has a mortgage of £150 000. At the end of each year 3% interest is added, then Selma pays £10 000.

a Show that at the end of the third year the amount owing is:

$£150\,000\times(1.03)^3 - 10\,000\times(1.03)^2 - 10\,000\times 1.03 - 10\,000$

b Find an expression for how much is owing at the end of the nth year.

c After how many years will the mortgage be paid off?

5 Rational functions and partial fractions

In this chapter you will learn how to:

- manipulate rational functions, including by using polynomial division with remainders
- use the factor theorem to find factors of the form $(ax + b)$
- decompose rational functions into a sum of algebraic fractions when the denominator contains distinct linear factors
- decompose rational functions into a sum of algebraic fractions when the denominator contains repeated linear factors.

Before you start...

GCSE	You should be able to add algebraic fractions.	1 Simplify into one fraction $\frac{1}{x}+\frac{1}{2+x}$.
Student Book 1, Chapter 3	You should be able to factorise quadratic expressions.	2 Factorise $6x^2+7x+2$.
Student Book 1, Chapter 4	You should be able to carry out polynomial division.	3 Simplify $(x^3-x^2-3x+6)\div(x+2)$.
Student Book 1, Chapter 4	You should be able to use the factor theorem.	4 Find a linear factor of x^3+x^2+5x-7.

What are rational functions?

A **rational function** is a fraction in which both the denominator and numerator are polynomials. These tend to have graphs with asymptotes, which makes them applicable to a wide range of real-world situations from economics to medicine - for example, the function $\frac{t}{t^2+1}$ is often used to model the amount of anaesthetic in a patient after time t.

This chapter will focus on applying algebraic techniques to change rational functions into forms that will allow more advanced manipulations later in the course. To do this, you will review a very important theorem about polynomials - the factor theorem.

Focus on...

Focus on ... Modelling 1 compares models using rational functions to those using exponential functions.

Section 1: An extension of the factor theorem

The version of the factor theorem that you met in Student Book 1 can be extended slightly to include factors where the coefficient of x is not 1.

Key point 5.1

If $f\left(\frac{b}{a}\right)=0$ then $(ax-b)$ is a factor of $f(x)$.

Rewind

In Student Book 1, Chapter 4, the factor theorem stated that if $f(b)=0$ then $(x-b)$ is a factor of $f(x)$.

WORKED EXAMPLE 5.1

a Show that $2x-1$ is a factor of $f(x)=2x^3-5x^2-14x+8$.

b Hence solve $2x^3-5x^2-14x+8=0$.

a $f\left(\frac{1}{2}\right)=2\times\frac{1}{8}-5\times\frac{1}{4}-14\times\frac{1}{2}+8$

$=\frac{1}{4}-\frac{5}{4}-7+8$

$=0$

Therefore $2x-1$ is a factor of $f(x)$.

By the factor theorem, if $f\left(\frac{1}{2}\right)=0$, then $2x-1$ is a factor of $f(x)$.

b

$$\begin{array}{r} x^2-2x-8 \\ 2x-1\overline{)2x^3-5x^2-14x+8} \\ \underline{2x^3-x^2} \\ -4x^2-14x+8 \\ \underline{-4x^2+2x} \\ -16x+8 \\ \underline{-16x+8} \\ 0 \end{array}$$

Use polynomial division to find the remaining factor.

So,

$2x^3-5x^2-14x+8=(2x-1)(x^2-2x-8)$

$=(2x-1)(x-4)(x+2)$

Factorise the quadratic.

$(2x-1)(x-4)(x+2)=0$

$x=\frac{1}{2},4,-2$

Solve as normal.

Tip

If you do not like long division you can write $2x^3-5x^2-14x+8$ as $(2x-1)(ax^2+bx+c)$, then multiply out and compare coefficients.

Rewind

You met polynomial division in Student Book 1, Chapter 4.

WORK IT OUT 5.1

If $f\left(-\frac{5}{2}\right)=0$, state a linear factor of the polynomial $f(x)$.

Which is the correct solution? Identify the errors made in the incorrect solutions.

Solution 1	Solution 2	Solution 3
By the factor theorem, $5x-2$ is a factor of $f(x)$.	$f\left(-\frac{5}{2}\right)=0 \Rightarrow 2x+5$ is a factor of $f(x)$.	By the factor theorem, $2x-5$ is a factor of $f(x)$.

The factor theorem also works in reverse, so if you know that $ax - b$ is a factor then $f\left(\frac{b}{a}\right) = 0$.

WORKED EXAMPLE 5.2

If $3x + 2$ is a factor of $3x^3 - 10x^2 + 4x + a$, find the value of a.

$$3\left(-\frac{2}{3}\right)^3 - 10\left(-\frac{2}{3}\right)^2 + 4\left(-\frac{2}{3}\right) + a = 0$$

$3x + 2$ can be written as $3x - (-2)$ so the factor theorem says that $f\left(\frac{-2}{3}\right) = 0$.

$$-8 + a = 0$$
$$a = 8$$

Solve the resulting equation.

EXERCISE 5A

1. Show that each expression has the given factor and hence factorise completely.
   - a i $2x^3 - 9x^2 + 7x + 6$ has factor $2x + 1$
     ii $3x^3 - 7x^2 + 5x - 1$ has factor $3x - 1$
   - b i $2x^3 + 7x^2 - 3x - 18$ has factor $2x - 3$
     ii $3x^3 - 5x^2 - 48x + 80$ has factor $3x - 5$

2. Show that $2x + 5$ is a factor of $4x^3 - 31x - 15$. Hence factorise it completely.

3. $5x - 2$ is a factor of $20x^3 + 22x^2 + ax - 4$.

   Find the value of a.

4. a $3x + 1$ is a factor of $f(x) = 12x^3 - 2x^2 + ax - 2$.

   Find the value of a.

   b Hence factorise $f(x)$.

5. a Show that $2x - a$ is a factor of $2x^3 - 5ax^2 + 4a^2x - a^3$.

   b Hence factorise $2x^3 - 5ax^2 + 4a^2x - a^3$.

6. Show that $x - a$ is a factor of $f(x) = x^3 - 6ax^2 + 11a^2x - 6a^3$. Hence solve $f(x) = 0$.

7. Prove that if $ax + b$ is a factor of $ax^2 + bx + c$ then $c = 0$.

8. Given that $ax + b$ is a factor of $x^2 + bx + a$, find an expression for b in terms of a.

9. a $2x + 1$ and $3x + 1$ are factors of $f(x) = 30x^4 + 67x^3 + 4x^2 + ax + b$.

   Find the values of a and b.

   b Hence solve $f(x) = 0$.

10. Show that $4x^2 - 1$ is a factor of $f(x) = 4x^4 + 4x^3 - 25x^2 - x + 6$ and hence solve $f(x) = 0$.

Section 2: Simplifying rational expressions

You need to be able to simplify and manipulate rational expressions, using your knowledge of algebra – in particular, always look to factorise expressions and simplify.

WORKED EXAMPLE 5.3

Simplify $\frac{x+1}{x^2-1} \div \frac{2}{x^2-3x+2}$.

$\frac{x+1}{x^2-1} \div \frac{2}{x^2-3x+2} \equiv \frac{x+1}{(x-1)(x+1)} \div \frac{2}{(x-2)(x-1)}$ — Always try to factorise first. Look out for the difference of two squares!

$\equiv \frac{1}{x-1} \div \frac{2}{(x-2)(x-1)}$ — Divide the top and bottom of the first fraction by the factor of $x+1$.

$\equiv \frac{1}{x-1} \times \frac{(x-2)(x-1)}{2}$ — To divide, flip the second fraction and multiply.

$\equiv \frac{1}{1} \times \frac{x-2}{2}$ — There is a factor of $x-1$ in the denominator of the first fraction and the denominator of the second fraction. Cancel these when the fractions are being multiplied.

$\equiv \frac{x-2}{2}$

You can use polynomial division to simplify rational functions in which the degree of the numerator is at least the degree of the denominator. This turns the rational function into the sum of a polynomial and a simpler rational function called the **remainder term**.

Tip

In Worked example 5.3 you used an identity sign to emphasise that the expression is just being simplified rather than solving it, but you will often see an equals sign used instead, and this is an acceptable alternative.

Key point 5.2

If $p(x)$ is a polynomial then

$$\frac{p(x)}{ax+b} \equiv q(x) + \frac{r}{ax+b}$$

$q(x)$ is also a polynomial and is called the quotient. r is the remainder.

WORKED EXAMPLE 5.4

Write $\frac{2x^2+5x+1}{2x-3}$ in the form $Ax+B+\frac{C}{2x-3}$.

$$\begin{array}{r} x+4 \\ 2x-3\overline{)2x^2+5x+1} \\ 2x^2-3x \\ \hline 8x+1 \\ 8x-12 \\ \hline 13 \end{array}$$

Use polynomial division.

So $\frac{2x^2+5x+1}{2x-3} = x+4+\frac{13}{2x-3}$ — The remainder found in polynomial division forms the numerator of the rational expression.

WORK IT OUT 5.2

Simplify $\frac{x^2-1}{x-1}$.

Which is the correct solution? Identify the errors made in the incorrect solutions.

Solution 1	Solution 2	Solution 3
$\frac{x^2-1}{x-1} \equiv \frac{(x-1)^2}{x-1}$ $\equiv \frac{x-1}{1}$ $\equiv x-1$	$\frac{x^2-1}{x-1} = \frac{(x-1)(x+1)}{x-1}$ $= \frac{x+1}{1}$ $= x+1$	Cancelling −1 from top and bottom: $\frac{x^2-1}{x-1} = \frac{x^2}{x} = x$

EXERCISE 5B

1 Simplify each expression.

a i $\frac{6x+9}{3}$ ii $\frac{10x+20}{5}$

b i $\frac{x-4}{2x-8}$ ii $\frac{x+3}{5x+15}$

c i $\frac{3x^2+4x}{x}$ ii $\frac{5x^2-7x}{x}$

d i $\frac{1-x}{x-1}$ ii $\frac{7-2x}{2x-7}$

e i $\frac{2x+4}{x^2-4}$ ii $\frac{4x+12}{x^2+6x+9}$

f i $\frac{x^2+2x+1}{x^2+5x+4}$ ii $\frac{x^2+5x+6}{x^2+7x+12}$

g i $\frac{6x^2+5x+1}{8x^2+6x+1}$ ii $\frac{12x^2-7x-10}{9x^2-4}$

2 Simplify each expression.

a i $4x \times \frac{1}{2}$ ii $12x \times \frac{1}{4}$

b i $\frac{x^2}{4} \times \frac{2}{x}$ ii $\frac{x^3}{3} \times \frac{15}{x}$

c i $\frac{1}{x^2} \times \frac{x}{2} \times \frac{x}{3}$ ii $\frac{x^2}{4} \times 2x \times \frac{5}{x^3}$

d i $\frac{2x+4}{5x} \times \frac{10x^2}{x+2}$ ii $\frac{2x+8}{x} \times \frac{3x^2}{3x+12}$

e i $\frac{x^2+7x+12}{x^2+2x-3} \times \frac{x+1}{x+4}$ ii $\frac{x^2-10x+21}{x^2-6x+5} \times \frac{x-5}{x-7}$

3 Simplify each expression.

a i $\frac{x}{2} \div \frac{x}{4}$ ii $\frac{3x}{4} \div \frac{9x}{20}$

b i $\frac{x}{x+1} \div \frac{x}{2}$ ii $\frac{3x}{x-2} \div \frac{5}{x-2}$

c i $\frac{(x^2-x)}{3x+6} \div \frac{x-1}{x+2}$ ii $\frac{x^2+5x}{5x+5} \div \frac{x}{5}$

d i $\frac{x^2+3x}{3x+2} \div (x+3)$ ii $\frac{5x}{7x-2} \div x$

4 Write each expression in the form $ax+b+\frac{r}{px+q}$.

a i $\frac{x^2}{2x+1}$ ii $\frac{x^2}{2x+3}$

b i $\frac{2x^2+3x+4}{2x+1}$ ii $\frac{5x^2+3x+4}{5x+3}$

5 Simplify $\frac{1}{x^2+7x+12} \div \frac{1}{x^2+8x+15}$.

6 Simplify $\left(\frac{1}{x-3}-\frac{1}{x}\right) \div \frac{6x}{x^2-9}$.

7 a Simplify $\frac{x^2+5x+6}{x+3}$.

b Hence or otherwise solve $\frac{x^2+5x+6}{x+3} = x^2+4x+4$.

8 Solve $\frac{x}{x+3} \div \frac{9}{2x+6} = \frac{8}{x}$.

9 Find the quotient and remainder when $4x+15$ is divided by $2x-5$.

10 Find the quotient and remainder when $2x^2-3x+4$ is divided by $2x-1$.

11 Write $\frac{2x^2-x-28}{x^2-2x-8}$ in the form $A+\frac{B}{x+C}$.

12 Show that $5x-2$ is a factor of $5x^3+18x^2+7x-6$ and hence simplify $\frac{5x^3+18x^2+7x-6}{5x^2+13x-6}$.

13 a Show that $2x+a$ is a factor of $2x^3+ax^2-2a^2x-a^3$.

b Hence simplify $\frac{(2x^3+ax^2-2a^2x-a^3)}{2x^2+3ax+a^2}$.

14 The remainder when x^2+ax+3 is divided by $x-2$ is 5. Find the value of a and the quotient.

15 When the polynomial $f(x)$ is divided by $x-2$ the quotient is the same as the remainder. Prove that $f(x)$ is divisible by $x-1$.

16 a A car travels at speed a km h^1 for 10 km then at speed b km h^1 for 20 km. Find and simplify an expression for the average speed during this journey.

b If the average speed equals the arithmetic mean of the two speeds, show that either the speeds in the two sections of the journey are equal or the speed in the second section is twice that of the first section.

Section 3: Partial fractions with distinct factors

You already know how to write a sum of two algebraic fractions as a single fraction, for example:

$$\frac{1}{x+1}+\frac{1}{x+2} \equiv \frac{x+2}{(x+1)(x+2)}+\frac{x+1}{(x+1)(x+2)}$$
$$\equiv \frac{2x+3}{(x+1)(x+2)}$$

However there are situations where you need to reverse this process. The method for doing this involves **partial fractions**.

To do this, you need to know the form the partial fractions will take.

Gateway to A level

For a reminder of adding algebraic fractions, see Gateway to A Level Section Y.

Fast forward

You will see that you need to do this with binomial expansions in Chapter 6 and integration in Chapter 11. You will also use it to help sum some series if you study Further Mathematics.

Key point 5.3

- $\dfrac{px+q}{(ax+b)(cx+d)}$ decomposes into $\dfrac{A}{ax+b}+\dfrac{B}{cx+d}$
- $\dfrac{px+q}{(ax+b)(cx+d)(ex+f)}$ decomposes into $\dfrac{A}{ax+b}+\dfrac{B}{cx+d}+\dfrac{C}{ex+f}$

Once you know the form you are aiming for, write it as an identity and multiply both sides by the denominator of the original fraction. You can then either compare coefficients or substitute in convenient values to find the values of A and B.

Rewind

Comparing coefficients was discussed in Student Book 1, Chapter 1.

WORKED EXAMPLE 5.5

Write $\dfrac{2x+3}{(x+1)(x+2)}$ in partial fractions.

$$\frac{2x+3}{(x+1)(x+2)} \equiv \frac{A}{x+1}+\frac{B}{x+2}$$

Write the expression in the correct form (given in Key point 5.3).

$$2x+3 \equiv A(x+2)+B(x+1)$$

Multiply both sides by $(x+1)(x+2)$ to remove fractions.

When $x=-2$:

$$2(-2)+3 = A(0)+B(-1)$$
$$-1=-B$$
$$B=1$$

As this is an identity, you can substitute in any value of x. Choosing $x=-2$ makes the coefficient of A zero, so you can find B.

When $x=-1$:

$$2(-1)+3 = A(1)+B(0)$$
$$A=1$$

Similarly, choosing $x=-1$ eliminates B so you can find A.

So $\dfrac{2x+3}{(x+1)(x+2)} \equiv \dfrac{1}{x+1}+\dfrac{1}{x+2}$

State the final answer.

You can only write in terms of partial fractions using this method if the degree of the numerator is less than the degree of the denominator. If this is not the case, you need to use algebraic manipulation to turn it into the required form.

WORKED EXAMPLE 5.6

a Show that $\frac{x^4-x^2+2}{x^3-x}$ can be written in the form $x+\frac{R}{x^3-x}$.

b Hence express $\frac{x^4-x^2+2}{x^3-x}$ in partial fractions.

a $\frac{x^4-x^2+2}{x^3-x}\equiv\frac{x(x^3-x)+2}{x^3-x}$ — Make a link between the numerator and denominator.

$\equiv\frac{x(x^3-x)}{x^3-x}+\frac{2}{x^3-x}$ — You can then split the expression into two fractions.

$\equiv x+\frac{2}{x^3-x}$

b $\frac{2}{x^3-x}\equiv\frac{2}{x(x-1)(x+1)}$ — First work with the fraction from part **a**. If you want to use partial fractions you must factorise the denominator.

$\frac{2}{x(x-1)(x+1)}\equiv\frac{A}{x}+\frac{B}{x-1}+\frac{C}{x+1}$ — Now use the form given in Key point 5.3.

$2\equiv A(x-1)(x+1)+Bx(x+1)+Cx(x-1)$ — Multiply by the denominator to remove fractions.

When $x=0$:

$2=A(-1)(1)+B(0)+C(0)$ — Substitute in $x=0$, to eliminate B and C.

$A=-2$

When $x=1$:

$2=A(0)+B(2)+C(0)$ — Substitute in $x=1$, to eliminate A and C.

$2=2B$

$B=1$

When $x=-1$

$2=A(0)+B(0)+C(-1)(-2)$ — Substitute in $x=-1$, to eliminate A and B.

$2=2C$

$C=1$

Therefore $\frac{2}{x(x-1)(x+1)}\equiv-\frac{2}{x}+\frac{1}{x-1}+\frac{1}{x+1}$

So $\frac{x^4-x^2+2}{x^3-x}\equiv x-\frac{2}{x}+\frac{1}{x-1}+\frac{1}{x+1}$ — Remember that the original expression had an x as well.

EXERCISE 5C

1. Write each expression in terms of partial fractions.

   a i $\frac{2x+2}{x(x+2)}$ ii $\frac{3}{x(x-3)}$ b i $\frac{3x+4}{(x+1)(x+2)}$ ii $\frac{5x+1}{(x-1)(x+2)}$

   c i $\frac{10-x}{x^2+x-12}$ ii $\frac{27-x}{x^2+x-30}$ d i $\frac{3x+2}{(2x-1)(x+3)}$ ii $\frac{x+12}{(3x-2)(2x+5)}$

2. Write each expression in terms of partial fractions.

   a i $\frac{6-4x}{x(x-3)(x-2)}$ ii $\frac{3x+4}{x(x+1)(x+2)}$ b i $\frac{13x+17}{(x-1)(x+1)(x+2)}$ ii $\frac{19x+55}{(x-3)(x+4)(x+1)}$

   c i $\frac{10}{(2x+1)(x+3)(x+1)}$ ii $\frac{49-4x}{(3x-1)(2x+3)(x+4)}$

3. Given that $\frac{A}{2x+5}+\frac{B}{5x+2}=\frac{1}{(2x+5)(5x+2)}$ find the values of A and B.

4. Express $\frac{1}{4x^2-1}$ in terms of partial fractions.

5. Simplify $\frac{3x+2}{x^3+3x^2+2x}$.

6. Decompose $\frac{2}{27x^3-3x}$ into partial fractions.

7. Write $\frac{1}{x^4-5x^2+4}$ in partial fractions.

8. a Show that $\frac{x^2-3x+5}{x-1}=x-2+\frac{3}{x-1}$.

   b Hence write $\frac{x^2-3x+5}{(x-1)(x-2)}$ in terms of partial fractions.

9. Expand in partial fractions $\frac{a}{x(x-a)}$ where a is a constant.

10. Expand in partial fractions $\frac{2x-3a}{x^2-3ax+2a^2}$ where a is a constant.

11. a Show that $\frac{x^2+7x+7}{x^2+7x+6}$ can be written in the form $1+\frac{R}{x^2+7x+6}$ where R is a constant to be determined.

    b Hence write $\frac{x^2+7x+7}{x^2+7x+6}$ in terms of partial fractions.

12. What happens if you try to write $\frac{x^2}{(x-1)(x-2)}$ in the form $\frac{A}{x-1}+\frac{B}{x-2}$?

Section 4: Partial fractions with a repeated factor

If there is a repeated factor in the denominator you need to use an alternative expression for the partial fractions.

Key point 5.4

$\frac{px+q}{(ax+b)(cx+d)^2}$ decomposes into $\frac{A}{ax+b}+\frac{B}{cx+d}+\frac{C}{(cx+d)^2}$

WORKED EXAMPLE 5.7

Express $\frac{11x-16}{(2x-1)^2(x+3)}$ in partial fractions.

$$\frac{11x-16}{(2x-1)^2(x+3)} \equiv \frac{A}{2x-1}+\frac{B}{(2x-1)^2}+\frac{C}{x+3}$$

Write the expression in the correct form (given in Key point 5.4).

$$11x-16 \equiv A(2x-1)(x+3)+B(x+3)+C(2x-1)^2$$

Multiply both sides by $(2x-1)^2(x+3)$ to eliminate fractions.

When $x=\frac{1}{2}$:

Choose $x=\frac{1}{2}$ so that the coefficients of A and C are zero.

$$11\times\frac{1}{2}-16 = A(0)+B\left(\frac{1}{2}+3\right)+C(0)^2$$

$$-\frac{21}{2}=\frac{7}{2}B$$

$$B=-3$$

When $x=-3$:

Choose $x=-3$ so that the coefficients of A and B are zero.

$$11(-3)-16 = A(0)+B(0)+C(2(-3)-1)^2$$

$$-49=49C$$

$$C=-1$$

When $x=0$:

There is no value of x that can make the coefficients of B and C zero, so you have to choose another simple alternative. $x=0$ often makes the arithmetic simple.

$$-16=A(-1)(3)+B(3)+C(-1)^2$$

$$-16=-3A+3B+C$$

$$-16=-3A+3(-3)-1$$

Substitute the values of B and C found above.

$$-6=-3A$$

$$A=2$$

So $\frac{11x-16}{(2x-1)^2(x+3)} \equiv \frac{2}{2x-1}-\frac{3}{(2x-1)^2}-\frac{1}{x+3}$

Write out the final answer.

EXERCISE 5D

1 Write each expression in terms of partial fractions.

a i $\frac{x}{(x+1)^2}$ ii $\frac{x}{(x-2)^2}$

b i $\frac{16}{x^2(x+4)}$ ii $\frac{1}{x^2(x-1)}$

c i $\frac{9x+9}{(x-1)(x+2)^2}$ ii $\frac{9x}{(x+1)(x-2)^2}$

d i $\frac{25}{(1-x)(2x+3)^2}$ ii $\frac{3x+17}{(2x-5)(x+1)^2}$

2 Split $\frac{4}{x^2(x-2)}$ into partial fractions.

3 Express $\frac{5x+1}{x^3+x^2}$ in partial fractions.

4 Write $\dfrac{1}{(x^2-4)(x-2)}$ in terms of partial fractions.

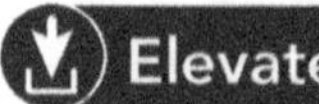

Elevate

See Support Sheet 5 for a further example of partial fractions with a repeated factor and for more practice questions.

5 **a** Use the factor theorem to show that $2x-1$ is a factor of $2x^3+3x^2-1$.

b Hence factorise $2x^3+3x^2-1$.

c Write $\dfrac{9x}{2x^3+3x^2-1}$ in partial fractions.

6 **a** Use the factor theorem to show that $3x+2$ is a factor of $12x^3-28x^2+3x+18$.

b Hence express $\dfrac{14x-47}{12x^3-28x^2+3x+18}$ in partial fractions.

7 **a** Write $\dfrac{x^4+5x^3+6x^2+2}{x^3+3x^2}$ in the form $Ax+B+\dfrac{C}{x^3+3x^2}$.

b Hence write $\dfrac{x^4+5x^3+6x^2+2}{x^3+3x^2}$ in partial fractions.

8 **a** Prove that if a function can be written as $\dfrac{A}{x-p}+\dfrac{B}{(x-p)^2}$ it can also be written as $\dfrac{Cx+d}{(x-p)^2}$.

b Write $\dfrac{1}{(x-1)(x-2)^2}$ in the form $\dfrac{A}{x-1}+\dfrac{Bx+C}{(x-2)^2}$.

9 Write $\dfrac{x(a+1)-a^2}{x(x-a)^2}$ in terms of partial fractions.

Checklist of learning and understanding

- The factor theorem can be used to check for factors of polynomials: if $f\left(\dfrac{b}{a}\right)=0$ then $(ax-b)$ is a factor of $f(x)$.
- A **rational function** is a fraction in which both the denominator and numerator are polynomials. In arithmetic they follow all the same rules as normal fractions.
- If the degree of the numerator is equal to or greater than the degree of the denominator you can use polynomial division to simplify the function.

  $\dfrac{p(x)}{ax+b}\equiv q(x)+\dfrac{r}{ax+b}$.
- If the rational function has a numerator with degree less than the denominator it can be decomposed into partial fractions.
  - $\dfrac{px+q}{(ax+b)(cx+d)}$ decomposes into $\dfrac{A}{ax+b}+\dfrac{B}{cx+d}$.
  - $\dfrac{px+q}{(ax+b)(cx+d)(ex+f)}$ decomposes into $\dfrac{A}{ax+b}+\dfrac{B}{cx+d}+\dfrac{C}{ex+f}$.
  - $\dfrac{px+q}{(ax+b)(cx+d)^2}$ decomposes into $\dfrac{A}{ax+b}+\dfrac{B}{cx+d}+\dfrac{C}{(cx+d)^2}$.

Mixed practice 5

1 What is the expression $\frac{x^2}{x^2-1} \div \left(\frac{4x}{x^2-2x+1}\right)$ equal to?
Choose from these options.

A $\frac{x^2-2x+1}{-4}$ B $\frac{4x^3}{(x-1)^3(x+1)}$ C $\frac{x(x-1)}{4(x+1)}$ D $-\frac{x}{4}$

2 **a** Simplify $\frac{x^3+1}{x+1}$.

b Hence find the prime factors of 1001.

3 **a** Show that $2x+3$ is a factor of $6x^3+27x^2+3x-36$.

b Factorise $6x^3+27x^2+3x-36$.

c Solve $6x^3+27x^2+3x-36=0$.

d Simplify $\frac{6x^3+27x^2+3x-36}{4x^2+22x+24}$.

4 Write $\frac{5}{x^2-x-6}$ in partial fractions.

5 Express $\frac{9x-9}{x^3-9x}$ in partial fractions.

6 Write $\frac{x}{18x^2-8}$ in terms of partial fractions.

7 Write $\frac{1}{x(x+2)^2}$ in partial fractions.

8 Decompose $\frac{20}{x^2-10}$ into partial fractions.

9 Express $\frac{16x}{(1-3x)(1+x)^2}$ in the form $\frac{A}{1-3x}+\frac{B}{1+x}+\frac{C}{(1+x)^2}$.

[© AQA 2014]

10 The polynomial $f(x)$ is defined by $f(x)=27x^3-9x+2$.

a Find the remainder when $f(x)$ is divided by $3x+1$.

b **i** Show that $f\left(-\frac{2}{3}\right)=0$.

ii Express $f(x)$ as a product of three linear factors.

iii Simplify
$\frac{27x^3-9x+2}{9x^2+3x-2}$.

[© AQA 2008]

11 What is the expression $\frac{2}{x^2-1}$ equal to?
Choose from these options.

A $\frac{1}{x-1}-\frac{1}{x+1}$ B $\frac{1}{x^2}$ C $\frac{1}{x-1}+\frac{1}{(x-1)^2}$ D $\frac{1}{x-1}\times\frac{1}{x+1}$

12 a Express $\frac{2x+3}{4x^2-1}$ in the form $\frac{A}{2x-1}+\frac{B}{2x+1}$, where A and B are integers.

b Express $\frac{12x^3-7x-6}{4x^2-1}$ in the form $Cx+\frac{D(2x+3)}{4x^2-1}$, where C and D are integers.

c Hence express $\frac{12x^3-7x-6}{4x^2-1}$ as the sum of a linear function and partial fractions.

[© AQA 2012 (adapted)]

13 a Simplify $f(x)=\frac{3x+3}{6x^2+13x+6}\times\frac{3x+2}{x^2+3x+2}$.

b Hence write $f(x)$ in terms of partial fractions

14 a Use the factor theorem to show that $3x+1$ is a factor of $9x^3-3x^2-5x-1$.

b Hence factorise $9x^3-3x^2-5x-1$.

c Solve $9x^3-3x^2-5x=1$.

d Write $\frac{4x+4}{9x^3-3x^2-5x-1}$ in partial fractions.

15 a Show that $2x+1$ is a factor of $f(x)=4x^3+4x^2-x-1$.

b Solve $f(x)=0$.

c Write $\frac{4x+1}{4x^3+4x^2-x-1}$ in partial fractions.

16 a Write $f(x)=\frac{2x^3+11x^2+20x+13}{x^2+5x+6}$ in the form $Ax+B+\frac{Cx+D}{x^2+5x+6}$.

b Hence write $f(x)$ in partial fractions.

17 a $f(x)=6x^4+35x^3+62x^2+ax+b$ is divisible by $2x+1$ and by $x+2$. Find the values of a and b.

b Fully factorise $f(x)$.

c Solve $f(x)=0$.

18 a Write $\frac{4}{(x+1)(x+5)}$ in terms of partial fractions.

b Hence write $\frac{16}{(x+1)^2(x+5)^2}$ in terms of partial fractions.

19 a Use polynomial division to simplify $\frac{u+4}{u-4}$.

b Hence or otherwise write $\frac{x^2+4}{x^2-4}$ in partial fractions.

20 Write $\frac{3x^2-a^2}{x^2(x-a)}$ in terms of partial fractions.

21 $\frac{2x^2+3x+k}{2x-1}$ can be written as $Ax+B+\frac{5}{2x-1}$.
Find the value of A, B and k.

22 The remainder when $x^2+5ax+b$ is divided by $x-a$ is $4a^2$. Write b in terms of a.

23 The remainder when a polynomial $f(x)$ is divided by $x-1$ is the same as the quotient. Find the value of $f(0)$.

24 $$\frac{A}{x+a}+\frac{B}{x+b}+\frac{C}{x+c}\equiv\frac{f(x)}{(x+a)(x+b)(x+c)}$$

Prove that if $f(x)$ is linear then $A+B+C=0$.

25 If two resistors with resistance R_1 and R_2 are connected in parallel the combined system has resistance R_T. These are related by the equation:

$$\frac{1}{R_T}=\frac{1}{R_1}+\frac{1}{R_2}$$

a Find and simplify an expression for R_T in terms of R_1 and R_2.

b Hence prove that $R_T < R_1$.

Elevate

See Extension Sheet 5 for more questions on continued fractions.

6 General binomial expansion

In this chapter you will learn how to:

- expand $(a+bx)^n$ where n is any rational power
- decide when a binomial expansion will converge
- use partial fractions to write expressions in the form required for the binomial expansion
- use binomial expansions to approximate functions.

Before you start...

GCSE	You should be able to simplify expressions with exponents.	1 Simplify $(8x^6)^{\frac{1}{3}}$.
Student Book 1, Chapter 9	You should be able to use binomial expansions for positive integers.	2 Expand $(3-2x)^4$.
Chapter 3	You should be able to write inequalities, using the modulus function.	3 Write $\lvert x-2\rvert < 3$ in the form $a < x < b$
Chapter 5	You should be able to write an expression in partial fractions.	4 Write $\frac{4-x}{x(x-2)}$ in the form $\frac{A}{x-2}+\frac{B}{x}$.

Extending the binomial theorem

In Student Book 1 the binomial expansion was simply a quick way to expand brackets. In this chapter it is extended to allow you to approximate many other functions with polynomials.

Section 1: The general binomial theorem

In Student Book 1 you met the binomial expansion when raising a sum of two terms to a positive integer power. For example:

$$(1+x)^n = {}^nC_0 + {}^nC_1x + {}^nC_2x^2 + {}^nC_3x^3 \ldots + {}^nC_nx^n$$

You also saw that the first three binomial coefficients could be written as:

$${}^nC_0 = 1,\ {}^nC_1 = n,\ {}^nC_2 = \frac{n(n-1)}{2!}$$

Continuing this pattern of the expressions for the coefficients, and using them in the same binomial expansion formula as before, gives a formula that also works for any rational number.

Tip

! is the symbol for factorial.
2! means 2×1
3! means $3 \times 2 \times 1$
In your formula book you will see this written as 2.1 and 3.2.1 etc.

Key point 6.1

If $|x|<1$

$(1+x)^n = 1+nx+\frac{n(n-1)}{2!}x^2+\frac{n(n-1)(n-2)}{3!}x^3\ldots$ for any rational value of n.

This will be given in your formula book.

Tip

The condition that $|x|<1$ is important. Although you can always create the polynomial on the right-hand side, the series will only converge when $|x|<1$, so the expansion is only valid if $|x|<1$.

Notice that this if n is not a positive whole number this is an infinitely long polynomial.

WORKED EXAMPLE 6.1

Find the first four terms in ascending powers of x in the binomial expansion of $\frac{1}{1-x}$ for $-1<x<1$.

$\frac{1}{1-x}=(1-x)^{-1}$ — Rewrite the fraction in the form $(1+x)^n$.

$=1+(-1)(-x)+\frac{(-1)(-1-1)}{2}(-x)^2+\frac{(-1)(-1-1)(-1-2)}{6}(-x)^3\ldots$ — Use Key point 6.1 with $n=-1$.

Be careful with powers of $-x$: $(-x)^2=x^2$.

$=1+x+x^2+x^3\ldots$

The formula in Key point 6.1 only works if the first number in the brackets is 1. If it is something else then you have to factorise the expression to turn the first number into 1. The range over which the expansion converges might also then change.

Rewind

The idea of a sequence converging was covered in Chapter 4, Section 1.

WORKED EXAMPLE 6.2

a Find the first three terms, in ascending powers of x, in the binomial expansion of $\sqrt{4+x}$.

b Find the values of x for which this expansion is valid.

a $\sqrt{4+x}=(4+x)^{\frac{1}{2}}$ — Rewrite the square root in the form $(a+x)^n$.

$=\left(4\left(1+\frac{x}{4}\right)\right)^{\frac{1}{2}}$ — Take out a factor of 4 inside the bracket to leave $(1+\ldots)^n$.

$=4^{\frac{1}{2}}\left(1+\frac{x}{4}\right)^{\frac{1}{2}}$ — Make sure the factor comes out as $4^{\frac{1}{2}}$.

$=2\left(1+\left(\frac{1}{2}\right)\left(\frac{x}{4}\right)+\frac{\frac{1}{2}\left(\frac{1}{2}-1\right)}{2}\left(\frac{x}{4}\right)^2\ldots\right)$ — Now use Key point 6.1 with $n=\frac{1}{2}$.

Be careful with powers of $\frac{x}{4}$: $\left(\frac{x}{4}\right)^2=\frac{x^2}{16}$.

$=2\left(1+\frac{1}{8}x-\frac{1}{128}x^2\ldots\right)$

$=2+\frac{1}{4}x-\frac{1}{64}x^2\ldots$ — Finally, multiply out the bracket.

Continues on next page

b $\left|\frac{x}{4}\right|<1$ — For $|x|<1$, use $\left|\frac{x}{4}\right|<1$.

$|x|<4$ — Multiply through by 4.

You could also write $-4<x<4$.

Binomial expansions such as these produce infinite polynomials so you might wonder how they can be useful. If x is small then very large powers of x are extremely small, so they can be neglected. You can therefore use the first few terms of the binomial expansions to approximate the original expression.

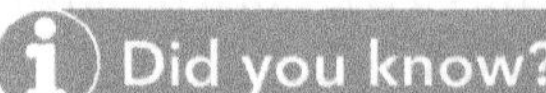

Did you know?

You might like to try finding $\sqrt{9.4}$ on your calculator and seeing how close this is to the binomial approximation. Your calculator uses a method very similar to the binomial approximation to work out square roots.

WORKED EXAMPLE 6.3

a Find the first two terms, in ascending powers of x, of the binomial expansion of $\sqrt{9+4x}$.
b Use this polynomial to approximate $\sqrt{9.4}$ to three decimal places.

a $\sqrt{9+4x}=(9+4x)^{\frac{1}{2}}$ — Rewrite the square root in the form $(a+x)^n$.

$=\left(9\left(1+\frac{4x}{9}\right)\right)^{\frac{1}{2}}$ — Take out a factor of 9 inside the bracket to leave $(1+\ldots)^n$.

$=9^{\frac{1}{2}}\left(1+\frac{4x}{9}\right)^{\frac{1}{2}}$ — Make sure the factor comes out as $9^{\frac{1}{2}}$.

$=3\left(1+\left(\frac{1}{2}\right)\left(\frac{4x}{9}\right)+\ldots\right)$ — Now use Key point 6.1 with $n=\frac{1}{2}$.

$=3\left(1+\frac{2}{9}x+\ldots\right)$

$=3+\frac{2x}{3}+\ldots$ — Finally, multiply out the bracket.

b Need x such that $\sqrt{9+4x}=\sqrt{9.4}$. — Find the value of x that will give $\sqrt{9.4}$.

$9+4x=9.4$

$4x=0.4$

$x=0.1$

$\therefore \sqrt{9.4}=3+\frac{2\times 0.1}{3}+\ldots$ — Substitute $x=0.1$ into $\sqrt{9+4x}=3+\frac{2x}{3}+\ldots$

≈ 3.066

EXERCISE 6A

1 Find the first three terms of the binomial expansion of each expression, stating the range of convergence.

a **i** $(1+x)^{-2}$ **ii** $(1+x)^{-3}$ **b** **i** $(1+x)^{\frac{1}{3}}$ **ii** $(1+x)^{\frac{1}{4}}$

c **i** $\sqrt{1-2x}$ **ii** $\sqrt{1-3x}$ **d** **i** $\dfrac{1}{4-x}$ **ii** $\dfrac{1}{5-x}$

2 Find the expansion of $\dfrac{1}{\left(1-\frac{1}{3}x\right)^3}$ in ascending powers of x up to and including the term in x^2.

Elevate

See Support Sheet 6 for a further example of an expansion when the constant term isn't 1, and for more practice questions.

3 Find the expansion of $\dfrac{1}{\sqrt[3]{8+x}}$ in ascending powers of x up to and including the term in x^3.

4 **a** Find the first three terms in the expansion of $\dfrac{1}{4x^2+4x+1}$.

b Find the values of x for which this expansion is valid.

5 **a** Use the binomial expansion to show that $\sqrt{1+\dfrac{x}{9}} = 1+\dfrac{x}{18}-\dfrac{x^2}{648}\ldots$

b State the range of values for which this expansion converges.

c Deduce the first three terms of the binomial expansion of $\sqrt{9+x}$.

d Use the first three terms of the expansion to find an approximation for $\sqrt{10}$ to four decimal places.

6 **a** Find the first four terms of $\sqrt{1-4x}$ in ascending powers of x.

b State the range of values for which this expansion is valid.

c Hence approximate:

i $\sqrt{96}$ to five decimal places

ii $\sqrt{0.006}$ to four decimal places.

7 The cubic term in the expansion of $(1+ax)^{-2}$ is $-256x^3$. Find the value of a.

8 Given that the expansion of $(1+ax)^n$ is $1-12x+90x^2+bx^3+\ldots$ find the value of b.

9 Given that the expansion of $(1+ax)^n$ is $1-x-\dfrac{x^2}{2}+bx^3+\ldots$ find the value of b.

10 Assume that $\sqrt{1+x} = a_0+a_1x+a_2x^2+\ldots$

By squaring both sides and comparing coefficients find the values of a_0, a_1 and a_2.

Does this agree with the binomial expansion of $\sqrt{1+x}$?

Section 2: Binomial expansions of compound expressions

Binomial expansions can be part of a larger expression.

WORKED EXAMPLE 6.4

Find the quadratic term in the binomial expansion of $\sqrt{\frac{1+x}{1-x}}$ if $|x|<1$.

$\sqrt{\frac{1+x}{1-x}} = (1+x)^{\frac{1}{2}}(1-x)^{-\frac{1}{2}}$

Use the laws of indices to create expressions of the form $(1+x)^n$.

$$= \left(1+\frac{1}{2}x+\frac{\left(\frac{1}{2}\right)\left(\frac{1}{2}-1\right)}{2}x^2 \ldots\right) \times \left(1+\left(-\frac{1}{2}\right)(-x)+\frac{\left(-\frac{1}{2}\right)\left(-\frac{1}{2}-1\right)}{2}(-x)^2 \ldots\right)$$

Expand both brackets: $n=\frac{1}{2}$ in the first bracket and $n=-\frac{1}{2}$ in the second.

$$= \left(1+\frac{1}{2}x-\frac{1}{8}x^2 \ldots\right)\left(1+\frac{1}{2}x+\frac{3}{8}x^2 \ldots\right)$$

Simplify each series before continuing.

Quadratic term:

$$1\times\frac{3}{8}x^2+\frac{1}{2}x\times\frac{1}{2}x+-\frac{1}{8}x^2\times 1$$

There are three ways to get an x^2 term.

$$=\frac{3}{8}x^2+\frac{1}{4}x^2-\frac{1}{8}x^2$$

$$=\frac{1}{2}x^2$$

Sometimes you have to use partial fractions to write an expression in a form where the binomial expansion is possible.

 **Rewind**

Partial fractions were covered in Chapter 5, Sections 3 and 4.

WORKED EXAMPLE 6.5

a Express $\frac{12x+5}{(2x+1)(3x+1)}$ in the form $\frac{A}{2x+1}+\frac{B}{3x+1}$.

b Hence find the first two terms in the expansion of $\frac{12x+5}{(2x+1)(3x+1)}$.

c State the range of values for which this expansion converges.

a $12x+5 \equiv A(3x+1)+B(2x+1)$

Apply the standard method for finding partial fractions.

Substituting in $x=-\frac{1}{2}$:

$-1=-\frac{1}{2}A$

$A=2$

Substituting in $x=-\frac{1}{3}$:

$1=\frac{1}{3}B$

$B=3$

Therefore:

$$\frac{12x+5}{(2x+1)(3x+1)} \equiv \frac{2}{2x+1}+\frac{3}{3x+1}$$

b $\frac{2}{2x+1}=2\times(1+2x)^{-1}$

$=2(1+(-1)(2x)+\ldots)$

$=2-4x+\ldots$

Find the binomial expansion of each term separately.

$\frac{3}{3x+1}=3\times(1+3x)^{-1}$

$=3(1+(-1)(3x)+\ldots)$

$=3-9x+\ldots$

So:

Combine the two expansions.

$$\frac{12x+5}{(2x+1)(3x+1)}=(2-4x+\ldots)+(3-9x+\ldots)$$

$$=5-13x+\ldots$$

c The first expansion converges if $|2x|<1$ so $|x|<\frac{1}{2}$.

The second expansion converges if $|3x|<1$ so $|x|<\frac{1}{3}$.

Consider the range of convergence of each term separately.

Therefore both will converge if $|x|<\frac{1}{3}$.

Think about which range provides the limiting factor:

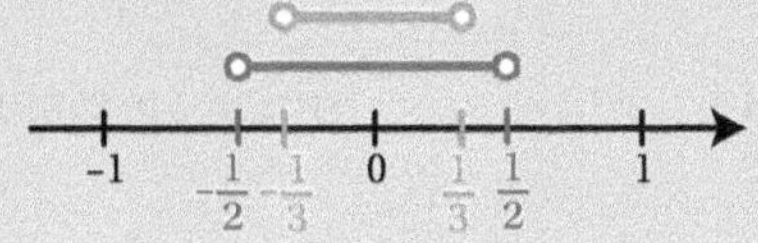

WORK IT OUT 6.1

Find the first three non-zero terms in ascending powers of n in the binomial expansion of $f(x)=\frac{1}{2+x}$.

Which is the correct solution? Identify the errors made in the incorrect solutions.

Solution A	Solution B	Solution C
Set $u=1+x$. Then you need the expansion of $\frac{1}{1+u}=(1+u)^{-1}$. $(1+u)^{-1}=1+(-1)(u)+\frac{(-1)(-1-1)}{2}(u)^2+\ldots$ $=1-u+u^2+\ldots$ Substituting back for u: $f(x)=1-(1+x)+(1+x)^2+\ldots$ $=1-1-x+1+2x+x^2+\ldots$ $=1+x+x^2+\ldots$	$f(x)=(2+x)^{-1}$ $=\left(2\left(1+\frac{x}{2}\right)\right)^{-1}$ $=\frac{1}{2}\left(1+\frac{x}{2}\right)^{-1}$ $=\frac{1}{2}(1+(-1)\left(\frac{x}{2}\right)+\frac{(-1)(-2)}{2}\left(\frac{x}{2}\right)^2$ $=\frac{1}{2}\left(1-\frac{x}{2}+\frac{x^2}{4}\right)$ $=\frac{1}{2}-\frac{x}{4}+\frac{x^2}{8}$	$f(x)=2\left(1+\frac{x}{2}\right)^{-1}$ $=2\left(1-\frac{x}{2}+\frac{(-1)(-2)}{2}\frac{x^2}{2}\right)$ $=2-x+x^2$

EXERCISE 6B

1. Find the first three terms of the expansion of $x(1-x)^{-2}$ in ascending powers of x.

2. Find the first three terms of the expansion of $\sqrt{x^2+2x^3}$ in ascending powers of x, stating the range over which the expansion converges.

3. Find the first three terms of the expansion of $\frac{1+x}{1-x}$.

4. Find the first three terms of the expansion of $\left(\frac{1+x}{1+2x}\right)^2$.

5. a Express $\frac{3x+2}{(x+1)(2x+1)}$ in the form $\frac{A}{x+1}+\frac{B}{2x+1}$.

   b Hence find the first three terms in the binomial expansion of $\frac{3x+2}{(x+1)(2x+1)}$.

   c Write down the set of values for which this expansion is valid.

6. a Express $\frac{5x+8}{(2-x)(x+1)^2}$ in partial fractions.

   b Find the first three terms in ascending powers of the binomial expansion of $\frac{5x+8}{(2-x)(x+1)^2}$.

   c Find the values of x for which this expansion converges.

7. Find the first three terms of the binomial expansion for $\frac{1}{(1+2x)(1+x)}$.

8 a Find the first two terms in ascending powers of the expansion of $\sqrt{\frac{1+5x}{1+12x}}$.

b Over what range of values is the expansion valid?

c By substituting in $x = 0.01$ find an approximation to $\sqrt{15}$ to two decimal places.

9 The first three non-zero terms of the expansion of $(1-x)(1+ax)^n$ are $1+x^2+bx^3$. Find the value of b.

10 Is it possible to find a binomial expansion for $\sqrt{x-1}$?

Checklist of learning and understanding

- If $|x|<1$
  $(1+x)^n = 1+nx+\frac{n(n-1)}{2!}x^2+\frac{n(n-1)(n-2)}{3!}x^3 \ldots$ for any rational value of n.
- You can use the first few terms of a binomial expansion to approximate the expression for small values of x.
- You can find a series expansion for rational functions by first decomposing them into partial fractions.

Mixed practice 6

1. Which expression has a binomial expansion valid for $|x| < 2$?

   Choose from these options.

   A $(1+x)^{-2}$ **B** $(1+2x)^3$ **C** $(2+x)^4$ **D** $\frac{2}{2+x}$

2. Find the first four terms in ascending order in the expansion of $\frac{1}{(1-2x)^2}$, stating the range of x-values for which the expansion converges.

3. Find the first three non-zero terms in ascending order in the expansion of $\frac{1}{\sqrt{4+x}}$, stating the range of x-values for which this is valid.

4. **a** Express $\frac{1}{1+2x+x^2}$ in the form $\frac{1}{(a+x)^n}$.

   b Hence find the first four terms in the expansion of $\frac{1}{1+2x+x^2}$.

5. Find the first three non-zero terms in ascending order in the expansion of $\sqrt{1-x}\sqrt{1+x}$.

6. Find the quadratic term in the expansion of $\sqrt{2+x^2}$.

   Choose from these options.

   A $\frac{1}{2}$ **B** $\frac{1}{2\sqrt{2}}$ **C** 2 **D** $\sqrt{2}$

7. Which expression has a binomial expansion with no linear term?

   Choose from these options.

   A $\frac{1+x}{(1-x)^2}$ **B** $\frac{1+2x}{(1+x)^2}$ **C** $\frac{1+x}{(1+2x)^2}$ **D** $\frac{1+x}{(1-2x)^2}$

8. **a** Find the first three terms of the binomial expansion of $\sqrt[3]{8+x}$.

   b Hence find an approximation for $\sqrt[3]{8100}$, to two decimal places, showing your reasoning.

9. **a** Write $\frac{3x+16}{(3-x)(x+2)^2}$ in partial fractions.

   b Hence find the first three terms of the expansion of $\frac{3x+16}{(3-x)(x+2)^2}$.

   c Write down the set of x-values for which this expansion is valid.

10. Find the first three terms in the binomial expansion of $\frac{1-x}{x^2-5x+6}$, stating the range of values for which it is valid.

11. The first three terms in the binomial expansion of $\frac{1}{(1+ax)^b}+\frac{1}{(1+bx)^a}$ are $2-6x+15x^2$. Find the values of a and b.

12. **a** **i** Find the binomial expansion of $(1+x)^{-\frac{1}{3}}$ up to and including the term in x^2.

   ii Hence find the binomial expansion of $\left(1+\frac{3}{4}x\right)^{-\frac{1}{3}}$ up to and including the term in x^2.

   b Hence show that $\sqrt[3]{\frac{256}{4+3x}} \approx a+bx+cx^2$ for small values of x, stating the values of the constants a, b and c.

[© AQA 2010]

13 **a** Find the binomial expansion of $(1+6x)^{\frac{2}{3}}$ up to and including the term in x^2.

b Find the binomial expansion of $(8+6x)^{\frac{2}{3}}$ up to and including the term in x^2.

c Use your answer from part **b** to find an estimate for $\sqrt[3]{100}$ in the form $\frac{a}{b}$, where a and b are integers.

[© AQA 2012]

14 **a** Find the first three non-zero terms of the binomial expansion of $\sqrt[3]{\frac{1+2x}{1-x}}$.

b By setting $x = 0.04$, find an approximation for $\sqrt[3]{9}$ to five decimal places.

15 Given that the expansion of $(1+ax)^n$ is $1-9x+54x^2+bx^3$, find the value of b.

16 Find the first three terms of the expansion of $\frac{1}{1+x+x^2}$.

17 **a** $f(x) = \frac{x}{1+x}$ can be written in the form $Ax + Bx^2 + Cx^3 + \ldots$.

Find the values of A, B and C and state the set of values of x for which this converges.

b Show that $f(x)$ can be written in the form $\frac{1}{1+\frac{1}{x}}$.

c $f(x)$ can be written in the form $P + \frac{Q}{x} + \frac{R}{x^2} + \ldots$.

Find the values of P, Q and R and state the set of values of x for which this converges.

d Use an appropriate expansion to approximate $\frac{100}{101}$ to four decimal places, showing your reasoning.

e Use an appropriate expansion to approximate $\frac{1}{51}$ to five significant figures, showing your reasoning.

18 In special relativity the energy of an object with mass m and speed v is given by:

$$E = \frac{mc^2}{\sqrt{1-\frac{v^2}{c^2}}}$$

where $c \approx 3\times10^8\ \text{m s}^{-1}$ is the speed of light.

a Find the first three non-zero terms of the binomial expansion, in increasing powers of v, stating the range over which it is valid.

Let E_2 be the expansion containing two terms and E_3 be the expansion containing three terms.

b By what percentage is E_3 bigger than E_2 if v is:

i 10% of the speed of light

ii 90% of the speed of light?

c Prove that $E > E_3 > E_2$.

See Extension Sheet 6 for a look at Babylonian multiplication.

PROOF 1

Proof of the sum of arithmetic series

You are going to demonstrate that, for an arithmetic series with first term a and common difference d, the sum of the first n terms is:

$$S_n = \frac{n}{2}[2a+(n-1)d]$$

Copy the proof and fill in the gaps.

$S_n = \underbrace{a+[a+d]+\ldots+[a+(n-2)d]+[a+(n-1)d]}_{n \text{ terms}}$ Write out the first couple of terms of the series and the last couple of terms.

$S_n = \underbrace{[\_\_\_\_\_\_]+[\_\_\_\_\_\_]+\ldots+[a+d]+a}_{n \text{ terms}}$ Writing these terms in the opposite order has no effect on the outcome.

$2S_n = \underbrace{[a+(n-1)d]+[\_\_\_\_\_\_]+\ldots+[\_\_\_\_\_\_]}_{n \text{ terms}}$ Adding the two expressions gives n identical terms.

$2S_n = \_\_(2a+(n-1)d)$ Collect like terms.

$S_n = \_\_(2a+(n-1)d)$ Dividing by 2 gives the final result.

Did you know?

Carl Friedrich Gauss (1777–1855) was amongst the most eminent mathematicians of the 19th century. His many contributions to mathematics included great strides in number theory, statistics and physics. He was a child prodigy and there is a famous legend about a lesson where his teacher was hoping to keep him quiet by asking him to add together all of the numbers from 1 to 100. The teacher was somewhat disappointed when he replied with the correct answer within seconds. It is believed that he applied a procedure similar to the one used in this proof.

PROOF 2

Proof of the sum of geometric series

You are going to demonstrate that, for a geometric series with first term a and common ratio r the sum of the first n terms is:

$$S_n = \frac{a(1-r^n)}{1-r}$$

Copy the proof and fill in the gaps.

$S_n = a + ar + ar^2 + \ldots + ar^{n-2} + ar^{n-1}$ — Write out the first few terms and the last few terms of the sum.

$rS_n = ar + ar^2 + ar^3 + \ldots + ar^{n-1} + ar^n$ — ____________

$rS_n - S_n =$ ______ — Subtracting removes all the terms in common between the two series.

$S_n(\_\_\_\_) = a(\_\_\_\_)$ — Factorise both sides.

$S_n = \frac{a(r^n-1)}{r-1}$ — Divide by $r-1$

QUESTIONS

1. This proof doesn't work when $r = 1$. At which stage does it break down? What is the formula for S_n when $r = 1$?

2. Does this proof work when $r = -1$? Can you find a simplified version of the formula in this case?

3. This proof appeals to the very important mathematical idea of self-similarity - looking to get similar structures in two different ways so that things cancel out. Use this idea to evaluate these expressions.

   a $\sqrt{1+\sqrt{1+\sqrt{1+\ldots}}}$ **b** $\frac{1}{1+\frac{2}{1+\frac{2}{1+\ldots}}}$

FOCUS ON... PROBLEM SOLVING 1

Trying small cases

In this section you will look at solving problems about sequences, but the ideas apply in other contexts too. In addition to thinking about trying small cases to spot patterns in sequences, you should also remember several other problem-solving ideas. In particular:

- introduce letters to represent unknowns
- look for things which stay the same
- persevere.

WORKED EXAMPLE 1

The terms of a sequence of positive integers $u_1, u_2, u_3, \ldots$ follow the rule:

$$u_n = \frac{1+u_{n-1}}{u_{n-2}} \text{ for } n \geqslant 3$$

Find $u_{2020} - u_{2000}$.

Let $u_1 = a$ and $u_2 = b$.

First thoughts are that 2020 is a very large number and you obviously can't work out all the terms on the way up to it. Also you are not told what u_1 and u_2 are, so perhaps it doesn't matter. You'll need to give them names to work with them though.

Then $u_3 = \frac{1+b}{a}$

$$u_4 = \frac{1+\frac{1+b}{a}}{b} = \frac{a+1+b}{ab}$$

Keep working out some more cases and hope you see something useful! It's worth simplifying these expressions to help anything useful stand out.

$$u_5 = \frac{1+\frac{a+1+b}{ab}}{\frac{1+b}{a}} = \frac{ab+a+b+1}{ab} \times \frac{a}{1+b} = \frac{(a+1)(b+1)}{ab} \times \frac{a}{1+b} = \frac{(a+1)}{b}$$

Persevere....

$$u_6 = \frac{1+\frac{a+1}{b}}{\frac{a+1+b}{ab}} = \frac{b+a+1}{b} \times \frac{ab}{a+1+b} = a$$

The algebra gets messy, but it works out quite nicely in the end!

Continues on next page

$$u_7 = \frac{1+a}{\left(\frac{a+1}{b}\right)}$$

This is interesting - you've got back to a followed by b.

$$= (1+a)\times\frac{b}{a+1}$$

$$= b$$

Importantly,

Since $u_1 = u_6$ and $u_2 = u_7$ and the sequence only depends on the previous two terms, it will continue to repeat every five terms.

$u_{2000} = u_{2005} = u_{2010} = u_{2015} = u_{2020}$

So $u_{2020} - u_{2000} = 0$

QUESTIONS

1 The sequence of integers $u_o, u_1, u_2 \ldots$ satisfies $u_0 = 1$ and $u_{n+1} = \dfrac{ku_n}{u_{n-1}}$ for $n \geqslant 1$ where k is a positive constant.

Given that $u_{2025} = 2025$, find the value of k.

2 Find a formula for the sum of the first n odd numbers.

3 An expression for the sum of the first n terms of an arithmetic sequence is given by $S_n = n^2 + 6n$.

Find an expression for the nth term of the sequence.

4 Rebecca can walk up stairs one at a time or two at a time. For example, to go up 5 stairs she might go 1, 2, 1, 1 or 2, 1, 2.

Her house has stairs with 10 steps. How many different ways can she go up the stairs?

FOCUS ON... MODELLING 1

Modelling with rational functions

You should already be familiar with exponential models, where the rate of change of a quantity is proportional to the quantity itself. In Student Book 1 they were used to model situations such as chemical reactions, population growth and cooling of a liquid. Here are some graphs resulting from exponential models.

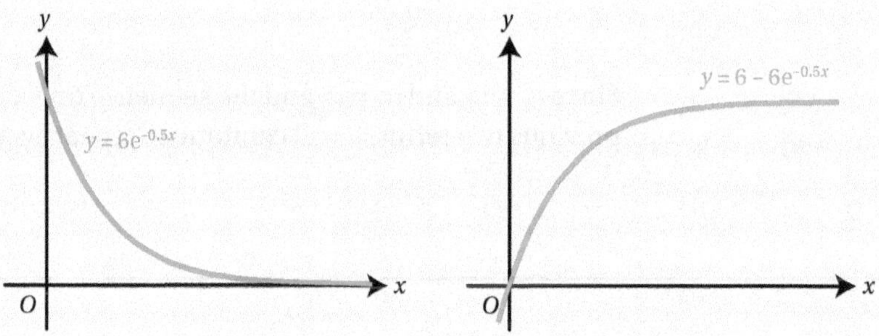

But exponential equations are not the only ones that result in graphs of this shape. Here are the graphs of some rational functions.

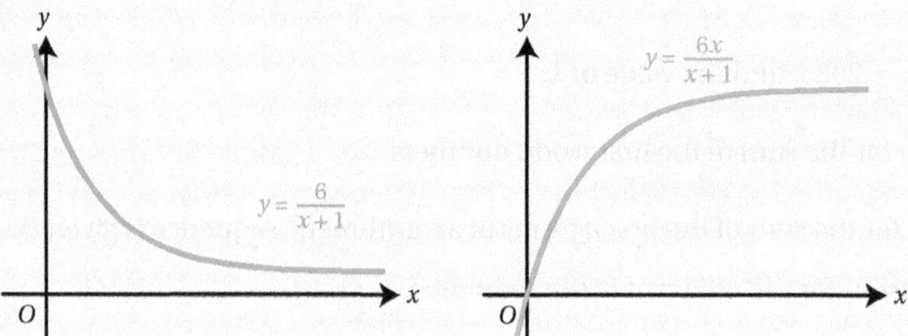

Notice that for a simple rational function, the rate of change is proportional to the **square** of its value.

For example, if $y = \frac{3}{x}$ then $\frac{dy}{dx} = -\frac{3}{x^2} = -\frac{1}{3}y^2$.

In this section you will meet an example of a model using a rational function and then learn a technique that allows you to decide whether an exponential or a rational model is a better fit for given data.

QUESTIONS

1 In biochemistry, the Michaelis-Menten model is used for the rate of reactions involving enzymes. It relates the reaction rate, v, to the concentration of the substrate, S:

$$v = \frac{v_0 S}{k+S}$$

where v_0 and k are constants that depend on the substances involved in the reaction.

a Given that $k = 8000$ and $v_0 = 0.6$ (in suitable units):

  i find the rate of reaction when the concentration is 10 000

  ii find the concentration for which the rate of reaction is 0.3

  iii sketch the graph of v against S.

b What does v_0 represent?

In Student Book 1, Chapter 8, you learnt a method for determining the parameters in an exponential model by using logarithms to turn it into an equation of a straight line. If $y = Ae^{kx}$, then $\ln y = kx + \ln A$.

You can use a similar idea for a rational function of the form of the Michaelis-Menten model. If $y = \frac{bx}{x+c}$ then

$$\frac{1}{y} = \frac{x+c}{bx} = \frac{1}{b} + \frac{c}{b}\frac{1}{x}$$

Hence the graph of $\frac{1}{y}$ against $\frac{1}{x}$ is a straight line with gradient $\frac{c}{b}$ and vertical axis intercept $\frac{1}{b}$.

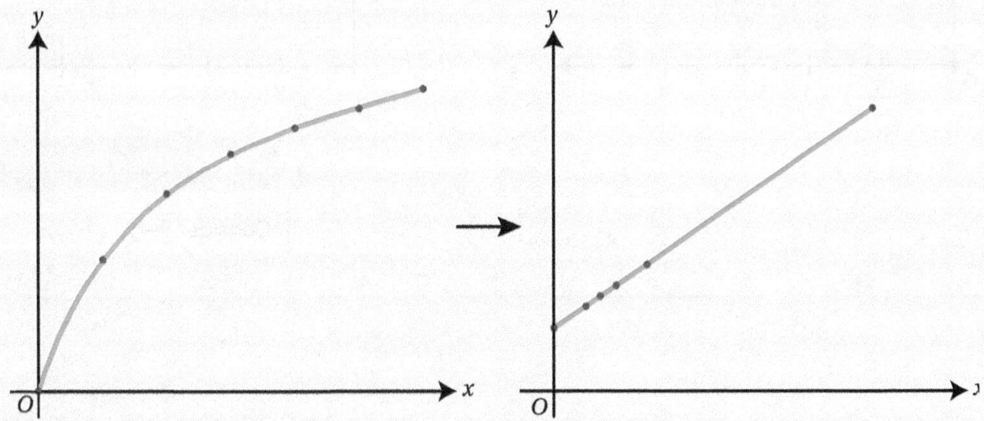

2 The data in the table can be modelled by an equation of the form $y = \frac{bx}{x+c}$.

x	y
0	0.0
1	1.0
2	1.3
3	1.5
4	1.6
5	1.7

Draw the graph of $\frac{1}{y}$ against $\frac{1}{x}$. Hence determine the values of the constants b and c.

The technique of transforming a graph into a straight line can also be used to decide whether a rational or an exponential function is a better model for a set of data.

3 Each graph shows a set of experimental data. Plot $\ln y$ against x and $\frac{1}{y}$ against $\frac{1}{x}$.

Hence decide whether an exponential or a rational function is a better model for the data.

a

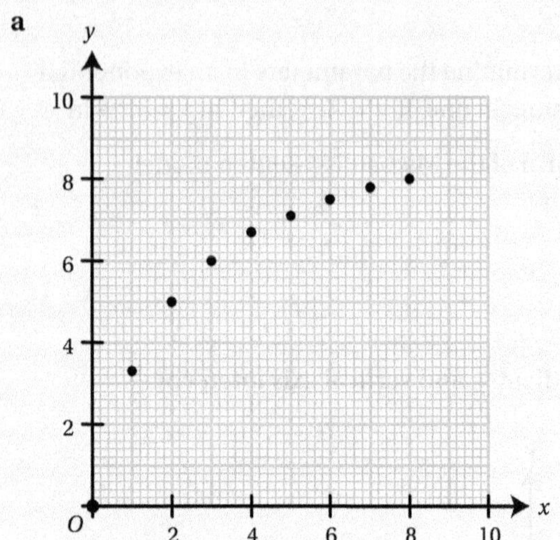

b

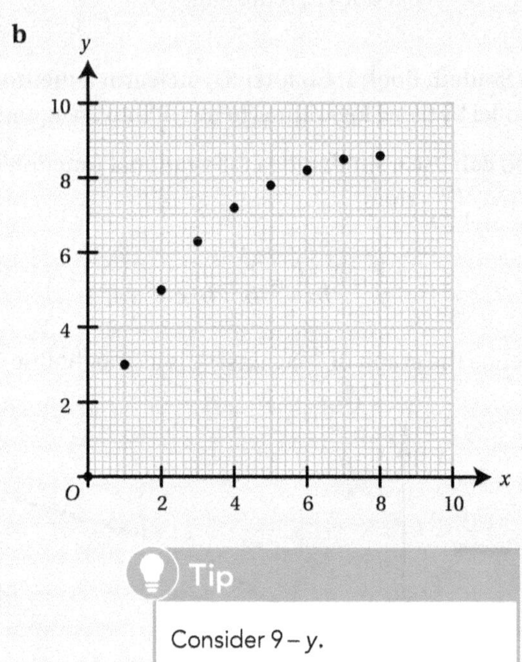

Tip

Consider $9 - y$.

CROSS-TOPIC REVIEW EXERCISE 1

1. Prove that if x is rational and xy is irrational, then y must be irrational.

2. The sum of the first n terms of an arithmetic sequence is given by $S_n = 3n + 2n^2$.

   Find the common difference of the sequence.

3. Find an expression for the mean value of the first n terms of a geometric series with first term a and common ratio r.

4. The function f is defined by $\text{f}(x) = (x-1)^2 + 3$.

   The function g is defined by $\text{g}(x) = ax + b$, where a and b are constants.

   Given that $\text{f}(\text{g}(x)) = 16x^2 - 16x + 7$, find the possible values of a and the corresponding values of b.

5. 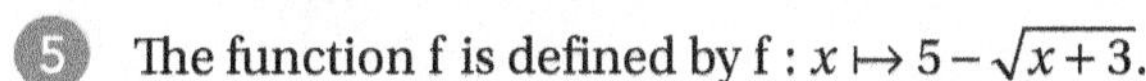
   The function f is defined by $\text{f} : x \mapsto 5 - \sqrt{x+3}$.

   a Find the largest possible domain of f and the corresponding range.

   b **i** State why the inverse function exists.

   ii Find the inverse function, stating its domain.

   c Find the set of values of x for which $\text{f}(x) = |\text{f}(x)|$.

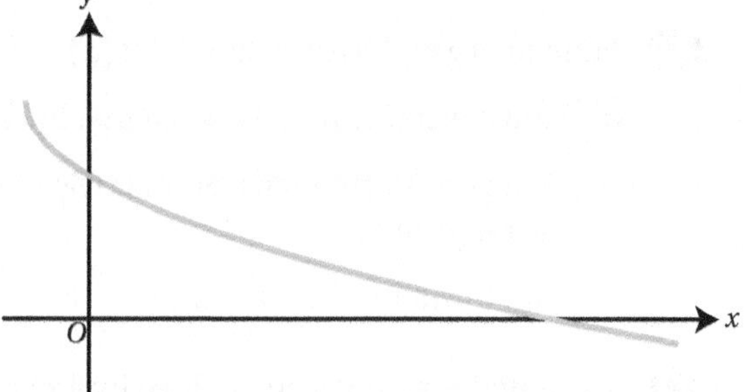

6. If $\text{f}(x) = |x|$ sketch $\text{f}'(x)$ for $-2 \leqslant x \leqslant 2$, $x \neq 0$.

7. **a** Sketch the graph of $y = |9 - 3x|$.

   b Solve the equation $|9 - 3x| = 6$.

   c Solve the inequality $|9 - 3x| > 6$.

8. The polynomial $\text{f}(x)$ defined by $\text{f}(x) = 15x^3 + 19x^2 - 4$.

   a **i** Find f(−1).

   ii Show that $(5x - 2)$ is a factor of $\text{f}(x)$.

   b Simplify

   $$\frac{15x^2 - 6x}{\text{f}(x)}$$

   giving your answer in a fully factorised form.

[© AQA 2010]

9 a Find the range of the function $f(x) = 2x^2 - 12x + 25$.

b Prove that there is no real value of x such that $x^2 + 10, 6x$ and $x^2 + 15$ are consecutive terms of an arithmetic sequence.

10 The sequence u_n is defined by $u_n = 0.5^n$.

a Find the exact value of $\sum_{0}^{10} u_r$.

b Find the exact value of $\sum_{0}^{10} \ln u_r$.

11 Find the range of values of x for which the series $x^2 - x + (x^2 - x)^2 + (x^2 - x)^3 \ldots$ converges.

12 Functions g and h are defined by $g(x) = \sqrt{x}$ and $h(x) = \frac{2x-3}{x+1}$ $(x \neq -1)$.

a Find constants A and B such that $h(x) = A + \frac{B}{x+1}$.

b Using transformations, or otherwise, sketch the graph of $y = h(x)$. Label the axis intercepts and state the range of h.

c Find the domain and range of $g \circ h$.

13 The functions f and g are defined by $f(x) = 3x + 1$ and $g(x) = ax^2 - x + 5$ respectively.

Find the value of a such that $f(g(x)) = 0$ has equal roots.

14 Daniel has a new kitten. In the first week the kitten eats 140 g of cat food. As the kitten grows, the amount of food it needs increases by 7% each week.

a Show that in the fourth week the kitten will need 172 g of cat food.

Daniel bought a 5 kg bag of cat food and wants to estimate how many full weeks the food will last.

b Show that the number of weeks, N, satisfies the inequality $1.07^N \leqslant 3.5$.

c Find the number of whole weeks the bag of food can be expected to last.

d Is this number likely to be an overestimate or an underestimate? Explain your answer.

15 It is given that $f(x) = \frac{7x-1}{(1+3x)(3-x)}$.

a Express $f(x)$ in the form $\frac{A}{3-x} + \frac{B}{1+3x}$, where A and B are integers.

b i Find the first three terms of the binomial expansion of $f(x)$ in the form $a + bx + cx^2$, where a, b and c are rational numbers.

ii State why the binomial expansion cannot be expected to give a good approximation to $f(x)$ at $x = 0.4$.

[© AQA 2013]

16 $\sum_{r=1}^{n} \ln\left(\frac{r+1}{r}\right) = 8$

Find the exact value of n.

17 This table shows the values and gradient of $f(x)$ at various points.

a

x	0	1	2	3	4
$f(x)$	4	2	3	4	6
$f'(x)$	7	9	−3	4	2

Is $f(x)$ a one-to-one function?

b Evaluate $f \circ f(3)$.

c The graph $y = g(x)$ is formed by translating the graph of $y = f(x)$ by a vector $\begin{pmatrix} 2 \\ 3 \end{pmatrix}$ and then reflecting it in the x-axis. Find $g'(2)$.

18 a Find the coordinates of the image of $P(x, y)$ after a reflection in the line $y = x$ followed by a reflection in the y-axis.

b $f(x)$ is a one-to-one function. The graph of $f(x)$ is rotated 90° anticlockwise.

Find the equation of the resulting graph.

19 a If the polynomial $f(x)$ is divided by $(x-a)^2$ the remainder is a linear function.

Explain why this statement can be written as $f(x) = (x-a)^2\, g(x) + mx + c$ where $g(x)$ is a polynomial.

b Find an expression for $f'(x)$ in terms of $g'(x)$ and $g(x)$.

c Hence show that the remainder when $f(x)$ is divided by $(x-a)^2$ is $f(a) + f'(a)(x-a)$.

d State the condition which must be satisfied if $(x-a)^2$ is to be a factor of $f(x)$.

7 Radian measure

In this chapter you will:

- learn about different units for measuring angles, called radians
- learn how to calculate certain special values of trigonometric functions in radians
- learn how to use trigonometric functions in modelling real-life situations
- learn how to solve geometric problems involving circles
- learn that trigonometric functions can be approximated by polynomials
- revise solving trigonometric equations.

Before you start...

Student Book 1, Chapter 10	You should be able to define trigonometric functions beyond acute angles, including exact values.	1 What is the exact value of $\sin 120°$?
Student Book 1, Chapter 10	You should be able to solve trigonometric equations.	2 Solve $\cos 2x = \frac{1}{2}$ for $0° < x < 360°$.
Student Book 1, Chapter 11	You should be able to use the sine and cosine rules.	3 Find the smallest angle in a triangle with sides 3 cm, 4 cm and 6 cm.
Chapter 3	You should be able to identify transformations of graphs.	4 The graph of $y = \cos x$ is translated 30° in the positive x direction and stretched vertically with the scale factor of 2. Find the equation of the new graph.
Chapter 6	You should be able to use the binomial expansion for negative and fractional powers.	5 Find the first three non-zero terms in the expansion of $\frac{2}{1-3x^2}$, in ascending powers of x.

What are radians?

Measuring angles is related to measuring lengths around the perimeter of the circle. This observation leads to the introduction of a new unit for measuring angles, the **radian**, which is a more useful unit of measurement than degrees in advanced mathematics.

Section 1: Introducing radian measure

The measure of 360° for a full turn may seem a little arbitrary – there are many other ways of measuring sizes of angles. In advanced mathematics, the most useful unit for measuring angles is the radian. This measure relates the size of the angle to the distance moved by a point around a circle.

Consider a circle with centre O and radius 1 (this is called the unit circle), and two points, A and B, on its circumference. As the line OA rotates into position OB, point A moves a distance equal to the length of the arc AB. The measure of the angle AOB in radians is simply this arc length.

If point A makes a full rotation around the circle, it will cover the distance equal to the length of the circumference of the circle. As the radius of the circle is 1, the length of the circumference is 2π. Hence a full turn measures 2π radians.

You can then deduce the sizes of other common angles in radians; for example, a right angle is one quarter of a full turn, so it measures $2\pi \div 4 = \frac{\pi}{2}$ radians. Although sizes of common angles measured in radians are often expressed as fractions of π, you can also use decimal approximations. Thus a right angle measures approximately 1.57 radians.

WORKED EXAMPLE 7.1

a Convert 75° to radians.
b Convert 2.5 radians to degrees.

a $\frac{75}{360} = \frac{5}{24}$ — What fraction of a full turn is 75°?

$\frac{5}{24} \times 2\pi = \frac{5\pi}{12}$ — Calculate the same fraction of 2π.

$\therefore 75° = \frac{5\pi}{12}$ radians
$(= 1.31 \text{ radians (3 s.f.)})$ — This is the exact answer. You can also find the decimal equivalent, to three significant figures.

b $\frac{2.5}{2\pi}$ — What fraction of a full turn is 2.5 radians?

$\frac{2.5}{2\pi} \times 360 = 143.24\ldots$ — Calculate the same fraction of 360°.

$\therefore 2.5 \text{ radians} = 143° \text{ (3 s.f.)}$

Key point 7.1

$360° = 2\pi$ radians

- To convert from degrees to radians, divide by 180 and multiply by π.
- To convert from radians to degrees, divide by π and multiply by 180.

Did you know?

There are many different measures of angle. One historical attempt was gradians, which split a right angle into 100 units.

WORKED EXAMPLE 7.2

Mark on the unit circle the points corresponding to these angles, measured in radians.

A π **B** $-\frac{\pi}{2}$ **C** $\frac{5\pi}{2}$ **D** $\frac{13\pi}{3}$

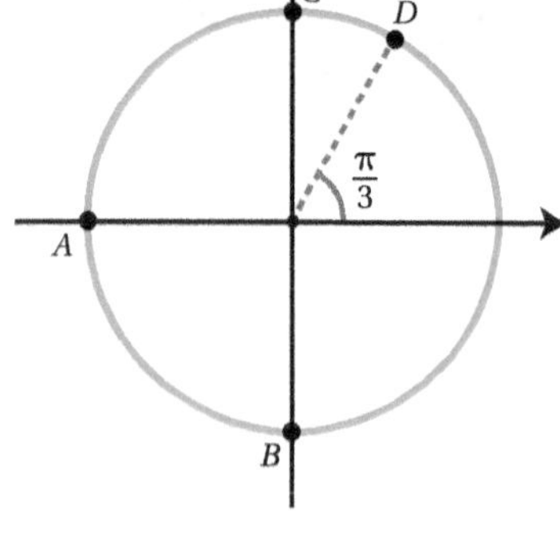

A: π radians is one half of a full turn.

B: $\frac{\pi}{2}$ is one quarter of the full turn and the minus sign represents clockwise rotation.

C: $\frac{5\pi}{2} = 2\pi + \frac{\pi}{2}$, so point C represents a full turn followed by another quarter of a turn.

D: $\frac{13\pi}{3} = 4\pi + \frac{\pi}{3}$, so point D represents two full turns followed by another 6th of a turn.

You can still find the values of trigonometric functions in radians, just as you could in degrees, but you'll need to make sure you change your calculator to radian mode.

Tip

Make sure you can change your calculator between degree and radian modes.

However, as well as getting values of trigonometric functions from your calculator you also need to recognise a few special angles for which you can find exact values. The method relies on properties of special right-angled triangles.

WORKED EXAMPLE 7.3

Find the exact values of $\sin\frac{\pi}{6}$, $\cos\frac{\pi}{6}$ and $\tan\frac{\pi}{6}$.

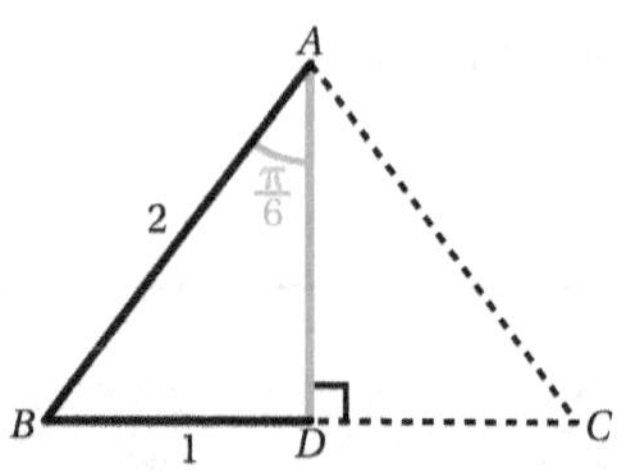

If a right-angled triangle has one angle of $\frac{\pi}{6}$ then the third angle is $\frac{\pi}{3}$, so this is half of an equilateral triangle.

You can choose any length for the side of the equilateral triangle. Let $AB = 2$; then $BD = 1$.

$AD^2 = 2^2 - 1^2 = 3$

$\therefore AD = \sqrt{3}$

Use Pythagoras' theorem to find AD.

$\sin\frac{\pi}{6} = \frac{o}{h} = \frac{1}{2}$

$\cos\frac{\pi}{6} = \frac{a}{h} = \frac{\sqrt{3}}{2}$

$\tan\frac{\pi}{6} = \frac{o}{a} = \frac{1}{\sqrt{3}}$

You can now use the definitions of sin, cos and tan in right-angled triangles.

The results for other values follow similarly.

Key point 7.2

Radians	0	$\frac{\pi}{6}$	$\frac{\pi}{4}$	$\frac{\pi}{3}$	$\frac{\pi}{2}$	π
Degrees	0	30	45	60	90	180
$\sin\theta$	0	$\frac{1}{2}$	$\frac{\sqrt{2}}{2}$	$\frac{\sqrt{3}}{2}$	1	0
$\cos\theta$	1	$\frac{\sqrt{3}}{2}$	$\frac{\sqrt{2}}{2}$	$\frac{1}{2}$	0	−1
$\tan\theta$	0	$\frac{1}{\sqrt{3}}$	1	$\sqrt{3}$	not defined	0

Gateway to A level

See Gateway to A Level Section Z for revision of working with exact values in degrees.

You can find corresponding values for angles greater than $\frac{\pi}{2}$, and derive other properties, by using the symmetries of the trigonometric graphs. You therefore need to recognise x-intercepts and turning points of these graphs, in radians.

Key point 7.3

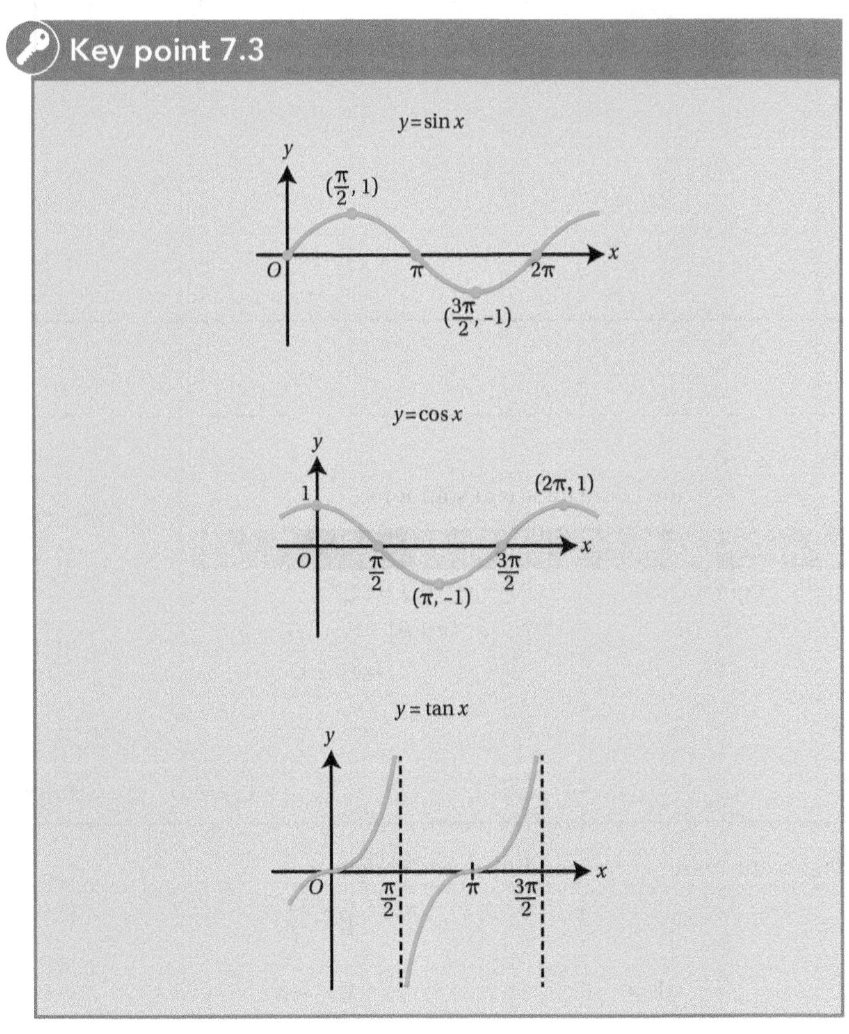

WORKED EXAMPLE 7.4

Given that $\sin\theta = 0.6$, find the values of:

a $\sin(\pi - \theta)$

b $\sin(\theta + \pi)$.

a

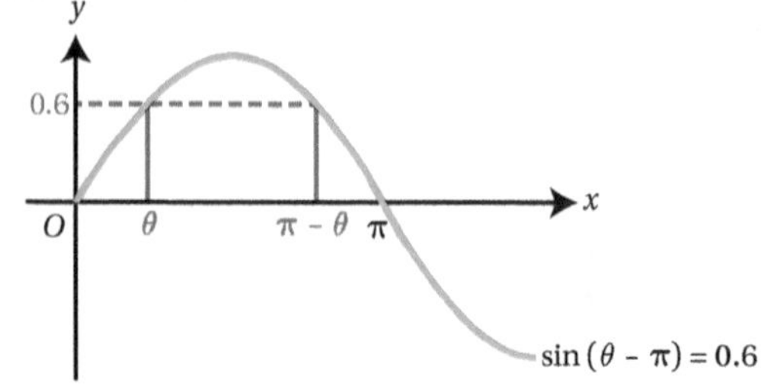

Label θ and $\pi - \theta$ on the horizontal axis.

so $\sin(\pi - \theta) = 0.6$.

The graph has the same height at the two points.

b

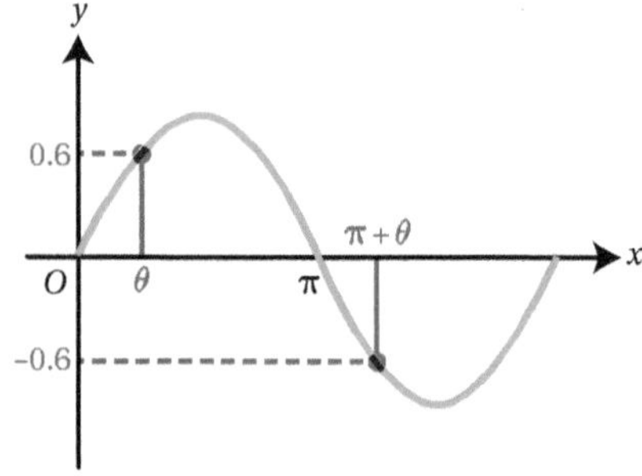

Label θ and $\theta + \pi$ on the horizontal axis.

So $\sin(\theta + \pi) = -0.6$.

WORK IT OUT 7.1

Given that $\cos\theta = 0.2$ find the value of $\cos(\pi + \theta)$.
Which is the correct solution? Identify the errors made in the incorrect solutions.

Solution 1	Solution 2	Solution 3
$\cos(\pi+\theta) = \cos\pi + \cos\theta$	From the graph,	$\theta = \cos^{-1} 0.2 = 1.37$
$= -1 + 0.2$	$\cos(\pi+\theta) = -\cos\theta$	So $\cos(\pi+\theta) = \cos 4.51$
$= -0.8$	$= -0.2$	$= -0.202$

EXERCISE 7A

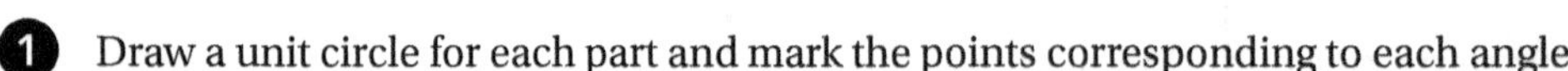

1 Draw a unit circle for each part and mark the points corresponding to each angle.

a i $\frac{\pi}{4}$ ii $\frac{\pi}{3}$ **b** i $\frac{4\pi}{3}$ ii $\frac{3\pi}{4}$

c i $-\frac{\pi}{3}$ ii $-\frac{\pi}{6}$ **d** i -2π ii -4π

2 Express each angle in radians, giving your answers in terms of π.

a **i** $135°$ **ii** $45°$ **b** **i** $90°$ **ii** $270°$

c **i** $120°$ **ii** $150°$ **d** **i** $50°$ **ii** $80°$

3 Express each angle in radians, correct to three decimal places.

a **i** $320°$ **ii** $20°$ **b** **i** $270°$ **ii** $90°$

c **i** $65°$ **ii** $145°$ **d** **i** $100°$ **ii** $83°$

4 Express each angle in degrees.

a **i** $\frac{\pi}{3}$ **ii** $\frac{\pi}{4}$ **b** **i** $\frac{5\pi}{6}$ **ii** $\frac{2\pi}{3}$

c **i** $\frac{3\pi}{2}$ **ii** $\frac{5\pi}{3}$ **d** **i** 1.22 **ii** 4.63

5 Sketch the graph of:

a **i** $y = \sin x$ for $-\frac{\pi}{2} \leqslant x \leqslant \frac{\pi}{2}$ **ii** $y = \sin x$ for $-\pi \leqslant x \leqslant 2\pi$

b **i** $y = \cos x$ for $-\frac{\pi}{2} \leqslant x \leqslant \frac{3\pi}{2}$ **ii** $y = \cos x$ for $-\pi \leqslant x \leqslant 2\pi$.

6 Given that $\sin\frac{\pi}{7} = 0.434$ find, without using a calculator, the value of:

a $\sin\frac{6\pi}{7}$ **b** $\sin\frac{29\pi}{7}$ **c** $\sin\frac{8\pi}{7}$ **d** $\sin\frac{13\pi}{7}$

7 Given that $\cos\frac{\pi}{5} = 0.809$ find, without using a calculator, the value of:

a $\cos\frac{4\pi}{5}$ **b** $\cos\frac{21\pi}{5}$ **c** $\cos\frac{9\pi}{5}$ **d** $\cos\frac{6\pi}{5}$

8 Given that $\tan\frac{\pi}{8} = 0.414$ find, without using a calculator, the value of:

a $\tan\frac{9\pi}{8}$ **b** $\tan\frac{7\pi}{8}$ **c** $\tan\frac{25\pi}{8}$ **d** $\tan\frac{15\pi}{8}$

9 Without using a calculator, find the exact values of:

a **i** $\cos\frac{3\pi}{4}$ **ii** $\cos\frac{5\pi}{4}$ **b** **i** $\sin\left(-\frac{\pi}{6}\right)$ **ii** $\sin\left(-\frac{\pi}{3}\right)$

c **i** $\tan\frac{3\pi}{4}$ **ii** $\tan\left(-\frac{\pi}{4}\right)$

10 Without using a calculator, evaluate each expression, simplifying as far as possible.

a $1 - \sin^2\frac{\pi}{6}$ **b** $\sin\frac{\pi}{4} + \sin\frac{\pi}{3}$ **c** $\cos\frac{\pi}{3} - \cos\frac{\pi}{6}$

11 Show that $\cos^2\frac{\pi}{6} - \sin^2\frac{\pi}{6} = \cos\frac{\pi}{3}$.

12 Show that $\left(1 + \tan\frac{\pi}{3}\right)^2 = 4 + 2\sqrt{3}$.

13 Simplify $\cos(\pi + x) + \cos(\pi - x)$.

14 Simplify the expression $\sin x + \sin\left(x + \frac{\pi}{2}\right) + \sin(x + \pi) + \sin\left(x + \frac{3\pi}{2}\right) + \sin(x + 2\pi)$.

Section 2: Inverse trigonometric functions and solving trigonometric equations

When solving trigonometric equations in Student Book 1, you used $\sin^{-1} x$ to solve equations of the form $\sin x = k$. Based on the work in Chapter 2 you can see that this is an example of an **inverse trigonometric function**.

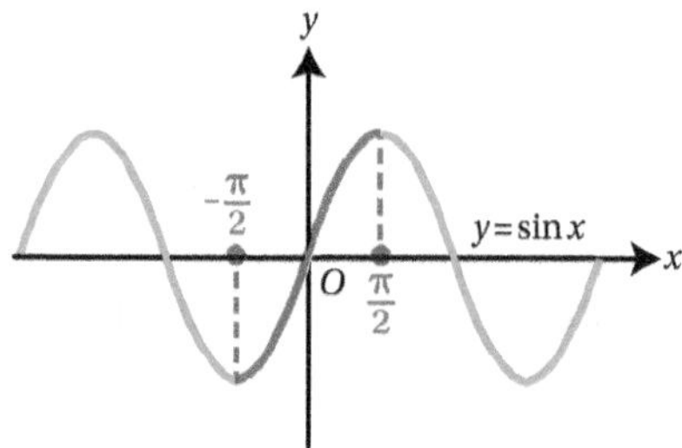

Since the original function needs to be one-to-one for the inverse function to exist, it is clear that you will need to consider the sine graph on a restricted domain.

From the graph you can see that a suitable domain is $-\frac{\pi}{2} \leqslant x \leqslant \frac{\pi}{2}$. (There are other options but this is chosen by convention.) You can now draw the graph of the inverse function as the reflection of the graph of the original function in the line $y = x$. From the graph you can identify the domain and range of the inverse function.

Key point 7.4

The inverse function of $f(x) = \sin x$ is $f^{-1}(x) = \arcsin x$.

Its domain is $[-1, 1]$ and its range is $\left[-\frac{\pi}{2}, \frac{\pi}{2}\right]$.

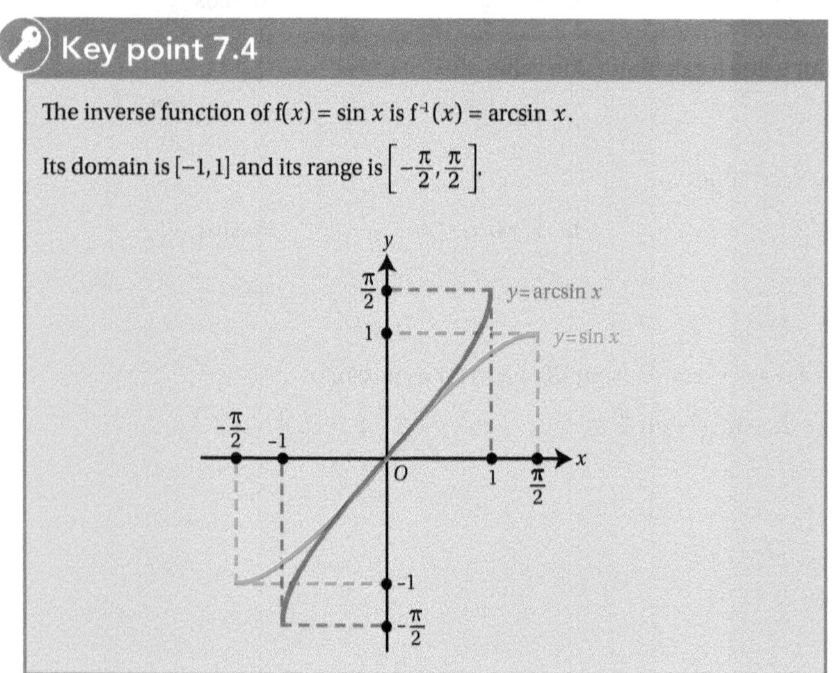

Rewind

You met inverse functions in Chapter 2, Section 4.

You can carry out a similar analysis for cosine and tangent functions to identify the domains on which their inverse functions are defined.

Key point 7.5

The inverse function of $f(x) = \cos x$ is $f^{-1}(x) = \arccos x$.

Its domain is $[-1, 1]$ and its range is $[0, \pi]$.

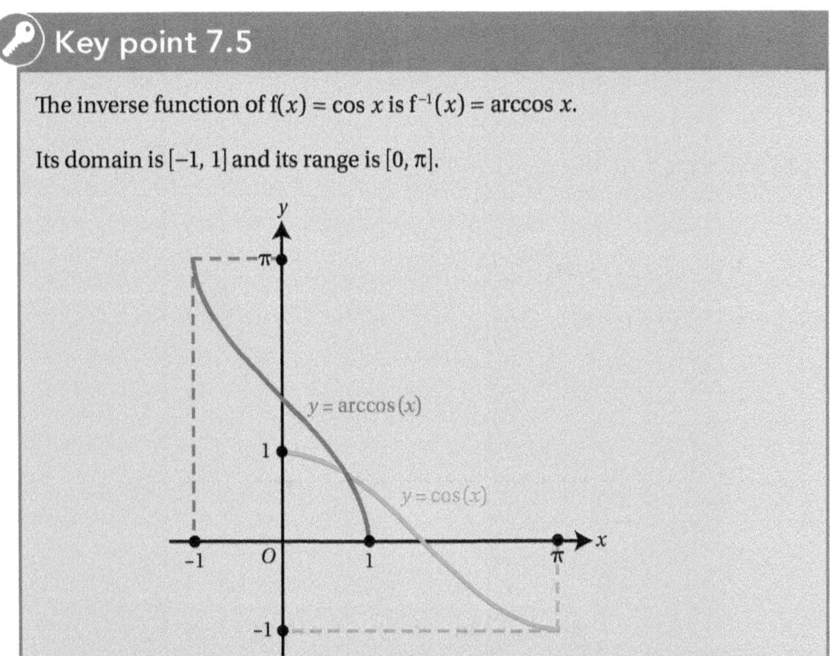

Key point 7.6

The inverse function of $f(x) = \tan x$ is $f^{-1}(x) = \arctan x$.

Its domain is $\mathbb{R}$ and its range is $\left(-\frac{\pi}{2}, \frac{\pi}{2}\right)$.

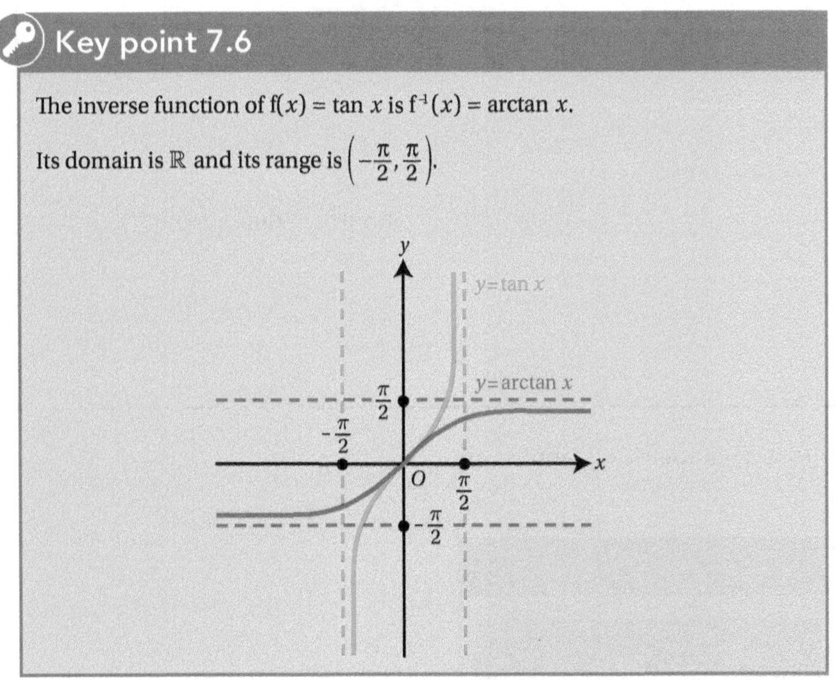

Tip

You will also see $\sin^{-1} x$, $\cos^{-1} x$ and $\tan^{-1} x$ used for the inverse sine, cosine and tangent functions.

You should remember what happens when you compose a function with its inverse: $f(f^{-1}(x)) = x$. Applying this to inverse trigonometric functions, you get:

$$\sin(\arcsin x) = x, \quad \cos(\arccos x) = x, \quad \tan(\arctan x) = x.$$

You need to be a little more careful when composing the other way round: for example, $\arcsin(\sin x) = x$ only when x is in the restricted domain used to define the arcsin function, i.e. $x \in \left[-\frac{\pi}{2}, \frac{\pi}{2}\right]$.

WORK IT OUT 7.2

Evaluate arctan 1.

Which is the correct solution? Identify the errors made in the incorrect solutions.

Solution 1	Solution 2	Solution 3
$\tan^{-1} 1 = \frac{\pi}{4}$	$\tan^{-1} 1 = \frac{1}{\tan 1} \approx 0.642$	$\tan^{-1} 1 = 45°$

WORKED EXAMPLE 7.5

Solve the equation $3 \arcsin 3x = \arcsin \frac{1}{2} + \arccos \frac{1}{2}$

$$3 \arcsin 3x = \arcsin \frac{1}{2} + \arccos \frac{1}{2}$$

Evaluate the arcsin terms on the RHS.

$$= \frac{\pi}{6} + \frac{\pi}{3}$$

$$= \frac{\pi}{2}$$

$$\therefore \arcsin 3x = \frac{\pi}{6}$$

Rearrange.

$$\sin(\arcsin 3x) = \sin \frac{\pi}{6}$$

Taking sin of both sides undoes the arcsin.

$$3x = \frac{1}{2}$$

$$x = \frac{1}{6}$$

As you learnt in Student Book 1, Chapter 10, trigonometric equations can have more than one solution.

Key point 7.7

Equation	First solution, θ_1	Second solution, θ_2	Further solutions
$\sin \theta = k$	$\sin^{-1} k$	$\pi - \theta_1$	Add or subtract multiples of 2π.
$\cos \theta = k$	$\cos^{-1} k$	$-\theta_1$	Add or subtract multiples of 2π.
$\tan \theta = k$	$\tan^{-1} k$		Add or subtract multiples of π.

WORKED EXAMPLE 7.6

Find the values of x between $-\pi$ and 2π for which $\cos x = \frac{\sqrt{2}}{2}$.

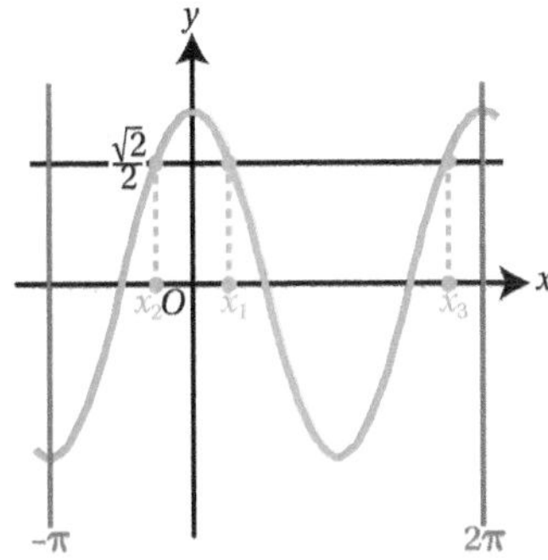

Sketch the graph.

Note there are three solutions.

Three solutions.

$x_1 = \arccos\frac{\sqrt{2}}{2} = \frac{\pi}{4}$

Use inverse cos to find the first solution.

$x_2 = -\frac{\pi}{4}$

Use the symmetry of the graph to find the other solutions.

$x_3 = 2\pi - \frac{\pi}{4} = \frac{7\pi}{4}$

$\therefore x = -\frac{\pi}{4}, \frac{\pi}{4}, \frac{7\pi}{4}$

WORK IT OUT 7.3

Solve $\tan 3x = \sqrt{3}$ for $0 < x < \frac{\pi}{2}$.

Which is the correct solution? Identify the errors made in the incorrect solutions.

Solution 1	Solution 2	Solution 3
$\tan x = \frac{\sqrt{3}}{3} = \frac{1}{\sqrt{3}}$ So $x = \tan^{-1}\frac{1}{\sqrt{3}}$ $= \frac{\pi}{6}$	$3x = \tan^{-1}\sqrt{3} = \frac{\pi}{3}$ $x = \frac{\pi}{9}$ Adding on π would take us outside of the required region so this is the only solution.	$3x = \tan^{-1}\sqrt{3} = \frac{\pi}{3}$ $3x = \frac{\pi}{3}, \frac{4\pi}{3}, \frac{7\pi}{3} \ldots$ $x = \frac{\pi}{9}, \frac{4\pi}{9}$ are the only solutions in the required range.

EXERCISE 7B

1 Use your calculator to evaluate these ratios, in radians, correct to three significant figures.

a **i** arccos 0.6 **ii** arcsin 0.2 **b** **i** arctan (−3) **ii** arcsin (−0.8)

2 Without using a calculator, find the exact value in radians.

a i $\arcsin\frac{1}{2}$ ii $\arccos\frac{\sqrt{3}}{2}$ b i $\arctan\left(-\sqrt{3}\right)$ ii $\arccos\left(-\frac{1}{\sqrt{2}}\right)$

c i $\arcsin(-1)$ ii $\arctan 1$

3 Without using a calculator, find the exact value.

a i $\arcsin\left(\sin\frac{\pi}{3}\right)$ ii $\arccos\left(\cos\frac{5\pi}{6}\right)$ b i $\arcsin\left(\sin\frac{2\pi}{3}\right)$ ii $\arccos\left(\cos(-3\pi)\right)$

c i $\arccos\left(\cos\frac{5\pi}{3}\right)$ ii $\arcsin\left(\sin\frac{7\pi}{4}\right)$ d i $\arctan\left(\tan\frac{7\pi}{4}\right)$ ii $\arctan\left(\tan\frac{11\pi}{6}\right)$

4 Without using a calculator, solve each equation.

a $\arcsin x = \frac{\pi}{3}$ b $\arccos 2x = \frac{5\pi}{6}$ c $\arctan(3x-1) = -\frac{\pi}{6}$

5 Without using a calculator, find the values of x between 0 and 2π for which:

a i $\cos x = \frac{\sqrt{3}}{2}$ ii $\cos x = \frac{\sqrt{2}}{2}$ b i $\cos x = -\frac{1}{2}$ ii $\cos x = -\frac{\sqrt{3}}{2}$

c i $\sin x = \frac{\sqrt{2}}{2}$ ii $\sin x = \frac{\sqrt{3}}{2}$ d i $\tan x = \frac{1}{\sqrt{3}}$ ii $\tan x = -1$.

6 Solve these equations in the given interval, giving your answers to three significant figures, where appropriate. Do not use graphs on your calculator.

a i $\cos t = \frac{4}{5}$ for $t \in [0, 4\pi]$ ii $\cos t = \frac{2}{3}$ for $t \in [0, 4\pi]$

b i $\sin\theta = -0.8$ for $\theta \in [-2\pi, 2\pi]$ ii $\sin\theta = -0.35$ for $\theta \in [-2\pi, 2\pi]$

c i $\tan\theta = -\frac{2}{3}$ or $-\pi \leqslant \theta \leqslant \pi$ ii $\tan\theta = -3$ for $-\pi \leqslant \theta \leqslant \pi$

d i $\cos\theta = 1$ for $\theta \in [0, 4\pi]$ ii $\cos\theta = 0$ for $\theta \in [0, 4\pi]$

7 Find the exact values of $x \in (-\pi, \pi)$ for which $2\sin x + 1 = 0$.

8 Sketch the graph of $y = 3\arcsin x$ for $-1 \leqslant x \leqslant 1$.

9 Find the exact solutions to $\tan 4x = \sqrt{3}$ for $0 \leqslant x \leqslant \pi$.

10 a If $3\cos x = \tan x$ show that $\sin x = \frac{-1+\sqrt{37}}{6}$.

b Find all solutions to $3\cos x = \tan x$ for $0 < x < 2\pi$.

> **Elevate**
>
> See Support Sheet 7 for a further example of solving equations in radians and for more practice questions.

11 Find the exact solutions of the equation $2\cos\theta\tan\theta = \tan\theta$ for $\theta \in [0, \pi]$.

12 Leo says that $\arccos(\cos x) = x$ for all x. Prove by counter example that this statement is false.

13 Find the exact solutions of the equation $\sin(x^2) = \frac{1}{2}$ for $-\pi < x < \pi$.

14 a Prove by a counter example that $\arctan x \neq \frac{\arcsin x}{\arccos x}$.

b Express $\arcsin x$ in terms of $\arccos x$.

c Solve the equation $2\arctan x = \arcsin x + \arccos x$.

Section 3: Modelling with trigonometric functions

Trigonometric functions can be used to model real-life situations that show periodic behaviour, for example, the height of the tide, or the motion of a point on a Ferris wheel. To do this, you need to consider trigonometric functions with different periods and amplitudes.

Using your knowledge of combining transformations from Chapter 3, you can adjust both the amplitude and period of $\sin x$ and $\cos x$.

Key point 7.8

The functions $y = a \sin bx$ and $y = a \cos bx$ have amplitude a and period $\frac{2\pi}{b}$.

Rewind

You know, from Student Book 1, Chapter 10, that $\sin x$ and $\cos x$ have period 2π and amplitude 1.

WORKED EXAMPLE 7.7

a Sketch the graph of $y = 4\cos\frac{x}{3}$ for $0 \leqslant x \leqslant 6\pi$.

b Write down the amplitude and the period of the function.

a Vertical stretch with scale factor 4

Horizontal stretch with scale factor $\frac{1}{3}$

Start with the graph of $y = \cos x$ and consider what transformations to apply to it.

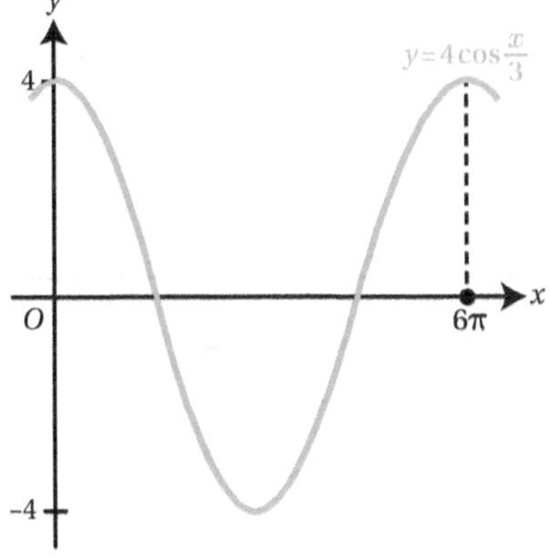

b Amplitude $= 4$

period $= \frac{2\pi}{\left(\frac{1}{3}\right)} = 6\pi$

As well as vertical and horizontal stretches, you can also apply translations to graphs. They will leave the period and the amplitude unchanged, but will change the positions of maximum and minimum points and the axes intercepts.

WORKED EXAMPLE 7.8

a Sketch the graph of $y = \sin\left(x - \frac{\pi}{3}\right) + 2$ for $x \in [0, 2\pi]$.

b Find the coordinates of the maximum and the minimum points on the graph.

a

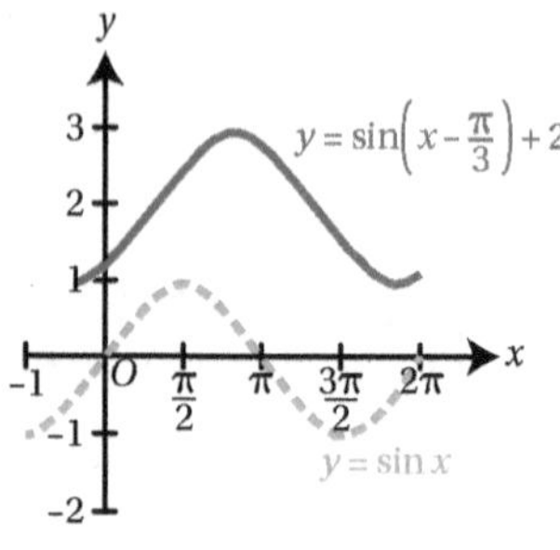

The equation is of the form $y = f\left(x - \frac{\pi}{3}\right) + 2$.

This represents a translation of $\frac{\pi}{3}$ units to the right and 2 units up.

b Maximum point:

$x = \frac{\pi}{2} + \frac{\pi}{3} = \frac{5\pi}{6}$

$y = 1 + 2 = 3$

So the maximum point is $\left(\frac{5\pi}{6}, 3\right)$.

The maximum and minimum points of $\sin x$ are $\left(\frac{\pi}{2}, 1\right)$ and $\left(\frac{3\pi}{2}, -1\right)$. You need to apply the same translation to these points.

Minimum point:

$x = \frac{3\pi}{2} + \frac{\pi}{3} = \frac{11\pi}{6}$

$y = -1 + 2 = 1$

So the minimum point is $\left(\frac{11\pi}{6}, 1\right)$

You can use your knowledge of transformations of graphs to find an equation of a function, given its graph.

WORKED EXAMPLE 7.9

The graph shown has the equation $y = a\sin(bx) + d$.

Find the values of a, b and d.

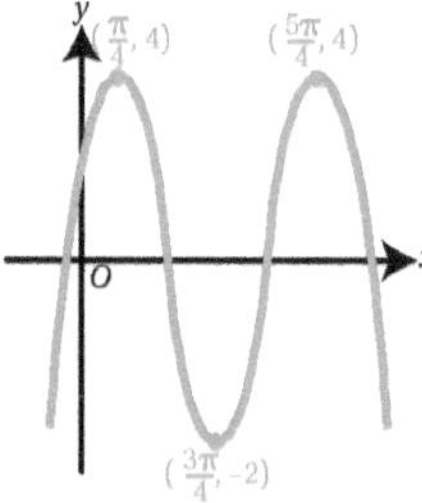

Amplitude $= \frac{4 - (-2)}{2} = 3$

$\therefore a = 3$

a is the amplitude, which is half the difference between the minimum and maximum values.

Period $= \frac{5\pi}{4} - \frac{\pi}{4} = \pi$

$\pi = \frac{2\pi}{b}$

$\therefore b = 2$

b is related to the period, which is the distance between the two consecutive maximum points. The formula is period $= \frac{2\pi}{b}$.

$d = \frac{4 + (-2)}{2}$

$\therefore d = 1$

d represents the vertical translation of the graph. It is the value halfway between the minimum and the maximum values.

You will now use these ideas to create mathematical models of periodic motion, such as motion around a circle, oscillation of a particle attached to the end of a spring, water waves or heights of tides. In practice, you would collect experimental data to sketch a graph and then use your knowledge of trigonometric functions to find its equation. You can then use the equation to do further calculations.

WORKED EXAMPLE 7.10

The height of water in a harbour is 16 m at high tide, and 10 m at low tide, 6 hours later. The graph shows how the height of water changes with time over 12 hours.

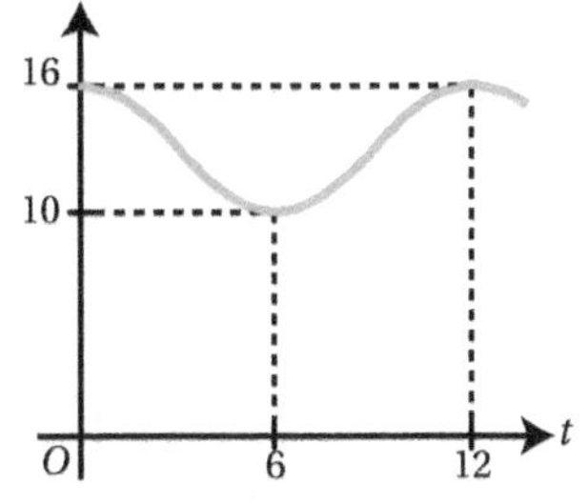

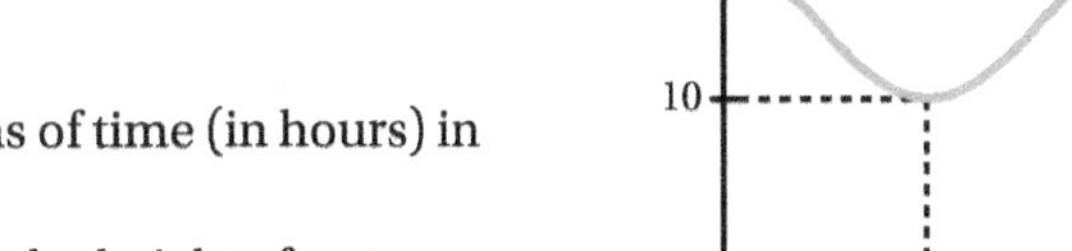

a Find the equation for height (in metres) in terms of time (in hours) in the form $h = m + a\cos bt$.

b Find the first two times after the high tide when the height of water is 12 m.

a $m = \frac{16+10}{2} = 13$ — m is the 'central value', half-way between minimum and maximum values.

amplitude $= \frac{16-10}{2} = 3$ — a is the amplitude, which is half the distance between the minimum and maximum values.

$\therefore a = 3$

period $= 12$ — The period is $\frac{2\pi}{b}$.

$12 = \frac{2\pi}{b}$

$\therefore b = \frac{\pi}{6}$

So $h = 13 + 3\cos\frac{\pi}{6}t$

b $13 + 3\cos\frac{\pi}{6}t = 12$ — Form an equation with $h = 12$.

$\cos\frac{\pi}{6}t = -\frac{1}{3}$ — Rearrange the equation into the form $\cos A = k$.

$\frac{\pi}{6}t = \arccos\left(-\frac{1}{3}\right)$ — The high tide is when $t = 0$, so you want the first two answers with $t > 0$.

$= 1.91$ or $2\pi - 1.91 = 4.37$ — Find two possible values of $\frac{\pi}{6}t$.

$t_1 = \frac{1.91\times 6}{\pi} = 3.64$ (3 s.f.) — Solve for t.

$t_2 = \frac{4.37\times 6}{\pi} = 8.34$ (3 s.f.)

The height of the water will be 12 m 3.64 hours and 8.34 hours after the high tide.

WORKED EXAMPLE 7.11

A point moves with constant speed around a circle of radius 2 cm, starting from the positive x-axis and taking 3 seconds to complete one full rotation. Let h be the height of the point above the x-axis.

Find an equation for h in terms of time.

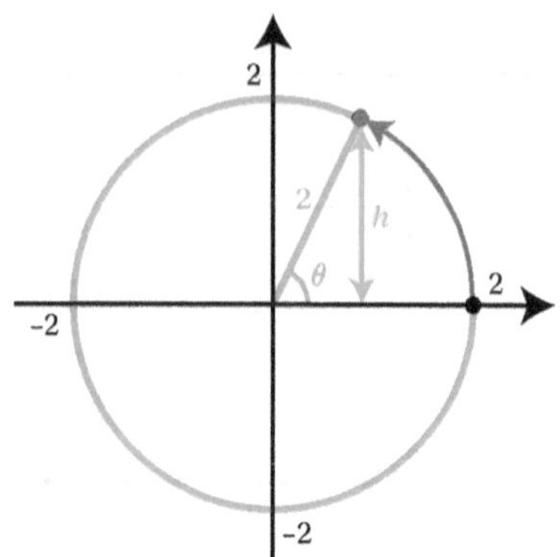

Draw a diagram; you can see that $h = 2\sin\theta$, where θ is the angle between the radius and the x-axis.

You now need to find how θ depends on time.

From the diagram: $h = 2\sin\theta$

Let t be the time, measured in seconds.

You have information about the values of θ at the start and 3 seconds later.

$\theta = 0$ when $t = 0$

$\theta = 2\pi$ and $t = 3$

So $\theta = \frac{2\pi}{3}t$.

As the point moves with constant speed, θ is directly proportional to t so $\theta = kt$.

Hence $h = 2\sin\left(\frac{2\pi}{3}t\right)$.

EXERCISE 7C

1 State the amplitude and the period of each function, where x is in radians.

a $f(x) = 3\sin 4x$ **b** $f(x) = \cos\frac{x}{2}$

c $f(x) = \cos 3x$ **d** $f(x) = 2\sin\pi x$

2 Sketch each graph, giving coordinates of maximum and minimum points.

a i $y = 2\cos\left(x - \frac{\pi}{3}\right)$ for $0 \leqslant x \leqslant 2\pi$ **ii** $y = 3\sin\left(x + \frac{\pi}{2}\right)$ for $0 \leqslant x \leqslant 2\pi$

b i $y = \sin 2x$ for $-\pi \leqslant x \leqslant \pi$ **ii** $y = \cos 3x$ for $0 \leqslant x \leqslant \pi$

c i $y = \tan\left(x - \frac{\pi}{2}\right)$ for $0 \leqslant x \leqslant \pi$ **ii** $y = \tan\left(x + \frac{\pi}{3}\right)$ for $0 \leqslant x \leqslant \pi$

d i $y = 3\cos x - 2$ for $0 \leqslant x \leqslant 4\pi$ **ii** $y = 2\sin x + 1$ for $-\pi \leqslant x \leqslant \pi$

3 The depth of water in a harbour varies during the day and is given by the equation $d = 16 + 7\sin\left(\frac{\pi}{6}t\right)$, where d is measured in metres and t in hours after midnight.

a Find the depth of the water at low and high tide.

b At what times does high tide occur?

4 A small ball is attached to one end of an elastic spring, and the other end of the spring is fixed to the ceiling. The ball is pulled down and released, and starts to oscillate vertically. The graph shows how the length of the spring, x cm, varies with time. The equation of the graph is $x = L + A\cos(5\pi t)$.

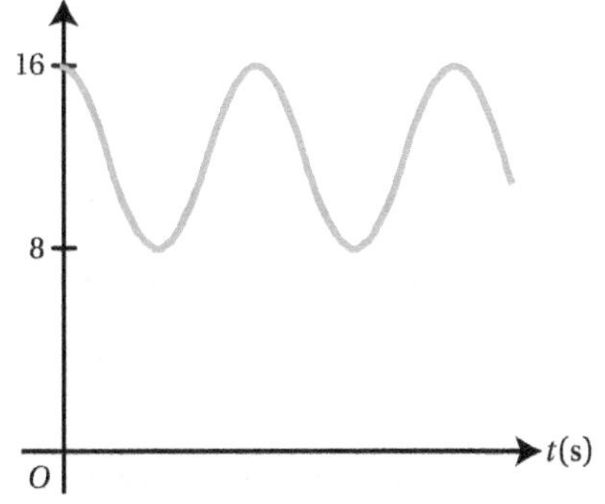

a Write down the amplitude of the oscillation.

b How long does it take for the ball to perform five complete oscillations?

5 This graph has equation $y = p\sin qx$ for $0 \leqslant x \leqslant 2\pi$. Find the values of p and q.

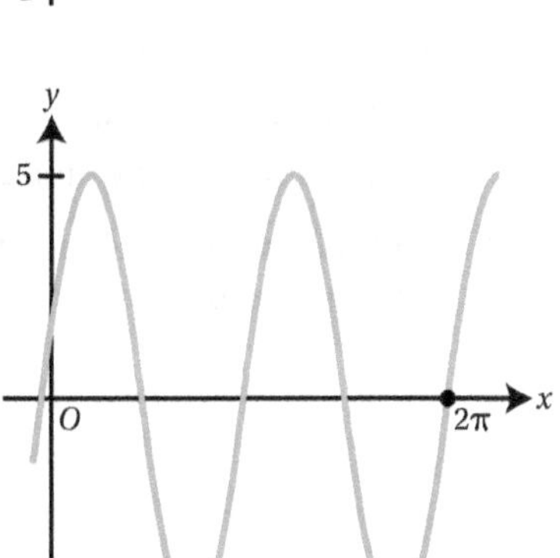

6 This graph has equation $y = a\cos(x - b)$ for $0° \leqslant x \leqslant 720°$. Find the values of a and b.

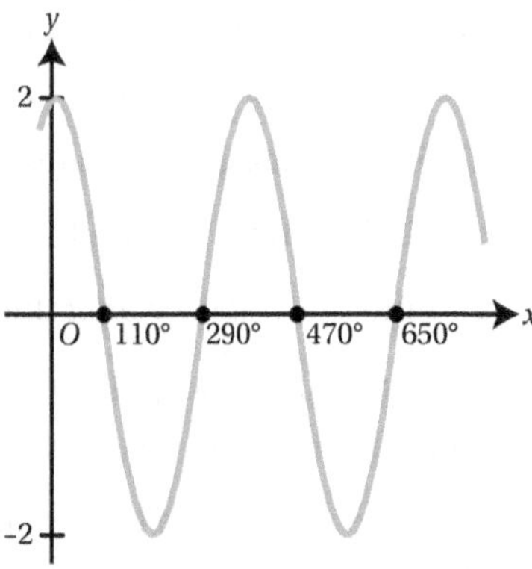

7 a On the same set of axes sketch the graphs of $y = 1 + \sin 2x$ and $y = 2\cos x$ for $0 \leqslant x \leqslant 2\pi$.

b Hence state the number of solutions of the equation $1 + \sin 2x = 2\cos x$ for $0 \leqslant x \leqslant 2\pi$.

c Write down the number of solutions of the equation $1 + \sin 2x = 2\cos x$ for $-2\pi \leqslant x \leqslant 6\pi$.

8 The graph shows the height of water below the level of a walkway as a function of time. The equation of the graph is of the form $y = a\cos bt + m$. Find the values of a, b and m.

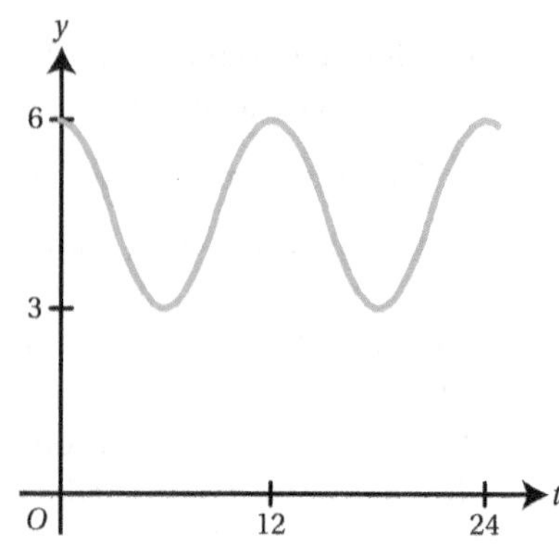

9 **a** Sketch the graph of $y = 2\cos(x + 60°)$ for $x \in [0°, 360°]$.

b Find the coordinates of the maximum and minimum points on the graph.

c Write down the coordinates of the maximum and minimum points on the graph of $y = 2\cos(x + 60°) - 1$ for $x \in [0°, 360°]$.

10 A point moves around a vertical circle of radius 5 cm, as shown in the diagram. It takes 10 s to complete one revolution.

a The height of the point above the x-axis is given by $h = a\sin kt$, where t is time measured in seconds. Find the values of a and k.

b Find the time during the first revolution when the point is 5 cm below the x-axis.

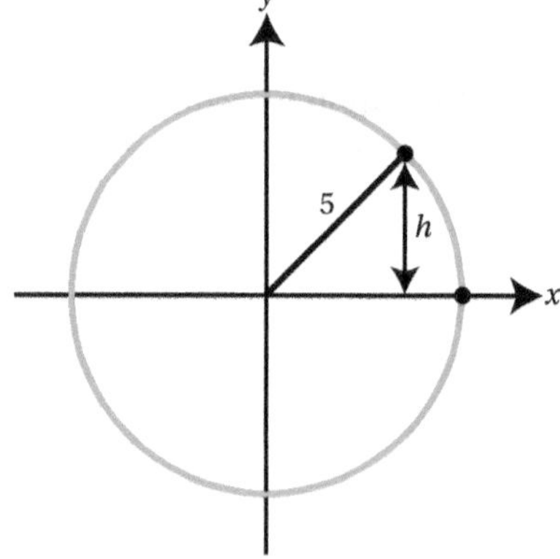

11 A ball is attached at its top and bottom to elastic strings, with each string attached at the far end and under tension. When the ball is pulled down and released, it starts moving up and down, so that the height of the ball above the ground is given by the equation $h = 120 - 10\cos 10t$, where h is measured in centimetres and t is time in seconds.

a Find the least and greatest height of the ball above ground.

b Find the time required to complete one full oscillation.

c Find the first time after the ball is released when it reaches the greatest height.

12 A Ferris wheel has radius 12 m and the centre of the wheel is 14 m above ground. The wheel takes 4 minutes to complete a full rotation.

Seats are attached to the circumference of the wheel. Let θ radians be the angle the radius connecting a seat to the centre of the wheel makes with the downward vertical, and let h be the height of the seat above ground.

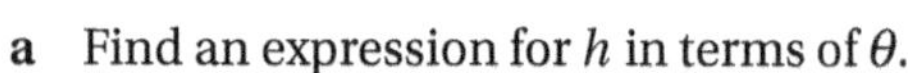

a Find an expression for h in terms of θ.

b Initially the seat is at the lowest point on the wheel. Assuming that the wheel rotates at constant speed, find an expression for θ in terms of t, where t is the time measured in minutes.

c Write down an expression for h in terms of t.
For how long is the seat more than 20 m above ground?

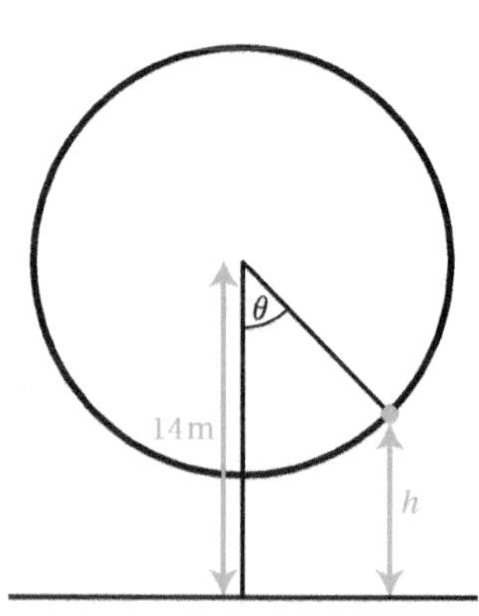

Section 4: Arcs and sectors

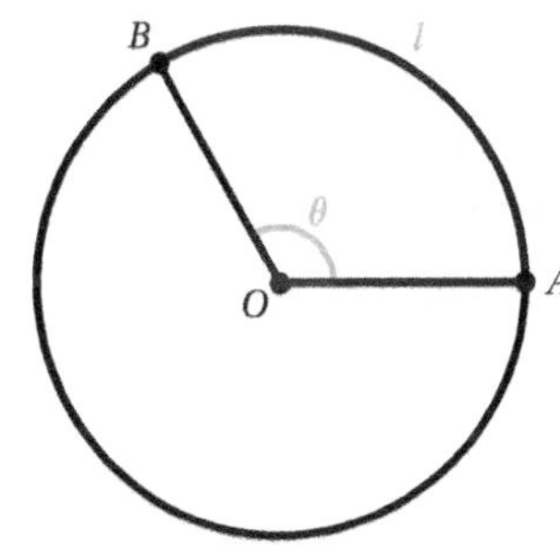

The diagram shows a circle with centre O and radius r, and points A and B on its circumference. The part of the circumference between points A and B is called an **arc** of the circle. You can see that there are in fact two such parts; the shorter one is called the **minor arc**, and the longer one the **major arc**. The minor arc AB **subtends** an angle θ at the centre of the circle.

The ratio of the length of the arc to the circumference of the whole circle is the same as the ratio of angle θ to the angle measuring a full turn. If l is the length of the arc, and angle θ is measured in radians, this means that $\frac{l}{2\pi r} = \frac{\theta}{2\pi}$. Rearranging this equation gives the formula for the arc length.

Key point 7.9

The length of an arc is

$$l = r\theta$$

where r is the radius of the circle and θ is the angle subtended at the centre, measured in radians.

WORKED EXAMPLE 7.12

Arc AB of a circle with radius 5 cm subtends an angle of 0.6 radians at the centre, as shown on the diagram.

a Find the length of the minor arc AB.

b Find the length of the major arc AB.

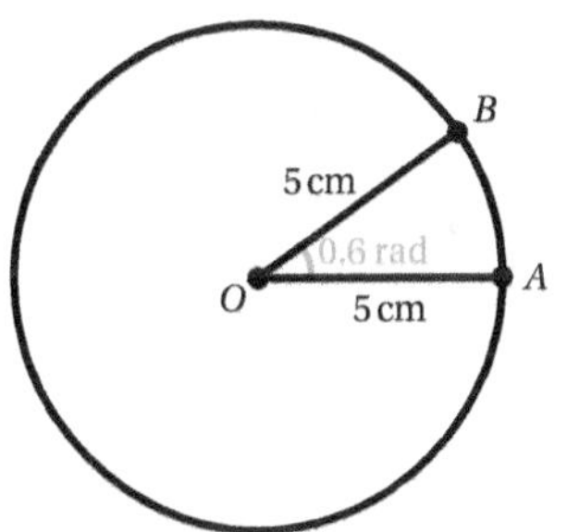

a $l = r\theta$
$= 5 \times 0.6$
$= 3$ cm

Use the formula for the length of an arc.

b $\theta_1 = 2\pi - 0.6$
$= 5.683$

The angle subtended by the major arc is equal to a full turn minus the smaller angle. A full turn is 2π radians.

$l = r\theta_1$
$= 5 \times 5.683$
$= 28.4$ cm (3 s.f.)

A **sector** is a part of a circle bounded by two radii and an arc. As with arcs, you can distinguish between a **minor sector** and a **major sector**.

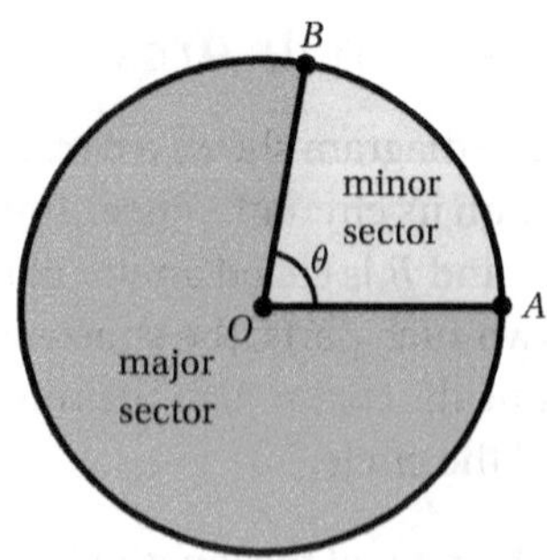

The ratio of the area of the sector to the area of the whole circle is the same as the ratio of angle θ to the angle measuring a full turn. If A is the area of the sector, and angle θ is measured in radians, this means that $\frac{A}{\pi r^2}=\frac{\theta}{2\pi}$. Rearranging this equation gives the formula for the area of the sector.

Key point 7.10

The area of a sector of a circle is

$$A=\frac{1}{2}r^2\theta$$

where r is the radius of the circle and θ is the angle subtended at the centre, measured in radians.

Common error

You can only use the formulae for arc length and area of sector in Key points 7.9 and 7.10 if the angle is in radians.

WORKED EXAMPLE 7.13

A sector of a circle has perimeter $p=12$ cm and angle at the centre $\theta=50°$. Find the area of the sector.

$\theta=50\times\frac{\pi}{180}=0.873$ radians

To use the formula for the area of a sector, the angle needs to be in radians.

$p=r\theta+2r$

You are given the perimeter, and you know the formula for it. Use this to find r.

$12=0.873r+2r$

$12=2.873r$

$r=\frac{12}{2.873}=4.18$ cm

$A=\frac{1}{2}(4.18)^2(0.873)$

$=7.63$ cm^2

Now use $A=\frac{1}{2}r^2\theta$.

EXERCISE 7D

1 Calculate the length of the minor arc subtending an angle of θ radians at the centre of the circle of radius r cm.

a $\theta=1.2,\ r=6.5$ **b** $\theta=0.4,\ r=4.5$

2 Points A and B lie on the circumference of a circle with centre O and radius r cm. Angle AOB is θ radians. Calculate the length of the major arc AB.

a $r=15,\ \theta=0.8$ **b** $r=1.4,\ \theta=1.4$

3 Points M and N lie on the circumference of a circle with centre O and radius r cm, and $\angle MON=\alpha$. Calculate the area of the minor sector MON.

a $r=5,\ \alpha=1.3$ **b** $r=0.4,\ \alpha=0.9$

4 Points A and B lie on the circumference of a circle with centre C and radius r cm. The size of the angle ACB is θ radians. Calculate the area of the major sector ACB.

a $r=13,\ \theta=0.8$ b $r=1.4,\ \theta=1.4$

5 Calculate the length of the minor arc AB in the diagram.

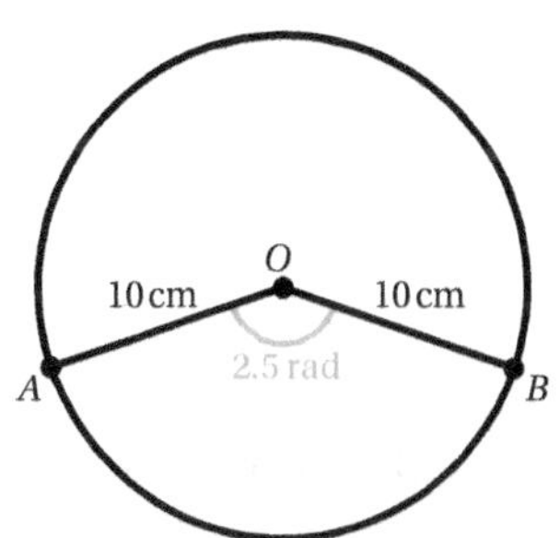

6 In the diagram, the radius of the circle is 8 cm and the length of the minor arc AB is 7.5 cm. Calculate the size of the angle AOB:

a in radians b in degrees.

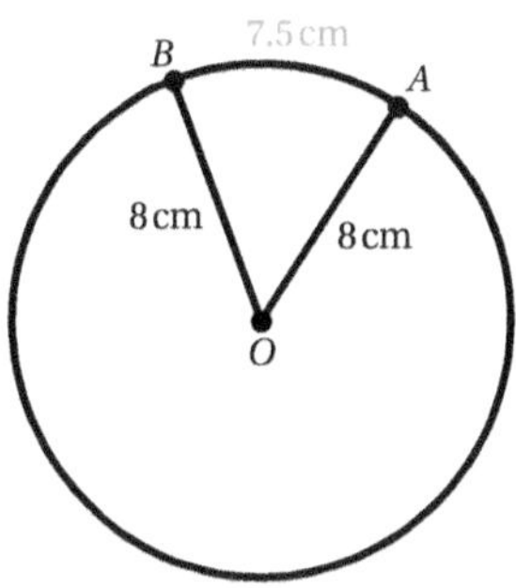

7 Points M and N lie on the circumference of a circle with centre C and radius 4 cm. The length of the major arc MN is 15 cm. Calculate the size of the smaller angle MCN.

8 Points P and Q lie on the circumference of the circle with centre O. The length of the minor arc PQ is 12 cm and $\angle POQ = 1.6$ radians. Find the radius of the circle.

9 A circle has centre O and radius 10 cm. Points A and B lie on the circumference of the circle so that the area of the minor sector AOB is 40 cm^2. Calculate the size of the angle AOB.

10 Points P and Q lie on the circumference of a circle with radius 21 cm. The area of the major sector POQ is 744 cm^2. Find the size of the smaller angle POQ in degrees.

11 In the diagram, the length of the major arc XY is 28 cm. Find the radius of the circle.

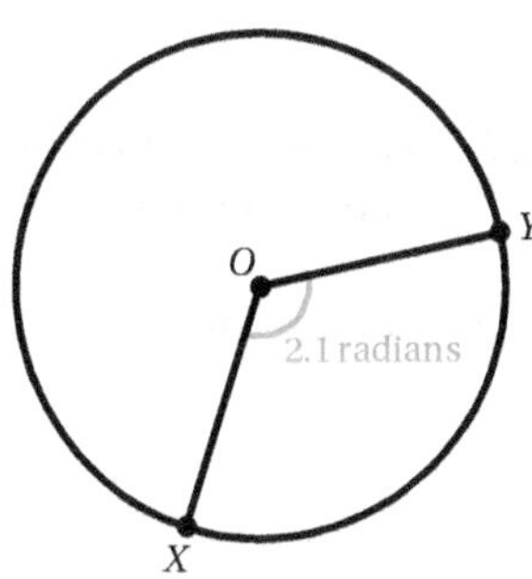

12 A sector of a circle with angle 1.2 radians has area 54 cm^2. Find the radius of the circle.

13 The perimeter of the sector shown in the diagram is 28 cm. Find its area.

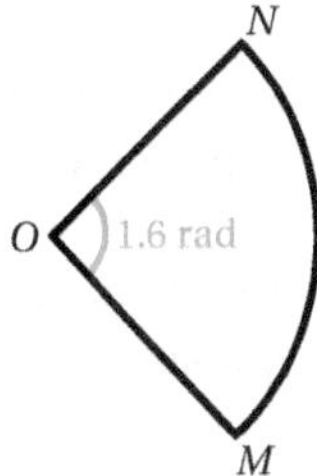

14 The diagram shows an equilateral triangle ABC with side $a = 5$ cm, and three arcs of circles with centres at the vertices of the triangle. Calculate the perimeter of the figure.

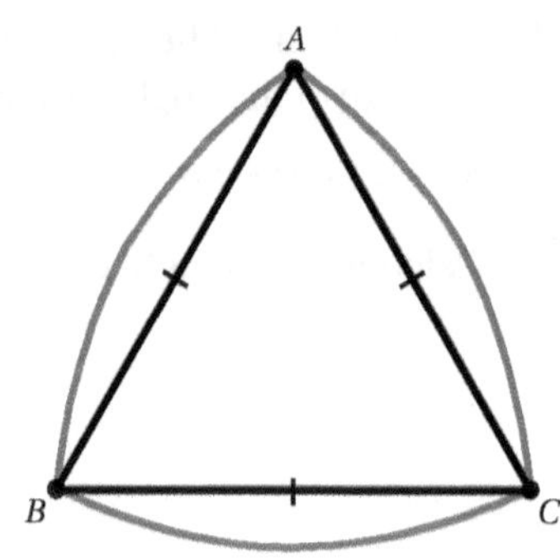

15 Calculate the perimeter of the figure shown in the diagram.

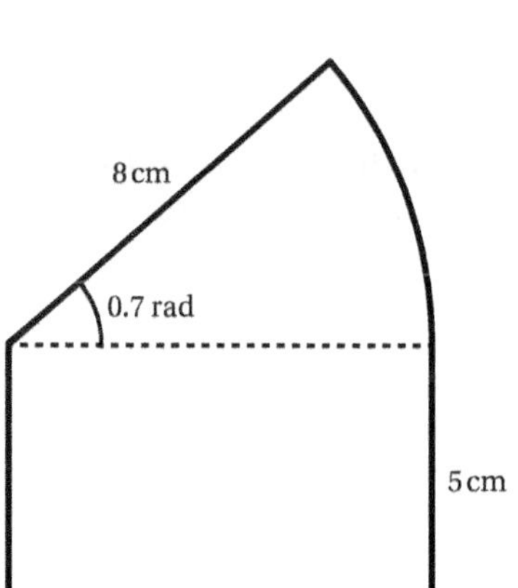

16 A sector of a circle with angle 162° has area 180 cm^2. Find the radius of the circle.

17 The diagram shows a triangle and the segment of a circle. Find the exact perimeter of the figure.

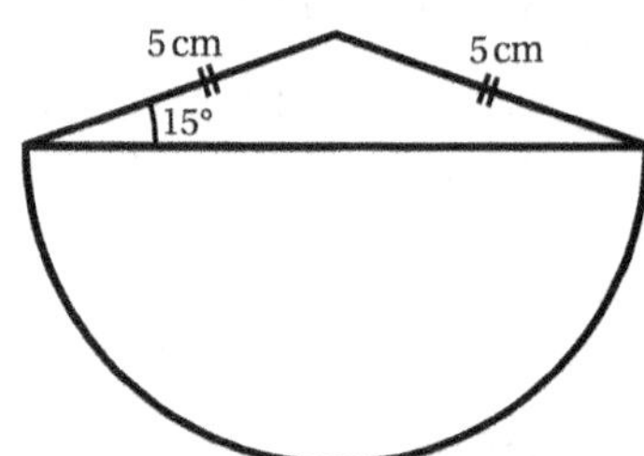

18 A sector of a circle has perimeter $p = 12$ cm and angle at the centre $\theta = 0.4$ radians. Find the radius of the circle.

19 Find the area of the shaded region.

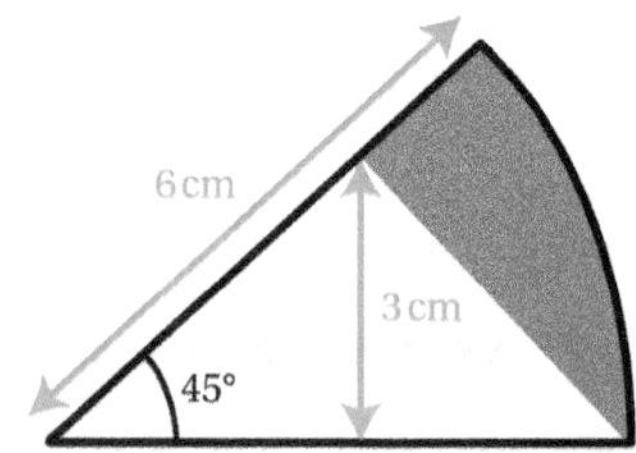

20 A sector of a circle has perimeter 7 cm and area 3 cm^2. Find the possible values of the radius of the circle.

21 Points P and Q lie on the circumference of a circle with centre O and radius 5 cm. The difference between the areas of the major sector POQ and the minor sector POQ is 15 cm^2. Find the size of the angle POQ.

22 A cone is made by rolling a piece of paper shown in the diagram.

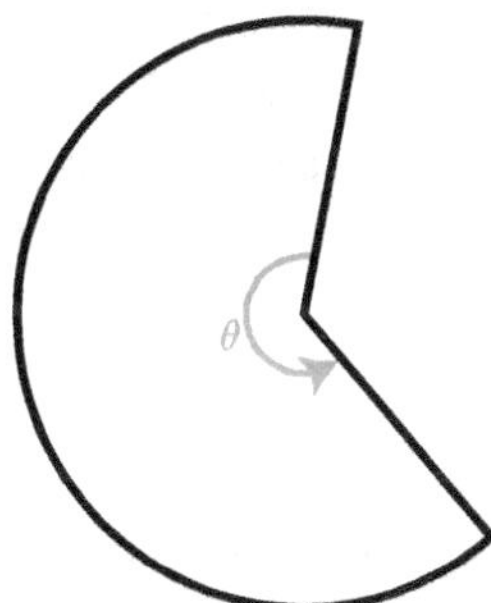

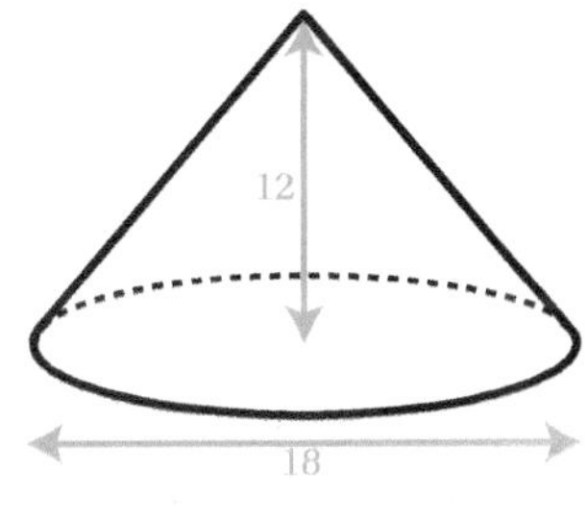

If the cone is to have height 12 cm and base diameter 18 cm, find the size of the angle marked θ.

Section 5: Triangles and circles

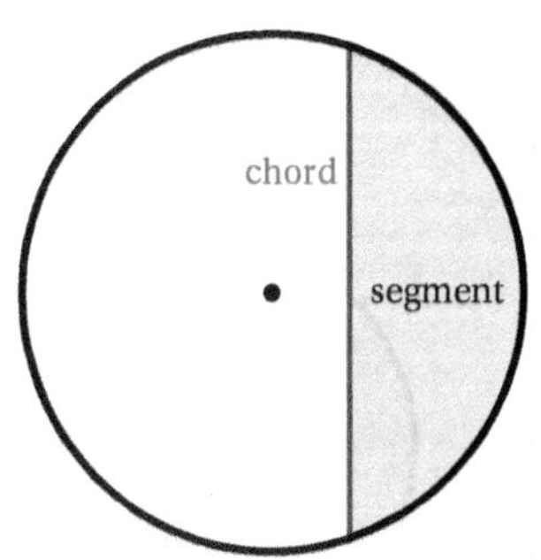

In this section you will look at two other important parts of circles: **chords** and **segments**. You will need to combine the results about arcs and sectors with the formulae for lengths and angles in triangles.

WORKED EXAMPLE 7.14

The diagram shows a sector of a circle of radius 7 cm and the angle at the centre is 0.8 radians. Find:

a the perimeter
b the area

of the shaded region.

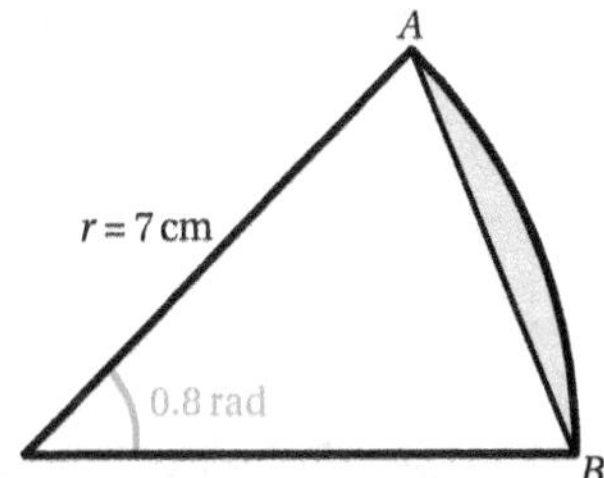

a $p = \text{arc} + \text{chord}$

The perimeter is made up of the arc AB and the chord AB.

$l = 7 \times 0.8 = 5.6\text{ cm}$

The formula for the length of the arc is $l = r\theta$.

Using the cosine rule:

$AB^2 = 7^2 + 7^2 - 2 \times 7 \times 7 \cos 0.8$

$AB^2 = 29.7$

$AB = \sqrt{29.7} = 5.45\text{ cm}$

$\therefore p = 5.6 + 5.45 = 11.1\text{ cm}$

The chord AB is the third side of the triangle ABC. As two sides and the angle between them are known, you can use the cosine rule. Remember that the angle is in radians.

b $A = \text{sector} - \text{triangle}$

If you subtract the area of triangle ABC, you are left with the area of the segment.

$\text{sector} = \frac{1}{2}(7^2 \times 0.8) = 19.6\text{ cm}^2$

Area of a sector $= \frac{1}{2}r^2\theta$.

$\text{triangle} = \frac{1}{2}(7 \times 7)\sin 0.8 = 17.58\text{ cm}^2$

Area of a triangle $= \frac{1}{2}ab\sin C$.

$\therefore A = 19.6 - 17.6 = 2.02\text{ cm}^2$

Worked example 7.15 shows how to solve more complex geometry problems, by splitting up the figure into basic shapes such as triangles and sectors.

Rewind

See Student Book 1, Chapter 11, if you need a reminder of the cosine rule and the formula for the area of the triangle.

WORKED EXAMPLE 7.15

The diagram shows two equal circles of radius 12 cm such that the centre of one circle is on the circumference of the other.

a Find the exact size of angle $P\hat{C_1}Q$ in radians.

b Calculate the exact area of the shaded region.

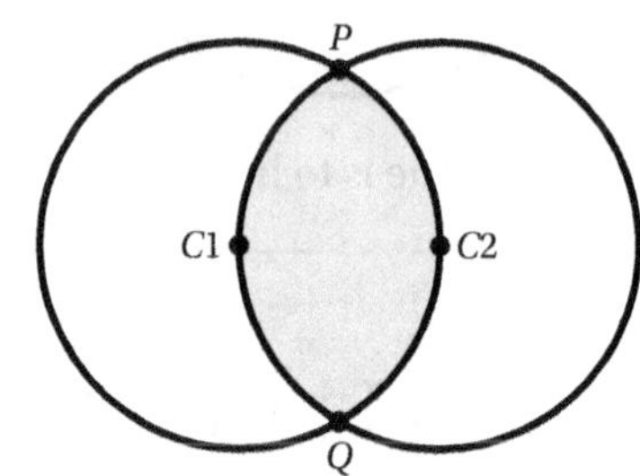

a

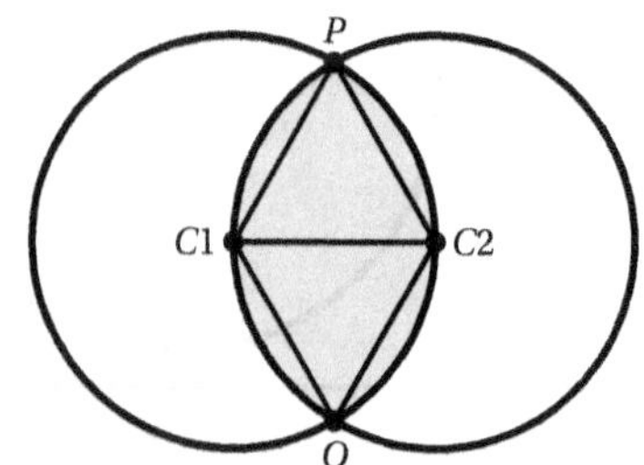

The only thing you know is the radius of the circle, so draw all the lengths that are equal to the radius.

$P\hat{C_1}C_2 = \frac{\pi}{3}$

$\therefore P\hat{C_1}Q = \frac{2\pi}{3}$

The lengths C_1P, C_2P and C_1C_2 are all equal to the radius of the circle. Therefore triangle PC_1C_2 is equilateral.

b

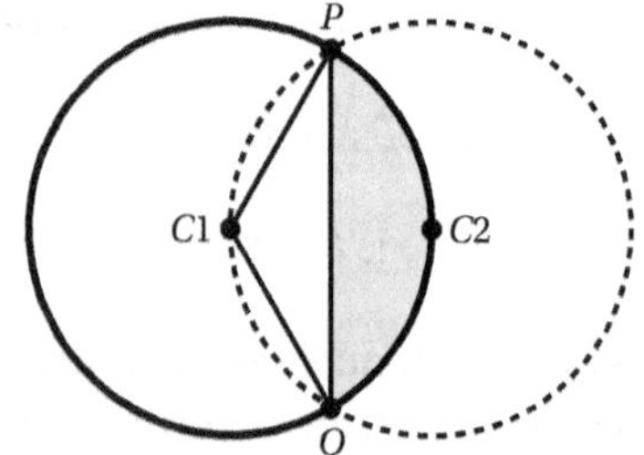

The shaded area is made up of two segments, each with angle at the centre $\frac{2\pi}{3}$ radians.

Area of one segment $= \frac{1}{2}(12^2)\left(\frac{2\pi}{3} - \sin\frac{2\pi}{3}\right)$

$= 72\left(\frac{2\pi}{3} - \frac{\sqrt{3}}{2}\right)$

$= 48\pi - 36\sqrt{3}$

You can find the area of one segment by using the formula. Remember to use the exact value of $\sin\frac{2\pi}{3}$.

$\therefore$ shaded area $= 96\pi - 72\sqrt{3}$ cm^2

The shaded area consists of two segments.

EXERCISE 7E

1 Find the length of the chord AB.

a **i**

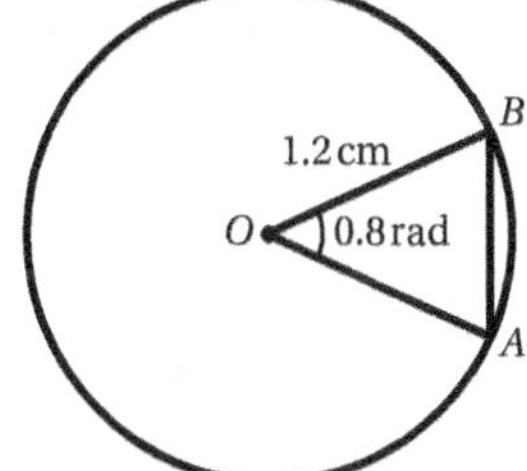

ii

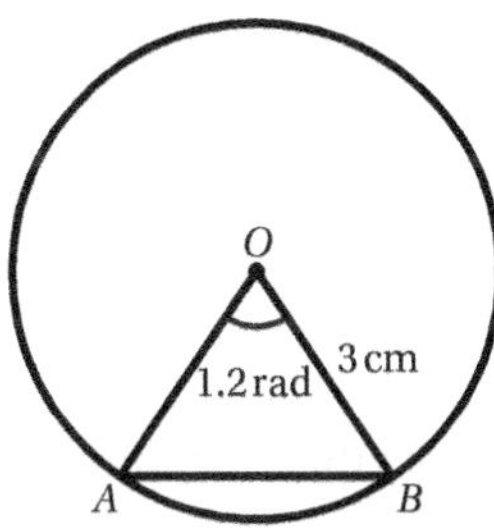

b **i**

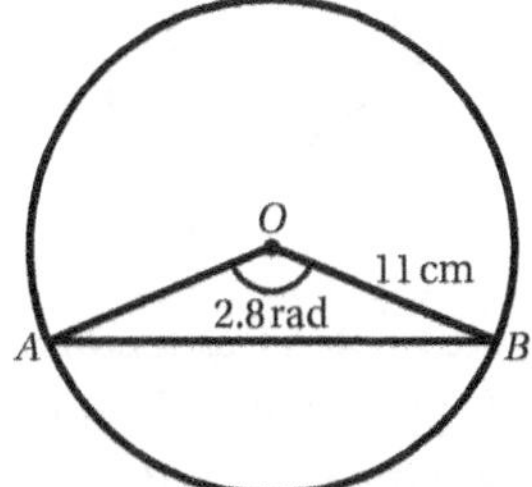

ii

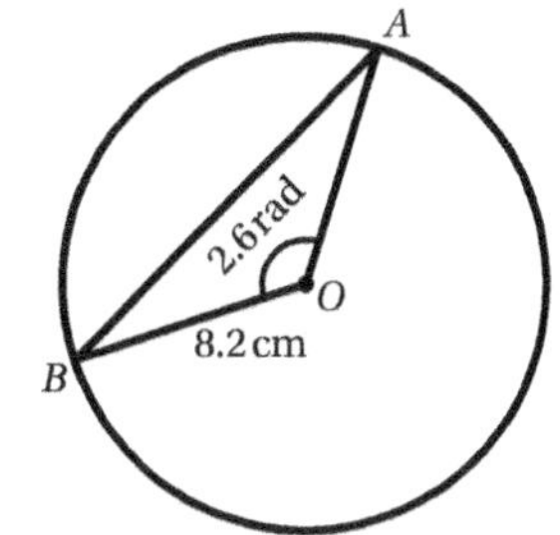

2 Find the perimeters of the minor segments from question 1.

3 Find the areas of minor segments from question 1.

4 A circle has centre O and radius 5 cm. Chord PQ subtends angle θ at the centre of the circle. Given that the area of the minor segment is $15\ \text{cm}^2$, show that $\sin\theta = \theta - 1.2$.

5 Two circles, with centres A and B, intersect at P and Q. The radii of the circles are 6 cm and 4 cm, and $PAQ = 45°$.

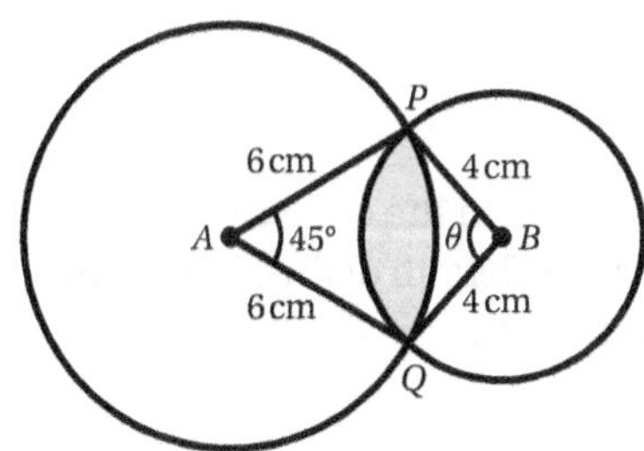

Elevate

See Extension Sheet 7 for some more challenging questions of this type.

a Show that $PQ = 6\sqrt{2-\sqrt{2}}$.

b Find the size of angle PBQ.

c Find the area of the shaded region.

Section 6: Small angle approximations

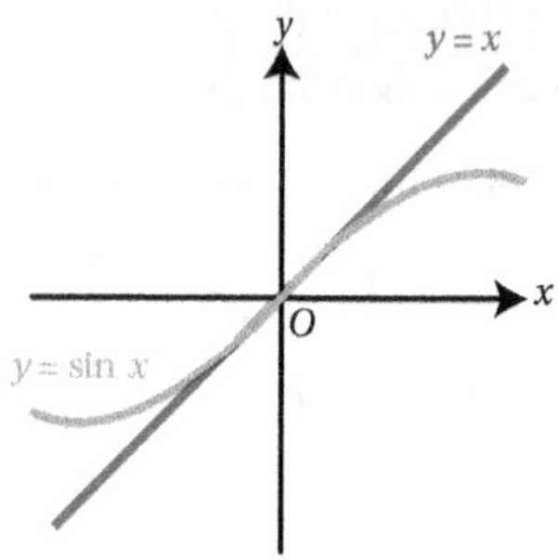

The diagram shows the graphs of $y = \sin x$ and $y = x$. As you can see, near the origin, the two graphs are very close to each other. This means that $\sin x \approx x$ for x close to zero.

PROOF 3

By considering the diagram, prove that, for small values of θ, $\sin\theta \approx \theta$.

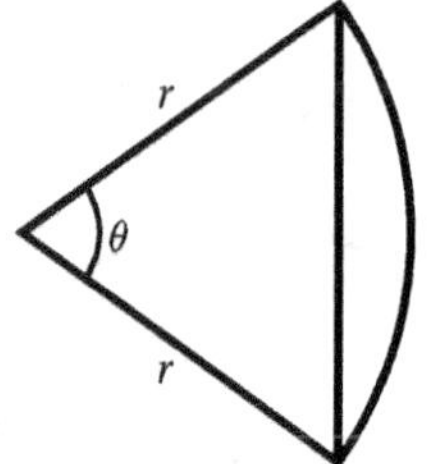

If the angle of the sector is θ radians then its area is $\frac{1}{2}r^2\theta$.

The area of the triangle is $\frac{1}{2}r^2\sin\theta$.

When θ is small, the sector and the triangle have nearly the same area.

Hence, $\frac{1}{2}r^2\sin\theta \approx \frac{1}{2}r^2\theta \Rightarrow \sin\theta \approx \theta$

For small values of θ, the sector and the triangle have approximately equal areas. So write down the expression for each.

For small values of x, the graph of $y = \cos x$ looks like a (negative) parabola. You can find its equations by looking at the sector and the triangle again.

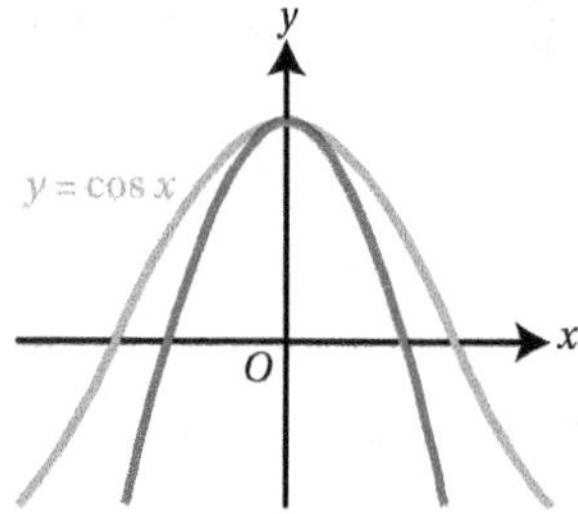

PROOF 4

Prove that, for small values of θ, $\cos\theta \approx 1 - \frac{1}{2}\theta^2$.

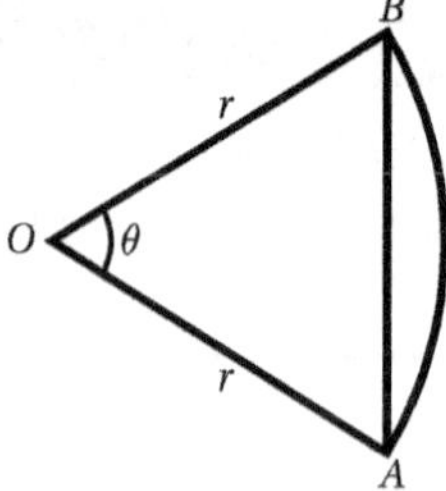

cos θ appears in the cosine rule, so compare the length of the arc and the straight line AB.

The length of the arc AB is $r\theta$.

The length of the line is $AB^2 = r^2 + r^2 - 2r^2\cos\theta$

When θ is small:

$$2r^2 - 2r^2\cos\theta \approx (r\theta)^2$$

$$\Rightarrow 2 - 2\cos\theta \approx \theta^2$$

$$\Rightarrow \cos\theta \approx 1 - \frac{1}{2}\theta^2$$

For small θ, the arc and the line have approximately equal lengths.

It is also possible to derive a similar approximation for $\tan\theta$. All three results, called the **small angle approximations**, are summarised in Key point 7.11.

Key point 7.11

For small θ, measured in radians:

- $\sin\theta \approx \theta$
- $\cos\theta \approx 1 - \frac{1}{2}\theta^2$
- $\tan\theta \approx \theta$

These will be given in your formula book.

Rewind

This is similar to using the binomial expansion to find approximate values of powers, as you learnt to do in Chapter 6.

You can use these results to find approximate values of trigonometric functions.

WORKED EXAMPLE 7.16

a Use a small angle approximation to estimate the value of $\cos 0.4$.
b Find the percentage error in your estimate.

a $\cos 0.4 \approx 1 - \frac{1}{2}(0.4^2)$ — Using $\cos\theta \approx 1 - \frac{1}{2}\theta^2$, as 0.4 is close to 0.

$= 1 - \frac{0.16}{2}$

$= 0.92$

b Percentage error $= \frac{|\cos 0.4 - 0.92|}{\cos 0.4} \times 100$ — Use a calculator to find the actual value.

$= 0.115\%$

Fast forward

The 'exact' value of $\cos 0.4$ you get from the calculator in Worked example 7.16 is in fact also an approximation. It is obtained from a generalisation of small angle approximations, called Maclaurin series, which you will study if you take the Further Mathematics course.

Just as with the binomial expansion, you can replace θ by another function, such as 3θ or x^2. You can also multiply two expansions together.

WORKED EXAMPLE 7.17

Assuming that x is sufficiently small that terms in x^3 and higher can be ignored, find an approximate expression for:

a $\cos 3x$ **b** $\cos 3x\,(1 + \sin 5x)$.

a $\cos 3x \approx 1 - \frac{1}{2}(3x)^2$ — Replace θ by $3x$ in $\cos\theta \approx 1 - \frac{1}{2}\theta^2$.

$= 1 - \frac{9}{2}x^2$

b $\cos 3x(1 + \sin 5x) \approx \left(1 - \frac{9}{2}x^2\right)(1 + 5x)$ — Use the result from **a** and $\sin\theta \approx \theta$.

$\approx 1 + 5x - \frac{9}{2}x^2$ — Expand the brackets, but only keep terms up to x^2.

You need to be a little more careful when dividing two approximate expressions – you may need to turn the quotient into a product before using the binomial expansion.

Rewind

See Chapter 6 for a reminder of the binomial expansion with negative powers.

WORKED EXAMPLE 7.18

Find an approximate expression for $\frac{\sin 4\theta}{1+\cos\theta}$ given that θ is small enough to neglect the terms in θ^3 and above.

$\frac{\sin 4\theta}{1+\cos\theta} \approx \frac{4\theta}{1+\left(1-\frac{1}{2}\theta^2\right)}$ Replace $\sin 4\theta$ and $\cos\theta$ by their small angle approximations.

$= \frac{8\theta}{4-\theta^2}$

$= 8\theta(4-\theta^2)^{-1}$ To complete the expansion, turn division into multiplication.

$= 8\theta \times 4^{-1}\left(1-\frac{\theta^2}{2}\right)^{-1}$ Now do the binomial expansion. Remember that the bracket needs to be in the form $(1+x)^{-1}$.

$\approx 2\theta\left(1+\frac{\theta^2}{2}+\ldots\right)$

$\approx 2\theta$ Expand the bracket and ignore any terms in θ^3 or higher.

EXERCISE 7F

1 Find the approximate value of each expression.

a i $\sin 0.2$ ii $\sin(-0.14)$ b i $\cos(-0.3)$ ii $\cos 0.2$

c i $\tan 0.12$ ii $\tan(-0.2)$

2 Find the small angle approximation for each expression.

a i $\sin 2\theta$ ii $\sin(-3x)$ b i $\cos 3x$ ii $\cos(-5\theta)$

c i $\tan(x^2)$ ii $\tan\left(\frac{\theta^2}{2}\right)$

3 Assuming θ is sufficiently small so that terms in θ^3 and above can be ignored, find an approximate expression for each expression.

a i $\cos 2\theta \cos 3\theta$ ii $\cos\frac{\theta}{2}\cos 4\theta$

b i $(1+2\sin\theta)\cos 2\theta$ ii $\cos\theta(1-\sin 2\theta)$

c i $(2+\sin\theta)(1-\tan 2\theta)$ ii $(3+\tan\theta)(\sin 3\theta-1)$

4 a Find a small angle approximation for $(1-\sin 2\theta)(1+3\tan\theta)$.

b Hence find an approximate value of $(1-\sin 0.4)(1+3\tan 0.2)$.

5 **a** Given that θ is sufficiently small, write down an approximate expression for $\cos\theta$ in the form $a + b\theta + c\theta^2$.

b Use your expression with $\theta = \frac{\pi}{6}$ to find an approximate value of π.

6 **a** Find an approximate expression for $(1 + \sin 2x)^3$ when x is sufficiently small so that the terms in x^3 and higher can be ignored.

b Find the percentage error when this approximation is used to estimate the value of:

i $(1 + \sin 0.4)^3$ **ii** $(1 + \sin 1.4)^3$.

7 Given that θ is close to zero, find an approximate expression for $\frac{\cos\theta}{1 - \sin 3\theta}$, ignoring powers of θ higher than 2.

8 Find an approximate expression for $\frac{\cos 2\theta}{1 + \tan 3\theta}$ when θ is close to zero.

9 Let θ be a small angle, measured in **degrees**. Use small angle approximations to find an approximate expression for $\sin\theta$ and $\cos\theta$.

10 **a** Find an approximate expression for $\sqrt{1 + \tan\theta}$ for small values of θ, including terms up and including θ^2.

b Hence find an approximate value of $\int_0^{\frac{\pi}{10}} \sqrt{1 + \tan\theta}\ d\theta$.

11 Given that x is close to zero so that terms in x^3 and higher can be ignored, find an approximate expression for $\frac{3 + \sin 2x}{(3 + \tan x)^2}$.

12 Find an approximate expression for $\frac{4 + \sin\theta}{\sqrt{3 + \cos\theta}}$ when θ is sufficiently small to ignore the terms in θ^3 and higher.

13 **a** For each of the three small angle approximations, find the largest value of θ for which the approximation error is less than 1%.

b **i** Show that using the small angle approximation for $\sin\theta$ and the identity $\cos\theta = \sqrt{1 - \sin^2\theta}$ gives the correct small angle approximation for $\cos\theta$.

ii Why is it not appropriate to use $\pm$ in front of the square root?

Checklist of learning and understanding

- The **radian** is defined in terms of the distance travelled around the circle, so that a full turn $= 2\pi$ radians.
  - To convert from degrees to radians, divide by 180 and multiply by π.
  - To convert from radians to degrees, divide by π and multiply by 180.
- For some real numbers the three functions have exact values, which you should learn.

Radians	0	$\frac{\pi}{6}$	$\frac{\pi}{4}$	$\frac{\pi}{3}$	$\frac{\pi}{2}$
Degrees	0	30	45	60	90
sin	0	$\frac{1}{2}$	$\frac{\sqrt{2}}{2}$	$\frac{\sqrt{3}}{2}$	1
cos	1	$\frac{\sqrt{3}}{2}$	$\frac{\sqrt{2}}{2}$	$\frac{1}{2}$	0
tan	0	$\frac{\sqrt{3}}{3}$	1	$\sqrt{3}$	undefined

- The functions $y = a\sin b(x+c)+d$ and $y = a\cos b(x+c)+d$ have:
  - amplitude a
  - central value d
  - minimum value $d-a$ and maximum value $d+a$
  - period $\frac{2\pi}{b}$
- Properties of inverse trigonometric functions:

Inverse function	Domain	Range
$\arcsin x$	$[-1, 1]$	$\left[-\frac{\pi}{2}, \frac{\pi}{2}\right]$
$\arccos x$	$[-1, 1]$	$[0, \pi]$
$\arctan x$	$\mathbb{R}$	$\left(-\frac{\pi}{2}, \frac{\pi}{2}\right)$

- Trigonometric equations can be solved in radians.

Equation	First solution, θ_1	Second solution, θ_2	Further solutions
$\sin\theta = k$	$\sin^{-1} k$	$\pi - \theta_1$	Add or subtract multiples of 2π.
$\cos\theta = k$	$\cos^{-1} k$	$-\theta_1$	Add or subtract multiples of 2π.
$\tan\theta = k$	$\tan^{-1} k$		Add or subtract multiples of π.

- If r is the radius of the circle and θ is the angle subtended at the centre, measured in radians, then the length of an arc is $l = r\theta$ and the area of a sector is $A = \frac{1}{2}r^2\theta$.
- For small θ, measured in radians:
  - $\sin\theta \approx \theta$
  - $\cos\theta \approx 1 - \frac{1}{2}\theta^2$
  - $\tan\theta \approx \theta$

Mixed practice 7

1 What is 152° in radians?
Choose from these options.

A 152π B $\frac{152}{\pi}$ C $\frac{38}{45}\pi$ D $\frac{45}{38}\pi$

2 The height of a wave (in metres) at a distance x metres from a buoy is modelled by the function $f(x) = 1.4\sin(3x - 0.1) - 0.6$.

a State the amplitude of the wave.

b Find the distance between consecutive peaks of the wave.

3 Solve the equation $5\sin^2\theta = 4\cos^2\theta$ for $-\pi \leqslant \theta \leqslant \pi$.

4 Use small angle approximations to estimate the value of $(1 - \sin 0.2)(1 + \cos 0.3)$.

5 In the diagram, $OABC$ is a rectangle with sides 7 cm and 2 cm. PQ is a straight line. AP and CQ are circular arcs, and $\angle AOP = \frac{\pi}{6}$ radians.

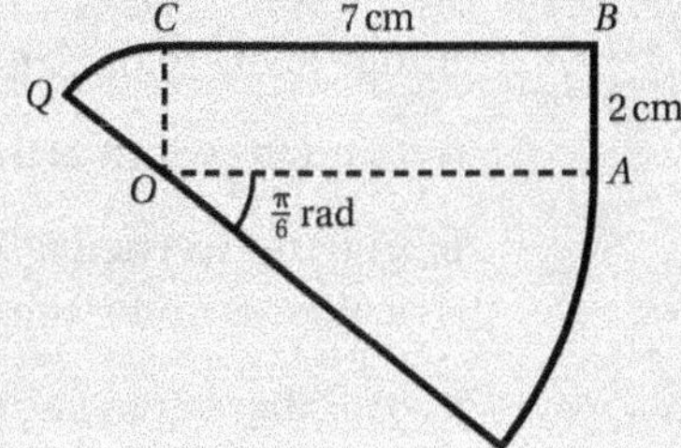

a Write down the size of $\angle COQ$.

b Find the area of the whole shape.

c Find the perimeter of the whole shape.

6 A sector has perimeter 36 cm and radius 10 cm. Find its area.

7 The diagram shows a circle with centre O and radius $r = 7$ cm. The chord PQ subtends angle $\theta = 1.4$ radians at the centre of the circle.

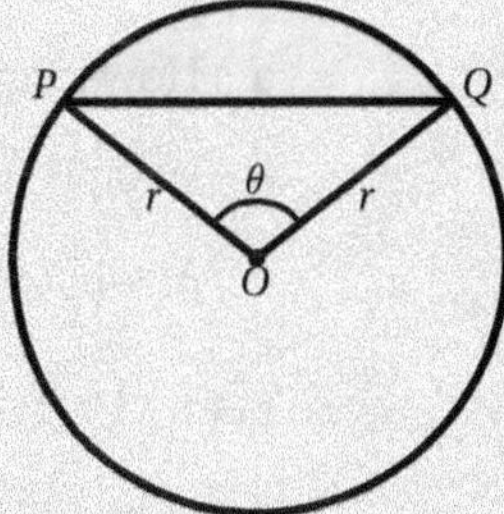

Find:

a the area of the shaded region

b the perimeter of the shaded region.

8 The diagram shows a sector OAB of a circle with centre O.

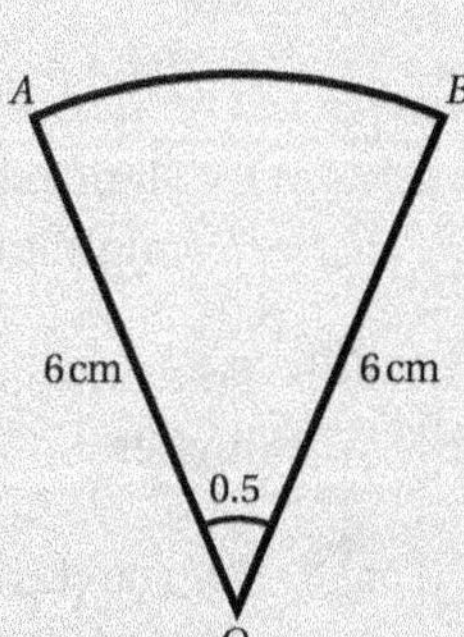

The radius of the circle is 6 cm and the angle $AOB = 0.5$ radians.

a Find the area of the sector OAB.

b i Find the length of the arc AB.

ii Hence show that the perimeter of the sector $OAB = k \times$ the length of the arc AB where k is an integer.

[© AQA 2011]

9 How many solutions for x are there to the equation $x = \arcsin\left(\sin\frac{2\pi}{3}\right)$?

Choose from these options.

A 0 **B** 1 **C** 2 **D** Infinitely many

10 The diagram shows the graph of the function $f(x)$ $a \sin bx$.
Find the values of a and b.

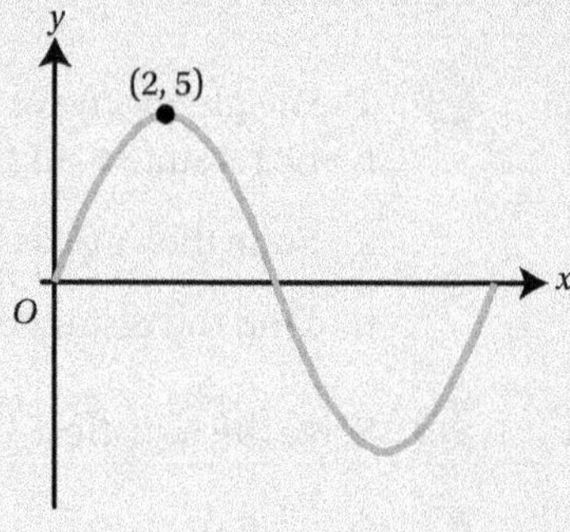

11 The shape of a small bridge can be modelled by the equation $y = 1.8 \sin \frac{x}{3}$, where y is the height of the bridge above water, and x is the distance from one river bank, both measured in metres.

a Find the width of the river.

b A barge has height 1.2 m above the water level. Find the maximum possible width of the barge so it can pass under the bridge.

c Another barge has width 3.5 m. What is the maximum possible height of the barge so it can pass under the bridge?

12 A runner is jogging around a level circular track. His distance north of the centre of the track, in metres, is given by $60 \cos 0.08\,t$ where t is measured in seconds.

a How long does is take the runner to complete one lap?

b What is the length of the track?

c At what speed is the runner jogging?

13 Let $f(x) = 3 \sin 2\left(x - \frac{\pi}{3}\right)$.

a State the period of the function.

b Find the coordinates of the points where the graph of $y = f(x)$, for $x \in [0, 2\pi]$ crosses the x-axis.

c Hence sketch the graph of, $y = f(x)$, for $x \in [0, 2\pi]$, showing the coordinates of the maximum and minimum points.

14 Find an approximate expression for $\frac{\cos\theta}{1+\tan\theta}$, assuming θ is small enough to ignore terms in θ^3 and higher.

15 Two circles have equal radius r and intersect at points S and T. The centres of the circles are A and B, and $\angle ASB = 90°$.

a Explain why $\angle SAT$ is also $90°$.

b Find the length AB in terms of r.

c Find the area of the sector AST.

d Find the area of the overlap of the two circles.

16 In the diagram, O is the centre of the circle and AT is the tangent to the circle at T.

If $OA = 8$ cm, and the circle has a radius of 4 cm, find the area of the shaded region.

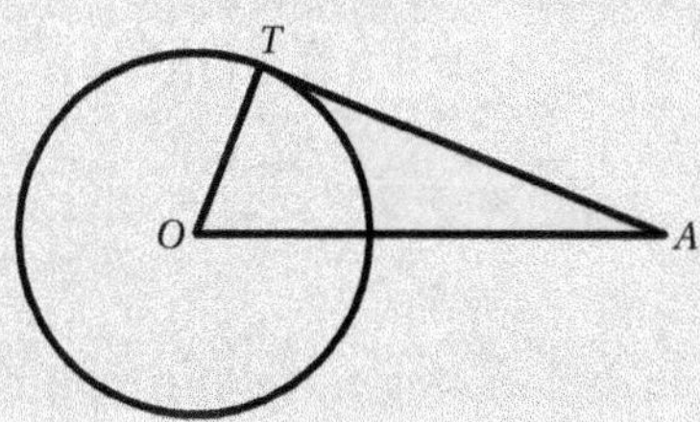

17 **a** Sketch the graph of $y = \cos^{-1} x$, where y is in radians. State the coordinates of the end points of the graph.

b Sketch the graph of $y = \pi - \cos^{-1} x$, where y is in radians. State the coordinates of the end points of the graph.

[© AQA 2013]

18 Find the exact values of $x \in [-\pi, \pi]$ satisfying the equation $2\cos\left(2x + \frac{\pi}{3}\right) = \sqrt{2}$.

19 Find the exact solutions to the equation $\cos\theta - 2\sin^2\theta + 2 = 0$ for $\theta \in [0, 2\pi]$.

20 **a** Given that $5 + \sin^2\theta = (5 + 3\cos\theta)\cos\theta$, show that $\cos\theta = \frac{3}{4}$.

b Hence solve the equation $5 + \sin^2 2x = (5 + 3\cos 2x)\cos 2x$ in the interval $0 < x < 2\pi$, giving your values of x in radians to three significant figures.

[© AQA 2013]

21 **a** Given that x is small enough so that the terms in x^3 and higher can be neglected, find an approximate expression for $\sqrt{4 + \sin 3x}$.

b Hence find an estimate for $\int_0^{\frac{\pi}{8}} \sqrt{4 + \sin 3x}\, dx$.

22 Two circular cogs are connected by a chain as shown in diagram A. The radii of the cogs are 3 cm and 8 cm and the distance between their centres is 25 cm.

Diagram B shows the quadrilateral O_1ABO_2. Line O_2P is drawn parallel to AB.

A

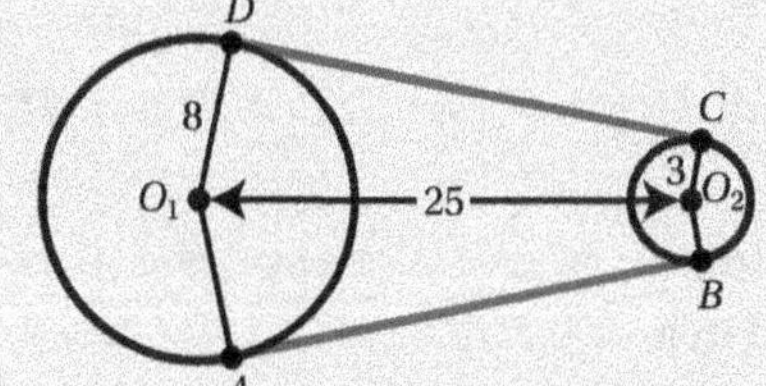

B

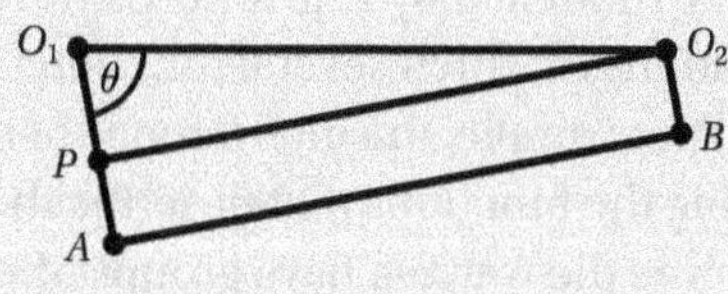

a Write down the size of O_1AB in radians, giving a reason for your answer.

b Explain why $PO_2 = AB$.

c Hence find the length AB.

d Find the size of the angle marked θ, giving your answer in radians correct to four significant figures.

e Calculate the length of the chain $ABCD$.

23 **a** Write down an expression for $\sin(\arcsin x)$.

b Show that $\sin(\arccos x) = \sqrt{1 - x^2}$.

c Hence solve the equation $\arcsin x = \arccos x$ for $0 \leqslant x \leqslant 1$.

8 Further trigonometry

In this chapter you will learn how to:

- work with trigonometric functions of sums and differences of two angles, for example, $\sin(A+B)$
- work with trigonometric functions of double angles, for example, $\sin 2A$
- work with sums of trigonometric functions, for example, $\sin A + \sin B$
- work with reciprocal trigonometric functions, for example, $\frac{1}{\sin x}$.

Before you start...

Student Book 1, Chapter 10	You should be able to use the identities $\sin^2 x + \cos^2 x \equiv 1$ and $\tan x \equiv \frac{\sin x}{\cos x}$.	1 Given that x is an acute angle with $\cos x = \frac{1}{3}$, find the exact value of: a $\sin x$ b $\tan x$.
Chapter 7; Student Book 1, Chapter 10	You should know and be able to use graphs of trigonometric functions, in degrees and radians.	2 State the coordinates of the minimum point on the graph of $y = 1 - 3\sin 2x$, for $x \in \left[0, \frac{\pi}{2}\right]$.
Chapter 7; Student Book 1, Chapter 10	You should be able to solve trigonometric equations in degrees and radians.	3 Solve each equation. a $\sin 3x = 2\cos 3x$, for $0° \leqslant x \leqslant 90°$. b $2\cos^2\theta - \sin\theta = 1$, for $\theta \in [-\pi, \pi]$.

Combining trigonometric functions

Periodic phenomena, such as water waves, sound waves or motion around a circle, can be modelled by sine and cosine functions. There are many situations in which several functions need to be combined together. For example, the interference of two waves can be modelled by adding the functions that describe their shapes. Some fairground rides, such as the waltzer, have groups of seats arranged in a circle that rotates on top of a larger rotating platform.

In this chapter you will learn how to simplify expressions involving sums and products of trigonometric functions.

Section 1: Compound angle identities

Use your calculator, working in radians, to find:

- $\sin 1.2$, $\sin 0.3$ and $\sin 1.5$
- $\cos 1.2$, $\cos 0.3$ and $\cos 1.5$.

Tip

You will also see these identities referred to as compound angle formulae.

There seems to be no obvious connection between the values of, for example, $\sin A$, $\sin B$ and $\sin(A+B)$.

In fact, though, there are formulae, called **compound angle identities**, to express $\sin(A+B)$ in terms of the functions of the individual angles.

Focus on...

See Focus on... Proof 1 for the proofs of these results.

Key point 8.1

- $\sin(A+B) \equiv \sin A \cos B + \cos A \sin B$
- $\sin(A-B) \equiv \sin A \cos B - \cos A \sin B$
- $\cos(A+B) \equiv \cos A \cos B - \sin A \sin B$
- $\cos(A-B) \equiv \cos A \cos B + \sin A \sin B$

These will be given in your formula book.

Common error

Notice the signs in the cosine identities: in the identity for the sum the minus sign is used, and in the identity for the difference the plus sign.

In practice, you only need to prove the identities for $\sin(A+B)$ and $\cos(A+B)$. Then you can use them to prove those for $A-B$.

PROOF 5

Use the identity $\cos(A+B) \equiv \cos A \cos B - \sin A \sin B$ to prove that $\cos(A-B) \equiv \cos A \cos B + \sin A \sin B$.

Replace B by $-B$ in the first identity:

$\cos(A+(-B)) \equiv \cos A \cos(-B) - \sin A \sin(-B)$

The only difference between the two expressions on the left-hand side is the sign of B, so replace B by $-B$.

$\cos(A-B) \equiv \cos A \cos B - \sin A(-\sin B)$

$\cos(A-B) \equiv \cos A \cos B + \sin A \sin B$

as required.

Now use the symmetries of the sin and cos graphs: $\cos(-x) = \cos x$ and $\sin(-x) = -\sin x$.

One of the simplest applications is to calculate exact values of trigonometric functions.

Rewind

You met exact values of trigonometric functions in Chapter 7, Section 1.

WORKED EXAMPLE 8.1

Find the exact values of:

a $\sin 75°$ **b** $\cos 15°$.

a $\sin 75° = \sin(30°+45°)$

$= \sin 30° \cos 45° + \cos 30° \sin 45°$

$= \frac{1}{2}\frac{\sqrt{2}}{2} + \frac{\sqrt{3}}{2}\frac{\sqrt{2}}{2}$

$= \frac{\sqrt{2}+\sqrt{6}}{4}$

You only know the exact values of the sin of a few angles: 0°, 30°, 45°, 60°, 90°. You can write 75° as a sum of two of those.

b $\cos 15° = \cos(45°-30°)$

$= \cos 45° \cos 30° + \sin 45° \sin 30°$

$= \frac{\sqrt{2}}{2}\frac{\sqrt{3}}{2} + \frac{\sqrt{2}}{2}\frac{1}{2}$

$= \frac{\sqrt{6}+\sqrt{2}}{4}$

Look for a way to make 15° from the special angles.

WORK IT OUT 8.1

Expand $\cos\left(x-\frac{\pi}{3}\right)$.

Which is the correct solution? Identify the errors made in the incorrect solutions.

Solution 1	Solution 2	Solution 3
$\cos\left(x-\frac{\pi}{3}\right)=\cos x-\cos\frac{\pi}{3}$ $=\cos x-\frac{1}{2}$	$\cos\left(x-\frac{\pi}{3}\right)=\cos x\cos\frac{\pi}{3}-\sin x\sin\frac{\pi}{3}$ $=\cos x\times\frac{1}{2}-\sin x\times\frac{\sqrt{3}}{2}$ $=\frac{1}{2}\cos x-\frac{\sqrt{3}}{2}\sin x$	$\cos\left(x-\frac{\pi}{3}\right)=\cos x\cos\frac{\pi}{3}+\sin x\sin\frac{\pi}{3}$ $=\cos x\times\frac{1}{2}+\sin x\times\frac{\sqrt{3}}{2}$ $=\frac{1}{2}\cos x+\frac{\sqrt{3}}{2}\sin x$

You can use the sine and cosine compound angle identities to derive new ones.

Key point 8.2

- $\tan(A+B)\equiv\frac{\tan A+\tan B}{1-\tan A\tan B}$
- $\tan(A-B)\equiv\frac{\tan A-\tan B}{1+\tan A\tan B}$

These will be given in your formula book.

WORKED EXAMPLE 8.2

Prove that $\tan(A+B)\equiv\frac{\tan A+\tan B}{1-\tan A\tan B}$.

$\tan(A+B)\equiv\frac{\sin(A+B)}{\cos(A+B)}$ — Express tan in terms of sin and cos.

$\equiv\frac{\sin A\cos B+\cos A\sin B}{\cos A\cos B-\sin A\sin B}$ — Use the identities for sin and cos.

$\equiv\frac{\frac{\sin A\cos B}{\cos A\cos B}+\frac{\cos A\sin B}{\cos A\cos B}}{\frac{\cos A\cos B}{\cos A\cos B}-\frac{\sin A\sin B}{\cos A\cos B}}$ — You want to express this in terms of tan. Looking at the top of the fraction, if you divide by $\cos A$ you will get $\tan A$ in the first term, and if you divide by $\cos B$ you will get $\tan B$ in the second term. So divide top and bottom by $\cos A\cos B$.

$\equiv\frac{\frac{\sin A}{\cos A}+\frac{\sin B}{\cos B}}{1-\frac{\sin A}{\cos A}\frac{\sin B}{\cos B}}$

$\equiv\frac{\tan A+\tan B}{1-\tan A\tan B}$ — Convert back to tan.

as required.

EXERCISE 8A

1 Without using a calculator, express each angle in the form $a \sin x + b \cos x$, giving exact values of a and b.

a $\sin\left(x+\frac{\pi}{3}\right)$ **b** $\sin\left(x-\frac{\pi}{4}\right)$ **c** $\cos\left(x+\frac{3\pi}{4}\right)$ **d** $\cos\left(x-\frac{3\pi}{2}\right)$

2 Without using a calculator, find the exact value of each expression.

a $\cos 75°$ **b** $\sin\frac{7\pi}{12}$ **c** $\tan 105°$ **d** $\sin\frac{\pi}{12}$

3 Without using a calculator, find the exact value of $\cos(A-B)$ for the angles shown in the diagrams.

a

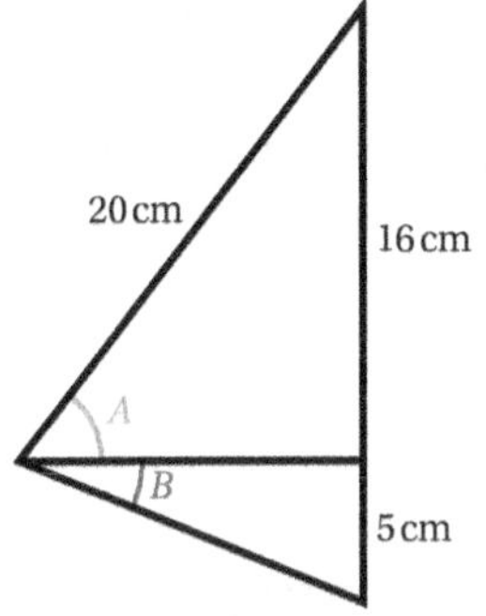

b

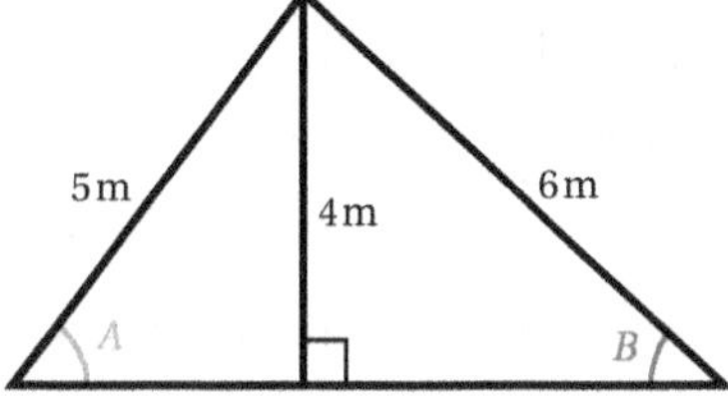

4 **a** Show that $\sin\left(x+\frac{\pi}{3}\right)+\sin\left(x-\frac{\pi}{3}\right)\equiv \sin x$.

b Simplify $\sin\left(x+\frac{\pi}{4}\right)+\cos\left(x+\frac{\pi}{4}\right)$.

5 **a** Express $\tan\left(\theta-\frac{\pi}{4}\right)$ in terms of $\tan\theta$.

b Given that $\tan\left(\theta-\frac{\pi}{4}\right)=6\tan\theta$, find two possible values of $\tan\theta$.

c Hence solve the equation $\tan\left(\theta-\frac{\pi}{4}\right)=6\tan\theta$ for $0<\theta<\pi$.

6 Write each of these expressions as a single trigonometric function, and hence find its maximum value and the smallest positive value of x for which it occurs.

a $\sin x \cos\frac{\pi}{4}+\cos x \sin\frac{\pi}{4}$ **b** $2\cos x \cos 25° + 2\sin x \sin 25°$

7 Assuming x is small enough so that the terms in x^3 and higher can be ignored, find an approximate expression for $\sin\left(\frac{\pi}{3}+x\right)$ in the form $a+bx+cx^2$.

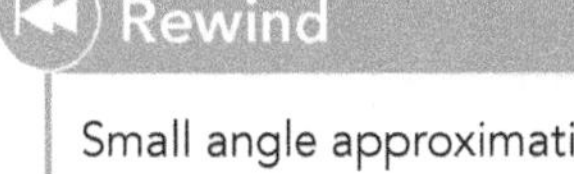

Small angle approximations for sin and cos were covered in Chapter 7, Section 6.

8 **a** Show that $\sin(A+B)+\sin(A-B)\equiv 2\sin A\cos I$.

b Hence solve the equation $\sin\left(x+\frac{\pi}{6}\right)+\sin\left(x-\frac{\pi}{6}\right)=3\cos x$ for $0\leqslant x\leqslant\pi$.

9 **a** Show that $\cos(x+y)+\cos(x-y)\equiv 2\cos x\cos y$.

b Hence solve the equation $\cos 3x+\cos x = 3\cos 2x$ for $x\in[0, 2\pi]$.

Section 2: Double angle identities

If you set $A = B$ in the compound angle identity for $\sin(A+B)$, you get an identity for $\sin 2A$. You can do the same with cos and tan giving a set of **double angle identities.**

Key point 8.3

- $\sin 2A \equiv 2\sin A\cos A$
- $\cos 2A \equiv \begin{cases} \cos^2 A - \sin^2 A \\ 2\cos^2 A - 1 \\ 1 - 2\sin^2 A \end{cases}$
- $\tan 2A \equiv \dfrac{2\tan A}{1-\tan^2 A}$

Tip

You can convert between the three different forms of the $\cos 2A$ identity by using $\sin^2 A + \cos^2 A \equiv 1$, but they are used so often that it is best to just remember all three.

A useful application of these identities is finding exact values of half-angles.

WORKED EXAMPLE 8.3

Using the exact value of $\cos 30°$, show that $\sin 15° = \sqrt{\dfrac{2-\sqrt{3}}{4}}$.

Using $\cos 2A \equiv 1 - 2\sin^2 A$:

$$\cos(2\times 15°) = 1 - 2\sin^2 15°$$
$$\cos 30° = 1 - 2\sin^2 15°$$

You know that $\cos 30° = \dfrac{\sqrt{3}}{2}$, so use a double angle formula to relate this to $\sin 15°$.

But $\cos 30° = \dfrac{\sqrt{3}}{2}$, so

$$\frac{\sqrt{3}}{2} = 1 - 2\sin^2 15°$$
$$\sqrt{3} = 2 - 4\sin^2 15°$$
$$4\sin^2 15° = 2 - \sqrt{3}$$
$$\sin^2 15° = \frac{2-\sqrt{3}}{4}$$

$\therefore \sin 15° = \sqrt{\dfrac{2-\sqrt{3}}{4}}$ $\quad(\sin 15° > 0)$

You have to choose between the positive and negative square root here. $\sin 15° > 0$ so take the positive square root.

Recognising the form of double angle identities can be useful in solving trigonometric equations.

WORKED EXAMPLE 8.4

Solve the equation $6\sin x\cos x = 1$ for $-\pi < x < \pi$.

$6\sin x\cos x = 1$

$2\sin x\cos x = \frac{1}{3}$

Write the left-hand side in terms of only one trig function.

$\sin 2x = \frac{1}{3}$

Rearrange so that the left-hand side is $2\sin x\cos x$, which is equal to $\sin 2x$.

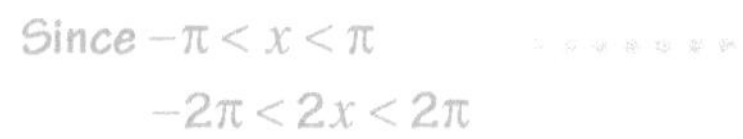

Follow the standard procedure and look at the graph to find that that there are four solutions in the given domain.

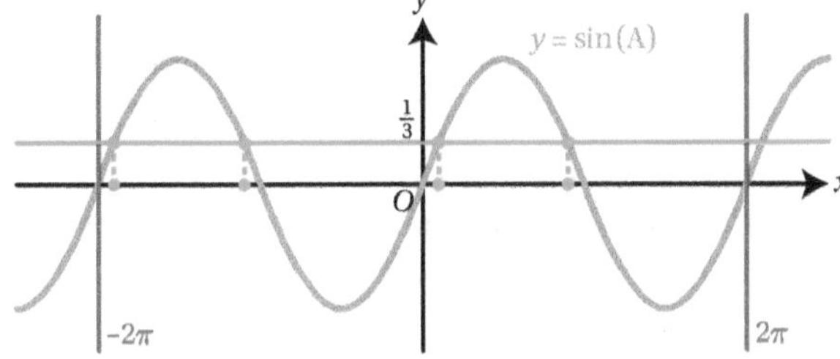

$\arcsin\left(\frac{1}{3}\right) = 0.3398$

$2x = 0.3398, 2.802, -5.943, -3.481$

$\therefore x = 0.170, 1.40, -2.97, -1.74$ (3 s.f.)

If an equation contains both $\cos 2\theta$ and $\cos\theta$, you can use identities to turn it into an equation involving only $\cos\theta$.

WORKED EXAMPLE 8.5

Find exact solutions of the equation $\cos 2x = \cos x$ for $0° \leqslant x \leqslant 360°$.

$\cos 2x = \cos x$

$2\cos^2 x - 1 = \cos x$

As before, write the equation in terms of only one trig function. (Note that $\cos 2x$ and $\cos x$ are not the same function!).

The left-hand side involves a double angle, so choose the $\cos 2x$ identity involving just $\cos x$.

$2\cos^2 x - \cos x - 1 = 0$

$(2\cos x + 1)(\cos x - 1) = 0$

$\cos x = -\frac{1}{2}$ or $\cos x = 1$

This is a quadratic in $\cos x$.

When $\cos x = -\frac{1}{2}$:

$x = 120°$ or $360 - 120 = 240°$

Solve each equation separately.

When $\cos x = 1$:

$x = 0°$ or $360°$

$\therefore x = 0, 120°, 240°, 360°$

Remember to list all the solutions at the end.

Although they are called the double angle identities, you can also use them with higher multiples.

Fast forward

In Chapter 11 you will use double angle identities to integrate some trigonometric functions.

WORKED EXAMPLE 8.6

Find an expression for $\cos 4x$ in terms of:

a $\cos 2x$ **b** $\cos x$.

a $\cos 4x \equiv \cos(2(2x))$
$\equiv 2\cos^2(2x) - 1$

$4x = 2\times(2x)$, so use one of the cos double angle identities.

Since you want an expression involving only cos it has to be $\cos 2A \equiv 2\cos^2 A - 1$.

b From part **a**:

$\cos(4x) \equiv 2\cos^2 2x - 1$
$\equiv 2(\cos 2x)^2 - 1$
$\equiv 2(2\cos^2 x - 1)^2 - 1$

Use the same double angle formula again to replace $\cos 2x$ in the answer in part **a**.

You can use a combination of double angle and compound angle identities to derive 'triple-angle identities'.

WORKED EXAMPLE 8.7

a Show that $\sin 3A \equiv 3\sin A - 4\sin^3 A$.
b Solve the equation $\sin 3x = \sin x$ for $x \in [0, 2\pi]$.

a $\sin 3A \equiv \sin(2A + A)$
$\equiv \sin 2A\cos A + \cos 2A\sin A$

If you write $3A = 2A + A$, you can use the $\sin(A+B)$ compound angle identity.

$\equiv (2\sin A\cos A)\cos A + (1 - 2\sin^2 A)\sin A$
$\equiv 2\sin A\cos^2 A + \sin A - 2\sin^3 A$
$\equiv 2\sin A(1 - \sin^2 A) + \sin A - 2\sin^3 A$
$\equiv 2\sin A - 2\sin^3 A + \sin A - 2\sin^3 A$
$\equiv 3\sin A - 4\sin^3 A$

You need an expression involving only single angles so use the double angle identities.

Since the answer only involves sin, use $\cos 2A \equiv 1 - 2\sin^2 A$.

You want only sin in the expression so replace $\cos^2 x$ with $1 - \sin^2 x$.

Continues on next page

b

$$\sin 3x = \sin x$$

$$3\sin x - 4\sin^3 x = \sin x$$

You need an equation in only one trig function, so use the result in **a**.

$$2\sin x - 4\sin^3 x = 0$$

$$2\sin x(1 - 2\sin^2 x) = 0$$

$$\sin x = 0 \text{ or } \sin^2 x = \frac{1}{2}$$

This is a cubic in $\sin x$ so rearrange to make the right-hand side equal to zero and then factorise.

When $\sin x = 0$:

$x = 0, \pi, 2\pi$

Solve these two equations in the usual way.

When $\sin^2 x = \frac{1}{2}$:

$$\sin x = \frac{1}{\sqrt{2}} \text{ or } -\frac{1}{\sqrt{2}}$$

$$x = \frac{\pi}{4}, \frac{3\pi}{4}, \frac{5\pi}{4}, \frac{7\pi}{4}$$

$$\therefore x = 0, \frac{\pi}{4}, \frac{3\pi}{3}, \pi, \frac{5\pi}{4}, \frac{7\pi}{4}, 2\pi$$

The double angle formulae allow you to simplify some complicated expressions involving functions and inverse functions.

Rewind

You met inverse trigonometric functions in Chapter 7, Section 2.

WORKED EXAMPLE 8.8

Let $x \in [-1, 1]$. Find an expression for:

a $\cos(2\arccos x)$ **b** $\sin(2\arccos x)$

in terms of x.

a

$$\cos(2\arccos x) = 2\cos^2(\arccos x) - 1$$
$$= 2(\cos(\arccos x))^2 - 1$$
$$= 2x^2 - 1$$

This is of the form $\cos 2A$, where $A = \arccos x$. So use the double angle identity involving only cos (as you don't want to introduce sin).

b

$$\sin(2\arccos x) = 2\sin(\arccos x)\cos(\arccos x)$$
$$= 2\sin(\arccos x)x \quad (1)$$

This is of the form $\sin 2A$ with $A = \arccos x$, so use the sin double angle identity this time.

$\cos(\arccos x) = x$

$$\sin^2(\arccos x) = 1 - \cos^2(\arccos x)$$
$$= 1 - (\cos(\arccos x))^2$$
$$= 1 - x^2$$

You need to convert $\sin(\arccos x)$ into an expression involving $\cos(\arccos x)$. Use $\sin^2 A \equiv 1 - \cos^2 A$, with $A = \arccos x$.

$\arccos x \in [0, \pi]$ so $\sin(\arccos x) > 0$

$$\therefore \sin(\arccos x) = \sqrt{1 - x^2}$$

When taking the square root, you have to choose between the positive and negative roots.

Substituting into (1):

$$\sin(2\arccos x) = 2x\sqrt{1 - x^2}$$

EXERCISE 8B

1 a i Given that $\cos\theta = -\frac{1}{4}$, find the exact value of $\cos 2\theta$.

ii Given that $\sin A = -\frac{2}{3}$, find the exact value of $\cos 2A$.

b i Given that $\sin x = \frac{1}{3}$ and $0 < x < \frac{\pi}{2}$, find the exact value of $\sin 2x$.

ii Given that $\sin x = \frac{3}{5}$ and $0 < x < \frac{\pi}{2}$, find the exact value of $\sin 2x$.

c i Given that $\sin x = \frac{3}{4}$ and $\frac{\pi}{2} < x < \pi$, find the exact value of $\sin 2x$.

ii Given that $\sin x = \frac{2}{3}$ and $\frac{\pi}{2} < x < \pi$, find the exact value of $\sin 2x$.

2 Without using a calculator, find the exact value of:

a i $\sin^2 22.5°$ ii $\cos^2 22.5°$ b i $\cos^2 75°$ ii $\sin^2 75°$.

3 Simplify each expression, using a double angle identity.

a $2\cos^2(3A) - 1$ b $4\sin 5x \cos 5x$ c $3 - 6\sin^2 \frac{b}{2}$ d $5\sin\frac{x}{3}\cos\frac{x}{3}$

4 Use double angle identities to solve each equation.

a $\sin 2x = 3\sin x$ for $x \in [0, 2\pi]$

b $\cos 2x - \sin^2 x = -2$ for $0° \leqslant x \leqslant 180°$

c $5\sin 2x = 3\cos x$ for $-\pi < x < \pi$

d $\tan 2x - \tan x = 0$ for $0° \leqslant x \leqslant 360°$

5 Prove each identity.

a $(\sin x + \cos x)^2 \equiv 1 + \sin 2x$

b $\cos^4\theta - \sin^4\theta \equiv \cos 2\theta$

c $\tan 2A - \tan A \equiv \frac{\tan A}{\cos 2A}$

d $\tan\alpha - \frac{1}{\tan\alpha} \equiv -\frac{2}{\tan 2\alpha}$

6 Find all the values of $\theta \in [-\pi, \pi]$ that satisfy the equation $\cos^2\theta + \cos 2\theta = 0$.

7 a Use the identity $\tan(A+B) \equiv \frac{\tan A + \tan B}{1 - \tan A \tan B}$ to show that $\tan 2A \equiv \frac{2\tan A}{1 - \tan^2 A}$.

b Show that $\tan 22.5° = \sqrt{2} - 1$.

8 Show that $\frac{1 - \cos 2\theta}{1 + \cos 2\theta} \equiv \tan^2\theta$.

9 a Given that $\tan\alpha \tan 2\alpha = 6$, find the possible values of $\tan\alpha$.

b Solve the equation $\tan\alpha \tan 2\alpha = 1$ for $\alpha \in (0, \pi)$, giving your answer in terms of π.

10 a Express $\cos 3A$ in terms $\cos A$.

b Express $\tan 3A$ in terms of $\tan A$.

11 Express $\cos 4\theta$ in terms of:

a $\cos\theta$ b $\sin\theta$.

12 a Show that:

i $\cos^2 \frac{1}{2}x \equiv \frac{1}{2}(1 + \cos x)$

ii $\sin^2\left(\frac{1}{2}x\right) \equiv \frac{1}{2}(1 - \cos x)$.

b Express $\tan^2 \frac{1}{2}x$ in terms of $\cos x$.

13 Given that $a \sin 4x = b \sin 2x$ and $0 < x < \frac{\pi}{2}$, express $\sin^2 x$ in terms of a and b.

14 Show that:

a $\tan(\arctan 1.2 + \arctan 0.5) = \frac{17}{4}$

b $\tan\left(2 \arctan \frac{1}{3}\right) = \frac{3}{4}$.

15 **a** Show that $\cos^2 \frac{1}{2} A \equiv \frac{1}{2}(1 + \cos A)$.

b Hence show that $\cos\left(\frac{1}{2} \arccos x\right) = \sqrt{\frac{1}{2}(1 + x)}$ for $-1 \leqslant x \leqslant 1$.

Section 3: Functions of the form $a \sin x + b \cos x$

In this section you will look at a useful method for dealing with sums of trigonometric functions.

Suppose you are trying to solve the equation $3 \sin x + 4 \cos x = 2$. You know that you need to try to write everything in terms of one trigonometric function so start by considering the identities met so far.

You cannot just replace $\sin x$ with a function of $\cos x$ (or $\cos x$ with a function of $\sin x$) as the only identity you have linking these is $\sin^2 x + \cos^2 x = 1$ and there is neither $\sin^2 x$ nor $\cos^2 x$ in the equation. So instead look for an identity involving both $\sin x$ and $\cos x$. From the list of compound angle identities there are several options – try one of the two that has a + sign.

Tip

Notice that $\cos(x - y) = \cos x \cos y + \sin x \sin y$ would work here as well as it also has a + sign.

$$\sin(x + y) \equiv \sin x \cos y + \cos x \sin y$$
$$\equiv \cos y \sin x + \sin y \cos x$$

Now, compare this with the equation:

$$3 \sin x + 4 \cos x = 2$$

You can see that the left-hand side of the equation would be the same as the compound angle expression if you could find y so that $\cos y = 3$ and $\sin y = 4$. This is not possible because $\sin^2 y + \cos^2 y \equiv 1$ so adjust the original identity by multiplying by a constant, R:

$$R \sin(x + y) \equiv (R \cos y) \sin x + (R \sin y) \cos x$$

Now you have:

$$R \cos y = 3 \text{ and } R \sin y = 4$$

which constitutes a pair of simultaneous equations in two unknowns (R and y).

To find R you can use $\sin^2 \theta + \cos^2 \theta \equiv 1$:

$$(R \sin y)^2 + (R \cos y)^2 = 4^2 + 3^2$$
$$R^2(\sin^2 y + \cos^2 y) = 16 + 9$$
$$R^2 = 25$$
$$\therefore\ R = 5 \quad (\text{by convention } R > 0)$$

To find y eliminate R by dividing one equation by the other:

$$\frac{R\sin y}{R\cos y} = \frac{4}{3}$$

$$\tan y = \frac{4}{3}$$

$$y = \tan^{-1}\frac{4}{3} = 0.927 \text{ (3 s.f.)}$$

So, you have seen that you can write $3\sin x + 4\cos x$ as $5\sin(x + 0.927)$ which allows you to rewrite the equation as:

$$5\sin(x+0.927) = 2$$

This is now a type of equation that you know how to solve.

You can use the same procedure with a cos compound angle identity, and you can apply it to differences as well as sums of trigonometric functions.

Key point 8.4

To write $a\sin x \pm b\cos x$ in the form $R\sin(x \pm \alpha)$ or $R\cos(x \pm \alpha)$:

1. Expand the brackets using a compound angle identity.
2. Equate coefficients of $\sin x$ and $\cos x$ to get equations for $R\sin\alpha$ and $R\cos\alpha$.
3. To get R, use $R^2 = a^2 + b^2$.
4. To get α, use $\tan\alpha \equiv \frac{\sin\alpha}{\cos\alpha}$.

Tip

You will usually be told which identity to use. However, if not, choose one that gives the same sign as the expression you have. In Worked example 8.9, $R\sin(x+\alpha)$ would also work as it gives a + sign.

WORKED EXAMPLE 8.9

a Write $3\sin x + 4\cos x$ in the form $R\cos(x-\alpha)$.

b Hence solve $3\sin x + 4\cos x = 2$ for $x \in [0, 2\pi]$

a $R\cos(x-\alpha) \equiv R(\cos x\cos\alpha + \sin x\sin\alpha)$
$\equiv (R\cos\alpha)\cos x + (R\sin\alpha)\sin x$
$\equiv (R\sin\alpha)\sin x + (R\cos\alpha)\cos x$

You are told to use $R\cos(x-\alpha)$, so first expand this using the compound angle identity.

Comparing this with $3\sin x + 4\cos x$:
$R\sin\alpha = 3$, $R\cos\alpha = 4$

Compare coefficients of $\sin x$ and $\cos x$.

$R^2(\cos^2\alpha + \sin^2\alpha) = 4^2 + 3^2$
$R^2 = 25$
$R = 5$

Find R by using $\sin^2 + \cos^2 \equiv 1$.

$\frac{R\sin\alpha}{R\cos\alpha} = \frac{3}{4}$

$\tan\alpha = \frac{3}{4}$

$\alpha = \tan^{-1}\frac{3}{4} = 0.644$ (3 s.f.)

Find α by using $\frac{\sin}{\cos} \equiv \tan$.

$\therefore 3\sin x + 4\cos x = 5\cos(x - 0.644)$

Continues on next page

b $3\sin x + 4\cos x = 2$

$5\cos(x - 0.644) = 2$

$\cos(x - 0.644) = \frac{2}{5}$

Use the answer from part **a**.

Since $0 < x < 2\pi$

$-0.664 < x - 0.664 < 2\pi - 0.664$

Now follow the standard procedure. Sketch the cos graph on the domain of $x - 0.664$.

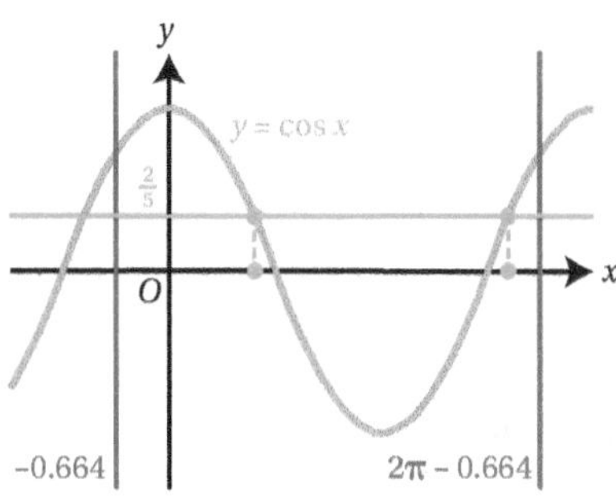

$\arccos\frac{2}{5} = 1.159$

$x - 0.644 = 1.159, 5.124$

$x = 1.80, 5.77$ (3 s.f.)

WORKED EXAMPLE 8.10

a Write $4\sin x - 6\cos x$ in the form $R\sin(x - \theta)$ where $R > 0$ and $\theta \in \left[0, \frac{\pi}{2}\right]$.

b Hence find the maximum value of $4\sin x - 6\cos x$ and the smallest positive value of x for which it occurs.

a $R\sin(x - \theta) \equiv R(\sin x\cos\theta - \cos x\sin\theta)$

$\equiv (R\cos\theta)\sin x - (R\sin\theta)\cos x$

Follow the standard procedure. So, start by expanding $R\sin(x - \theta)$.

Comparing this with $4\sin x - 6\cos x$:

$R\cos\theta = 4$, $R\sin\theta = 6$

Compare coefficients of $\sin x$ and $\cos x$.

$R^2 = 4^2 + 6^2 = 52$

$R = \sqrt{52} = 2\sqrt{13}$

Find R by using $\sin^2 + \cos^2 \equiv 1$.

$\tan\theta = \frac{R\sin\theta}{R\cos\theta} = \frac{6}{4} = \frac{3}{2}$

$\theta = \arctan\frac{3}{2} = 0.983$ (3 s.f.)

Find α by using $\frac{\sin}{\cos} \equiv \tan$.

$\therefore\ 4\sin x - 6\cos x = 2\sqrt{13}\sin(x - 0.983)$

b Maximum value $= 2\sqrt{13}$

Since sin has a maximum value of 1, the maximum value of this function must be R.

This value occurs when $\sin(x - 0.983) = 1$

$x - 0.983 = \frac{\pi}{2}$

$x = 0.983 + \frac{\pi}{2} = 2.55$ (3 s.f.)

Solve $\sin(x - 0.983) = 1$ for the smallest positive value of x.

EXERCISE 8C

1 Express in the form $R\sin(x+\alpha)$ where $R>0$ and $0<\alpha<\frac{\pi}{2}$:

i $4\sin x+6\cos x$ ii $\cos x+3\sin x$.

2 Express in the form $R\sin(\theta-a)$ where $R>0$ and $0<a<90°$:

i $2\sin\theta-2\cos\theta$ ii $\sin\theta-\sqrt{3}\cos\theta$.

3 Express in the form $R\cos(x+\theta)$ where $R>0$ and $0<\theta<\frac{\pi}{2}$:

i $\sqrt{6}\cos x-\sqrt{2}\sin x$ ii $5\cos x-5\sin x$.

4 Express in the form $R\cos(x-\theta)$ where $R>0$ and $0<\theta<90°$:

i $7\cos x+6\sin x$ ii $5\sin x+12\cos x$.

5 a Express $5\sin x+12\cos x$ in the form $R\sin(x+\theta)$, where $R>0$ and $0<\theta<\frac{\pi}{2}$.

b Hence give details of two successive transformations that transform the graph of $y=\sin x$ into the graph of $y=5\sin x+12\cos x$.

6 a Express $3\sin x-7\cos x$ in the form $R\sin(x-\theta)$, where $R>0$ and $0<\theta<\frac{\pi}{2}$.

b Hence find the range of the function $f(x)=3\sin x-7\cos x$.

Elevate

See Support Sheet 8 for a further example of solving equations of the form $a\sin x+b\cos x$ and for more practice questions.

7 a Express $4\cos x-5\sin x$ in the form $R\cos(x+\alpha)$, where $R>0$ and $0<\alpha<\frac{\pi}{2}$.

b Hence find the smallest positive value of x for which $4\cos x-5\sin x=0$.

8 a Express $\sqrt{3}\sin x+\cos x$ in the form $R\cos(x-\theta)$, where $R>0$ and $0<\theta<\frac{\pi}{2}$.

b Hence find the coordinates of the minimum and maximum points on the graph of $y=\sqrt{3}\sin x+\cos x$ for $x\in[0,2\pi]$.

9 Find, to 3 significant figures, all values of x in the interval $[0, 2\pi]$ for which $2\sin x-\cos x=2$.

10 Solve the equation $\sin 2x+\cos 2x=1$ for $-\pi\leqslant x\leqslant\pi$.

Section 4: Reciprocal trigonometric functions

Since $\tan x\equiv\frac{\sin x}{\cos x}$ you can do all the calculations you need with just sine and cosine. However, having the notation for $\tan x$ can simplify many expressions.

Similarly, since expressions of the form $\frac{\cos x}{\sin x}\left(\equiv\frac{1}{\tan x}\right)$, $\frac{1}{\sin x}$ and $\frac{1}{\cos x}$ often occur, it seems sensible to have notations for these as well.

Using the graphs of sine, cosine and tangent and what you know about reciprocal transformations, you can draw the graphs of $y=\sec x$, $y=\operatorname{cosec} x$ and $y=\cot x$.

Key point 8.5

The **reciprocal trigonometric functions** are:

- secant: $\sec x \equiv \dfrac{1}{\cos x}$
- cosecant: $\operatorname{cosec} x \equiv \dfrac{1}{\sin x}$
- cotangent: $\cot x \equiv \dfrac{1}{\tan x}$

Common error

It is easy to think that $\sec x \equiv \dfrac{1}{\sin x}$ and $\operatorname{cosec} x \equiv \dfrac{1}{\cos x}$. Be careful – it's the other way round!

$y = \sec x$ has domain $x \in \mathbb{R}$, $x \neq \dfrac{n\pi}{2}$ for integer n (it has vertical asymptotes at every multiple of $\dfrac{\pi}{2}$ corresponding to $\cos x = 0$).

It has range $(-\infty, -1] \cup [1, \infty)$.

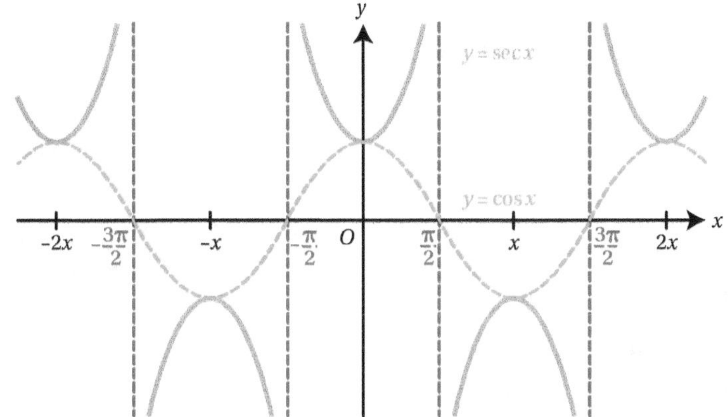

$y = \operatorname{cosec} x$ has domain $x \in \mathbb{R}$, $x \neq n\pi$ for integer n (it has vertical asymptotes at every multiple of π corresponding to $\sin x = 0$).

It has range $(-\infty, -1] \cup [1, \infty)$.

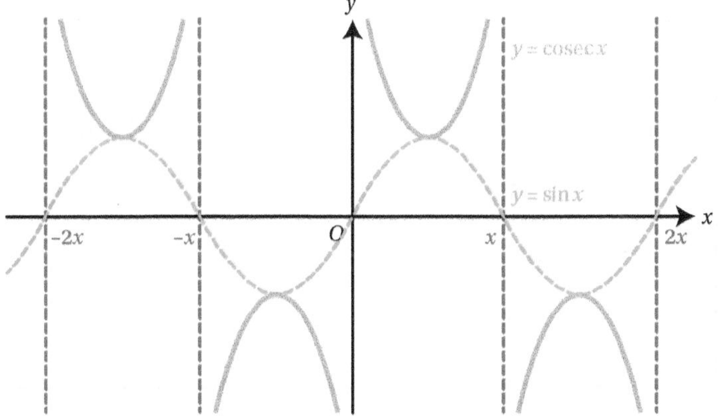

$y = \cot x$ has domain $x \in \mathbb{R}$, $x \neq n\pi$ for integer n (it has vertical asymptotes at every multiple of π corresponding to $\tan x = 0$).

It has range $\mathbb{R}$.

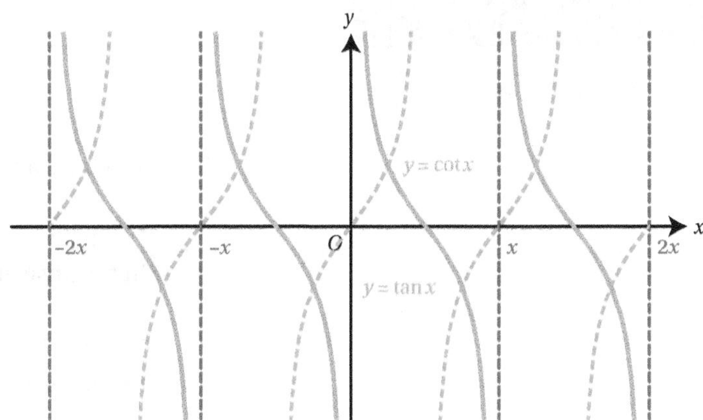

Perhaps the most common usage of these functions is in the identities that can be deduced from the familiar $\sin^2 x + \cos^2 x = 1$ by dividing through by $\cos^2 x$ and $\sin^2 x$ respectively (see Key point 8.6).

Key point 8.6

- $\sec^2 x \equiv 1+\tan^2 x$
- $\text{cosec}^2 x \equiv 1+\cot^2 x$

Tip

The inverse trigonometric functions follow the same conventions as the normal trigonometric functions so $\sec^2 x$ means $(\sec x)^2$.

WORKED EXAMPLE 8.11

Solve the equation $2\tan^2 x+\frac{3}{\cos x}=0$ for $-\pi < x < \pi$.

$2\tan^2 x+\frac{3}{\cos x}=0$

$2\tan^2 x+3\sec x=0$

$\frac{1}{\cos x} \equiv \sec x$ so $\frac{3}{\cos x} \equiv 3\sec x$.

$2(\sec^2 x-1)+3\sec x=0$

$2\sec^2 x+3\sec x-2=0$

As always you want only one trig function in the equation. Use $\tan^2 x \equiv \sec^2 x - 1$ to achieve this.

$(2\sec x-1)(\sec x+2)=0$

$\sec x=\frac{1}{2}$ or $\sec x=-2$

$\frac{1}{\cos x}=\frac{1}{2}$ or $\frac{1}{\cos x}=-2$

You do not know how to find inverse sec, so express sec x in terms of cos x.

$\cos x=2$ or $\cos x=-\frac{1}{2}$

When $\cos x=2$: no solutions.

Finally solve each equation separately.

When $\cos x=-\frac{1}{2}$:

$\arccos\left(-\frac{1}{2}\right)=\frac{2\pi}{3}$

$x=\frac{2\pi}{3}$ or $-\frac{2\pi}{3}$

WORKED EXAMPLE 8.12

Show that $\sec 2\theta \equiv \frac{\sec^2\theta}{2-\sec^2\theta}$.

$\sec 2\theta \equiv \frac{1}{\cos 2\theta}$

You have a formula for $\cos 2\theta$, so start by introducing cos.

$\equiv \frac{1}{2\cos^2\theta-1}$

Use the version that only contains cos as only sec is wanted in the answer: $\cos 2\theta \equiv 2\cos^2\theta-1$.

$\equiv \frac{1}{\frac{2}{\sec^2\theta}-1}$

You need $\sec\theta$ in the answer so change back from $\cos\theta$ using $\cos\theta \equiv \frac{1}{\sec\theta}$.

$\equiv \frac{\sec^2\theta}{2-\sec^2\theta}$

Finally, simplify by multiplying top and bottom of the fraction by $\sec^2\theta$.

EXERCISE 8D

1 Giving your answers to four significant figures, find the value of:

a **i** $\sec 1.2$ **ii** $\operatorname{cosec} 2.4$ **b** **i** $\cot(-3.5)$ **ii** $\sec(-0.6)$

c **i** $\operatorname{cosec}\frac{2\pi}{5}$ **ii** $\cot\frac{4\pi}{3}$.

2 Find the exact value of:

a **i** $\operatorname{cosec}\frac{\pi}{3}$ **ii** $\operatorname{cosec}\frac{\pi}{4}$ **b** **i** $\sec\frac{3\pi}{4}$ **ii** $\sec\frac{5\pi}{6}$

c **i** $\cot\left(-\frac{\pi}{4}\right)$ **ii** $\cot\left(-\frac{2\pi}{3}\right)$ **d** **i** $\operatorname{cosec}\frac{3\pi}{2}$ **ii** $\cot\frac{\pi}{2}$.

3 Find the values of $\operatorname{cosec} A$ and $\sec B$.

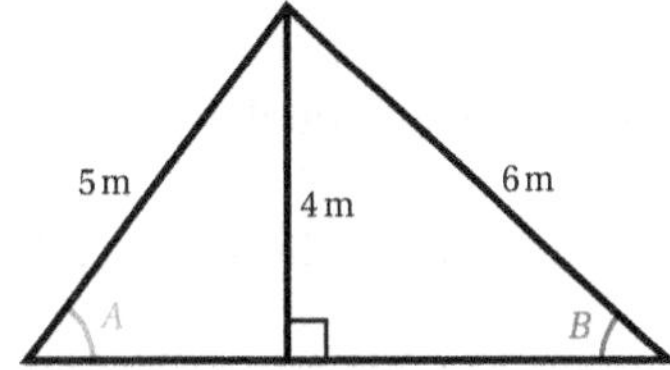

4 Solve these equations for $x \in [0, 2\pi]$, giving your answers to 3 s.f.

a **i** $\sec x = 2$ **ii** $\sec x = 3$ **b** **i** $\operatorname{cosec} x = 1.5$ **ii** $\operatorname{cosec} x = 2.7$

c **i** $\cot x = 5$ **ii** $\cot x = 0.5$ **d** **i** $\sec 2x = 3$ **ii** $\operatorname{cosec} 2x = 4$.

5 Find the exact solution of each equation for $-\pi \leqslant \theta \leqslant \pi$.

a **i** $\operatorname{cosec}\theta = -2$ **ii** $\operatorname{cosec}\theta = -1$ **b** **i** $\cot\theta = \sqrt{3}$ **ii** $\cot\theta = 1$

c **i** $\sec\theta = 1$ **ii** $\sec\theta = -\frac{2}{\sqrt{3}}$ **d** **i** $\cot\theta = 0$ **ii** $\cot\theta = -1$

6 **a** **i** Given that $\tan\theta = \frac{4}{3}$ and $0° < \theta < 90°$, find the exact value of $\sec\theta$.

ii Given that $\tan\theta = \frac{2}{5}$ and $0° < \theta < 90°$, find the exact value of $\sec\theta$.

b **i** Given that $\operatorname{cosec}\theta = 5$ and $\theta \in \left[0, \frac{\pi}{2}\right]$, find the exact value of $\cot\theta$.

ii Given that $\operatorname{cosec}\theta = 3$ and $\theta \in \left[0, \frac{\pi}{2}\right]$, find the exact value of $\cot\theta$.

c **i** Given that $\cot\theta = 3$ and $\pi < \theta < \frac{3\pi}{2}$, find the exact value of $\sin\theta$.

ii Given that $\cot\theta = \frac{1}{2}$ and $\pi < \theta < \frac{3\pi}{2}$, find the exact value of $\sin\theta$.

d **i** Given that $\sin\theta = \frac{\sqrt{2}}{3}$, find the possible values of $\sec\theta$.

ii Given that $\sin\theta = -\frac{1}{2}$, find the possible values of $\sec\theta$.

7 Prove that $\sin^2\theta + \cot^2\theta\sin^2\theta \equiv 1$.

8 Solve the equation $3\operatorname{cosec}^2 x = 4$ for $0° < x < 360°$.

9 Find, to three significant figures where appropriate, the values of x in the interval $-\pi < x < \pi$ for which $\cot x = 3\cos x$.

10 Show that $\tan x + \cot x \equiv \sec x \operatorname{cosec} x$.

11 Prove that $\operatorname{cosec} 2x \equiv \frac{\sec x \operatorname{cosec} x}{2}$.

12 Prove that $\cot 2x \equiv \frac{\cot^2 x - 1}{2 \cot x}$.

13 **a** Given that $\sec^2 x - 3\tan x + 1 = 0$, show that $\tan^2 x - 3\tan x + 2 = 0$.

b Hence solve the equation $\sec^2 x - 3\tan x + 1 = 0$ for $x \in [0, 2\pi]$.

14 **a** Show that $\frac{\cot^2 \theta}{1+\cot^2 \theta} \equiv \cos^2 \theta$.

b Hence solve the equation $\frac{\cot^2 \theta}{1+\cot^2 \theta} = 3\sin 2\theta$ for $\theta \in [-\pi, \pi]$.

15 **a** Prove that $\frac{\sin\theta}{1-\cos\theta} - \frac{\sin\theta}{1+\cos\theta} \equiv 2\cot\theta$.

b Hence solve the equation $\frac{\sin 2x}{1-\cos 2x} - \frac{\sin 2x}{1+\cos 2x} = 3$ for $0° < x < 180°$.

16 Given that θ is small enough to neglect the terms in θ^3 and above, find an approximate expression for $\sec 3\theta$.

17 Solve the equation $\cot^2 2\theta = \operatorname{cosec} 2\theta + 1$ for $\theta \in [-\pi, \pi]$.

18 Find the inverse function of $\sec x$ in terms of the arccosine function.

Checklist of learning and understanding

- **Compound angle identities** (be careful to get the signs right!):
  - $\sin(A \pm B) \equiv \sin A \cos B \pm \cos A \sin B$
  - $\cos(A \pm B) \equiv \cos A \cos B \mp \sin A \sin B$
  - $\tan(A \pm B) \equiv \frac{\tan A \pm \tan B}{1 \mp \tan A \tan B}$
- **Double angle identities:**
  - $\sin 2A \equiv 2\sin A \cos A$
  - $\cos 2A \equiv \begin{cases} \cos^2 A - \sin^2 A \\ 2\cos^2 A - 1 \\ 1 - 2\sin^2 A \end{cases}$
  - $\tan 2A \equiv \frac{2\tan A}{1-\tan^2 A}$
- One particular application is to write $a\sin x \pm b\cos x$ in the form $R\sin(x \pm \alpha)$ or $R\cos(x \pm \alpha)$:
  - Expand the brackets, using a compound angle identity.
  - Equate coefficients of $\sin x$ and $\cos x$ to get equations for $R\sin\alpha$ and $R\cos\alpha$.
  - To get R, use $R^2 = a^2 + b^2$.
  - To get α, use $\tan\alpha \equiv \frac{\sin\alpha}{\cos\alpha}$.
- Reciprocal trigonometric functions are defined by:
  - $\sec x \equiv \frac{1}{\cos x}$
  - $\operatorname{cosec} x \equiv \frac{1}{\sin x}$
  - $\cot x \equiv \frac{1}{\tan x}$.
- Two identities for reciprocal trigonometric functions can be derived from $\sin^2 x + \cos^2 x = 1$:
  - $\sec^2 x \equiv 1 + \tan^2 x$
  - $\operatorname{cosec}^2 x \equiv 1 + \cot^2 x$.

Mixed practice 8

1. What is the maximum value of $8\sin x - 6\cos x$?

   Choose from these options.

   A 2   **B** 8   **C** 10   **D** 14

2. **a** Use the identity for $\cos(A+B)$ to prove that $\cos 2\theta \equiv 1 - 2\sin^2\theta$.

   b Find the exact solutions of the equation $\cos 2\theta = \sin\theta$ for $0 \leqslant \theta \leqslant 2\pi$.

3. **a** Write $\cos\left(x+\frac{\pi}{3}\right)$ in the form $a\cos x + b\sin x$.

   b Hence find the exact value of $x \in [-2\pi, 2\pi]$ for which $\cos\left(x+\frac{\pi}{3}\right) = \cos\left(x-\frac{\pi}{3}\right)$.

4. The circle shown in the diagram has centre O and radius r.

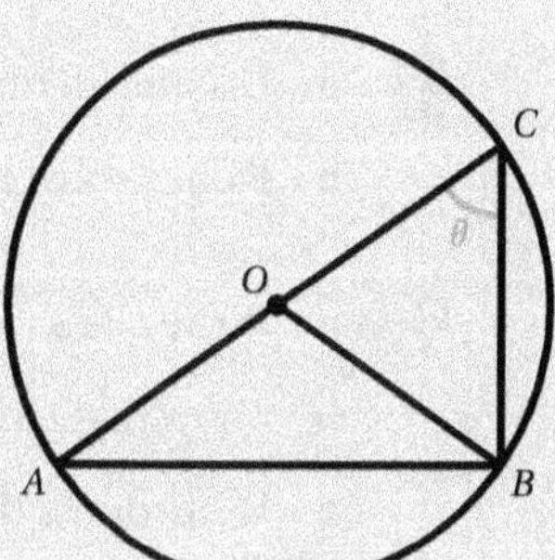

   a Write down the lengths of AB and BC in terms of r and θ.

   b Write down an expression for the area of the triangle ABC.

   c Write down an expression for the area of the triangle OBC.

   d Hence find the ratio of the two areas in the form $\frac{\text{area}(OBC)}{\text{area}(ABC)} = k$ where $k \in \mathbb{Q}$.

5. **a** Use the identity for $\tan(A+B)$ to show that $\tan 2A \equiv \frac{2\tan A}{1-\tan^2 A}$.

   b Write down the value of $\tan 135°$.

   c Hence find the exact value of $\tan 67.5°$.

6. A water wave has the profile shown in the graph, where y represents the height of the wave, in metres, and x is the horizontal distance, also in metres.

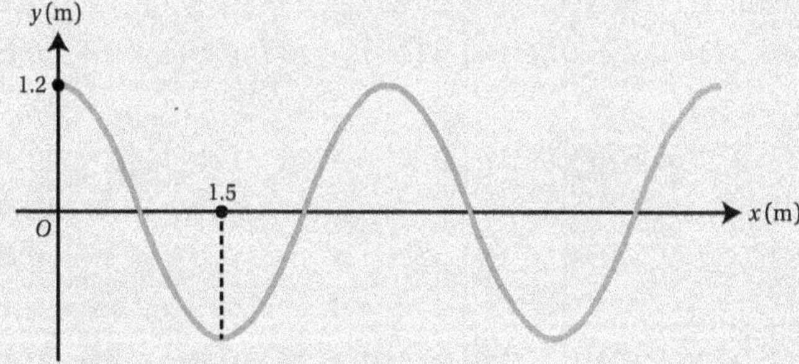

   a Given that the equation of the wave can be written as $y_1 = a\cos px$, find the values of a and p.

   b A second wave has the profile given by the equation $y_2 = 0.9\sin\left(\frac{2\pi}{3}x\right)$. Write down the amplitude and the period of the second wave.

   When the two waves combine a new wave is formed, with the profile given by $y = y_1 + y_2$.

   c Write the equation for y in the form $R\sin(x+\alpha)$, where $R > 0$ and $0 < \alpha < \frac{\pi}{2}$.

   d State the amplitude and the period of the combined wave.

   e Find the smallest positive value of x for which the height of the combined wave is zero.

   f Find the first **two** positive values of x for which the height of the combined wave is 1.3 m.

7 a Express $\sin x - 3\cos x$ in the form $R\sin(x-\alpha)$, where $R>0$ and $0° < \alpha < 90°$, giving your value of α to the nearest $0.1°$.

b Hence find the values of x in the interval $0° < x < 360°$ for which

$\sin x - 3\cos x + 2 = 0$

giving your values of x to the nearest degree.

[© AQA 2013]

8 Simplify $\dfrac{1}{\operatorname{cosec}\theta + \cot\theta} - \dfrac{1}{\operatorname{cosec}\theta - \cot\theta}$.

Choose from these options.

A $-2\cos\theta$ B $-\dfrac{2}{\tan\theta}$ C $-2\sec\theta$ D $-\dfrac{2}{\sin\theta}$

9 a Use the identity for $\cos(A+B)$ to show that $\cos 2\theta \equiv 2\cos^2\theta - 1$.

b Hence solve the equation $\dfrac{\sin\theta}{1+\cos\theta} = 3\cot\dfrac{\theta}{2}$ for $\theta \in (0, 2\pi)$.

10 a Write $t^3 - 3t^2 - 3t + 1$ as a product of a linear and a quadratic factor.

b Show that $\tan 3A \equiv \dfrac{3\tan A - \tan^3 A}{1 - 3\tan^2 A}$.

c Write down the exact value of $\tan 45°$.

d Hence find the exact value of $\tan 15°$ and $\tan 75°$.

11 By forming and solving a quadratic equation, solve the equation

$8\sec x - 2\sec^2 x = \tan^2 x - 2$

in the interval $0 < x < 2\pi$, giving the values of x in radians to three significant figures.

[© AQA 2013]

12 a Show that the equation

$$\frac{1}{1+\cos\theta} + \frac{1}{1-\cos\theta} = 32$$

can be written in the form

$\operatorname{cosec}^2\theta = 16$

b Hence, or otherwise, solve the equation

$$\frac{1}{1+\cos(2x-0.6)} + \frac{1}{1-\cos(2x-0.6)} = 32$$

giving all values of x in radians to two decimal places in the interval $0 < x < \pi$.

[© AQA 2012]

13 a Express $\sqrt{15}\sin(2x) + \sqrt{5}\cos(2x)$ in the form $R\sin(2x+\alpha)$.

b The function f is defined by $f(x) = \dfrac{2}{5 + \sqrt{15}\sin(2x) + \sqrt{5}\cos(2x)}$.

Using your answer to part **a**, find:

i the maximum value of $f(x)$, giving your answer in the form $p + q\sqrt{5}$ where $p, q \in \mathbb{Q}$

ii the smallest value of x for which this maximum occurs, giving your answer exactly, in terms of π.

14 **a** Write down an expression for $\sin(\arcsin x)$.

b Show that $\cos(\arcsin x) = \sqrt{1-x^2}$.

c Hence find an expression for $\sin(2 \arcsin x)$.

15 **a** **i** Expand $\left(x-\sqrt{10}\right)^2$.

ii Hence show that $x^2 + 10 \geqslant 2\sqrt{10}\, x$.

iii State the value of x for which the equality holds.

A picture of height 3 m hangs on the wall so that the lower end of the picture is 2 m above the floor. An observer sits on the floor at a distance d m from the wall, and sees the picture at an angle θ, as shown in the diagram.

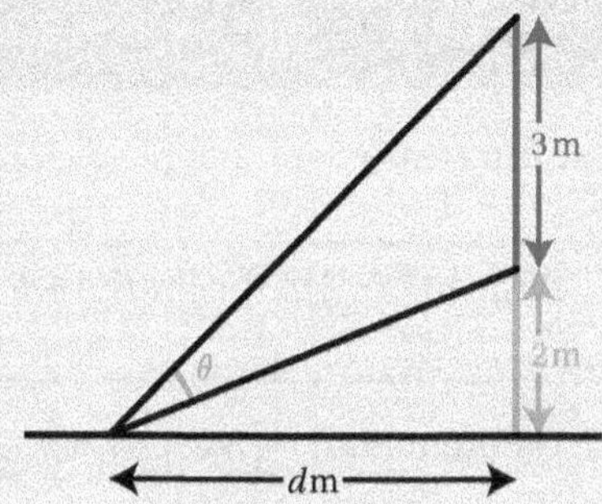

b Show that $\tan\theta = \dfrac{3d}{d^2+10}$.

c Use the answer to part **a** to show that $\tan\theta \leqslant \dfrac{3}{20}\sqrt{10}$.

d Hence find the value of d that gives the largest possible value of $\tan\theta$.

e State the distance the observer should stand from the wall in order to see the picture at the largest possible angle, and find the value of that angle.

Elevate

See Extension Sheet 8 for a selection of more challenging problems.

9 Calculus of exponential and trigonometric functions

In this chapter you will learn how to:

- differentiate e^x, $\ln x$, $\sin x$, $\cos x$ and $\tan x$
- integrate e^x, $\frac{1}{x}$, $\sin x$ and $\cos x$
- review applications of differentiation to find tangents, normals and stationary points
- review applications of integration to find the equation of a curve and areas.

Before you start…

Student Book 1, Chapter 7	You should be able to use rules of indices and logarithms.	1 Write $\ln(3x^4)$ in the form $A + B\ln x$.
Student Book 1, Chapter 12	You should be able to differentiate x^n.	2 Differentiate $y = \left(3x - \frac{1}{x}\right)\left(x + \frac{2}{3x}\right)$.
Student Book 1, Chapter 14	You should be able to integrate x^n for $n \neq -1$.	3 Find the exact value of $\int_1^2 \frac{x^3+3}{2x^2}\,dx$.
Chapters 7, 8	You should be able to use compound angle formulae and small angle approximations.	4 Use small angle approximations to find the approximate value of $\cos\left(\frac{\pi}{3} + \frac{\pi}{100}\right)$.

Extending differentiation and integration

You have already seen many situations that can be modelled using trigonometric, exponential and logarithm functions. For example, in Student Book 1, Chapter 8, you studied exponential models for population growth, and in Chapter 7 of this book you saw how to use sine and cosine functions to model periodic motion, such as motion around a circle, the oscillation of a particle attached to the end of a spring, or the heights of tides.

Often, you need to find the rate of change of quantities in such models; for example, at what rate is the population increasing after three years? Finding rates of change involves differentiation. Integration reverses this process – given the rate of change, you can find the equation for the original quantity.

So far you have learnt how to differentiate and integrate functions of the form ax^n. In this section rules of differentiation and integration will be extended to include a wider variety of functions.

Section 1: Differentiation

You already know, from Student Book 1, that the rate of growth of an exponential function is proportional to the value of the function. In particular, for $y = e^x$ the rate of growth equals the y value. Key point 9.1 gives this in terms of differentiation.

Rewind

In fact you already know, from Student Book 1, Chapter 8, that if $y = e^{kx}$ then $\frac{dy}{dx} = ke^{kx}$. You will see how this arises in the next chapter.

Key point 9.1

- If $y = e^x$ then $\frac{dy}{dx} = e^x$

The derivative of the natural logarithm function follows from the fact that $y = \ln x$ is the inverse function of $y = e^x$.

Rewind

You learnt about inverse functions and their graphs in Chapter 2, Section 4.

Key point 9.2

If $y = \ln x$ then $\frac{dy}{dx} = \frac{1}{x}$

PROOF 6

Let point A on the graph of $y = e^x$ have y-coordinate a.

Let point B be the point on the graph of $y = \ln x$, which is the reflection of A in the line $y = x$.

The reflection swaps x and y coordinates, so B has x-coordinate a.

From Chapter 2, since they are inverse functions, you know how the graphs of $y = e^x$ and $y = \ln x$ are related. So you can use the gradient of the first graph to find the gradient of the second graph.

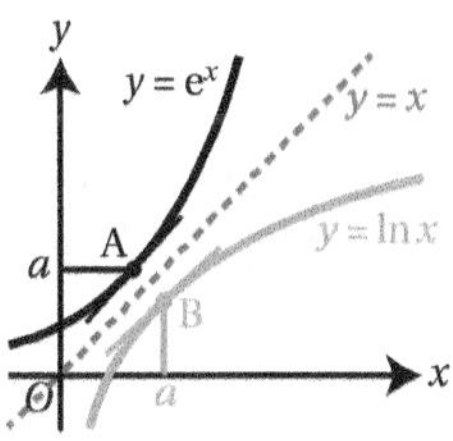

The gradient of $y = e^x$ at A is a.

You know that the gradient of $y = e^x$ at A equals the y-value.

The gradient of the reflected line:

To see what happens to the gradient on reflection in $y = x$, consider reflecting a triangle.

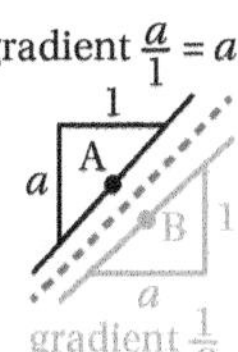

This shows that, if a line with gradient m is reflected in $y = x$, the gradient of the reflected line is $\frac{1}{m}$.

The gradient of the graph $y = \ln x$ at B is $\frac{1}{a}$.

This is 1 divided by the x-coordinate of B, so the gradient of $y = \ln x$ is $\frac{1}{x}$.

The rules for differentiating sums and constant multiples of expressions still apply. Sometimes you will need to simplify an expression before differentiating.

WORKED EXAMPLE 9.1

Differentiate $y = e^{x+3} - 3\ln(4x^2)$.

$$y = e^3 e^x - 3(\ln 4 + 2\ln x)$$
$$= e^3 e^x - 3\ln 4 - 6\ln x$$

You can only differentiate e^x and $\ln x$, so you need to use rules of indices and logarithms to simplify the expression.

$$\frac{dy}{dx} = e^3 e^x - \frac{6}{x}$$

e^x is multiplied by a constant (e^3).
$3\ln 4$ is a constant, so its derivative is 0.
The derivative of $6\ln x$ is $6 \times \frac{1}{x}$.

To differentiate $\sin x$ and $\cos x$ you need to remember differentiation from first principles, which you met in Student Book 1, Chapter 12. You also need to use compound angle formulae and small angle approximations.

Rewind

You can recap compound angle formulae in Chapter 8, Section 1, and small angle approximations in Chapter 7, Section 6.

WORKED EXAMPLE 9.2

Use differentiation from first principles to prove that the derivative of $\sin x$ is $\cos x$, where x is measured in radians.

Let $f(x) = \sin x$

$$f'(x) = \lim_{h \to 0}\left\{\frac{\sin(x+h) - \sin x}{h}\right\}$$

$f'(x) = \lim_{h \to 0}\left\{\frac{f(x+h) - f(x)}{h}\right\}$

$$= \lim_{h \to 0}\left\{\frac{\sin x\cos h + \sin h\cos x - \sin x}{h}\right\}$$

Use $\sin(A+B) = \sin A\cos B + \sin B\cos A$.

$$\approx \lim_{h \to 0}\left\{\frac{\sin x\left(1 - \frac{1}{2}h^2\right) + (h)\cos x - \sin x}{h}\right\}$$

Use small angle approximations:
$\cos h \approx 1 - \frac{1}{2}h^2$
$\sin h \approx h$

$$= \lim_{h \to 0}\left\{\frac{-\frac{1}{2}h^2\sin x + h\cos x}{h}\right\}$$

$$= \lim_{h \to 0}\left\{-\frac{1}{2}h\sin x + \cos x\right\}$$

$$= \cos x$$

Let h tend to 0.

Differentiating $\cos x$ follows the same method. You can also prove the result for $\tan x$ in this way.

Key point 9.3

- If $y = \sin x$ then $\frac{dy}{dx} = \cos x$
- If $y = \cos x$ then $\frac{dy}{dx} = -\sin x$
- If $y = \tan x$ then $\frac{dy}{dx} = \sec^2 x$

Fast forward

You can also prove the result for $\tan x$ using the quotient rule, which you will meet in Chapter 10.

Tip

These formulae only apply if x is in radians.

Worked examples 9.3 and 9.4 should help you remember two applications of differentiation: finding equations of tangents and normals and finding stationary points.

WORKED EXAMPLE 9.3

Find the equations of the tangent and the normal to the graph of the function $f(x) = \cos x + e^x$ at the point $x = 0$.

Give your answer in the form $ax + by + c = 0$, where a, b, c are integers.

$f'(x) = -\sin x + e^x$

$\therefore f'(0) = -\sin 0 + e^0 = 1$

You need the gradient, which is $f'(0)$.

When $x = 0$,

$y = f(0) = \cos 0 + e^0$

$= 1 + 1$

$= 2$

To find the equation of a straight line you also need coordinates of one point.

The tangent passes through the point on the graph where $x = 0$. Its y-coordinate is $f(0)$.

Tangent:

$y - y_1 = m(x - x_1)$

$y - 2 = 1(x - 0)$

$y = x + 2$

$y - x - 2 = 0$

Put all the information into the equation of a line.

Normal:

$y - 2 = -1(x - 0)$

$y = -x + 2$

$y + x - 2 = 0$

The gradient of the normal is $-\frac{1}{m}$ and it passes through the same point.

WORKED EXAMPLE 9.4

Find the coordinates of the stationary point on the graph of $y = \ln x + \frac{1}{x^2}$ and determine its nature.

$y = \ln x + x^{-2}$

$\frac{dy}{dx} = \frac{1}{x} - 2x^{-3}$ ...... First find $\frac{dy}{dx}$.

For stationary points, $\frac{dy}{dx} = 0$:

$\frac{1}{x} - \frac{2}{x^3} = 0$ ...... Rewrite $2x^{-3}$ as $\frac{2}{x^3}$ and multiply through by x^3.

$x^2 - 2 = 0$

$\therefore x = \sqrt{2}$ (as $x > 0$) ...... $\ln x$ is only defined for $x > 0$.

$y = \ln\sqrt{2} + \frac{1}{2}$ ...... Find the y-value.

The stationary point is $\left(\sqrt{2}, \frac{1}{2}\ln 2 + \frac{1}{2}\right)$ ...... Use laws of logs to rewrite the y-coordinate.

$\frac{d^2y}{dx^2} = -\frac{1}{x^2} + 6x^{-4}$ ...... To determine the nature of the stationary point, evaluate the second derivative at $x = \sqrt{2}$.

$= -\frac{1}{(\sqrt{2})^2} + \frac{6}{(\sqrt{2})^4}$

$= -\frac{1}{2} + \frac{6}{4} = 1 > 0$

The stationary point is a minimum.

EXERCISE 9A

1 Differentiate each function.

a i $y = 3e^x$ ii $y = \frac{2e^x}{5}$ b i $y = -2\ln x$ ii $y = \frac{1}{3}\ln x$

c i $y = \frac{\ln x}{5} - 3x + 4e^x$ ii $y = 4 - \frac{e^x}{2} + 3\ln x$ d i $y = 3\sin x$ ii $y = 2\cos x$

e i $y = 2x - 5\cos x$ ii $y = \tan x + 5$ f i $y = \frac{\sin x + 2\cos x}{5}$ ii $y = \frac{1}{2}\tan x - \frac{1}{3}\sin x$

2 Simplify each function and then differentiate it.

a i $y = \ln x^3$ ii $y = 2\ln x^5$ b i $y = 3\ln 2x$ ii $y = \ln 5x$

c i $y = e^{x+3}$ ii $y = e^{x-3}$ d i $y = e^{2\ln x}$ ii $y = e^{3\ln x + 2}$

e i $y = \frac{3\sin x - \cos x}{\cos x}$ ii $y = \frac{2\cos x + 4\sin x}{\cos x}$ f i $y = \frac{e^{2x} - 2e^x}{e^x}$ ii $y = \frac{4e^x - e^{2x}}{2e^x}$

3 Find the exact value of the gradient of the graph of $f(x) = \frac{1}{2}e^x - 7\ln x$ at the point $x = \ln 4$.

4 Find the exact value of the gradient of the graph of $f(x) = e^x - \frac{\ln x}{2}$ when $x = \ln 3$.

5 Find the value of x where the gradient of $f(x) = 5 - 2e^x$ is –6.

6 Find the value of x where the gradient of $g(x) = x^2 - 12\ln x$ is 2.

7 Find the rate of change of $f(x) = \sin x + x^2$ at the point $x = \frac{\pi}{2}$.

8 Find the rate of change of $g(x) = \frac{1}{4}\tan x - 3\cos x - x^3$ at the point $x = \frac{\pi}{6}$.

9 Find the equation of the tangent to the curve $y = e^x + x$ that is parallel to $y = 3x$.

10 Find the equation of the tangent to the curve $y = 5\sin x$ that is parallel to $2x - y = 6$.

11 Given $h(x) = \sin x + \cos x, 0 \leqslant x < 2\pi$, find the values of x for which $h'(x) = 0$.

12 Find the equations of the tangent and the normal to the graph of $y = 3\tan x - 2\sqrt{2}\sin x$ at $x = \frac{\pi}{4}$. Give all the coefficients in an exact form.

13 Given that $y = \frac{1}{4}\tan x + \frac{1}{x^2}, 0 < x \leqslant 2\pi$, solve the equation $\frac{dy}{dx} = 1 - \frac{2}{x^3}$.

14 Find and classify the stationary points on the curve $y = \sin x + 4\cos x$ in the interval $0 < x < 2\pi$.

15 Show that the function $f(x) = \ln x + \frac{1}{x^k}$ has a stationary point with y-coordinate $\frac{\ln k + 1}{k}$.

16 Find the range of the function $f: x \mapsto e^x - 4x + 2$.

17 Find and classify the stationary points of:

a $y = \ln x - \sqrt{x}$ **b** $y = 2e^x - 5x$.

18 The volume V of water, in millions of litres, in a tidal lake is controlled by a dam. It can be modelled by $V = 60\cos t + 100$, where t is the time, in days, after the dam is first opened.

a What is the smallest volume of the lake?

b A hydroelectric plant produces an amount of electricity proportional to the rate of flow of water. Find the time, in the first 6 days, when the plant is producing maximum electricity.

Section 2: Integration

All the differentiation results from Section 1 can be reversed to integrate several new functions.

Key point 9.4

- $\int e^x \, dx = e^x + c$
- $\int \frac{1}{x} \, dx = \ln|x| + c$
- $\int \sin x \, dx = -\cos x + c$
- $\int \cos x \, dx = \sin x + c$

Note the modulus sign in $\int \frac{1}{x} \, dx = \ln|x| + c$. This is because the x in $\ln x$ cannot be negative, but the x in $\frac{1}{x}$ can. Taking the modulus means that you can find the area between the x-axis and the negative part of the curve $y = \frac{1}{x}$. (You can see that this is the same as the corresponding region on the positive x-axis.)

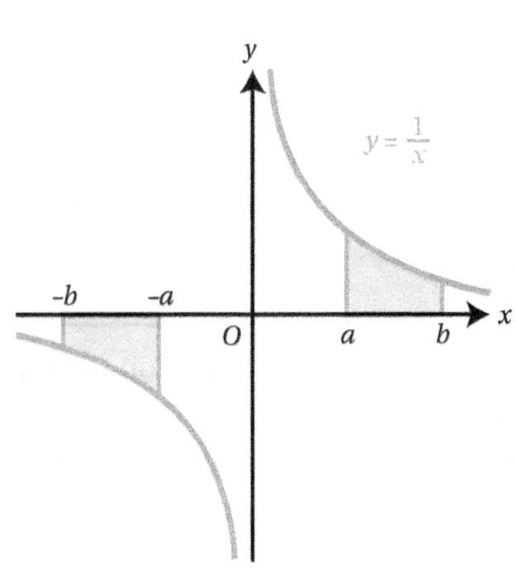

You can combine these facts with rules of integration you already know. Sometimes you will need to rewrite an expression in a different form before you can integrate it.

WORKED EXAMPLE 9.5

Find $\int \frac{5x-3x^3}{2x^2}\,dx$.

$\int \frac{5x-3x^3}{2x^2}\,dx = \int \left(\frac{5}{2x} - \frac{3}{2}x\right)dx$ — You don't know how to integrate a quotient, but you can split the fraction and integrate each term separately.

$= \frac{5}{2}\ln x - \frac{3}{4}x^2 + c$

WORK IT OUT 9.1

Decide which statement is correct. Identify the errors made in the incorrect statements.

Statement 1	Statement 2	Statement 3	Statement 4
$y=\frac{1}{x}$ $\Rightarrow \frac{dy}{dx}=\ln x$	$\int\frac{1}{x^2}\,dx=\ln x^2+c$	$y=\ln 3x$ $\Rightarrow \frac{dy}{dx}=\frac{1}{3x}$	$y=\ln 3x$ $\Rightarrow \frac{dy}{dx}=\frac{1}{x}$

Remember that integration is the reverse of differentiation, so if you know the derivative of a function you can use integration to find the function itself.

WORKED EXAMPLE 9.6

A curve passes through the point $(-1, 3)$ and its gradient is given by $\frac{dy}{dx}=3e^x-1$. Find the equation of the curve.

$y=\int(3e^x-1)\,dx$ — Integrate $\frac{dy}{dx}$ to get y.

$=3e^x-x+c$ — Don't forget the constant of integration.

Using $x=-1, y=3$: — Use given values of x and y to find the constant of integration.

$3=3e^{-1}-(-1)+c$

$c=2-\frac{3}{e}$

$\therefore y=3e^x-x+2-\frac{3}{e}$ — Write the full equation.

You can also use integration to find the area between the curve and the x-axis. This involves evaluating **definite integrals**. It is always a good idea to draw a diagram to make sure you are finding the correct area.

Rewind

You know from Student Book 1, Chapter 14, that when a curve is below the x-axis the integral is negative.

WORKED EXAMPLE 9.7

Find the exact area enclosed between the x-axis, the curve $y = \sin x$ and the lines $x = 0$ and $x = \frac{\pi}{3}$.

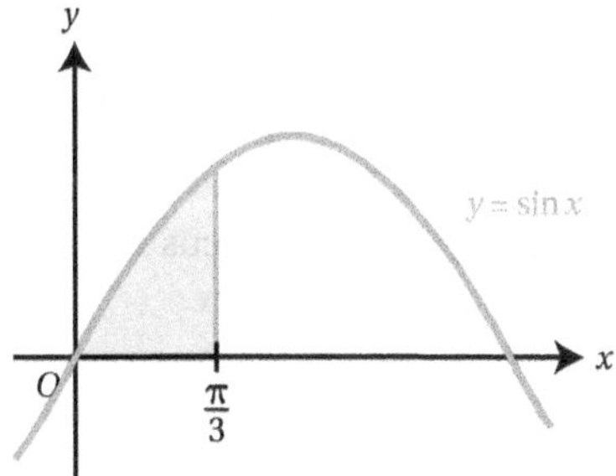

Sketch the graph and identify the required area.

$A = \int_0^{\frac{\pi}{3}} \sin x \, dx = [-\cos x]_0^{\frac{\pi}{3}}$

Integrate and write in square brackets.

$= \left(-\cos\frac{\pi}{3}\right) - (-\cos 0)$

$= \left(-\frac{1}{2}\right) - (-1) = \frac{1}{2}$

Evaluate the integrated expression at the upper and lower limits and subtract the value at the lower limit from the value at the upper limit.

WORK IT OUT 9.2

Three students are considering the problem, 'Evaluate $\int_{-2}^{6} \frac{1}{x} \, dx$.'

Discuss their solutions. What is the value of the area shaded in the second solution?

Solution 1	Solution 2	Solution 3
$\int_{-2}^{6} \frac{1}{x} \, dx = [\ln \lvert x \rvert]_{-2}^{6}$ $= \ln \lvert 6 \rvert - \ln \lvert -2 \rvert$ $= \ln 6 - \ln 2$ $= \ln 3$	This is the area between the graph and the x-axis: y -2 O 6 x You need to calculate the two parts separately and add them up: $\int_{-2}^{0} \frac{1}{x} \, dx + \int_{0}^{6} \frac{1}{x} \, dx$ But $\int_{-2}^{0} \frac{1}{x} \, dx = \ln 0 - \ln \lvert -2 \rvert$ And $\ln 0$ is not defined, so there is no answer.	You are trying to integrate between −2 and 6, but this includes $x = 0$, which is not in the domain of $\ln x$. So there is no answer.

EXERCISE 9B

1 Find each integral.

a i $\int 5e^x \, dx$ ii $\int 9e^x \, dx$ b i $\int \frac{2e^x}{5} \, dx$ ii $\int \frac{7e^x}{11} \, dx$

c i $\int \frac{e^x + 3x}{2} \, dx$ ii $\int \frac{e^x + x^3}{5} \, dx$ d i $\int 3 \cos x \, dx$ ii $\int 4 \sin x \, dx$

e i $\int \frac{\sin x - 2\cos x}{2} \, dx$ ii $\int \frac{2\cos x - \sin x}{3} \, dx$ f i $\int (\sqrt{x} + \sin x) \, dx$ ii $\int \left(\cos x - \frac{1}{\sqrt{x}} \right) dx$

2 Find each integral.

a i $\int \frac{2}{x} \, dx$ ii $\int \frac{3}{x} \, dx$ b i $\int \frac{1}{2x} \, dx$ ii $\int \frac{1}{3x} \, dx$

c i $\int \frac{5}{2x} \, dx$ ii $\int \frac{2}{3x} \, dx$ d i $\int \frac{x^2 - 1}{x} \, dx$ ii $\int \frac{x^3 + 5}{x} \, dx$

e i $\int \frac{3x + 2}{x^2} \, dx$ ii $\int \frac{3x - 5x^2}{x^3} \, dx$ f i $\int \frac{2\sqrt{x} + 3x}{x\sqrt{x}} \, dx$ ii $\int \frac{x^2 - 4\sqrt{x}}{x\sqrt{x}} \, dx$

3 Find the exact value of each definite integral.

a i $\int_0^2 3e^x \, dx$ ii $\int_1^3 2e^x \, dx$ b i $\int_0^{\ln 3} e^x \, dx$ ii $\int_0^{\ln 5} 2e^x \, dx$

c i $\int_1^{\ln 2} (3e^x + 2) \, dx$ ii $\int_1^{\ln 3} (4 - 3e^x) \, dx$ d i $\int_1^3 \frac{3}{2x} \, dx$ ii $\int_2^5 \frac{4}{3x} \, dx$

e i $\int_e^{e^2} \frac{2}{x} \, dx$ ii $\int_1^{e^3} \frac{3}{2x} \, dx$ f i $\int_0^{\frac{\pi}{2}} \cos x \, dx$ ii $\int_{\pi}^{2\pi} \sin x \, dx$

g i $\int_{-\pi}^{\pi} (3\sin x - 4\cos x) \, dx$ ii $\int_0^{2\pi} (2\cos x - \sin x) \, dx$

4 Find the exact value of the area enclosed by the curve $y = \frac{2}{3x}$, the x-axis and the lines $x = 2$ and $x = 6$. Give your answer in the form $a \ln b$.

5 Find the area enclosed by the curve $y = 3 \sin x$, the x-axis and the line $x = \frac{\pi}{3}$.

6 Find the exact value of $\int_0^{\frac{\pi}{3}} (\sin x + 2\cos x) \, dx$.

7 Find the exact value of $\int_1^{e^5} \frac{3}{x} \, dx$.

8 a Evaluate $\int_{-9}^{-3} \frac{2}{x} \, dx$.

b State the value of the area between the graph of $y = \frac{1}{x}$, the x-axis and the lines $x = -9$ and $x = -3$.

9 Find the equation of the curve given that $\frac{dy}{dx} = \cos x + \sin x$ and the curve passes through the point $(\pi, 1)$.

10 Find $\int \frac{(\sin x + 1)(\sin x - 1)}{2\cos x} \, dx$.

11 The derivative of the function $f(x)$ is $\frac{1}{2x}$.

a Find an expression for all possible functions $f(x)$.

b If the curve $y = f(x)$ passes through the point $(2, 7)$ find the equation of the curve.

Elevate

See Support Sheet 9 for a further example of finding the equation of a curve and for more practice questions.

12 The diagram shows part of the curve $y = e^x - 5\sin x$.

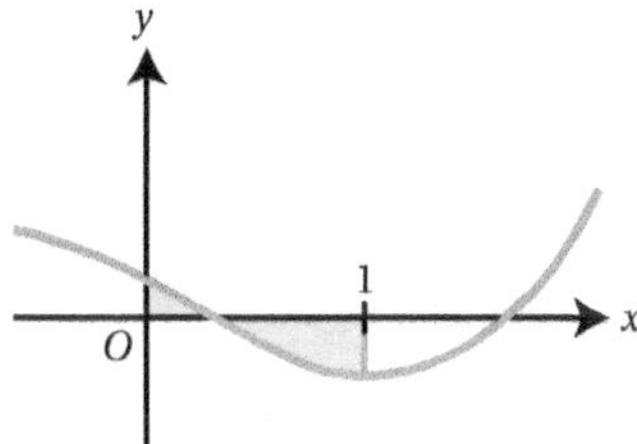

Find the area of the shaded region, enclosed between the curve, the x-axis and the lines $x = 0$ and $x = 1$.

Fast forward

Areas of regions that are partly above and partly below the x-axis are discussed in more detail in Chapter 12, Section 4.

13 The diagram shows a part of the curve with equation $y = \frac{4}{x} + x - 5$.

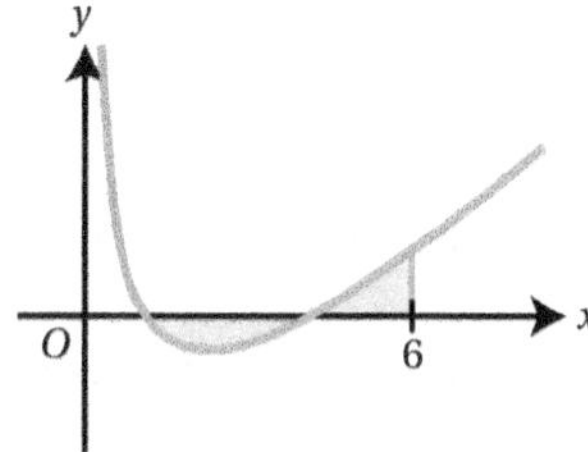

a Show that the curve crosses the x-axis at $x = 1$ and $x = 4$.

b Find the exact value of the area of the shaded region. Give your answer in the form $p - 4\ln q$, where p and q are rational numbers.

14 Show that the value of the integral $\int_k^{2k} \frac{1}{x}\,dx$ is independent of k.

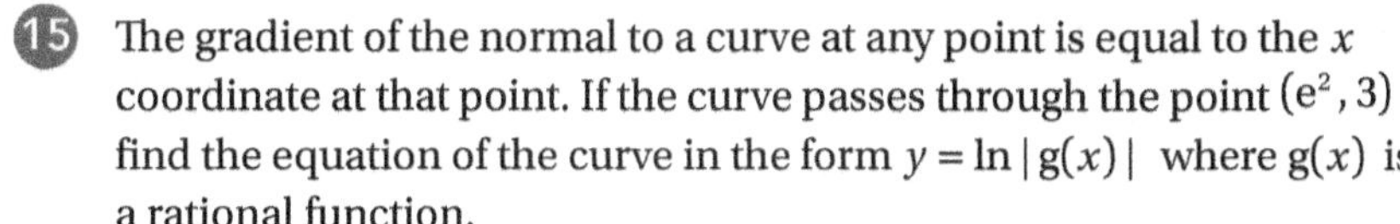

15 The gradient of the normal to a curve at any point is equal to the x coordinate at that point. If the curve passes through the point $(e^2, 3)$ find the equation of the curve in the form $y = \ln|g(x)|$ where $g(x)$ is a rational function.

Checklist of learning and understanding

- The derivatives of basic trigonometric, exponential and logarithm functions are:
  - $\frac{d}{dx}(\sin x) = \cos x$
  - $\frac{d}{dx}(e^x) = e^x$
  - $\frac{d}{dx}(\cos x) = -\sin x$
  - $\frac{d}{dx}(\ln x) = \frac{1}{x}$
  - $\frac{d}{dx}(\tan x) = \sec^2 x$
- The results for $\sin x$ and $\cos x$ can be derived using differentiation from first principles.
- The integrals of basic trigonometric, exponential and logarithm functions are:
  - $\int \sin x\,dx = -\cos x + c$
  - $\int e^x\,dx = e^x + c$
  - $\int \cos x\,dx = \sin x + c$
  - $\int \frac{1}{x}\,dx = \ln|x| + c$
- The rules of differentiation and integration can be combined to find:
  - rates of change
  - equations of tangents and normals
  - stationary points
  - the equation of a curve with a given gradient
  - the area between a curve and the x-axis.

Mixed practice 9

1 $\int_1^a \frac{1}{x}\,dx = 4$

Find a. Choose from these options.

A $a=4$ B $a=\sqrt{3}$ C $a=\ln 3$ D $a=e^4$

2 Find the equation of the tangent to the curve $y=e^x+2\sin x$ at the point where $x=\frac{\pi}{2}$.

3 If $f'(x)=\sin x$ and $f\left(\frac{\pi}{3}\right)=0$ find $f(x)$

4 i Find the exact value of the gradient of the graph of $f(x)=\frac{1}{2}e^x-7\ln x$ at the point $x=\ln 4$.

ii Find the exact value of the gradient of the graph $f(x)=e^x-\frac{\ln x}{2}$ when $x=\ln 3$.

5 Find the exact value of the integral $\int_0^{\pi}(e^x+\sin x+1)\,dx$

6 The diagram shows the curve with equation $y=x-7+\frac{10}{x}$, which crosses the x-axis at 2 and 5.

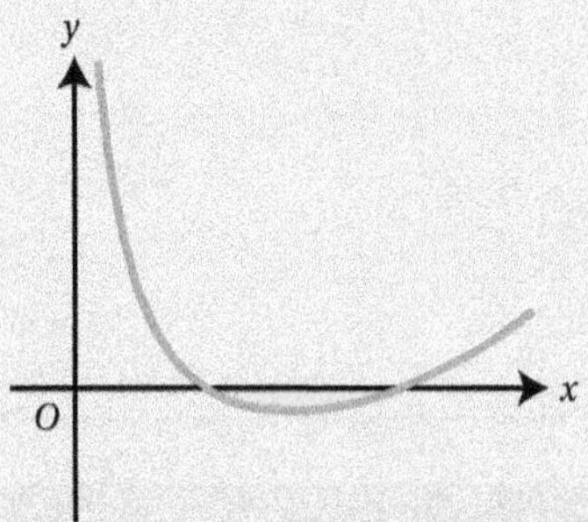

Find the exact value of the shaded area.

7 Find $\frac{dy}{dx}$ when $y=e^{3x}+\ln x$.

[© AQA 2013]

8 A curve has equation $y=e^{2x}-10e^x+12x$.

i Find $\frac{dy}{dx}$. ii Find $\frac{d^2y}{dx^2}$.

[© AQA 2006]

9 Find the x-coordinate of the stationary point on the curve $y=8\sin x+\tan x$, for $0<x<\pi$.

Choose from these options.

A $x=30°$ B $x=60°$ C $x=120°$ D $x=150°$

10 Find the indefinite integral $\int \frac{1+x^2\sqrt{x}}{x}\,dx$.

11 Find and classify the stationary points on the curve $y=3\sin x+4\cos x$ in the interval $0<x<2\pi$.

12 Find and classify the stationary points on the curve $y = \tan x - \frac{4x}{3}$ for $-\pi < x < \pi$.

Give only the x-coordinates, and leave your answers in terms of π.

13 Find the equation of the normal to the curve $y = 3e^x$ at the point $x = \ln 2$.

Give your answer in the form $x + ky = p + \ln q$ where k, p and q are integers.

14 The population P of bacteria, in thousands, at a time t, in hours, is modelled by $P = 10 + e^t - 3t, t \geqslant 0$.

a **i** Find the initial population of bacteria.

ii At what time does the number of bacteria reach 14 million?

b **i** Find $\frac{dP}{dt}$.

ii Find the time at which the bacteria are growing at a rate of 6 million per hour.

c **i** Find $\frac{d^2P}{dt^2}$ and explain the physical significance of this quantity.

ii Find the minimum number of bacteria, justifying that it is a minimum.

15 The diagram shows the graphs of $y = \sin x$ and $y = \sqrt{3} \cos x$.

Find the exact area of the shaded region.

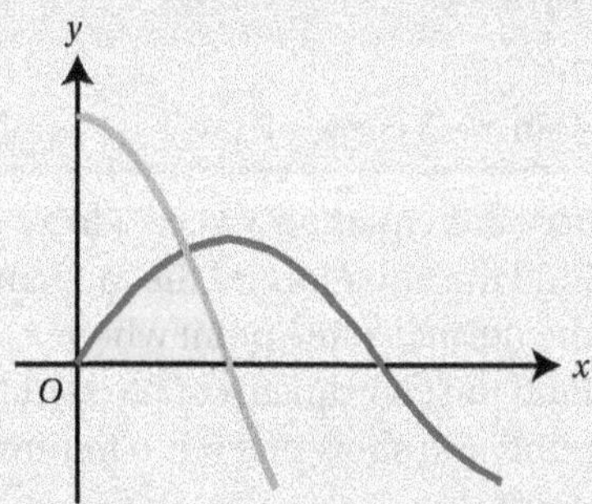

Elevate

See Extension Sheet 9 for a selection of more challenging problems.

16 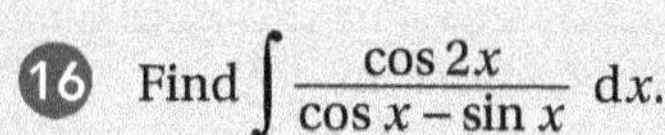
Find $\int \frac{\cos 2x}{\cos x - \sin x} \, dx$.

10 Further differentiation

In this chapter you will learn how to:

- use the chain rule to differentiate composite functions
- differentiate products and quotients of functions
- work with implicit functions and their derivatives
- differentiate inverse functions.

Before you start...

Chapter 9; Student Book 1, Chapter 12	You should be able to differentiate the functions x^n, $\sin x$, $\cos x$, $\tan x$, e^x, $\ln x$.	1 Differentiate these expressions. a $2x^3 - 3\sqrt{x}$ b $5\ln x + \frac{1}{3x^3}$ c $5e^x$ d $4\sin x - 3\cos x + 2\tan x$
Chapter 9; Student Book 1, Chapter 13	You should be able to use differentiation to find the equations of tangents, normals and stationary points.	2 A curve has equation $y = x - 2\ln x$. a Find the equations of the tangent and the normal at the point where $x = 1$. b Find the coordinates of the stationary point and show that it is a minimum point.
Student Book 1, Chapter 10	You should know basic trigonometric identities.	3 Simplify these expressions. a $3\sin^2 x + 3\cos^2 x$ b $\frac{3\sin x}{4\cos x}$
Chapter 8	You should know the definitions of reciprocal trigonometric functions.	4 Write these in terms of $\sin x$ and $\cos x$. a $\sec x \tan x$ b $\frac{\operatorname{cosec} x}{\cot x}$
Chapter 5	You should be able to simplify expressions involving fractions and surds.	5 Simplify each expression. a $\frac{x - \frac{1}{x+1}}{\frac{2}{x+1} - 3}$ b $\frac{\sqrt{x-1} + \frac{1}{\sqrt{x-1}}}{x-1}$
GCSE; Student Book 1, Chapter 7	You should be able to change the subject of a formula.	6 Make y the subject of each formula. a $x = e^{2y-1}$ b $2x + 3xy = (x-2)y$

Differentiating more complex functions

In Chapter 9 you learnt how to differentiate a variety of functions. You can also differentiate sums, differences and constant multiples of those functions, for example, $y = 3\sin x + e^x - 3x^2$.

The next question to ask is: How can you differentiate products and quotients of functions, for example, $y = x^2 e^x$ or $\frac{\sin x}{x}$?

You know from Student Book 1, Chapter 12, that you cannot just differentiate the separate components of a product (or quotient) and then multiply (or divide) the results, so you need a new rule for these situations.

Before looking at products and quotients you will learn how to differentiate composite functions, such as $\sin 3x$ or e^{x^2}. You will also learn how to differentiate equations such as $x^2 + y^2 = 5$ without making y the subject, and apply this method to differentiating inverse functions.

Section 1: The chain rule

To differentiate $y = (3x^2 + 5x + 2)^2$ you could expand the brackets and differentiate term by term. But often this is either too difficult or not possible; for example, $\sqrt{3x^2 + 5x + 2}$ cannot be 'expanded' and $(3x^2 + 5x + 2)^7$ would lead to a very long expression.

What about functions such as $y = \sin 3x$ or $y = e^{x^2}$? You can differentiate $y = \sin x$ and $y = e^x$, but you have no rules, so far, telling you what to do when x is replaced by $3x$ or x^2.

These functions may seem quite different but they do have something in common – they are all **composite functions**.

- $y = (3x^2 + 5x + 2)^7$ $\quad$ $y = u^7$ where $u(x) = 3x^2 + 5x + 2$
- $y = \sin 3x$ $\quad$ $y = \sin u$ where $u(x) = 3x$
- $y = e^{x^2}$ $\quad$ $y = e^u$ where $u(x) = x^2$

To differentiate a composite function, think about how a change in x affects y. If x changes by Δx this causes u to change by Δu which in turn causes y to change by Δy. If you multiply the two rates of change you can see that:

$$\frac{\Delta y}{\Delta u} \times \frac{\Delta u}{\Delta x} = \frac{\Delta y}{\Delta x}$$

This leads to the rule for differentiating composite functions.

Key point 10.1

The **chain rule** states that if $y = f(u)$ where $u = g(x)$, so that $y = f(g(x))$, then

$$\frac{dy}{dx} = \frac{dy}{du} \times \frac{du}{dx}$$

WORKED EXAMPLE 10.1

Differentiate $y=(3x^2+5x+2)^7$.

$y=u^7$ where $u=3x^2+5x+2$ — This is a composite function.

$\frac{dy}{dx}=\frac{dy}{du}\times\frac{du}{dx}$ — Use the chain rule.

$=7u^6\times(6x+5)$

$=7(3x^2+5x+2)^6(6x+5)$ — Write the answer in terms of x. There is no need to expand the brackets.

WORKED EXAMPLE 10.2

Differentiate $y=\sin 3x$.

$y=\sin u$ where $u=3x$ — This is a composite function.

$\frac{dy}{dx}=\frac{dy}{du}\times\frac{du}{dx}$ — Use the chain rule.

$=\cos u\times(3)$

$=3\cos 3x$ — Write the answer in terms of x and rearrange.

Rewind

For a reminder of how to differentiate $\sin x$, $\cos x$ and $\tan x$ see Chapter 9, Section 1.

WORKED EXAMPLE 10.3

Differentiate $y=e^{x^2}$.

$y=e^u$ where $u=x^2$ — This is a composite function.

$\frac{dy}{dx}=\frac{dy}{du}\times\frac{du}{dx}$ — Use the chain rule.

$=e^u\times 2x$

$=2xe^{x^2}$ — Write the answer in terms of x and rearrange.

Tip

Once you get used to using the chain rule, you won't need to write the 'inner function' u each time. In particular, since the derivative of kx is just k, you can now immediately recognise that:

- $\frac{d}{dx}(\sin kx)=k\cos kx$
- $\frac{d}{dx}(\cos kx)=-k\sin kx$
- $\frac{d}{dx}(\tan kx)=k\sec^2 kx$

Sometimes you need to apply the chain rule more than once.

WORKED EXAMPLE 10.4

Differentiate $y = \cos^3(\ln 2x)$.

$y = (\cos(\ln 2x))^3$ — Remember that $\cos^3 A$ means $(\cos A)^3$.

$y = u^3$ where $u = \cos v$ and $v = \ln 2x$ — This is a composite of three functions.

$\frac{dy}{dx} = \frac{dy}{du} \times \frac{du}{dv} \times \frac{dv}{dx}$ — Use the chain rule with three derivatives.

$= 3u^2 \times (-\sin v) \times 2 \times \frac{1}{2x}$

$= 3(\cos(\ln 2x))^2 \times (-\sin(\ln 2x)) \times \frac{1}{x}$ — Write everything in terms of x and simplify.

$= -\frac{3}{x}\cos^2(\ln 2x)\sin(\ln 2x)$

You can use the chain rule to differentiate reciprocal trigonometric functions.

WORKED EXAMPLE 10.5

Show that $\frac{d}{dx}(\sec x) = \sec x \tan x$.

$y = \sec x = (\cos x)^{-1}$ — Express $\sec x$ in terms of $\cos x$, as you know how to differentiate that.

$y = u^{-1}$ where $u = \cos x$ — This is a composite function...

$\frac{dy}{dx} = -u^{-2} \times (-\sin x)$

$= -(\cos x)^{-2}(-\sin x)$ — ...so apply the chain rule.

$= \frac{\sin x}{\cos^2 x}$

$= \frac{1}{\cos x}\frac{\sin x}{\cos x}$ — The answer contains $\tan x$, which is $\frac{\sin x}{\cos x}$.

$= \sec x \tan x$

The proofs for the other two reciprocal trigonometric functions follow the same pattern, giving the results shown in Key point 10.2.

Key point 10.2

- $y = \sec x \qquad \frac{dy}{dx} = \sec x \tan x$
- $y = \operatorname{cosec} x \qquad \frac{dy}{dx} = -\operatorname{cosec} x \cot x$
- $y = \cot x \qquad \frac{dy}{dx} = -\operatorname{cosec}^2 x$

These will be given in your formula book.

EXERCISE 10A

1 Use the chain rule to differentiate each expression with respect to x.

a i $(3x+4)^5$ ii $(5x+4)^7$ **b** i $(5-x)^{-4}$ ii $(1-x)^{-7}$

c i $\sqrt{3x-2}$ ii $\sqrt{x+1}$ **d** i $\frac{1}{3-x}$ ii $\frac{1}{(2x+3)^2}$

e i e^{5x-3} ii e^{10x+1} **f** i e^{1-2x} ii e^{4-3x}

g i $\sin 4x$ ii $\sin \pi x$ **h** i $\cos 2\pi x$ ii $\cos 3x$

i i $\tan 5x$ ii $\tan \frac{\pi}{4} x$ **j** i $\cos(1-4x)$ ii $\sin(2-x)$

k i $\sec 4x$ ii $\sec(2x+1)$ **l** i $\cot 3x$ ii $\operatorname{cosec} 5x$

m i $\ln(5x+2)$ ii $\ln(x-4)$ **n** i $\ln(5-x)$ ii $\ln(3-2x)$

2 Use the chain rule to differentiate each expression with respect to x.

a i $(x^2-3x+1)^7$ ii $(x^3+1)^5$ **b** i e^{x^2-2x} ii e^{4-x^3}

c i $(2e^x+1)^{-3}$ ii $(2-5e^x)^{-4}$ **d** i $\sin(3x^2+1)$ ii $\cos(x^2+2x)$

e i $\cos^3 x$ ii $\sin^4 x$ **f** i $\ln(2x-5x^3)$ ii $\ln(4x^2-1)$

g i $(4\ln x-1)^4$ ii $(\ln x+3)^{-5}$ **h** i $\sqrt{3x^2+1}$ ii $\sqrt{5-2x^2}$

3 Differentiate each expression, using the chain rule twice.

a i $\sec^2 3x$ ii $\tan^2 2x$ **b** i $e^{\sin^2 3x}$ ii $e^{(\ln 2x)^2}$

c i $(1-2\sin^2 2x)^2$ ii $(4\cos 3x+1)^2$ **d** i $\ln(1-3\cos 2x)$ ii $\ln(2-\cos 5x)$

4 Find the equation of the tangent to the graph of $y=(3x+5)^2$ at the point where $x=2$.

5 Find the equation of the normal to the curve $y=\frac{1}{\sqrt{4x^2+1}}$ at the point where $x=\sqrt{2}$.

6 For what values of x does the function $f: x \mapsto \ln(x^2-35)$ have a gradient of 1?

7 Find the coordinates of the stationary points on the curve with equation $y = (3x^2 - 6)^3$.

8 Find the coordinates of the stationary points on the graph of $y = (x^3 - 1)^5$.

Rewind

You met kinematics in Student Book 1, Chapter 16.

9 A particle moves in a straight line with displacement at time t from O given by $s = 6\cos\left(\frac{t}{3}\right)$ m.

Find, at time $t = 2\pi$ seconds:

a the velocity of the particle

b the acceleration of the particle.

10 Find the exact coordinates of stationary points on the curve $y = e^{\sin x}$ for $x \in [0, 2\pi]$.

11 Find the exact coordinates of the stationary point on the curve $y = \frac{4}{x^2 - 12x}$.

12 A population of bacteria increases exponentially so that the number of bacteria, N, after t minutes is given by $N = 45\,e^{0.4t}$.

Find the size of the population at the moment when its rate of increase is 198 bacteria per minute.

13 Given that $f(x) = \text{cosec}^2 x$:

a find $f'(x)$

b solve the equation $f'(x) = 2\,f(x)$ for $-\pi < x < \pi$

14 A non-uniform chain hangs from two posts. Its height h above the ground satisfies the equation

$h = e^x + \frac{1}{e^{2x}}, -1 \leqslant x \leqslant 2.$

The left post is positioned at $x = -1$, and the right post is positioned at $x = 2$.

a State, with reasons, which post is taller.

b Show that the minimum height occurs when $x = \frac{1}{3}\ln 2$.

c Find the exact value of the minimum height of the chain.

15 **a** Solve the equation $\sin 2x = \sin x$ for $0 \leqslant x \leqslant 2\pi$, giving your answers in terms of π.

b Find the coordinates of the stationary points of the curve $y = \sin 2x - \sin x$ for $0 \leqslant x \leqslant 2\pi$, giving your answers correct to three significant figures.

c Hence sketch the curve $y = \sin 2x - \sin x$ for $0 \leqslant x \leqslant 2\pi$.

Did you know?

Many people think that a chain fixed at both ends will hang as a parabola, but it can be proved that it hangs in the shape of the curve in question 14, called a catenary. The proof requires techniques from a mathematical area called differential geometry.

Section 2: The product rule

There is also a rule for differentiating products, such as $y = x^2 \cos x$ or $y = x \ln x$.

Key point 10.3

The **product rule** states that if $y = u(x)v(x)$ then

$$\frac{dy}{dx} = \frac{du}{dx}v + u\frac{dv}{dx}$$

WORKED EXAMPLE 10.6

Differentiate: **a** $y = x^2 \cos x$ **b** $y = x \ln x$.

a Let $u = x^2$ and $v = \cos x$.

It doesn't matter which function you call u and which v.

Then $\frac{du}{dx} = 2x$ and $\frac{dv}{dx} = -\sin x$

So $\frac{dy}{dx} = \frac{du}{dx}v + u\frac{dv}{dx}$

Applying the product rule.

$= 2x \cos x + x^2(-\sin x)$

$= 2x \cos x - x^2 \sin x$

b Let $u = x$ and $v = \ln x$.

y is a product of two functions, so use the product rule.

Then $\frac{du}{dx} = 1$ and $\frac{dv}{dx} = \frac{1}{x}$

So $\frac{dy}{dx} = \frac{du}{dx}v + u\frac{dv}{dx}$

$= (1)(\ln x) + (x)\left(\frac{1}{x}\right)$

$= \ln x + 1$

WORK IT OUT 10.1

Differentiate $y = \sin(x^2 + 3x)$.

Which is the correct solution?

Identify the errors made in the incorrect solutions.

Solution 1	Solution 2	Solution 3
$u = \sin, v = x^2 + 3x$ $\frac{du}{dx} = \cos, \frac{dv}{dx} = 2x + 3$ $\frac{dy}{dx} = \frac{du}{dx}v + u\frac{dv}{dx}$ $= \cos(x^2 + 3x) + \sin(2x + 3)$	$\frac{dy}{dx} = (2x + 3)\cos(x^2 + 3x)$	$\frac{dy}{dx} = \cos(2x + 3)$

When differentiating a more complicated product, you may need to use the chain rule as well as the product rule. When the function involves powers, the two terms in the product rule often have a common factor.

WORKED EXAMPLE 10.7

Differentiate $y = x^4(3x^2 - 5)^5$ and factorise your answer.

Let $u = x^4$ and $v = (3x^2 - 5)^5$ — This is a product, so use the product rule.

$\frac{du}{dx} = 4x^3$

$\frac{dv}{dx} = 5(3x^2 - 5)^4(6x)$

$= 30x(3x^2 - 5)^4$ — $v(x)$ is a composite function, so use the chain rule.

$\frac{dy}{dx} = \frac{du}{dx}v + u\frac{dv}{dx}$ — Now apply the product rule.

$= 4x^3(3x^2 - 5)^5 + x^4 \times 30x(3x^2 - 5)^4$

$= 2x^3(3x^2 - 5)^4[2(3x^2 - 5) + 15x^2]$ — You are asked to factorise the answer, so instead of expanding the brackets look for common factors.

$= 2x^3(3x^2 - 5)^4(6x^2 - 10 + 15x^2)$

$= 2x^3(3x^2 - 5)^4(21x^2 - 10)$

EXERCISE 10B

1. Differentiate each function. Use the product rule.

   a i $y = x^2 \cos x$ ii $y = x^{-1} \sin x$ b i $y = x^{-2} \ln x$ ii $y = x \ln x$

   c i $y = x^3\sqrt{2x+1}$ ii $y = x^{-1}\sqrt{4x}$ d i $e^{2x} \tan x$ ii $e^{x+1} \sec 3x$

2. Find $f'(x)$ and fully factorise your answer.

   a i $f(x) = (x+1)^4(x-2)^5$ ii $f(x) = (x-3)^7(x+5)^4$

   b i $f(x) = (2x-1)^4(1-3x)^3$ ii $f(x) = (1-x)^5(4x+1)^2$

3. Differentiate $y = (3x^2 - x + 2)e^{2x}$, giving your answer in the form $P(x)e^{2x}$ where $P(x)$ is a polynomial.

4. Given that $f(x) = x^2 e^{3x}$, find $f''(x)$ in the form $(ax^2 + bx + c)e^{3x}$.

5. Find the x-coordinates of the stationary points on the curve $y = (2x+1)^5 e^{-2x}$.

6. Find the exact values of the x-coordinates of the stationary points on the curve $y = (3x+1)^5(3-x)^3$.

7. Find the derivative of $\sin(x e^x)$ with respect to x.

8. a Given that $f(x) = x \ln x$ find $f'(x)$.   b Hence find $\int \ln x \, dx$.

9. Find the exact coordinates of the minimum point of the curve $y = e^{-x} \cos x$, $0 \leqslant x \leqslant \pi$.

10 Given that $f(x) = x^2\sqrt{1+x}$ show that $f'(x) = \dfrac{x(a+bx)}{2\sqrt{1+x}}$ where a and b are constants to be found.

11 **a** Write $y = x^x$ in the form $y = e^{f(x)}$.

b Hence or otherwise find $\dfrac{dy}{dx}$.

c Find the exact coordinates of the stationary points of the curve $y = x^x$.

12 **a** If $a < b$ and p, q are positive integers find the x-coordinate of the stationary point of the curve $y = (x-a)^p(x-b)^q$ in the domain $a < x < b$.

b Sketch the graph in the case when $p = 2$ and $q = 3$.

c By considering the graph or otherwise, determine a condition involving p and/or q to determine when this stationary point is a maximum.

Section 3: Quotient rule

To differentiate a quotient such as $y = \dfrac{x^2-4x+12}{(x-3)^2}$ you can express it as:

$$y = (x^2-4x+12)(x-3)^{-2}$$

and use a combination of the chain rule and the product rule. However, there is a shortcut that allows you to differentiate quotients directly.

Key point 10.4

The **quotient rule** states that if $y = \dfrac{u(x)}{v(x)}$ then

$$\frac{dy}{dx} = \frac{\frac{du}{dx}v - u\frac{dv}{dx}}{v^2}$$

This will be given in your formula book.

WORKED EXAMPLE 10.8

Differentiate $y = \dfrac{x^2-4x+12}{(x-3)^2}$. Use the quotient rule and simplify as far as possible.

Let $u = x^2-4x+12$ and $v = (x-3)^2$.

$$\frac{dy}{dx} = \frac{\frac{du}{dx}v - u\frac{dv}{dx}}{v^2}$$

Use the quotient rule, making sure to get u and v the right way round.

$$= \frac{(2x-4)(x-3)^2 - (x^2-4x+12)2(x-3)}{[(x-3)^2]^2}$$

Use the chain rule to differentiate v.

$$= \frac{(2x-4)(x-3) - (x^2-4x+12)2}{(x-3)^3}$$

Notice that a factor of $(x-3)$ can be cancelled.

$$= \frac{2x^2-10x+12-2x^2+8x-24}{(x-3)^3}$$

$$= \frac{-2x-12}{(x-3)^3}$$

In Chapter 9, it was stated that the derivative of $\tan x$ is $\sec^2 x$. Worked example 10.9 shows how you can use the quotient rule, together with the derivatives of $\sin x$ and $\cos x$, to prove this result.

WORKED EXAMPLE 10.9

Prove that $\frac{d}{dx}(\tan x) = \sec^2 x$.

$\tan x \equiv \frac{\sin x}{\cos x}$, $u = \sin x$, $v = \cos x$ — You know how to differentiate $\sin x$ and $\cos x$, so use $\tan x \equiv \frac{\sin x}{\cos x}$.

$\frac{dy}{dx} = \frac{\frac{du}{dx}v - u\frac{dv}{dx}}{v^2}$ — Use the quotient rule.

$= \frac{\cos x \cos x - \sin x(-\sin x)}{(\cos x)^2}$

$= \frac{\cos^2 x + \sin^2 x}{\cos^2 x}$

$= \frac{1}{\cos^2 x}$ — Notice that $\sin^2 x + \cos^2 x \equiv 1$

$= \sec^2 x$

WORK IT OUT 10.2

Differentiate $y = \frac{3}{\tan^2 x}$.

Which is the correct solution? Identify the errors made in the incorrect solutions.

Solution 1	Solution 2	Solution 3
$u = 3, v = \tan^2 x$ $\frac{du}{dx} = 0, \frac{dv}{dx} = 2\tan x \sec^2 x$ So: $\frac{dy}{dx} = \frac{0(\tan^2 x) - 3(2\tan x \sec^2 x)}{(\tan^2 x)^2}$ $= \frac{-6\tan x \sec^2 x}{(\tan x)^4}$ $= -\frac{6\sec^2 x}{\tan^3 x}$	$y = \frac{3}{\tan^2 x} = 3\tan^{-2} x$ $u = 3, v = \tan^{-2} x$ $\frac{du}{dx} = 0, \frac{dv}{dx} = -2\tan^{-3} x \sec^2 x$ So: $\frac{dy}{dx} = 0(\tan^{-2} x)$ $+3(-2\tan^{-3} x \sec^2 x)$ $= -6\tan^{-3} x \sec^2 x$	$y = \frac{3}{\tan^2 x} = 3\cot^2 x$ $\frac{dy}{dx} = 3 \times 2\cot x(-\mathrm{cosec}^2 x)$ $= -6\cot x\,\mathrm{cosec}^2 x$

The quotient rule, just like the product rule, often leads to a long expression. If you are required to simplify it, you may need to work with fractions and roots, as in Worked example 10.10.

Tip

You only need to use the quotient rule when differentiating a quotient of two functions. If a quotient is made up of a constant and a function you can do it more simply. For example, $\frac{3}{1+x^2}$ is just $3(1+x^2)^{-1}$, which you can differentiate by using the chain rule.

WORKED EXAMPLE 10.10

Differentiate $\frac{x}{\sqrt{x+1}}$ giving your answer in the form $\frac{x+c}{k\sqrt{(x+1)^p}}$ where $c, k, p \in \mathbb{N}$.

$y = \frac{x}{\sqrt{x+1}}$, $u = x$, — Use the quotient rule.

$v = \sqrt{x+1} = (x+1)^{\frac{1}{2}}$

$$\frac{dy}{dx} = \frac{\frac{du}{dx}v - u\frac{dv}{dx}}{v^2}$$

$$= \frac{1\times(x+1)^{\frac{1}{2}} - x\times\frac{1}{2}(x+1)^{-\frac{1}{2}}}{\left((x+1)^{\frac{1}{2}}\right)^2}$$

$$= \frac{\sqrt{x+1} - \frac{x}{2\sqrt{x+1}}}{x+1}$$ — As you want a square root in the answer, turn the fractional powers back into roots.

$$= \frac{2(x+1)-x}{2(x+1)\sqrt{x+1}}$$ — Remove fractions within fractions by multiplying top and bottom by $2\sqrt{x+1}$.

$$= \frac{x+2}{2\sqrt{(x+1)^3}}$$ — Notice that $a\sqrt{a} = a^{\frac{3}{2}} = \sqrt{a^3}$.

Elevate

See Support Sheet 10 for a further example of combining the chain rule with the product or quotient rule and some more practice questions.

EXERCISE 10C

1. Differentiate each equation. Use the quotient rule.
   - **a** **i** $y = \frac{x-1}{x+1}$ **ii** $y = \frac{x+2}{x-3}$
   - **b** **i** $y = \frac{\sqrt{2x+1}}{x}$ **ii** $y = \frac{x^2}{\sqrt{x-1}}$
   - **c** **i** $y = \frac{1-2x}{x^2+2}$ **ii** $y = \frac{4-x^2}{1+x}$
   - **d** **i** $y = \frac{\ln 3x}{x}$ **ii** $y = \frac{\ln 2x}{x^2}$
2. Find the equation of the normal to the curve $y = \frac{\sin x}{x}$ at the point where $x = \frac{\pi}{2}$, giving your answer in the form $y = mx + c$, where m and c are exact.
3. Find the coordinates of the stationary points on the graph of $y = \frac{x^2}{2x-1}$.
4. The graph of $y = \frac{x-a}{x+2}$ has gradient 1 at the point $(a, 0)$ and $a \neq -2$. Find the value of a.
5. Find the exact coordinates of the stationary point on the curve $y = \frac{\ln x}{x}$ and determine its nature.

6 Find the range of values of x for which the function $f(x)=\frac{x^2}{1-x}$ is increasing.

7 Given that $y=\frac{x^2}{\sqrt{x+1}}$ show that $\frac{dy}{dx}=\frac{x(ax+b)}{2(x+1)^p}$, stating clearly the value of the constants a, b and p.

8 Show that if the curve $y=f(x)$ has a maximum stationary point at $x=a$ then the curve $y=\frac{1}{f(x)}$ has a minimum stationary point at $x=a$ as long as $f(a)\neq 0$.

Section 4: Implicit differentiation

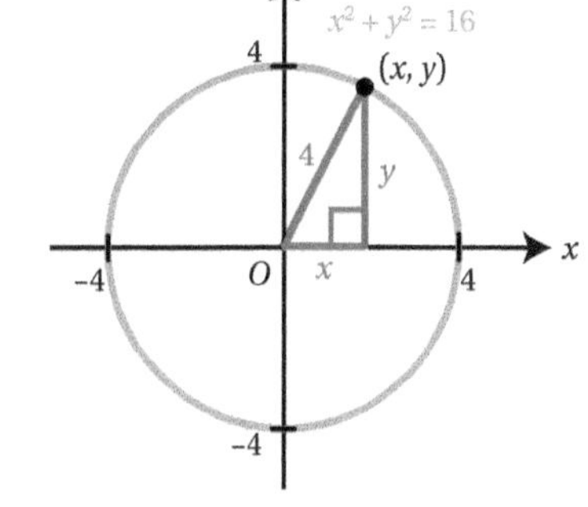

The functions you have differentiated so far have always been of the form $y=f(x)$. From time to time you will come across functions that are written differently. For example, the equation of the circle shown in the diagram is $x^2+y^2=16$. Such functions are said to be **implicit** whereas those in the form $y=f(x)$ are said to be **explicit**.

The gradient of the tangent at any point on the circle is still given by $\frac{dy}{dx}$. Rather than attempting to rearrange the equation to make it explicit, you can just differentiate term by term with respect to x:

$$\frac{d}{dx}(x^2)+\frac{d}{dx}(y^2)=\frac{d}{dx}(16)$$

You need to take care when you come to y^2 as it is not a function of x - here you will need the chain rule:

$$\frac{d(y^2)}{dx}=\frac{d(y^2)}{dy}\times\frac{dy}{dx}=2y\frac{dy}{dx}$$

This will generally be the case when differentiating terms involving y.

Key point 10.5

When differentiating implicitly, you need to use:

$$\frac{d}{dx}[f(y)]=\frac{d}{dy}[f(y)]\times\frac{dy}{dx}$$

WORKED EXAMPLE 10.11

Find an expression for $\frac{dy}{dx}$ for the circle $x^2+y^2=16$.

$$\frac{d}{dx}(x^2)+\frac{d}{dx}(y^2)=\frac{d}{dx}(16)$$

Differentiate each term.

$$2x+2y\frac{dy}{dx}=0$$

Use the chain rule on the term involving y.

$$2y\frac{dy}{dx}=-2x$$

$$\frac{dy}{dx}=-\frac{x}{y}$$

Now rearrange the equation to find $\frac{dy}{dx}$.

Notice that the expression for $\frac{dy}{dx}$ will often be in terms of both x and y.

Sometimes you may need to use the product rule as well as the chain rule in this process of implicit differentiation.

WORKED EXAMPLE 10.12

Find an expression for $\frac{dy}{dx}$ if $e^x + x\sin y = \cos 2y$.

$\frac{d}{dx}(e^x) + \frac{d}{dx}(x\sin y) = \frac{d}{dx}(\cos 2y)$ — Differentiate term by term, using the chain rule on all terms involving y.

$e^x + \left(1 \times \sin y + x \times \cos y \frac{dy}{dx}\right) = -2\sin 2y \frac{dy}{dx}$ — $x\sin y$ is a product, so use the product rule together with the chain rule on all terms involving y.

$x\cos y \frac{dy}{dx} + 2\sin 2y \frac{dy}{dx} = -e^x - \sin y$ — Group the terms involving $\frac{dy}{dx}$.

$(x\cos y + 2\sin 2y)\frac{dy}{dx} = -e^x - \sin y$

$$\frac{dy}{dx} = \frac{-e^x - \sin y}{x\cos y + 2\sin 2y}$$

If you are only interested in the gradient at a particular point, or are given the gradient and need to find the x-and y-coordinates, you can substitute the given value into the differentiated equation without rearranging it.

WORKED EXAMPLE 10.13

For the curve with equation $x^2 + y^2 - xy = 3$:

a find the gradient at the point (1, 3)

b find the coordinates of the point where the gradient is 1.

$\frac{d}{dx}(x^2) + \frac{d}{dx}(y^2) - \frac{d}{dx}(xy) = \frac{d}{dx}(3)$ — Differentiate each term with respect to x. The term xy will require the product rule.

$2x + 2y\frac{dy}{dx} - \left(1 \times y + x \times \frac{dy}{dx}\right) = 0$ — Use the chain rule on terms involving y.

$(2y - x)\frac{dy}{dx} + (2x - y) = 0$ — Group terms involving $\frac{dy}{dx}$.

a When $x = 1$ and $y = 3$: — Substitute in the numbers before rearranging.

$(6 - 1)\frac{dy}{dx} + (2 - 3) = 0$

$\frac{dy}{dx} = \frac{1}{5}$

Continues on next page

b When $\frac{dy}{dx} = 1$:

Put in the given value of the gradient.

$$(2y - x)(1) + (2x - y) = 0$$

$$x + y = 0$$

$$y = -x$$

This is a second equation relating x and y. You can solve it simultaneously with the original equation.

Substitute into $x^2 + y^2 - xy = 3$:

$$x^2 + (-x)^2 - x(-x) = 3$$

$$3x^2 = 3$$

$$x = \pm 1$$

Usin $y = -x$, the coordinates are $(1, -1)$ and $(-1, 1)$

Remember to find both x- and y-coordinates.

You can also use the method in part **b** to find stationary points.

WORKED EXAMPLE 10.14

Find the coordinates of the stationary points on the curve $y^3 + 3xy^2 - x^3 = 27$.

$$\frac{d}{dx}(y^3) + \frac{d}{dx}(3xy^2) - \frac{d}{dx}(x^3) = \frac{d}{dx}(27)$$

Differentiate each term with respect to x but be aware that the term $3xy^2$ will require the product rule.

$$3y^2\frac{dy}{dx} + \left(3 \times y^2 + 3x \times 2y\frac{dy}{dx}\right) - 3x^2 = 0$$

Use the chain rule on all terms involving y.

$$3y^2\frac{dy}{dx} + 6xy\frac{dy}{dx} + 3y^2 - 3x^2 = 0$$

For stationary points, $\frac{dy}{dx} = 0$

You know the value of $\frac{dy}{dx}$.

$$3y^2 - 3x^2 = 0$$

$$(y - x)(y + x) = 0$$

$$y = x \text{ or } y = -x$$

When $x = y$: $x^3 + 3x \times x^2 - x^3 = 27$

You have found a relationship between x and y at the stationary points but to find the points substitute back into the original function.

$$3x^3 = 27$$

$$x^3 = 9$$

$$x = \sqrt[3]{9}$$

$\therefore (\sqrt[3]{9}, \sqrt[3]{9})$ is a stationary point.

When $x = -y$: $(-x)^3 + 3x(-x)^2 - x^3 = 27$

$$-x^3 + 3x^3 - x^3 = 27$$

$$x^3 = 27$$

$$x = 3$$

$\therefore (3, -3)$ is a stationary point.

One application of implicit differentiation is to differentiate exponential functions with base other than e.

Key point 10.6

$$\frac{d}{dx}(a^x) = a^x \ln a$$

PROOF 7

Let $y = a^x$

Then $\ln y = x \ln a$ — As this is an exponential expression, take ln of both sides.

$\frac{d}{dx}(\ln y) = \frac{d}{dx}(x \ln a)$ — Now use implicit differentiation.

$\frac{1}{y}\frac{dy}{dx} = \ln a$ — Remember that $\ln a$ is a constant.

$\frac{dy}{dx} = y \ln a$

$= a^x \ln a$

Rewind

This confirms that the gradient of any exponential function is proportional to the y-value, as you learnt in Student Book 1, Chapter 8.

EXERCISE 10D

1 Find the gradient of each curve at the given point.

a i $x^2 + 3y^2 = 7$ at $(2, -1)$ ii $2x^3 - y^3 = -6$ at $(1, 2)$

b i $\cos x + \sin y = 0$ at $(0, \pi)$ ii $\tan x + \tan y = 2$ at $\left(\frac{\pi}{4}, \frac{\pi}{4}\right)$

c i $x^2 + 3xy + y^2 = 20$ at $(2, 2)$ ii $3x^2 - xy^2 + 3y = 21$ at $(-1, 3)$

d i $xe^y + ye^x = 2e$ at $(1, 1)$ ii $x \ln y - \frac{x}{y} = 2$ at $(-1, 1)$

2 Find $\frac{dy}{dx}$ in terms of x and y.

a i $3x^2 - y^3 = 15$ ii $x^4 + 3y^2 = 20$

b i $xy^2 - 4x^2y = 6$ ii $y^2 - xy = 7$

c i $\frac{x+y}{x-y} = 2y$ ii $\frac{y^2}{xy+1} = 1$

d i $xe^y - 4 \ln y = x^2$ ii $3x \sin y + 2 \cos y = \sin x$

3 Find the coordinates of stationary points on the curves given by each implicit equation.

a $-x^2 + 3xy + y^2 = 13$ b $2x^2 - xy + y^2 = 28$

4 Find the exact value of the gradient at the given point.

a i $y = 3^x$ at $(1, 3)$ ii $y = 5^x$ at $(2, 25)$

b i $y = \left(\frac{1}{2}\right)^x$ when $x = -2$ ii $y = \left(\frac{1}{3}\right)^x$ when $x = -1$

c i $y = 2^{3x}$ when $x = -1$ ii $y = 4^{2x}$ when $x = \frac{1}{4}$

d i $y = 3^{3-x}$ when $x = 2$ ii $y = 5^{1-x}$ when $x = 2$

5 Find the gradient of the curve $x^2 + y^2 = 5$ at the point $(-2, 1)$.

6 A curve has equation $3x^2 + 5y^3 = 22$.

a Show that $\frac{dy}{dx} = -\frac{2x}{5y^2}$.

b Find the equation of the tangent to the curve at the point $(3, -1)$.

7 A curve has equation $3x^2 - y^2 = 8$. Point A has coordinates $(-2, 2)$.

a Show that point A lies on the curve.

b Find the equation of the normal to the curve at A.

8 Find the gradient of the curve with equation $x^2 - 3xy + y^2 + 1 = 0$ at the point $(1, 2)$.

9 Find the equation of the tangent to the curve with equation $3x^2 + xy - y^2 = -3$ at the point $(2, 5)$.

10 Find the equation of the tangent to the curve with equation $4x^2 - 3xy - y^2 = 25$ at the point $(2, -3)$.

11 A curve has implicit equation $x2^y = \ln y$. Find an expression for $\frac{dy}{dx}$ in terms of x and y.

12 Find the coordinates of the stationary point on the curve given by $e^x + ye^{-x} = 2e^2$.

13 The line L is tangent to the curve C which has the equation $y^2 = x^3$ when $x = 4$ and $y > 0$.

a Find the equation of L.

b Show that L meets C again at the point P with an x-coordinate which satisfies the equation $x^3 - 9x^2 + 24x - 16 = 0$.

c Find the coordinates of the point P.

Section 5: Differentiating inverse functions

Another application of the chain rule is to differentiate inverse functions. If $y = f(x)$ then $x = f^{-1}(y)$. The derivative of f is $\frac{dy}{dx}$ and the derivative of f^{-1} is $\frac{dx}{dy}$. But the chain rule says that:

$$\frac{dy}{dx} \times \frac{dx}{dy} = \frac{dy}{dy} = 1$$

In Chapter 9, the result for the derivative of $\ln x$ was just stated. You can prove it by using the fact that $\ln x$ is the inverse function of e^x.

Key point 10.7

The derivative of the inverse function is:

$$\frac{dx}{dy} = \frac{1}{\left(\frac{dy}{dx}\right)}$$

WORKED EXAMPLE 10.15

Use the derivative of e^x to prove that the derivative of $\ln x$ is $\frac{1}{x}$.

Let $y = \ln x$.

Then $x = e^y$.

Rewrite in terms of e^y (the inverse function of ln).

$\Rightarrow \frac{dx}{dy} = e^y$

Now differentiate with respect to y.

$\Rightarrow \frac{dy}{dx} = \frac{1}{e^y}$

Use the inverse function rule.

But $e^y = x$, so

The answer needs to be in terms of x.

$\frac{dy}{dx} = \frac{1}{x}$, as required.

Fast forward

If you study Further Mathematics you will use this same method to differentiate inverse trigonometric functions.

EXERCISE 10E

1 If you prefer, you can differentiate inverse functions by using implicit differentiation. For example, given that $y = \sqrt{x}$:

a express x in terms of y

b use implicit differentiation to find $\frac{dy}{dx}$; write your answer in terms of x.

2 If $x = \ln 4y$, find $\frac{dy}{dx}$ in terms of x.

3 If $x = ye^{2y}$, find the exact value of $\frac{dy}{dx}$ at the point where $y = \ln 3$, giving your answer in the form $\frac{1}{a(1+\ln a)}$, where a is an integer.

4 Given that $f(x) = x^3 + 3x + 2$:

a by considering $f'(x)$, prove that $f(x)$ has an inverse function

b find the gradient of the graph of $y = f^{-1}(x)$ at the point where $x = 2$.

5 a Given that $f(x) = \log_a x$ write down an expression for $f^{-1}(x)$.

b Hence prove that $\frac{d}{dx}(\log_a x) = \frac{1}{x \ln a}$.

6 A function is defined by $f(x) = x + \cos x$.

a Prove that f is an increasing function.

b Show that the point $\left(\frac{\pi}{4}, \frac{\pi+2\sqrt{2}}{4}\right)$ lies on the graph of $y = f(x)$.

c Find the exact value of the gradient of the graph of $y = f^{-1}(x)$ at the point $x = \frac{\pi+2\sqrt{2}}{4}$.

7 Given that $y = \arcsin x$:

a find $\frac{dx}{dy}$ in terms of y

b hence find $\frac{dy}{dx}$ in terms of x.

8 Given that $f(x)$ is an increasing function, prove that $f^{-1}(x)$ is also an increasing function.

Checklist of learning and understanding

- The **chain rule** is used to differentiate composite functions.
  If $y = f(u)$ where $u = g(x)$ then:
  $$\frac{dy}{dx} = \frac{dy}{du} \times \frac{du}{dx}$$
- The **product rule** is used to differentiate two functions multiplied together.
  If $y = u(x)\, v(x)$ then:
  $$\frac{dy}{dx} = \frac{du}{dx} v + u \frac{dv}{dx}$$
- The **quotient rule** is used to differentiate one function divided by another.
  If $y = \frac{u(x)}{v(x)}$ then:
  $$\frac{dy}{dx} = \frac{\frac{du}{dx} v - u \frac{dv}{dx}}{v^2}$$
- The derivatives of the **reciprocal trigonometric functions** are:
  - $\frac{d}{dx}(\sec x) = \sec x \tan x$
  - $\frac{d}{dx}(\operatorname{cosec} x) = -\operatorname{cosec} x \cot x$
  - $\frac{d}{dx}(\cot x) = -\operatorname{cosec}^2 x$
- The derivative of an **exponential function** is:
  $$\frac{d}{dx}(a^x) = a^x \ln a$$
- To differentiate functions given **implicitly**:
  - differentiate each term
  - use the chain rule for any term containing y:
    $$\frac{d}{dx}[f(y)] = \frac{d}{dy}[f(y)] \times \frac{dy}{dx}$$
- Differentiation of **inverse functions**:
  $$\frac{dx}{dy} = \frac{1}{\left(\frac{dy}{dx}\right)}$$

Mixed practice 10

1 $y=\frac{3x^2-6x-4}{(x-1)^2}$; $\frac{dy}{dx}=\frac{k}{(x-1)^3}$

Find the value of the constant k. Choose from these options.

A −2 **B** 1 **C** 10 **D** 14

2 $y=\ln(1+e^x)$

Find $\frac{dy}{dx}$ at the point where $x=-\ln 2$. Choose from these options.

A $\frac{1}{3}$ **B** −1 **C** $\frac{2}{3}$ **D** −2

3 **a** **Find $\frac{dy}{dx}$ when:**

i $y=e^{5x}$ **ii** $y=\sqrt{3x+2}$

b Hence find $\frac{dy}{dx}$ for $y=e^{5x}\sqrt{3x+2}$, giving your answer in the form $\frac{e^{5x}(ax+b)}{c\sqrt{3x+2}}$, where a, b and c are integers to be found.

4 Find the exact value of the gradient of the curve with equation $y=\frac{1}{4-x^2}$ when $x=\frac{1}{2}$.

5 A curve has equation $x^2+3y^2-2xy=22$.

a Show that the point $P(1, 3)$ lies on the curve.

b Find the gradient of the curve at P.

6 The curve C has equation $y=\frac{x}{\sqrt{x^2-5}}$. The tangent to the curve at the point $x = 3$ crosses the coordinate axes at points A and B. Find the area of the triangle AOB.

7 Find the exact gradient of the curve with equation $y=3^x$ at the point where $x=2$.

8 A curve has equation $y=x^3\ln x$.

a Find $\frac{dy}{dx}$.

b **i** Find an equation of the tangent to the curve $y=x^3\ln x$ at the point on the curve where $x=e$.

ii This tangent intersects the x-axis at the point A. Find the exact value of the x-coordinate of the point A.

[© AQA 2012]

9 The graph of $y=xe^{-kx}$ has a stationary point when $x=\frac{2}{5}$. Find the value of k.

10 Find the exact coordinates of the stationary point on the curve with equation $ye^x=3x-6$.

11 **a** Find the value of k so that $a^x=e^{kx}$.

b Hence show that the derivative of a^x is $a^x\ln a$.

Rewind

Compare the derivation in Question 10 with the one using implicit differentiation shown in Proof 7.

12 A particle moves in a straight line with velocity given by $v = (e^{2t} - 13e^{t} + 15t + 20)$ m s^{-1}, where $0 \leqslant t \leqslant 2$. Find, as an exact value, the minimum value of the velocity during the motion, fully justifying your answer.

13 At time t seconds, the displacement of a particle from O is $s = 10(t^2 - 3)e^{-t}$ m.

a Find the value of t when the particle is instantaneously at rest.

b Find the particle's maximum speed for $0 \leqslant t \leqslant 10$, fully justifying your answer.

14 A curve has equation $y = \dfrac{x^2}{1-2x}$.

a Write down the equation of the vertical asymptote of the curve.

b Use differentiation to find the coordinates of the stationary points on the curve.

c Determine the nature of the stationary points.

d Sketch the graph of $y = \dfrac{x^2}{1-2x}$.

15 a Prove that the derivative of $\operatorname{cosec} x$ is $-\operatorname{cosec} x \cot x$.

b Find the equation of the tangent to the graph of $y = \operatorname{cosec} x$ at the point where $x = \dfrac{\pi}{6}$.

16 A function is defined by $g(x) = 3x + \ln(2x)$ for $x > 0$.

a Find $g'(x)$ and hence prove that $g(x)$ has an inverse function.

b Find the gradient of $g^{-1}(x)$ at the point $(3, 1)$.

17 A curve is defined by the equation $9x^2 - 6xy + 4y^2 = 3$.

Find the coordinates of the two stationary points of this curve.

[© AQA 2012]

18 A curve is given by the implicit equation $x^2 - xy + y^2 = 12$.

a Find the coordinates of the stationary points on the curve.

b Show that, at the stationary points, $(x-2y)\dfrac{d^2y}{dx^2} = 2$.

c Hence determine the nature of the stationary points.

19 a Let $y = \arctan x$. By first expressing x in terms of y, prove that $\dfrac{dy}{dx} = \dfrac{1}{1+x^2}$.

b Find the equation of the normal to the curve $y = \arctan(3x)$ at the point where $x = \dfrac{1}{\sqrt{3}}$.

20 a Given that $y = \dfrac{x+1}{\sqrt{x+2}}$, find $\dfrac{dy}{dx}$. Give your answer in the form $\dfrac{Ax+B}{2(x+2)^{\frac{3}{2}}}$ where A and B are integers.

The diagram shows a part of the graph of the curve $y = \dfrac{x}{(x+2)^{\frac{3}{2}}}$.

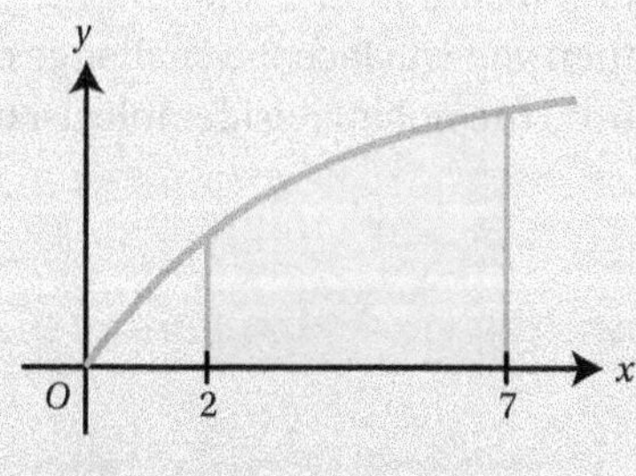

b Find the coordinates of the stationary point on the curve.

c Find the shaded area enclosed by the curve, the x-axis and the lines $x = 2$ and $x = 7$.

Elevate

See Extension Sheet 10 for questions on some properties of e that use differentiation.

11 Further integration techniques

In this chapter you will learn how to integrate using:

- known derivatives
- the chain rule in reverse
- a change of variable (substitution)
- the product rule in reverse (integration by parts)
- trigonometric identities
- the separation of a fraction into two fractions.

Before you start...

Chapter 9	You should be able to differentiate and integrate polynomial, exponential and trigonometric functions.	1 Find: a $\int\left(4x^2+\frac{3}{x}\right)dx$ b $\int 5\sin x\,dx$. 2 Given that $y=4\,e^x$, find: a $\frac{dy}{dx}$ b $\int_0^1 y\,dx$.
Chapter 10	You should be able to use the chain rule for differentiation.	3 Differentiate: a $\sin 4x$ b $\ln(x^2+1)$.
Chapter 8	You should be able to use double angle formulae.	4 Given that $\cos 2A = 0.28$, find the possible values of $\cos A$.
Chapter 2, Chapter 5	You should be able to split an expression into partial fractions.	5 Write $\frac{36}{(x-1)(x+2)^2}$ in partial fractions.

Integrating more complex functions

Having extended the range of functions you can differentiate, you now need to do the same for integration. In some cases you will be able to use the results from Chapter 10 directly, but in many others you will require new techniques. In this chapter you will look at each of these in turn and then you will face the challenge of selecting the appropriate technique from the not inconsiderable list of options you have built up.

Section 1: Reversing standard derivatives

You already know how to integrate many functions, by reversing the corresponding differentiation results.

- $\int x^n \, dx = \frac{1}{n+1} x^{n+1} + c$
- $\int e^x \, dx = e^x + c$
- $\int \frac{1}{x} \, dx = \ln|x| + c$
- $\int \sin x \, dx = -\cos x + c$
- $\int \cos x \, dx = \sin x + c$

In Chapter 10 you differentiated sec x, cosec x and cot x. You can now reverse these standard derivatives, too, and add these results to your list.

- $\int \sec^2 x \, dx = \tan x + c$
- $\int \sec x \tan x \, dx = \sec x + c$
- $\int \operatorname{cosec} x \cot x \, dx = -\operatorname{cosec} x + c$
- $\int \operatorname{cosec}^2 x \, dx = -\cot x + c$

Thinking about reversing the chain rule for differentiation allows you to go one step further and deal with integrals such as $\int \cos 2x \, dx$.

You know that the answer must include sin $2x$ as sin is the integral of cos, but this is not the final answer. If you differentiate sin $2x$:
$\frac{d}{dx} \sin 2x = 2\cos 2x$.

Since you do not want the 2 in front of cos $2x$, divide by 2 (or multiply by $\frac{1}{2}$).

$$\therefore \int \cos 2x \, dx = \frac{1}{2} \sin 2x + c$$

WORKED EXAMPLE 11.1

Find $\int (7x-3)^4 \, dx$.

$$\int (7x-3)^4 \, dx = \frac{1}{7} \times \frac{1}{5} (7x-3)^5$$

$$= \frac{1}{35}(7x-3)^5 + c$$

$()^4$ integrates to $\frac{1}{5}()^5$ but, differentiating back, the chain rule will also give a factor of 7 (the derivative of $7x-3$). Remove this by multiplying by $\frac{1}{7}$.

WORKED EXAMPLE 11.2

Find $\int e^{4x+5}\,dx$.

$\int e^{4x+5}\,dx = \frac{1}{4} \times e^{4x+5} + c$

$e^{(\,)}$ integrates to $e^{(\,)}$ but, differentiating back, the chain rule will also give a factor of 4 (the derivative of $4x+5$). Remove this by multiplying by $\frac{1}{4}$.

You may notice a pattern here: you always end up integrating the function and then dividing by the coefficient of x. This is indeed a general rule when the 'inside' function is of the form $(ax+b)$.

Key point 11.1

$$\int f(ax+b)\,dx = \frac{1}{a}F(ax+b) + c$$

where $F(x)$ is the integral of $f(x)$.

Tip

Note that this rule only applies when the 'inside' function is of the form $(ax+b)$.

WORKED EXAMPLE 11.3

Find $\int \frac{1}{5-3x}\,dx$.

$\int \frac{1}{5-3x}\,dx = \frac{1}{-3}\ln|5-3x| + c$

$= -\frac{1}{3}\ln|5-3x| + c$

Integrate $\frac{1}{(\,)}$ to $\ln|\;|$ and divide by the coefficient of x.

WORK IT OUT 11.1

Find $\int \frac{1}{3x}\,dx$. Which is the correct solution? Identify the errors made in the incorrect solutions.

Solution A	Solution B	Solution C
$\int \frac{1}{3x}\,dx = \int 3x^{-1}\,dx$ $= 3\ln\lvert x\rvert + c$	$\int \frac{1}{3x}\,dx = \frac{1}{3}\ln\lvert 3x\rvert + c$	$\int \frac{1}{3x}\,dx = \frac{1}{3}\int \frac{1}{x}\,dx$ $= \frac{1}{3}\ln\lvert x\rvert + c$

WORKED EXAMPLE 11.4

Find $\int 3\sec^2(5x-2)\,dx$.

$\int 3\sec^2(5x-2)\,dx = 3 \times \frac{1}{5}\tan(5x-2) + c$

$= \frac{3}{5}\tan(5x-2) + c$

Integrate $\sec^2(\,)$ to $\tan(\,)$ and divide by the coefficient of x.

EXERCISE 11A

1 Find each indefinite integral.

a **i** $\int 5(x+3)^4\,dx$ **ii** $\int (x-2)^5\,dx$ **b** **i** $\int (4x-5)^7\,dx$ **ii** $\int \left(\frac{1}{8}x+1\right)^3 dx$

c **i** $\int 4\left(3-\frac{1}{2}x\right)^6 dx$ **ii** $\int (4-x)^8\,dx$ **d** **i** $\int \sqrt{2x-1}\,dx$ **ii** $\int 7(2-5x)^{\frac{3}{4}}\,dx$

e **i** $\int \frac{1}{\sqrt[4]{2+\frac{x}{3}}}\,dx$ **ii** $\int \frac{6}{(4-3x)^2}\,dx$

2 Work out each integral.

a **i** $\int 3e^{3x}\,dx$ **ii** $\int e^{2x+5}\,dx$ **b** **i** $\int 4e^{\frac{2x-1}{3}}\,dx$ **ii** $\int e^{\frac{1}{2}x}\,dx$

c **i** $\int -6e^{-3x}\,dx$ **ii** $\int \frac{1}{e^{4x}}\,dx$ **d** **i** $\int \frac{-2}{e^{\frac{x}{4}}}\,dx$ **ii** $\int e^{-\frac{2}{3}x}\,dx$

3 Find each integral.

a **i** $\int \frac{1}{x+4}\,dx$ **ii** $\int \frac{5}{5x-2}\,dx$ **b** **i** $\int \frac{2}{3x+4}\,dx$ **ii** $\int \frac{-8}{2x-5}\,dx$

c **i** $\int \frac{-3}{1-4x}\,dx$ **ii** $\int \frac{1}{7-2x}\,dx$ **d** **i** $\int \left(1-\frac{3}{5-x}\right)dx$ **ii** $\int \left(3+\frac{1}{3-x}\right)dx$

4 Integrate each expression.

a $\int -\operatorname{cosec} x \cot x\,dx$ **b** $\int 3\sec^2 3x\,dx$ **c** $\int \sin(2-3x)\,dx$

d $\int \operatorname{cosec}^2\left(\frac{1}{4}x\right)dx$ **e** $\int 2\cos 4x\,dx$ **f** $\int \sec\frac{x}{2}\tan\frac{x}{2}\,dx$

5 Find the exact value of $\int_0^{\frac{\pi}{3}} 2\sin(5x)\,dx$.

6 Find the exact area enclosed by the graph of $y = 3e^{-2x}$, the x-axis and the lines $x = 1$ and $x = 4$.

7 Find the area enclosed by the x-axis and the curve with equation $y = 9-(2x-5)^2$.

8 Given that $0 < a < 1$ and the area between the x-axis, the lines $x = a^2$, $x = a$ and the graph of $y = \frac{1}{1-x}$ is 0.4, find the value of a correct to three significant figures.

Section 2: Integration by substitution

The shortcut for reversing the chain rule (Key point 11.1) works only when the derivative of the 'inside' function is a constant. This is because a constant factor can 'move through the integral sign', for example:

$$\begin{aligned}\int \cos 2x\,dx &= \int \frac{1}{2}\times 2\cos 2x\,dx\\ &= \frac{1}{2}\int 2\cos 2x\,dx\\ &= \frac{1}{2}\sin 2x + c\end{aligned}$$

Fast forward

Another method for integrating products is integration by parts, which is the reverse of the product rule. You will meet this in Section 3.

This cannot be done with a variable: $\int x \sin x \, dx$ is not the same as $x \int \sin x \, dx$. So you need a different rule for integrating a product of two functions. In some cases this can be achieved by extending the principle of reversing the chain rule, leading to the method of **integration by substitution.**

Reversing the chain rule

 **Rewind**

See Chapter 10, Section 1, for a reminder of the chain rule for differentiation.

When using the chain rule to differentiate a composite function, you differentiate the outer function and multiply this by the derivative of the inner function; for example:

$$\frac{d}{dx}(\sin(x^2+2)) = \cos(x^2+2) \times 2x$$

You can think of this as using a substitution $u = x^2 + 2$, and then $\frac{dy}{dx} = \frac{dy}{du} \times \frac{du}{dx}$.

Look now at $\int x \cos(x^2+2) \, dx$. Since $\cos(x^2+2)$ is a composite function it can be written as $\cos u$, where $u = x^2 + 2$. Thus the integral becomes $\int x \cos u \, dx$.

Tip

When making a substitution of this type, only replace the inner function with u. Any other instances of x will cancel out when changing dx to du.

You know how to integrate $\cos u$, but you need to integrate with respect to u, so you should have du instead of dx. Although du and dx are not the same thing, they are related because $u = x^2 + 2 \Rightarrow \frac{du}{dx} = 2x$.

You can then rearrange this to make x the subject so that it can be replaced in the integral:

$$x = \frac{1}{2}\frac{du}{dx}$$

Substituting all of this into the integral gives:

$$\int x \cos(x^2+2) \, dx = \int \frac{1}{2}\frac{du}{dx}\cos(u) \, dx$$
$$= \frac{1}{2}\int \cos u \frac{du}{dx} \, dx$$

It follows from the chain rule that $\int \cos u \frac{du}{dx} \, dx = \int \cos u \, du$ so:

$$\int x \cos(x^2+2) \, dx = \frac{1}{2}\int \cos u \, du$$
$$= \frac{1}{2}\sin u + c$$
$$= \frac{1}{2}\sin(x^2+2) + c$$

Notice that having found an answer in terms of u, you need to put it back in terms of x.

In practice, this method can be shorted by appreciating that $\frac{du}{dx}$ can be split up just like a fraction, so $\frac{du}{dx} = 2x$ gives $dx = \frac{1}{2x}du$.

This is illustrated in Worked examples 11.5 and 11.6.

WORKED EXAMPLE 11.5

Find each integral. **a** $\int \sin^5 x \cos x \, dx$ **b** $\int x^2 e^{x^3+4} \, dx$

a Let $u = \sin x$.

Then $\frac{du}{dx} = \cos x \Rightarrow dx = \frac{1}{\cos x} du$

Think of $\sin^5 x$ as $(\sin x)^5$; therefore the inner function is $\sin x$.

$\int (\sin x)^5 \cos x \, dx = \int u^5 \cos x \frac{1}{\cos x} du$

Make the substitution.

$= \int u^5 \, du$

$= \frac{1}{6} u^6 + c$

$= \frac{1}{6} \sin^6 x + c$

Write the answer in terms of x.

b Let $u = x^3 + 4$.

Then $\frac{du}{dx} = 3x^2 \Rightarrow dx = \frac{1}{3x^2} du$

e^{x^3+4} is a composite function with inner function $x^3 + 4$.

$\int x^2 e^{x^3+4} \, dx = \int x^2 e^u \frac{1}{3x^2} du$

Make the substitution.

$= \int \frac{1}{3} e^u \, du$

$= \frac{1}{3} e^u + c$

$= \frac{1}{3} e^{x^3+4} + c$

Write the answer in terms of x.

When limits are given, you must ensure they are changed too. In this case there is no need to change back to the original variable at the end.

WORKED EXAMPLE 11.6

Evaluate $\int_0^1 \frac{x-3}{x^2-6x+7} dx$, giving your answer in the form $a \ln p$.

Let $u = x^2 - 6x + 7$.

The 'inner' function is $x^2 - 6x + 7$.

Then $\frac{du}{dx} = 2x - 6 \Rightarrow dx = \frac{1}{2x-6} du$

Limits: $x = 0 \Rightarrow u = 7$, $x = 1 \Rightarrow u = 2$

Write the limits in terms of u.

$\int_0^1 \frac{x-3}{x^2-6x+7} dx = \int_7^2 \frac{x-3}{u} \frac{1}{2x-6} du$

Make the substitution.

$= \int_7^2 \frac{1}{2u} du$

Simplify: $2x - 6 = 2(x-3)$

$= \left[\frac{1}{2} \ln |u| \right]_7^2$

$= \frac{1}{2} (\ln 2 - \ln 7)$

$= \frac{1}{2} \ln \left(\frac{2}{7} \right)$

This particular case of substitution, in which the top of the fraction is the derivative of the bottom, is definitely worth remembering:

Key point 11.2

$$\int \frac{f'(x)}{f(x)}\,dx = \ln|f(x)| + c$$

> **Tip**
>
> You do not have to make a substitution if you can see that the expression is of the form $\int f'(x)[f(x)]^n\,dx$, i.e. with both a function and its derivative present. You can just go straight to the answer, using the reverse chain rule: $\frac{1}{n+1}[f(x)]^{n+1} + c$

You can use this result to integrate some trigonometric functions.

WORKED EXAMPLE 11.7

Show that $\int \tan x\,dx = \ln(\sec x) + c$.

$\int \tan x\,dx = \int \frac{\sin x}{\cos x}\,dx$ — If $f(x) = \cos x$ then $f'(x) = -\sin x$.

$= -\int \frac{-\sin x}{\cos x}\,dx$ — So you can write the integral in the form $\int \frac{f'(x)}{f(x)}$.

$= -\ln(\cos x) + c$ — Use the result of Key point 11.2.

$= \ln\left(\frac{1}{\cos x}\right) + c$

$= \ln(\sec x) + c$ — The answer is in terms of $\sec x$, so remember that $\sec = \frac{1}{\cos}$ and use $-\ln a = \ln(a^{-1})$.

EXERCISE 11B

1 Use either a suitable substitution or the reverse chain rule to find each integral.

a i $\int 2x(x^2+3)^3\,dx$ ii $\int 2x(x^2-1)^5\,dx$

b i $\int (2x-5)(3x^2-15x+4)^4\,dx$ ii $\int (x^2+2x)(x^3+3x^2-5)^3\,dx$

c i $\int \frac{2x}{x^2+3}\,dx$ ii $\int \frac{3x^2-4}{x^3-4x+5}\,dx$

d i $\int \frac{x+4}{x^2+8x-3}\,dx$ ii $\int \frac{x^2+2x-5}{x^3+3x^2-15x+1}\,dx$

e i $\int \frac{x}{\sqrt{x^2+2}}\,dx$ ii $\int \frac{x^2}{(x^3-4)^2}\,dx$

f i $\int 4\cos^5 x \sin x\,dx$ ii $\int \cos 2x \sin^3 2x\,dx$

g i $\int \tan^3 x \sec^2 x\,dx$ ii $\int \cot^4 x\,\mathrm{cosec}^2 x\,dx$

h i $\int 3xe^{3x^2-1}\,dx$ ii $\int 3x\,e^{x^2}dx$

i i $\int \frac{e^{2x+3}}{e^{2x+3}+4}\,dx$ ii $\int \frac{\cos x}{3+4\sin x}\,dx$

2 Find $\int \sin x\, e^{\cos x}\,dx$.

3 Find the exact value of $\int_0^2 (2x+1)e^{x^2+x-1}\,dx$, showing all your working.

4 Show that $\int_2^5 \frac{2x}{x^2-1}\,dx = \ln k$ where k is an integer to be found.

5 Find $\int \frac{\cos 3x}{\sin^5 3x}\,dx$.

6 Find $\int \operatorname{cosec}^5 2x \cot 2x\,dx$.

General substitution

In all the examples so far, after the substitution, the part of the integral that was still in terms of x cancelled with a similar term coming from $\frac{du}{dx}$. For example, in Worked example 11.5 part b,

$\int x^2 e^{x^3+4}\,dx = \int x^2 e^u \frac{1}{3x^2}\,du = \int \frac{1}{3} e^u\,du.$

This will always happen when one part of the expression to be integrated is an exact multiple of the derivative of the inner function.

In some cases the remaining x-terms will not cancel and you will have to express x in terms of u. The full method of substitution will then be as summarised in Key point 11.3.

Key point 11.3

Integration by substitution:

1. Select a substitution (if not already given).
2. Differentiate the substitution and write dx in terms of du.
3. Replace dx by the expression in **2**, and replace any obvious occurrences of u.
4. Change the limits from x to u.
5. Simplify as far as possible.
6. If any terms with x remain, write them in terms of u.
7. Do the new integral in terms of u.
8. Write the answer in terms of x.

Tip

If you are not told which substitution to use, look for a composite function and take u = 'inner' function.

WORKED EXAMPLE 11.8

Find $\int x\sqrt{4x-1}\,dx$, using the substitution $u = 4x-1$.

$u = 4x-1$

$\frac{du}{dx} = 4 \Rightarrow \quad dx = \frac{1}{4}du$

Differentiate the substitution and write dx in terms of du (step 2).

$\int x\sqrt{4x-1}\,dx = \int x\sqrt{u}\,\frac{1}{4}\,du = \int \frac{1}{4}x\sqrt{u}\,du$

Replace parts that you have expressions for, and simplify if possible (steps 3 to 5).

$= \int \frac{1}{4}\frac{u+1}{4}u^{\frac{1}{2}}\,du$

There is still an x remaining, so replace it by using $u = 4x-1 \Rightarrow x = \frac{u+1}{4}$ (step 6).

$= \frac{1}{16}\int \left(u^{\frac{3}{2}} + u^{\frac{1}{2}}\right)du$

$= \frac{1}{16}\left(\frac{2}{5}u^{\frac{5}{2}} + \frac{2}{3}u^{\frac{3}{2}}\right) + c$

Now everything is in terms of u so you can integrate (step 7). Remember that $\sqrt{u} = u^{\frac{1}{2}}$.

$= \frac{1}{8}\left(\frac{1}{5}(\sqrt{4x-1})^5 + \frac{1}{3}(\sqrt{4x-1})^3\right) + c$

Write the answer in terms of x, using $u = 4x-1$ (step 8).

A substitution can be given as x in terms of u^2, rather than u in terms x.

WORKED EXAMPLE 11.9

Use the substitution $x = u^2$ (with $u > 0$) to evaluate $\int_9^{25} \frac{1}{x - 2\sqrt{x}}\,dx$.

$\frac{dx}{du} = 2u \Rightarrow dx = 2u\,du$ — Differentiate the substitution. Note: now you are using $\frac{dx}{du}$.

$x = u^2 = 9 \Rightarrow u = 3$
$x = u^2 = 25 \Rightarrow u = 5$ — Find limits for u. This involves solving an equation.

$\int_9^{25} \frac{1}{x - 2\sqrt{x}}\,dx = \int_3^5 \frac{1}{u^2 - 2u} 2u\,du$ — Replace dx by $2u\,du$ and x by u^2. Use the fact that $\sqrt{u^2} = u$.

$= \int_3^5 \frac{2}{u-2}\,du$ — Simplify if possible.

$= \left[2\ln|u-2|\right]_3^5$
$= 2\ln 3 - 2\ln 1$
$= \ln 9$ — Now everything is in terms of u so you can integrate. Remember the modulus sign with the ln.

EXERCISE 11C

1 Find each integral. Use the given substitution.

a **i** $\int x\sqrt{x+1}\,dx,\ u = x+1$ **ii** $\int x^2\sqrt{x-2}\,dx,\ u = x-2$

b **i** $\int 2x(x-5)^7\,dx,\ u = x-5$ **ii** $\int x(x+3)^5\,dx,\ u = x+3$

2 Use the given substitution to find each integral.

a **i** $\int \frac{1}{x+\sqrt{x}}\,dx,\ x = u^2$ **ii** $\int \frac{1}{3\sqrt{x}+4x}\,dx,\ x = u^2$

b **i** $\int \frac{1}{x\ln x}\,dx,\ x = e^u$ **ii** $\int \frac{1}{x(\ln x)^3}\,dx,\ x = e^u$

3 Find each integral. Use an appropriate substitution.

a **i** $\int x(2x-1)^4\,dx$ **ii** $\int 9x(3x+2)^5\,dx$ **b** **i** $\int x\sqrt{x-3}\,dx$ **ii** $\int (x+1)\sqrt{5x-6}\,dx$

c **i** $\int \frac{x^2}{\sqrt{x-5}}\,dx$ **ii** $\int \frac{4(x+5)}{(2x-3)^3}\,dx$

4 Use the given substitution to evaluate each definite integral.

a **i** $\int_1^3 4x(2x+1)^3\,dx,\ u = 2x+1$ **ii** $\int_0^1 6x(3x-2)^4\,dx,\ u = 3x-2$

b **i** $\int_0^{\frac{\pi}{2}} \cos x \sin^5 x\,dx,\ u = \sin x$ **ii** $\int_0^{\frac{\pi}{4}} \sec^2 x \tan^2 x\,dx,\ u = \tan x$

c **i** $\int_2^3 \left(\frac{x}{4-x}\right)^2 dx,\ u = 4-x$ **ii** $\int_1^3 \frac{x^3}{(x+2)^2}\,dx,\ u = x+2$

5 Use the substitution $u = x - 2$ to find $\int \frac{x}{\sqrt{x-2}}\,dx$.

6 **a** Show that $(x-1)$ is a factor of $x^3 - 1$. **b** Find $\int \frac{2x^2 - x - 1}{x^3 - 1}\,dx$

7 Use the substitution $x = e^u$ to find $\int \frac{\sec^2(\ln(x^2))}{2x}\,dx$.

8 Show that $\int_1^3 \frac{(2x-3)\sqrt{x^2-3x+3}}{x^2-3x+3}\,dx = a\sqrt{b} + c$ where a, b and c are integers to be found.

9 By using the substitution $u = e^x$, find the exact value of $\int_0^{\frac{1}{2}\ln 3} \frac{1}{e^x + e^{-x}}\,dx$.

Section 3: Integration by parts

In Section 2, you saw cases in which products of functions can be integrated by using the reverse chain rule or a substitution. But you still cannot work out integrals such as $\int x \sin x\,dx$ or $\int x^2 e^x\,dx$.

In order to integrate these, return to the product rule for differentiation:

$$\frac{d}{dx}(uv) = u\frac{dv}{dx} + v\frac{du}{dx}$$

Integrating with respect to x you get:

$$uv = \int u\frac{dv}{dx}\,dx + \int v\frac{du}{dx}\,dx$$

$$\Rightarrow \int u\frac{dv}{dx}\,dx = uv - \int v\frac{du}{dx}\,dx$$

Tip

This working shows that integration by parts can be thought of as the 'reverse product rule'. In practice, though, while the idea of integrating using the 'reverse chain rule' is used, the 'reverse product rule' is not used; the parts formula is used instead.

Key point 11.4

The **integration by parts** formula is:

$$\int u\frac{dv}{dx}\,dx = uv - \int v\frac{du}{dx}\,dx$$

This will be given in your formula book.

WORKED EXAMPLE 11.10

Find $\int x \sin x\,dx$.

$u = x$ and $\frac{dv}{dx} = \sin x$ — This is a product to which the reverse chain rule cannot be applied, so try integration by parts.

$\Rightarrow \frac{du}{dx} = 1$ and $v = -\cos x$

$\int u\frac{dv}{dx}\,dx = uv - \int v\frac{du}{dx}\,dx$ — Apply the formula.

$\int x \sin x\,dx = x(-\cos x) - \int (-\cos x)1\,dx$

$= -x\cos x + \int \cos x\,dx$

$= -x\cos x + \sin x + c$

In Worked example 11.10, taking u to be x worked well because $\frac{du}{dx}$ is a constant, so after applying the formula the resulting integral was just $\int \cos x$. You could only do this because $\frac{dv}{dx} = \sin x$ is easy to integrate, so you could find v. In some examples this is not possible.

WORKED EXAMPLE 11.11

Find $\int x \ln x \, dx$.

$u = \ln x, \ \frac{dv}{dx} = x$

This is a product, so use integration by parts. You cannot take $u = x$ because you don't know how to integrate $\frac{dv}{dx} = \ln x$, so try choosing them the other way round.

$\Rightarrow \frac{du}{dx} = \frac{1}{x}, \ v = \frac{1}{2}x^2$

$\int x \ln x \, dx = \ln x \times \frac{1}{2}x^2 - \int \frac{1}{2}x^2 \frac{1}{x} \, dx$

Apply the formula.

$= \frac{1}{2}x^2 \ln x - \int \frac{1}{2}x \, dx$

Always simplify before integrating.

$= \frac{1}{2}x^2 \ln x - \frac{1}{4}x^2 + c$

The strategy for choosing which function is u and which is $\frac{dv}{dx}$ is summarised in Key point 11.5.

Key point 11.5

When using integration by parts for $\int x^n \, f(x) \, dx$, choose $u = x^n$ in all cases except when $f(x) = \ln x$.

You can even use this strategy to integrate $\ln x$ by itself.

WORKED EXAMPLE 11.12

Use integration by parts to find $\int \ln x \, dx$.

$\int \ln x \, dx = \int 1 \times \ln x \, dx$

The trick is to write $\ln x$ as a product of $\ln x$ and 1 so that you can use integration by parts.

$u = \ln x$ and $\frac{dv}{dx} = 1$

As suggested in Key point 12.5, let $u = \ln x$.

$\Rightarrow \frac{du}{dx} = \frac{1}{x}$ and $v = x$

$\int u \frac{dv}{dx} \, dx = uv - \int \frac{du}{dx} v \, dx$

Apply the formula.

$\int 1 \times \ln x \, dx = (\ln x)x - \int \frac{1}{x} x \, dx$

$= x \ln x - \int 1 \, dx$

$= x \ln x - x + c$

EXERCISE 11D

1 Find each integral. Use integration by parts.

a i $\int x\cos 2x\,dx$ ii $\int x\sin\left(\frac{x}{2}\right)dx$ b i $\int 4xe^{-2x}\,dx$ ii $\int xe^{4x}\,dx$

c i $\int 2x\ln 5x\,dx$ ii $\int x\ln x\,dx$ d i $\int \frac{x^3}{2}\ln x\,dx$ ii $\int 3x^5\ln 2x\,dx$

2 Use integration by parts to evaluate $\int_0^{\frac{\pi}{2}} x\sin x\,dx$.

3 Find the exact value of $\int_1^{e} 3x^2\ln(2x)\,dx$.

Repeated integration by parts

It may be necessary to use integration by parts more than once. As long as the integrals are becoming simpler each time, you are on the right track!

WORKED EXAMPLE 11.13

Find the exact value of $\int_0^{\ln 2} x^2e^x\,dx$

$u = x^2$ and $\frac{dv}{dx} = e^x$

$\Rightarrow \frac{du}{dx} = 2x$ and $v = e^x$

This is a product to which you cannot apply the reverse chain rule, so try integration by parts.

Choose u to be the polynomial.

$\int u\frac{dv}{dx}\,dx = uv - \int \frac{du}{dx}v\,dx$

Apply the formula.

$\int_0^{\ln 2} x^2e^x\,dx = \left[x^2e^x\right]_0^{\ln 2} - \int_0^{\ln 2} 2x\,e^x\,dx$

Put in the limits on the uv part straight away.

$u = 2x$ and $\frac{dv}{dx} = e^x$

$\Rightarrow \frac{du}{dx} = 2$ and $v = e^x$

You have to integrate a product again, so use integration by parts for the second time. Choose u to be the polynomial again. (Notice that if you were to change and choose $u = e^x$ and $\frac{dv}{dx} = 2x$ you would end up back where you started!)

So,

$\int_0^{\ln 2} 2xe^x\,dx = \left[2xe^x\right]_0^{\ln 2} - \int_0^{\ln 2} 2e^x\,dx$

$= \left[2xe^x\right]_0^{\ln 2} - \left[2e^x\right]_0^{\ln 2}$

Apply the formula again and use the limits.

Therefore,

$\int_0^{\ln 2} x^2e^x\,dx$

$= \left[x^2e^x\right]_0^{\ln 2} - \left\{\left[2xe^x\right]_0^{\ln 2} - \left[2e^x\right]_0^{\ln 2}\right\}$

$= ((\ln 2)^2e^{\ln 2} - 0) - (2\ln 2e^{\ln 2} - 0) + (2e^{\ln 2} - 2)$

$= 2(\ln 2)^2 - 4\ln 2 + 2$

Put both integrals together, making sure to keep track of negative signs by using brackets appropriately.

Common error

When using integration by parts twice, remember to apply the negative sign in front of the integral to *all* of the second application of the parts formula.

EXERCISE 11E

1 Use integration by parts twice to find each integral.

a $\int x^2 \cos 3x \, dx$ b $\int x^2 \sin 2x \, dx$

c $\int \frac{1}{4} x^2 e^{\frac{x}{4}} \, dx$ d $\int x^2 (x+2)^5 \, dx$

2 Use integration by parts to find each integral.

a $\int 2 \ln(3x) \, dx$ b $\int \ln\left(\frac{1}{x}\right) dx$

3 Evaluate each interval exactly.

a $\int_0^{\frac{\pi}{2}} x \cos x \, dx$ b $\int_1^2 \frac{\ln x}{x^2} \, dx$

4 When using the integration by parts formula, you start with $\frac{dv}{dx}$ and find v.

Why not include a constant of integration when you do this?

Try a few examples adding $+c$ to v and see what happens.

5 Find $\int 2xe^{-3x} \, dx$.

6 Evaluate $\int_1^e x^5 \ln x \, dx$.

7 Evaluate $\int_1^e (\ln x)^2 \, dx$

8 a Show that $\int \tan x \, dx = \ln |\sec x| + c$.

. b Hence find $\int \frac{x}{\cos^2 x} \, dx$

9 Use the substitution $\sqrt{x+1} = u$ to find the exact value of $\int_{-1}^{3} \frac{1}{2} e^{\sqrt{x+1}} \, dx$.

10 Let $I = \int e^x \cos x \, dx$.

a Use integration by parts twice to show that $I = e^x \cos x + (e^x \sin x - I) + c$.

b Hence find $\int e^x \cos x \, dx$.

Section 4: Using trigonometric identities in integration

The most common use of trigonometric identities in integration is integrating the squares of trigonometric functions.

WORKED EXAMPLE 11.14

Find $\int \tan^2 x \, dx$.

$\int \tan^2 x \, dx = \int (\sec^2 x - 1) \, dx$

$= \tan x - x + c$

You can integrate $\sec^2 x$, so use $\tan^2 + 1 \equiv \sec^2 x$.

In Section 2 you integrated expressions such as $\sin^2 x \cos x$ by using the substitution $u = \sin x$; the derivative $\frac{du}{dx}$ cancels with the cos x. But if you try to integrate just $\int \sin^2 x \, dx$ the same substitution does not work. You could try rewriting it as $\int (1 - \cos^2 x) \, dx$, but you don't know how to do this either.

The trick is to notice that $\sin^2 x$ appears in one of the versions of the double angle formulae for $\cos 2x$: $\cos 2x \equiv 1 - 2\sin^2 x$, and you know how to integrate $\cos 2x$. This leads to a method for integrating both $\sin^2 x$ and $\cos^2 x$, which you need to remember.

Key point 11.6

To integrate $\sin^2 x$ use $\cos 2x \equiv 1 - 2\sin^2 x$.

To integrate $\cos^2 x$ use $\cos 2x \equiv 2\cos^2 x - 1$.

Rewind

Double angle identities were covered in Chapter 8, Section 2.

WORKED EXAMPLE 11.15

Find $\int \sin^2 x \, dx$.

$\cos 2x \equiv 1 - 2\sin^2 x$

$\Rightarrow \sin^2 x \equiv \frac{1}{2}(1 - \cos 2x)$

You can find an alternative expression for $\sin^2 x$ by using a double angle identity.

$\therefore \int \sin^2 x \, dx = \int \frac{1}{2}(1 - \cos 2x) \, dx$

$= \int \left(\frac{1}{2} - \frac{1}{2}\cos 2x\right) dx$

$= \frac{1}{2}x - \frac{1}{2} \times \frac{1}{2}\sin 2x + c$

$= \frac{1}{2}x - \frac{1}{4}\sin 2x + c$

Remember to divide by the coefficient of x when integrating $\cos 2x$.

Sometimes trigonometric identities turn out to be useful in integrals that do not appear to involve trigonometry.

WORKED EXAMPLE 11.16

Use the substitution $x = 3\sin\theta$ to find $\int_0^3 \sqrt{9-x^2}\,dx$.

$$9-x^2 = 9-(3\sin\theta)^2$$
$$= 9-9\sin^2\theta$$
$$= 9\cos^2\theta$$

Make the substitution. Simplify each part separately first.

Use $1-\sin^2\theta = \cos^2\theta$

$$\frac{dx}{d\theta} = 3\cos\theta \Rightarrow dx = 3\cos\theta\, d\theta$$

Differentiate the substitution.

Limits:

$$x = 0 \Rightarrow \sin\theta = 0 \Rightarrow \theta = 0$$
$$x = 3 \Rightarrow \sin\theta = 1 \Rightarrow \theta = \frac{\pi}{2}$$

Find the limits for θ.

So:

$$\int_0^3 \sqrt{9-x^2}\,dx = \int_0^{\frac{\pi}{2}} \sqrt{9\cos^2\theta} \times 3\cos\theta\, d\theta$$
$$= \int_0^{\frac{\pi}{2}} 3\cos\theta \times 3\cos\theta\, d\theta$$
$$= \int_0^{\frac{\pi}{2}} 9\cos^2\theta\, d\theta$$
$$= \int_0^{\frac{\pi}{2}} \frac{9}{2}(\cos 2\theta + 1)\, d\theta$$
$$= \left[\frac{9}{4}\sin 2\theta + \frac{9}{2}\theta\right]_0^{\frac{\pi}{2}}$$
$$= \left(\frac{9}{4}\sin\pi + \frac{9\pi}{4}\right) - \left(\frac{9}{4}\sin 0 + 0\right)$$
$$= \frac{9\pi}{4}$$

Simplify fully before attempting to integrate.

Now use the method from Key point 11.6: $\cos 2\theta \equiv 2\cos^2\theta - 1 \Rightarrow \cos^2\theta \equiv \frac{\cos 2\theta + 1}{2}$.

Remember to divide by 2 when integrating $\cos 2\theta$.

In integrals of this type, you may need to use trigonometric identities again to write the answer in terms of x at the end.

Tip

The substitution in Worked example 11.16 is useful in other integrals involving expressions like $\sqrt{a^2-x^2}$.

WORKED EXAMPLE 11.17

a Show that $\frac{\sec\theta}{\tan^2\theta} \equiv \operatorname{cosec}\theta\cot\theta$.

b Use the substitution $x = \sec\theta$ to find $\int (x^2-1)^{-\frac{3}{2}}\,dx$.

a
$$\frac{\sec\theta}{\tan^2\theta} \equiv \frac{1}{\cos\theta}\frac{1}{\left(\frac{\sin\theta}{\cos\theta}\right)^2}$$
$$\equiv \frac{1}{\cos\theta}\frac{\cos^2\theta}{\sin^2\theta}$$
$$\equiv \frac{1}{\sin\theta}\frac{\cos\theta}{\sin\theta}$$
$$\equiv \operatorname{cosec}\theta\cot\theta$$

There is no obvious identity linking the functions on the left to the ones on the right. So write everything in terms of sin and cos.

b
$$x = \sec\theta$$
$$\Rightarrow \frac{dx}{d\theta} = \sec\theta\tan\theta$$

Differentiate the substitution and express dx in terms of $d\theta$.

$$dx = \sec\theta\tan\theta\,d\theta$$
$$\int (x^2-1)^{-\frac{3}{2}}\,dx = \int (\sec^2\theta - 1)^{-\frac{3}{2}}\sec\theta\tan\theta\,d\theta$$

Replace those parts that you already have expressions for.

$$= \int (\tan^2\theta)^{-\frac{3}{2}}\sec\theta\tan\theta\,d\theta$$
$$= \int (\tan\theta)^{-2}\sec\theta\,d\theta$$

There are no instances of x remaining, so you can integrate.

Notice that $\sec^2\theta - 1 \equiv \tan\theta$

$$= \int \operatorname{cosec}\theta\cot\theta\,d\theta$$
$$= -\operatorname{cosec}\theta + c$$

Using the result from part **a**, you now have a standard derivative (Section 11.1).

$$x = \sec\theta \Rightarrow \cos\theta = \frac{1}{x}$$
$$\Rightarrow \sin\theta = \sqrt{1-\cos^2\theta}$$
$$= \sqrt{1-\frac{1}{x^2}}$$
$$= \sqrt{\frac{x^2-1}{x^2}}$$
$$\Rightarrow \operatorname{cosec}\theta = \frac{1}{\sin\theta}$$
$$= \sqrt{\frac{x^2}{x^2-1}}$$
$$\therefore \int (x^2-1)^{-\frac{3}{2}}\,dx = -\sqrt{\frac{x^2}{x^2-1}} + c$$

To express the answer in terms of x you need to link $\sin\theta$ to $\cos\theta$.

WORK IT OUT 11.2

Three students integrate $\cos x \sin x$ in three different ways.

Which is the correct solution? Identify the errors made in the incorrect solutions.

Amara uses reverse chain rule with $u = \sin x$:	Ben uses reverse chain rule with $u = \cos x$:	Carlos uses a double angle formula:
$\frac{du}{dx} = \cos x$, so $\int \cos x \sin x \, dx = \int u \, du$ $= \frac{1}{2}\sin^2 x + c$	$\frac{du}{dx} = -\sin x$, so $\int \cos x \sin x \, dx = \int -u \, du$ $= -\frac{1}{2}\cos^2 x + c$	$\int \cos x \sin x \, dx = \int \frac{1}{2}\sin 2x \, dx$ $= -\frac{1}{4}\cos 2x + c$

EXERCISE 11F

1 Simplify to get a standard integral, and then integrate:

a $\int \frac{\tan 3x}{\cos 3x} \, dx$ b $\int \frac{1}{\sin^2 x} \, dx$ c $\int (\sin 5x \cos x - \cos 5x \sin x) \, dx$

d $\int \frac{3 - \cos 2x}{\sin^2 2x} \, dx$ e $\int \frac{\cos 2x}{\cos x + \sin x} \, dx$

2 Use trigonometric identities before using a substitution (or reversing the chain rule) to integrate:

a $\int \cos^3 x \sin^2 x \, dx$ b $\int \frac{\cos^3 x}{\sin^2 x} \, dx$ c $\int \sin x \cos x \, e^{\cos 2x} \, dx$

d $\int \tan^4 3x + \tan^6 3x \, dx$ e $\int \frac{\sin 2x \cos 2x}{\sqrt{1 + \cos 4x}} \, dx$

3 Find each integral:

a i $\int 2\cos^2 x \, dx$ ii $\int \cos^2 3x \, dx$ b i $\int 2\tan^2\left(\frac{x}{2}\right) dx$ ii $\int \tan^2 3x \, dx$

4 Find the exact value of each of the following.

a i $\int_0^{\pi} \sin^2 2x \, dx$ ii $\int_{\frac{\pi}{4}}^{\pi} (1 + \cos 2x)^2 \, dx$

b i $\int_0^{\frac{\pi}{4}} (\tan x - 1)^2 \, dx$ ii $\int_0^{2\pi} \tan^2\left(\frac{x}{6}\right) dx$

5 Use the given substitution to find each of the following.

a i $\int_0^{\frac{\pi}{6}} \frac{1 + \sin\theta}{\cos\theta} \, d\theta$, $u = \sin\theta$ ii $\int_0^{\frac{\pi}{2}} \frac{\sin 2\theta}{1 + \cos\theta} \, d\theta$, $u = 1 + \cos\theta$

b i $\int_0^{\frac{1}{3}} \sqrt{4 - 9x^2} \, dx$, $x = \frac{2}{3}\sin\theta$ ii $\int_{\frac{1}{4}}^{\frac{1}{2}} \sqrt{1 - 4x^2} \, dx$, $x = \frac{1}{2}\cos\theta$

c i $\int \sqrt{1 - x^2} \, dx$, $x = \cos\theta$ ii $\int \sqrt{36 - x^2} \, dx$, $x = 6\sin\theta$

6 Find $\int \sin^2\left(\frac{x}{3}\right)dx$.

7 a Show that $\tan^3 x = \tan x \sec^2 x - \tan x$.

b Hence find $\int \tan^3 x\, dx$.

8 Given that $\int_0^{\frac{\pi}{12}} \tan^2(kx)\,dx = \frac{4-\pi}{12}$ for some positive k,
find the least value of k that satisfies the equation.

9 a Use the formula for $\cos(A+B)$ to show that $\cos 2x \equiv 2\cos^2 x - 1$.

b Hence find $\int \cos 2x \sin x\, dx$

10 Use the substitution $x = \sin^2 u$ to find $\int \frac{1}{\sqrt{x(1-x)}}dx$.

11 Use the substitution $x = \tan\theta$ to find $\int \frac{1}{1+x^2}dx$

12 a Show that $\sin^3\theta = \sin\theta - \sin\theta\cos^2\theta$.

b Hence find the exact value of $\int_0^{3\pi} \sin^3\left(\frac{x}{3}\right)dx$.

13 The diagram shows a part of the circle with centre at the origin and radius 5.

a Write the equation of the curve in the form $y = f(x)$.

b Use integration to prove that the shaded area is $\frac{25\pi}{4}$.

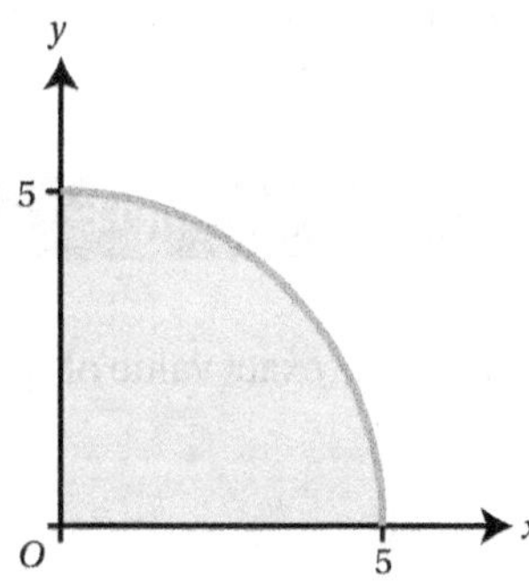

Section 5: Integrating rational functions

You can already use Key point 11.4 to work out integrals such as $\int \frac{x}{x^2-1}dx$ where the derivative of the denominator is a multiple of the numerator. You can do this with a substitution or directly with the reverse chain rule.

If the numerator or the denominator is changed just a little, this may not work any more.

For example, using $u = x^2 - 1$ in $\int \frac{x-2}{x^2-1}dx$ gives $\int \frac{1}{u} \times \frac{x-2}{2x}du$, so the xs do not cancel.

An alternative method is to split the fraction into two. You can do this either by splitting the numerator, or by using partial fractions. You will look at partial fractions first.

Rewind

See Chapter 5, Sections 3 and 4, for a reminder of partial fractions.

WORKED EXAMPLE 11.18

Use partial fractions to find $\int \frac{2}{x^2-1}\,dx$ in the form $\ln(f(x))+c$.

$\frac{2}{x^2-1} = \frac{A}{x-1} + \frac{B}{x+1}$

$\Rightarrow 2 = A(x+1) + B(x-1)$

$\Rightarrow \begin{cases} A+B=0 \\ A-B=2 \end{cases}$ — Comparing coefficients of x^1 and x^0.

$\Rightarrow A=1, B=-1$

$\int \frac{2}{x^2-1}\,dx = \int \left(\frac{1}{x-1} - \frac{1}{x+1}\right) dx$

$= \ln(x-1) - \ln(x+1) + c$

$= \ln\left(\frac{x-1}{x+1}\right) + c$ — Use $\ln a - \ln b = \ln\left(\frac{a}{b}\right)$.

Remember that you can also use partial fractions with a non-linear denominator.

WORKED EXAMPLE 11.19

Find the exact value of $\int_{-1}^{0} \frac{5x+4}{(x-1)(x+2)^2}\,dx$.

$\frac{5x+4}{(x-1)(x+2)^2} = \frac{A}{x-1} + \frac{B}{x+2} + \frac{C}{(x+2)^2}$ — There are three fractions.

$\Rightarrow 5x+4 = A(x+2)^2 + B(x-1)(x+2) + C(x-1)$

$x=1$: $\quad 9 = A(9) \Rightarrow A = 1$

$x=-2$: $\quad -6 = C(-3) \Rightarrow C = 2$

x^2 term: $\quad Ax^2 + Bx^2 = 0 \Rightarrow B = -A = -1$

$\int_{-1}^{0} \left(\frac{1}{x-1} - \frac{1}{x+2} + \frac{2}{(x+2)^2}\right) dx$ — Remember the modulus signs when integrating $\frac{1}{x}$. To integrate $\frac{2}{(x+2)^2}$ write it as $2(x+2)^{-1}$.

$= \left[\ln|x-1| - \ln|x+2| - \frac{2}{x+2}\right]_{-1}^{0}$

$= [\ln 1 - \ln 2 - 1] - [\ln 2 - \ln 1 - 2]$

$= 1 - 2\ln 2$

Before starting the complicated partial fraction process, it is worth checking whether you can simplify the fraction.

WORKED EXAMPLE 11.20

Find $\int \frac{x+4}{12-5x-2x^2}\,\mathrm{d}x$.

$$\int \frac{x+4}{12-5x-2x^2}\,\mathrm{d}x = \int \frac{x+4}{(3-2x)(x+4)}\,\mathrm{d}x$$

It is generally a good idea to check whether polynomials factorise.

$$= \int \frac{1}{3-2x}\,\mathrm{d}x$$

$$= -\frac{1}{2}\ln|3-2x|+c$$

You now have a standard integral, just remember to divide by the coefficient of x.

You will now look at improper fractions.

You can integrate these by splitting them into a polynomial plus a proper fraction, or by splitting a numerator so that one part cancels with the denominator.

Rewind

See Chapter 5, Section 2, for a reminder of division of improper rational functions.

WORKED EXAMPLE 11.21

Find:

a $\int \frac{x^2+5}{x+2}\,\mathrm{d}x$ **b** $\int \frac{x+2}{x-1}\,\mathrm{d}x$.

a $$\frac{x^2+5}{x+2} = x-2+\frac{9}{x+2}$$

Polynomial division gives quotient $x-2$ and remainder 9.

$$\therefore \int \frac{x^2+5}{x+2}\,\mathrm{d}x = \int \left(x-2+\frac{9}{x+2}\right)\mathrm{d}x$$

$$= \frac{1}{2}x^2-2x+9\ln|x+2|+c$$

b $$\frac{x+2}{x-1} = \frac{(x-1)+3}{x-1}$$

$$= 1+\frac{3}{x-1}$$

In this case you can perform polynomial division informally by splitting the numerator in a convenient way.

$$\therefore \int \frac{x+2}{x-1}\,\mathrm{d}x = \int \left(1+\frac{3}{x-1}\right)\mathrm{d}x$$

$$= x+3\ln|x-1|+c$$

EXERCISE 11G

1 State whether each integral can be done by using a substitution (or reversing the chain rule). For those that can, carry out the integration.

a $\int \frac{3x}{x^2-4}\,dx$ b $\int \frac{4x-6}{x^2-3x+1}\,dx$ c $\int \frac{4}{x^2-4}\,dx$ d $\int \frac{5x}{x^2+1}\,dx$

e $\int \frac{4x^2+12x-18}{x^3-9x}\,dx$ f $\int \frac{x^2-3}{x^3-9x}\,dx$ g $\int \frac{12x^2}{x^3+2}\,dx$

2 By first simplifying, find each integral.

a $\int \frac{(4x^2-9)^2}{(2x+3)^2}\,dx$ b $\int \frac{x+3}{6-13x-5x^2}\,dx$ c $\int \frac{x^2}{3x^3-3x^2}\,dx$ d $\int \frac{x^2-3x+2}{x^2-2x}\,dx$

3 Find each integral by splitting into partial fractions.

a i $\int \frac{5x-29}{(x-3)(x-10)}\,dx$ ii $\int \frac{x-7}{(x+1)(x-3)}\,dx$

b i $\int \frac{1}{x^2-1}\,dx$ ii $\int \frac{x}{x^2-1}\,dx$

c i $\int \frac{1-2x}{(x-2)(1-x)}\,dx$ ii $\int \frac{3x-1}{(1-x)(1+x)}\,dx$

d i $\int \frac{3(2x^2+2x+3)}{x^3+3x^2}\,dx$ ii $\int \frac{4x^2-5x+2}{x^3-2x^2}\,dx$

e i $\int \frac{6x+2}{(x-1)^2(x+3)}\,dx$ ii $\int \frac{-(2x+5)}{(x+1)^2(x-2)}\,dx$

4 Find each integral by first performing polynomial division (or splitting the numerator).

a i $\int \frac{x+1}{x+2}\,dx$ ii $\int \frac{2x+3}{x-1}\,dx$ b i $\int \frac{x^2+2}{x-3}\,dx$ ii $\int \frac{x^2+2x-1}{x+5}\,dx$

5 Find the remaining integrals from question **1**.

6 a Write $\frac{5}{x^2+x-6}$ as a sum of partial fractions.

b Hence find $\int \frac{5}{x^2+x-6}\,dx$ giving your answer in the form $\ln(\text{f}(x))+c$.

7 Find the exact value of $\int_0^1 \frac{4}{x^2-4}\,dx$.

8 Find the exact value of $\int_2^5 \frac{x-1}{x+2}\,dx$.

9 a Split $\frac{5-x}{2+x-x^2}$ into partial fractions.

b Given that $\int_0^1 \frac{5-x}{2+x-x^2}\,dx = \ln k$, find the value of k.

10 Find the exact value of $\int_3^4 \frac{8-3x}{x^3-4x^2+4x}\,dx$. Give your answer in the form $\ln p+\frac{1}{q}$, where p and q are rational numbers.

Elevate

See Support Sheet 11 for further examples of integrating rational functions and for more practice questions.

Checklist of learning and understanding

- When using **integration by substitution** remember to:
  - differentiate the substitution and express dx in terms of du
  - simplify the resulting expression
  - find the limits for u.
- Two special cases of integration by substitution should be remembered:
  - $\int f(ax+b)\,dx = \frac{1}{a}F(ax+b)+c$ where $F(x)$ is the integral of $f(x)$
  - $\int \frac{f'(x)}{f(x)}\,dx = \ln|f(x)|+c$
- **Integration by parts** can be used for some products: $\int u\frac{dv}{dx}\,dx = uv - \int \frac{du}{dx}v\,dx$
  - In integrals of the form $\int x^n f(x)\,dx$, take $u = x^n$ unless $f(x) = \ln x$.
  - You may need to do integration by parts more than once.
- Some integrals can be simplified by using a trigonometric identity. The particularly useful ones to remember are:
  - for $\int \sin^2 x\,dx$ and $\int \cos^2 x\,dx$ use a rearrangement of the double angle formula:
    $\cos 2x \equiv 2\cos^2 x - 1 \equiv 1 - 2\sin^2 x$
  - for $\int \sqrt{a^2-x^2}\,dx$ use the substitution $x = a\sin\theta$.
- Some rational fractions can be integrated by splitting them into **partial fractions**. Look out for improper fractions where you may be able to split the numerator

Mixed practice 11

1. Find $\int 6\sec^2 x \tan^2 x \,dx$.

   Choose from these options.

   A $3\tan^2 x + c$ **B** $3\sec^2 x + c$ **C** $2\tan^3 x + c$ **D** $2\sec^3 x + c$

2. Find the exact value of $\int_0^{\pi} \cos^2 3x\,dx$.

3. Use integration by parts to find $\int x\cos 2x\,dx$.

4. Given that $\int_0^m \frac{1}{3x+1}\,dx = 1$ calculate, to three significant figures, the value of m.

5. Find the exact value of $\int_0^{\frac{\pi}{12}} \frac{1}{\cos^2 4x}\,dx$.

6. Find each integral.

   a $\int \frac{1}{1-3x}\,dx$ **b** $\int \frac{1}{(2x+3)^2}\,dx$

7. By using integration by parts, find $\int xe^{6x}\,dx$.

   [© AQA 2012]

8. **a** Given that $y = 4x^3 - 6x + 1$, find $\frac{dy}{dx}$.

   b Hence find $\int_2^3 \frac{2x^2-1}{4x^3-6x+1}\,dx$, giving your answer in the form $p\ln q$, where p and q are rational numbers.

   [© AQA 2012]

9. Find an expression in terms of a and b for $\int_{\ln a}^{\ln b} xe^x\,dx$.

   Choose from these options.

   A $\ln\left(\frac{b^b}{a^a}\right) + a - b$ **B** $b\ln b - a\ln a$ **C** $e^b - e^a$ **D** $be^b - ae^a$

10. Find $\int \ln x\,dx$.

11. **a** Simplify $\frac{e^{-4x}+3e^{-2x}}{e^{-4x}-9}$.

    b Hence find $\int \frac{e^{-4x}+3e^{-2x}}{e^{-4x}-9}\,dx$.

12. Use the substitution $x = 2\sin u$ to find $\int \frac{1}{\sqrt{4-x^2}}\,dx$.

13. **a** Write $\frac{x+5}{(x-1)(x+2)}$ in partial fractions.

    b Hence find, in the form $\ln k$, the exact value of $\int_5^7 \frac{x+5}{(x-1)(x+2)}\,dx$.

14. Find $\int \frac{1}{x\ln x}\,dx$.

15. Using the substitution $u = \frac{1}{2}x - 1$, or otherwise, find $\int \frac{x}{\sqrt{\frac{1}{2}x-1}}\,dx$.

16 **a** Given that $\frac{9x^2-6x+5}{(3x-1)(x-1)}$ can be written in the form $3+\frac{A}{3x-1}+\frac{B}{x-1}$, where A and B are integers, find the values of A and B.

b Hence or otherwise, find $\int \frac{9x^2-6x+5}{(3x-1)(x-1)}\,dx$.

[© AQA 2006]

17 Use the substitution $u = 3 - x^3$ to find the exact value of $\int_0^1 \frac{x^5}{3-x^3}\,dx$.

[© AQA 2014]

18 A particle P moves in a straight line with acceleration given by

$$a = \frac{2t+3}{t+1} \text{ m s}^{-2}$$

When $t = 0$ P is at rest.

Find the distance P travels in the first 3 seconds of its motion.

19 Find $\int \frac{x+3}{(x-2)^2}\,dx$.

20 Given that $\int_{-a}^{a} \frac{2}{1-x^2}\,dx = 2$, find the exact value of a.

21 **a** Use a suitable substitution to find $\int_{-3}^{3} \sqrt{9-x^2}\,dx$.

b The ellipse shown in the diagram has equation $\frac{x^2}{9}+\frac{y^2}{4}=1$.

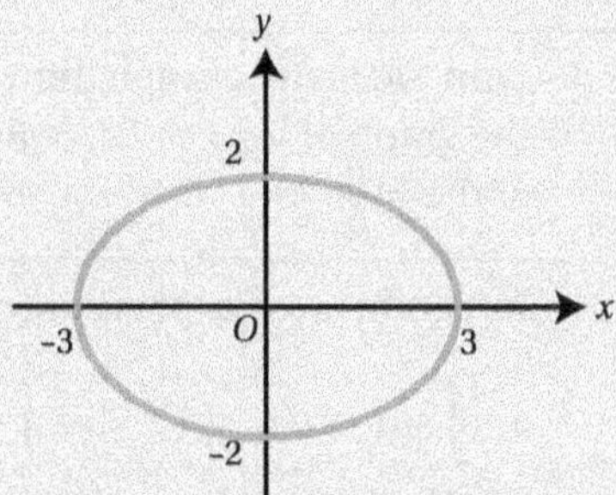

Use integration to prove that the area enclosed by the ellipse is 6π.

22 Let $I = \int \frac{\sin x}{\sin x + \cos x}\,dx$ and $J = \int \frac{\cos x}{\sin x + \cos x}\,dx$.

a Find $I + J$.

b By using a substitution $u = \sin x + \cos x$ find $J - I$.

c Hence find $\int \frac{\sin x}{\sin x + \cos x}\,dx$.

Elevate

See Extension Sheet 11 for a selection of more challenging problems.

12 Further applications of calculus

In this chapter you will learn how to:

- use the second derivative to determine the shape of a curve
- use a parameter to describe curves
- calculate rates of change of related quantities
- find the area between two curves, or between a curve and the y-axis.

Before you start...

Chapter 10	You should be able to find the first and second derivatives of various functions including using the chain, product and quotient rules.	1 Differentiate each function. a $y = e^{2x} \sin 3x$ b $y = \frac{\ln(x^2+1)}{x^2+1}$
Chapter 8	You should be able to use trigonometric identities to simplify expressions and solve equations.	2 Solve the equation $\cos 2x = 3 \sin x$ for $0 \leqslant x \leqslant 2\pi$.
Student Book 1, Chapter 14	You should be able to find the area between a curve and the x-axis.	3 Find the area between the x-axis and the graph of $y = \cos 2x$, between $x = 0$ and $x = \frac{\pi}{6}$.
Chapter 11	You should be able to integrate various functions, use substitution and integration by parts.	4 Integrate these expressions. a $\int \sin^2 x \, dx$ b $\int \frac{x}{x^2-2} dx$ c $\int x \cos 3x \, dx$

More uses of differentiation and integration

In this chapter you will use some of the ideas you have already met:

- from differentiation to establish further properties of curves and to look at rate of change calculations that involve more than one variable
- from integration to calculate more complex areas between curves.

You will be introduced to a new way of describing the equation of a curve, which has applications in mechanics and engineering, for example, in circular motion and projectile motion.

Elevate

See Extension Sheet 12 for some uses of differentiation in economics.

Section 1: Properties of curves

You already know that the first derivative tells you whether a function is increasing or decreasing. But an increasing function can have two different shapes:

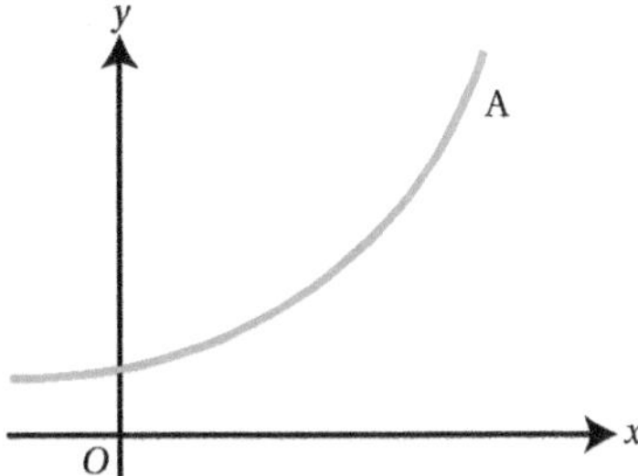

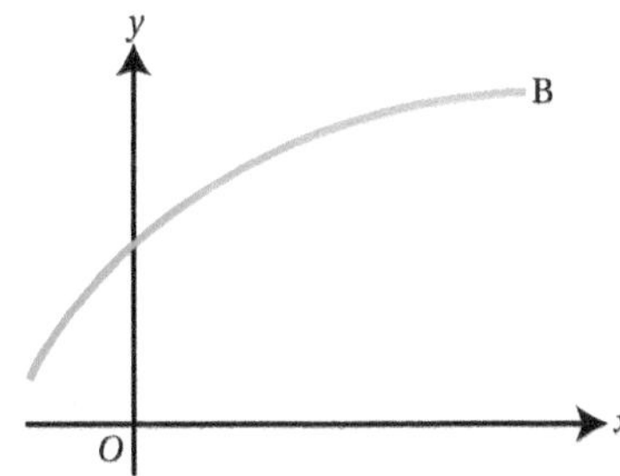

Similarly, a decreasing function can do so in two ways:

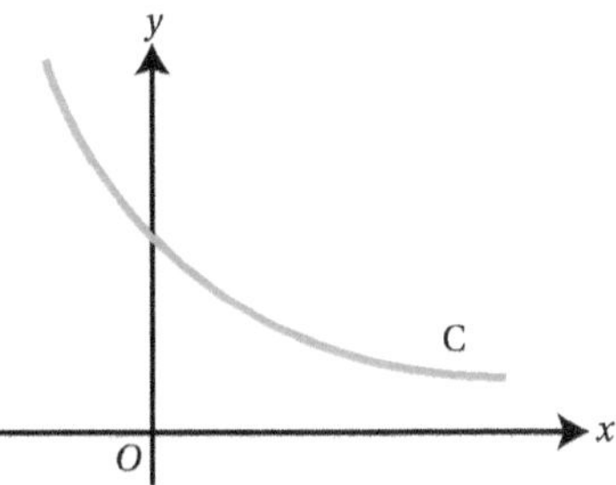

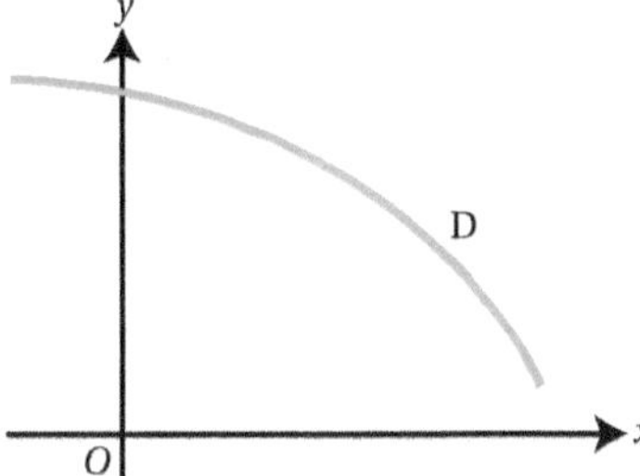

The curves A and C both curve upwards; they are called **convex** curves.

The curves B and D, which both curve downwards, are called **concave** curves.

The two convex curves have something in common: Their gradients are increasing (curve C has negative gradient, which becomes less negative as you move to the right). So the rate of change of gradient is positive. But this rate of change is measured by the second derivative, so

$$\frac{d^2y}{dx^2} > 0$$

for curves A and C. Similarly for curves B and D, the second derivative is negative.

Key point 12.1

A curve that curves upwards is called **convex** and has $\frac{d^2y}{dx^2} > 0$.

A curve that curves downwards is called **concave** and has $\frac{d^2y}{dx^2} < 0$.

WORKED EXAMPLE 12.1

Find the values of x corresponding to the convex sections of the curve $y = x^4 - 6x^3 + 3x - 2$.

$\frac{dy}{dx} = 4x^3 - 18x^2 + 3$

$\frac{d^2y}{dx^2} = 12x^2 - 36x$

Find the second derivative.

$\therefore 12x^2 - 36x > 0$

The curve is convex when $\frac{d^2y}{dx^2} > 0$.

$12x(x-3) > 0$

This is a quadratic inequality, so factorise and sketch the graph.

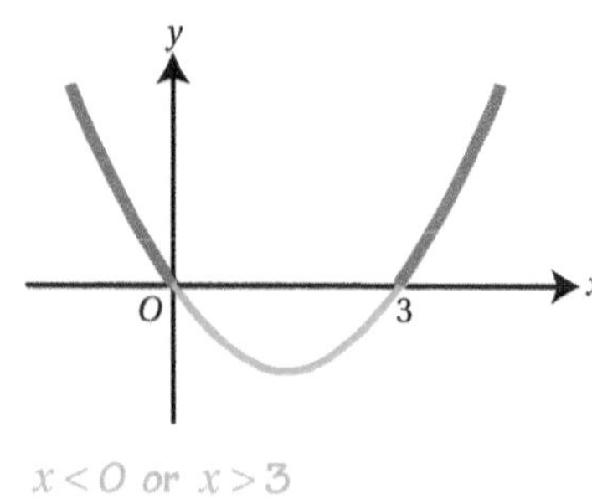

$x < 0$ or $x > 3$

When a curve changes from convex to concave (or vice versa) the second derivative changes sign. So there is a point where $\frac{d^2y}{dx^2} = 0$. This is called a **point of inflection.**

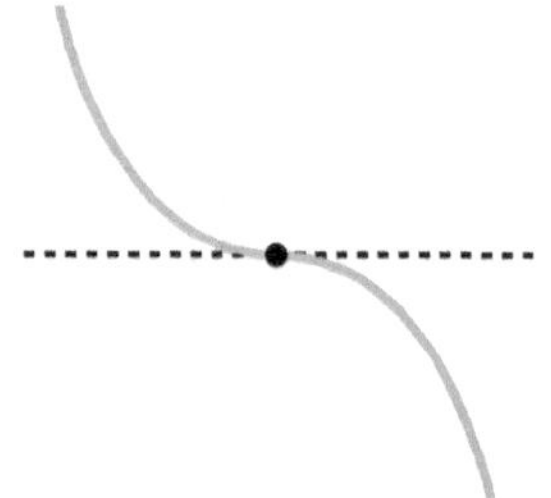

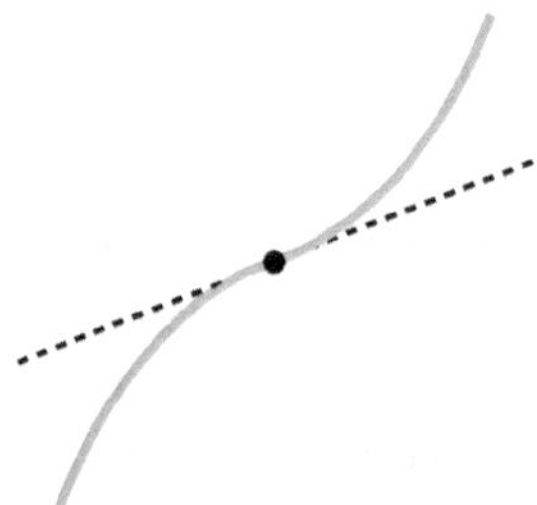

The diagrams show that a point of inflection can be horizontal (so it is a stationary point) or non-horizontal. Notice that if you try to draw a tangent at a point of inflection, it will cross the curve.

Key point 12.2

For a **point of inflection:**

$$\frac{d^2y}{dx^2} = 0$$

WORKED EXAMPLE 12.2

The curve $y = 3\cos x - x$ has two points of inflection for $x \in [0, 2\pi]$. Find their coordinates.

$\frac{dy}{dx} = -3\sin x - 1$ — Find the second derivative.

$\frac{d^2y}{dx^2} = -3\cos x$

$\therefore -3\cos x = 0$ — The points of inflection satisfy $\frac{d^2y}{dx^2} = 0$.

$x = \frac{\pi}{2}, \frac{3\pi}{2}$

The points of inflection are $\left(\frac{\pi}{2}, -\frac{\pi}{2}\right)$ and $\left(\frac{3\pi}{2}, -\frac{3\pi}{2}\right)$. — Remember to find the y-coordinates as well.

WORKED EXAMPLE 12.3

The curve $y = x^3(6 - x)$ has two points of inflection. Find their coordinates and determine whether they are stationary or non-stationary. Hence sketch the curve.

$y = 6x^3 - x^4$ — Rather than using the product rule, expand the brackets before differentiating.

$\frac{dy}{dx} = 18x^2 - 4x^3$

$\frac{d^2y}{dx^2} = 36x - 12x^2$

$\therefore 12x(3 - x) = 0$ — Points of inflection have $\frac{d^2y}{dx^2} = 0$.

$x = 0$ or 3

When $x = 0$, $\frac{dy}{dx} = 0$ and $y = 0$. So $(0, 0)$ is a stationary point of inflection. — Stationary points have $\frac{dy}{dx} = 0$.

When $x = 3$, $\frac{dy}{dx} = 54$ and $y = 81$. So $(3, 81)$ is a non-stationary point of inflection.

x-intercepts: $x^3(6 - x) = 0 \Rightarrow x = 0$ or 6 — To sketch the curve, you also need the x-intercepts and the stationary points.

Stationary points:

$18x^2 - 4x^3 = 2x^2(9 - 2x) = 0 \Rightarrow x = 0$ or $\frac{9}{2}$

When $x = \frac{9}{2}$, $\frac{d^2y}{dx^2} = -81 < 0$, so this is a maximum point. — You already know that $(0, 0)$ is a point of inflection.

Use the second derivative to find out about the other stationary point.

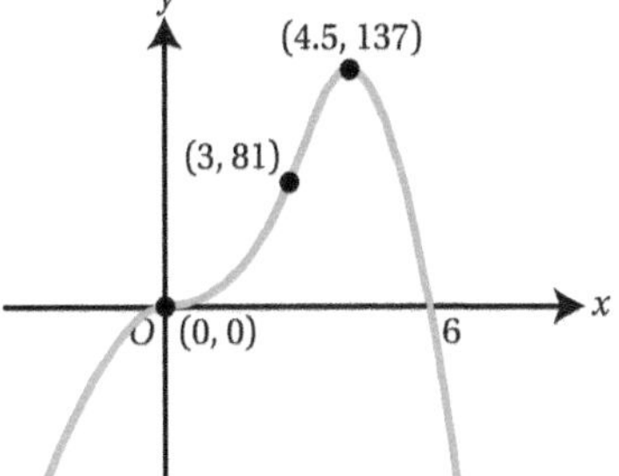

The curve has a point of inflection at the origin and a maximum point. The second point of inflection is between those two, where the curve changes from convex to concave.

Be careful not to assume that when a stationary point has $\frac{d^2y}{dx^2} = 0$, the point is always a point of inflection.

The point could be a local maximum, a local minimum or a point of inflection. To establish which it is, consider the gradient on either side of the stationary point.

> **Common error**
>
> At a point of inflection $\frac{d^2y}{dx^2} = 0$, but just because $\frac{d^2y}{dx^2} = 0$ at a particular point it does not follow that it is a point of inflection.

WORKED EXAMPLE 12.4

a Find the coordinates of the stationary point of $f(x) = x^4$.
b Determine the nature of this stationary point.

a $f'(x) = 4x^3$ — Stationary points have $f'(x) = 0$.

For stationary points $f'(x) = 0$:

$4x^3 = 0$

$x = 0$

$f(0) = 0$ — Find the y-coordinate.

Therefore, the stationary point is $(0, 0)$.

b Find the nature of this point. — Use $f''(x)$ to determine its nature.

$f''(x) = 12x^2$

$f''(0) = 0$ — When $f''(x) = 0$, no conclusion can be made.

So, examine $f'(x)$: — Check the gradient on either side instead.

$f'(-1) = 4(-1)^3 = -4 < 0$

$f'(1) = 4(1)^3 = 4 > 0$

The gradient goes from negative to positive, so this is a minimum point.

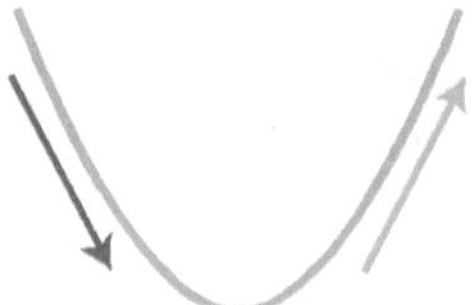

$\therefore (0, 0)$ is a minimum.

WORK IT OUT 12.1

Show that the function $f(x) = x^2 \sin x$ has a stationary point at the origin, and determine its nature.

Which is the correct solution? Identify the errors made in the incorrect solutions.

<table>
<tr><th>Solution 1</th><th>Solution 2</th><th>Solution 3</th></tr>
<tr><td colspan="3">$f'(x) = 2x \sin x + x^2 \cos x$
$f'(0) = 0$, so there is a stationary point at $x = 0$.
$f''(x) = 2 \sin x + 2x \cos x + 2x \cos x - x^2 \sin x$
$= 2 \sin x + 4x \cos x - x^2 \sin x$</td></tr>
<tr><td>$f''(0) = 0$

So this is a point of inflection.</td><td>$f''(0) = 0$

Test the derivative on either side:
$f'(-1) = 2.2 > 0$
$f'(1) = 2.2 > 0$

The curve goes from increasing to increasing, so this is a point of inflection.</td><td>$f''(0) = 0$

Test the second derivative on either side:
$f''(-1) = -3.0 < 0$
$f''(1) = 3.0 > 0$

The curve changes from concave to convex, so this is a point of inflection.</td></tr>
</table>

EXERCISE 12A

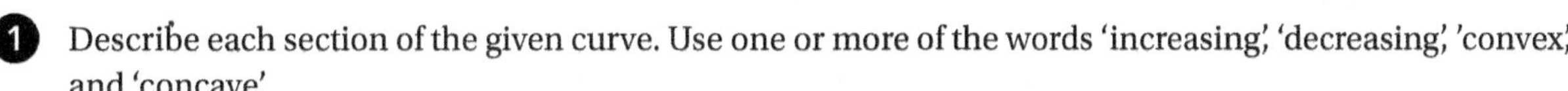

1 Describe each section of the given curve. Use one or more of the words 'increasing', 'decreasing', 'convex', and 'concave'.

a

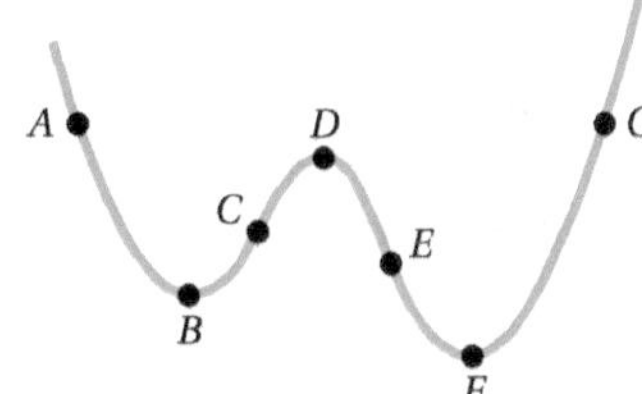

i A to B

ii C to E

iii F to G

b

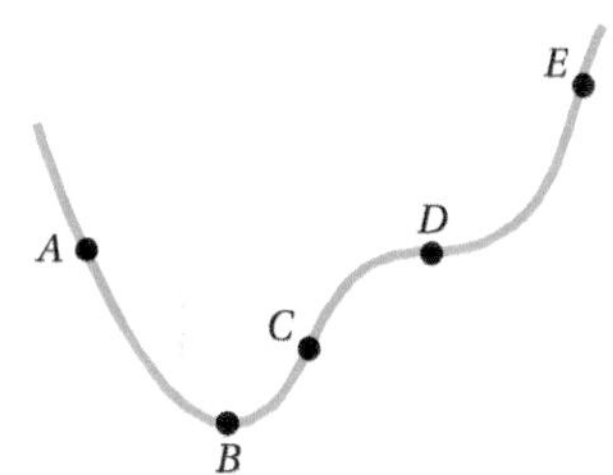

i B to D

ii C to D

iii D to E

2 Mark these points on each curve.

A, local maximum point B, stationary point of inflection C, non-stationary point of inflection

a

b

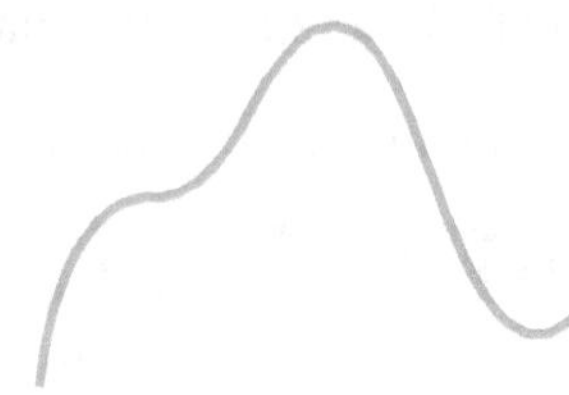

3 Mark one or more points on the curve $y = f(x)$ that satisfy the given conditions.

A: $f'(x) = 0$ and $f''(x) > 0$

B: $f'(x) < 0$ and $f''(x) = 0$

C: $f'(x) = 0$ and $f''(x) = 0$

D: $f'(x) = 0$ and $f''(x) < 0$

E: $f'(x) > 0$ and $f''(x) = 0$

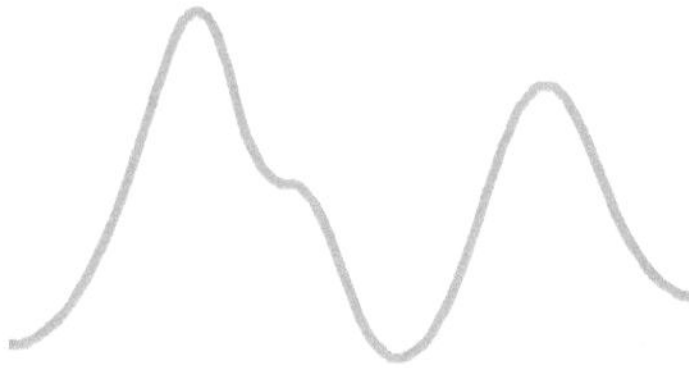

4 Find the coordinates of the point of inflection on the curve $y = e^x - x^2$.

5 The curve $y = x^4 - 6x^2 + 7x + 2$ has two points of inflection. Find their coordinates.

6 Show that all points of inflection on the curve $y = \sin x$ lie on the x-axis.

7 Find the set of values of x corresponding to the convex sections of the curve $y = x^4 + 2x^3 - 36x^2 + 5$.

8 Find the coordinates of the points of inflection on the curve $y = 2\cos x + x$ for $0 \leqslant x \leqslant 2\pi$.
Justify carefully that these points are points of inflection.

9 Show that the graph of $y = \tan^3 x$ has only one stationary point for $x \in \left(-\frac{\pi}{2}, \frac{\pi}{2}\right)$, and that this is a point of inflection.

10 Find the x-coordinates of the stationary points on the graph of $y = x^5 - 5x^4$ and determine their nature.

11 The curve $y = x^3 + ax^2 + bx + c$ has a stationary point of inflection. Show that $a^2 = 3b$.

12 The curve $y = 4x + x^2 - ax^3$ is concave for $x < 3$. Find the value of a.

13 Find the set of values of x corresponding to the concave section of the curve $y = x^2e^{-x}$.

14 The graph shows $y = f'(x)$.

On a copy of the diagram:

a mark points corresponding to a local minimum of $f(x)$ as A

b mark points corresponding to a local maximum of $f(x)$ as B

c mark points corresponding to a point of inflection of $f(x)$ as C.

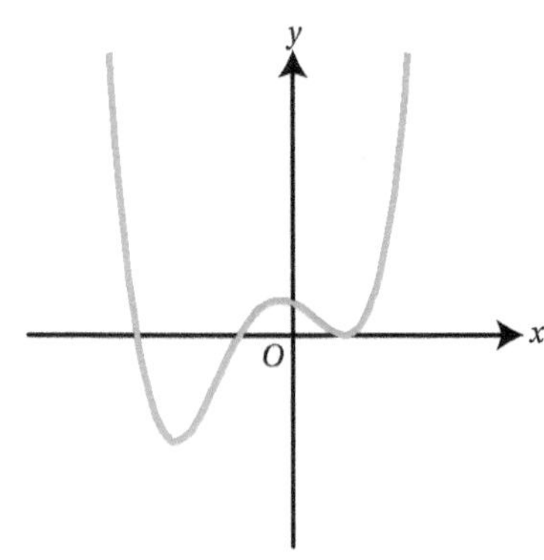

Section 2: Parametric equations

Rewind

For a reminder of implicit functions and how to differentiate them, see Chapter 10, Section 4.

You should by now be very familiar with the idea of using equations to represent graphs and using differentiation to find their gradients.

You also know that some curves cannot be represented by equations of the form $y = f(x)$. For example, a circle with radius 5 centred at the origin has an implicit equation $x^2 + y^2 = 25$.

But there is another way to describe points on this circle, using the angle from the horizontal. Let P be a point on the circle and let θ be the angle OP makes with the x-axis.

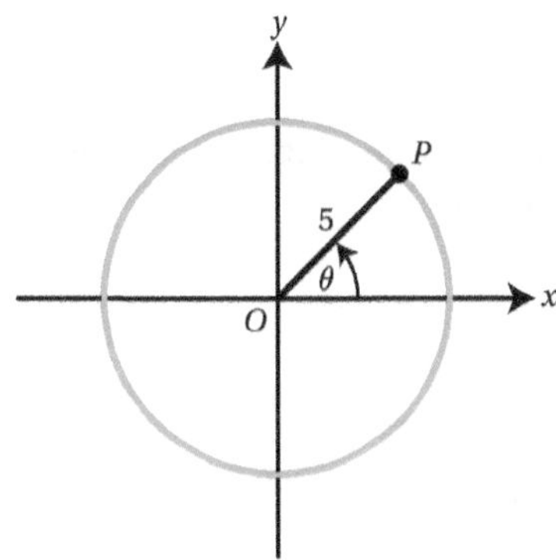

Then the coordinates of P are $x = 5\cos\theta,\ y = 5\sin\theta$.

Both x and y have been expressed in terms of a third variable, θ.

These are called **parametric equations** and θ is called the **parameter**.

You can check that these coordinates satisfy the original equation of the circle:

$$\begin{aligned} x^2 + y^2 &= 25\cos^2\theta + 25\sin^2\theta \\ &= 25(\sin^2\theta + \cos^2\theta) \\ \Rightarrow x^2 + y^2 &= 25 \end{aligned}$$

The last equation, involving just x and y, is called a **Cartesian equation**.

Each value of the parameter corresponds to a single point on the curve.

Worked example 12.5 illustrates how to use parametric equations.

WORKED EXAMPLE 12.5

A curve has parametric equations $x = 3t^2 - 9,\ y = t^3 - 3t$.

a By filling in the table, mark the points corresponding to the given parameter values. Hence sketch the curve.

Point	A	B	C	D	E
t	−2	−1	0	1	2
x					
y					

b Find the parameter value(s) at the points:
i (21.72, 23.168) **ii** (0, 0).

a

Point	A	B	C	D	E
t	−2	−1	0	1	2
x	3	−6	−9	−6	3
y	−2	2	0	−2	2

Use the equations to find x and y for each value of t.

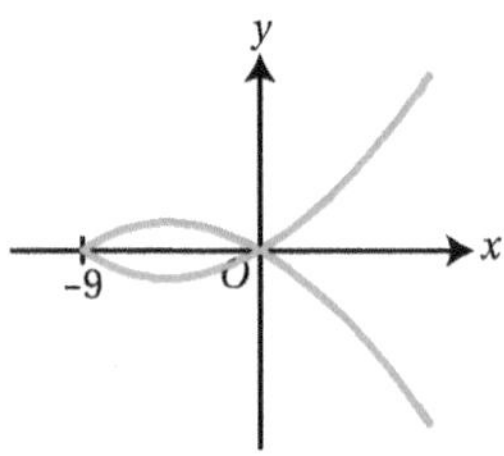

Sketch the curve.

b i $3t^2 - 9 = 21.72$

$t^2 = 10.24$

$t = \pm 3.2$

Substitute $x = 21.72$ into the equation for x and solve.

When $t = -3.2,\ y = -23.168$

When $t = 3.2,\ y = 23.168$

So the value of t is 3.2.

There are two possible values of x, so you need to check which one gives the correct value of y.

ii $3t^2 - 9 = 0$

$t^2 = 3$

$t = \pm\sqrt{3}$

Substitute $x = 0$ into the equation for x and solve.

When $t = -\sqrt{3},\ y = -3\sqrt{3} - 3(-\sqrt{3}) = 0$

When $t = \sqrt{3},\ y = 3\sqrt{3} - 3\sqrt{3} = 0$

So the possible values of t are $\sqrt{3}$ or $-\sqrt{3}$.

Although each parameter value gives a single point, there can be several parameter values giving the same point.

Notice that $y = 0$ also when $t = 0$, but $x \neq 0$ there.

You can sometimes convert parametric equations into a Cartesian equation by eliminating the parameter. The two most common ways of doing this are either by substitution, or by using a trigonometric identity.

Common error

When you are given x- and y-coordinates and want to find t at that point, don't assume that the first value of t you get is correct – check it works in both the x and y equation.

WORKED EXAMPLE 12.6

Find the Cartesian equation of each curve: **a** $x = t^2 - 1,\ y = t + 2$ **b** $x = \tan\theta,\ y = 2\cos\theta$.

a $y = t + 2 \Rightarrow t = y - 2$

$x = (y-2)^2 - 1$

$\Rightarrow x = y^2 - 4y + 3$

Make t the subject of the y-equation and substitute it into the x-equation.

b $\cos\theta = \frac{y}{2} \Rightarrow \sec\theta = \frac{2}{y}$

There is a trigonometric identity relating tan and sec, so write $\sec\theta$ in terms of y.

$\sec^2\theta - \tan^2\theta = 1$

$\left(\frac{2}{y}\right)^2 - x^2 = 1$

$\Rightarrow \frac{4}{y^2} - x^2 = 1$

The Cartesian equation does not need to be in the form $y = f(x)$; any equation containing just x and y is fine.

In the first example of parametric equations of a circle, the parameter represented the angle between the radius and the x-axis. In other contexts the parameter can represent distance along the curve, or time taken to travel to that point; sometimes the parameter has no obvious meaning, but in all cases it determines where you are on the curve.

Fast forward

Kinematics in two dimensions and projectiles are covered in more detail in Chapter 17.

One important application of parametric equations is in studying motion in two dimensions.

WORKED EXAMPLE 12.7

A ball is thrown upwards at an angle, from the point (0, 0). The x-axis represents ground level. The coordinates of the ball are given by $x = 3.2t,\ y = 10.6t - 4.9t^2$, where t represents the time in seconds.

a Find the coordinates of the ball 2 seconds after projection.

b Find the two times when the ball is 2.7 m above ground.

c Show that the path of the ball is a parabola, and find its Cartesian equation, giving the coefficients to 2 significant figures.

a When $t = 2$:

$x = 3.2 \times 2 = 6.4$ and $y = 10.6 \times 2 - 4.9 \times 4 = 1.6$

The coordinates are (6.4, 1.6).

Substitute in $t = 2$.

b $10.6t - 4.9t^2 = 2.7$

$4.9t^2 - 10.6t + 2.7 = 0$

$t = 0.295$ and 1.87

The two times are 0.295 s and 1.87 s.

The height is measured by the y-coordinate.

c $x = 3.2t \Rightarrow t = \frac{x}{3.2}$

$y = 10.6\left(\frac{x}{3.2}\right) - 4.9\left(\frac{x}{3.2}\right)^2 \Rightarrow y = 3.3x - 0.48x^2$

As the equation is quadratic, the path of the ball is a parabola.

Rearrange the first equation to make t the subject and substitute into the second.

EXERCISE 12B

1 By creating a table of values sketch the curves given by these parametric equations.

a i $x=2t^2,\ y=3t$ for $t\in[-2,\ 2]$ ii $x=t^2,\ y=5t$ for $t\in[-4,\ 4]$

b i $x=3\sin t,\ y=5\cos t$ for $t\in[0,\ 2\pi]$ ii $x=4\cos t,\ y=2\sin t$ for $t\in[0,\ 2\pi]$

c i $x=\sin 2t,\ y=\sin t$ for $t\in[0,2\pi]$ ii $x=\cos t,\ y=\sin 2t$ for $t\in[0,\ 2\pi]$

2 Find the parameter value at the given point on each curve.

a i $x=3t^2,\ y=2t$; point $(3,\ -2)$ ii $x=5t^2,\ y=2-t$; point $(20,\ 0)$

b i $x=2t^2,\ y=t^3$; point $(8,\ -8)$ ii $x=2t^3,\ y=5t^2$; point $(-54,\ 45)$

c i $x=5\sin\theta,\ y=5\cos\theta$; point $(3,\ -4)$ ii $x=2\cos\theta,\ y=6\sin\theta$; point $\left(1,\ 3\sqrt{3}\right)$

3 Find a Cartesian equation for each parametric curve.

a i $x=3t^2,\ y=2t$ ii $x=5t^2,\ y=2-t$

b i $x=2t^2,\ y=t^3$ ii $x=2t^3,\ y=5t^2$

c i $x=5\sin\theta,\ y=5\cos\theta$ ii $x=2\cos\theta,\ y=6\sin\theta$

d i $x=2\cos 2\theta,\ y=\cos\theta$ ii $x=\cos 2\theta,\ y=3\sin\theta$

e i $x=\tan\theta,\ y=\sec\theta$ ii $x=3\sec\theta,\ y=2\tan\theta$

4 A ball is projected from point O on the horizontal ground and moves in the vertical plane. The ball follows a parabolic path with parametric equations $x=3.5t,\ y=12t-4.9t^2$, where t is the time after the projection. The units on the x- and y-axes are metres, and time is measured in seconds.

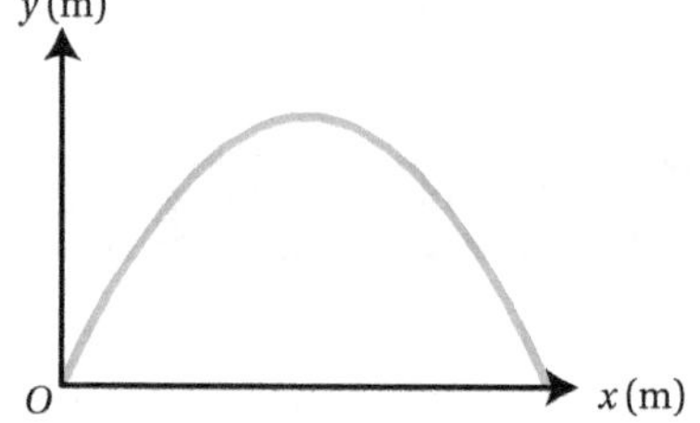

a Find the height of the ball above ground 2 seconds after projection.

b Find the distance from O of the point where the ball hits the ground.

5 A straight line has parametric equations $x=1+\frac{\sqrt{3}}{2}t,\ y=2+\frac{1}{2}t$.

a Find the y-intercept of the line.

b Find the distance between the points on the line with $t=4$ and $t=6$. What does the parameter t represent?

c Find the Cartesian equation of the line in the form $ax+by=c$.

Differentiating parametric equations

When a curve is given parametrically you can find its gradient, $\frac{dy}{dx}$, by using the chain rule to take account of the parameter t.

> **Tip**
>
> To differentiate when there are three (or more) variables involved, always use the chain rule. You will see this in the next section as well with connected rates of change.

Key point 12.3

$$\frac{dy}{dx}=\frac{dy}{dt}\times\frac{dt}{dx}$$

WORKED EXAMPLE 12.8

A curve has parametric equations $x = \tan t,\ y = \sin 2t$. Find the gradient of the curve:

a at the point where $t = \frac{\pi}{4}$

b at the point $\left(\sqrt{3}, \frac{\sqrt{3}}{2}\right)$.

a $\frac{dy}{dt} = 2\cos 2t$

$\frac{dx}{dt} = \sec^2 t$

Find $\frac{dy}{dt}$ and $\frac{dx}{dt}$.

$\frac{dy}{dx} = \frac{dy}{dt} \times \frac{dt}{dx}$

$= 2\cos 2t \times \frac{1}{\sec^2 t}$

$= 2\cos 2t \times \cos^2 t$

$= 2\cos\frac{\pi}{2} \times \cos^2\frac{\pi}{4} = 0$

Use the chain rule. Remember that $\frac{dt}{dx} = \frac{1}{\frac{dx}{dt}}$

b To find t:

$\tan t = \sqrt{3} \Rightarrow t = \frac{\pi}{3}, \frac{4\pi}{3}, \ldots$

$\sin 2t = \frac{\sqrt{3}}{2} \Rightarrow 2t = \frac{\pi}{3}, \frac{2\pi}{3}, \ldots$

$t = \frac{\pi}{6}, \frac{\pi}{3}, \ldots$

The formula for the gradient is in terms of t, so you need to find the value of t first.

$\therefore t = \frac{\pi}{3}$ at this point.

You need the value of t that satisfies both equations.

$\frac{dy}{dx} = 2\cos\frac{2\pi}{3} \times \cos^2\frac{\pi}{3} = -\frac{1}{4}$

Use the equation from part **a**.

WORK IT OUT 12.2

Find the equation of the tangent to the curve with parametric equations $x = 2t^2,\ y = t^3$ at the point $(18, -27)$. Which is the correct solution? Identify the errors made in the incorrect solutions.

Solution 1	Solution 2	Solution 3
$\frac{dy}{dx} = \frac{3t^2}{4t} = \frac{3}{4}t$ Equation of the tangent: $y+27 = \frac{3}{4}t(x-18)$ $\Leftrightarrow y = \frac{3}{4}tx - \frac{27}{2}t - 54$	$\frac{dy}{dx} = \frac{3t^2}{4t} = \frac{3}{4}t$ When $x = 18,\ t^2 = 9 \Rightarrow t = \pm 3$ But $y = -27$, so $t = -3$ When $t = -3$, $\frac{dy}{dx} = -\frac{9}{4}$ Tangent: $y+27 = -\frac{9}{4}(x-18)$ $\Leftrightarrow 9x + 4y = 54$	$\frac{dy}{dx} = \frac{3t^2}{4t} = \frac{3t}{4}$ When $x = 18,\ t^2 = 9,\ t = 3$ Then $\frac{dy}{dx} = \frac{9}{4}$ Tangent: $y+27 = \frac{9}{4}(x-18)$ $\Leftrightarrow 9x + 4y = 270$

Curves given by parametric equations can have both horizontal and vertical tangents. You can find these by looking at the values of $\frac{dx}{dt}$ and $\frac{dy}{dt}$.

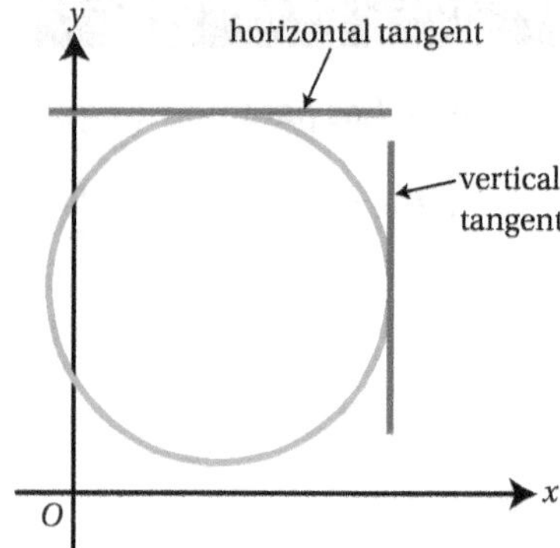

Key point 12.4

At a point where the tangent is parallel to the x-axis, $\frac{dy}{dt} = 0$.

At a point where the tangent is parallel to the y-axis, $\frac{dx}{dt} = 0$.

WORKED EXAMPLE 12.9

A curve has parametric equations $x = 2\sin t$, $y = e^{-2t}$ for $0 \leqslant t \leqslant \pi$.

a Find the equation of the tangent parallel to the y-axis.
b Show that the curve has no tangents parallel to the x-axis.

a $\frac{dx}{dt} = 0$:

If a tangent is parallel to the y-axis then $\frac{dx}{dt} = 0$.

$2\cos t = 0 \Rightarrow t = \frac{\pi}{2}$

$\therefore x = 2\sin\frac{\pi}{2} = 2$

This tangent is vertical, so it is of the form $x = k$.

The equation of the tangent is $x = 2$.

b $\frac{dy}{dt} = -2e^{-2t}$

If a tangent is parallel to the x-axis then $\frac{dy}{dt} = 0$.

But $-2e^{-2t} < 0$ for all t.

The exponential function is always positive.

Hence $\frac{dy}{dt} \neq 0$, so there are no horizontal tangents.

You an use parametric equations to describe some interesting curves and to prove properties of their tangents and normals. In many of these proofs you need to work with the general value of the parameter.

Did you know?

The curve in Worked example 12.10 is called an astroid. A ladder sliding down a wall is always tangent to this curve.

WORKED EXAMPLE 12.10

A curve has parametric equations $x = 5\cos^3 t,\ y = 5\sin^3 t$, for $0 \leqslant t < 2\pi$. P is a point of the curve.

a Find, in terms of t, the equation of the tangent to the curve at P.

The tangent at a point P meets the x-axis at A and the y-axis at B.

b Prove that the length of AB does not depend on the position of P on the curve.

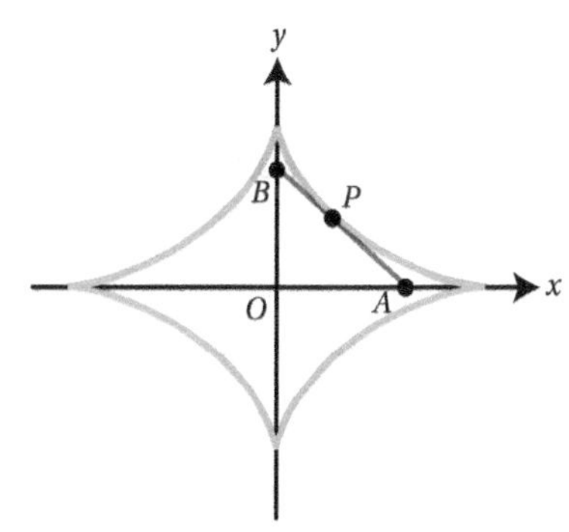

a $\frac{dy}{dt} = 15\sin^2 t\cos t,$

$\frac{dx}{dt} = -15\cos^2 t\sin t$

First find $\frac{dy}{dt}$ and $\frac{dx}{dt}$. Treat $5\cos^3 t$ as $5(\cos t)^3$ and use the chain rule, and likewise for $5\sin^3 t$.

$\frac{dy}{dx} = \frac{dy}{dt} \times \frac{dt}{dx}$

$= 15\sin^2 t\cos t \times \frac{1}{-15\cos^2 t\sin t}$

$= \frac{15\sin^2 t\cos t}{-15\cos^2 t\sin t}$

$= -\tan t$

Then use the chain rule to find $\frac{dy}{dx}$.

Equation of the tangent:

$y - 5\sin^3 t = -\tan t\,(x - 5\cos^3 t)$

The equation of the tangent is $y - y_1 = m(x - x_1)$.

b At A, $y = 0$:

Set $y = 0$ and $x = 0$ to find axis intercepts.

$-5\sin^3 t = -\tan t\,(x - 5\cos^3 t)$

$5\sin^2 t\cos t = x - 5\cos^3 t$

Divide both sides by $-\tan t$.

$x = 5\sin^2 t\cos t + 5\cos^3 t$

$= 5\cos t\,(\sin^2 t + \cos^2 t)$

$= 5\cos t$

Factorise and use $\sin^2 t + \cos^2 t = 1$.

At B, $x = 0$:

$y - 5\sin^3 t = -\tan t(-5\cos^3 t)$

$y = 5\sin t\cos^2 t + 5\sin^3 t$

$= 5\sin t(\cos^2 t + \sin^2 t)$

$= 5\sin t$

The distance between $A(5\cos t, 0)$ and $B(0, 5\sin t)$ is:

Now you have the coordinates of A and B so you can find the distance between them.

$\sqrt{(5\cos t)^2 + (5\sin t)^2} = \sqrt{25} = 5$

This length does not depend on t, so it is the same for every position of P.

EXERCISE 12C

1 Find an expression for $\frac{dy}{dx}$ in terms of the parameter (t or θ) for these parametric curves.

a i $x = 3t^2,\ y = 2t$ ii $x = 5t^2,\ y = 2 - t$

b i $x = 2\cos 2\theta,\ y = \cos\theta$ ii $x = \cos 2\theta,\ y = 3\sin\theta$

c i $x = \tan\theta,\ y = \sec\theta$ ii $x = 3\sec\theta,\ y = 2\tan\theta$

2 Find the gradient of each curve at the given point. (You need to find the parameter value first.)

a i $x = 3t^2,\ y = 2t$; point $(3, -2)$ ii $x = 5t^2,\ y = 2 - t$; point $(20, 0)$

b i $x = 2t^2,\ y = t^3$; point $(8, -8)$ ii $x = 2t^3,\ y = 5t^2$; point $(-54, 45)$

c i $x = 5\sin\theta,\ y = 5\cos\theta$; point $(3, -4)$ ii $x = 2\cos\theta,\ y = 6\sin\theta$; point $\left(1, 3\sqrt{3}\right)$

3 A curve has parametric equations $x = 3t^2,\ y = t - 2$.
Find the equation of the normal to the curve at the point where $t = 1$.

4 A curve has parametric equations $x = t + \sin t,\ y = \cos t$.
Find the equation of the tangent to the curve at the point where $t = \pi$.

Elevate

See Support Sheet 12 for a further example of finding tangents and normals with parametric equations and for more practice questions.

5 The tangent to the curve with parametric equations $x = t^2,\ y = e^{-t}$ at the point $\left(1, \frac{1}{e}\right)$ crosses the coordinate axes at the points M and N.
Find the exact area of triangle OMN.

6 a Find the equation of the normal to the curve with equation $x = 4t,\ y = \frac{4}{t}$ at the point $(16, 1)$.

b Find the coordinates of the point where the normal crosses the curve again.

7 A parabola has parametric equations $x = 3t^2,\ y = 6t$.
The normal to the parabola at the point where $t = -1$ crosses the parabola again at the point Q. Find the coordinates of Q.

8 Prove that the curve with parametric equations $x = t^3 - 3t^2 + 7t,\ y = 5t^2 + 1$ has no tangents parallel to the y-axis.

9 Let P be the point on the curve $x = t^2,\ y = \frac{1}{t}$ with coordinates $\left(p^2, \frac{1}{p}\right)$.

The tangent to the curve at P meets the x-axis at point A and the y-axis at point B. Prove that $AP = 2BP$.

10 Point $Q\,(aq^2, 2aq)$ lies on the parabola with parametric equations $x = at^2,\ y = 2at$.

a Let M be the point where the normal to the parabola at Q crosses the x-axis.
Find, in terms of a and q, the coordinates of M.

b N is the perpendicular from Q to the x-axis. Prove that the distance $MN = 2a$.

Integrating parametric equations

When the equation of a curve is given in a parametric form, you can still use integration to find the area between the curve and the x-axis.

Key point 12.5

The area under the curve $(x(t),\ y(t))$, between points $x = a$ and $x = b$, is given by:

$$\int_{t_1}^{t_2} y \frac{\mathrm{d}x}{\mathrm{d}t}\,\mathrm{d}t$$

where t_1 and t_2 are the parameter values at the points on the curve where $x = a$ and $x = b$.

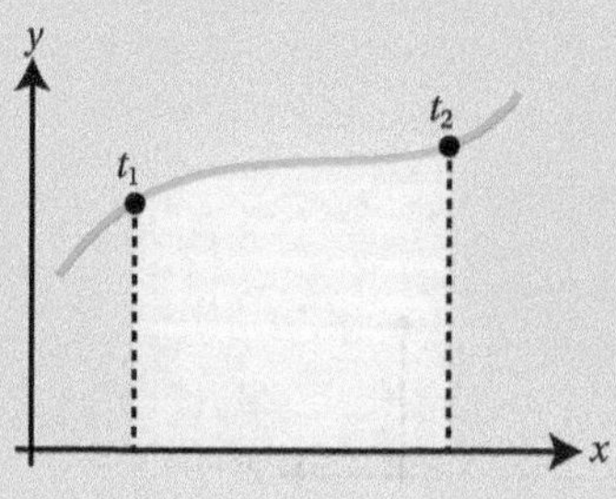

Rewind

This is very much like the idea used in integration by substitution, in Chapter 11, Section 2, to change from x to u.

WORKED EXAMPLE 12.11

The diagram shows a part of the curve with parametric equations $x = t^2,\ y = 3 - t$. Points A and B have x-coordinates 0 and 9.

a Find the values of the parameter at A and B.

b Find the area bounded by the curve, the x-axis and the y-axis.

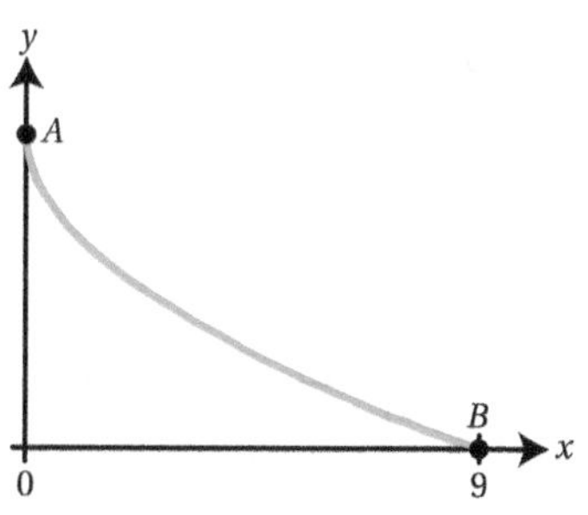

a At A: $x = 0 \Rightarrow t = 0$

At B: $y = 0 \Rightarrow t = 3$

Notice that there are two possible values of t that give $x = 9$, but only one of them is consistent with $y = 0$.

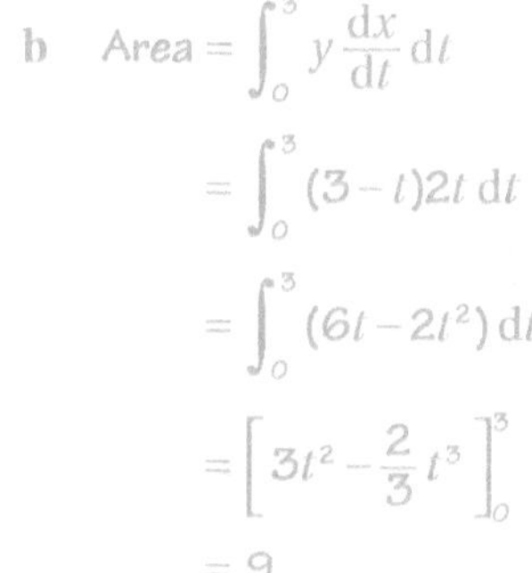

Use the formula for the parametric integration, with the parameter values at A and B.

EXERCISE 12D

1 Find the area of the shaded region for each curve, with the given parametric equations.

a **i** $x = t^2 + 2,\ y = t + 3$ 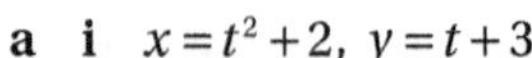

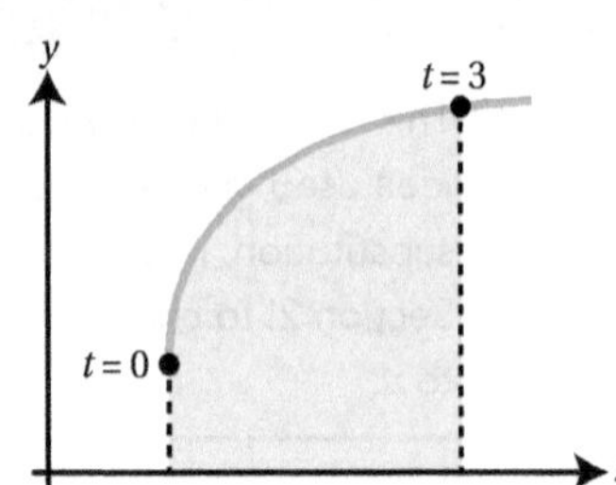

ii $x = t^2,\ y = t + 5$

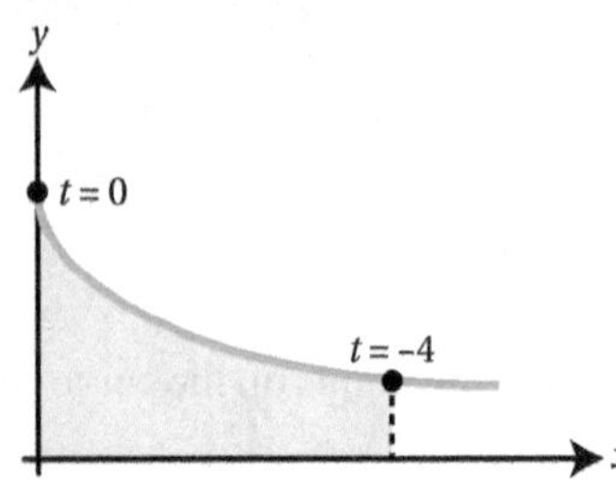

b **i** $x = 1 + t,\ y = 1 - \frac{1}{t}$

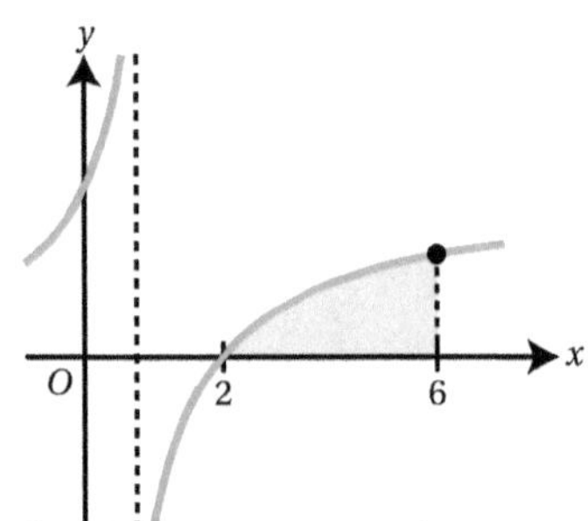

ii $x = 2t^2,\ y = 1 - \frac{1}{t^2}$

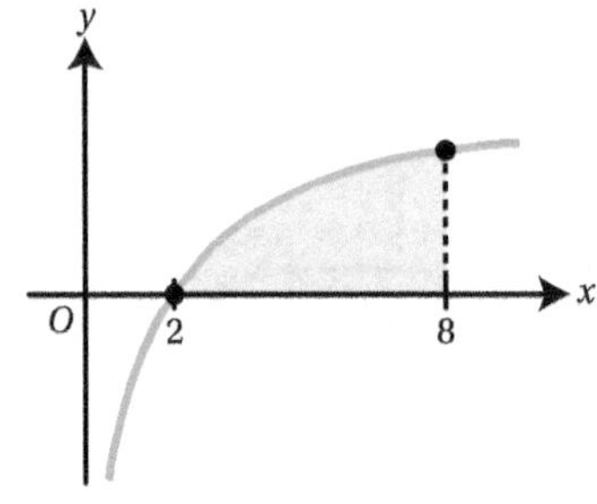

c **i** $x = 3t,\ y = \sin t$

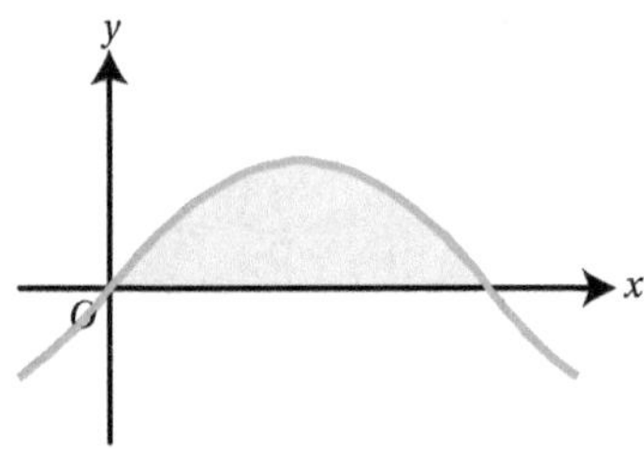

ii $x = 4t,\ y = \sec^2 t - 2$

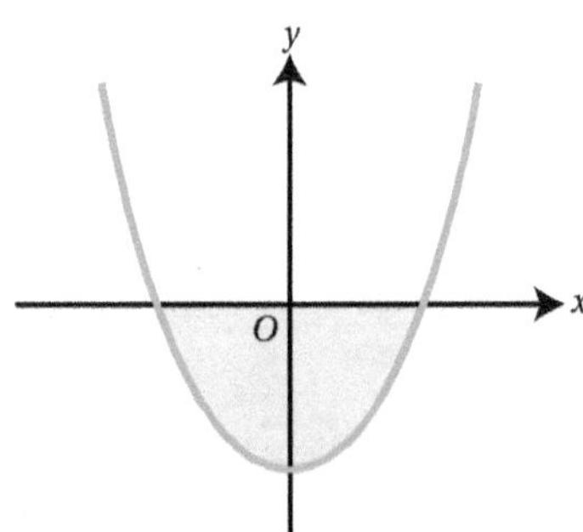

2 **a** Find the coordinates of the points where the curve with parametric equations $x = (t + 2)^2,\ y = t^2 - t - 2$ crosses the x-axis.

b Find the area enclosed by the curve and the x-axis.

3 A curve is defined by the parametric equations $x = t + e^{-t},\ y = 1 - e^{-t}$.

a Find the exact values of t at the points where $y = 0$ and $y = \frac{1}{2}$.

b Find the area bounded by the curve, the x-axis and the line $x = \frac{1}{2} + \ln 2$.

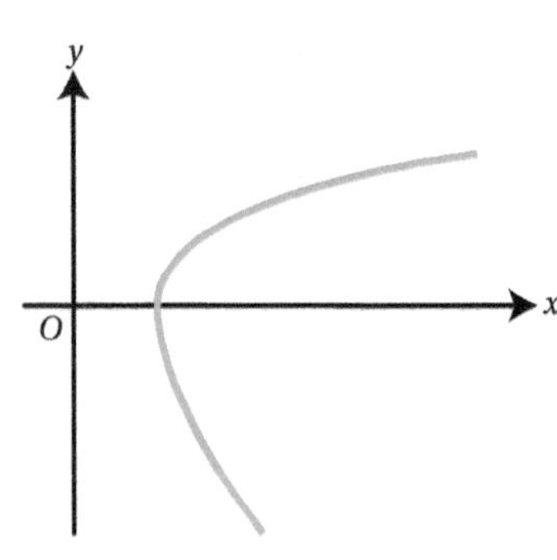

4 The graph shows the curve with parametric equations:

$x = t - \sin t,\ y = 1 - \cos t$ for $0 \leqslant t \leqslant 2\pi$.

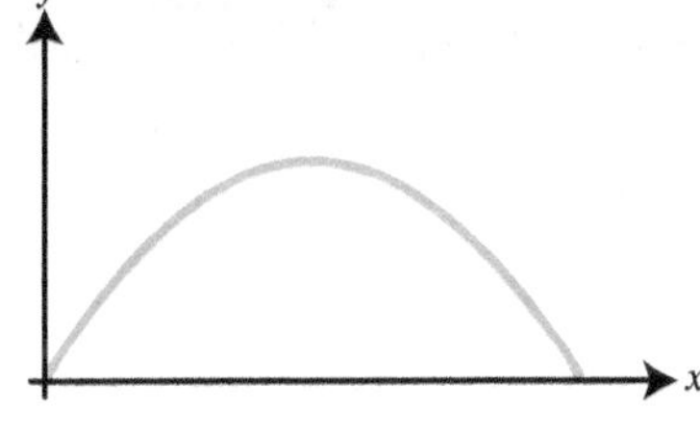

a Show that the shaded area is given by:

$$A = \int_0^{2\pi} (1 - \cos t)^2 \, dt$$

b Hence find the exact value of A.

5 Point $P\left(\frac{10}{3}, \frac{8}{3}\right)$ lies on the curve with parametric equations

$x = t + \frac{1}{t},\ y = t - \frac{1}{t}.$

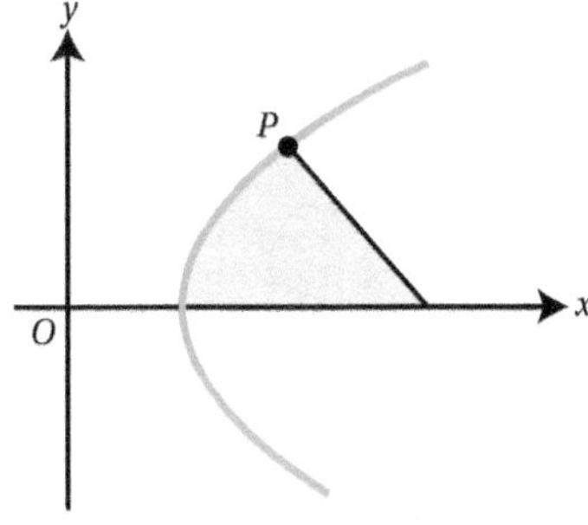

a Find the parameter value at P.

b The diagram shows a part of the curve and the normal at P. Find the area of the shaded region.

c Find the Cartesian equation of the curve.

Section 3: Connected rates of change

So far you have only looked at functions of one variable; for example, you can use differentiation to find how quickly the velocity of an object changes with time.

In this section, you will use the chain rule to solve problems where there are several related variables.

Consider, for example, an inflating balloon. You can control the amount (V) of gas in the balloon, but you may want to know how fast the radius (r) is increasing. These are two different rates of change, but intuitively they are linked – the faster the gas fills the balloon the faster the radius will increase.

The two derivatives need to be linked: $\frac{dV}{dt}$ and $\frac{dr}{dt}$. You can do this by using the chain rule and the geometric context.

Rewind

You met this idea of writing rates of change in terms of derivatives in Student Book 1, Chapter 12.

Key point 12.6

To link variables A and B via their rates of change, use the chain rule:

$$\frac{dA}{dt} = \frac{dA}{dB} \times \frac{dB}{dt}$$

Tip

In these practical applications, the rate of change is nearly always with respect to time, t. For example the rate of change of volume, V, is $\frac{dV}{dt}$.

WORKED EXAMPLE 12.12

A spherical balloon is being inflated with air at a rate of 200 cm^3 per minute. At what rate is the radius increasing when the radius is 8 cm?

V = volume of air in balloon in cm^3
r = radius of balloon in cm
t = time in minutes

When solving problems in context you should always start by defining the variables.

$\frac{dV}{dt} = 200\ cm^3$ per minute

The balloon being inflated at a rate of 200 cm^3 per minute means you know the rate of change of volume, $\frac{dV}{dt}$.

Since the balloon is spherical:

$V = \frac{4}{3}\pi r^3$ so $\frac{dV}{dr} = 4\pi r^2$

Use the geometric context and differentiate to find $\frac{dV}{dr}$.

$$\frac{dr}{dt} = \frac{dr}{dV} \times \frac{dV}{dt}$$

You want the rate of change of the radius, $\frac{dr}{dt}$. Relate this to the two derivatives you already have using the chain rule.

$$= \frac{1}{4\pi r^2} \times 200$$

$$= \frac{1}{4\pi \times 8^2} \times 200$$

Substitute into the chain rule. Remember that $\frac{dr}{dV} = \frac{1}{\frac{dV}{dr}}$.

$$= 0.249 \text{ (3 s.f.)}$$

So the radius is increasing at about 0.249 cm per minute.

You want the rate of change of the radius when the radius is 8 cm.

Sometimes you need to combine the chain rule with the product or the quotient rule.

WORKED EXAMPLE 12.13

The length of a rectangle is increasing at a constant rate of 3 cm s^{-1}, and the width is decreasing at a constant rate of 2 cm s^{-1}. Find the rate of change of the area of the rectangle at the instant when the length is 18 cm and the width is 10 cm.

x = length in cm
y = width in cm
A = area in cm^2

Start by defining the variables.

$\frac{dx}{dt} = 3\ cm\ s^{-1}$

Write the given information in terms of derivatives. The rate of change of x is 3 cm s^{-1}.

$\frac{dy}{dt} = -2\ cm\ s^{-1}$

As the width is decreasing, the rate of change will be negative.

$A = xy$

$$\Rightarrow \frac{dA}{dt} = \frac{dx}{dt}y + x\frac{dy}{dt}$$

$$= 3y - 2x$$

Use the product rule to differentiate A. You want $\frac{dA}{dt}$.

When $x = 18$ and $y = 10$

$\frac{dA}{dt} = 3(10) - 2(18) = -6$

The area is decreasing at the rate of 6 $cm^2\ s^{-1}$.

Use the given values of x and y.

WORKED EXAMPLE 12.14

As a conical ice stalactite melts, the rate of decrease of height, h, is 1 cm h^{-1} and the rate of decrease of the radius of the base, r, is 0.1 cm h^{-1}.
At what rate is the volume of the stalactite decreasing when the height is 30 cm and the base radius is 4 cm?

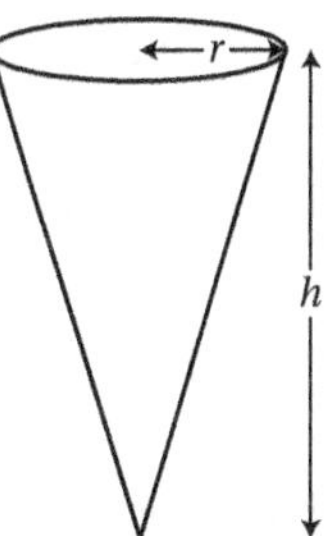

$\frac{dh}{dt} = -1\text{ cm h}^{-1}$

$\frac{dr}{dt} = -0.1\text{ cm h}^{-1}$

Write the given information in terms of derivatives.

Remember that decrease means a negative derivative.

$V = \frac{1}{3}\pi r^2 h$

Use the geometric context to relate the variables.

$$\frac{dV}{dt} = \frac{d}{dt}\left(\frac{1}{3}\pi r^2\right)h + \left(\frac{1}{3}\pi r^2\right)\frac{dh}{dt}$$
$$= \frac{2}{3}\pi r\frac{dr}{dt}h + \frac{1}{3}\pi r^2\frac{dh}{dt}$$

You want $\frac{dV}{dt}$ so differentiate both sides with respect to t. This requires the product rule and the chain rule.

$$\frac{dV}{dt} = \frac{2}{3}\pi \times 4 \times (-0.1) \times 30 + \frac{1}{3}\pi \times 4^2 \times (-1)$$
$$= -41.9\text{ cm}^3\text{ h}^{-1}$$

Substitute in the given values.

The volume is decreasing at $41.9\text{ cm}^3\text{ h}^{-1}$.

EXERCISE 12E

1 In each case, find an expression for $\frac{dz}{dx}$ in terms of x.

a **i** $z = 4y^2,\ y = 3x^2$ **ii** $z = y^2,\ y = x^3 + 1$

b **i** $z = \cos y,\ y = 3x^2$ **ii** $z = \tan y,\ y = x^2 + 1$

2 **a** **i** Given that $z = y^2 + 1$ and $\frac{dy}{dx} = 5$, find $\frac{dz}{dx}$ when $y = 5$.

ii Given that $z = 2y^3$ and $\frac{dy}{dx} = -2$, find $\frac{dz}{dx}$ when $y = 1$.

b **i** If $w = \sin x$ and $\frac{dw}{dt} = -3$, find $\frac{dx}{dt}$ when $x = \frac{\pi}{3}$.

ii If $P = \tan h$ and $\frac{dP}{dx} = 2$, find $\frac{dh}{dx}$ when $h = \frac{\pi}{4}$.

c **i** Given that $V = 12r^3$, $\frac{dr}{dt} = 1$ and $\frac{dV}{dt} = 4$, find the possible values of r.

ii Given that $H = 3S^{-2}$, find the value of S for which $\frac{dH}{dx} = 3$ and $\frac{dS}{dx} = 4$.

3 **a** **i** Given that $A = xy$, find $\frac{dA}{dt}$ when $x = 4,\ y = 5,\ \frac{dx}{dt} = 2$ and $\frac{dy}{dt} = 3$.

ii Given that $S = ab$, find $\frac{dS}{dt}$ when $a = 12,\ b = 5,\ \frac{da}{dt} = -2$ and $\frac{db}{dt} = 4$.

b **i** Given that $V = 3r^2h$, find $\frac{dV}{dt}$ when $r = 3,\ h = 2,\ \frac{dr}{dt} = 2$ and $\frac{dh}{dt} = -1$.

ii Given that $N = kx^4$, find $\frac{dN}{dt}$ when $x = 2,\ k = 5,\ \frac{dk}{dt} = 1,\ \frac{dx}{dt} = 1$.

c **i** Given that $m = \frac{S}{n}$ and that $S = 100,\ \frac{dS}{dt} = 20,\ N = 50,\ \frac{dN}{dt} = 4$, find $\frac{dm}{dt}$.

ii Given that $\rho = \frac{m}{V}$ and that $m = 24,\ \frac{dm}{dt} = 2,\ V = 120,\ \frac{dV}{dt} = 6$, find $\frac{d\rho}{dt}$.

4 A circular stain is spreading so that the radius is increasing at the constant rate of $1.5\ \text{cm s}^{-1}$.
Find the rate of increase of the area when the radius is 12 cm.

5 The area of a square is increasing at the constant rate of $50\ \text{cm}^2\ \text{s}^{-1}$.
Find the rate of increase of the side of the square when the length of the side is 12.5 cm.

6 A spherical ball is inflated so that its radius increases at the rate of 3 cm per second.
Find the rate of change of the volume when the radius is 8 cm.

7 A rectangle has length a and width b. Both the length and the width are increasing at a constant rate of 3 cm per second. Find the rate of increase of the area of the rectangle at the instant when $a = 10$ cm and $b = 15$ cm.

8 The surface area of a closed cylinder is given by $A = 2\pi r^2 + 2\pi rh$, where h is the height and r is the radius of the base. At the time when the surface area is increasing at the rate of $20\pi\ \text{cm}^2\ \text{s}^{-1}$, the radius is 4 cm, the height is 1 cm and is decreasing at the rate of $2\ \text{cm s}^{-1}$. Find the rate of change of radius at this time.

9 The radius of a cone is increasing at the rate of $0.5\ \text{cm s}^{-1}$ and its height is decreasing at the rate of $0.3\ \text{cm s}^{-1}$. Find the rate of change of the volume of the cone at the instant when the radius and 20 cm and the height is 30 cm.

10 A point is moving in the plane so that its coordinates are both functions of time, $(x(t), y(t))$. When the coordinates of the point are $(5, 7)$, the x-coordinate is increasing at the rate of 16 units per second and the y-coordinate is increasing at the rate of 12 units per second. At what rate is the distance of the point from the origin changing at this instant?

Section 4 : More complicated areas

Area above and below the x-axis

You know from Student Book 1, Chapter 14, that when a region is below the x-axis, the value of the definite integral will be negative. Therefore, when there are parts of the curve above and below the axis, to calculate the total area you must separate out the two sections.

WORKED EXAMPLE 12.15

a Find $\int_1^4 x^2 - 4x + 3\,dx$.

b Sketch the graph of $y = x^2 - 4x + 3$. Hence find the area of the region enclosed between the x-axis, the curve $y = x^2 - 4x + 3$ and the lines $x = 1$ and $x = 4$.

a

$$\int_1^4 x^2 - 4x + 3\,dx = \left[\frac{1}{3}x^3 - 2x^2 + 3x\right]_1^4$$
$$= \left(\frac{1}{3}(4)^3 - 2(4)^2 + 3(4)\right) - \left(\frac{1}{3}(1)^3 - 2(1)^2 + 3(1)\right)$$
$$= \left(\frac{4}{3}\right) - \left(\frac{4}{3}\right)$$
$$= 0$$

Integrate and evaluate at the upper and lower limits.

b

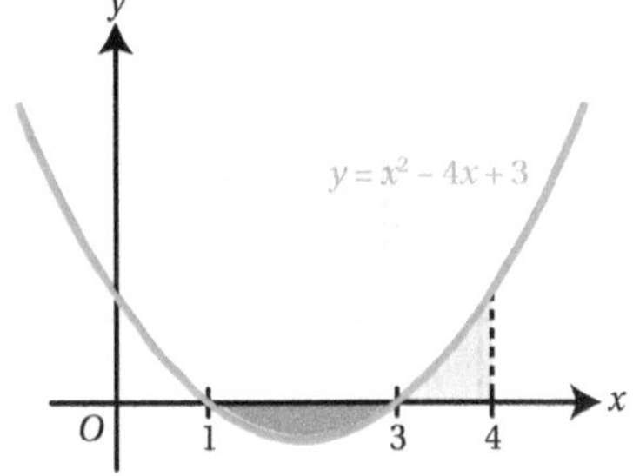

You can see that the required area is made up of two parts, one above the x-axis and one below it, so evaluate each of them separately.

$$\int_1^3 x^2 - 4x + 3\,dx = \left[\frac{1}{3}x^3 - 2x^2 + 3x\right]_1^3$$
$$= (0) - \left(\frac{4}{3}\right)$$
$$= -\frac{4}{3}$$

The integral for the part of the curve below the axis is negative, but the area must be positive.

$\therefore$ Area below the axis is $\frac{4}{3}$.

$$\int_3^4 x^2 - 4x + 3\,dx = \left[\frac{1}{3}x^3 - 2x^2 + 3x\right]_3^4$$
$$= \left(\frac{4}{3}\right) - (0)$$
$$= \frac{4}{3}$$

The area of the part above the axis is found as normal.

$\therefore$ Area above the axis is $\frac{4}{3}$.

Total area $= \frac{4}{3} + \frac{4}{3} = \frac{8}{3}$

Now add the two areas together.

You can interpret the fact that the integral in Worked example 12.15 was zero as meaning that the area above the axis is exactly cancelled by the area below the axis.

Common error

When you are asked to find an area, it is very important to sketch the graph first (if it is not given).
Don't just integrate in one go between the x-limits of the region to find its area unless you are sure the curve is always totally above or totally below the x-axis. If part of the region is above the x-axis and part is below, you must find the area of each separately.

Area between two curves

The area A in the diagram is bounded by two curves with equations $y = \text{f}(x)$ and $y = \text{g}(x)$.

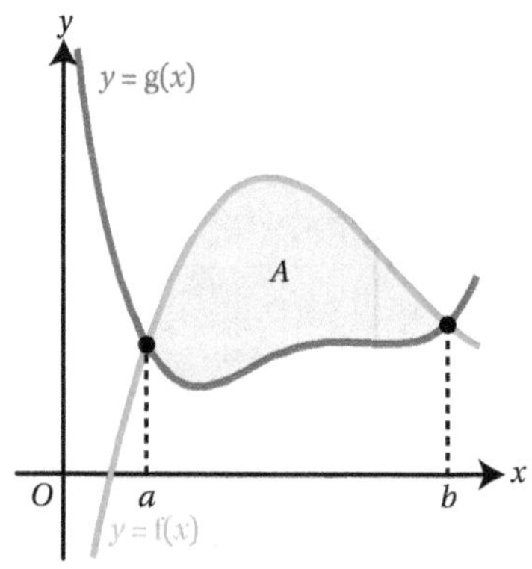

You can find the area by taking the area bounded by the upper curve $\text{f}(x)$ and the x-axis and subtracting the area bounded by the lower curve $\text{g}(x)$ and the x-axis, i.e.

$$A = \int_a^b \text{f}(x)\,\text{d}x - \int_a^b \text{g}(x)\,\text{d}x$$

You can do the subtraction before integrating so that you only have to integrate one expression instead of two. This gives an alternative formula for the area.

Key point 12.7

The area A bounded by two curves with equations $y = \text{f}(x)$ and $y = \text{g}(x)$, where curve $\text{g}(x)$ is below curve $\text{f}(x)$, is given by:

$$A = \int_a^b (\text{f}(x) - \text{g}(x))\,\text{d}x$$

where a and b are the x-coordinates of the intersection points of the two curves.

Tip

Make sure you subtract the lower curve from the upper curve when using this method.

WORKED EXAMPLE 12.16

Find the exact area enclosed between the curves $y = \frac{1}{2}x + \frac{15}{2}$ and $y = x^2 - 6x + 13$.

For the intersection:

$$x^2 - 6x + 13 = \frac{1}{2}x + \frac{15}{2}$$

First you need to find the x-coordinates of the intersections.

$$2x^2 - 12x + 26 = x + 15$$
$$2x^2 - 13x + 11 = 0$$
$$(2x - 11)(x - 1) = 0$$
$$x = \frac{11}{2}, 1$$

Continues on next page

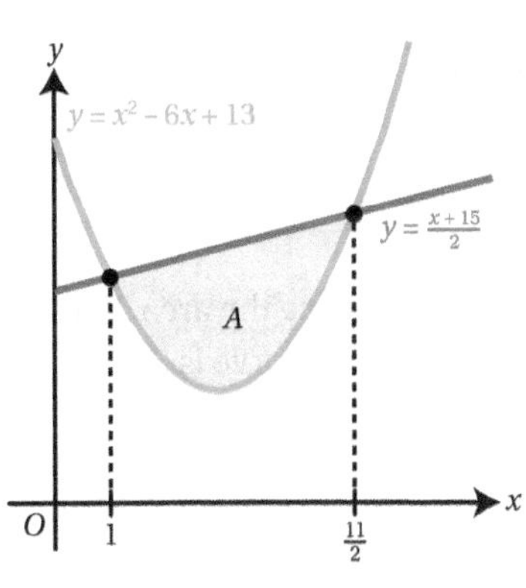

It may help to do a rough sketch to see the relative positions of the two curves.

$$A = \int_1^{\frac{11}{2}} \left(\frac{1}{2}x + \frac{15}{2}\right) - (x^2 - 6x + 13)\,dx$$

Subtract the lower curve from the upper curve before integrating.

$$= \int_1^{\frac{11}{2}} -x^2 + \frac{13}{2}x - \frac{11}{2}\,dx$$

$$= \left[-\frac{1}{3}x^3 + \frac{13}{4}x^2 - \frac{11}{2}x\right]_1^{\frac{11}{2}}$$

$$= \left(\frac{605}{48}\right) - \left(-\frac{31}{12}\right) = \frac{729}{48}$$

Subtracting the two equations before integrating is particularly useful when one of the curves is partly below the x-axis. As long as $f(x)$ is always above $g(x)$ then the expression being integrated, $f(x) - g(x)$, is always positive, so you do not have to worry about the signs of $f(x)$ and $g(x)$ themselves.

Rewind

In Student Book 1, Chapter 14, you will have found an area such as that in Worked example 12.16 by subtracting the area under the curve from the area of a trapezium. This new method is more direct.

WORKED EXAMPLE 12.17

Find the area bounded by the curves $y = e^x - 3$ and $y = 5 - 7e^{-x}$.

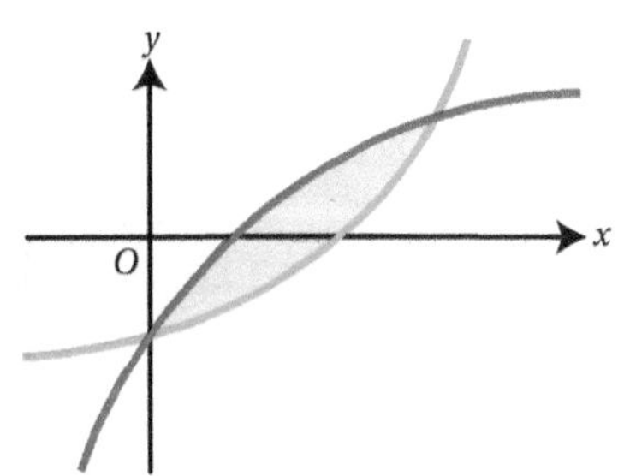

Sketch the graph to see the relative position of the two curves.

Intersections:

$$e^x - 3 = 5 - 7e^{-x}$$

$$e^{2x} - 3e^x = 5e^x - 7$$

Find the intersection points; start by multiplying through by e^x.

Continues on next page

$$e^{2x} - 8e^x + 7 = 0$$
$$(e^x - 1)(e^x - 7) = 0$$
$$\therefore x = 0 \text{ or } \ln 7$$

This is a disguised quadratic which can be factorised.

$$A = \int_0^{\ln 7} (5 - 7e^{-x}) - (e^x - 3)\,dx$$

Write down the integral representing the area. You can see from the graph that the top curve is $5 - 7e^{-x}$.

$$= \int_0^{\ln 7} 8 - 7e^{-x} - e^x\,dx$$

Simplify before integrating.

$$= \left[8x + 7e^{-x} - e^x\right]_0^{\ln 7}$$
$$= (8\ln 7 + 1 - 7) - (0 + 7 - 1)$$

Use $e^{\ln 7} = 7$.

$$= 8\ln 7 - 12$$

WORK IT OUT 12.3

Find the area enclosed between the line $y = 2x$ and the curve $y = x^2 - 3x$.

Which is the correct solution? Identify the errors made in the incorrect solutions.

Solution 1

Sketching the graph shows that part of the area is below the x-axis, so split it up into two parts.

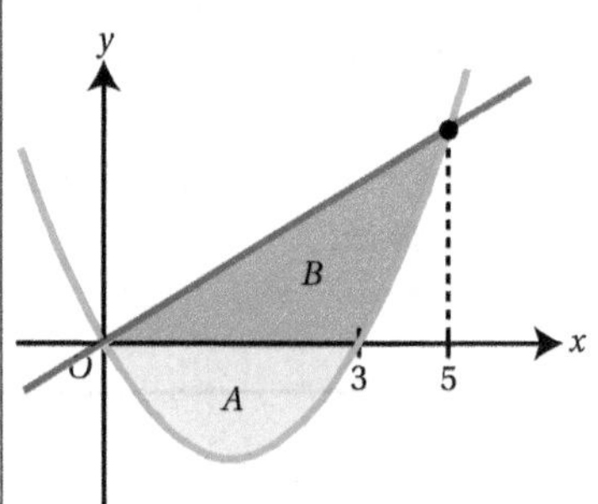

$$A = \int_0^3 x^2 - 3x\,dx$$
$$= \left[\frac{1}{3}x^3 - \frac{3}{2}x^2\right]_0^3 = -4.5$$
$$B = \int_0^5 2x\,dx$$
$$= \left[x^2\right]_0^5 = 25$$

Total area $= 25 + 4.5 = 29.5$

Solution 2

The intersection points are at $x = 0$ and $x = 5$.

$$\int_0^5 x^2 - 3x - 2x\,dx$$
$$= \left[\frac{1}{3}x^3 - \frac{5}{2}x^2\right]_0^5$$
$$= -\frac{125}{6}$$

Solution 3

The intersection points are at $x = 0$ and $x = 5$.

$$\int_0^5 2x - (x^2 - 3x)\,dx$$
$$= \left[\frac{5x^2}{2} - \frac{x^3}{3}\right]_0^5$$
$$= 20.8 \text{ (3 s.f.)}$$

Area between a curve and the y-axis

How can you find the area of the shaded region A in this diagram, bounded by the curve and the y-axis?

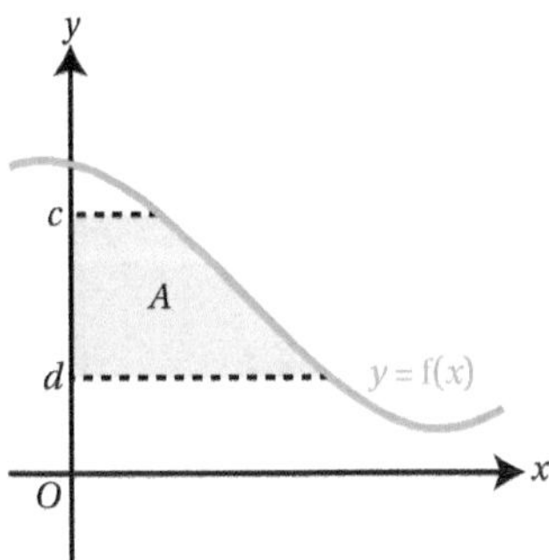

You know how to find the area between the curve and the x-axis, so one possible strategy is to draw vertical lines to create the region labelled A_1. The required area is equal to the area of A_1 minus the red rectangle plus the blue rectangle.

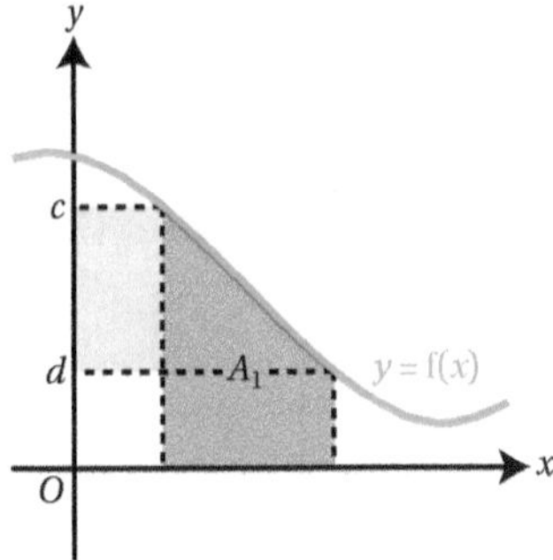

But there is a quicker way. You can treat x as a function of y, effectively reflecting the whole diagram in the line $y = x$.

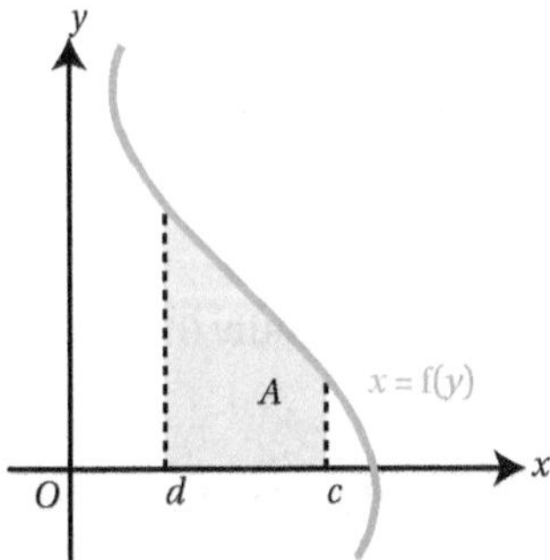

Key point 12.8

The area bounded by the curve $y = \text{f}(x)$, the y-axis and the lines $y = c$ and $y = d$ is given by:

$$\int_c^d \text{g}(y)\,\text{d}y$$

where $\text{g}(y)$ is the expression for x in terms of y, i.e. $x = \text{g}(y)$.

Rewind

From Chapter 2, Section 4, you may have realised that this is related to inverse functions.

WORKED EXAMPLE 12.18

The curve shown has equation $y = 2\sqrt{x-1}$.
Find the area of the shaded region.

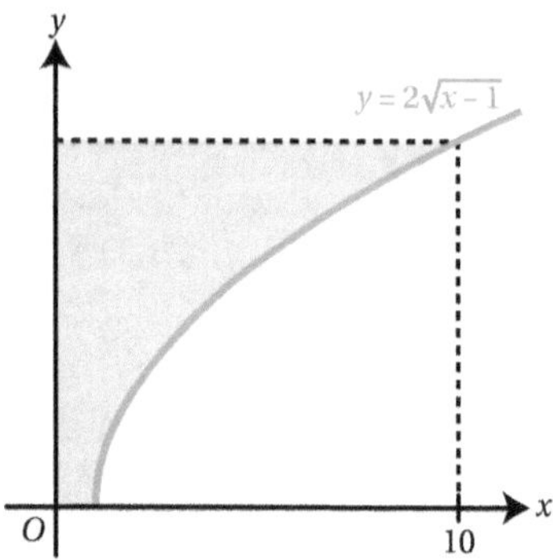

$x - 1 = \left(\frac{y}{2}\right)^2$ — Express x in terms of y.

$\Rightarrow x = \frac{y^2}{4} + 1$

When $x = 1$, $y = 2\sqrt{1-1} = 0$ — Find the limits on the y-axis.

When $x = 10$, $y = 2\sqrt{10-1} = 6$

$$\text{Area} = \int_0^6 \left(\frac{y^2}{4} + 1\right) dy$$

$$= \left[\frac{y^3}{12} + y\right]_0^6 = (18 + 6) - (0) = 24$$

WORKED EXAMPLE 12.19

Find the exact area enclosed by the graph of $y = \ln(x+3)$, the y-axis and the line $y = 1$.

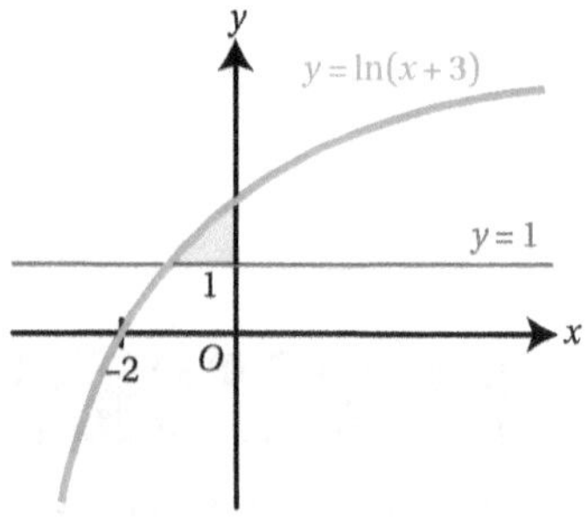

Sketch the graph and identify the area required.

The graph crosses the y-axis at $y = \ln(0+3)$.

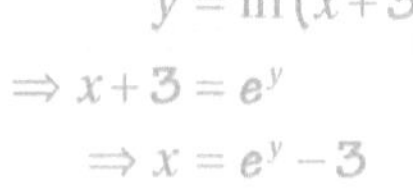

$y = \ln(x+3)$

$\Rightarrow x + 3 = e^y$

$\Rightarrow x = e^y - 3$

The area between a curve and the y-axis is given by $\int_a^b x \, dy$, so express x in terms of y...

Continues on next page

$$\int_1^{\ln 3} (e^y - 3)\,dy = \left[e^y - 3y\right]_1^{\ln 3}$$ ... and then evaluate the definite integral.

$$= (e^{\ln 3} - 3\ln 3) - (e^1 - 3)$$

$$= (3 - 3\ln 3) - (e - 3)$$ Remember that $e^{\ln k} = k$.

$$= 6 - e - 3\ln 3$$ Notice that this value is negative; this is because the shaded area is to the left of the y-axis.

So area $= 3\ln 3 + e - 6$

EXERCISE 12F

1 Find the area of each shaded region. You need to find the intersection points of the two curves first.

a **i**

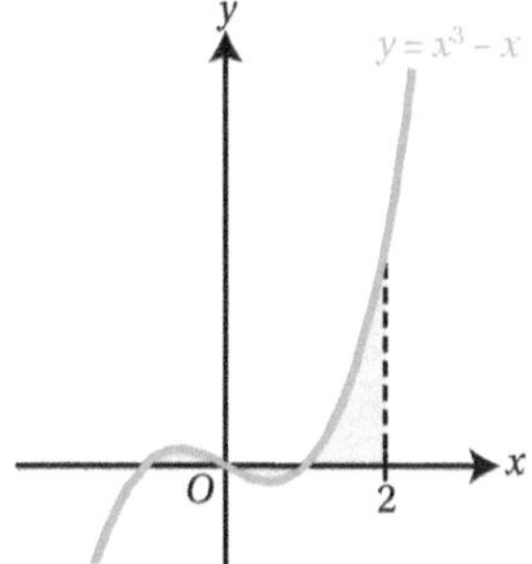

ii

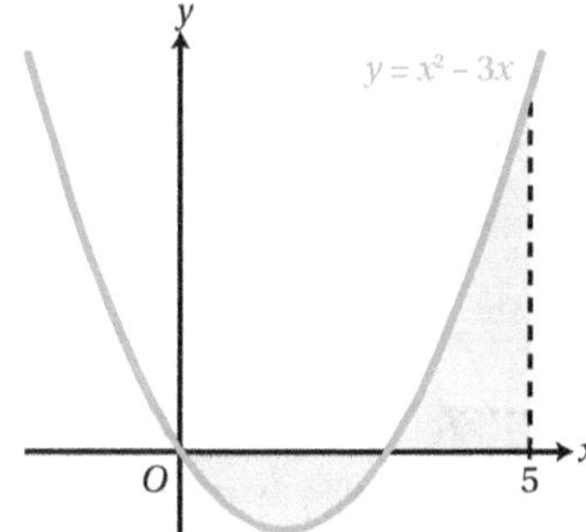

b **i**

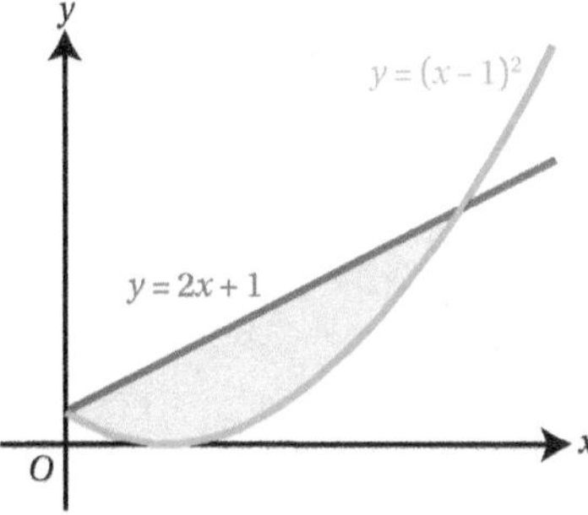

ii

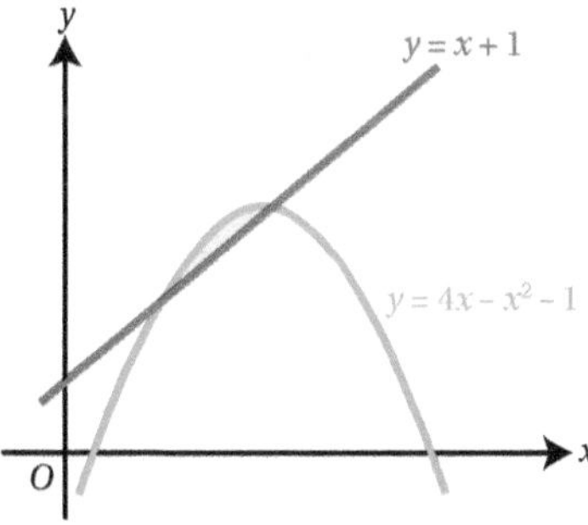

c **i**

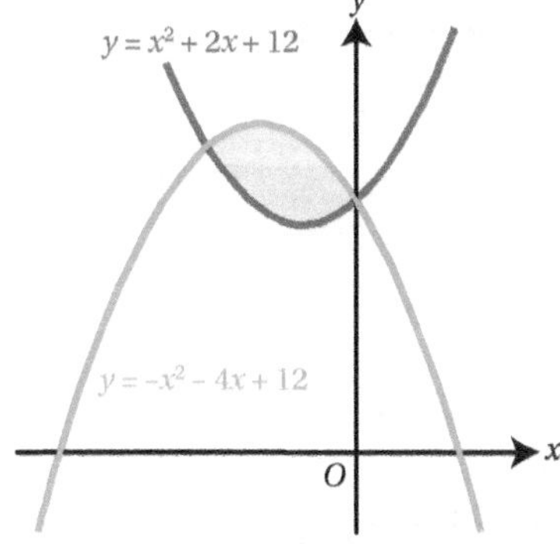

ii

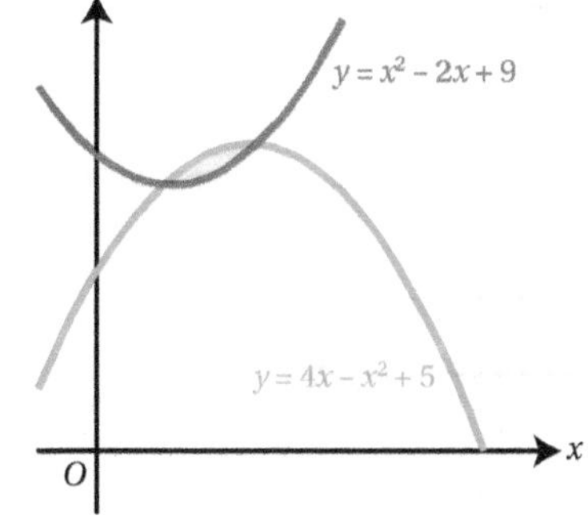

d **i**

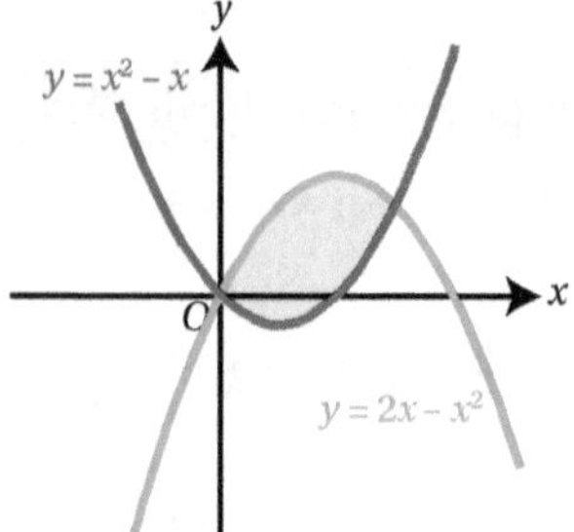

ii

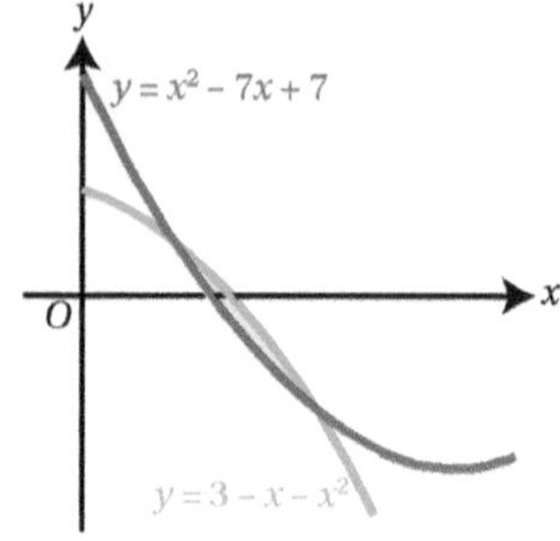

e **i**

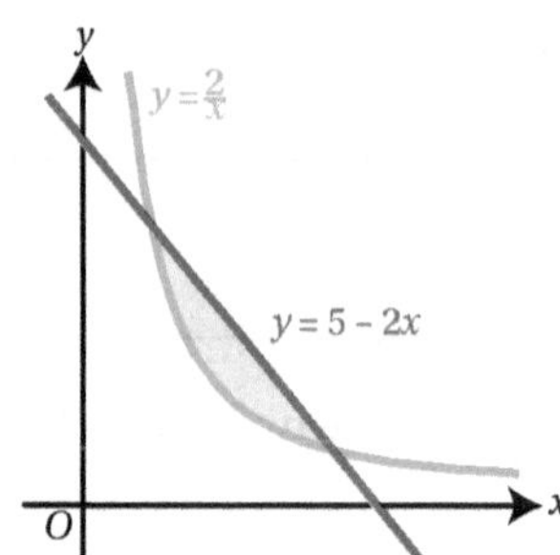

ii

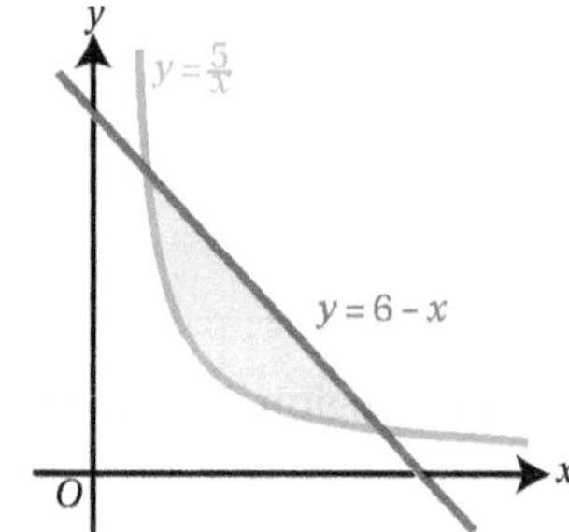

2 Find the areas of the shaded regions.

a **i**

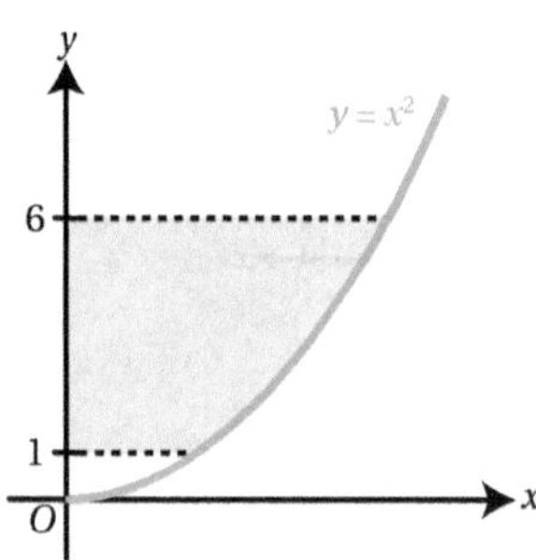

ii

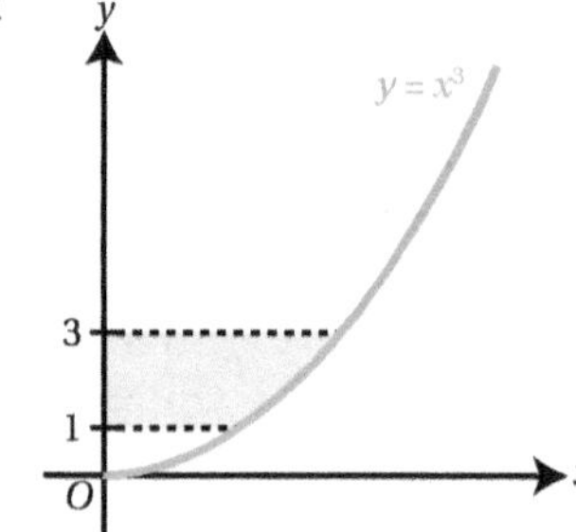

b **i**

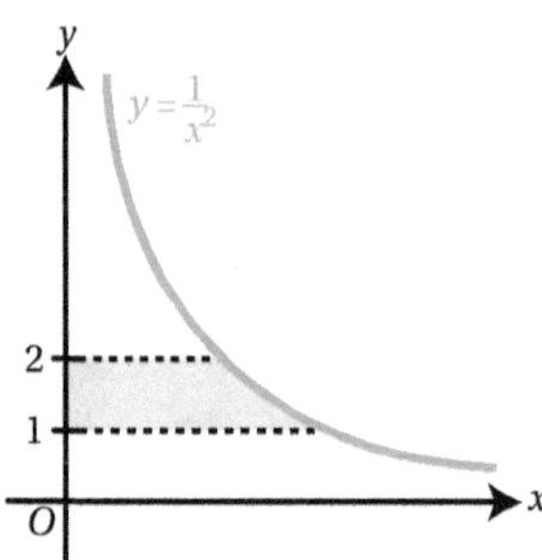

ii

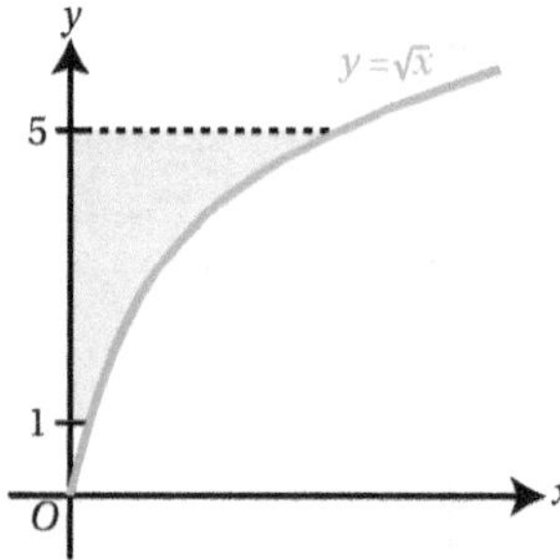

c **i**

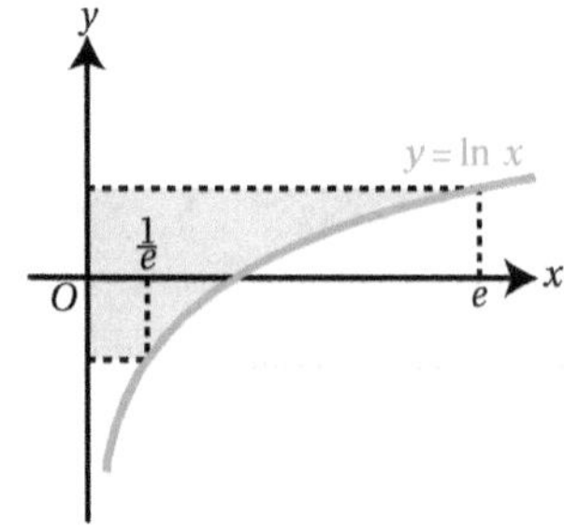

ii

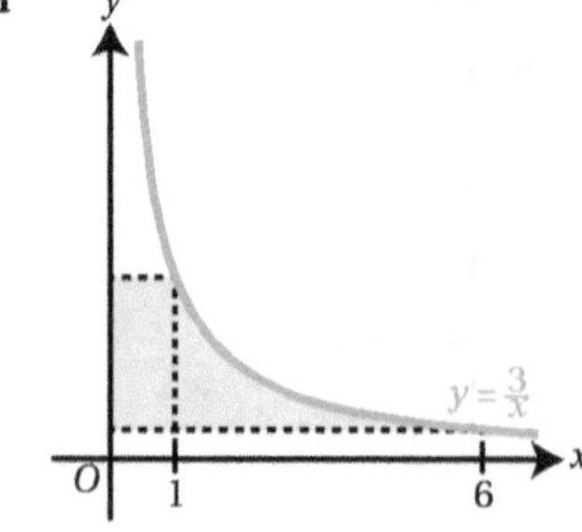

3 The diagram shows part of the graph of $y = x^2 - 6x + 8$. The shaded region is enclosed by the curve, the x-axis and the lines $x = 0$ and $x = 5$.

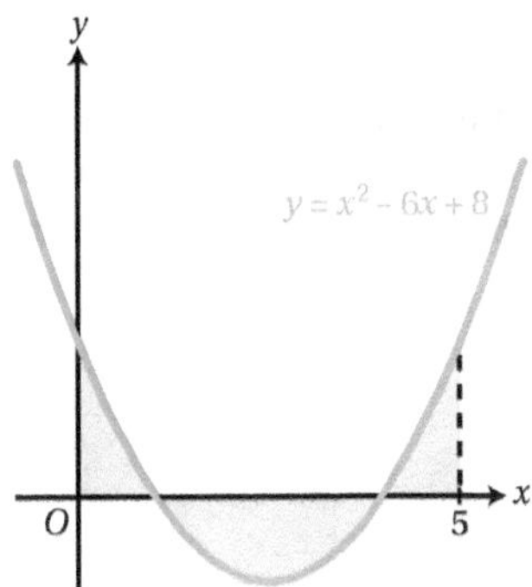

Find the area of the shaded region.

4 Find the area enclosed between the graphs of $y = x^2 + x - 2$ and $y = x + 2$.

5 The diagram shows the graphs of $y = e^x$ and $y = x^2$. Find the exact value of the area of the shaded region.

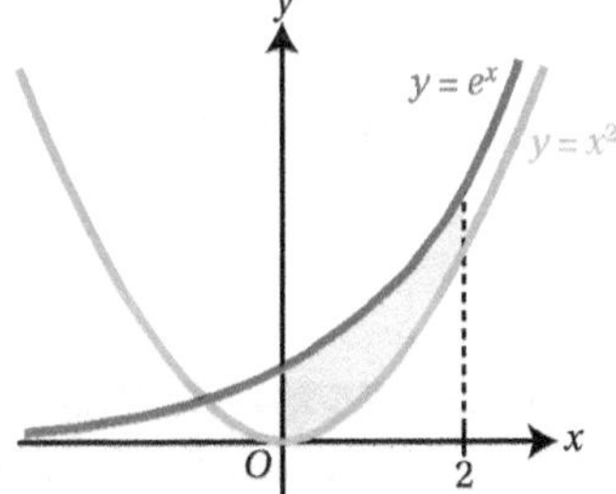

6 Show that the area of the shaded region is $\frac{9}{2}$.

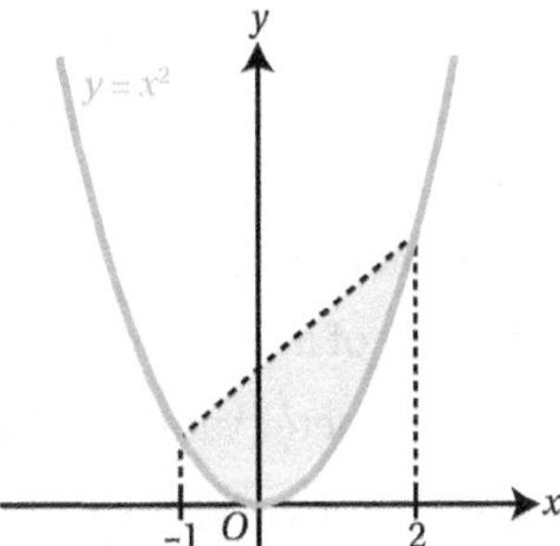

7 The diagram shows the curve $y = \sqrt{x}$.

If the shaded area is 504 find the value of a.

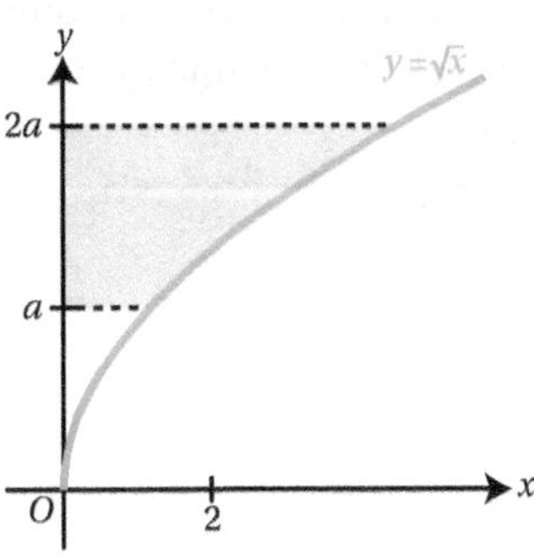

8 Find the exact value of the area enclosed by the graph of $y = \ln(x+1)$, the line $y = 2$ and the y-axis.

9 Find the exact value of the area between the graphs of $y = \frac{5}{x}$ and $y = 6 - x$.

10 Find the total area enclosed between the graphs of $y = x(x-4)^2$ and $y = 4x - x^2$.

11 The area enclosed between the curve $y = x^2$ and the line $y = mx$ is $10\frac{2}{3}$. Find the value of m if $m > 0$.

12 The diagram shows the graph of $y = x^2$.

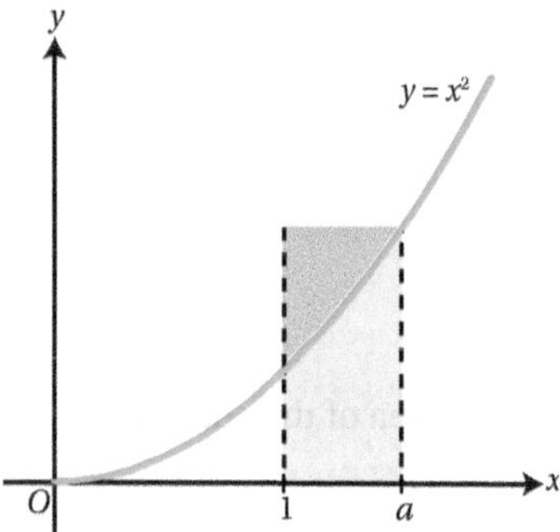

Find the value of a for which the areas of the red region and the blue region are equal.

Checklist of learning and understanding

- A convex curve has $\frac{d^2y}{dx^2} > 0$; a concave curve has $\frac{d^2y}{dx^2} < 0$.
- At a point of inflection, $\frac{d^2y}{dx^2} = 0$ and the curve changes from convex to concave, or vice versa.
- Parametric equations are a way of describing a curve where both x and y coordinates are given in terms of a parameter (usually called t or θ). Each parameter value corresponds to a single point on the curve.
- The gradient of a curve given in parametric form is $\frac{dy}{dx} = \frac{dy}{dt} \times \frac{dt}{dx}$.
- The area between the x-axis and a part of a curve with parametric equations $(x(t), y(t))$ is given by $\int_{t_1}^{t_2} y \frac{dx}{dt}\, dt$, where t_1 and t_2 are the parameter values at the end-points.
- The chain rule can be used to connect rates of change of two related variables; If u depends on t and y depends on u, then $\frac{dy}{dt} = \frac{dy}{du} \times \frac{du}{dt}$. You often need the geometric context of the question to work out how y depends on u.
- The area enclosed between two curves with equations $y = f(x)$ and $y = g(x)$ is given by $\int_a^b (f(x) - g(x))\,dx$, where a and b are x-coordinates of the intersection points.
- The area between a curve, the y-axis and the lines $y = c$ and $y = d$ is given by $\int_c^d g(y)\,dy$, where $x = g(y)$ is the expression for x in terms of y.

Mixed practice 12

1 A curve is defined by the parametric equations

$x = \sin 2t, y = \cos 3t \quad$ for $t \in \left[0, \frac{\pi}{2}\right]$

Find the value of the parameter t at the point $\left(\frac{1}{2}, -\frac{1}{\sqrt{2}}\right)$. Choose from these options.

A $\frac{\pi}{6}$ **B** $\frac{\pi}{4}$ **C** $\frac{\pi}{12}$ **D** $\frac{5\pi}{12}$

2 Find the coordinates of the point of inflection on the graph of $y = \frac{x^3}{6} - x^2 + x$.

3 A curve has parametric equations $x = 3t^2,\ y = 2t - t^3$.

a Show that the point $P\,(3, 1)$ lies on the curve, and the find the value of t at this point.

b Find the equation of the tangent to the curve at P.

4 The diagram shows the graph of $y = (x-3)^2$ with a horizontal line drawn through its y-intercept. Find the exact value of the area of the shaded region.

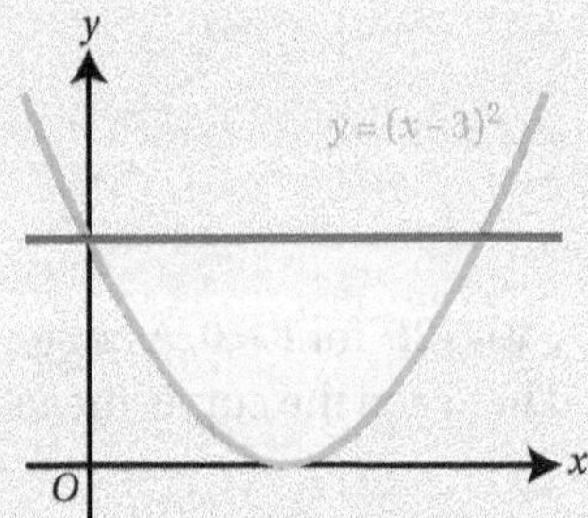

5 **a** Find $\int_0^3 (x^2 - 1)\,dx$.

b The diagram shows the graph of $y = x^2 - 1$. The shaded region is bounded by the curve, the x-axis and the lines $x = 0$ and $x = 3$.

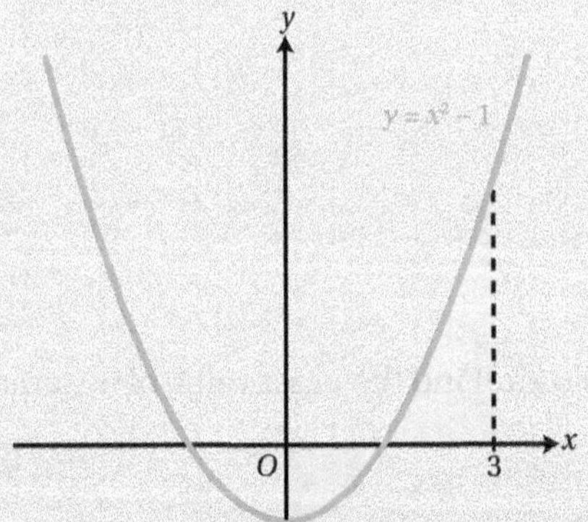

Find the area of the shaded region.

6 A curve is defined by the parametric equations $x = \frac{t^2}{2} + 1,\ y = \frac{4}{t} - 1$.

a Find the gradient at the point on the curve where $t = 2$.

b Find a Cartesian equation of the curve.

[© AQA 2014]

7 For the curve $y = \mathrm{f}(x)$, $\mathrm{f}''(a) = 0$.

Which statement is true?

A There is a point of inflection at $x = a$.

B There is a point of inflection at $x = a$ if $\mathrm{f}'(a) = 0$ as well.

C There isn't a point of inflection at $x = a$ if $\mathrm{f}'(a) \neq 0$.

D More information is required to determine whether there is a point of inflection at $x = a$.

8 The diagram shows the graph of $y = \sqrt{x}$.

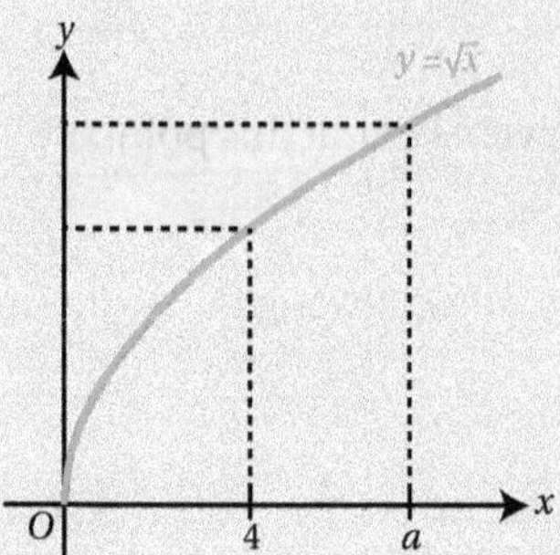

The area of the shaded region is 39 units. Find the value of a.

9 The curve shown in the diagram has parametric equations $x = 8t^3$, $y = 12t^2$ for $t \geqslant 0$. A tangent to the curve is drawn at the point (64, 48). Find the shaded area enclosed between the curve, the tangent and the y-axis.

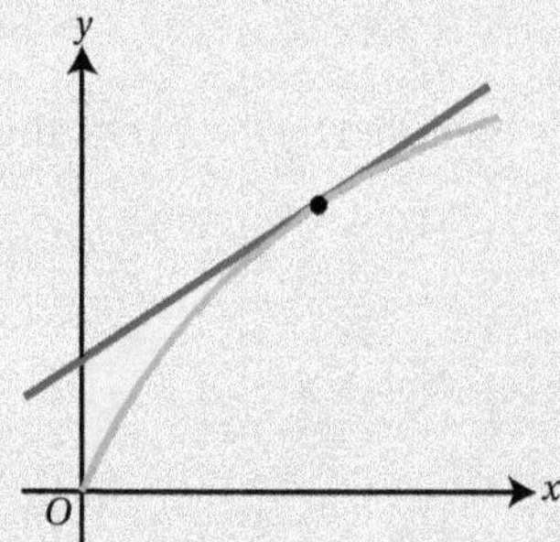

10 **a** Solve the equation $\sin x = \sin 2x$ for $0 \leqslant x \leqslant \pi$.

b The diagrams shows the curves $y = \sin x$ and $y = \sin 2x$. Find the exact value of the area of the shaded region.

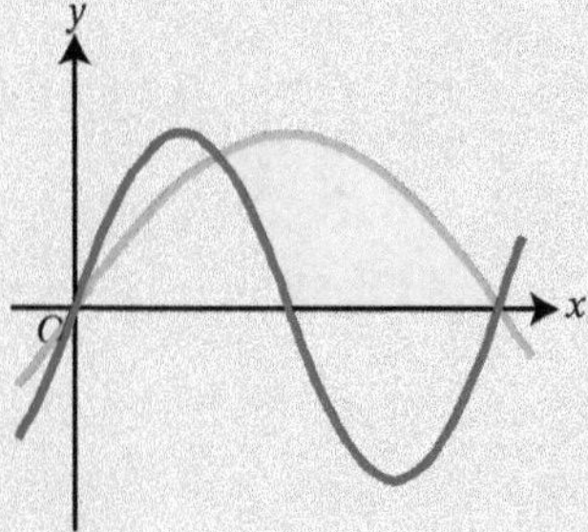

11 Consider the graph of $y = x\sin 2x$ for $x \in [0, 2\pi]$.

a Show that the x-coordinates of the points of inflection satisfy $\tan 2x = \frac{1}{x}$.

b Use graphs to find the number of points of inflection on the graph.

12 **a** Find the coordinates of the stationary points on the graph of $y = x^4 - x^5$ and determine their nature.

b Prove that the graph has one non-horizontal point of inflection.

13 Prove that the function $f(x) = x^3 \cos x$ has a stationary point of inflection at the origin.

14 The diagram shows a part of the graph of $y = x^n$ for $n > 1$.

The red area is three times larger than the blue area. Find the value of n.

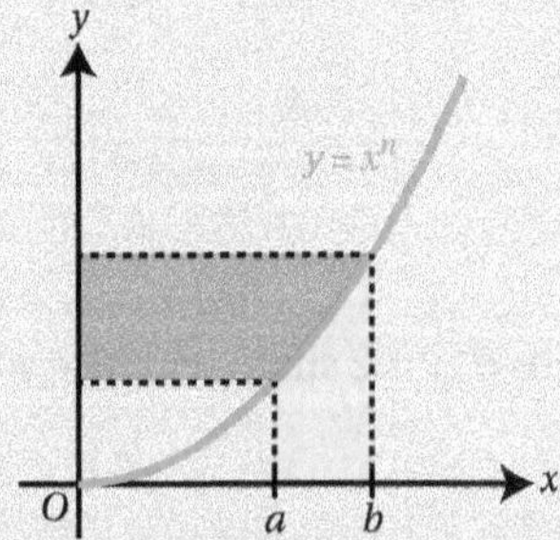

15 The diagram shows an isosceles right-angled triangle of side 100 cm. Point D is moving along the side AB towards point B so that the area of the trapezium $DBCE$ is decreasing at the constant rate of $18\ \text{cm}^2\ \text{s}^{-1}$.

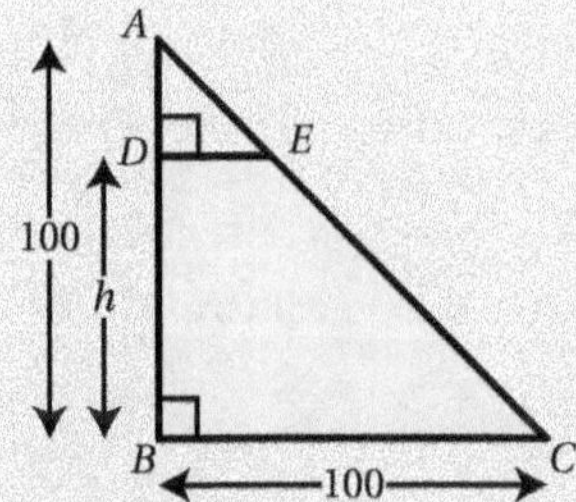

Let $BD = h$.

a Write down an expression for the area of the trapezium $DBCE$ in terms of h.

b Show that $\frac{dh}{dt} = \frac{18}{h - 100}$.

Initially point D is at vertex A.

c Given that $h = 100 - k\sqrt{t}$, find the value of k.

16 A curve is defined by the parametric equations $x = 8e^{-2t} - 4$, $y = 2e^{2t} + 4$.

a Find $\frac{dy}{dx}$ in terms of t.

b The point P, where $t = \ln 2$, lies on the curve.

i Find the gradient of the curve at P.

ii Find the coordinates of P.

iii The normal at P crosses the x-axis at the point Q. Find the coordinates of Q.

c Find the Cartesian equation of the curve in the form $xy + 4y - 4x = k$, where k is an integer.

[© AQA 2013]

17 A particle P moves in a straight line with velocity given by $v = 3e^{-t}(\cos t - \sin t)\ \text{m s}^{-1}\ 0 \leqslant t \leqslant 4$.

a Find the times when P is instantaneously at rest.

b Find the total distance travelled in the first π seconds of motion, giving your answer to three significant figures.

18 Show that the area of the shaded region in the diagram is $\frac{9}{2}$.

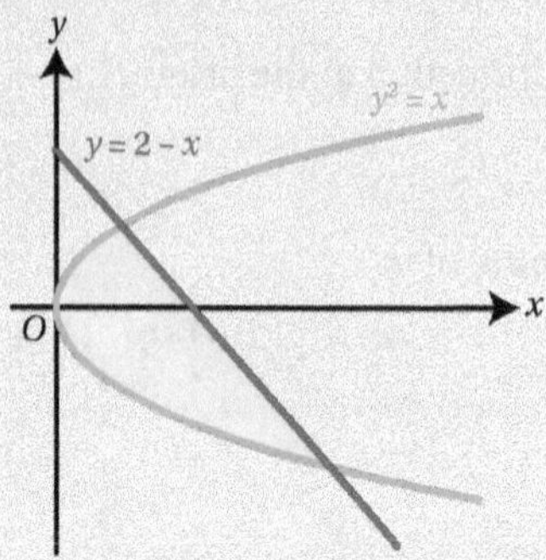

19 The ellipse shown in the diagram has parametric equations $x = 5\cos\theta$, $y = 2\sin\theta$, with $\theta \in [0, 2\pi)$.

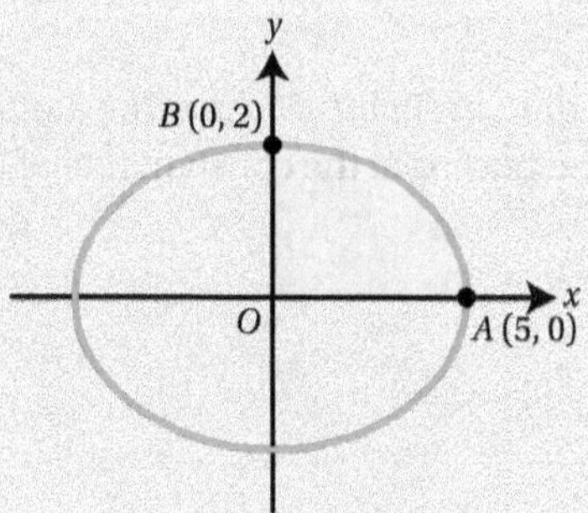

a State the values of θ at the points marked A and B.

b Find the area of the shaded region, and hence state the total area enclosed by the ellipse.

20 The graph shows $y = f'(x)$.

On a sketch of this graph:

a mark points corresponding to a local minimum of $f(x)$ as A

b mark points corresponding to a local maximum of $f(x)$ as B

c mark points corresponding to a point of inflection of $f(x)$ as C.

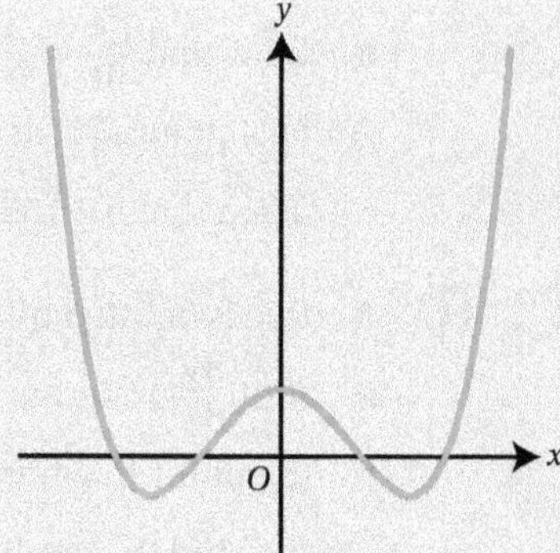

21 a Show that $5a^2 + 4ax - x^2 = (5a - x)(x + a)$.

b Find the coordinates of the points of intersection of the graphs $y = 5a^2 + 4ax - x^2$ and $y = x^2 - a^2$.

c Find the area enclosed between these two graphs.

d Show that the fraction of this area above the axis is independent of a and state the value that this fraction takes.

13 Differential equations

In this chapter you will learn how to:

- solve differential equations of the form $\frac{dy}{dx} = f(x)g(y)$
- write differential equations in a variety of contexts
- interpret a solution of a differential equation and decide whether it is realistic in the given context.

Before you start…

Student Book 1, Chapter 7	You should be able to rearrange expressions involving exponents and logarithms.	1 Given that $\ln(v-3) = t + \ln 5$, write v in terms of t.
Student Book 1, Chapter 18	You should be able to draw force diagrams and find net forces.	2 An object of weight 35 N falls under gravity. The magnitude of the air resistance is 8 N. Find the net force on the object.
Chapter 10	You should be able to write equations involving related rates of change.	3 The rate of change of the radius, r, of a sphere is $5\sqrt{r}$. Find an expression for the rate of change of volume.
Chapter 11	You should be able to integrate, using partial fractions, and simplify the answer, using laws of logs.	4 Integrate and simplify $\int \frac{8}{4-x^2}\,dx$
Chapter 11	You should be able to use integration by substitution and by parts.	5 Integrate a $\int \frac{4x}{x^2+3}\,dx$ b $\int x^2 \ln x\,dx$
Chapter 11	You should be able to integrate, using trigonometric identities.	6 Find $\int \tan^2 2x\,dx$.

Modelling using differential equations

In Student Book 1, Chapter 15, you met problems involving velocity as the rate of change of displacement, and acceleration as the rate of change of velocity. Many natural processes can be modelled by equations involving the rate of change of some variable, such as population growth and cooling of bodies. Newton's second law actually states that force is equal to the rate of change of momentum. To get from these rates of change to find the underlying variable involves solving differential equations.

In this chapter you will look at forming differential equations with an emphasis on real-world applications and at a method for solving a particular type of differential equation.

Section 1: Introduction to differential equations

To solve a **differential equation** you go from an equation involving derivatives to one without. You have done this already for the case where the equation can be written in the form $\frac{dy}{dx} = f(x)$.

As an example, consider the differential equation $\frac{dy}{dx} = 3x^2$.
To solve this differential equation all that you need is integration:
$y = \int 3x^2 \, dx = x^3 + c.$

Because of the constant of integration you find that there is not just one solution to the differential equation. It could be any one of a **family of solutions.** All the curves have the same gradient function, so they have the same shape. The solution $y = x^3 + c$ is called the **general solution** to the differential equation.

$y = x^3 + 4$, $y = x^3 + 2$, $y = x^3$, $y = x^3 - 2$

You may also be told that the curve passes through a point, e.g. $(0, 2)$. This is called an **initial condition** and it allows you to find the constant of integration and so narrow down the general solution to the **particular solution** – in this case $y = x^3 + 2$.

Sometimes you will need to rearrange the equation before integrating.

WORKED EXAMPLE 13.1

Find the particular solution of the differential equation $(x^2 - 2)\frac{dy}{dx} = 3x$, given that $y = \ln 2$ when $x = 2$.

$$\frac{dy}{dx} = \frac{3x}{x^2 - 2}$$

Write the equation in the form $\frac{dy}{dx} = f(x)$.

$$y = \int \frac{3x}{x^2 - 2} \, dx$$

Then integrate.

Let $u = x^2 - 2$

$$\frac{du}{dx} = 2x \Rightarrow dx = \frac{1}{2x} du$$

Since the numerator is a multiple of the derivative of the denominator, a substitution will work.

$$y = \int \frac{3x}{u} \frac{1}{2x} du$$

Make the substitution, and simplify as much as possible.

$$= \int \frac{3}{2u} du$$

$$= \frac{3}{2} \ln u + c$$

Remember $+c$; this is the general solution.

$$= \frac{3}{2} \ln (x^2 - 2) + c$$

When $x = 2$, $y = \ln 2$

Use the given condition to find c.

$$\frac{3}{2} \ln 2 + c = \ln 2$$

$$c = -\frac{1}{2} \ln 2$$

So $y = \frac{3}{2} \ln (x^2 - 2) - \frac{1}{2} \ln 2$

This value of c gives the particular solution.

EXERCISE 13A

1 Find the general solution of each differential equation.

a **i** $\frac{dy}{dx} = 3\sin 2x$ **ii** $\frac{dy}{dx} = 4\cos\left(\frac{x}{3}\right)$ **b** **i** $3\frac{dy}{dx} - 2e^{2x} = 0$ **ii** $4e^{\frac{x}{2}} - \frac{dy}{dx} = 0$

c **i** $\cos^2 x\frac{dy}{dx} = 3$ **ii** $\cot^2 x\frac{dy}{dx} = 1$ **d** **i** $x^3\frac{dy}{dx} = \ln x$ **ii** $\cos^2 x\frac{dy}{dx} = \sin x$

2 Find the particular solution of each differential equation.

a **i** $\frac{dy}{dx} = \frac{2}{\sqrt{3x+9}}$, $y = 2$ when $x = 0$ **ii** $\frac{dy}{dx} = \frac{1}{\sqrt{4-x}}$, $y = 1$ when $x = 3$

b **i** $(x^2+1)\frac{dy}{dx} = 2x$, $y = 0$ when $x = 1$ **ii** $2x\frac{dy}{dx} = x^2+1$, $y = 1$ when $x = 1$

c **i** $\frac{1}{2}e^{3x}\frac{dy}{dx} = 3$, $y = 0$ when $x = 0$ **ii** $e^{2x-1}\frac{dy}{dx} = 4$, $y = 0$ when $x = \frac{1}{2}$

d **i** $\sec x\frac{dy}{dx} = \sin^3 x$, $y = \frac{13}{64}$ when $x = \frac{\pi}{3}$ **ii** $\cos^3 x\frac{dy}{dx} = \sin x$, $y = 5$ when $x = \frac{\pi}{4}$

Section 2: Separable differential equations

In Section 1 you looked at differential equations where $\frac{dy}{dx}$ depends just on x. But there are situations where the gradient depends on y, or on both variables, for example, $\frac{dy}{dx} = x^2 y$.

You can't solve this equation by simply integration as the right-hand side contains y but if the y is moved to the left-hand side first, to give

$$\frac{1}{y}\frac{dy}{dx} = x^2$$

and you can then integrate both sides of the equation with respect to x.

$$\int \frac{1}{y}\frac{dy}{dx}\,dx = \int x^2\,dx$$

However, $\int \frac{1}{y}\frac{dy}{dx}\,dx = \int \frac{1}{y}\,dy$ so the equation becomes:

$$\int \frac{1}{y}\,dy = \int x^2\,dx \qquad (*)$$

$$\Rightarrow \ln y = \frac{x^3}{3} + c$$

$$\Rightarrow y = e^{\frac{x^3}{3}+c} = Ae^{\frac{x^3}{3}}$$

Key point 13.1

To solve a differential equation that can be written in the form $\frac{dy}{dx} = f(x)\,g(y)$:

- get all of the xs on one side and all of the ys on the other side by multiplication or division
- separate $\frac{dy}{dx}$ as if it were a fraction
- integrate both sides.

Just as with integration by substitution, you get the same results from just splitting up $\frac{dy}{dx}$ as if it were a fraction when you separate the xs and ys to different sides; this leads to the equation marked *. This method of solving differential equations is called **separation of variables**.

WORKED EXAMPLE 13.2

Solve the equation $\frac{dy}{dx} = (1.2 + 0.4x)y$ given that $y = 32$ when $x = 0$.

$\frac{1}{y}\frac{dy}{dx} = 1.2 + 0.4x$ — Get the y term onto the left-hand side

$\int \frac{1}{y}\,dy = \int (1.2 + 0.4x)\,dx$ — Separate $\frac{dy}{dx}$ and integrate.

$\ln y = 1.2x + 0.2x^2 + c$ — Since the difference of two constants is just another constant, you only need $+c$ on one side.

When $x = 0$, $y = 32$: — Use the initial condition to find c.

$\ln 32 = 0 + 0 + c$

$c = \ln 32$

$\therefore \ln y = 1.2x + 0.2x^2 + \ln 32$

$y = e^{1.2x + 0.2x^2 + \ln 32}$

$y = 32e^{1.2x + 0.2x^2}$ — Using $e^{\ln 32} = 32$.

Sometimes you need to factorise the equation first to get it into the correct form.

WORKED EXAMPLE 13.3

Show that the general solution to the differential equation $\frac{dy}{dx} = xy - x$ can be written as $y = 1 + Ae^{\frac{x^2}{2}}$ if $y > 1$.

$\frac{dy}{dx} = x(y-1)$ — You can use separation of variables if you can write the equation in the form $\frac{dy}{dx} = f(x)\,g(y)$.

$\int \frac{1}{y-1}\,dy = \int x\,dx$ — Separate the variables: divide by $y - 1$ and multiply by dx. Then integrate.

$\ln|y-1| = \frac{x^2}{2} + c$

$|y-1| = e^{\frac{x^2}{2}+c}$

But since $y - 1 > 0$

$y - 1 = e^{\frac{x^2}{2}+c} = e^{\frac{x^2}{2}}e^c = Ae^{\frac{x^2}{2}}$ — Since e^c is a constant, relabel it as A.

$\therefore y = 1 + Ae^{\frac{x^2}{2}}$

WORK IT OUT 13.1

Find the general solution of the differential equation $\frac{dy}{dx} = xy$.

Which is the correct solution? Identify the errors made in the incorrect solutions.

Solution 1	Solution 2	Solution 3
$\frac{dy}{dx} = xy$	$\frac{dy}{dx} = xy$	$\frac{dy}{dx} = xy$
$y = \int xy \, dx$	$\frac{1}{y}\frac{dy}{dx} = x$	$\frac{1}{y}\frac{dy}{dx} = x$
$= y\int x \, dx$	$\int \frac{1}{y} \, dy = \int x \, dx$	$\int \frac{1}{y} \, dy = \int x \, dx$
$= y\frac{x^2}{2} + c$	$\ln y = \frac{x^2}{2} + c$	$\ln y = \frac{x^2}{2} + c$
$y\left(1 - \frac{x^2}{2}\right) = c \Rightarrow y = \frac{c}{1 - \frac{x^2}{2}}$	$y = Ae^{\frac{x^2}{2}}$	$y = e^{\frac{x^2}{2}} + c$

If $\frac{dy}{dx}$ depends just on y you can use the fact that $\frac{dx}{dy} = \frac{1}{\frac{dy}{dx}}$.

Key point 13.2

If $\frac{dy}{dx} = f(y)$ then $x = \int \frac{1}{f(y)} \, dy$.

WORKED EXAMPLE 13.4

Newton's law of cooling states that the rate of change of temperature of a cooling body is proportional to the difference between the body's temperature and the temperature of the surroundings.

A cup of coffee cools in the room where the air temperature is 21 °C. The temperature of the coffee, θ °C, satisfies the differential equation

$\frac{d\theta}{dt} = -0.054(\theta - 21)$ where t is the time, measured in minutes.

The initial temperature of the coffee is 94 °C. Find the temperature of the coffee after six minutes.

$$\frac{dt}{d\theta} = \frac{1}{-0.054(\theta - 21)}$$

You want to integrate with respect to θ so take the reciprocal of both sides...

So:

$$t = \int \frac{1}{-0.054(\theta - 21)} \, d\theta$$

... and integrate.

$$= -\frac{1}{0.054} \ln(\theta - 21) + c$$

Continues on next page

When $t = 0, \theta = 94$:

$0 = -\frac{1}{0.054}\ln(94-21)+c$ — Use the initial conditions to find c.

$\Rightarrow c = \frac{\ln 73}{0.054} \approx 79.5$

When $t = 6$: — Substitute in the given value of t before rearranging.

$6 = -\frac{1}{0.054}\ln(\theta-21)+79.5$

$\ln(\theta-21) = 73.5 \times 0.054 = 3.97$

$\theta = 21 + e^{3.97} = 73.8\,°\text{C}$

EXERCISE 13B

1 Find the particular solution of each differential equation, giving your answer in the form $y = f(x)$ simplified as far as possible.

a **i** $\frac{dy}{dx} = \frac{2x^2}{3y}$, $y = 0$ when $x = 0$ **ii** $\frac{dy}{dx} = 4xy^2$, $y = 1$ when $x = 0$

b **i** $\frac{dy}{dx} = \frac{4y}{x}$, $y = 2$ when $x = 1$ **ii** $\frac{dy}{dx} = -3x^2y$, $y = 3$ when $x = 0$

2 Find the particular solution of each differential equation. You do not need to give the equation for y explicitly.

a **i** $\frac{dy}{dx} = \frac{\sin x}{\cos y}$, $y = 0$ when $x = \frac{\pi}{3}$ **ii** $\frac{dy}{dx} = \frac{\sec^2 x}{\sec^2 y}$, $y = 0$ when $x = \frac{\pi}{3}$

b **i** $\frac{dy}{dx} = x^2y$, $y = 1$ when $x = 0$ **ii** $\frac{dy}{dx} = \frac{y^2}{x}$, $y = 1$ when $x = 1$

c **i** $\frac{dy}{dx} = 2e^{x+2y}$, $y = 0$ when $x = 0$ **ii** $\frac{dy}{dx} = e^{x-y}$, $y = 2$ when $x = 0$

3 Find the general solution of each differential equation, giving your answer in the form $y = f(x)$ simplified as far as possible.

a **i** $2y\frac{dy}{dx} = 3x^2$ **ii** $\frac{1}{y^2}\frac{dy}{dx} = 2x$ **b** **i** $x\frac{dy}{dx} = \sec y$ **ii** $(x-2)\frac{dy}{dx} = \cos^2 y$

c **i** $(x-1)\frac{dy}{dx} = x(y+3)$ **ii** $\frac{(1-x^2)\,dy}{dx} = xy + y$

4 Solve the differential equation $\frac{dy}{dx} = 2y(1-x)$ given that when $x = 1$, $y = 1$. Give your answer in the form $y = f(x)$.

5 The function $H(t)$ satisfies the differential equation $t\frac{dH}{dt} = H$. When $t = 1$, $H = 2$. Find the value of H when $t = 5$.

6 Given that $\frac{dN}{dt} = -kN$, where k is a positive constant, show that $N = Ae^{-kt}$.

7 Find the general solution of the differential equation $x\frac{dy}{dx} + 4 = y^2$, giving your answer in the form $y = f(x)$.

8 Given that $\frac{dy}{dx} = \sqrt{\frac{1-y^2}{1-x^2}}$ and that $y = \frac{\sqrt{3}}{2}$ when $x = \frac{1}{2}$, show that $2y = x\sqrt{k} + \sqrt{1-x^2}$, where k is a constant to be found.

Elevate

See Support Sheet 13 for a further example of solving a separable differential equation and for more practice questions.

Section 3: Modelling with differential equations

Now that you can solve various differential equations, you can look at how such equations arise in a variety of contexts. There are many situations where you know (or it is reasonable to assume) what the rate of change of a quantity depends on. You can then write down and solve a differential equation to find out how the actual quantity behaves.

Rewind

You met the idea of the rate of change of a quantity in practical applications being $\frac{\mathrm{d}}{\mathrm{d}t}$ in Chapter 12, Section 3. You have also met specific rates of change, such as acceleration $a = \frac{\mathrm{d}v}{\mathrm{d}t}$, in Student Book 1, in mechanics.

WORKED EXAMPLE 13.5

A skydiver of mass 60 kg jumps out of an airplane with zero initial velocity. The air resistance is proportional to velocity and may be modelled as $R = 0.8v$.

Write and solve a differential equation to find an expression for the velocity of the skydiver in terms of time. Give your answer in terms of g.

Forces on the skydiver:

Always start by drawing a force diagram! The weight acts downwards and the air resistance upwards.

$$F = ma$$

$$60\frac{\mathrm{d}v}{\mathrm{d}t} = 60g - 0.8v$$

Write Newton's second law equation, taking downwards as positive (because that's the direction of motion).

$$\frac{\mathrm{d}t}{\mathrm{d}v} = \frac{60}{60g - 0.8v}$$

Since the right-hand side is in terms of v, re-write the equation in terms of $\frac{\mathrm{d}t}{\mathrm{d}v}$.

$$t = \int \frac{60}{60g - 0.8v}\,\mathrm{d}v$$

$$= \frac{60}{-0.8}\ln|60g - 0.8v| + c$$

$$= -75\ln|60g - 0.8v| + c$$

Now integrate with respect to v.

When $t = 0, v = 0$:

$$0 = -75\ln 60g + c$$

$$\Rightarrow c = 75\ln 60g$$

Now use the initial condition.

So:

$$t = -75\ln|60g - 0.8v| + 75\ln 60g$$

You need to express v in terms of t.

$$\ln|60g - 0.8v| = \ln 60g - \frac{t}{75}$$

$$60g - 0.8v = 60g\,\mathrm{e}^{-\frac{t}{75}}$$

$$v = 75g - 75g\,\mathrm{e}^{-\frac{t}{75}}$$

Here you have used:

If $\ln A = \ln B - C$ then $A = \mathrm{e}^{\ln B - C} = B\mathrm{e}^{-C}$

Notice that on the penultimate line of Worked example 13.5 the modulus sign was removed. This is only justified if $588 - 0.8v \geqslant 0$ i.e., $v \leqslant 735$. Since initially $v = 0$, this is certainly true for some initial part of the motion. But you should ask whether eventually v becomes larger than 735; if it does then the equation needs to be changed. Looking at the final solution, $e^{-\frac{t}{75}}$ is always positive, so v is in fact always less than 735 and the solution is valid for all t.

Focus on...

There are a number of examples of modelling real life situations with differential equations in Focus on... Modelling 2.

Did you know?

In fact, $e^{-\frac{t}{75}}$ decreases and tends to zero, so v increases towards 735 without ever reaching it.

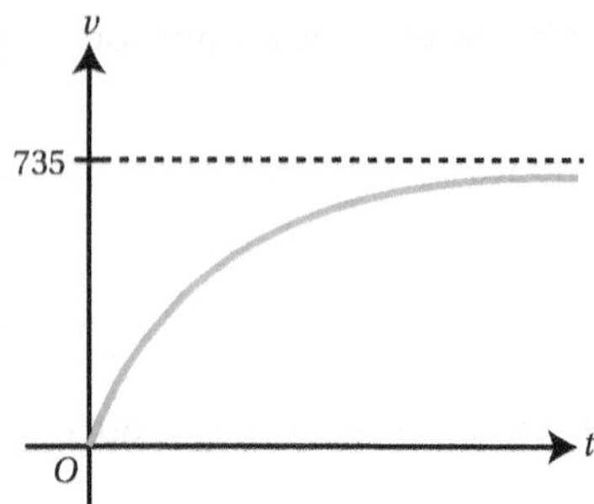

This maximum velocity is called the **terminal velocity**. Of course, the object might hit the ground before getting close to this velocity.

Often you have a model where some constants are unknown. You can find them by using experimental or observational information.

WORKED EXAMPLE 13.6

In a simple model of a population of bacteria, the growth rate is assumed to be proportional to the number of bacteria.

a Let N thousand be the number of bacteria after t minutes. Initially there are 2000 bacteria and this number increases to 10 000 after 12 minutes. Write and solve a differential equation to find the number of bacteria after t minutes.

b Comment on one limitation of this model.

a $\frac{dN}{dt} = kN$

The rate of growth is $\frac{dN}{dt}$.

'Proportional to N' means you can write it as kN for some constant k.

$\frac{dt}{dN} = \frac{1}{kN}$

$\Rightarrow t = \frac{1}{k}\ln N + c$

You have information about t and N, but not about $\frac{dN}{dt}$. So you need to solve the differential equation before you can put in the numbers.

Continues on next page

When $t = 0, N = 2000$:

To find the constants c and k, use the two given conditions: first use $N = 2000$ initially (i.e. when $t = 0$).

$$0 = \frac{1}{k}\ln 2000 + c$$

$$\Rightarrow c = -\frac{1}{k}\ln 2000$$

When $t = 12, N = 10\,000$:

A second equation is needed to find k so next use the condition that $N = 10\,000$ when $t = 12$.

$$12 = \frac{1}{k}\ln 10\,000 - \frac{1}{k}\ln 2000$$

$$12 = \frac{1}{k}\ln 5$$

Using $\ln A - \ln B = \ln\left(\frac{A}{B}\right)$.

$$k = \frac{\ln 5}{12}$$

$$\therefore c = -\frac{12\ln 2000}{\ln 5}$$

So:

$$t = \frac{12}{\ln 5}\ln N - \frac{12\ln 2000}{\ln 5}$$

$$t\ln 5 = 12\ln N - 12\ln 2000$$

$$t\ln 5 = 12\ln\left(\frac{N}{2000}\right)$$

$$\frac{N}{2000} = e^{\frac{t\ln 5}{12}}$$

$$N = 2000e^{\left(\frac{\ln 5}{12}\right)t}$$

Put k and c back in and rearrange the equation to get N in terms of t.

Check that you can follow all the steps!

b This model predicts unlimited population growth, which is not realistic.

Exponential growth models often work initially, but need to be adapted for longer time periods.

Sometimes a problem has several variables and you need to use the geometric context and related rates of change to produce a single differential equation.

Rewind

Remember, from Chapter 12, Section 3, that in connected rate of change problems you need to start by expressing all the information given in terms of derivatives. Then use the chain rule to link these derivatives.

WORKED EXAMPLE 13.7

A cylindrical tank with cross-sectional area 5 m^2 and height 4 m is initially filled with water. The water leaks out of the tank out of a small hole at the bottom at the rate of $0.08\sqrt{h}$ m^3 s^{-1}, where h m is the height of water in the tank after t seconds.

a Find an equation for $\frac{dh}{dt}$ in terms of h.

b Hence find how long it takes for the tank to empty.

Continues on next page

a Let V be the volume of the tank.

Define any variables not already defined in the question.

$V = 5h$

So, $\frac{\mathrm{d}V}{\mathrm{d}h} = 5$

Express all the given information in terms of derivatives. Use here that volume of cylinder = cross-sectional area × height

Also, $\frac{\mathrm{d}V}{\mathrm{d}t} = -0.08\sqrt{h}$

Water leaks out at the rate of $0.08\sqrt{h}$ m^3 s^{-1}. This tells you the rate of change of volume. It is negative as the volume is decreasing.

$$\frac{\mathrm{d}h}{\mathrm{d}t} = \frac{\mathrm{d}h}{\mathrm{d}V} \times \frac{\mathrm{d}V}{\mathrm{d}t}$$

You want to find $\frac{\mathrm{d}h}{\mathrm{d}t}$. Use the chain rule to express this in terms of V as this appears in the previous expressions.

$$= \frac{1}{5} \times (-0.08\sqrt{h})$$

$$= -0.016\sqrt{h}$$

$\frac{\mathrm{d}V}{\mathrm{d}h} = 5 \Rightarrow \frac{\mathrm{d}h}{\mathrm{d}V} = \frac{1}{5}$

b $\frac{\mathrm{d}t}{\mathrm{d}h} = \frac{1}{-0.016\sqrt{h}} = -62.5h^{-\frac{1}{2}}$

Solve the equation by integrating $\frac{\mathrm{d}t}{\mathrm{d}h}$.

$$\Rightarrow t = -62.5\int h^{-\frac{1}{2}}\,\mathrm{d}h$$

$$t = -125h^{\frac{1}{2}} + c$$

When $t = 0, h = 4$:

Initially the tank is full.

$$0 = -125\sqrt{4} + c$$

$$\Rightarrow c = 250$$

When $h = 0$:

The tank is empty when $h = 0$.

$$t = -125\sqrt{h} + 250$$

$$= 250$$

So the tank is empty after 250 seconds.

EXERCISE 13C

1 Write differential equations to describe each situation. You do not need to solve the equations.

a **i** A population increases at a rate equal to five times the size of the population (N).

ii The mass of a substance (M) decreases at a rate equal to three times the current mass.

b **i** The rate of change of velocity is directly proportional to the velocity and inversely proportional to the square root of time.

ii A population size increases at a rate proportional to the square root of the population size (N) and to the cube root of time.

c i The area of a circular stain increases at a rate proportional to the square root of the radius. Find an equation for the rate of change of radius with respect to time.

ii The volume of a sphere decreases at a constant rate of $0.8\ \text{m}^3\ \text{s}^{-1}$. Find an equation for the rate of decrease of the radius.

2 In a simple model of a population of bacteria, the rate of growth is proportional to the size of the population.

When the population size (N) is 5000 bacteria, the rate of growth is 1000 bacteria/minute.

a Show that $\frac{dN}{dt} = 0.2N$.

b Initially there were 700 bacteria. Solve the differential equation and predict the number of bacteria after 20 minutes.

3 The mass of a radioactive substance (M g) decays at a rate proportional to the mass. Initially the mass is 12 g and it decays at the rate of $2.4\ \text{g s}^{-1}$.

a Show that $\frac{dM}{dt} = -\frac{1}{5}M$.

b Find the amount of time it takes for the mass to halve.

4 A particle of mass 1.2 kg is moving with speed $8\ \text{m s}^{-1}$ in a straight line on a horizontal table. A resistance force is applied to the particle in the direction of motion. The magnitude of the force is proportional to the square of the speed, so that $|F| = 0.3v^2$.

a Show that $\frac{dv}{dt} = -0.25v^2$.

b Find an expression for the velocity in terms of time, and hence find how long it takes for the speed to decrease to below $2\ \text{m s}^{-1}$.

5 The population of the fish in a lake, N thousand, can be modelled by the differential equation

$$\frac{dN}{dt} = (0.8 - 0.14t)N$$

where t is the time, in years, since the fish were first introduced into the lake. Initially there were 2000 fish.

a Show that the population initially increases and find when it starts to decrease.

b Find the expression for N in terms of t.

c Hence find the maximum population of the fish in the lake.

d What does this model predict about the size of the population in the long term?

6 In Economics, there is a model for how the demand (Q) for a product depends on its price (P). It states that the rate of change of Q with respect to P is proportional to Q but inversely proportional to P.

a Explain how this model leads to the differential equation

$$\frac{1}{Q}\frac{dQ}{dP} = \frac{\varepsilon}{P}$$

where ε is a negative constant (called *elasticity*).

b Find the general solution of this differential equation.

c Sketch the graph of Q against P, and describe how demand depends on price, in each case.

i $\varepsilon = -1$ ii $\varepsilon = 0$

7 Newton's law of cooling states that the rate of change of temperature of the body is proportional to the difference in the temperature between the body and its surroundings.

A bottle of milk is taken out of the fridge, which is at the temperature of 5 °C, and placed on the table in the kitchen, where the temperature is 19 °C. Initially, the milk is warming up at the rate of 4.2 °C per minute.

a Show that $\frac{d\theta}{dt} = 0.3(19 - \theta)$, where θ °C is the temperature of the milk and t is time in minutes since the milk was taken out of the fridge.

The temperature of the milk is measured to the nearest degree.

b Solve the differential equation and hence find how long, to the nearest minute, it takes for the temperature of the milk to reach the kitchen temperature.

8 A particle of mass 3 kg is pulled through liquid by means of a light inextensible string. The tension in the string is 12 N. The resistance force is proportional to the velocity, and equals $2.4v$. The particle starts from rest.

a Find an expression for the velocity after time t.

b Describe the velocity of the particle for large values of t.

9 An inverted cone has base radius 4 cm and height 10 cm.

The cone is filled with water at a constant rate of 80 cm^3 s^{-1}.

a Show that the height of water (h) satisfies the differential equation

$$\pi h^2 \frac{dh}{dt} = 500.$$

b Given that the cone is initially empty, find how long it takes to fill it.

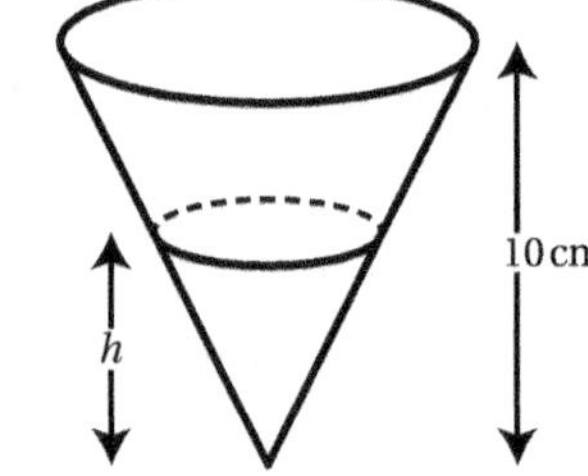

10 A particle of mass 0.4 kg is moving at the speed of 10 m s^{-1} when it enters a viscous liquid at the point B. Inside the liquid the resistance force is proportional to the velocity, and is initially equal to 1.2 N.

a Show that the velocity of the particle satisfies the differential equation $\frac{dv}{dt} = -0.3v$.

b Find an expression for the velocity of the particle t seconds after entering the liquid.

c Find the displacement of the particle from point B, and describe what happens to the displacement for large values of t.

Checklist of learning and understanding

- A **differential equation** is an equation for the derivative of a function. To solve a differential equation means to find an expression for the function itself.
- Some differential equations can be solved by **separation of variables.** Write the equation in the form $g(y)\frac{dy}{dx} = f(x)$ and integrate both sides.
- **Initial conditions** can be used to find the constant of integration.
- A differential equation often describes the rate of change of a quantity – this is the derivative with respect to time.
- The rate of change is often proportional to one of the variables. You may need to use given information to find the constant of proportionality.
- Sometimes a problem involves more than one variable and you need to use related rates of change to write a differential equation.

Mixed practice 13

1 A car depreciates at a rate proportional to its value, V, at time t years after it was made. $k > 0$ is a constant.

Find a possible differential equation to model the value of the car.

Choose from these options.

A $\frac{dV}{dt} = kV$ **B** $\frac{dV}{dt} = \frac{k}{V}$ **C** $\frac{dV}{dt} = -kV$ **D** $\frac{dV}{dt} = -\frac{k}{V}$

2 A cake is placed in the oven. Its temperature, T °C is modelled by the differential equation $\frac{dT}{dt} = k(150 - T)$ where k is a positive constant and t is the time, in minutes, after it was put in the oven. What is the maximum temperature the cake can reach?

Choose from these options.

A $150k$ °C **B** 150 °C **C** k °C **D** There is no maximum.

3 Solve the differential equation $\frac{dy}{dx} = 2xy - 6x$ given that $y = 4$ when $x = 0$. Give your answer in the form $y = f(x)$.

4 Find the particular solution of the differential equation $\frac{dy}{dx} = \cos x \cos^2 y$ such that $y = \frac{\pi}{4}$ when $x = \frac{\pi}{6}$.

5 In a chemical reaction, the amount of the reactant (M g) follows the differential equation $\frac{dM}{dt} = 3 - 0.5M$.

Initially there was 2 g of the reactant.

Show that $M = A - Be^{-\frac{t}{2}}$, where A and B are constants to be found.

6 **a** Find $\int x\sqrt{x^2+3}\,dx$.

b Solve the differential equation

$$\frac{dy}{dx} = \frac{x\sqrt{x^2+3}}{e^{2y}}$$

given that $y = 0$ when $x = 1$. Give your answer in the form $y = f(x)$.

[© AQA 2013]

7 A population of fish initially contains 250 fish and increases at the rate of 10 fish per month. Let N be the number of fish after t months.

In a simple model of population growth the rate of increase is directly proportional to the population size.

a Show that $\frac{dN}{dt} = 0.04N$.

b Solve the differential equation and find how long it takes for the number of fish to reach 1000.

c Comment on the long-term suitability of this model.

An improved model takes into account seasonal variation:

$$\frac{dN}{dt} = 0.04N\left(1 + 2.5\cos\left(\frac{\pi t}{6}\right)\right)$$

d Given that initially there are 250 fish, find an expression for the size of the population after t months.

8 A model for a relationship between the price (P) of a commodity and the demand (Q) for the commodity states that

$$\frac{dQ}{dP} = \frac{\varepsilon Q}{P}$$

where ε is the elasticity.

a Find the general solution to this differential equation.

b Describe the relationship between the price and demand when $\varepsilon = -\frac{1}{2}$.

c For most commodities, $\varepsilon \leqslant 0$. Suggest, with a reason, what sort of commodity might have $\varepsilon > 0$.

9 A cylindrical tank has radius 2 m and height 12 m. The tank is being filled with water so that $\frac{dV}{dt} = \frac{\pi}{5}(10-h)^2$. Initially the tank is empty.

a Show that $20\frac{dh}{dt} = (10-h)^2$ and hence find an expression for the height of water in the tank at time t, giving your answer in the form $h = A - \frac{B}{t+2}$.

b Hence or otherwise, determine whether the tank will ever completely fill with water.

10 A ball of mass 300 g falls vertically downwards. When the velocity of the ball is v m s^{-1} the air resistance has magnitude $1.47v$.

a Show that $\frac{dv}{dt} = 4.9(2-v)$.

b The ball falls from rest.

Show that $v = C(1-e^{-kt})$, stating the values of constants C and k.

c Hence describe the motion of the ball.

11 A giant snowball is melting. The snowball can be modelled as a sphere whose surface area is decreasing at a constant rate with respect to time. The surface area is A cm^2 at time t days after it begins to melt.

a Write down a differential equation in terms of the variables A and t and a constant k, where $k > 0$, to model the melting snowball.

b **i** Initially the radius of the snowball is 60 cm, and 9 days later, the radius has halved.

Show that $A = 1200\pi(12-t)$.

(You may assume that the surface area of a sphere is given by $A = 4\pi r^2$, where r is the radius.)

ii Use this model to find the number of days that it takes the snowball to melt completely.

[© AQA 2011]

12 Consider this model of population growth:

$$\frac{dN}{dt} = 1.2N - 0.4N^2$$

where N thousand is the population size at time t months.

a Suggest what the term $-0.4N^2$ could represent.

b Given that initially $N = 1.5$, solve the differential equation. Give your answer in the form $N = \frac{a}{1+e^{bt}}$, where a and b are constants.

c Hence describe what happens to the population in the long term.

13 Variables x and y satisfy the differential equation $\frac{dy}{dx} = e^{x-ey}$.

Given that $y = \frac{1}{e}$ when $x = 0$, show that $x - ey + 1 = 0$.

14 A particle moves in a straight line. Its acceleration depends on the displacement:

$$\frac{dv}{dt} = -8e^{-4x}.$$

a Find an expression for $\frac{dv}{dx}$ in terms of x and v.

b Initially the particle is at the origin and its speed is $2\ \mathrm{m\,s^{-1}}$.

Show that $v = 2e^{-2x}$.

c Find expressions for the displacement and velocity in terms of time.

Elevate

See Extension Sheet 13 for questions on some other types of differential equation.

14 Numerical solution of equations

In this chapter you will learn how to:

- work with equations that cannot be solved by algebraic rearrangement
- find an interval that contains a root of an equation, and how to check that a given solution is correct to a specified degree of accuracy (the sign-change method)
- approximate a part of the curve by a tangent, and use this to find an improved guess for a solution (the Newton–Raphson method)
- create a sequence that converges to a root of an equation (fixed point iteration)
- identify situations in which the above methods fail to find a solution.

Before you start...

Chapter 4	You should be able to use the term-to-term rule to generate a sequence.	1 Find the first four terms of the sequence defined by $x_{n+1} = 5x_n - 2x_n^2, x_1 = 1$.
Chapter 2; Chapter 8	You should be able to rearrange equations involving polynomials, fractions, exponentials, logarithms and trigonometric functions.	2 Rearrange each equation into the required form. a $x = 3\ln(x+2)$ into $x = \mathrm{e}^{kx} - C$ b $x = 2\sqrt{x^2 - 3}$ into $x = \frac{1}{2}\sqrt{x^2 + k}$ c $x\cos x - 3\sin x = 0$ into $x = \arctan\left(\frac{x}{a}\right)$
Chapter 9	You should be able to differentiate a variety of functions.	3 Differentiate each equation. a $y = 3x^2 \tan x$ b $y = \frac{\ln x}{x^2}$ c $y = \mathrm{e}^{3x^2} - \ln(2x)$

What is a numerical solution?

Throughout your studies of mathematics you have learnt to solve various types of equations: linear, quadratic, exponential, trigonometric. You can solve some equations by algebraic rearrangement, for example:

$$2x - 3 = 2$$
$$x = \frac{2+3}{2} = 2.5$$

Sometimes you can write solutions in an exact form, such as:

$$x^2 - 1 = 4$$
$$x = \pm\sqrt{5}$$

or

$$\sin 2x = \frac{1}{2} \quad (\text{for } 0 \leqslant x \leqslant \pi)$$
$$x = \frac{\pi}{12}, \frac{5\pi}{12}.$$

Alternatively, you can use your calculator to find approximate solutions; for example:

$$\sin x = \frac{1}{3} \quad \left(\text{for } 0 \leqslant x \leqslant \frac{\pi}{2}\right)$$

$$x = 0.340 \text{ (3 s.f.)}$$

Tip

You can write the solution of $\sin x = \frac{1}{3}$ in an exact form: it is just $x = \arcsin\left(\frac{1}{3}\right)$.

But there are some equations that you can't solve by any combination of algebraic rearrangement.

For example, how would you solve the equation $3 \sin x = 2x$? However you try rearranging it, you can't get it into the form $x =$ a number.

If you have a graphical calculator or graphing software you can use them to find approximate solutions by drawing the graphs of $y = 3 \sin x$ and $y = 2x$ and finding their intersection.

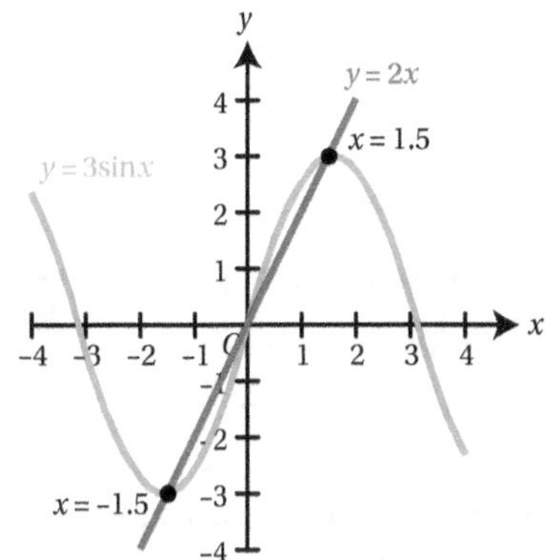

Your calculator might have a 'solver' function that allows it to find approximate solutions without drawing the graph. But how does it do that?

In this chapter you will learn several different methods for solving equations numerically (this means that a solution is not found by algebraic manipulation, but by evaluating certain expressions to find increasingly accurate approximations). You should be aware that these methods can only give approximate solutions, but you can choose the level of accuracy. You will also see that in some situations these methods fail to find an answer. You will learn how to recognise such situations.

Section 1: Locating roots of a function

The simplest numerical methods involve finding an interval in which the solution lies, and then improving accuracy by making this interval smaller.

For example, consider the equation $3 \sin x = 2x$. It is easier to see how the method works if you rearrange the equation into the form 'equation = 0':

$$3 \sin x - 2x = 0$$

A sensible way to start is to try to some values of x and see if any of them give an answer close to zero.

x	0	1	2	3
$3 \sin x - 2x$	0	0.524	−1.27	−5.58

From the table you can see that $x = 0$ is a solution. But you can also see that the value of the expression $3 \sin x - 2x$ changes sign between $x = 1$ and $x = 2$: it is positive when $x = 1$ but negative when $x = 2$. This means that it must equal zero somewhere between these two x-values.

You can see this if you draw the graph of $y = 3\sin x - 2x$. If the y-coordinate changes from positive to negative, then the graph must cross the x-axis.

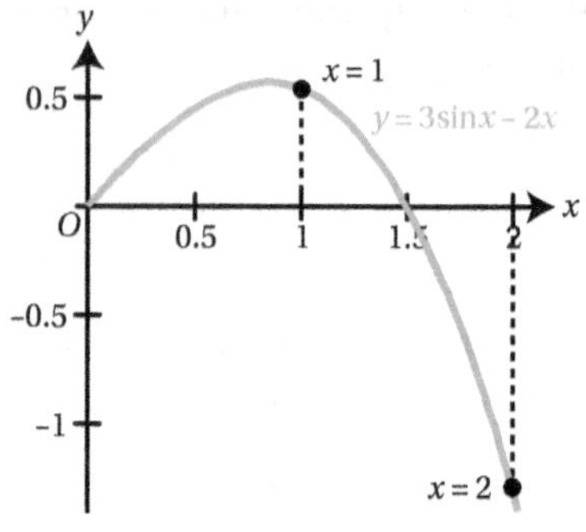

You can try to locate this solution more accurately by looking for a smaller interval. For example:

x	1	1.5	2
$3\sin x - 2x$	0.524	0.007 52	−1.27

This tells you that the solution is between 1.5 and 2. You can continue finding smaller and smaller intervals to find the solution to any degree of accuracy you want.

Tip

This method produces only one solution. The equation $3\sin x - 2x = 0$ also has a negative solution, between −2 and −1.5. In general, with numerical methods you can never be sure that you have found all the solutions.

Key point 14.1

The sign-change rule

If $f(x)$ is a **continuous function** and a and b are numbers such that $f(x)$ changes sign between a and b, then the equation $f(x) = 0$ has a root between a and b.

Common error

Note that the converse is not true: if $f(x) = 0$ has a root between a and b then $f(x)$ does not necessarily change sign between a and b. This is because there could be more than one root in the interval.

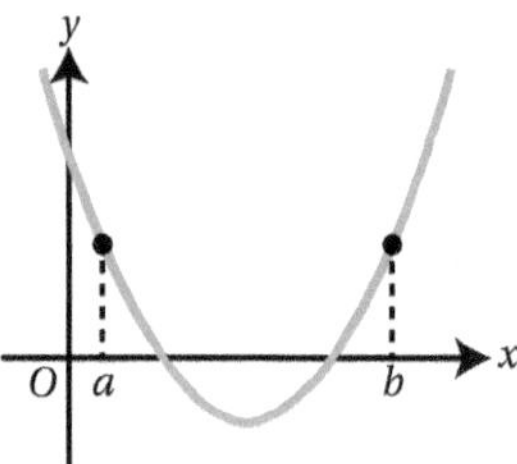

If the graph of $y = f(x)$ is not a continuous line (for example, when it has a vertical asymptote) then it is possible for $f(x)$ to change sign without the graph crossing the x-axis.

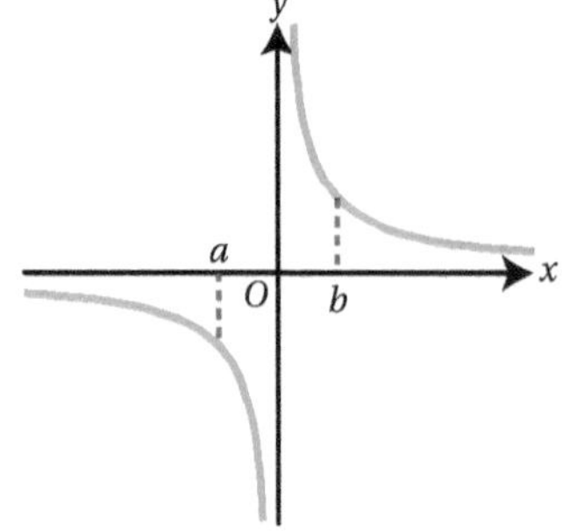

WORKED EXAMPLE 14.1

Let $f(x) = \frac{5}{x} - 2x + x^2$.

a Show that the equation $f(x) = 0$ has a solution between –1 and –2.

b **i** Evaluate $f(-0.5)$ and $f(0.5)$.

ii Explain why in this case, the change of sign does not imply that $f(x) = 0$ has another solution between –0.5 and 0.5.

a $f(-1) = \frac{5}{-1} - 2(-1) + (-1)^2 = -2 < 0$ — Show clearly which values of x you are using.

$f(-2) = \frac{5}{-2} - 2(-2) + (-2)^2 = 5.5 > 0$

There is a change of sign, so there is a root between $x = -2$ and $x = -1$. — State the conclusion.

b i $f(-0.5) = \frac{5}{-0.5} - 2(-0.5) + (-0.5)^2$ — Show clearly which values of x you are using.

$= -8.75$

$f(0.5) = \frac{5}{0.5} - 2(0.5) + (0.5)^2 = 9.25$

ii $f(x)$ has a vertical asymptote at $x = 0$, so the change of sign does not imply that the graph crosses the x-axis. — Think about the graph: if it has an asymptote then it does not necessarily cross the x-axis.

$\frac{1}{x}$ is not defined for $x = 0$, so the graph will have an asymptote there.

In practice, rather than knowing that a solution lies in a particular interval, you want to express a solution you already have to a specified degree of accuracy, such as one decimal place. Consider, as an example, the equation $\ln x = \frac{x}{4}$. By trying some values of x you can find that this equation has a solution around 8.6. But how can you check that this is in fact correct to one decimal place?

The solution, when rounded to one decimal place, equals 8.6. But the numbers that round to 8.6 are between 8.55 and 8.65, so you should look for a change of sign between those two values.

WORKED EXAMPLE 14.2

Show that the equation $\ln x = \frac{x}{4}$ has a root $x = 8.6$, correct to one decimal place.

Let $f(x) = \ln x - \frac{x}{4}$. — First write the equation in the form $f(x) = 0$.

$f(8.55) = \ln 8.55 - \frac{8.55}{4} = 0.008\,43 > 0$ — Then look for a change of sign.

$f(8.65) = \ln 8.65 - \frac{8.65}{4} = -0.004\,94 < 0$

There is a change of sign, so there is a root between 8.55 and 8.65.

So this root is 8.6 to 1 d.p.

You should note that you are looking at the degree of accuracy of the x-value here, not the y-value. For the equation from Worked example 14.2, $f(8.62) = 0.000\,91$ which equals 0 to two decimal places. However, the root is not is not equal to 8.62 to 2 d.p. Look at this table of values.

x	8.605	8.615	8.625
$f(x) = \ln x - \frac{x}{4}$	0.001 09	−0.000 254	−0.001 59

The change of sign actually occurs between 8.605 and 8.615 so the root is 8.61 correct to 2 d.p.

WORK IT OUT 14.1

Find the smallest positive root of $\tan x = x$ correct to two decimal places.

Which is the correct solution? Identify the errors made in the incorrect solutions.

Solution A	Solution B	Solution C
$f(x) = \tan x - x$ $f(1.565) = \tan 1.565 - 1.565$ $= 170.956\,121\,8 > 0$ $f(1.575) = \tan 1.575 - 1.575$ $= -239.460\,787\,2 < 0$	$f(x) = \tan x - x$ $f(4.48) = \tan 4.48 - 4.48$ $= -0.254\,613\,297 < 0$ $f(4.50) = \tan 4.50 - 4.50$ $= 0.137\,332\,054\,6 > 0$	$f(x) = \tan x - x$ $f(4.485) = \tan 4.485 - 4.485$ $= -0.1633 < 0$ $f(4.495) = \tan 4.495 - 4.495$ $= 0.0324 > 0$
As there is a change of sign, there is a root between 1.565 and 1.575.	As there is a change of sign, there is a root between 4.48 and 4.50.	As there is a change of sign, there is a root between 4.485 and 4.495.
So the root is 1.57 to 2 d.p.	So the root is 4.49 to 2 d.p.	So the root is 4.49 to 2 d.p.

Focus on...

Focus on... Problem solving 2 compares a numerical and analytical solution to the same problem.

EXERCISE 14A

1 Classify these equations as 'can find exact solutions' or 'cannot rearrange algebraically'.

a $x^2 - 4x = 7$ b $e^{-x} = 4x$ c $\sin x = 3\tan x$ d $e^{4x} = 5$

e $\tan x = 3x^2$ f $3\ln 4x = 4$ g $e^{5x} = 3\ln x$ h $x^3 - 4x - 4 = 0$

2 Each equation has a root between −3 and 3. In each case, find two integers between which the root lies.

a $x^3 - 4x - 4 = 0$ b $2\sin x - 4x + 1 = 0$ c $25\ln x = 3x^2$ d $\cos(3 - x) + x^3 = -5$

3 For each equation, show that there is a root in the given interval.

a i $x - 5\ln x = 0$, between 1 and 2 ii $3e^{2x} - 4x^2 = 0$, between −1 and 0

b i $\cos 2x = \sqrt{x}$, between 0 and 1 ii $3\tan x = 5x^3$, between −1 and −0.5

4 For each equation, show that the given root is correct to the stated degree of accuracy.

a **i** $x^5 - 3x^2 + 1 = 0,\ x = 0.6$ (1 d.p.) **ii** $x^3 - 3x + 4 = 0,\ x = -2.2$ (1 d.p.)

b **i** $4\sin x - e^{-2x} = 0,\ x = 3.14$ (3 s.f.) **ii** $2\ln x - 3\cos x = 0,\ x = 1.36$ (3 s.f.)

c **i** $3\tan x = 5x^3,\ x = 1.31$ (2 d.p.) **ii** $2x = e^{-x},\ x = 0.35$ (2 d.p.)

d **i** $3e^x = x^4,\ x = 6.20$ (3 s.f.) **ii** $3\cos x = \ln x,\ x = 5.30$ (3 s.f.)

5 **a** Show that the equation $x^3 - 3x - 1 = 0$ has a solution between 1 and 2.

b Show that this solution equals 1.9 correct to one decimal place.

6 The equation $\ln\left(\frac{x}{3}\right) - \frac{x^2}{4} + 2 = 0$ has two solutions.

a Show that one of the solutions equals 0.425 correct to three significant figures.

b The other solution lies between positive integers k and $k+1$. Find the value of k.

7 A function is defined by $f(x) = \frac{x^2+2}{2x-5}$.

a Show that the equation $f(x) = 0$ has no solutions.

b **i** Evaluate f(2) and f(3).

ii Alicia says that the change of sign implies that the equation $f(x) = 0$ has a root between 2 and 3. Explain why she is wrong.

8 Let $g(x) = \cos 8x$.

a **i** Sketch the graph of $y = g(x)$ for $0 \leqslant x \leqslant \frac{\pi}{2}$.

ii State the number of solutions of the equation $g(x) = 0$ between 0 and $\frac{\pi}{4}$.

b **i** State the values of g(0) and $g\left(\frac{\pi}{4}\right)$.

ii George says, 'There is no change of sign between 0 and $\frac{\pi}{4}$ so the equation $g(x) = 0$ has no roots in this interval.' Use your graph to explain why George's reasoning is incorrect.

iii Use the change of sign method (without referring to the graph) to show that the equation $g(x) = 0$ has two roots between 0 and $\frac{\pi}{4}$.

Section 2: The Newton–Raphson method

The sign change rule allows you to show that a root of an equation lies in a certain interval, but how do you know which interval to try? In the first example, with $f(x) = 3\sin x - 2x$, the table of integer values of x showed that there is a solution between 1 and 2. You could then look at $x = 1.1, 1.2, \ldots$ to locate the change of sign more accurately.

x	1	1.1	1.2	1.3	1.4	1.5
$3\sin x - 2x$	0.524	0.007 52	0.396	0.291	0.156	−0.007

This shows that the root is between 1.4 and 1.5. You can continue like this until you make the interval as small as you like.

If you wanted to locate the root to three or four decimal places, this could take quite a long time. In some applications solutions of equations are required to even higher accuracy, and then this method becomes unfeasible, even with a fast computer.

The numbers in the table suggest that the root may be much closer to 1.5 than to 1.4, so maybe the next search should not start from 1.41 but from 1.48 or 1.49 – but how do you know which number? It would be good to have a method to tell you which number to try next. There are in fact many such methods. In this section you will meet the **Newton–Raphson method**, which uses the tangent to the graph of $f(x)$ to suggest where to look for the root.

The diagram shows the section of the graph of $y = 3\sin x - 2x$ for $1.3 \leqslant x \leqslant 1.6$. You can see that it crosses the x-axis between 1.4 and 1.5. The diagram also shows the tangent to the graph at $x = 1.4$. Near this point, the tangent follows the graph closely, so it will cross the x-axis near the root of the equation.

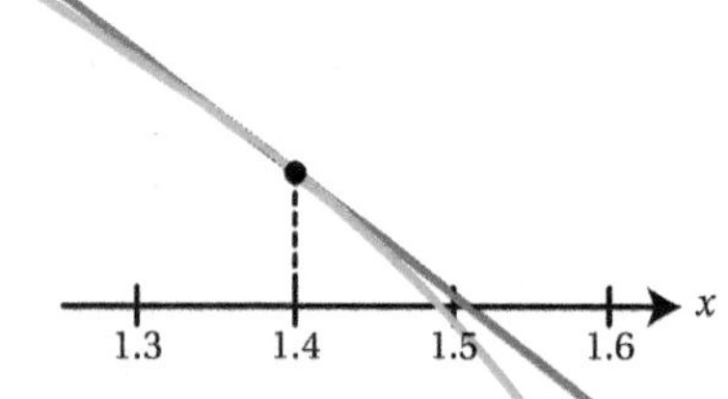

Use a graph plotter to draw the graph and the tangent and find where the tangent crosses the x-axis. You will find that the value is $x = 1.504\,925\ldots$

You can now repeat the same procedure by drawing a tangent at this new value of x; this new tangent crosses the x-axis at $x = 1.495\,85\ldots$

You can check that this gives the correct root to three decimal places:

$$\left.\begin{array}{l} f(1.4955) = 0.000\,499\,7 > 0 \\ f(1.4965) = -0.001\,276 < 0 \end{array}\right\} \Rightarrow x = 1.496 \text{ (3 d.p.)}$$

Repeating one more time gives the root correct to seven decimal places – this would have taken a very long time with the original sign-change method!

So far you have used a graph plotter to draw tangents and find their x-intercepts. But there is a formula to calculate the x-intercept of the tangent, without having to rely on the graph. You can find the equation of a tangent at any given point, and then use this equation to find where the tangent crosses the x-axis.

Key point 14.2

Newton–Raphson method

A sequence of approximations to the root of the equation $f(x) = 0$ is given by:

$$x_{n+1} = x_n - \frac{f(x_n)}{f'(x_n)}$$

This will be given in your formula book.

 Rewind

You met equations of tangents in Student Book 1, Chapter 13.

PROOF 8

The tangent to the graph at $x = x_0$ has gradient $m = \mathrm{f}'(x_0)$ and it passes through the point (x_0, y_0) where $y_0 = \mathrm{f}(x_0)$.

Therefore its equation is:

$$y - y_0 = m(x - x_0)$$
$$\Leftrightarrow y - \mathrm{f}(x_0) = \mathrm{f}'(x_0)(x - x_0)$$

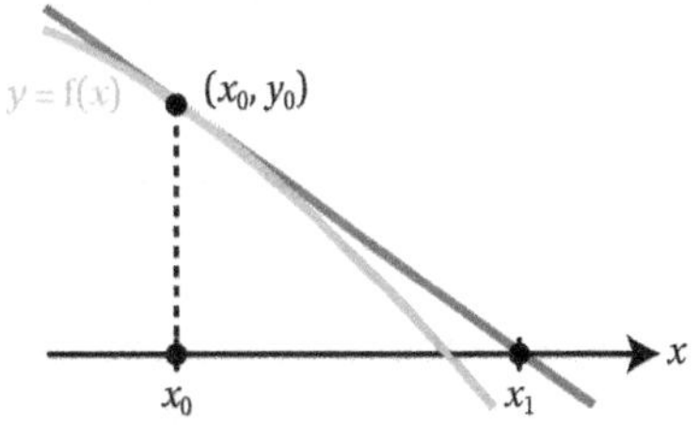

To find where this tangent crosses the x-axis set $y = 0$:

$$0 - \mathrm{f}(x_0) = \mathrm{f}'(x_0)(x - x_0)$$
$$\Leftrightarrow x - x_0 = -\frac{\mathrm{f}(x_0)}{\mathrm{f}'(x_0)}$$
$$\Leftrightarrow x = x_0 - \frac{\mathrm{f}(x_0)}{\mathrm{f}'(x_0)}$$

This is point x_1 as shown on the diagram.

You can repeat this process to find x_{n+1} from x_n.

WORKED EXAMPLE 14.3

a Show that the equation $e^{2x} - 10x = 0$ has a root between 1 and 2.
b Starting from $x_0 = 1.5$, use three repetitions of the Newton-Raphson method to find a better approximation for the root.
c Show that this approximation gives the root correct to three decimal places.

a Let $\mathrm{f}(x) = e^{2x} - 10x$
$\mathrm{f}(1) = e^2 - 10 = -2.61 < 0$
$\mathrm{f}(2) = e^4 - 20 = 34.6 > 0$

There is a change of sign, so there is a root between 1 and 2.

Use the sign change rule.

b $\mathrm{f}'(x) = 2e^{2x} - 10$

To use the Newton-Raphson formula you need $\mathrm{f}'(x)$.

$$x_1 = x_0 - \frac{e^{2x_0} - 10x_0}{2e^{2x_0} - 10} = 1.331\,44\ldots$$

Call the next approximation x_1.

$$x_2 = x_1 - \frac{e^{2x_1} - 10x_1}{2e^{2x_1} - 10} = 1.276\,655\ldots$$

Use x_1 to find x_2.

$x_3 = 1.271\,367\ldots.$

Use x_2 to find x_3.

c $\mathrm{f}(1.2705) = e^{2\times1.2705} - 12.705 = -0.0126\ldots < 0$
$\mathrm{f}(1.2715) = e^{2\times1.2715} - 12.715 = 0.0027\ldots > 0$

There is a change of sign, so there is a root between 1.2705 and 1.2715.
So the root is 1.271 (3 d.p.).

To show that the root equals 1.271 to 3 d.p. you need to show that it is between 1.2705 and 1.2715.

Repeating the Newton-Raphson calculation several times, as shown in Worked example 14.3, involves using the same formula with different numbers. Every time the number you need to put into the equation is the answer from the previous calculation. Most calculators have an ANS button that you can use to carry out such repetitive calculations. Here is how you can use it for Worked example 14.3.

 **Elevate**

See Support Sheet 14 for a further example of using the Newton-Raphson method and for more practice questions.

Start by entering the starting value of x_0 (in this case 1.5) and press the =/EXE button. Then type in ANS $- \frac{e^{2ANS} - 10ANS}{2e^{2ANS} - 10}$ and press the =/EXE button again.

The answer is the value of x_1. Pressing the =/EXE button repeatedly gives x_2, x_3 and so on. This way you can quickly generate as many approximations to the root as you like.

EXERCISE 14B

1 For each equation, use technology to sketch the graph. Draw the tangent at x_0 and find the next approximation to the root.

a i $x^2 - 10\ln x = 0,\ x_0 = 4$ ii $x - 2\cos x = 0,\ x_0 = 2$

b i $x^3 - 2x^2 - 1 = 0,\ x_0 = 2.5$ ii $5x - \frac{1}{2}x^4 - 3 = 0,\ x_0 = 1$

2 For each equation, use the Newton-Raphson method, with the given starting value, to find the root correct to three significant figures. Use the change of sign method to show that your root is correct to 3 s.f.

a i $x^4 - 3x + 1 = 0,\ x_0 = 1.5$ ii $3x - x^3 + 1 = 0,\ x_0 = 0$

b i $\sin\left(\frac{x}{2}\right) - x + 1 = 0,\ x_0 = 0$ ii $e^{0.2x} - 3\sqrt{x} = 0,\ x_0 = 11$

3 The equation $1 - x^2 + 2x^3 = 0$ has a root near –0.5. Using this value as the first approximation, use the Newton-Raphson method to find the next approximation to the root.

4 a Show that the equation $2x^2 - 1 = \frac{3}{x}$ can be written as $2x^3 - x - 3 = 0$.

b Given that the equation has a root near 1.5, use the Newton-Raphson method to find the next two approximations.

5 The equation $\sin x = \frac{2}{x}$ has a root near 6.5. Use the Newton-Raphson method to find the next two approximations to the root, giving your answers to five significant figures.

6 a Show that the equation $2 - x + \frac{4}{x} = 0$ has a root between 3 and 4.

b Using $x_0 = 3$ as the starting value, use the Newton-Raphson method to find the next approximation to the root. Give your answer correct to two decimal places.

c Show that the approximation from part **b** is correct to one decimal place, but not to two decimal places.

7 a Show that the equation $e^{-x} - 0.2x = 0$ has a root between 1 and 2.

b Use the Newton-Raphson method to find this root correct to three decimal places, and show that the root you found is correct to 3 d.p.

8 The equation $\cos 3x = \ln(x+1)$ has a root between 0 and 1. Use the Newton-Raphson method to find this root correct to three significant figures, and show that the solution you found is correct to 3 s.f.

Section 3: Limitations of the Newton–Raphson method

The Newton-Raphson method can find a root to a high degree of accuracy very quickly, so it seems to be a very good method. Unfortunately, there are some situations when it does not work.

For example, try to find a root of the equation $x^3 + x^2 - 0.2 = 0$. The sign-change check confirms that this equation has a root between 0 and 1.

The Newton-Raphson iteration formula is:

$$x_{n+1} = x_n - \frac{x_n^3 + x_n^2 - 0.2}{3x_n^2 + 2x_n}$$

If you start from $x_0 = 0$ you get $x_1 = 0 - \frac{-0.2}{0}$. But you cannot divide by zero, so it is impossible to find x_1. You could try changing the starting point slightly, for example $x_0 = 0.1$. Try it: the sequence moves away from the root but then comes back again.

Key point 14.3

The Newton-Raphson method does not work if the starting value is a stationary point of $f(x)$.

If the root is close to a stationary point, the sequence may initially move away from the root.

WORKED EXAMPLE 14.4

a Show that the equation $x^4 - 4x^3 - 7.5x^2 + 50x - 55 = 0$ has a root between 2 and 3.

b Explain why $x_0 = 2.5$ is not a suitable starting point for a Newton-Raphson iteration to find this root.

c Use the starting value $x_0 = 2.6$ to find the root correct to three decimal places.

a Let $f(x) = x^4 - 4x^3 - 7.5x^2 + 50x - 55$

Then

$f(2) = -1 < 0$

$f(3) = 0.5 > 0$

There is a change of sign, so $f(x) = 0$ has a root between 2 and 3.

Use the sign change rule.

b $f'(x) = 4x^3 - 12x^2 - 15x + 50$

$f'(2.5) = 0$

so there would be division by zero in the Newton-Raphson formula.

The Newton-Raphson method fails when $f'(x_0) = 0$, so find $f'(2.5)$ to check.

c $x_0 = 2.6$

$$x_{n+1} = x_n - \frac{x_n^4 - 4x_n^3 - 7.5x_n^2 + 50x_n - 55}{4x_n^3 - 12x_n^2 - 15x_n + 50}$$

Use the Newton-Raphson formula with $x_0 = 2.6$.

The sequence converges to 2.866 (3 d.p.).

Continue the sequence until you can see what the limit is. You are not being asked to write down all the iterations.

Another situation where the Newton–Raphson method can fail is if the function has an asymptote so that x_1 falls outside of the domain of the function.

WORKED EXAMPLE 14.5

a The equation $\frac{\ln x + 0.2}{x} = 0$ has a root between 0.5 and 2. Explain why a Newton–Raphson iteration from $x_0 = 1.5$ fails to find this root.

b Starting from $x_0 = 1$ carry out three iterations of the Newton–Raphson method, showing the values correct to three decimal places.

a $f(x) = \frac{\ln x + 0.2}{x}$

In order to use the Newton-Raphson formula first find $f'(x)$. You need to use the quotient rule.

$f'(x) = \frac{\frac{1}{x}x - 1(\ln x + 0.2)}{x^2}$

$= \frac{0.8 - \ln x}{x^2}$

$x_0 = 1.5$

Use the formula with $x_0 = 1.5$ to find x_1.

$x_1 = 1.5 - \frac{f(1.5)}{f'(1.5)} = -0.8$

$x_2 = -0.8 - \frac{f(-0.8)}{f'(-0.8)}$

In order to find x_2 you need to use the formula with $x_1 = -0.8$. But $f(x)$ has a vertical asymptote at $x = 0$ and is not defined for $x < 0$, so this is impossible.

But $f(x)$ is not defined for $x < 0$ so it is not possible to find x_2.

b $x_0 = 1$

$x_1 = 1 - \frac{f(1)}{f'(1)} = 0.750$

$x_2 = 0.810$

$x_3 = 0.819$

You can use graphing software to verify the result in part **a** of Worked example 14.5: you can see that the tangent at $x_0 = 1.5$ crosses the x-axis to the left of the vertical asymptote.

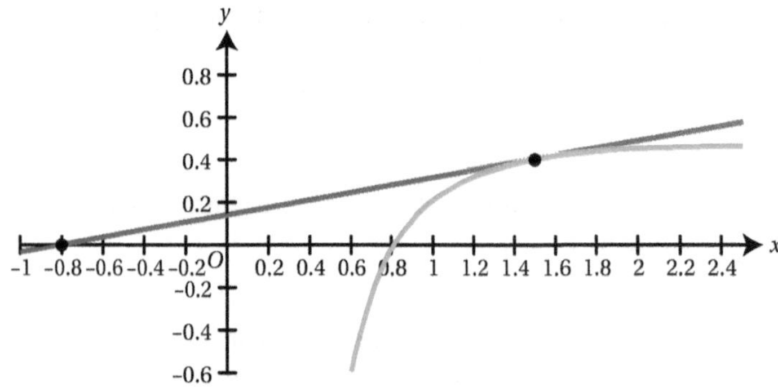

EXERCISE 14C

1. For each equation carry out one iteration of the Newton–Raphson method starting with the given value of x_0. In each case sketch the graph (using technology) to explain why x_1 is not a better approximation to the root than x_0.

   a i $2x^3 - 5x + 2,\ x_0 = 1$ ii $\sqrt{x} - 0.2x^3 - 0.5 = 0,\ x_0 = 1$

   b i $\tan x - x - 1 = 0,\ x_0 = 0.5$ ii $\frac{\ln 2x + x}{x} = 0,\ x_0 = 1$

2. Let $f(x) = 10x^3 - 5x^2 - 1$.

   a Find the x-coordinates of the stationary points of $f(x)$.

   b Show that the equation $f(x) = 0$ has a root between $\frac{1}{3}$ and 1.

   c The Newton–Raphson formula with $x_0 = 0.35$ is used to find x_1.
   Explain why x_1 may not be an improved approximation for the root.

3. The diagram shows the curve with equation $f(x) = (x-2)e^{-x} + 1$.

   Find the x-coordinate of the stationary point of $f(x)$ and hence explain, with an aid of a diagram, why a Newton–Raphson iteration with $x_0 = 3.5$ will not converge to the root of $f(x) = 0$.

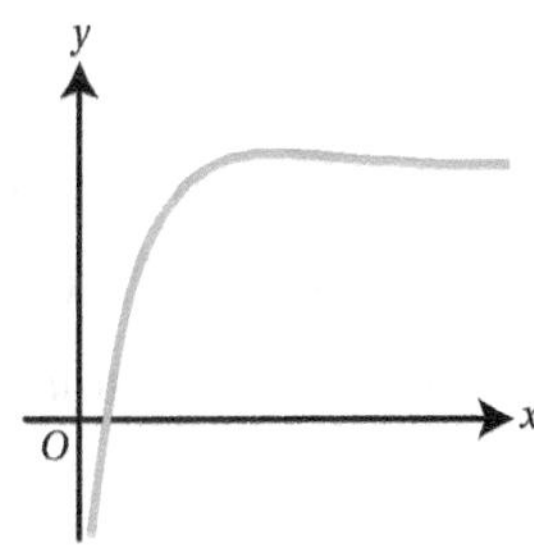

4. Let $f(x) = \ln x - \tan x + 2$.

   a Write down the values of x between 0 and 10 for which $f(x)$ is not defined.

   The equation $f(x) = 0$ has a root between 0 and 1.

   b Taking $x_0 = 0.7$, use the Newton–Raphson method to find x_1.

   c Explain why Newton–Raphson iteration cannot be continued to find a better approximation to the root.

5. The diagram shows the curve with equation $f(x) = 3x^2 - x^3 - 2$.

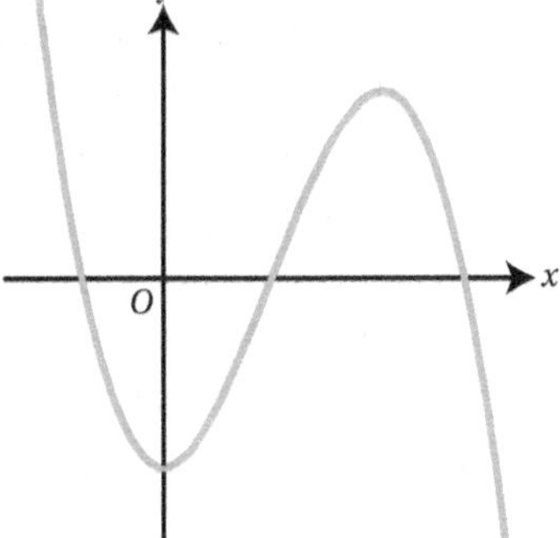

   a Find the coordinates of the turning points of $f(x)$.

   The curve crosses the x-axis at $x = \alpha,\ x = \beta$ and $x = \gamma$, with $\alpha < \beta < \gamma$.

   The Newton–Raphson method is to be used to find the roots of $f(x) = 0$, with $x_0 = k$.

   b To which root, if any, do the successive approximations converge when:

   i $k = -1$ ii $k = 2$?

   c Write down the range of values of k for which the Newton–Raphson iteration converges to γ.

6. The function $f(x) = 2x^3 - 6x + 1$ has three zeros $(a < b < c)$ between −2 and 2.

   a For each zero, find two integers between which it lies.

   b Find the exact coordinates of the stationary points of $f(x)$ and sketch its graph.

   c Find the coordinates of the point where the tangent to the graph at $x = 0.9$ crosses the x-axis.

   The Newton–Raphson method is to be used to find root $x = b$.

   d Explain why the iteration with $x_0 = 0.9$ does not converge to b.

   e Use an iteration with $x_0 = 0.5$ to find b correct to three significant figures.

7 Let $f(x) = \ln x - 0.5x^2 + 2$.

a Find the coordinates of the stationary point on the graph $y = f(x)$ and show that this is a maximum point.

b Show that $f(x)$ has no points of inflection. Hence sketch the graph of $y = f(x)$.

The equation $f(x) = 0$ has two roots.

c Show that one of the roots is between 0 and 1 and find two integers between which the other root lies.

The Newton–Raphson iteration is to be used to find the roots.

d Explain why the starting value $x_0 = 0.5$ cannot be used to find the smaller root.

e State the range of values of x_0 for which the Newton–Raphson iteration converges to the larger root.

8 Each equation has a root near the given starting value x_0. In each case:

i explain why starting with x_0 does not lead to a better approximation to the root

ii use technology to investigate for which starting values Newton–Raphson method works.

a $\sin(2x) + \cos(3x) = 0$, take $x_0 = 1.25$ to find the root between 1 and 2.

b $\sqrt{x} - \cos(3x) = 0$, $x_0 = 1$

c $2 - \operatorname{cosec} x = 0$, take $x_0 = 1$ to find the root between 0 and 1.

d $\tan x - x = 0$, take $x_0 = 4$ to find the root between 4 and 5.

9 A function is defined by $f(x) = \dfrac{x^4 - 10x^2 + 25}{x^2 - 1}$.

a Show that, if α is any root of the equation $f(x) = 0$, then $f'(\alpha) = 0$.

The graph of $y = f(x)$ is shown.

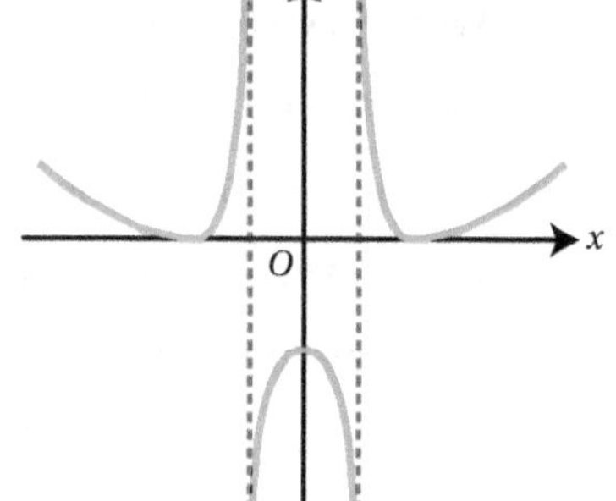

The Newton–Raphson formula with a positive value of x_0 is used to find x_1.

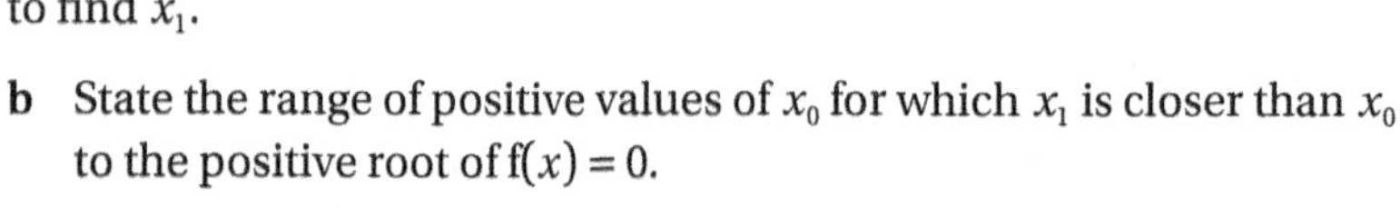

b State the range of positive values of x_0 for which x_1 is closer than x_0 to the positive root of $f(x) = 0$.

c Use a diagram to show that there is a positive value of x_0 such that $x_1 < 0$, but subsequent approximations converge to the positive root of $f(x) = 0$.

10 **a** Sketch the graphs of $y = \tan x$ and $y = 2x$ for $0 < x < \frac{1}{2}\pi$. Hence show that the equation $\tan x - 2x = 0$ has a solution, α, in that interval.

b Consider the Newton-Raphson iteration with the starting value x_0.

Show that the iteration converges to α when $x_0 = 1.1$ but not when $x_0 = 0.9$.

c There is a value k such that the iteration with $k < x_0 < \frac{1}{2}\pi$ converges to α. Show the value k on your sketch and draw the tangent to the curve at $x = k$.

d When $x_0 < k$, the Newton-Raphson iteration may not converge to α. Describe two different cases that can arise.

Section 4: Fixed-point iteration

You saw in Section 3 that the Newton–Raphson method does not always work. There are alternative methods you can use in such situations. In this section you will learn about **fixed-point iteration**, which also involves creating a sequence that gets closer and closer to the root, but in a different way from the Newton–Raphson method.

> **Did you know?**
>
> The term, fixed-point iteration, refers to the fact that the solution is a fixed point of the function $g(x)$ – it is the value where the output equals the input.

To use fixed-point iteration the equation needs to be rearranged into the form $x = g(x)$. Suppose you have a starting guess x_1. If $x_1 = g(x_1)$ you have a found a solution of the equation. Otherwise you are looking for an improved guess. Since you want x to equal $g(x)$, it makes sense to try $x_2 = g(x_1)$.

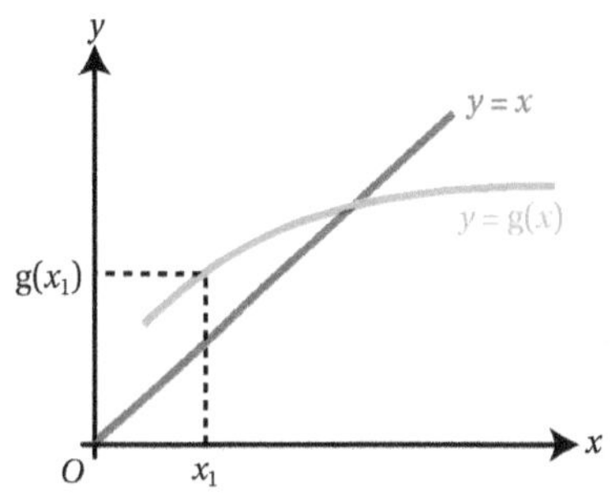

You can see why this works by looking at the graph. The solution of the equation $x = g(x)$ is the intersection of the graphs $y = x$ and $y = g(x)$. Starting from the point x_1 on the x-axis you can find x_2 on the y-axis by using the graph $y = g(x)$. To see whether x_2 is closer to the solution than x_1 was, you need to find x_2 on the x-axis. You can do this by reflecting it in the line $y = x$.

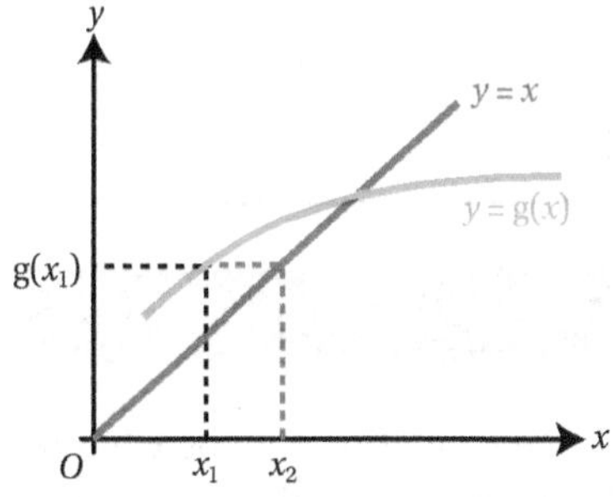

In the example shown in the diagram, x_2 is closer to the solution than x_1. If you now repeat the same process you can hope to get closer and closer to the solution. In other words, you follow this sequence:

- start with x_1
- $x_2 = g(x_1)$
- $x_3 = g(x_2)$ …

You can write the general rule for the sequence as $x_{n+1} = g(x_n)$. This sequence may **converge** to a **limit**, meaning that the terms of the sequence get closer and closer to a certain number α. If this is the case, both x_n and x_{n+1} will get closer to α so the sequence equation becomes $\alpha = g(\alpha)$ and so α is a solution of the equation $x = g(x)$.

> **Rewind**
>
> Sequence rules and convergence were covered in Chapter 4.

In Worked example 14.6 the sequence increased towards the limit. But it is also possible for a sequence to oscillate above and below the limiting value.

Key point 14.4

Fixed-point iteration

To solve an equation in the form $x = g(x)$:

- using a starting guess x_1, generate a sequence $x_{n+1} = g(x_n)$
- if this sequence converges to a limit, then this limit is a solution of the equation.

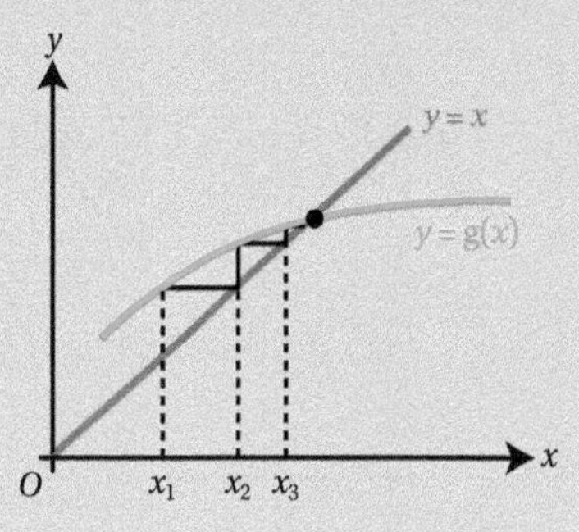

WORKED EXAMPLE 14.6

The equation $x = \ln x + 3$ has a solution between 4 and 5.

a Use fixed-point iteration with the starting value $x_1 = 4.5$ to find the first five approximations to the solution. Give the values correct to five significant figures.

b Show that this solution is correct to two decimal places.

a Sequence: $x_{n+1} = \ln x_n + 3$

$x_1 = 4.5$

$x_2 = \ln 4.5 + 3 = 4.5041$

$x_3 = 4.5050$

$x_4 = 4.5052$

$x_5 = 4.5052$

Start by inputting '4.5' into a calculator and use ln (ANS) + 3 to generate subsequent values.

b The solution appears to be 4.51 (2 d.p.).

$x = \ln x + 3 \Leftrightarrow \ln x + 3 - x = 0$

$\ln 4.505 + 3 - 4.505 = 1.88 \times 10^{-4} > 0$

$\ln 4.515 + 3 - 4.515 = -7.59 \times 10^{-3} < 0$

There is a sign change, so the solution is between 4.505 and 4.515, so it equals 4.51 (2 d.p.).

To check that the solution is 4.51 correct to 2 d.p. rearrange the equation into the form $f(x) = 0$ and look for a sign change between 4.505 and 4.515.

WORKED EXAMPLE 14.7

The equation $x = \cos x$ has a root between 0 and 1. The sequence $x_{n+1} = \cos(x_n)$ with $x_1 = 0.5$ is used to find an approximation to this root.

a Draw a graph to illustrate the first three approximations.

b Find the root correct to 3 d.p.

a

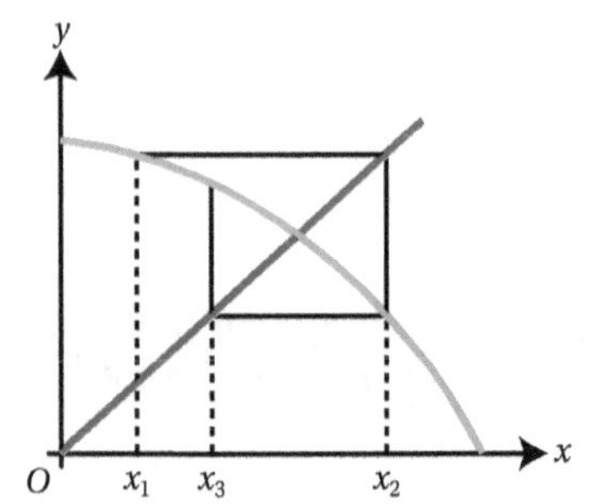

Draw the curve $y = \cos x$ and the line $y = x$. Start with $x_1 = 0.5$ on the x-axis. Find the corresponding point on the curve and reflect it in the line $y = x$ to find x_2. Repeat to find x_3.

b $x_2 = \cos 0.5 = 0.8776$

$x_3 = \cos 0.8776 = 0.6390$

$x_4 = 0.8027 \ldots$

Start with 0.5. use cos (ANS) to generate the sequence.

The sequence converges to $x = 0.739$ (3 d.p.)

Continue until the third decimal place stops changing.

Key point 14.5

The graphs showing fixed-point iteration are often called **staircase** and **cobweb diagrams**.

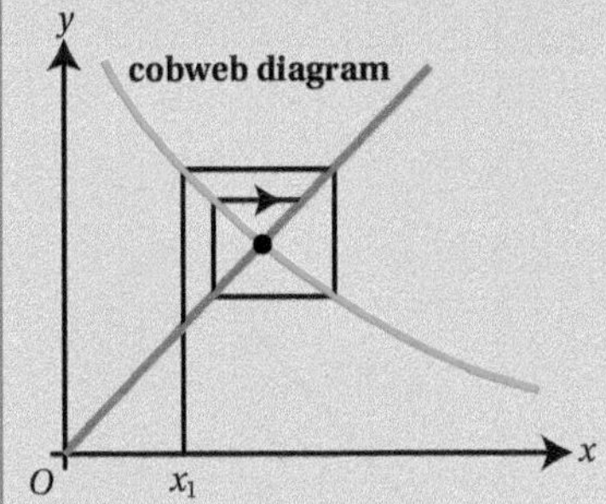

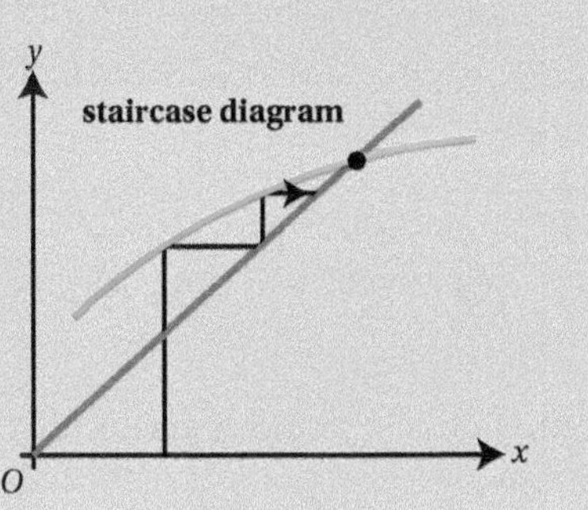

EXERCISE 14D

1 Use fixed-point iteration with the given starting value to find the first five approximations to the roots of each equation. Give your answers to three decimal places.

a **i** $x = 2\ln(x+2),\ x_1 = 3$ **ii** $x = 3 - e^{-x},\ x_1 = 4$

b **i** $x = \frac{3}{x+4} - 1,\ x_1 = -1$ **ii** $x = \cos(2x-1),\ x_1 = 1$

c **i** $\ln x - x^2 + 2 = x,\ x_1 = 0.5$ **ii** $1.5x\sin(x+1) = x,\ x_1 = 1$

2 Use fixed-point iteration with the given starting value to find an approximate solution correct to two decimal places.

a **i** $x = e^{-\frac{x}{2}},\ x_1 = 0$ **ii** $x = \cos\left(\frac{x}{3}\right),\ x_1 = 0$

b **i** $x = \ln x + 3,\ x_1 = 4$ **ii** $x = \tan\left(\frac{x}{2}\right) + 0.2,\ x_1 = 1$

3 Use fixed-point iteration to solve each equation. Draw a graph and use technology to investigate these questions.

i Does the limit depend on the starting point?

ii Does starting on different sides of the root give a different limit?

iii If there is more than one root, which one does the sequence converge to? Does it matter where you start?

a $x = \arctan x + 1$ **b** $x = \sqrt[3]{x+2}$ **c** $x = \ln(x+2)$ **d** $x = e^{x-2}$

4 The equation $x = \frac{13 - e^{-x}}{5}$ has a root between 2 and 3. Use fixed-point iteration to find this root correct to two decimal places.

5 **a** Use the iterative formula $x_{n+1} = \cos\left(\frac{x_n}{3}\right)$ with $x_1 = 0.5$ to find x_5. Give your answer correct to four decimal places.

b This value of x_5 is an approximation to the root of the equation $f(x) = 0$. Write down an expression for $f(x)$ and suggest the value of the root correct to two decimal places.

6 The equation $x = \arcsin\left(\frac{x+3}{5}\right)$ has a root between 0 and 1. Use fixed-point iteration to find this root correct to three decimal places. Show the first three approximations correct to five decimal places.

7 **a** Use the iterative formula $x_{n+1} = \frac{2x_n^3 - 5}{6x_n^2 + 3}$ to find x_4 and x_5 correct to four decimal places with $x_0 = 0$.

b The sequence $\{x_n\}$ converges to the root of the equation $ax^3 + bx + c = 0$. Find the values of a, b and c.

8 The diagram shows the graphs of $y = x$ and $y = g(x)$. The iteration $x_{n+1} = g(x_n)$ converges to the root of the equation $x = g(x)$. The value x_1 is shown on the diagram.

The values of x_2 and x_3 correspond to the points marked A and B. Which of the two values corresponds to the point A?

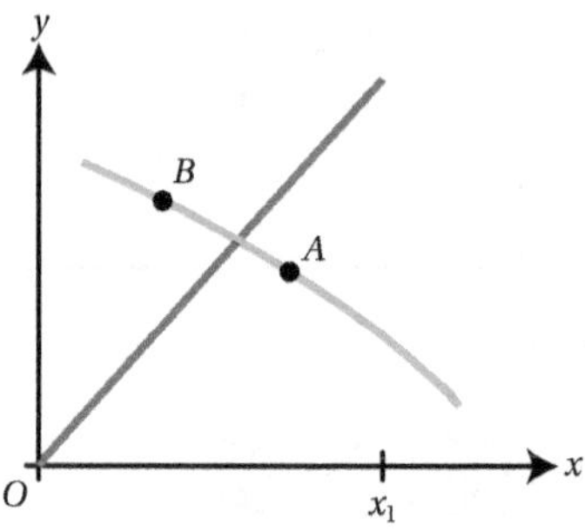

9 **a** Show that the equation $x = \frac{1}{3}\left(x + \frac{5}{x}\right)$ has a root between 1 and 2.

b Use fixed-point iteration with $x_1 = 1.5$ to find the next three approximations to the root.

c Show that your value of x_4 gives the root correct to two decimal places.

10 The diagram shows the graphs of $y = x$ and $y = 6 - \frac{4}{x}$.

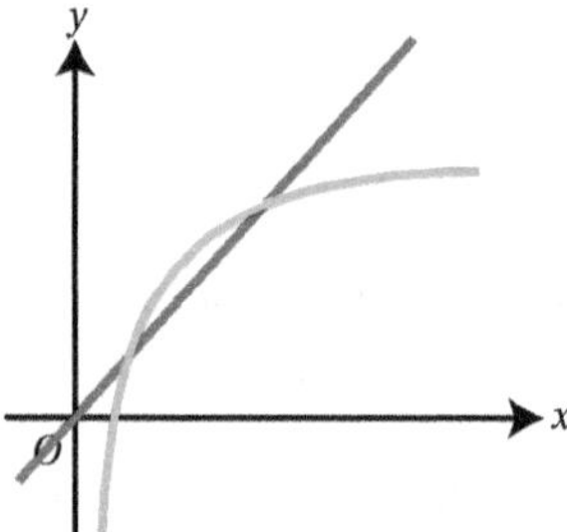

a Use the diagram to show that the iteration $x_{n+1} = 6 - \frac{4}{x_n}$ converges to a root of the equation $x = 6 - \frac{4}{x}$.

b Use the iteration from part **a** to find this root correct to two decimal places.

c By rewriting the equation in the form $ax^2 + bx + c = 0$, find the exact value of the root. Hence find the percentage error in your approximation.

11 The diagram shows graphs of $y = x$ and $y = \frac{1}{2}\left(x + \frac{10}{x}\right)$.

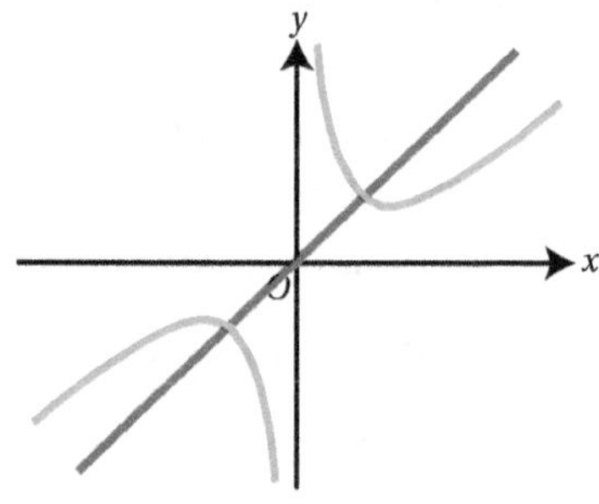

a Show on the diagram that the iteration $x_{n+1} = \frac{1}{2}\left(x_n + \frac{10}{x_n}\right)$ with $x_1 = 2$ converges.

b Prove that the iteration converges to $\sqrt{10}$.

Section 5: Limitations of fixed-point iteration and alternative rearrangements

Fixed-point iteration seems a little easier to implement than the Newton-Raphson method, as there is no need to differentiate or rearrange the equation in any other way. It also works in some situations when the Newton-Raphson method does not. However, there are other situations when fixed-point iteration does not work because the sequence fails to converge to the root.

For example, try solving the equation $x = e^x - 2$ for $x > 0$ by using the iterative formula $x_{n+1} = e^{x_n} - 2$. Sketching graphs and trying some integer values of x shows that there is a root between $x = 1$ and $x = 2$.

Starting the iteration at $x_1 = 1.5$ gives the sequence $x_1 = 1.5,\ x_2 = 2.482,\ x_3 = 9.965, \ldots$

The sequence is clearly getting further away from the root: the sequence **diverges**.

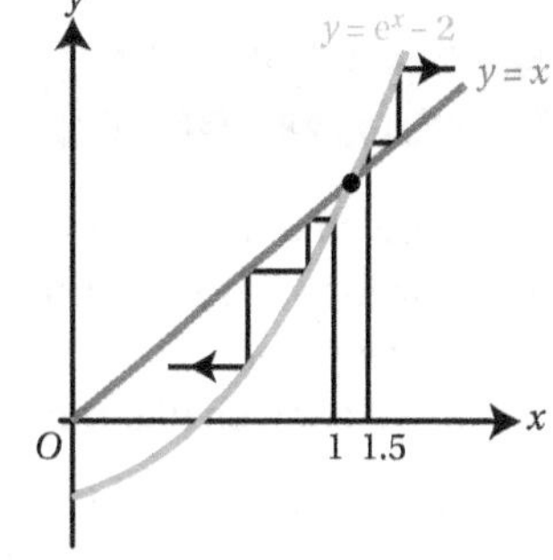

You can also try starting the sequence below the root, at $x_1 = 1$; the resulting sequence is $x_1 = 1,\ x_2 = 0.718,\ x_3 = 0.0509, \ldots$

Again, the sequence does not seem to be approaching the root. You can see this on the graph by drawing the staircase diagram.

If you continue the second sequence it actually converges to the other root of the equation:

$$x_1 = 1, x_2 = 0.718, x_3 = 0.0509, x_4 = -0.948,$$
$$x_5 = -1.61, x_6 = -1.83, x_7 = -1.84, x_8 = -1.84, \ldots$$

But is there any way fixed-point iteration can be used to find the first root?

You can write the equation $x = e^x - 2$ in a different way:

$$\begin{aligned} x &= e^x - 2 \\ \Leftrightarrow e^x &= x + 2 \\ \Leftrightarrow\ x &= \ln(x+2) \end{aligned}$$

The sequence based on this rearrangement, $x_{n+1} = \ln(x+2)$, starting at $x_1 = 1.5$, is:

$$x_1 = 1.5,\ x_2 = 1.25,\ x_3 = 1.18,\ x_4 = 1.16,\ x_5 = 1.15,\ x_6 = 1.15 \ldots$$

Looking at the graph confirms that this sequence does indeed converge to the root between 1 and 2.

Interestingly, starting this iteration near the other root ($x = -1.84$) leads either to a divergent sequence or to one converging to the positive root. Starting at $x_1 = -1.5$ gives the sequence $-0.693,\ 0.268,\ 0.819,\ 1.04,\ 1,\ 11,\ 1.13,\ 1.14,\ 1.15,\ 1.15 \ldots$

Starting with $x_1 = -1.9$ you get $x_2 = -2.30$ and then the sequence cannot be continued further because $\ln(x+2)$ is not defined for $x \leqslant -2$.

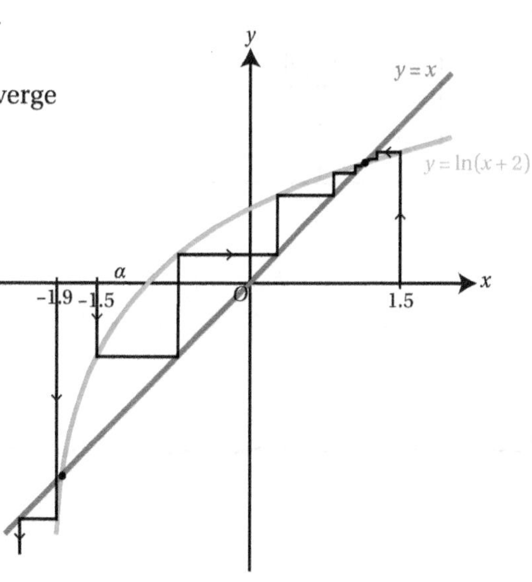

Key point 14.6

- Some rearrangements of the equation lead to **convergent** sequences, and others to **divergent** sequences.
- A divergent sequence does not find the required root.
- If an equation has more than one root, different rearrangements may converge to different roots.

WORKED EXAMPLE 14.8

a Show that the equation $x = 2\cos x$ has a root between 0 and 2.

b Starting with $x_1 = 1$ find the next four terms of the sequence $x_{n+1} = 2\cos x_n$ and describe the behaviour of the sequence.

c Use a cobweb diagram to illustrate the behaviour of the sequence.

d Show that the equation $x = 2\cos x$ can be written as $x = \cos^{-1}\left(\frac{x}{2}\right)$. Use this rearrangement to find an approximate solution of the equation correct to three decimal places.

a Write $f(x) = 2\cos x - x$.

To use the change of sign method, rewrite the equation in the form $f(x) = 0$.

$f(0) = 2 > 0$

$f(2) = -2.83 < 0$

There is a change of sign, so there is a root between 0 and 2.

b $x_2 = 2\cos 1 = 1.08$

$x_3 = 2\cos 1.08 = 0.942$

$x_4 = 2\cos 0.942 = 1.18$

$x_5 = 2\cos 1.18 = 0.77$

The sequence seems to diverge.

The terms of the sequence are getting further away from each other.

c

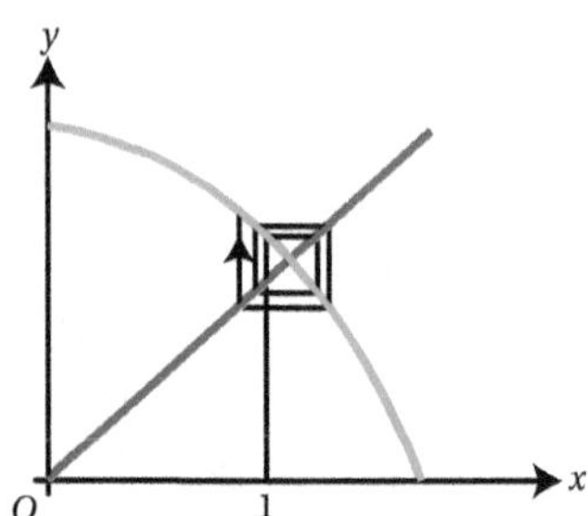

d $x = 2\cos x$

Make sure you show all the steps in a 'show that ...', question.

$\Leftrightarrow \cos x = \frac{x}{2}$

$\Leftrightarrow x = \cos^{-1}\left(\frac{x}{2}\right)$

Continues on next page

$x_{n+1} = \cos^{-1}\left(\frac{x_n}{2}\right)$

Use this new arrangement to form a sequence. It makes sense to start at $x_1 = 1$ again.

$x_1 = 1$
$x_2 = 1.047$
$x_3 = 1.020$
$x_4 = 1.036$
$\vdots$
$x_{12} = 1.029\,945\ldots$
$x_{13} = 1.029\,82\ldots$
$x_{14} = 1.029\,89\ldots$
$\therefore x = 1.030$ (3 d.p.)

You are not asked to show all the approximations, so just keep going until the third decimal place stops changing. In this question you are also not asked to demonstrate that the root is correct to 3 d.p.

Is there any way you can tell whether a sequence from a particular rearrangement would converge without trying it? Looking at some staircase diagrams suggests that the answer has something to do with the gradient of the graph of $g(x)$ at the point where it crosses the line $y = x$.

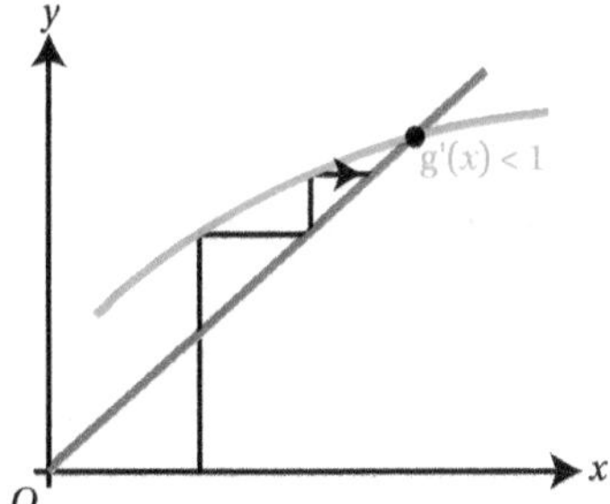

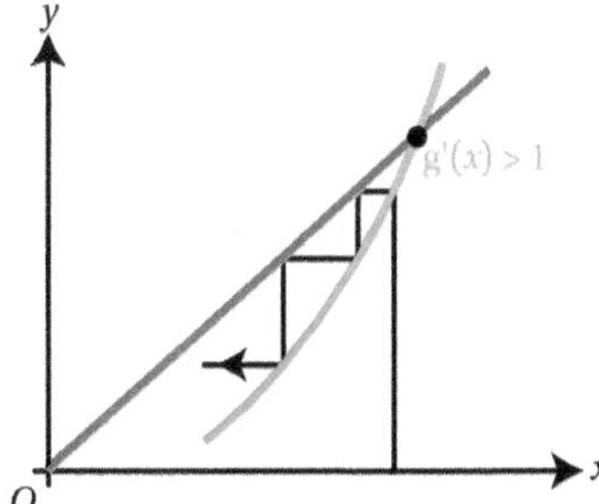

In the first diagram the gradient of the graph of $g(x)$ is smaller than the gradient of $y = x$ and the sequence seems to converge. In the second diagram, where the sequence diverges, the graph of $g(x)$ is steeper than the line $y = x$. This suggests that the sequence converges when the gradient of $g(x)$ is smaller than one and diverges when it is greater than one.

It is less obvious what happens when the gradient of $g(x)$ is negative so that the iteration produces a cobweb diagram. It turns out that it depends on whether the gradient of $g(x)$ is smaller or greater than -1.

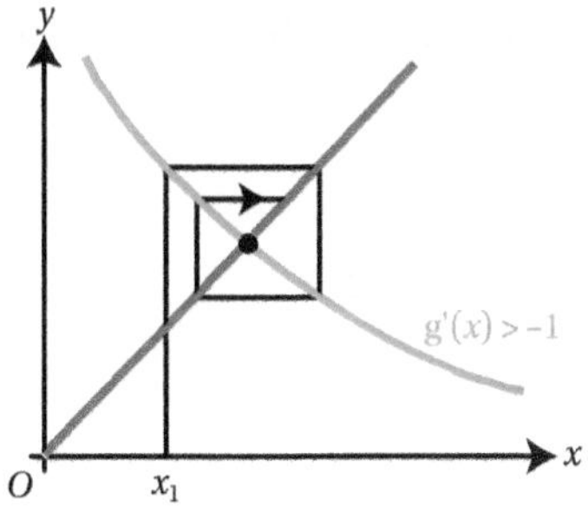

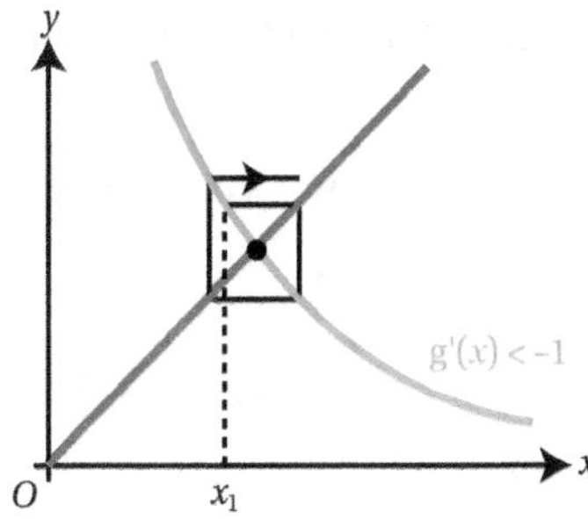

Both cases are summarised in Key point 14.7.

Key point 14.7

Fixed-point iteration $x_{n+1} = g(x_n)$:

- converges if $|g'(x)| < 1$ near the root
- diverges if $|g'(x)| > 1$ near the root.

If $g'(x) = 1$ or -1 it is impossible to tell whether the sequence will converge without trying it.

WORKED EXAMPLE 14.9

The equation $x^3 + 2x + 7 = 0$ has a root near -1.6. Two possible rearrangements of this equation are $x = -\dfrac{7}{x^2 + 2}$ and $x = -\sqrt[3]{2x+7}$.

Determine which rearrangement will produce a sequence that converges to the root.

Write the sequence as $x_{n+1} = g(x_n)$ — The iteration converges if $|g'(x)| < 1$ near the root.

When $g(x) = -\dfrac{7}{x^2+2}$: — Differentiate g.

$$g'(x) = \frac{14x}{(x^2+2)^2}$$

$$g'(-1.6) = -\frac{22.4}{4.56^2} < -1$$ — The root is near -1.6, so evaluate the derivative at this point.

$|g'(x)| > 1$ near the root, so this iteration will not converge.

When $g(x) = -\sqrt[3]{2x+7}$:

$$g'(x) = -\frac{2}{3}(2x+7)^{-\frac{2}{3}}$$

$$g'(-1.6) = -\frac{2}{3}(1.8)^{-\frac{2}{3}} > -1$$

$|g'(x)| < 1$ near the root, so this iteration will converge.

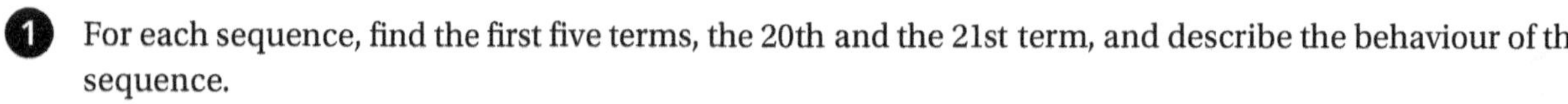

EXERCISE 14E

1 For each sequence, find the first five terms, the 20th and the 21st term, and describe the behaviour of the sequence.

a $x_{n+1} = 3x_n^2 - 1,\ x_1 = 0$

b $x_{n+1} = 5 - \frac{1}{2}x_n^2,\ x_1 = 1$

c $x_{n+1} = 2 - \dfrac{x_n^2}{2} + \dfrac{x_n^3}{5},\ x_1 = 0$

d $x_{n+1} = \sin\left(\frac{1}{2}x_n\right) - 0.2,\ x_1 = 2$

e $x_{n+1} = 3 - x_n,\ x_1 = 2$

f $x_{n+1} = 3\ln x_n,\ x_1 = 2$

g $x_{n+1} = 3\ln x_n,\ x_1 = 1.5$

2 Each diagram shows the graphs of $y = x$ and $y = \mathrm{g}(x)$ and the starting value x_1. Describe what happens to the iteration defined by $x_{n+1} = \mathrm{g}(x_n)$.

a

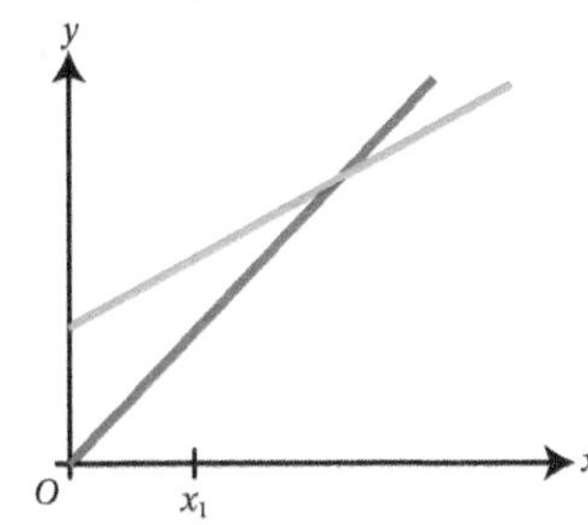

b

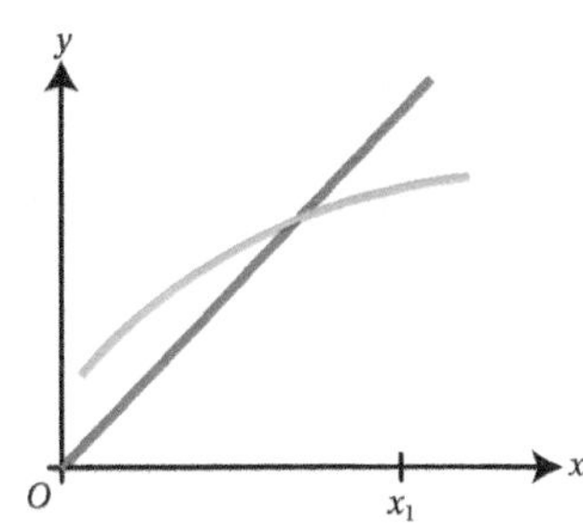

c

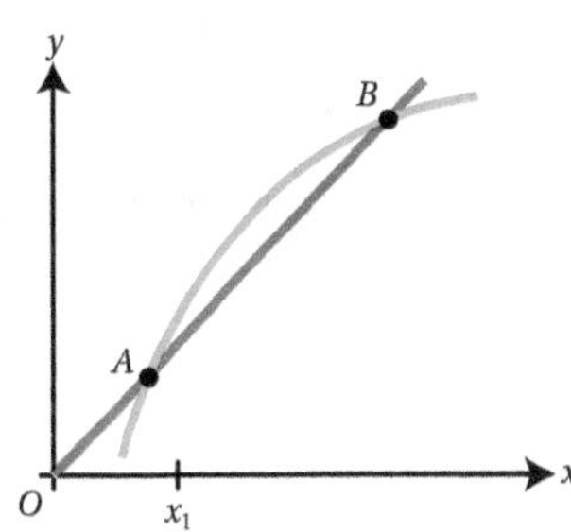

d

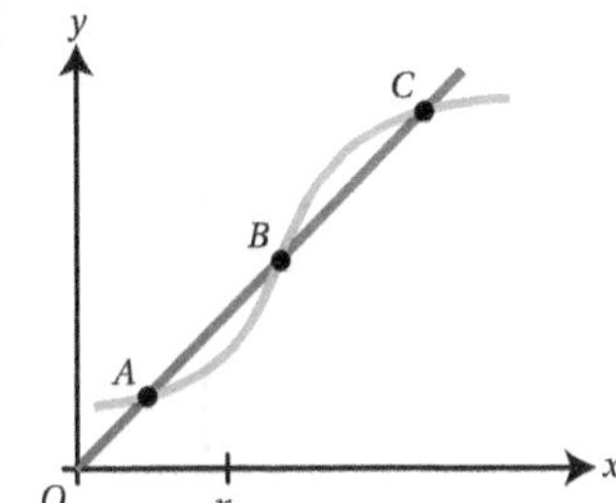

e

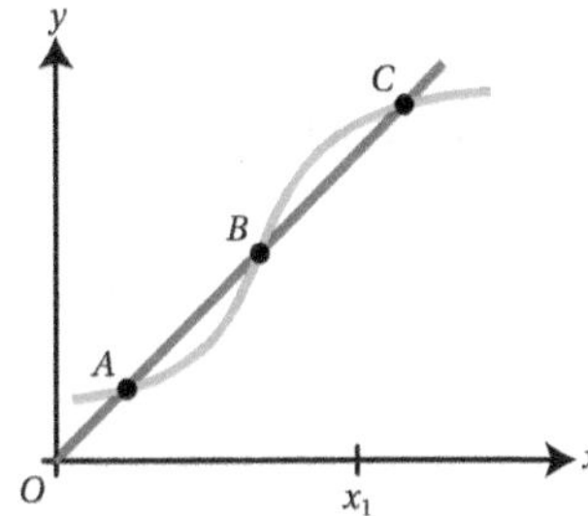

f

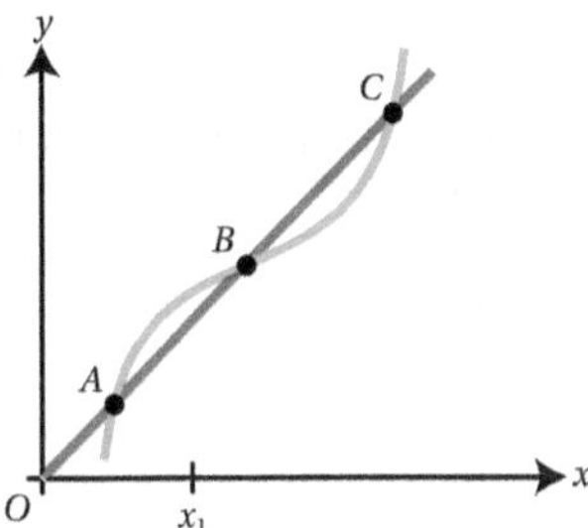

g

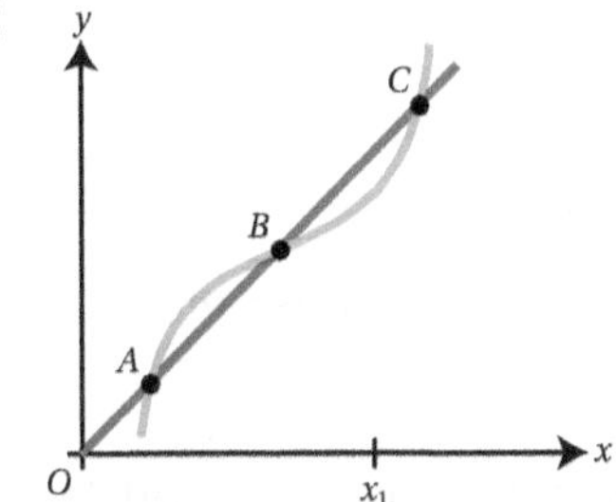

3 Find an alternative rearrangement for each equation.

a **i** $x = 3\sin x$ **ii** $x = 5\tan x$ **b** **i** $x = \ln(x-3)$ **ii** $x = \mathrm{e}^{x-2}$

c **i** $x = 3\sqrt{x-1}$ **ii** $x = 2\sqrt[3]{x+5}$ **d** **i** $x = 3x^2 - 1$ **ii** $x = \dfrac{x^2+1}{7}$

4 Find the missing constants to rearrange each equation into the given equivalent form.

a **i** $2x^3 - 4x + 1 = 0 \Leftrightarrow x = \sqrt[3]{ax+b}$ **ii** $x - \dfrac{1}{2}x^4 - 3 = 0 \Leftrightarrow x = \sqrt[4]{ax+b}$

b **i** $x^3 + x^2 - 4 = 0 \Leftrightarrow x = \dfrac{a}{x} - bx^2$ **ii** $2x^3 - 2x^2 + 1 = 0 \Leftrightarrow x = \dfrac{ax^3+b}{x}$

c **i** $2x^3 - x + 5 = 0 \Leftrightarrow x = \sqrt{a + \dfrac{b}{x}}$ **ii** $x^2 - 3x^3 + 2 = 0 \Leftrightarrow x = \sqrt{cx + \dfrac{d}{x}}$

d **i** $x^3 - 3x^2 + 5x + 2 = 0 \Leftrightarrow x = \dfrac{ax^3+b}{3x-5}$ **ii** $2x^3 + x^2 - 3x + 1 = 0 \Leftrightarrow x = \dfrac{2x^3+1}{a-bx}$

5 Each equation has more than one root. For each root, use technology to find a rearrangement that converges to it.

a $x = 2 + 3\ln(x-2)$ **b** $3\arctan(x-2) = x - 1$ **c** $x^3 - 3x^2 - 5x + 3 = 0$

6 The diagram shows the graph of a function $g(x)$ and the line $y = x$. The equation $g(x) = x$ has three roots, marked α, β and γ. Which of the roots does the iteration $x_{n+1} = g(x_n)$ converge to when:

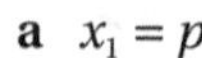

a $x_1 = p$ **b** $x_1 = q$?

Justify your answer by showing several iterations on the diagram.

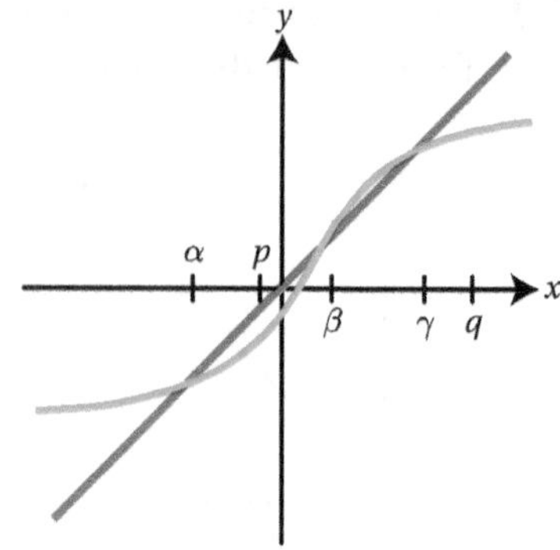

7 The iterative formula $x_{n+1} = f(x_n)$ is used to find an approximate solution of the equation $x = f(x)$. The graphs of $y = f(x)$ and $y = x$ are shown.

Use the diagram to determine the behaviour of the sequence $\{x_n\}$ when:

a $x_1 = a$ **b** $x_1 = b$ **c** $x_1 = c$.

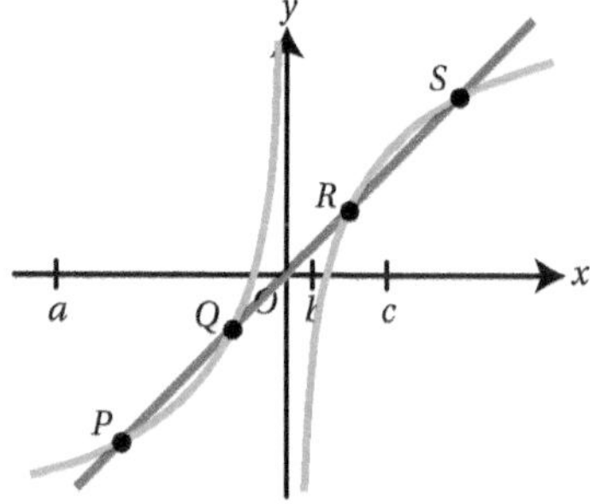

8 The Newton–Raphson iteration for solving $f(x) = 0$ is $x_{n+1} = x_n - \frac{f(x_n)}{f'(x_n)}$.

This can also be considered as a fixed-point iteration for solving $x = g(x)$.

a Express $g(x)$ in terms of $f(x)$.

Let a be the solution of the equation, and assume that $f'(x) \neq 0$ for x near a.

b Find $g'(x)$, and hence prove that the Newton–Raphson method always converges when the starting value is sufficiently close to a.

9 **a** Show that the equation $x = 2 - e^{-\frac{x}{2}}$ has a root between −4 and −3, and another between 1 and 2.

b Let $g(x) = 2 - e^{-\frac{x}{2}}$ and define the sequence $x_{n+1} = g(x_n)$ with $x_1 = 0$. Solve the inequality $g'(x) > 1$ and hence determine to which of the two roots the sequence $\{x_n\}$ converges.

10 The function $f(x)$ is defined by $f(x) = x^4 + 3x^3 - 15x + 1$. The equation $f(x) = 0$ has two positive real roots, $\alpha < \beta$.

a Show that α is between 0 and 1, and find two consecutive integers between which β lies.

b The rearrangement $x = \sqrt[3]{\frac{-x^4 + 15x - 1}{3}}$ is used to find an approximate root of the equation $f(x) = 0$. **Without** carrying out the iteration, determine to which of the two roots the sequence $x_{n+1} = \sqrt[3]{\frac{-4x_n^4 + 15x_n - 1}{3}}$ converges.

c Show that an alternative rearrangement is $x = \frac{x^4 + c}{k + nx^2}$ and find the constants c, k and n. Use this rearrangement to find the root α correct to three decimal places.

11 **a** Find, in terms of k, the roots of the equation $x = kx(1-x)$.

Let $g(x) = kx(1-x)$ where $k > 1$.

b Find $g'(\alpha)$, where α is the non-zero root if the equation $x = g(x)$.

c The iterative formula $x_{n+1} = kx_n(1-x_n)$ with $x_1 = 0.5$ is used to find approximate roots of the equation $x = g(x)$. Find the set of values of k for which the iteration converges to the non-zero root.

Checklist of learning and understanding

- Some equations can only be solved by finding numerical approximations. Numerical methods are methods that tell you how to find an improved approximation.
- When an equation is written in the form $f(x) = 0$ you can show that it has a root (solution) between $x = a$ and $x = b$ by showing that $f(a)$ and $f(b)$ have different signs.
- You can use the change-of-sign method to check the accuracy of the solution by 'unrounding' the number.
- Both methods involve creating a sequence x_n which **converges** to the root of the equation. The first term of the sequence needs to be chosen from an interval that contains the root.
- The **Newton–Raphson** method works by approximating the curve by its tangent. The equation needs to be written in the form $f(x) = 0$. The resulting sequence is given by:

$$x_{n+1} = x_n - \frac{f(x_n)}{f'(x_n)}$$

- The Newton–Raphson method fails to converge if the starting point is close to a stationary point of $f(x)$ (where $f'(x) = 0$).
- **Fixed-point iteration** requires the equation written in the form $x = g(x)$. The iteration sequence is given by:

$x_{n+1} = g(x_n)$

  - This iteration can be represented graphically as a **staircase** or a **cobweb** diagram.

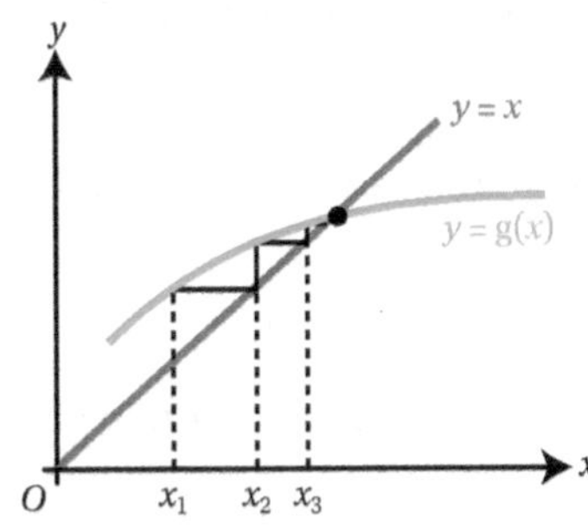

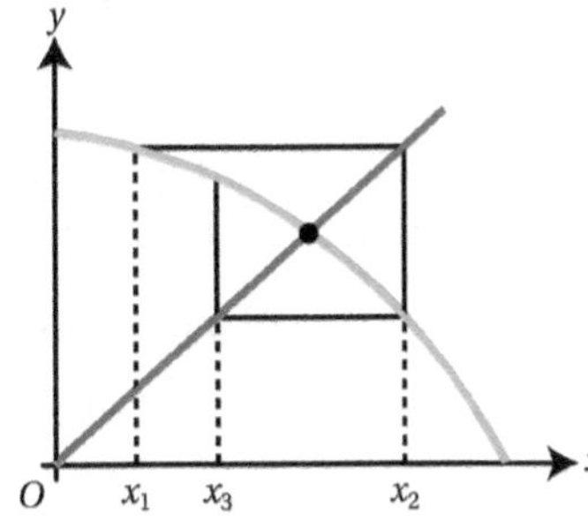

  - Sometimes the sequence **diverges**, meaning that the terms get further away from the required root. To get a convergent sequence you may need to rearrange the equation.
  - If the equation has several roots, different rearrangements may converge to different roots.
  - The iteration converges if $|g'(x)| < 1$ near the root and diverges if $|g'(x)| > 1$.

Mixed practice 14

In this exercise you will apply numerical solutions of equations in various contexts where the resulting equation cannot be solved exactly.

1. The recurrence relation $x_{n+1} = \sqrt[3]{1-\sqrt{x_n}}$ is used to try to find the root of $x^3 + \sqrt{x} - 1 = 0$.
   If $x_1 = 0.5$, how can the sequence $x_1, x_2, x_3 ...$ be demonstrated?

   Choose from these options.

   A cobweb diagram **B** staircase increasing diagram
   C staircase decreasing diagram **D** divergence diagram

2. $f(x) = 3\ln x - x$. $f(x) = 0$ has a roots α_1 and α_2 where $1 < \alpha_1 < 2$ and $4 < \alpha_2 < 5$.

   Using the Newton–Raphson method, which initial values, x_0, will fail to find either of these roots?

   Choose from these options.

   A $x_0 = 0.5$ **B** $x_0 = 3$ **C** $x_0 = 5.5$ **D** $x_0 = 6$

3. The graph of $y = \sin\left(\frac{x}{2}\right) - x^2$ has a stationary point between 0 and 1.

   a Show that the x-coordinate of the stationary point satisfies $x = \frac{1}{4}\cos\left(\frac{x}{2}\right)$.

   b Use fixed-point iteration with a suitable starting point to find the x-coordinate of the stationary point correct to three decimal places.

4. The diagram shows a sector of a circle with radius 5 cm. The angle at the centre is θ. The area of the shaded region equals 30π cm^2.

   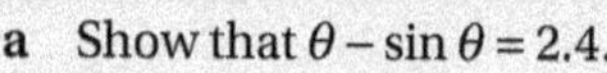

   a Show that $\theta - \sin\theta = 2.4$.

   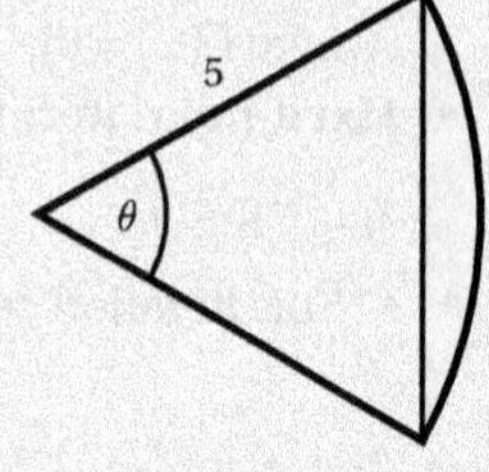

   b Show that the equation in part **a** has a root between 2 and 3.

   c Use the Newton–Raphson method with a suitable starting point to find the value of θ correct to two decimal places.

5. **a** Sketch the graphs of $y = e^{x-1}$ and $y = \ln(x+2)$ on the same set of axes. State the number of solutions of the equation $e^{x-1} = \ln(x+2)$.

   b Show that the equation in part **a** has a solution between 1 and 2.

   c Use an iteration of the form $x_{n+1} = A + \ln(\ln(x_n + B))$ to find this solution correct to two decimal places.

6. The diagram shows the graph of $y = x^2 + 1$. The shaded area is enclosed by the curve, the coordinate axes and the line $x = a$.

   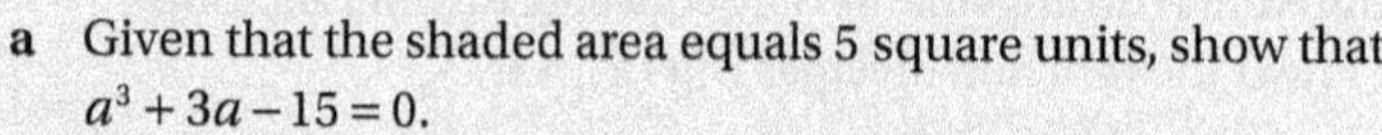

   a Given that the shaded area equals 5 square units, show that $a^3 + 3a - 15 = 0$.

   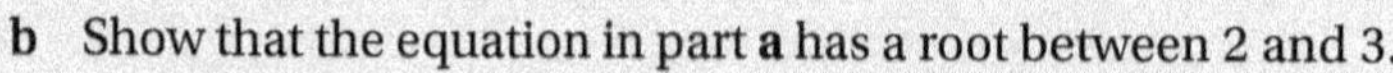

   b Show that the equation in part **a** has a root between 2 and 3.

   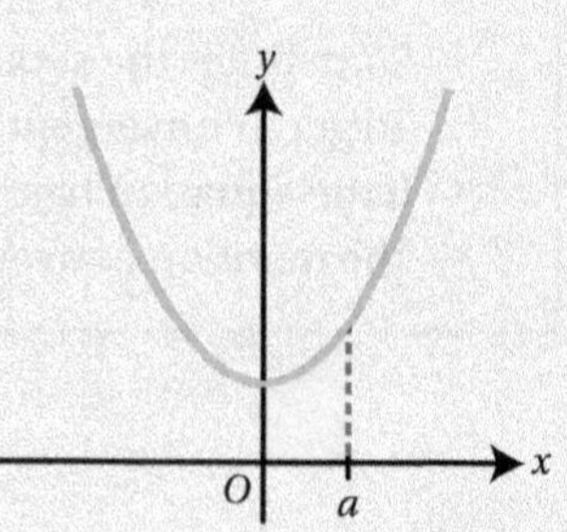

   c Use the Newton–Raphson method to find the value of a correct to three decimal places.

7 The equation

$x^3 - x^2 + 4x - 900 = 0$

has exactly one root, α.

Taking $x_1 = 10$ as a first approximation to α, use the Newton-Raphson method to find a second approximation, x_2, to α. Give your answer to four significant figures.

[© AQA 2013]

8 Let $\mathrm{f}(x) = \dfrac{x^3 + 4x + 8}{x^3 + 1}$ for $x < -1$.

a Show that the equation $\mathrm{f}(x) = 0$ has a root between -2 and -1.

b Starting from $x_0 = -2$ use the Newton-Raphson method to find x_1.

c Explain why this iteration cannot be continued to find the root.

d Use the Newton-Raphson method, with $x_0 = -1.5$, to find the root correct to three decimal places.

9 A curve is defined by $y = \dfrac{\sin x}{x^2}$ for $x > 0$.

a Show that the x-coordinate of any stationary point on the curve satisfies the equation $x = 2\tan x$.

One of the roots of this equation is between 4 and 5.

b By considering the derivative of $2\tan x$ prove that the iteration $x_{n+1} = 2\tan x_n$ does not converge to this root.

c Find an alternative rearrangement of the equation $x = 2\tan x$ and use it to find the x-coordinate of the stationary point on the curve $y = \dfrac{\sin x}{x^2}$ between $x = 4$ and $x = 5$.

Give your answer correct to three decimal places.

10 **a** The equation $\mathrm{e}^{-x} - 2 + \sqrt{x} = 0$ has a single root, α.

Show that α lies between 3 and 4.

b Use the recurrence relation $x_{n+1} = (2 - \mathrm{e}^{-x_n})^2$, with $x_1 = 3.5$, to find x_2 and x_3, giving your answers to three decimal places.

c The diagram shows parts of the graphs of $y = (2 - \mathrm{e}^{-x})^2$ and $y = x$, and a position of x_1.

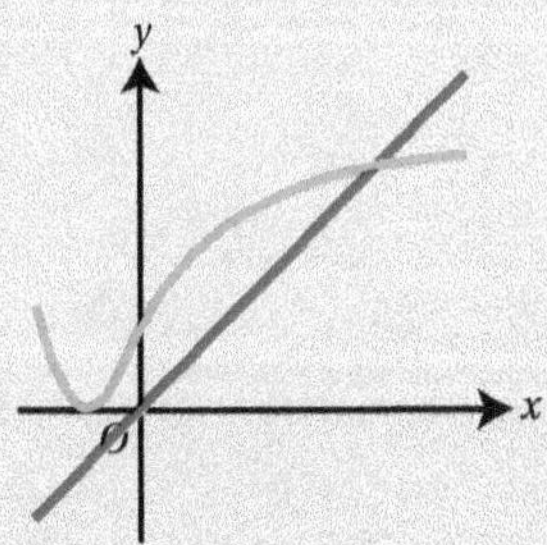

On a copy of the diagram, draw a staircase or cobweb diagram to show how convergence takes place, indicating the positions of x_2 and x_3 on the x-axis.

[© AQA 2013]

11 A ball is thrown vertically upwards from the roof of a house. Its height h metres above the ground after time t seconds is $h = 15t - 4.9t^2 + 6.5e^{-t}$.

a Write down the height of the ball above ground level when it was released.

b i Show that one possible iterative formula for finding an approximation to the maximum height reached by the ball is $t_{n+1} = \ln\left(\dfrac{6.5}{15 - 9.8t_n}\right)$.

ii Show that this iteration will not converge to a positive root for any $t \geqslant 0$.

iii Using the formula in part **b i**, with $t_0 = 1$, find the root to which this iteration converges.

iv Find a second iterative formula for finding the maximum height.

v Given that the root is near $t = 1$, use this second formula to find the root to two decimal places.

c Given that the time taken for the ball to hit the ground is between 2.5 and 3.5 seconds, use the Newton–Raphson method to find this time, correct to two decimal places.

Elevate

See Extension Sheet 14 for a look at Euler's method for finding approximate solutions to differential equations.

15 Numerical integration

In this chapter you will learn how to:

- understand why definite integration is connected to area under a curve
- approximate integrals that cannot be found exactly
- establish whether these approximations are overestimates or underestimates.

Before you start...

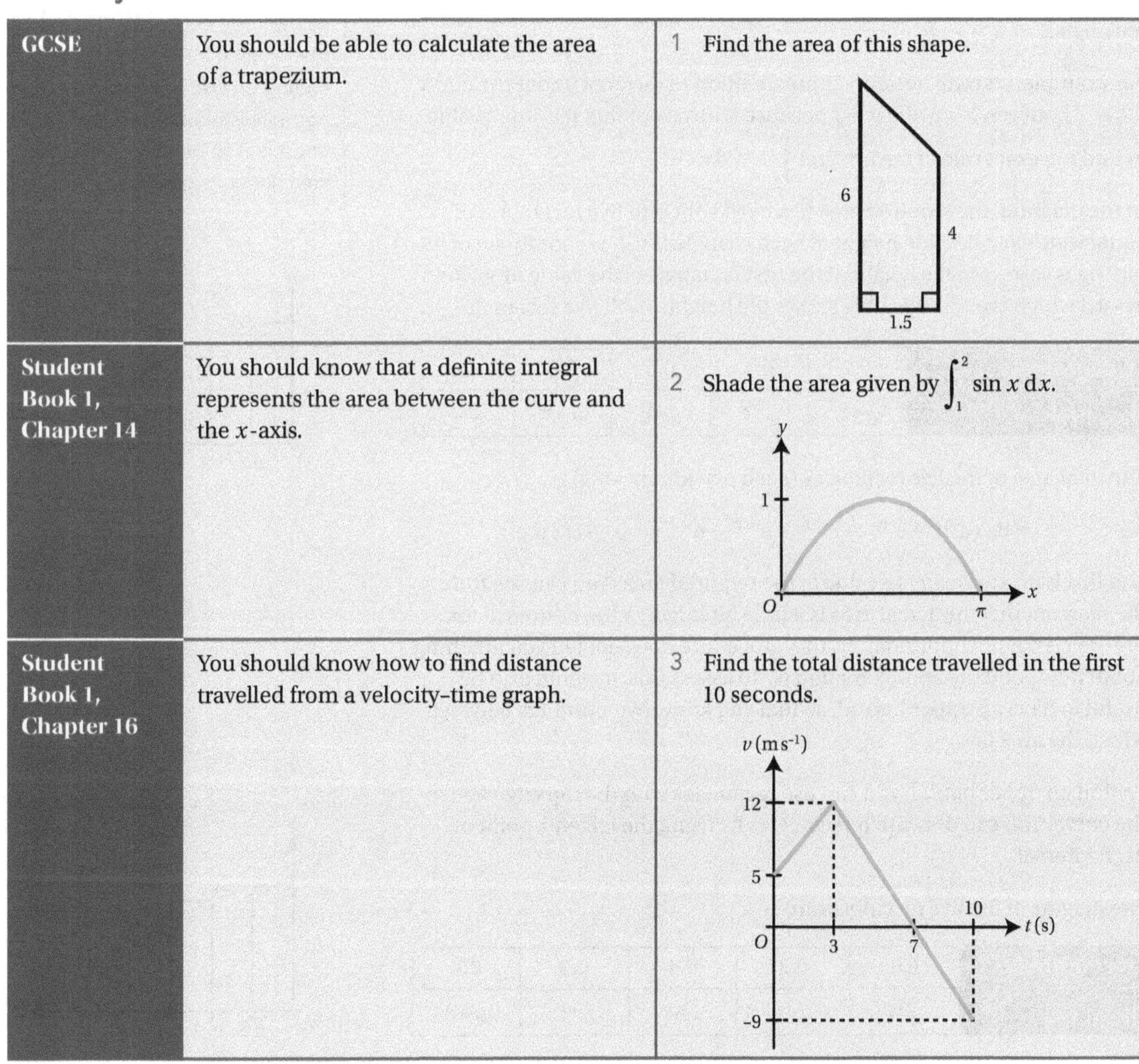

GCSE	You should be able to calculate the area of a trapezium.	1 Find the area of this shape.
Student Book 1, Chapter 14	You should know that a definite integral represents the area between the curve and the x-axis.	2 Shade the area given by $\int_1^2 \sin x \, dx$.
Student Book 1, Chapter 16	You should know how to find distance travelled from a velocity–time graph.	3 Find the total distance travelled in the first 10 seconds.

An approximation to definite integration

Using definite integration to find the area between a curve and the x-axis has many applications, for example, in Mechanics (finding distance travelled from a velocity-time graph) and Statistics (where the area under the normal distribution curve represents probability).

Since you already know many different integration methods, you may think that in most cases the area under the curve can be found exactly. It turns out that in many real-world problems in engineering and finance, for example, it is actually not possible to integrate the function exactly (or that the integration method is very complicated). In such cases you can find an approximation for the area, using one of the methods from this chapter.

Section 1: Integration as the limit of a sum

The simplest way to estimate the value of an area is to split it into rectangles.

For example, you will need an approximation to the area under the curve $y = e^{-x^2}$ between $x = 0$ and $x = 1$ because it turns out that it is impossible to find the exact vale of the integral $\int_0^1 e^{-x^2}\,dx$.

Fast forward

You will see in Chapter 21 that the function e^{-x^2} is used when finding probabilities from a normal distribution, so it is important to be able to find the area under its graph.

In the diagram, the required area has been split into five rectangles of equal width, $h = 0.2$. The height of each triangle is the y-coordinate of its top-right vertex; so the height of the first rectangle is the value of y when $x = 0.2$ which is $e^{-0.2^2}$. The table shows the heights of all five rectangles.

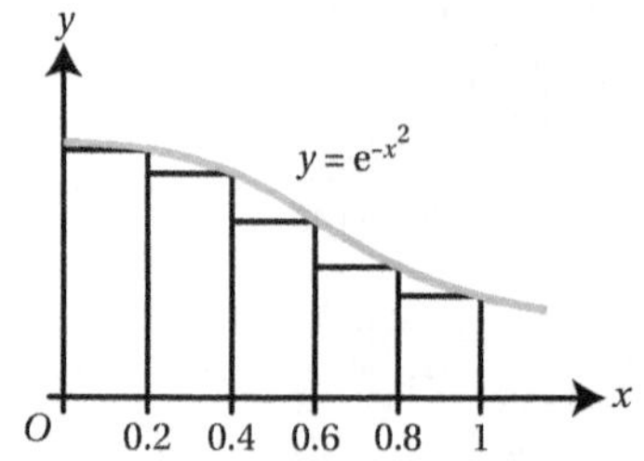

x-coordinate	0.2	0.4	0.6	0.8	1.0
height $y = e^{-x^2}$	$e^{-0.2^2}$	$e^{-0.4^2}$	$e^{-0.6^2}$	$e^{-0.8^2}$	$e^{-1.0^2}$

The total area of the five rectangles (each of width $h = 0.2$) is

$$0.2(e^{-0.2^2} + e^{-0.4^2} + e^{-0.6^2} + e^{-0.8^2} + e^{-1.0^2}) = 0.681 \text{ (3 d.p.)}$$

and this is an approximate value of the required area. You can see from the diagram that the actual area is a little bit larger; a **lower bound** for the area is 0.681. Unfortunately, this procedure does not tell you anything about how good the approximation is. To assess this, it would also be useful to have an **upper bound**, so that you know two numbers between which the area lies.

To find an upper bound, you can use rectangles with the top edge above the curve. You can draw such rectangles by using the left end-point of each interval.

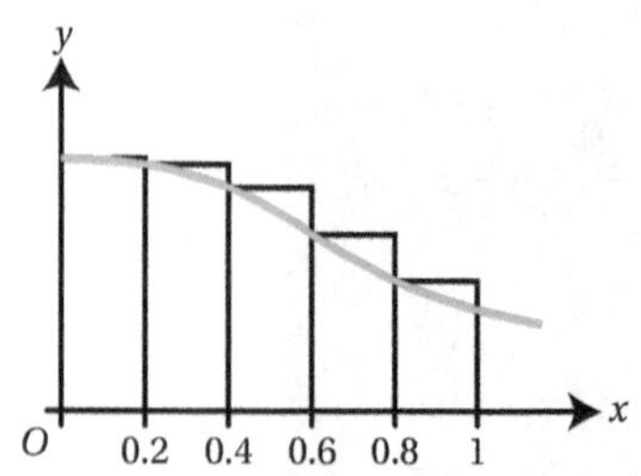

The heights of the five rectangles are:

x-coordinate	0.0	0.2	0.4	0.6	0.8
height $y = e^{-x^2}$	$e^{-0.0^2}$	$e^{-0.2^2}$	$e^{-0.4^2}$	$e^{-0.6^2}$	$e^{-0.8^2}$

So the total area is

$$0.2(e^{-0.0^2}+e^{-0.2^2}+e^{-0.4^2}+e^{-0.6^2}+e^{-0.8^2}) = 0.808\ (3\ \text{d.p.})$$

You can therefore say that the required area under the curve is certainly between 0.681 and 0.808, and you can write $0.681 < \int_0^1 e^{-x^2}\,dx < 0.808$.

Key point 15.1

You can find upper and lower bounds for the area under a curve by using rectangles that lie above and below the curve. The actual area lies between the lower and the upper bounds.

WORKED EXAMPLE 15.1

The diagram shows a part of the graph of $y = \cos(x^2)$ where x is in radians.

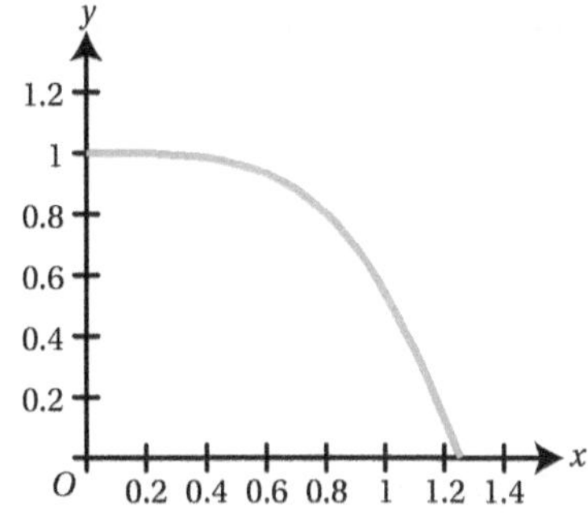

Using six rectangles of equal width, find a lower bound for the value of $\int_0^{1.2} \cos(x^2)\,dx$.

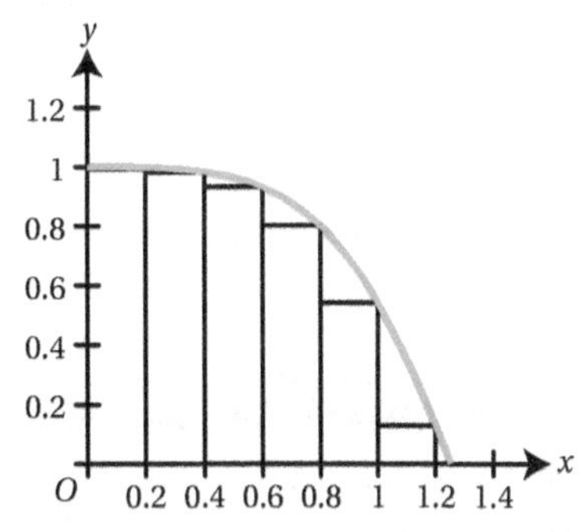

From the graph you can see that for the lower bound (rectangles below the curve) you need to use x-coordinates at the right end-point of each interval.

x	0.2	0.4	0.6	0.8	1.0	1.2
y	0.999	0.987	0.936	0.802	0.540	0.130

The width of each rectangle is $\frac{1.2}{6} = 0.2$.

Area $= 0.2(0.999 + 0.987 + \ldots) = 0.879$

So $\int_0^{2} \cos(x^2)\,dx > 0.879$

Each rectangle has width 0.2 and height equal to the y-coordinate.

WORKED EXAMPLE 15.2

A part of the curve with equation $y = \ln(x^2 + 1)$ is shown in the diagram.

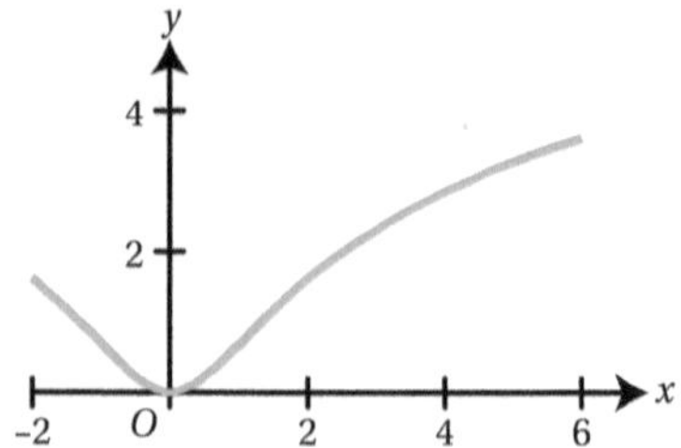

Use four rectangles of equal width to find an upper bound for $\int_1^5 \ln(x^2+1)\,dx$.

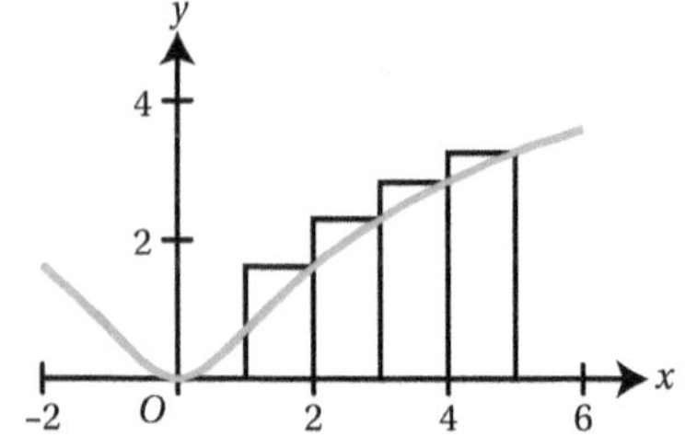

From the graph you can see that for the upper bound you now need to use x-coordinates at the right end-point of each interval.

x	2	3	4	5
y	1.609	2.303	2.833	3.258

The width of each rectangle is $\frac{5-1}{4} = 1$.

Area $= 1(1.609 + 2.303 + \ldots) = 10$ (2 s.f.)

So $\int_1^5 \ln(x^2+1)\,dx < 10$

Each rectangle has width 1 and height equal to the y-coordinate.

WORK IT OUT 15.1

Three students tried to find an upper bound for the value of $\int_2^5 \operatorname{cosec}\sqrt{x+1}\,dx$, using six rectangles.

Which solution is correct? Identify the errors in the incorrect solutions.

Solution 1	Solution 2	Solution 3
Using $x = 2, 2.5, 3, 3.5, 4, 4.5$: $0.5(1.01+1.05+1.10+1.17+1.27$ $+1.40) = 3.50$ So $\int_2^5 \operatorname{cosec}\sqrt{x+1}\,dx < 3.5$	Using $x = 2.5, 3, 3.5, 4, 4.5, 5$: $0.5(1.05+1.10+1.17+1.27$ $+1.40+1.57) = 3.78$ So $\int_2^5 \operatorname{cosec}\sqrt{x+1}\,dx < 3.8$	Using $x = 2, 2.5, 3, 3.5, 4, 4.5, 5$: $0.5(1.01+1.05+1.10+1.17$ $+1.27+1.40+1.57) = 4.29$ So $\int_2^5 \operatorname{cosec}\sqrt{x+1}\,dx < 4.3$

To get a more accurate approximation for the area, the upper and lower bounds need to be closer to each other. You can do this by using more rectangles of smaller width. For the original example, estimating

Tip

Always use the graph to decide which rectangles to use.

$\int_0^1 e^{-x^2}\,dx$, you can use a spreadsheet to calculate areas using more and more rectangles. Here are some of the results.

Number of rectangles	Width	Lower bound	Upper bound
5	0.2	0.681	0.808
10	0.1	0.715	0.778
20	0.05	0.731	0.762
50	0.02	0.740	0.753
100	0.01	0.744	0.750

You can use a graphical calculator or graphing software to check that the actual area is 0.747 (3 d.p.).

Key point 15.2

As the number of rectangles increases, the upper and the lower bounds approach a **limit**, which is the actual value of the definite integral.

Using rectangles of height y_i and width δx

$$\lim_{n\to\infty} = \sum_{i=1}^{n} y_i\,\delta x = \int_a^b y\,dx$$

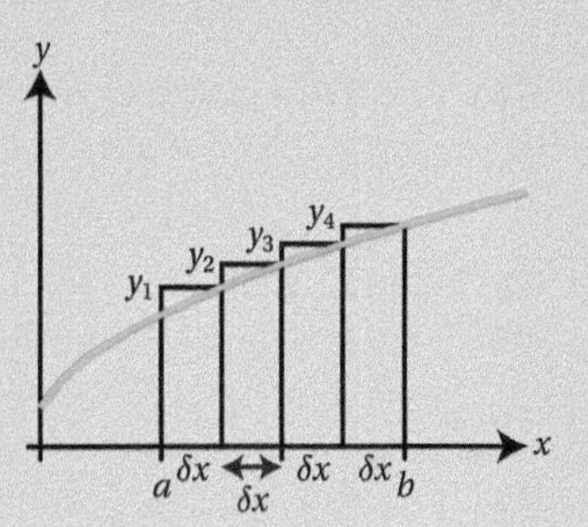

EXERCISE 15A

1 Use five rectangles to find upper and lower bounds for the value of each integral. Use technology to draw the graph first.

a i $\int_0^5 e^{-x^2}\,dx$ ii $\int_0^2 \frac{1}{x^3+1}\,dx$ b i $\int_0^2 \sin\sqrt{x}\,dx$ ii $\int_3^4 \ln(x^3-2)\,dx$

c i $\int_0^2 (e^{\sqrt{x}}-1)\,dx$ ii $\int_0^1 \tan(x^2)\,dx$

2 Use a spreadsheet to find upper and lower bounds for each integral, using:

i 10 rectangles ii 20 rectangles iii 40 rectangles.

In each case find the difference between the upper and lower bounds.

How does the difference decrease when the number of rectangles doubles?

a $\int_0^{\frac{\pi}{2}} \cos(\sin x)\,dx$ b $\int_2^3 \ln(\sin x)\,dx$ c $\int_0^2 e^{\frac{x^3}{10}}\,dx$ d $\int_2^5 \frac{1}{\sqrt{x}+1}\,dx$

3 The diagram shows part of the curve with equation $y = \sin(\ln x)$.

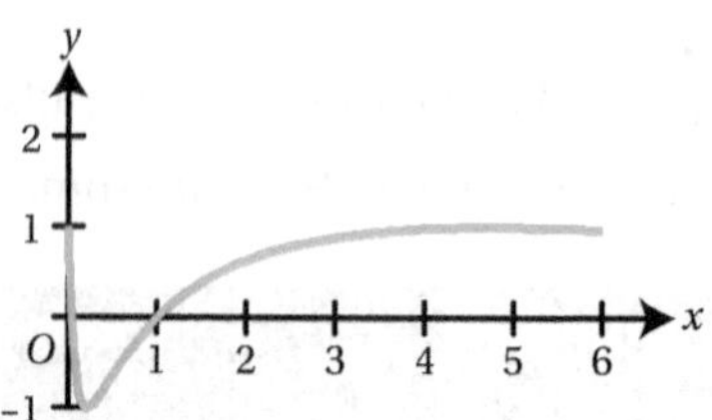

Using six rectangles of equal width, find upper and lower bounds for $\int_1^4 \sin(\ln x)\,dx$.

4 The diagram shows the graph of $y = 3e^{-\sqrt{x}}$.

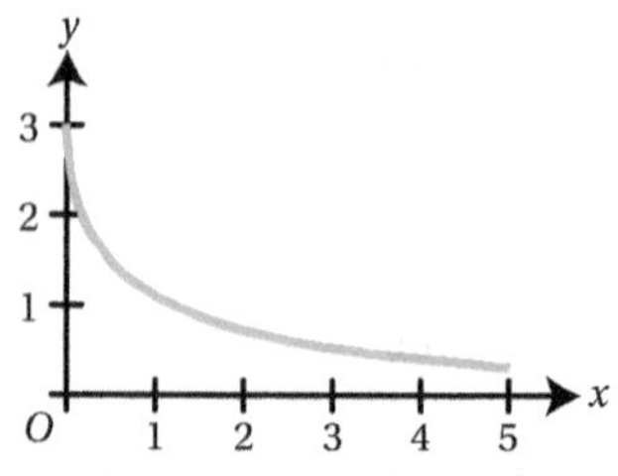

a Use five rectangles of equal width to find upper and lower bounds for $\int_0^4 3e^{-\sqrt{x}}\,dx$.

b How could the difference between the upper and lower bounds be reduced?

5 a Sketch the graph of $y = \sec x$ for $x \in \left[0, \frac{\pi}{2}\right)$.

b Use four rectangles of equal width to find an upper bound for $\int_0^1 \sec x\,dx$.

c If 20 rectangles were used, would the upper bound increase or decrease?

6 n rectangles of width δx are used to estimate the area under the graph of $y = \cos^2 x$ between $x = 0$ and $x = \frac{\pi}{2}$.

Find $\lim_{n\to\infty} \sum_{i=1}^{n} y_i\,\delta x$

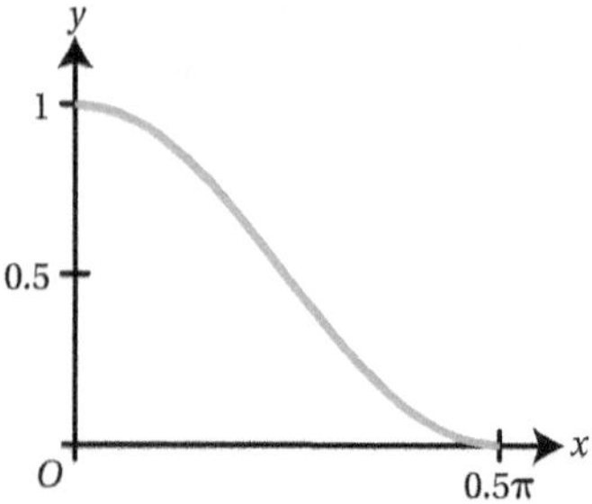

7 The diagram shows a part of the graph $y = \sqrt{\sin x}$.

a State the exact coordinates of the maximum point on the curve.

b In the part of the graph shown in the diagram, use four rectangles of equal width to find a lower bound for $\int_0^{\pi} \sqrt{\sin x}\,dx$.

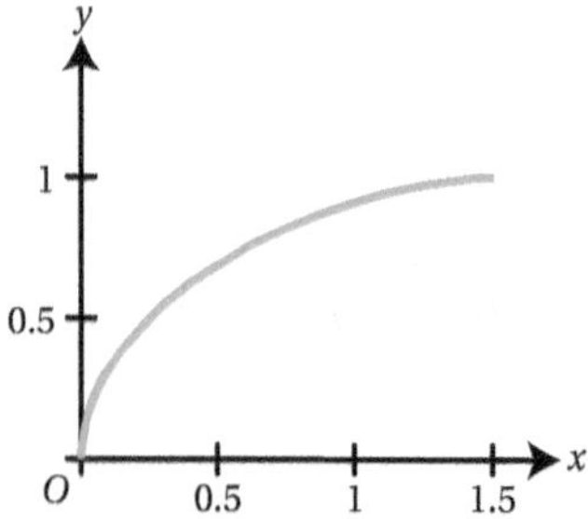

Section 2: The trapezium rule

You saw in Section 1 that you can approximate a definite integral by using sums of areas of rectangles. However, you may need many rectangles to achieve high accuracy. You can improve this method by replacing each rectangle by a trapezium that connects the end-points of the interval. Each trapezium has an area somewhere between the upper and the lower rectangles.

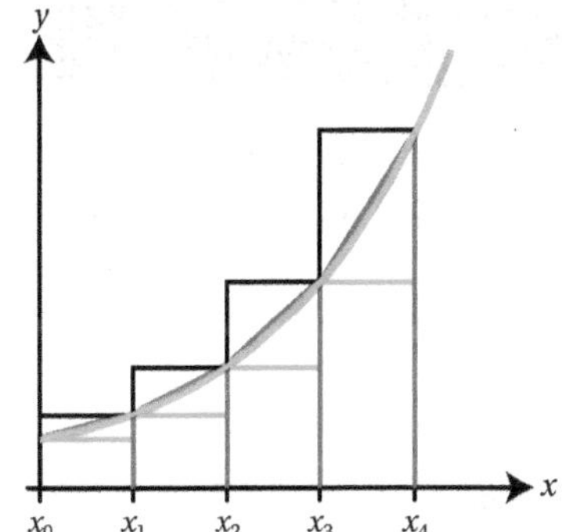

The area of a trapezium is given by $\frac{(a+b)h}{2}$, where a and b are the parallel sides and h is the height. Looking at the first trapezium, its height is equal to the width of the interval (h) and the parallel sides have lengths equal to the y-coordinates of the end-points.

If there are n intervals, you can label the x-coordinates $x_0, x_1, x_2 \dots x_n$ and the corresponding y-coordinates $y_0, y_1, y_2 \dots y_n$. The areas of the trapezia are then:

$$\frac{(y_0+y_1)h}{2}, \frac{(y_1+y_2)h}{2} \quad \dots \quad \frac{(y_{n-1}+y_n)h}{2}$$

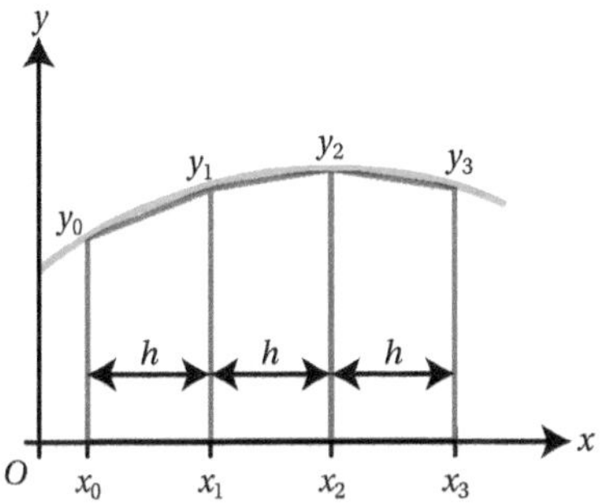

When you add these together, each y-coordinate appears twice, except for the first and the last one.

Key point 15.3

The area under a curve using n equal intervals with end-points $x_0, x_1, \dots x_n$ is given by the **trapezium rule**:

$$\int_a^b f(x)\,dx \approx \frac{h}{2}[y_0 + y_n + 2(y_1 + y_2 + \dots + y_{n-1})]$$

where $y_i = f(x_i)$ and $h = \frac{b-a}{n}$.

This will be given in your formula book.

Tip

It is a good idea to set out the x- and y-values in a table. The formula then says $\frac{h}{2}$ times (first + last + twice the rest), where h is the difference between the x-coordinates.

You may be asked to find an approximation by using a certain number of intervals, or by using a certain number of **ordinates**. The ordinates are the y-values you substitute into the formula.

Common error

n ordinates means $n-1$ intervals. So, for example, if you use the trapezium rule with five ordinates between 0 and 1, this means four intervals each of width 0.25, **not** five intervals of width 0.2.

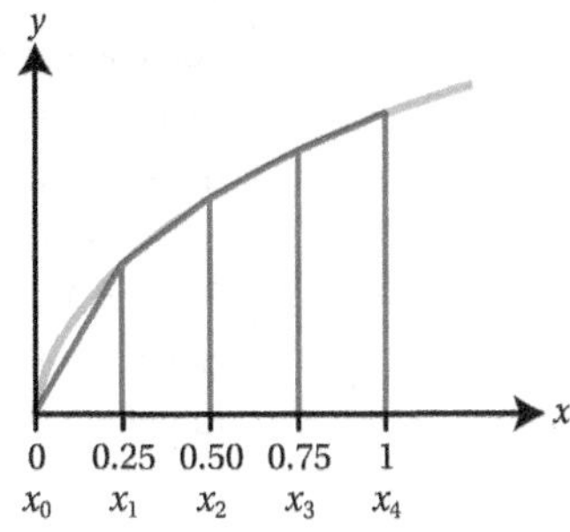

WORKED EXAMPLE 15.3

Use the trapezium rule with five equal intervals to find an approximate value of $\int_0^1 e^{-x^2}\, dx$. Give your answer correct to three decimal places.

$h = \frac{1-0}{5} = 0.2$

Divide the interval from 0 to 1 into five equal parts.

x	0	0.2	0.4	0.6	0.8	1.0
y	1	0.9608	0.8521	0.6977	0.5273	0.3679

The x-values start from 0 and go up in steps of 0.2 until they reach 1.

Note that since there are five intervals, there should be six x-values.

The y-values are calculated from $y = e^{-x^2}$. Since you want the answer correct to 3 d.p., record the y-values to 4 d.p. and round at the end.

$$\int_0^1 e^{-x^2}\, dx \approx \frac{0.2}{2}[1 + 0.3679 + 2(0.9608 + 0.8521 + 0.6977 + 0.5273)]$$
$$= 0.1(1.3679 + 2(3.0379))$$
$$= 0.744\,37$$
$$= 0.744 \text{ (3 d.p.)}$$

Use the formula. You should show the numbers used the in the calculation.

Tip

You may have a TABLE function on your calculator which will produce the table of values. Alternatively, you can save the six numbers in memory to use in the trapezium rule calculation.

Or if the expression for $f(x)$ is short, you can just type in the whole sum as $e^{-0^2} + e^{-1^2} + 2(e^{-0.2^2} + e^{-0.4^2} + e^{-0.6^2} + e^{-0.8^2})$.

Did you know?

Your calculator may have a function for finding approximate values of definite integrals. It probably uses the trapezium rule.

The actual value of the integral in Worked example 15.3 is 0.747 (3 d.p.), so the approximation is within 0.003 of the correct value. In Section 1, 100 rectangles were needed to get this close. This demonstrates that the trapezium rule, although slightly more complicated, is a lot more accurate than using rectangles.

In this case the approximation is a slight underestimate. Looking at the graph explains why this is - most of the trapezia lie below the curve.

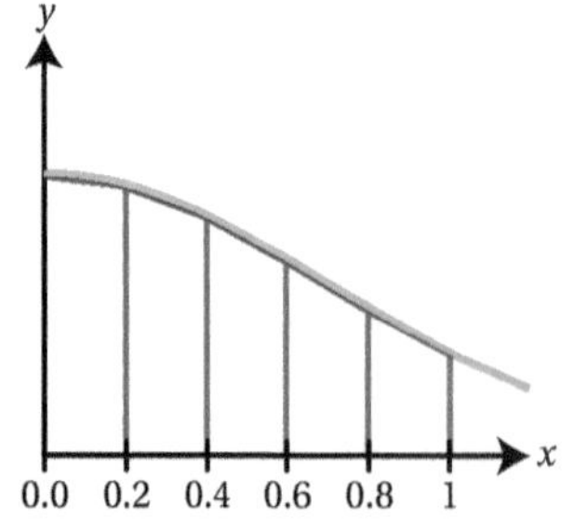

Rewind

You can see from the graph that the trapezium rule gives an underestimate when the function is concave – see Chapter 12, Section 1, for a reminder of the definition of concave and convex functions.

Key point 15.4

The trapezium rule will:

- underestimate the area when the curve is concave
- overestimate the area when the curve is convex.

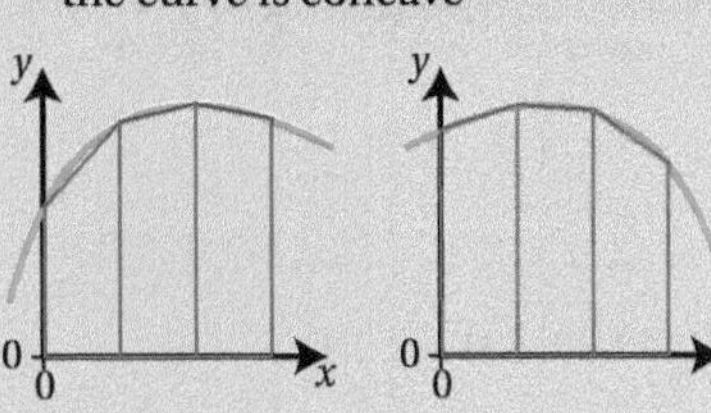

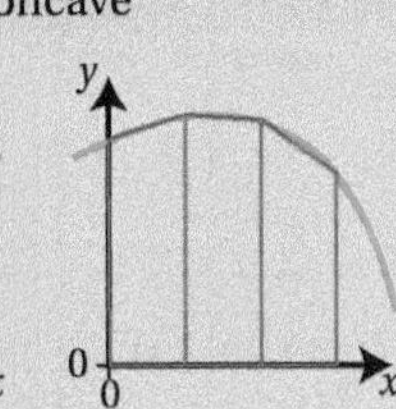
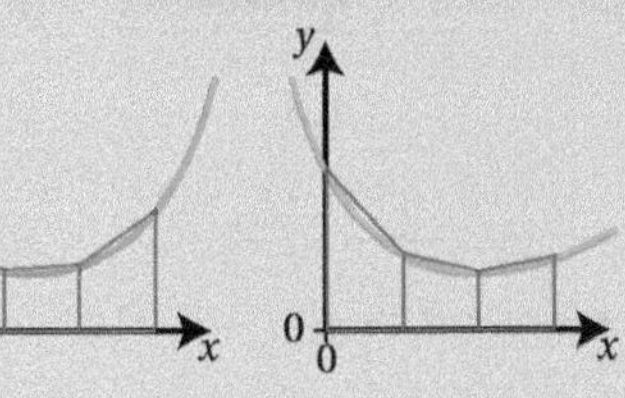

WORKED EXAMPLE 15.4

The diagram shows the graph of $y = \tan^3 x$.

a Use the trapezium rule with five ordinates to estimate the value of $\int_0^{\frac{\pi}{4}} \tan^3 x \, dx$, giving your answer correct to three significant figures.

b Explain whether your answer is an overestimate or an underestimate.

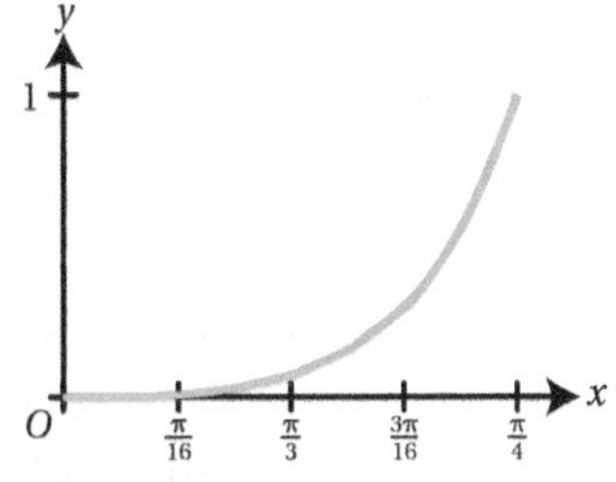

a $h = \frac{\frac{\pi}{4}}{4} = \frac{\pi}{16}$

Divide the interval from 0 to $\frac{\pi}{4}$ into four equal parts.

x	0	$\frac{\pi}{16}$	$\frac{\pi}{8}$	$\frac{3\pi}{16}$	$\frac{\pi}{4}$
y	0	0.007 87	0.071 07	0.298 32	1

The x-coordinates start from 0 and increase in steps of $\frac{\pi}{16}$. The y-coordinates are found using $y = \tan^3 x$.

$$\text{Area} \approx \frac{\pi}{32}[0 + 1 + 2(0.007\,87 + 0.071\,07 + 0.298\,32)]$$

Use the trapezium rule formula.

$$= \frac{\pi}{32}(1 + 2(0.377\,26)) = 0.172\,25$$

$$\therefore \int_0^{\frac{\pi}{4}} \tan^3 x \, dx \approx 0.172 \text{ (3 d.p.)}$$

b Since the function is convex between $x = 0$ and $x = \frac{\pi}{4}$ the trapezia are above the curve.

The explanation should refer to the shape of the graph.

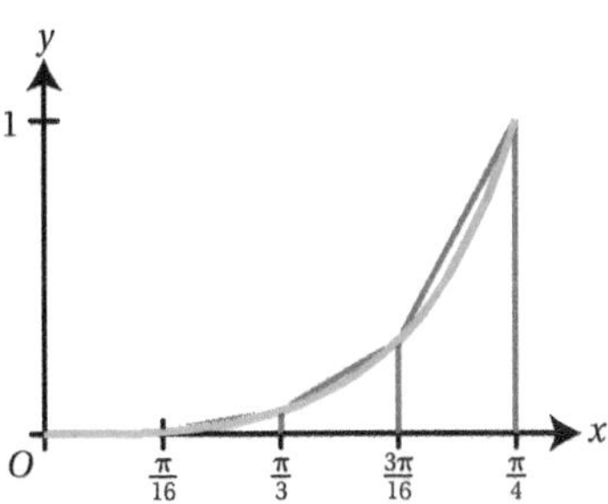

Hence this approximation is an overestimate.

WORK IT OUT 15.2

Three students were asked to estimate the value of $\int_1^2 (\ln x)^2 \, dx$ using trapezium rule with five strips.

Which solution is correct? Identify the errors in the incorrect solutions.

Solution 1

1	1.25	1.5	1.75	2
0	0.05	0.16	0.31	0.48

$0.125(0.48+2\times0.53)=0.192$

Solution 2

1	1.2	1.4	1.6	1.8	2
0	0.03	0.11	0.22	0.345	0.48

$0.1(0+0.03+0.11+\ldots)=0.119$

Solution 3

1	1.2	1.4	1.6	1.8	2
0	0.03	0.11	0.22	0.345	0.48

$0.1(0.48+1.43)=0.191$

EXERCISE 15B

1. Use the trapezium rule, with the given number of intervals, to find the approximate value of each integral. Compare your answer to the upper and lower bounds found in question 1 of Exercise 15A.

   a i $\int_0^5 e^{-x^2} \, dx$, five intervals　　ii $\int_0^2 \frac{1}{x^3+1} \, dx$, four intervals

   b i $\int_0^2 \sin\sqrt{x} \, dx$, five intervals　　ii $\int_3^4 \ln(x^3-2) \, dx$, four intervals

   c i $\int_0^2 (e^{\sqrt{x}}-1) \, dx$, six intervals　　ii $\int_0^1 \tan(x^2) \, dx$, three intervals

2. For each integral from question 1:

   i use technology to find its value, correct to eight decimal places

   ii use a spreadsheet to calculate trapezium rule approximations, using 2, 4, 8 and 16 intervals

   iii find the percentage error in each estimate. How do the percentage errors decrease when you double the number of intervals?

3. For each integral, find either its exact value where possible, or an approximation using six trapezia.

   a $\int_3^6 \ln\sqrt{x-2} \, dx$　　b $\int_1^4 \frac{1}{\sqrt{x+1}} \, dx$　　c $\int_0^\pi \sin^2 x \, dx$

   d $\int_0^\pi \sin(x^2) \, dx$　　e $\int_2^3 \ln(x^3) \, dx$　　f $\int_1^4 \operatorname{cosec}\left(\frac{x}{2}\right) dx$

4. a Sketch the graph of $y=3\ln(x-1)$.

   b Use the trapezium rule with six ordinates to estimate the value of $\int_2^4 3\ln(x-1) \, dx$. Give your answer to two decimal places.

   c Explain whether your answer is an overestimate or an underestimate.

5. a Use the trapezium rule with four intervals to find an approximate value of $\int_4^5 e^{\sqrt{5-x}} \, dx$.

   b Describe how you could obtain a more accurate approximation.

6 The diagram shows a part of the graph of $y = \cos(x^2)$.

The graph crosses the x-axis at the point where $x = a$.

a Find the exact value of a.

b Use the trapezium rule with five ordinates to find an approximation for $\int_0^a \cos(x^2)\,dx$.

c Is your approximation an overestimate or an underestimate? Explain your answer.

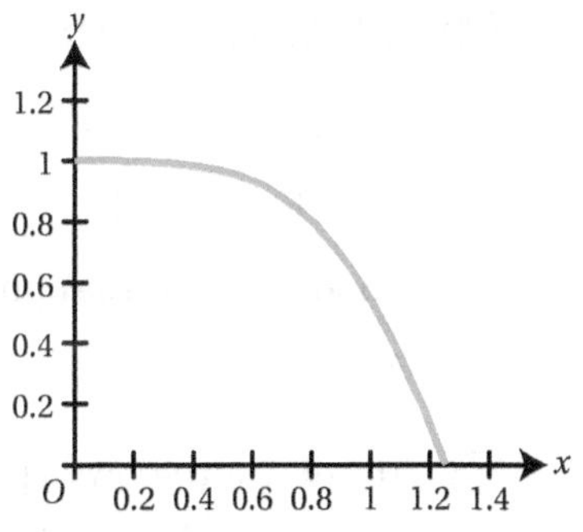

7 A particle moves in a straight line with the velocity given by $v = e^{\sqrt{t}}$, where v is measured in m s^{-1} and t in seconds. Use the trapezium rule with six strips to find the approximate distance travelled by the particle in the first 3 seconds.

Elevate

See Support Sheet 15 for a further example of using the trapezium rule and for more practice questions.

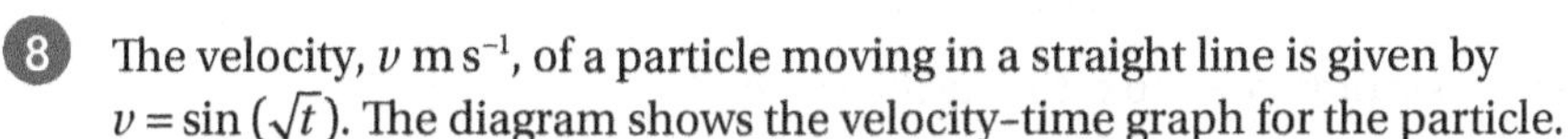

8 The velocity, $v\ \text{m s}^{-1}$, of a particle moving in a straight line is given by $v = \sin(\sqrt{t})$. The diagram shows the velocity–time graph for the particle.

a The particle changes direction when $t = p$ and $t = q$ (with $p < q$). Find the exact values of p and q.

b Use the trapezium rule, with eight intervals, to estimate the total distance travelled by the particle during the first q seconds.

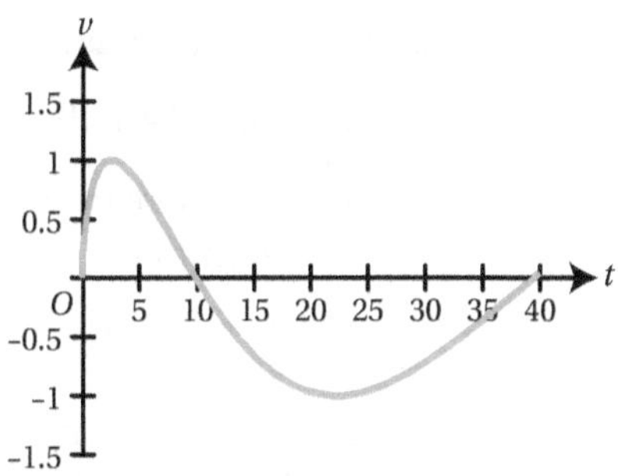

Checklist of learning and understanding

- Some definite integrals cannot be evaluated exactly. In those cases it is possible to use rectangles to find upper and lower bounds.
  - An **upper bound** for an area is a number that is larger than the area; a **lower bound** is a number that is smaller than the area.
  - As the rectangles get smaller, the upper and lower bounds get closer to each other. The actual area is the **limit** of the sum of the rectangles.
- The **trapezium rule** is a way of using trapezia to estimate the area. You need fewer trapezia than rectangles to achieve the same accuracy.
- Using n equal intervals with end-points $x_0, x_1, \dots x_n$, the area under the curve is:

$$\int_a^b f(x)\,dx \approx \frac{h}{2}[y_0 + y_n + 2(y_1 + y_2 + \dots + y_{n-1})]$$

where $y_i = f(x_i)$ and $h = \frac{b-a}{n}$

- The trapezium rule will:
  - underestimate the area when the curve is concave
  - overestimate the area when the curve is convex.

Mixed practice 15

1 The trapezium rule with seven ordinates is used to find an estimate for $\int_4^6 f(x)\,dx$.

What is the width of each interval?

Choose from these options.

A 2 B $\frac{4}{7}$ C $\frac{2}{7}$ D $\frac{1}{3}$

2 The diagram shows a part of the graph of $y = \frac{5}{\sqrt{x+2}+1}$.

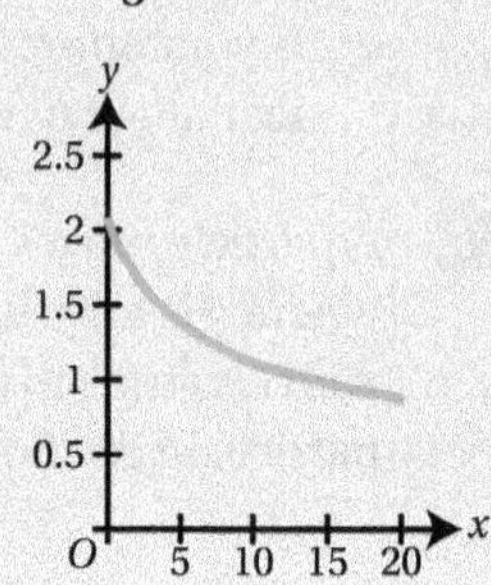

Using five rectangles of equal width, find a lower bound for the value of $\int_0^{15} \frac{5}{\sqrt{x+2}+1}\,dx$. Give your answer correct to one decimal place.

3 The diagram shows a part of the curve with equation $y = \ln(x^2+1)$.

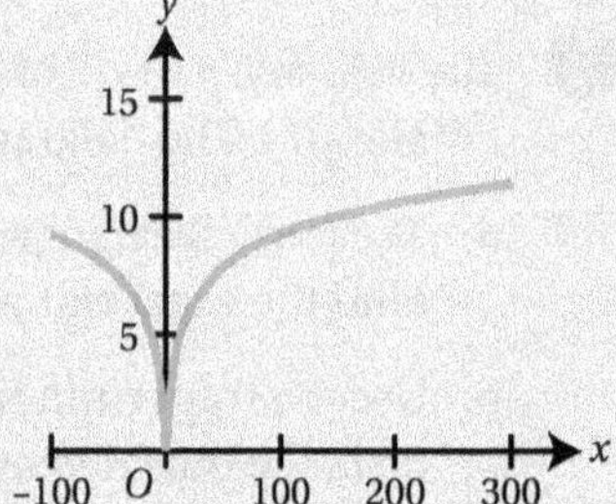

a Use the trapezium rule with six ordinates, to estimate the value of $\int_0^{200} \ln(x^2+1)\,dx$.

b State, with a reason, whether your answer is an underestimate or an overestimate.

c Explain how you could find a more accurate estimate.

4 A curve C, defined for $0 < x < 2\pi$ by the equation $y = \sin x$, where x is in radians, is sketched here. The region bounded by the curve C, the x-axis from 0 to 2 and the line $x = 2$ is shaded.

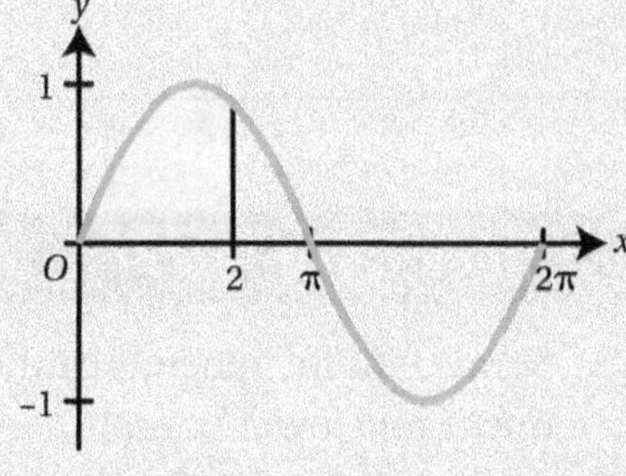

The area of the shaded region is given by $\int_0^2 \sin x\,dx$, where x is in radians.

Use the trapezium rule with five ordinates (four strips) to find an approximate value for the area of the shaded region, giving your answer to three significant figures.

[© AQA 2011]

5 **a** Use the trapezium rule with five ordinates (four strips) to find an approximate value for

$$\int_0^4 \frac{2^x}{x+1}\,dx$$

giving your answer to three significant figures.

b State how you could obtain a better approximation to the value of the integral using the trapezium rule.

[© AQA 2012]

6 Use six rectangles of equal width to find lower and upper bounds L and U such that

$$L < \int_2^5 (\ln x)^2\,dx < U.$$

7 The trapezium rule with four equal intervals is used to find an estimate for $\int_1^5 f(x)\,dx$.

If f is a decreasing function and $f''(x) > 0$, what will happen to the value obtained by the trapezium rule if 8 intervals are used?

Choose from these options.

A Increase **B** Decrease **C** Stay the same

D Depends on the particular function f.

8 The diagram shows a part of the curve with equation $y = \ln(7x - x^2 - 9)$. Use the trapezium rule with six strips of equal width to estimate the area enclosed between the curve and the x-axis.

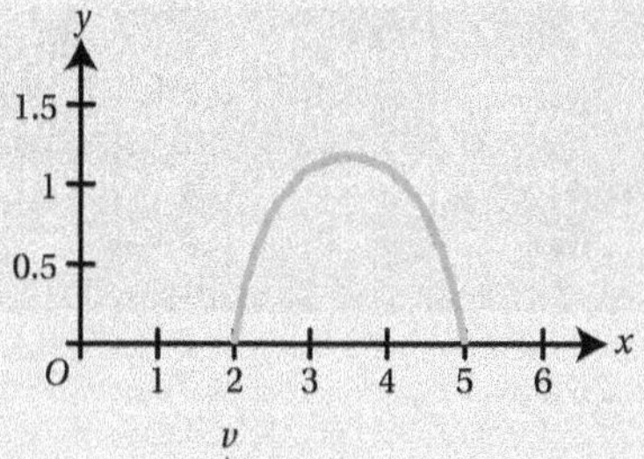

9 A particle moves in a straight line so that its velocity is given by the equation $v = \sin t \sin 3t$.

a Find the values of $t \in [0, 2\pi]$ when the velocity is zero.

b Use the trapezium rule with strips of width $\frac{\pi}{6}$ to find an approximate value of the distance travelled by the particle from $t = 0$ to $t = \pi$.

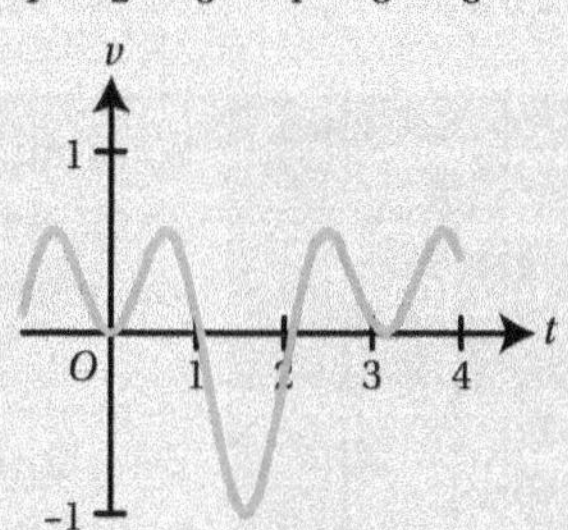

10 A part of the graph of $y = \ln\left(\frac{4x+2}{x+2}\right)$ is shown in the diagram.

Use four rectangles of equal width to find a rational number K such that $\int_0^{20} \ln\left(\frac{4x+2}{x+2}\right) dx < \ln K$.

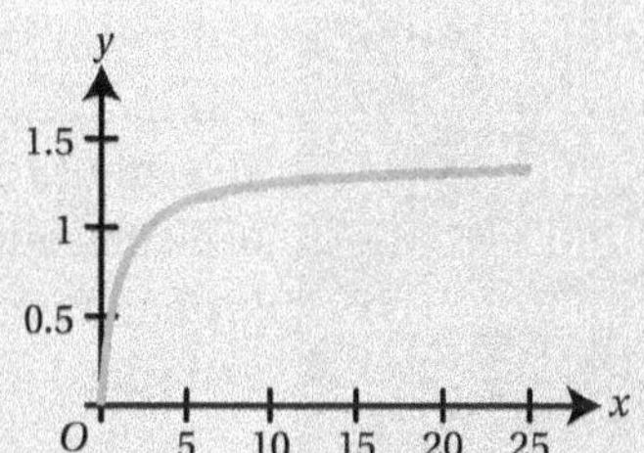

11 A curve has equation $y = \sqrt{x^4 + 1}$.

a Sketch the curve, showing the coordinates of any stationary points.

b Use four rectangles of equal width to find a lower bound for $\int_{-1}^{1} \sqrt{x^4 + 1}\,dx$.

12 The diagram shows the velocity–time graph for a particle moving in a straight line.
The velocity of the particle is measured at 10-second intervals and the results are recorded in the table.

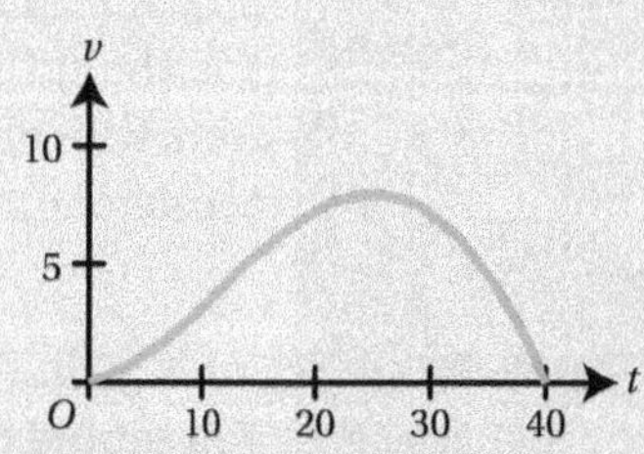

t (s)	0	10	20	30	40
v (m s^{-1})	0.0	3.1	7.2	7.1	0.0

Estimate the average speed of the particle during the 40 seconds.

Elevate

See Extension Sheet 15 for a look at how approximation methods were used in the development of calculus.

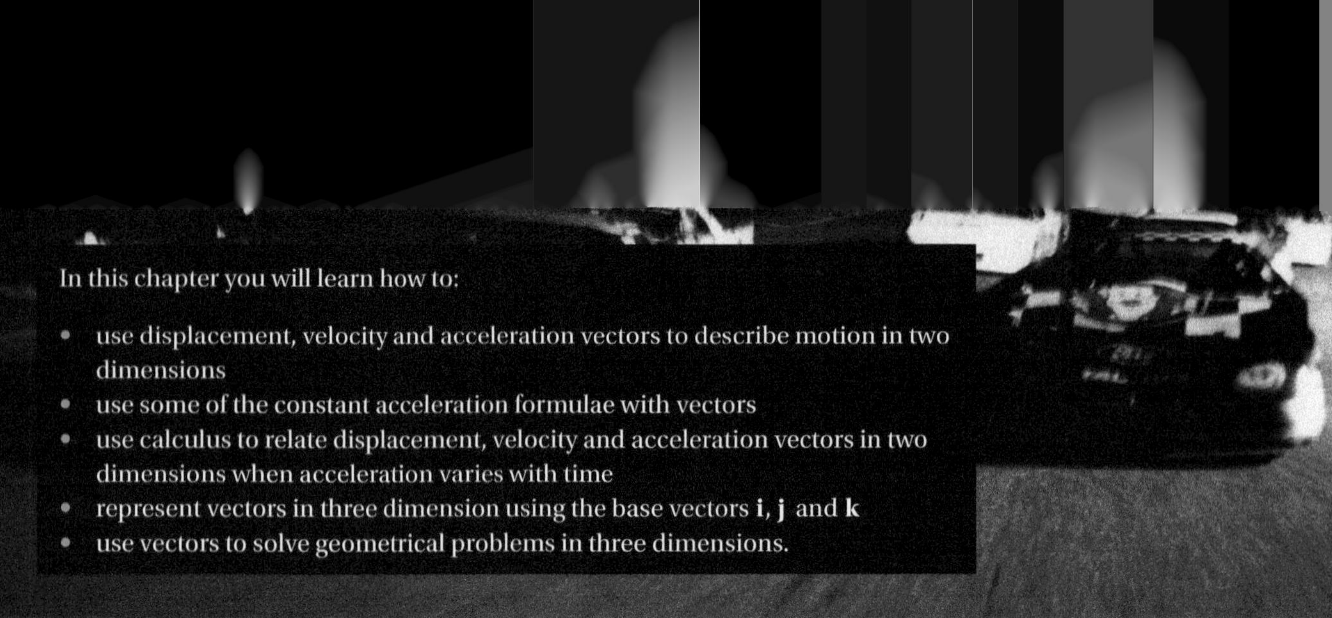

In this chapter you will learn how to:

- use displacement, velocity and acceleration vectors to describe motion in two dimensions
- use some of the constant acceleration formulae with vectors
- use calculus to relate displacement, velocity and acceleration vectors in two dimensions when acceleration varies with time
- represent vectors in three dimension using the base vectors **i**, **j** and **k**
- use vectors to solve geometrical problems in three dimensions.

Before you start...

Student Book 1, Chapter 15	You should be able to link displacement vectors to coordinates and perform operations with vectors.	1 Consider the points $A(2, 5)$, $B(-1, 3)$ and $C(7, -2)$. Let $\mathbf{p} = \overrightarrow{AB}$ and $\mathbf{q} = \overrightarrow{BC}$. Write in column vector form: a $\mathbf{p}$ b $\mathbf{q} - \mathbf{p}$ c $4\mathbf{q}$ d $\overrightarrow{AC}$.
Student Book 1, Chapter 15	You should be able to find the magnitude and direction of a vector.	2 Find the magnitude and direction of the vector $\begin{pmatrix} -3 \\ 2 \end{pmatrix}$.
Student Book 1, Chapter 16	You should understand the concepts of displacement and distance, instantaneous and average velocity and speed and acceleration.	3 In the diagram, positive displacement is measured to the right. A B C 120 m 180 m A particle takes 3 seconds to travel from B to C and another 7 seconds to travel from C to A. Find: a the average velocity b the average speed for the whole journey.
Student Book 1, Chapter 16	You should be able to use calculus to work with displacement, velocity and acceleration in one dimension.	4 A particle moves in a straight line with velocity $v = 2\,\mathrm{e}^t - t^2$. Find: a the acceleration when $t = 3$. b an expression for the displacement from the starting position.
Student Book 1, Chapter 17	You should be able to use constant acceleration formulae in one dimension.	5 A particle accelerates uniformly from 3 m s^{-1} to 7 m s^{-1} while covering a distance of 60 m in a straight line. Find the acceleration.
Chapter 12	You should be able to work with curves defined parametrically.	6 Find the Cartesian equation of the curve with parametric equations $x = 1 - 2t^2$, $y = 1 + t$.

Why do you need to use vectors to describe motion?

In Student Book 1, you studied motion in a straight line. You saw how displacement, velocity and acceleration are related through differentiation and integration:

$$v = \frac{dx}{dt}, \quad x = \int v\,dt$$

$$a = \frac{dv}{dt}, \quad v = \int a\,dt$$

In the special case when the acceleration is constant, you can use the constant acceleration equations:

$$v = u + at, \quad v^2 = u^2 + 2as, \quad s = ut + \frac{1}{2}at^2, \quad s = vt - \frac{1}{2}at^2, \quad s = \frac{1}{2}(u+v)t$$

But the real world has three dimensions, and objects do not always move in a straight line. You need to be able to describe positions and motion in a plane (such as a car moving around a race track) or in space (for example, flight paths of aeroplanes). This requires the use of vectors to describe displacement, velocity and acceleration.

Rewind

Vectors were introduced in Student Book 1, Chapter 15. Remember that you can also write the vector $\begin{pmatrix} 2 \\ 3 \end{pmatrix}$ as $2\mathbf{i} + 3\mathbf{j}$.

Section 1: Describing motion in two dimensions

When a particle moves in two dimensions, the displacement, velocity and acceleration are vectors. The distance and speed are still scalars.

WORKED EXAMPLE 16.1

Points A, B and C have position vectors $\begin{pmatrix} 3 \\ -1 \end{pmatrix}$, $\begin{pmatrix} 1 \\ 5 \end{pmatrix}$ and $\begin{pmatrix} -2 \\ 1 \end{pmatrix}$, where the distance is measured in metres. A particle travels in a straight line between each pair of points. It takes 5 seconds to travel from A to C and then a further 3 seconds to travel from C to B. Find:

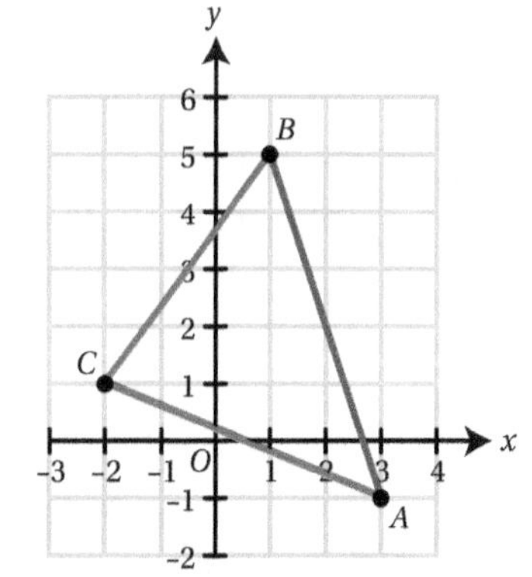

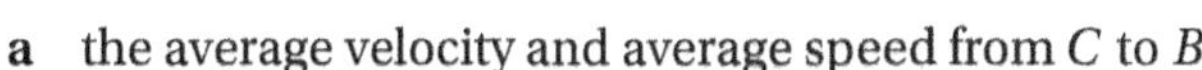

a the average velocity and average speed from C to B
b the final displacement of the particle from A
c the average velocity for the whole journey
d the average speed for the whole journey.

a $\overrightarrow{CB} = \mathbf{b} - \mathbf{c} = \begin{pmatrix} 1 \\ 5 \end{pmatrix} - \begin{pmatrix} -2 \\ 1 \end{pmatrix} = \begin{pmatrix} 3 \\ 4 \end{pmatrix}$

First find the displacement from C to B. This is the difference between the position vectors.

Remember: $\overrightarrow{CB} = \mathbf{b} - \mathbf{c}$

Continues on next page

$\text{Average velocity} = \begin{pmatrix} 3 \\ 4 \end{pmatrix} \div 3$

$= \begin{pmatrix} 1 \\ 1.33 \end{pmatrix} \text{m s}^{-1}$

Then use average velocity $= \dfrac{\text{displacement}}{\text{time}}$.

$CB = \sqrt{3^2 + 4^2} = 5$

Use Pythagoras' theorem to find the distance CB.

$\text{Speed} = \dfrac{5}{3} = 1.67 \text{ m s}^{-1}$

Then use average speed $= \dfrac{\text{distance}}{\text{time}}$.

b $\overrightarrow{AB} = \begin{pmatrix} 1 \\ 5 \end{pmatrix} - \begin{pmatrix} 3 \\ -1 \end{pmatrix}$

$= \begin{pmatrix} -2 \\ 6 \end{pmatrix}$

The displacement from A to B is the difference between their position vectors. It is highlighted blue in the diagram.

c Total time $= 5 + 3 = 8$ s

Average velocity $= \dfrac{\text{final displacement}}{\text{time}}$

$\text{Average velocity} = \begin{pmatrix} -2 \\ 6 \end{pmatrix} \div 8$

$= \begin{pmatrix} -0.25 \\ 0.75 \end{pmatrix} \text{m s}^{-1}$

d $\overrightarrow{AC} = \begin{pmatrix} -2 \\ 1 \end{pmatrix} - \begin{pmatrix} 3 \\ -1 \end{pmatrix}$

$= \begin{pmatrix} -5 \\ 2 \end{pmatrix}$

To get the total distance travelled, add the distance from A to C to the distance from C to B. This is highlighted red in the diagram.

$AC = \sqrt{5^2 + 2^2} = \sqrt{29}$

$\overrightarrow{CB} = \begin{pmatrix} 1 \\ 5 \end{pmatrix} - \begin{pmatrix} -2 \\ 1 \end{pmatrix}$

$= \begin{pmatrix} 3 \\ 4 \end{pmatrix}$

$CB = \sqrt{3^2 + 4^2} = \sqrt{25}$

Total distance $= \sqrt{29} + \sqrt{25} = 10.4$ m

$\therefore$ average speed $= \dfrac{10.4}{8} = 1.3 \text{ m s}^{-1}$

Average speed $= \dfrac{\text{total distance}}{\text{time}}$

You have already found that the total time is 8 seconds.

Common error

Notice that in Worked example 16.1 the average speed is **not** the magnitude of the average velocity vector. This is because the particle changes direction during the motion.

You can calculate average acceleration by considering the change in velocity.

WORKED EXAMPLE 16.2

A particle moves in a plane. It passes point A with velocity $(3\mathbf{i}-2\mathbf{j})$ m s^{-1} and passes point B 4 seconds later with velocity $(2\mathbf{i}+5\mathbf{j})$ m s^{-1}.

Find the magnitude and direction of the average acceleration of the particle between A and B.

Average acceleration $= ((2\underline{i}+5\underline{j})-(3\underline{i}-2\underline{j}))\div 4$

$= \left(-\frac{1}{4}\underline{i}+\frac{7}{4}\underline{j}\right)$ m s^{-2}

First find the acceleration vector using average acceleration $= \frac{\text{change in velocicty}}{\text{time}}$.

The magnitude is $|\underline{a}| = \sqrt{\left(\frac{1}{4}\right)^2+\left(\frac{7}{4}\right)^2}$

$= 1.77$ m s^2

Then find its magnitude.

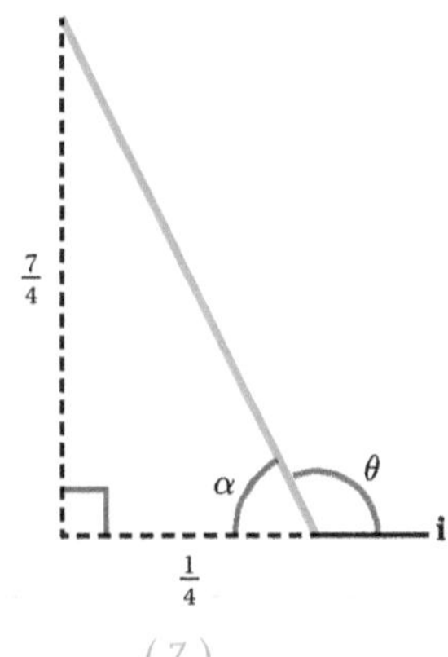

You can use any angle to define the direction as long as you clearly state where you are measuring the angle from. The angle above (or anti-clockwise from) the vector **i** is the most common choice. Draw a diagram to make sure you find the angle you intended!

$\tan\alpha = \frac{\left(\frac{7}{4}\right)}{\left(\frac{1}{4}\right)} = 7$

$\alpha = 81.7$

$\therefore \theta = 180 - 81.7$

$= 98.1°$

The direction is 98.1° above the vector $\underline{i}$.

Acceleration may cause a change in the direction of the velocity as well as its magnitude (speed). This means that the object will not necessarily move in a straight line. If you know how the displacement vector varies with time, you can sometimes find the Cartesian equation of the object's path.

Tip

The path an object follows is also called a **trajectory**.

Rewind

The displacement vector gives the parametric equations of the object's path, with the parameter being time. See Chapter 12, Section 2, for a reminder of parametric equations.

WORKED EXAMPLE 16.3

A particle moves in a plane. At time t the particle is at a point and its displacement from the origin O is given by the position vector $\overrightarrow{OP} = \begin{pmatrix} t+2 \\ 1-t^2 \end{pmatrix}$.

a Prove that the particle moves along a parabola.
b Sketch the parabola.

a The coordinates of P are:
$x = t+2,\ y = 1-t^2$

The two components of the position vector give the x- and y-coordinates of the particle's position.

$t = x-2$
$\therefore y = 1-(x-2)^2$
$y = -x^2+4x-3$

You want to find y in terms of x, so to eliminate t, substitute t from the first equation into the second equation.

Hence the path of the particle is a parabola.

b

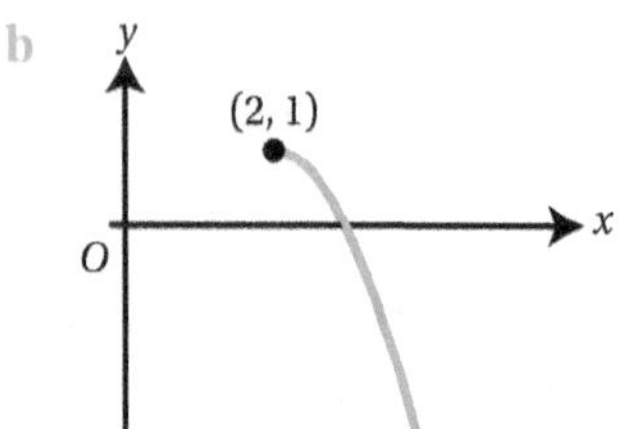

Note that, since $t > 0$, the object covers only a part of the parabola with $x > 2$.

You can also look at two particles moving in a plane and ask questions about the distance between them, and whether they ever meet. To do this you need to work with position vectors.

Key point 16.1

- If a particle starts at the point with position vector $\mathbf{r}_0$ and moves with constant velocity $\mathbf{v}$, its position vector at time t is $\mathbf{r} = \mathbf{r}_0 + \mathbf{v}t$.
- Two particles, A and B, meet if $\mathbf{r}_A = \mathbf{r}_B$ for the same value of t.

Rewind

In Worked example 16.4, part **b** is an example of proof by contradiction – see Chapter 1 for a reminder.

WORKED EXAMPLE 16.4

Two particles, A and B, move in the same plane. A starts from the origin and moves with constant velocity $\mathbf{v}_A = (3\mathbf{i} - 2\mathbf{j})\ \text{m s}^{-1}$.

a Write down the position vector of A in terms of t.

Particle B starts from the point with position vector $(\mathbf{i} - 5\mathbf{j})$ m and moves with constant velocity $\mathbf{v}_B = (\mathbf{i} + 3\mathbf{j})\ \text{m s}^{-1}$.

b Prove that A and B never meet.
c Find the minimum distance between the two particles.

Continues on next page

a $\mathbf{r}_A = (3\mathbf{i} - 2\mathbf{j})t$

$= 3t\mathbf{i} - 2t\mathbf{j}$

Use $\mathbf{r} = \mathbf{r}_0 + \mathbf{v}t$ with $\mathbf{r}_0 = 0$ as A starts at the origin.

b Position vector of B is

$\mathbf{r}_B = (\mathbf{i} - 5\mathbf{j}) + (\mathbf{i} + 3\mathbf{j})t$

$= (t + 1)\mathbf{i} + (3t - 5)\mathbf{j}$

This time $\mathbf{r}_0 = \mathbf{i} - 5\mathbf{j}$.

The particles meet when $\mathbf{r}_A = \mathbf{r}_B$:

You want to show that A and B are never in the same place at the same time. So try to find the value of t when the two displacements are equal and show that this is impossible.

$3t\mathbf{i} - 2t\mathbf{j} = (t + 1)\mathbf{i} + (3t - 5)\mathbf{j}$

$$\Rightarrow \begin{cases} 3t = t + 1 \\ -2t = 3t - 5 \end{cases}$$

If two vectors are equal then both components have to be equal. So you need a value of t that works in both equations.

From the first equation:

$2t = 1 \Rightarrow t = \frac{1}{2}$

Check in the second equation:

$-2\left(\frac{1}{2}\right) = -1$

$3\left(\frac{1}{2}\right) - 5 = -\frac{7}{2}$

Hence $\mathbf{r}_A \neq \mathbf{r}_B$ for all t so the particles never meet.

There is not a value of t which makes the two position vectors equal.

c At time t,

$\overrightarrow{AB} = \mathbf{r}_B - \mathbf{r}_A$

$= (t + 1)\mathbf{i} + (3t - 5)\mathbf{j} - (3t\mathbf{i} - 2t\mathbf{j})$

$= (1 - 2t)\mathbf{i} + (5t - 5)\mathbf{j}$

The distance between the particles is the magnitude of the displacement between them, which is found by subtracting the two position vectors.

The distance between A and B is:

$AB = \sqrt{(1 - 2t)^2 + (5t - 5)^2}$

$AB^2 = (1 - 4t + 4t^2) + (25t^2 - 50t + 25)$

$= 29t^2 - 54t + 26$

This expression has a minimum value when its square has a minimum value; so look at AB^2 to avoid having to work with the square root.

Let $y = 29t^2 - 54t + 26$

Then $\frac{dy}{dt} = 58t - 54 = 0$

$\Rightarrow t = \frac{27}{29}$

You could complete the square to find the minimum value, but it is not easy with these numbers so differentiate instead.

The minimum value of y is:

Don't forget that this is the minimum value for AB^2.

$29\left(\frac{27}{29}\right)^2 - 54\left(\frac{27}{29}\right) + 26 = 0.862$

Hence the minimum distance AB is

$\sqrt{0.862} = 0.928$ m

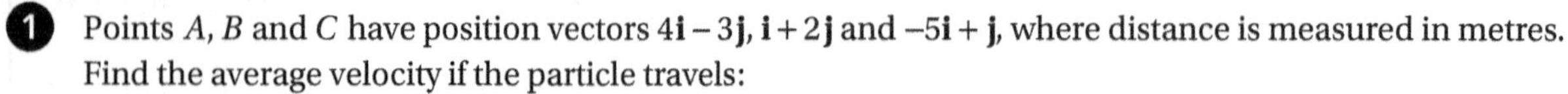

EXERCISE 16A

1 Points A, B and C have position vectors $4\mathbf{i} - 3\mathbf{j}$, $\mathbf{i} + 2\mathbf{j}$ and $-5\mathbf{i} + \mathbf{j}$, where distance is measured in metres. Find the average velocity if the particle travels:

a **i** from A to B in 3 seconds **ii** from A to C in 4 seconds

b **i** from C to B in 5 seconds **ii** from B to A in 4 seconds

c **i** from A to B in 3 seconds and then from B to C in 5 seconds

ii from C to A in 7 seconds and then from A to B in 4 seconds.

2 Find the average acceleration vector, and the magnitude of average acceleration in each case.

a **i** The velocity changes from $\begin{pmatrix} 6 \\ -2 \end{pmatrix}$ m s^{-1} to $\begin{pmatrix} 8 \\ 3 \end{pmatrix}$ m s^{-1} in 10 seconds.

ii The velocity changes from $\begin{pmatrix} -3 \\ 5 \end{pmatrix}$ m s^{-1} to $\begin{pmatrix} 1 \\ 10 \end{pmatrix}$ m s^{-1} in 8 seconds.

b **i** A particle accelerates from rest to $(4\mathbf{i} - 2\mathbf{j})$ m s^{-1} in 5 seconds.

ii A particle accelerates from rest to $(-3\mathbf{i} + 4\mathbf{j})$ m s^{-1} in 10 seconds.

3 Three points have coordinates $A(3, 5)$, $B(12, 7)$ and $C(8, 0)$.

a A particle travels in a straight line from A to B in 6 seconds.

Find its average velocity and average speed.

b Another particle travels in a straight line from B to C in 8 seconds and then in a straight line from C to B in 5 seconds.

Find its average velocity and average speed.

4 A particle moves in the plane so that its displacement from the origin at time t is given by the vector $\begin{pmatrix} t - 3 \\ 2 + t^2 \end{pmatrix}$.

a Find the particle's distance from the origin when $t = 2$.

b Find the Cartesian equation of the particle's trajectory.

5 **a** An object's velocity changes from $(5\mathbf{i} - 2\mathbf{j})$ m s^{-1} to $(3\mathbf{i} + 4\mathbf{j})$ m s^{-1} in 3 seconds.

Find the magnitude of its average acceleration.

b The object then moves for another 10 seconds with average acceleration $(-\mathbf{i} + 0.5\mathbf{j})$ m s^{-2}.

Find its direction of motion at the end of the 10 seconds.

6 A particle travels in a straight line from point P, with coordinates $(-4, 7)$, to point Q with coordinates $(3, -2)$. The journey takes 12 seconds and the distance is measured in metres.

a Find the average speed of the particle.

The particle then takes a further 7 seconds to travel in a straight line to point R with coordinates $(2, 5)$.

b Find the displacement from P to R.

c Find the average velocity of the particle for the whole journey.

d Find the average speed for the whole journey from P to R.
Explain why this is not equal to the magnitude of the average velocity.

7 Two particles, A and B, move in the plane. A has constant velocity $\begin{pmatrix} -3 \\ 1 \end{pmatrix}$ m s^{-1} and its initial displacement from the origin is $\begin{pmatrix} 14 \\ 0 \end{pmatrix}$ m. B starts from the origin and moves with constant velocity $\begin{pmatrix} 4 \\ 1 \end{pmatrix}$ m s^{-1}.

Show that the two particles meet and find the position vector of the meeting point.

8 A particle moves in a plane so that this displacement from the origin at time $t \geqslant 0$ is given by the vector $(t-1)\mathbf{i} + (6 + 4t - t^2)\mathbf{j}$ m.

a Find the distance of the particle from the origin when $t = 3$.

b Sketch the trajectory of the particle.

9 An object moves with a constant velocity $(-2\mathbf{i} + \mathbf{j})$ m s^{-1}. Its initial displacement from the origin is $(3\mathbf{i} - 4\mathbf{j})$ m.

a Find the Cartesian equation of the particle's trajectory.

b Find the minimum distance of the particle from the origin.

10 A particle moves in the plane so that its displacement from the origin at time t seconds is $(4\cos(2t)\mathbf{i} + 2\sin(2t)\mathbf{j})$ m.

Find the maximum distance of the particle from the origin.

Section 2: Constant acceleration equations

When a particle moves with constant acceleration, you can use formulae analogous to those for one-dimensional motion.

Key point 16.2

Constant acceleration formulae in two dimensions:

- $\mathbf{v} = \mathbf{u} + \mathbf{a}t$
- $\mathbf{s} = \mathbf{u}t + \frac{1}{2}\mathbf{a}t^2$
- $\mathbf{s} = \mathbf{v}t - \frac{1}{2}\mathbf{a}t^2$
- $\mathbf{s} = \frac{1}{2}(\mathbf{u} + \mathbf{v})t$

These will be given in your formula book.

Rewind

See Student Book 1, Chapter 17, for a reminder of the constant acceleration formulae.

Notice that these formulae give displacement ($\mathbf{s}$) rather than position vector ($\mathbf{r}$). To find the position vector at time t, you need to add on the particle's initial position vector $\mathbf{r}_0$ which gives $\mathbf{r} = \mathbf{r}_0 + \mathbf{s}$. This is the same idea as in Key point 16.1, except that now the velocity isn't constant so you need an equation of motion for $\mathbf{s}$ rather than just $\mathbf{v}t$.

Fast forward

The list in Key point 16.2 does not contain the vector version of the formula $v^2 = u^2 + 2as$. If you study Further Mathematics, you will meet a way of multiplying vectors (called the scalar product) that enables you to extend this formula to two dimensions as well.

WORKED EXAMPLE 16.5

A particle starts with initial velocity $(3\mathbf{i} - \mathbf{j})$ m s^{-1} and moves with constant acceleration. After 5 seconds its velocity is $(1.5\mathbf{i} + 2\mathbf{j})$ m s^{-1}. Find:

a the displacement from its initial position

b the distance from the initial position at this time.

a $\underline{u} = 3\underline{i} - \underline{j}$

$\underline{v} = 1.5\underline{i} + 2\underline{j}$

$t = 5$

$\underline{s} = ?$

Write down what you know and what you want.

$\underline{s} = \frac{1}{2}(\underline{u} + \underline{v})t$

$= \frac{1}{2}((3\underline{i} - \underline{j}) + (1.5\underline{i} + 2\underline{j})) \times 5$

$= 2.5(4.5\underline{i} + \underline{j})$

$= 11.25\underline{i} + 2.5\underline{j}$

Use $\mathbf{s} = \frac{1}{2}(\mathbf{u} + \mathbf{v})t$.

b Distance from starting position:

$\sqrt{11.25^2 + 2.5^2} = 11.5$ m

Distance is the magnitude of the displacement.

Since the particle does not move in a straight line, the distance from the starting point (which is measured in a straight line) is not the same as distance travelled (which is along a curve).

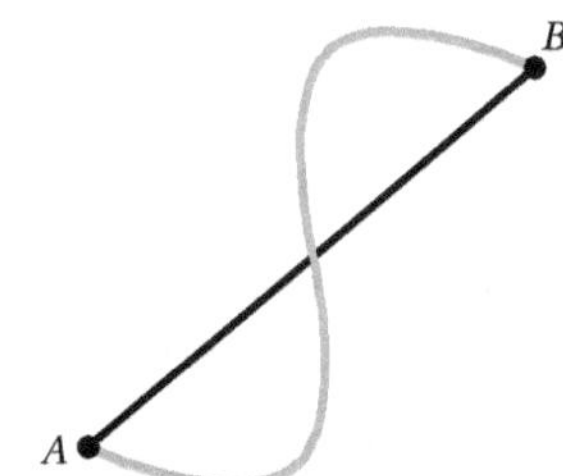

You need to be a little careful when solving equations with vectors. If you are comparing two sides of a vector equation, corresponding components need to be equal.

WORKED EXAMPLE 16.6

A particle moves with constant acceleration $\begin{pmatrix} -1.5 \\ 3 \end{pmatrix}$ m s^{-2}. It is initially at the origin and its initial velocity is $\begin{pmatrix} 2 \\ 5 \end{pmatrix}$ m s^{-1}.

Find the time when the particle is at the point with position vector $\begin{pmatrix} 1 \\ 4 \end{pmatrix}$ m.

$\underline{u} = \begin{pmatrix} 2 \\ 5 \end{pmatrix}$

$\underline{a} = \begin{pmatrix} -1.5 \\ 3 \end{pmatrix}$

$\underline{s} = \begin{pmatrix} 1 \\ 4 \end{pmatrix}$

$t = ?$

Write down what you know and what you want.

As the particle starts at the origin, its position vector at time t gives its displacement.

Continues on next page

$\mathbf{s} = \mathbf{u}t + \frac{1}{2}\mathbf{a}t^2$

$\begin{pmatrix} 1 \\ 4 \end{pmatrix} = \begin{pmatrix} 2 \\ 5 \end{pmatrix} t + \frac{1}{2}\begin{pmatrix} -1.5 \\ 3 \end{pmatrix} t^2$

Use $\mathbf{s} = \mathbf{u}t + \frac{1}{2}\mathbf{a}t^2$

First component:

$1 = 2t - 0.75t^2$

$0.75t^2 - 2t + 1 = 0$

$t = 2$ or $\frac{2}{3}$

The first component of the vectors gives a quadratic equation for t. This will give two possible values.

Second component:

when $t = 2$: $5t + \frac{1}{2}(3)t^2 = 16$

when $t = \frac{2}{3}$: $5t + \frac{1}{2}3t^2 = 4$

You need to check for which of these values of t the second component equals 4.

The particle is at $\begin{pmatrix} 1 \\ 4 \end{pmatrix}$ m when $t = \frac{2}{3}$ seconds.

WORK IT OUT 16.1

A particle moves with constant acceleration, starting from the origin. Its position vector at time t is given by $\mathbf{r} = (t^2 - 3t)\mathbf{i} + (2t^2 - 15t)\mathbf{j}$.

How many times does the particle pass through the origin during the subsequent motion?

Which is the correct solution? Identify the errors made in the incorrect solutions.

Solution 1	Solution 2	Solution 3
At the origin the position vector is zero.	At the origin both components of the position vector are zero.	At the origin both components of the position vector are zero.
When $t^2 - 3t = 0$	When $t^2 - 3t = 0$	When $t^2 - 3t = 0$
$t = 0$ or 3	$t = 0$ or 3	$t = 0$ or 3
So the particle passes through the origin once, when $t = 3$.	When $2t^2 - 15t = 0$	When $2t^2 - 15t = 0$
	$t = 0$ or 7.5	$t = 0$ or 7.5
	So the particle passes through the origin twice, when $t = 3$ and 7.5	So the particle does not pass through the origin again.

You saw in Student Book 1, Chapter 18, that Newton's second law still applies in two dimensions: $\mathbf{F} = m\mathbf{a}$, where $\mathbf{F}$ and $\mathbf{a}$ are vectors and m is a scalar. This means that the particle accelerates in the direction of the net force. You can now use this in conjunction with the constant acceleration formulae for vectors.

WORKED EXAMPLE 16.7

A particle of mass 2.4 kg moves under the action of a constant force, $\mathbf{F}$ N. When $t = 0$ the particle is at the point with position vector $\begin{pmatrix} 3 \\ -1 \end{pmatrix}$ moving with velocity $\begin{pmatrix} -2 \\ 5 \end{pmatrix}$ m s^{-1} and when $t = 4$ seconds its position vector is $\begin{pmatrix} 19 \\ -5 \end{pmatrix}$ m. Find the vector $\mathbf{F}$.

$\underline{u} = \begin{pmatrix} -2 \\ 5 \end{pmatrix}$

$\underline{s} = \begin{pmatrix} 19 \\ -5 \end{pmatrix} - \begin{pmatrix} 3 \\ -1 \end{pmatrix} = \begin{pmatrix} 16 \\ -4 \end{pmatrix}$

$t = 4$

$\underline{a} = ?$

Since the force is constant the acceleration is also constant, so you can use the constant acceleration equations.

The displacement, **s**, is the change in position vector.

$$\underline{s} = \underline{u}t + \frac{1}{2}\underline{a}t^2$$

$$\begin{pmatrix} 16 \\ -4 \end{pmatrix} = 4\begin{pmatrix} -2 \\ 5 \end{pmatrix} + \frac{1}{2}(4^2)\underline{a}$$

$$8\underline{a} = \begin{pmatrix} 16 \\ -4 \end{pmatrix} - \begin{pmatrix} -8 \\ 20 \end{pmatrix}$$

Rearrange to find **a**

$$= \begin{pmatrix} 24 \\ 16 \end{pmatrix}$$

$$\underline{a} = \begin{pmatrix} 3 \\ 2 \end{pmatrix} \text{ m s}^{-2}$$

$$\underline{F} = m\underline{a}$$

Now use $\mathbf{F} = m\mathbf{a}$.

$$= 2.4\begin{pmatrix} 3 \\ 2 \end{pmatrix}$$

$$= \begin{pmatrix} 7.2 \\ 4.8 \end{pmatrix} \text{ N}$$

EXERCISE 16B

1 In each question the particle moves with constant acceleration. Time is measured in seconds and displacement in metres.

a i $\mathbf{u} = \begin{pmatrix} 3 \\ -1 \end{pmatrix}$, $\mathbf{a} = \begin{pmatrix} -0.6 \\ 0.7 \end{pmatrix}$, find $\mathbf{v}$ when $t = 4$

ii $\mathbf{u} = 4\mathbf{i} + 2\mathbf{j}$, $\mathbf{a} = 1.2\mathbf{i} - 0.6\mathbf{j}$, find $\mathbf{v}$ when $t = 7$

b i $\mathbf{u} = -2\mathbf{i} + 0.5\mathbf{j}$, $\mathbf{a} = 0.3\mathbf{i} - 0.8\mathbf{j}$, find $\mathbf{s}$ when $t = 5$

ii $\mathbf{u} = \begin{pmatrix} 3 \\ -1 \end{pmatrix}$, $\mathbf{a} = \begin{pmatrix} -0.6 \\ 0.7 \end{pmatrix}$, find $\mathbf{s}$ when $t = 3$

c i $\mathbf{v} = 2\mathbf{i} + 5\mathbf{j}$, $\mathbf{s} = -2\mathbf{i} + \mathbf{j}$, $t = 4$, find $\mathbf{u}$.

ii $\mathbf{v} = \begin{pmatrix} -1 \\ 3 \end{pmatrix}$, $\mathbf{s} = \begin{pmatrix} 2 \\ 2 \end{pmatrix}$, $t = 6$, find $\mathbf{u}$.

d i $\mathbf{a} = \begin{pmatrix} -1 \\ 1 \end{pmatrix}$, $\mathbf{u} = \begin{pmatrix} 2 \\ 3 \end{pmatrix}$, $\mathbf{s} = \begin{pmatrix} 2 \\ 8 \end{pmatrix}$, find t.

ii $\mathbf{a} = \begin{pmatrix} 2 \\ 4 \end{pmatrix}$, $\mathbf{u} = \begin{pmatrix} 2 \\ -9 \end{pmatrix}$, $\mathbf{s} = \begin{pmatrix} 48 \\ 18 \end{pmatrix}$, find t.

2 An object of mass m kg moves under the action of the force $\mathbf{F}$ N. The object is initially at rest. Find the speed of the object at time t seconds, in each case.

a i $m = 6, t = 5, \mathbf{F} = 24\mathbf{i} + 6\mathbf{j}$

ii $m = 2, t = 10, \mathbf{F} = 6\mathbf{i} + 10\mathbf{j}$

b i $m = 0.2, t = 7, \mathbf{F} = 6\mathbf{i} - 2\mathbf{j}$

ii $m = 0.5, t = 5, \mathbf{F} = -3\mathbf{i} + 9\mathbf{j}$

3 An object moves with constant acceleration $\begin{pmatrix} 0.6 \\ -0.4 \end{pmatrix}$ m s^{-2} and initial velocity $\begin{pmatrix} 3.5 \\ 2.4 \end{pmatrix}$ m s^{-1}.

Find its velocity and the displacement from the initial position after 7 seconds.

4 A particle moves with constant acceleration $(3\mathbf{i} - \mathbf{j})$ m s^{-2}. It is initially at the the point with position vector $(\mathbf{i} + 4\mathbf{j})$ m and its velocity is $(2\mathbf{i} + 5\mathbf{j})$ m s^{-1}.

a Find the distance of the particle from the origin after 3 seconds.

b Find the direction of motion of the particle at this time.

5 A particle passes the origin with velocity $(2\mathbf{i} + 5\mathbf{j})$ m s^{-1} and moves with constant acceleration.

a Given that 7 seconds later its velocity is $(-12\mathbf{i} + 15.5\mathbf{j})$ m s^{-1}, find the acceleration.

b Find the time when the particle's displacement from the origin is $(-8\mathbf{i} + 32\mathbf{j})$ m.

6 An object moves with constant acceleration. When $t = 0$ s it has velocity $\begin{pmatrix} -1 \\ 3 \end{pmatrix}$ m s^{-1}.
When $t = 5$ s its displacement from the initial position is $\begin{pmatrix} 5 \\ 7 \end{pmatrix}$ m.

Find the magnitude of the acceleration.

7 A particle moves with constant acceleration. Its initial velocity is $(3\mathbf{i} - 2\mathbf{j})$ m s^{-1}.
8 seconds later, its displacement from the initial position is $(-44\mathbf{i} + 20\mathbf{j})$ m.
Find its direction of motion at this time.

8 An object moves with constant acceleration and initial velocity $5\mathbf{j}$ m s^{-1}.
When its displacement from the initial position is $(12.5\mathbf{i} + 5\mathbf{j})$ m, its velocity is $(5\mathbf{i} - 3\mathbf{j})$ m s^{-1}.

Find the magnitude of the acceleration.

9 A particle moves with constant acceleration $\begin{pmatrix} 3.8 \\ 2.2 \end{pmatrix}$ m s^{-2}. Given that its initial velocity is $\begin{pmatrix} -1 \\ 2 \end{pmatrix}$ m s^{-1},

find the time when its displacement from the initial position is $\begin{pmatrix} 180 \\ 130 \end{pmatrix}$ m.

10 A particle of mass 2.5 kg is subjected to a constant force $\mathbf{F} = (1.2\mathbf{i} + 0.9\mathbf{j})$ N. The initial velocity of the particle is $(0.6\mathbf{i} - 1.3\mathbf{j})$ m s^{-1}.

Find the velocity of the particle after 5 seconds.

11 A particle starts with initial velocity $6\mathbf{j}$ m s^{-1} and moves with constant acceleration $0.5\mathbf{i}$ m s^{-2}.

Prove that the speed of the particle increases with time.

12 A particle moves with constant acceleration $\mathbf{a} = (-2\mathbf{i} + \mathbf{j})$ m s^{-2}. When $t = 0$ s the particle is at rest, at the point with the position vector $(5\mathbf{i} + 3\mathbf{j})$ m.

Find the shortest distance of the particle from the origin during the subsequent motion.

Section 3: Calculus with vectors

When the acceleration is not constant you need to use differentiation and integration to find expressions for displacement and velocity. In Student Book 1, Chapter 15, you learnt how to do that for motion in one dimension. The same principles apply to two-dimensional motion: differentiating the displacement equation gives the velocity equation, and differentiating the velocity equation gives the acceleration equation. The only difference is that those quantities are now represented by vectors.

Key point 16.3

To differentiate or integrate a vector, differentiate or integrate each component separately.

WORKED EXAMPLE 16.8

A particle moves in two dimensions. Its position vector, measured in metres, varies with time (measured in seconds) as $\mathbf{r} = \begin{pmatrix} 2t^2 - 4 \\ 1 - t^3 \end{pmatrix}$.

Find the speed of the particle when $t = 3$ s.

$\underline{v} = \frac{d\underline{r}}{dt} = \begin{pmatrix} 4t \\ -3t^2 \end{pmatrix}$

To find the velocity, differentiate the displacement vector. Do this by differentiating each component separately.

When $t = 3$:

$\underline{v} = \begin{pmatrix} 12 \\ -27 \end{pmatrix}$

$\text{Speed} = |\underline{v}|$

$= \sqrt{12^2 + 27^2}$

$= 29.5 \text{ m s}^{-1}$

Speed is the magnitude of velocity.

When using integration with vectors, the constant of integration will also be a vector.

WORKED EXAMPLE 16.9

A particle moves with acceleration $\mathbf{a} = \begin{pmatrix} 2t+1 \\ 2\sin t \end{pmatrix}$ m s^{-2}. The initial velocity is $\begin{pmatrix} -1 \\ 3 \end{pmatrix}$ m s^{-1}.

Find an expression for the velocity at time t.

$\underline{v} = \int \underline{a}\, dt$

$= \int \begin{pmatrix} 2t+1 \\ 2\sin t \end{pmatrix} dt$

$= \begin{pmatrix} t^2+t \\ -2\cos t \end{pmatrix} + \underline{c}$

To find the velocity, integrate the acceleration vector. Do this by integrating each component separately.

When $t = 0$, $\underline{v} = \begin{pmatrix} -1 \\ 3 \end{pmatrix}$:

Use the initial velocity to find c.

$\begin{pmatrix} -1 \\ 3 \end{pmatrix} = \begin{pmatrix} 0 \\ -2 \end{pmatrix} + \underline{c}$

$\Rightarrow \underline{c} = \begin{pmatrix} -1 \\ 5 \end{pmatrix}$

So $\underline{v} = \begin{pmatrix} t^2+t-1 \\ -2\cos t+5 \end{pmatrix}$ m s^{-1}

You can include the constant within the existing vector.

Remember that, for two vectors to be equal, corresponding components need to be equal.

WORKED EXAMPLE 16.10

A particle starts from point P with velocity $(3\mathbf{i}+\mathbf{j})$ m s^{-1}. Its acceleration is given by $\mathbf{a} = (-t\mathbf{i}+2t\mathbf{j})$ m s^{-2}.

Show that the particle never returns to P.

$\underline{v} = \int (-t\underline{i}+2t\underline{j})\, dt$

$= -\frac{1}{2}t^2\underline{i}+t^2\underline{j}+\underline{c}$

You need to find the displacement vector from P and show that this is never 0 for $t > 0$.

First integrate $\mathbf{a}$ to find $\mathbf{v}$.

Continues on next page

$\underline{v} = 3\underline{i} + \underline{j}$ when $t = 0$, so $\underline{c} = 3\underline{i} + \underline{j}$.

$\therefore \underline{v} = \left(-\frac{1}{2}t^2 + 3\right)\underline{i} + (t^2 + 1)\underline{j}$

Use the initial velocity to find **c**.

$\underline{s} = \int\left[\left(-\frac{1}{2}t^2 + 3\right)\underline{i} + (t^2 + 1)\underline{j}\right]dt$

$= \left(-\frac{1}{6}t^3 + 3t\right)\underline{i} + \left(\frac{1}{3}t^3 + t\right)\underline{j} + \underline{c}$

Now integrate the velocity to find the displacement vector. Initially **s** = 0, so **c** = 0.

If $\underline{s} = 0$ then the $\underline{i}$ component of the displacement is 0:

$-\frac{1}{6}t^3 + 3t = 0$

$\frac{1}{6}t(-t^2 + 18) = 0$

$t = 0$ or $\sqrt{18}$

Now check if there is any value of t (other than $t = 0$) when **s** = 0. First find the time at which the **i** component is 0.

When $t = \sqrt{18}$ the $\underline{j}$ component of displacement is:

$\frac{1}{3}t^3 + t = 19.7 \neq 0$

Hence $\underline{s} \neq 0$ for $t > 0$, so the particle does not return to the starting point.

$t = 0$ is the starting point so use $t = \sqrt{18}$ and check whether the **j** component is 0.

WORK IT OUT 16.2

A particle starts from rest and moves with acceleration $\mathbf{a} = \begin{pmatrix} 3\sin t \\ -5\cos t \end{pmatrix}$ m s^{-2}. Find an expression for the velocity of the particle.

Which is the correct solution? Identify the errors made in the incorrect solutions.

Solution 1	Solution 2	Solution 3
$\mathbf{v} = \int\begin{pmatrix} 3\sin t \\ -5\cos t \end{pmatrix}dt$ $= \begin{pmatrix} -3\cos t \\ -5\sin t \end{pmatrix} + c$ Initially at rest means that the speed is zero. When $t = 0$: $\mathbf{v} = \begin{pmatrix} -3 \\ 0 \end{pmatrix}$ $\Rightarrow$ speed $= \sqrt{(-3)^2 + 0^2} = 3$ $\Rightarrow c = -3$ $\mathbf{v} = \begin{pmatrix} -3\cos t - 3 \\ -5\sin t - 3 \end{pmatrix}$ m s^{-1}	$\mathbf{v} = \int\begin{pmatrix} 3\sin t \\ -5\cos t \end{pmatrix}dt$ $= \begin{pmatrix} -3\cos t \\ -5\sin t \end{pmatrix} + \mathbf{c}$ Initially at rest: $\mathbf{v}(0) = \begin{pmatrix} 0 \\ 0 \end{pmatrix} = \begin{pmatrix} -3 \\ 0 \end{pmatrix} + \mathbf{c}$ $\Rightarrow \mathbf{c} = \begin{pmatrix} 3 \\ 0 \end{pmatrix}$ $\therefore \mathbf{v} = \begin{pmatrix} -3\cos t + 3 \\ -5\sin t \end{pmatrix}$ m s^{-1}	$\mathbf{v} = \int\begin{pmatrix} 3\sin t \\ -5\cos t \end{pmatrix}dt$ $= \begin{pmatrix} -3\cos t \\ -5\sin t \end{pmatrix} + \mathbf{c}$ Initially at rest, so $\mathbf{c} = \mathbf{0}$ $\therefore \mathbf{v} = \begin{pmatrix} -3\cos t \\ -5\sin t \end{pmatrix}$ m s^{-1}

When the force is variable (a function of t) you can still use $\mathbf{F} = m\mathbf{a}$, but because the acceleration will now be variable, you need to use calculus rather than the constant acceleration formulae.

WORKED EXAMPLE 16.11

A particle of mass 0.5 kg starts from rest and moves under the action of the force $((4t)\mathbf{i} + (2t-2)\mathbf{j})$ N.

Find:

a the speed after 3 seconds.

b the direction of motion of the particle after 3 seconds.

a $\underline{F} = m\underline{a}$

$(4t)\underline{i} + (2t-2)\underline{j} = 0.5\underline{a}$

$\Rightarrow \underline{a} = (8t)\underline{i} + (4t-4)\underline{j}\text{ m s}^{-2}$

$\underline{v} = \int (8t)\underline{i} + (4t-4)\underline{j}\,dt$

$= (4t^2)\underline{i} + (2t^2 - 4t)\underline{j} + \underline{c}$

Find the velocity by integrating the acceleration. Use $\mathbf{F} = m\mathbf{a}$ first.

When $t = 0$, $\underline{v} = 0$ so $\underline{c} = 0$

Remember to find **c**.

$\therefore \underline{v} = ((4t^2)\underline{i} + (2t^2 - 4t)\underline{j})\text{ m s}^{-1}$

When $t = 3$ s:

$\underline{v} = (36\underline{i} + 6\underline{j})\text{ m s}^{-1}$

You can now use the given value of t to find the velocity vector.

The speed is:

$|\underline{v}| = \sqrt{36^2 + 6^2} = 36.5\text{ m s}^{-1}$

Speed is the magnitude of the velocity vector.

b

$v = 36\mathbf{i} + 6\mathbf{j}$

6

θ

36

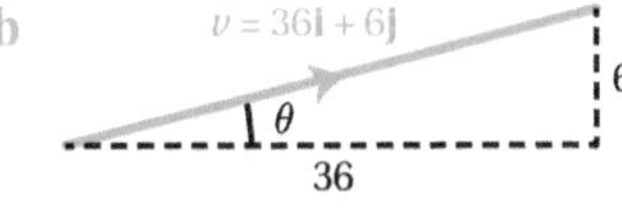

The direction is given by the angle with the vector **i**.

It is always a good idea to draw a diagram to see which angle you are looking for.

$\tan\theta = \frac{6}{36}$

$\theta = \tan^{-1}\left(\frac{6}{36}\right) = 9.46°$

The direction of motion is 9.46° above the horizontal.

EXERCISE 16C

1 For a particle with the given position vector, find expressions for the velocity and acceleration vectors. Also find the speed when $t = 3$ s.

a **i** $\mathbf{r} = (3t - \sin t)\mathbf{i} + (t - t^2)\mathbf{j}$ m

ii $\mathbf{r} = (e^{2t} - t)\mathbf{i} + (t^2 + e^{2t})\mathbf{j}$ m

b **i** $\mathbf{r} = \begin{pmatrix} 4\cos 3t \\ 3\sin 2t \end{pmatrix}$ m

ii $\mathbf{r} = \begin{pmatrix} 3\ln(t+1) - t \\ t^2 + \ln(t+1) \end{pmatrix}$ m

2 For a particle moving with the given acceleration, find expressions for the velocity and position vectors.

The particle is initially at the origin, and the initial velocity is given in each question part. Also find the distance from the starting point when $t = 3$.

a **i** $\mathbf{a} = (3 - t^2)\mathbf{i} + 2t\mathbf{j}\,\text{m s}^{-2}$, $\mathbf{v}(0) = 2\mathbf{i} + 5\mathbf{j}\,\text{m s}^{-1}$ **ii** $\mathbf{a} = (t+1)\mathbf{i} + 3\mathbf{j}\,\text{m s}^{-2}$, $\mathbf{v}(0) = \mathbf{i} - 2\mathbf{j}\,\text{m s}^{-1}$

b **i** $\mathbf{a} = \begin{pmatrix} 3\,\text{e}^{t} \\ 2\,\text{e}^{-t} \end{pmatrix}\text{m s}^{-2}$, $\mathbf{v}(0) = \begin{pmatrix} -1 \\ 1 \end{pmatrix}\text{m s}^{-1}$ **ii** $\mathbf{a} = \begin{pmatrix} 3\sin 2t \\ 3\cos 2t \end{pmatrix}\text{m s}^{-2}$, $\mathbf{v}(0) = \begin{pmatrix} 3 \\ 0 \end{pmatrix}\text{m s}^{-1}$

c **i** $\mathbf{a} = 2\cos(t)\mathbf{i} + 3\sin(t)\mathbf{j}\,\text{m s}^{-2}$, $\mathbf{v}(0) = \mathbf{0}\,\text{m s}^{-1}$ **ii** $\mathbf{a} = \text{e}^{2t}\mathbf{i} + 3\text{e}^{t}\mathbf{j}\,\text{m s}^{-2}$, $\mathbf{v}(0) = \mathbf{0}\,\text{m s}^{-1}$

3 A particle moves in a plane with position vector given by $\mathbf{r} = \text{e}^{2t}\mathbf{i} + (t-1)\mathbf{j}$.

a Find an expression for the velocity of the particle at time t.

b Find the speed of the particle when $t = 5$.

4 A particle moves in a plane, starting from rest. Its acceleration varies according to the equation

$\mathbf{a} = \begin{pmatrix} 6t \\ \cos 2t \end{pmatrix}\text{m s}^{-2}$.

a Find an expression for the velocity of the particle at time t.

b Find the displacement from the initial position after 3 seconds.

5 The velocity of a particle, in m s^{-1}, moving in a plane is given by $\mathbf{v} = (3 - \sin(2t))\mathbf{i} + 2\cos(2t)\mathbf{j}$.

a Find the initial speed of the particle.

b Find the magnitude of the acceleration when $t = 12$.

c Find an expression for the displacement from the initial position after t seconds.

6 A particle starts from rest and moves with acceleration $((2 + \text{e}^{-2t})\mathbf{i} + 4\text{e}^{-2t}\mathbf{j})\,\text{m s}^{-2}$. Find its distance from the initial position after 1.2 seconds.

7 The velocity of a particle moving in a plane is given by $\mathbf{v} = \begin{pmatrix} 2 - 3t^2 \\ 4t - 1 \end{pmatrix}\text{m s}^{-1}$.

Show that the particle never returns to its initial position.

8 For a particle moving in two dimensions, the displacement vector from the starting point is given by

$\mathbf{s} = \begin{pmatrix} 3t^3 - 4t \\ t^4 - 2t^3 + t \end{pmatrix}$

a The components of the displacement vector give parametric equations of the trajectory of the particle, $x = x(t)$, $y = y(t)$. Use parametric differentiation to find the gradient of the tangent to this curve, $\dfrac{\text{d}y}{\text{d}x}$, when $t = 3$.

b Find the velocity vector when $t = 3$. What do you notice?

9 A particle of mass 2 kg moves under the action of the force $\mathbf{F} = (24\cos(2t)\mathbf{i} - 24\sin(2t)\mathbf{j})\,\text{N}$. Its initial velocity is $\mathbf{v} = 6\mathbf{j}\,\text{m s}^{-1}$.

a Show that the speed of the particle is constant.

b By considering the x and y components of the displacement vector, show that the particle moves in a circle.

10 A particle moves in the plane, from the initial position $\begin{pmatrix} 5 \\ 0 \end{pmatrix}$ m. Its velocity, $\mathbf{v}$ m s^{-1}, at time t s is given by the equation: $\mathbf{v} = \begin{pmatrix} -8t \\ 2 \end{pmatrix}$.

Find the time when the particle is closest to the origin, and find this minimum distance.

Section 4: Vectors in three dimensions

In the preceding sections you learnt how to use vector equations to represent motion in two dimensions. Now vector methods will be extended to enable you to describe positions and various types of motion in the three-dimensional world.

To represent positions and displacements in three-dimensional space, you need three base vectors, all perpendicular to each other. They are conventionally called **i**, **j**, **k**.

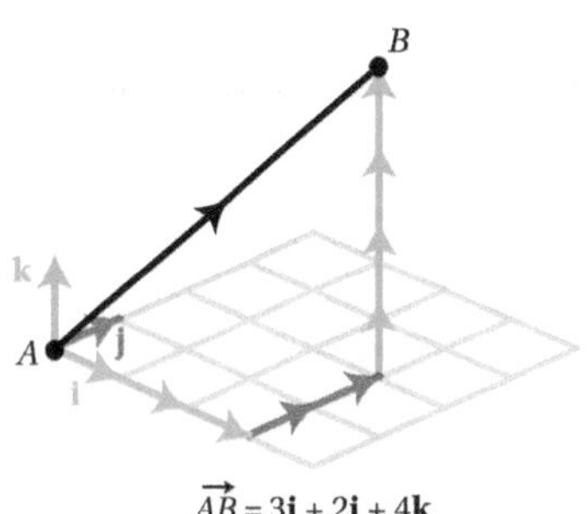

$\overrightarrow{AB} = 3\mathbf{i} + 2\mathbf{j} + 4\mathbf{k}$

You can also show the components in a column vector: $\overrightarrow{AB} = \begin{pmatrix} 3 \\ 2 \\ 4 \end{pmatrix}$.

Each point in a three-dimensional space can be represented by a position vector, which equals its displacement from the origin. The displacement from one point to another is the difference between their position vectors.

WORKED EXAMPLE 16.12

Points A and B have coordinates $(3, -1, 2)$ and $(5, 0, 3)$, respectively. Write as column vectors:

a the position vectors of A and B

b the displacement vector $\overrightarrow{AB}$.

a $\underline{a} = \begin{pmatrix} 3 \\ -1 \\ 2 \end{pmatrix}, \underline{b} = \begin{pmatrix} 5 \\ 0 \\ 3 \end{pmatrix}$

The components of the position vectors are the coordinates of the point.

b $\overrightarrow{AB} = \underline{b} - \underline{a}$

$= \begin{pmatrix} 5 \\ 0 \\ 3 \end{pmatrix} - \begin{pmatrix} 3 \\ -1 \\ 2 \end{pmatrix}$

$= \begin{pmatrix} 2 \\ 1 \\ 1 \end{pmatrix}$

Relate $\overrightarrow{AB}$ to the position vectors **a** and **b** exactly as you would in two dimensions.

Did you know?

Although our space is three dimensional, it turns out that many situations can be modelled as motion in two dimensions. For example, it is possible to prove that the orbit of a planet lies in a plane, so two-dimensional vectors are sufficient to describe it.

The formula for the magnitude of a three-dimensional vector is analogous to the two-dimensional one.

Key point 16.4

- The magnitude (modulus) of a vector $\mathbf{a} = \begin{pmatrix} a_1 \\ a_2 \\ a_3 \end{pmatrix}$ is $|\mathbf{a}| = \sqrt{a_1^2 + a_2^2 + a_3^2}$.
- The distance between points with position vectors **a** and **b** is $|\mathbf{b} - \mathbf{a}|$.

WORKED EXAMPLE 16.13

Points A and B have position vectors $\mathbf{a} = 2\mathbf{i} - \mathbf{j} + 5\mathbf{k}$ and $\mathbf{b} = 5\mathbf{i} + 2\mathbf{j} + 3\mathbf{k}$. Find the exact distance AB.

$\overrightarrow{AB} = \underline{b} - \underline{a}$
$= (5\underline{i} + 2\underline{j} + 3\underline{k}) - (2\underline{i} - \underline{j} + 5\underline{k})$
$= 3\underline{i} + 3\underline{j} - 2\underline{k}$

The distance is the magnitude of the displacement vector, so you need to find $\overrightarrow{AB}$ first.

$|\overrightarrow{AB}| = \sqrt{3^2 + 3^2 + 2^2}$
$= \sqrt{22}$

Now use the formula for the magnitude.

Remember that you can use vector addition and subtraction to combine displacements.

WORKED EXAMPLE 16.14

The diagram shows points M, N, P, Q such that $\overrightarrow{MN} = 3\mathbf{i} - 2\mathbf{j} + 6\mathbf{k}$, $\overrightarrow{NP} = \mathbf{i} + \mathbf{j} - 3\mathbf{k}$ and $\overrightarrow{MQ} = -2\mathbf{j} + 5\mathbf{k}$.

Write each vector in component form.

a $\overrightarrow{MP}$ **b** $\overrightarrow{PM}$ **c** $\overrightarrow{PQ}$.

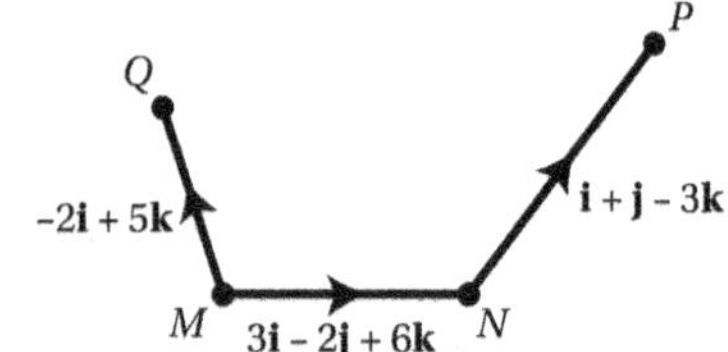

a $\overrightarrow{MP} = \overrightarrow{MN} + \overrightarrow{NP}$
$= (3\underline{i} - 2\underline{j} + 6\underline{k}) + (\underline{i} + \underline{j} - 3\underline{k})$
$= 4\underline{i} - \underline{j} + 3\underline{k}$

You can get from M to P via N.

b $\overrightarrow{PM} = -\overrightarrow{MP}$
$= -4\underline{i} + \underline{j} - 3\underline{k}$

Going from P to M is the reverse of going from M to P.

c $\overrightarrow{PQ} = \overrightarrow{PM} + \overrightarrow{MQ}$
$= (-4\underline{i} + \underline{j} - 3\underline{k}) + (-2\underline{j} + 5\underline{k})$
$= -4\underline{i} - \underline{j} + 2\underline{k}$

You can get from P to Q via M, using the answers from previous parts.

EXERCISE 16D

1 Write each vector in column vector notation (in three dimensions).

a i $4\mathbf{i}$ ii $-5\mathbf{j}$

b i $3\mathbf{i}+\mathbf{k}$ ii $2\mathbf{j}-\mathbf{k}$

2 Let $\mathbf{a}=\begin{pmatrix}7\\1\\12\end{pmatrix}$, $\mathbf{b}=\begin{pmatrix}5\\-2\\3\end{pmatrix}$ and $\mathbf{c}=\begin{pmatrix}1\\1\\2\end{pmatrix}$. Find each vector.

a i $3\mathbf{a}$ ii $4\mathbf{b}$ b i $\mathbf{a}-\mathbf{b}$ ii $\mathbf{b}+\mathbf{c}$

c i $2\mathbf{b}+\mathbf{c}$ ii $\mathbf{a}-2\mathbf{b}$ d i $\mathbf{a}+\mathbf{b}-2\mathbf{c}$ ii $3\mathbf{a}-\mathbf{b}+\mathbf{c}$

3 Let $\mathbf{a}=\mathbf{i}+2\mathbf{j}$, $\mathbf{b}=\mathbf{i}-\mathbf{k}$ and $\mathbf{c}=2\mathbf{i}-\mathbf{j}+3\mathbf{k}$. Find each vector.

a i $-5\mathbf{b}$ ii $4\mathbf{a}$

b i $\mathbf{c}-\mathbf{a}$ ii $\mathbf{a}-\mathbf{b}$

c i $\mathbf{a}-\mathbf{b}+2\mathbf{c}$ ii $4\mathbf{c}-3\mathbf{b}$

4 Find the magnitude of each vector in three dimensions.

$\mathbf{a}=\begin{pmatrix}4\\1\\2\end{pmatrix}$, $\mathbf{b}=\begin{pmatrix}1\\-1\\0\end{pmatrix}$, $\mathbf{c}=2\mathbf{i}-4\mathbf{j}+\mathbf{k}$, $\mathbf{d}=\mathbf{j}-\mathbf{k}$

5 Find the distance between each pair of points in three dimensions.

a i $A(1, 0, 2)$ and $B(2, 3, 5)$ ii $C(2, 1, 7)$ and $D(1, 2, 1)$

b i $P(3, -1, -5)$ and $Q(-1, -4, 2)$ ii $M(0, 0, 2)$ and $N(0, -3, 0)$

6 Find the distance between the points with the given position vectors.

a $\mathbf{a}=2\mathbf{i}+4\mathbf{j}-2\mathbf{k}$ and $\mathbf{b}=\mathbf{i}-2\mathbf{j}-6\mathbf{k}$

b $\mathbf{a}=\begin{pmatrix}3\\7\\-2\end{pmatrix}$ and $\mathbf{b}=\begin{pmatrix}1\\-2\\-5\end{pmatrix}$

c $\mathbf{a}=\begin{pmatrix}2\\0\\-2\end{pmatrix}$ and $\mathbf{b}=\begin{pmatrix}0\\0\\5\end{pmatrix}$

d $\mathbf{a}=\mathbf{i}+\mathbf{j}$ and $\mathbf{b}=\mathbf{j}-\mathbf{k}$

7 Given that $\mathbf{a}=4\mathbf{i}-2\mathbf{j}+\mathbf{k}$, find the vector $\mathbf{b}$ such that:

a $\mathbf{a}+\mathbf{b}$ is the zero vector b $2\mathbf{a}+3\mathbf{b}$ is the zero vector

c $\mathbf{a}-\mathbf{b}=\mathbf{j}$ d $\mathbf{a}+2\mathbf{b}=3\mathbf{i}$.

8 Given that $\mathbf{a}=\begin{pmatrix}-1\\1\\2\end{pmatrix}$ and $\mathbf{b}=\begin{pmatrix}5\\3\\3\end{pmatrix}$ find vector $\mathbf{x}$ such that $3\mathbf{a}+4\mathbf{x}=\mathbf{b}$.

9 Given that $\mathbf{a}=3\mathbf{i}-2\mathbf{j}+5\mathbf{k}$, $\mathbf{b}=\mathbf{i}-\mathbf{j}+2\mathbf{k}$ and $\mathbf{c}=\mathbf{i}+\mathbf{k}$, find the value of the scalar t such that $\mathbf{a}+t\mathbf{b}=\mathbf{c}$.

10 Find the possible values of the constant c such that the vector $\begin{pmatrix}2c\\c\\-c\end{pmatrix}$ has magnitude 12.

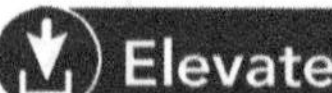

See Support Sheet 16 for a further example of three-dimensional vectors and for more practice questions.

11 Let $\mathbf{a}=\begin{pmatrix}-2\\0\\-1\end{pmatrix}$ and $\mathbf{b}=\begin{pmatrix}2\\-1\\2\end{pmatrix}$. Find the possible values of λ such that $|\mathbf{a}+\lambda\mathbf{b}|=5\sqrt{2}$.

12 Points A and B are such that $\overrightarrow{OA}=\begin{pmatrix}-1\\-6\\13\end{pmatrix}$ and $\overrightarrow{OB}=\begin{pmatrix}1\\-2\\4\end{pmatrix}+t\begin{pmatrix}2\\1\\-5\end{pmatrix}$ where O is the origin. Find the possible values of t such that $AB=3$.

13 Points P and Q have position vectors $\mathbf{p}=\mathbf{i}+\mathbf{j}+3\mathbf{k}$ and $\mathbf{q}=(2+t)\mathbf{i}+(1-t)\mathbf{j}+(1+t)\mathbf{k}$. Find the value of t for which the distance PQ is minimum possible and find this minimum distance.

Section 5: Solving geometrical problems

This chapter finishes with a review of how you can use vector methods to solve geometrical problems. You have already used these results:

- the position vector of the midpoint of line segment AB is $\frac{1}{2}(\mathbf{a}+\mathbf{b})$
- if vectors $\mathbf{a}$ and $\mathbf{b}$ are parallel then there is a scalar k so that $\mathbf{b}=k\mathbf{a}$
- the unit vector in the same direction as $\mathbf{a}$ is $\hat{\mathbf{a}}=\frac{1}{|\mathbf{a}|}\mathbf{a}$.

Rewind

These results were introduced in Student Book 1, Chapter 15.

WORKED EXAMPLE 16.15

Points A, B, C, and D have position vectors $\mathbf{a}=\begin{pmatrix}3\\-1\\1\end{pmatrix}$, $\mathbf{b}=\begin{pmatrix}5\\0\\3\end{pmatrix}$, $\mathbf{c}=\begin{pmatrix}7\\8\\-3\end{pmatrix}$, $\mathbf{d}=\begin{pmatrix}4\\3\\-2\end{pmatrix}$.

Point E is the midpoint of BC.

a Find the position vector of E.

b Show that $ABED$ is a parallelogram.

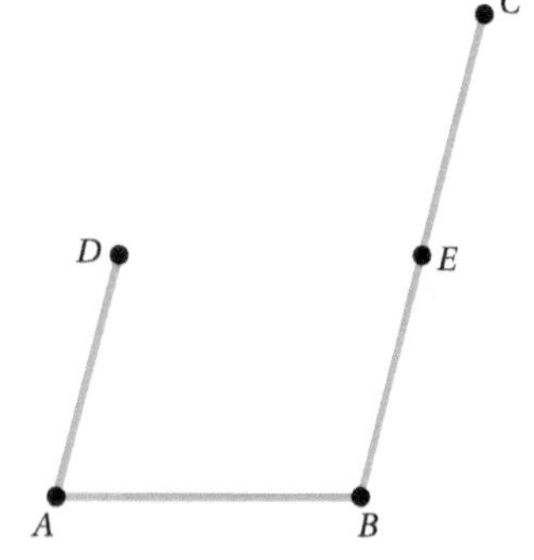

Draw a diagram to show what is going on.

Continues on next page

a

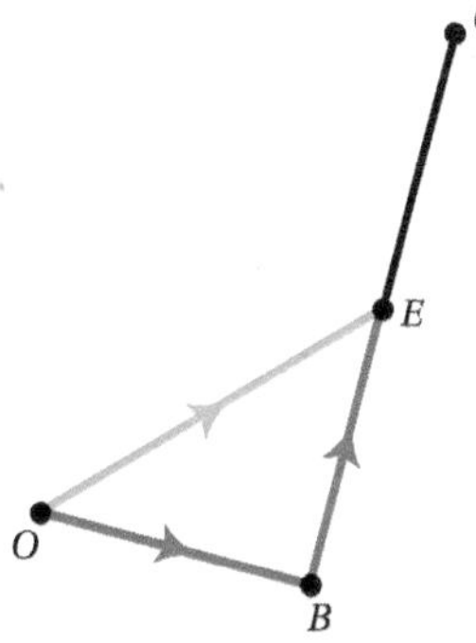

As you are given position vectors, it may help to show the origin on the diagram. For this part, you only need to look at points B, C and E.

Tip

Vector diagrams do not have to be accurate or to scale to be useful. A two-dimensional sketch of a three-dimensional situation is often enough to show you what's going on.

$$\overrightarrow{OE} = \overrightarrow{OB} + \overrightarrow{BE}$$
$$= \overrightarrow{OB} + \frac{1}{2}\overrightarrow{BC}$$
$$= \underline{b} + \frac{1}{2}(\underline{c} - \underline{b})$$
$$= \frac{1}{2}\underline{b} + \frac{1}{2}\underline{c}$$
$$= \begin{pmatrix} 2.5 \\ 0 \\ 1.5 \end{pmatrix} + \begin{pmatrix} 3.5 \\ 4 \\ -1.5 \end{pmatrix}$$
$$= \begin{pmatrix} 6 \\ 4 \\ 0 \end{pmatrix}$$

b $\overrightarrow{AD} = \underline{d} - \underline{a}$

In a parallelogram, opposite sides are equal and parallel, which means that the vectors corresponding to those sides are equal. So you need to show that $\overrightarrow{AD} = \overrightarrow{BE}$.

$$= \begin{pmatrix} 4 \\ 3 \\ -2 \end{pmatrix} - \begin{pmatrix} 3 \\ -1 \\ 1 \end{pmatrix}$$
$$= \begin{pmatrix} 1 \\ 4 \\ -3 \end{pmatrix}$$

$$\overrightarrow{BE} = \underline{e} - \underline{b}$$
$$= \begin{pmatrix} 6 \\ 4 \\ 0 \end{pmatrix} - \begin{pmatrix} 5 \\ 0 \\ 3 \end{pmatrix}$$
$$= \begin{pmatrix} 1 \\ 4 \\ -3 \end{pmatrix}$$

$\overrightarrow{AD} = \overrightarrow{BE}$

$\therefore$ $ABED$ is a parallelogram.

WORKED EXAMPLE 16.16

Given vectors $\mathbf{a} = \begin{pmatrix} 1 \\ 2 \\ 7 \end{pmatrix}$, $\mathbf{b} = \begin{pmatrix} -3 \\ 4 \\ 2 \end{pmatrix}$ and $\mathbf{c} = \begin{pmatrix} -2 \\ p \\ q \end{pmatrix}$

a Find the values of p and q such that $\mathbf{c}$ is parallel to $\mathbf{a}$.

b Find the value of scalar k such that $\mathbf{a} + k\mathbf{b}$ is parallel to vector $\begin{pmatrix} 0 \\ 10 \\ 23 \end{pmatrix}$.

a Write $\underline{c} = t\underline{a}$ for some scalar t.

If two vectors are parallel you can write $\mathbf{v}_1 = t\mathbf{v}_2$.

Then:

$$\begin{pmatrix} -2 \\ p \\ q \end{pmatrix} = t\begin{pmatrix} 1 \\ 2 \\ 7 \end{pmatrix} = \begin{pmatrix} t \\ 2t \\ 7t \end{pmatrix}$$

$$\begin{cases} -2 = t \\ p = 2t \\ q = 7t \end{cases}$$

If two vectors are equal then all their corresponding components are equal.

$\therefore p = -4, q = -14$

b $\underline{a} + k\underline{b} = \begin{pmatrix} 1 \\ 2 \\ 7 \end{pmatrix} + \begin{pmatrix} -3k \\ 4k \\ 2k \end{pmatrix} = \begin{pmatrix} 1-3k \\ 2+4k \\ 7+2k \end{pmatrix}$

Write vector $\mathbf{a} + k\mathbf{b}$ in terms of k and then use $\mathbf{a} + k\mathbf{b} = t\begin{pmatrix} 0 \\ 10 \\ 23 \end{pmatrix}$.

Parallel to $\begin{pmatrix} 0 \\ 10 \\ 23 \end{pmatrix}$:

$$\begin{pmatrix} 1-3k \\ 2+4k \\ 7+2k \end{pmatrix} = t\begin{pmatrix} 0 \\ 10 \\ 23 \end{pmatrix}$$

$$\begin{cases} 1-3k = 0 \\ 2+4k = 10t \\ 7+2k = 23t \end{cases}$$

$1-3k = 0 \Rightarrow k = \frac{1}{3}$

You can find k from the first equation, but you need to check that all three equations are satisfied.

$2+4\left(\frac{1}{3}\right) = 10t \Rightarrow t = \frac{1}{3}$

Check in the third equation:

$\text{LHS} = 7+2\left(\frac{1}{3}\right) = \frac{23}{3}$

$\text{RHS} = 23\left(\frac{1}{3}\right) = \frac{23}{2}$

$\therefore k = \frac{1}{3}$

WORKED EXAMPLE 16.17

a Find the unit vector in the same direction as $\mathbf{a} = \begin{pmatrix} 2 \\ -2 \\ 1 \end{pmatrix}$.

b Find a vector of magnitude 5 parallel to **a**.

a Let the required unit vector be $\hat{\underline{a}}$.

$|\underline{a}| = \sqrt{2^2 + 2^2 + 1^2} = 3$

Find the magnitude of **a**.

$\hat{\underline{a}} = \frac{\underline{a}}{|\underline{a}|}$

$= \begin{pmatrix} 2 \\ -2 \\ 1 \end{pmatrix} \div 3 = \begin{pmatrix} \frac{2}{3} \\ -\frac{2}{3} \\ \frac{1}{3} \end{pmatrix}$

Then divide **a** by its magnitude to produce a vector in the same direction as **a** but of length 1.

b Let $\underline{b}$ be parallel to $\underline{a}$ and $|\underline{b}| = 5$.

Then

$\underline{b} = 5\hat{\underline{a}} = \begin{pmatrix} \frac{10}{3} \\ -\frac{10}{3} \\ \frac{5}{3} \end{pmatrix}$

To get a vector of magnitude 5, multiply the unit vector by 5.

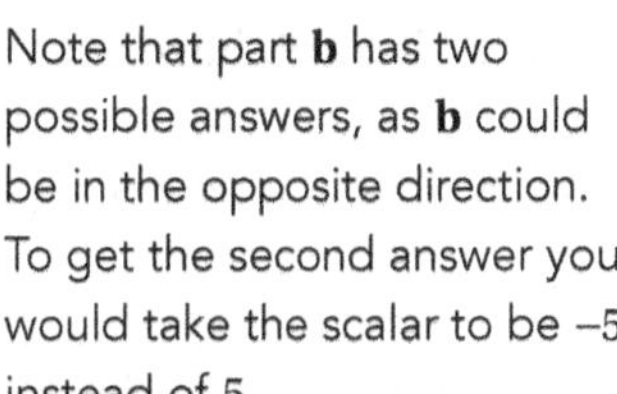

Note that part **b** has two possible answers, as **b** could be in the opposite direction. To get the second answer you would take the scalar to be –5 instead of 5.

The midpoint is just a special case of dividing a line segment in a given ratio.

WORKED EXAMPLE 16.18

Points A and B have position vectors **a** and **b**. Find, in terms of **a** and **b**, the position vector of the point M on AB such that $AM : MB = 2 : 3$.

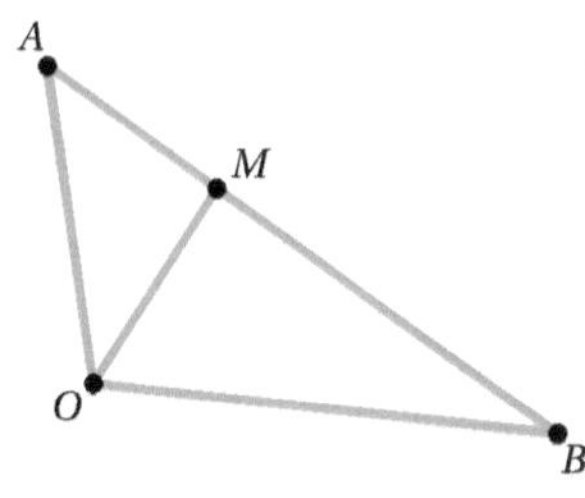

Always start by drawing a diagram. Since the question is about position vectors, include the origin.

$\overrightarrow{OM} = \overrightarrow{OA} + \overrightarrow{AM}$

You can get from O to M via either A or B.

$= \overrightarrow{OA} + \frac{2}{5}\overrightarrow{AB}$

$AM : MB = 2 : 3$ means that $\overrightarrow{AM} = \frac{2}{5}\overrightarrow{AB}$

$= \underline{a} + \frac{2}{5}(\underline{b} - \underline{a})$

$= \frac{3}{5}\underline{a} + \frac{2}{5}\underline{b}$

EXERCISE 16E

1 a i Find a unit vector parallel to $\begin{pmatrix} 2 \\ 2 \\ 1 \end{pmatrix}$.

ii Find a unit vector parallel to $6\mathbf{i}+6\mathbf{j}-3\mathbf{k}$.

b i Find a unit vector in the same direction as $\mathbf{i}+\mathbf{j}+\mathbf{k}$.

ii Find a unit vector in the same direction as $\begin{pmatrix} 4 \\ -1 \\ 2\sqrt{2} \end{pmatrix}$.

2 Points A and B have position vectors $\overrightarrow{OA}=\begin{pmatrix} 3 \\ 1 \\ -2 \end{pmatrix}$ and $\overrightarrow{OB}=\begin{pmatrix} 4 \\ -2 \\ 5 \end{pmatrix}$.

a Write $\overrightarrow{AB}$ as a column vector.

b Find the position vector of the midpoint of AB.

3 Points A, B and C have position vectors $\mathbf{a}=\begin{pmatrix} 2 \\ -1 \\ 4 \end{pmatrix}$, $\mathbf{b}=\begin{pmatrix} 5 \\ 1 \\ 2 \end{pmatrix}$ and $\mathbf{c}=\begin{pmatrix} 3 \\ 1 \\ 4 \end{pmatrix}$. Find the position vector of point D such that $ABCD$ is a parallelogram.

4 Given that $\mathbf{a}=\begin{pmatrix} 2 \\ 0 \\ 2 \end{pmatrix}$ and $\mathbf{b}=\begin{pmatrix} 3 \\ 1 \\ 3 \end{pmatrix}$ find the value of the scalar p such that $\mathbf{a}+p\mathbf{b}$ is parallel to the vector $\begin{pmatrix} 3 \\ 2 \\ 3 \end{pmatrix}$.

5 Given that $\mathbf{x}=2\mathbf{i}+3\mathbf{j}+\mathbf{k}$ and $\mathbf{y}=4\mathbf{i}+\mathbf{j}+2\mathbf{k}$ find the value of the scalar λ such that $\lambda\mathbf{x}+\mathbf{y}$ is parallel to vector $\mathbf{j}$.

6 Points A and B have position vectors $\mathbf{a}=\begin{pmatrix} 2 \\ 2 \\ 1 \end{pmatrix}$ and $\mathbf{b}=\begin{pmatrix} 1 \\ -1 \\ 3 \end{pmatrix}$. Point C lies on the line segment AB so that $AC : BC = 2 : 3$. Find the position vector of C.

7 Points P and Q have position vectors $\mathbf{p}=2\mathbf{i}-\mathbf{j}-3\mathbf{k}$ and $\mathbf{q}=\mathbf{i}+4\mathbf{j}-\mathbf{k}$.

a Find the position vector of the midpoint M of PQ.

b Point R lies on the line PQ such that $QR = QM$. Find the coordinates of R.

8 Given that $\mathbf{a}=\mathbf{i}-\mathbf{j}+3\mathbf{k}$ and $\mathbf{b}=2q\mathbf{i}+\mathbf{j}+q\mathbf{k}$ find the values of scalars p and q such that $p\mathbf{a}+\mathbf{b}$ is parallel to vector $\mathbf{i}+\mathbf{j}+2\mathbf{k}$.

9 a Find a vector of magnitude 6 parallel to $\begin{pmatrix} 4 \\ -1 \\ 1 \end{pmatrix}$.

b Find a vector of magnitude 3 in the same direction as $2\mathbf{i}-\mathbf{j}+\mathbf{k}$.

10 Points A and B have position vectors $\mathbf{a}$ and $\mathbf{b}$. Point M lies on AB and $AM : MB = p : q$. Express the position vector of M in terms of $\mathbf{a}$, $\mathbf{b}$, p and q.

11 In the diagram, O is the origin and points A and B have position vectors $\mathbf{a}$ and $\mathbf{b}$. P, Q and R are points on OA, OB and AB such that $OP : PA = 1 : 4$, $OQ : QB = 3 : 2$ and $AB : BR = 5 : 1$.

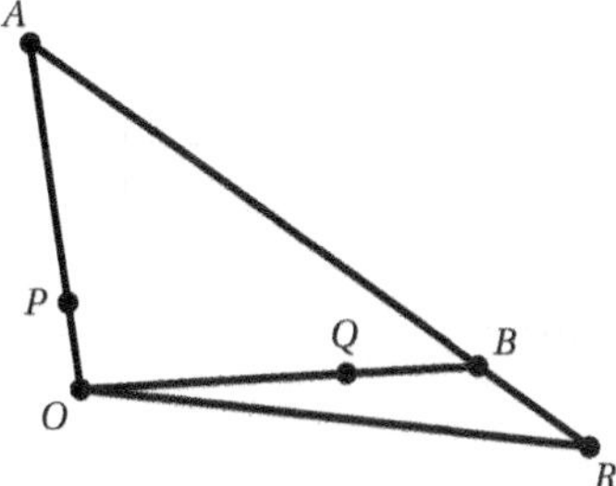

Prove that:

a PQR is a straight line

b Q is the midpoint of PR.

Checklist of learning and understanding

- Constant acceleration formulae in two dimensions:
  - $\mathbf{v} = \mathbf{u} + \mathbf{a}t$
  - $\mathbf{s} = \mathbf{u}t + \frac{1}{2}\mathbf{a}t^2$
  - $\mathbf{s} = \mathbf{v}t - \frac{1}{2}\mathbf{a}t^2$
  - $\mathbf{s} = \frac{1}{2}(\mathbf{u} + \mathbf{v})t$
- To differentiate or integrate a vector, differentiate or integrate each component separately.
- Vectors in three dimensions can be expressed in terms of **base vectors i, j, k** using **components.**
- The **magnitude** of a vector can be calculated using the components of the vector $|\mathbf{a}| = \sqrt{a_1 + a_2 + a_3}$.
- The **distance** between the points with position vectors **a** and **b** is given by $|\mathbf{b} - \mathbf{a}|$.
- The unit vector in the same direction as **a** is $\hat{\mathbf{a}} = \frac{1}{|\mathbf{a}|}\mathbf{a}$.

Mixed practice 16

1. The point A has position vector $\begin{pmatrix} 2 \\ -3 \\ -1 \end{pmatrix}$ and the point B has position vector $\begin{pmatrix} 1 \\ -2 \\ 5 \end{pmatrix}$. Find the vector $\overrightarrow{AB}$.

   Choose from these options.

   A $\begin{pmatrix} 1 \\ -1 \\ -6 \end{pmatrix}$ **B** $\begin{pmatrix} -1 \\ 1 \\ 6 \end{pmatrix}$ **C** $\begin{pmatrix} 3 \\ -5 \\ 4 \end{pmatrix}$ **D** $\begin{pmatrix} 1 \\ -5 \\ 4 \end{pmatrix}$

2. A particle of mass 6 kg moves with constant acceleration $(1.6\mathbf{i} - 0.3\mathbf{j})$ m s^{-2}.

   a Find the magnitude of the net force acting on the particle.

   When $t = 0$ the particle has velocity $(-2\mathbf{i} + 2.5\mathbf{j})$ m s^{-1}.

   b Find the speed and the direction of motion of the particle 5 seconds later.

3. Points A and B have position vectors $\mathbf{a} = \begin{pmatrix} 4 \\ 1 \\ 2 \end{pmatrix}$ and $\mathbf{b} = \begin{pmatrix} 2 \\ -1 \\ 3 \end{pmatrix}$. C is the midpoint of AB.

   Find the exact distance AC.

4. Points A, B and C have position vectors $\mathbf{a} = \mathbf{i} + 3\mathbf{j} - 4\mathbf{k}$, $\mathbf{b} = 3\mathbf{i} + 2\mathbf{j} + 2\mathbf{k}$ and $\mathbf{c} = -3\mathbf{i} + 3\mathbf{j} + 4\mathbf{k}$.

   a Find the position vector of the point D such that $ABCD$ is a parallelogram.

   b Prove that $ABCD$ is a rhombus.

5. A particle moves in the plane so that its position vector at time t seconds is $(3.2 - t^2)\mathbf{i} + (-4.6 + 0.2t^3)\mathbf{j}$ m. Find the speed of the particle when $t = 2.5$ s.

6. Points A, B and C have position vectors $\mathbf{a} = -7\mathbf{i} + 11\mathbf{j} + 9\mathbf{k}$, $\mathbf{b} = 13\mathbf{i} - 4\mathbf{j} + 14\mathbf{k}$ and $\mathbf{c} = 3\mathbf{i} + \mathbf{j} + 4\mathbf{k}$.

   a Prove that the triangle ABC is isosceles.

   b Find the position vector of point D such that the four points form a rhombus.

7. A particle moves with constant acceleration between the points A and B. At A, it has velocity $(4\mathbf{i} + 2\mathbf{j})$ m s^{-1}. At B, it has velocity $(7\mathbf{i} + 6\mathbf{j})$ m s^{-1}. It takes 10 seconds to move from A to B.

   a Find the acceleration of the particle.

   b Find the distance between A and B.

   c Find the average velocity as the particle moves from A to B.

   [© AQA 2015]

8. A particle moves with velocity vector $\mathbf{v} = (1 - t)\mathbf{i} + (3t - 2)\mathbf{j}$.

   Find the time at which the particle is moving parallel to the vector $\mathbf{i} + \mathbf{j}$.

   Choose from these options.

   A $t = 1$ **B** $t = \frac{2}{3}$ **C** $t = \frac{4}{3}$ **D** $t = \frac{3}{4}$

9 A particle of mass 0.3 kg starts from rest and moves under the action of a constant force $(6\mathbf{i} - 2\mathbf{j})$ N. Find how long it takes to reach the speed of 12 m s^{-1}.

10 A helicopter is initially hovering above the helipad. It sets off with constant acceleration $(0.3\mathbf{i} + 1.2\mathbf{j})$ m s^{-2}, where the unit vectors $\mathbf{i}$ and $\mathbf{j}$ are directed east and north, respectively. The helicopter is modelled as a particle moving in two dimensions.

a Find the bearing on which the helicopter is travelling.

b Find the time at which the helicopter is 300 m from its initial position.

c Explain in everyday language the meaning of the modelling assumption that the helicopter moves in two dimensions.

11 Points P and Q have position vectors $\mathbf{p} = 4\mathbf{i} - \mathbf{j} + 11\mathbf{k}$ and $\mathbf{q} = 3\mathbf{j} - \mathbf{k}$. S is the point on the line segment PQ such that $PS : SQ = 3 : 2$. Find the exact distance of S from the origin.

12 A particle of mass 2 kg moves in the plane under the action of the force $\mathbf{F} = (20\sin(2t)\,\mathbf{i} + 30\cos(t)\,\mathbf{j})$ N. The particle is initially at rest at the origin.

Find the direction of motion of the particle after 5 seconds.

13 In this question, vectors $\mathbf{i}$ and $\mathbf{j}$ point due east and north, respectively.

A port is located at the origin. One ship starts from the port and moves with velocity $\mathbf{v}_1 = (3\mathbf{i} + 4\mathbf{j})$ km h^{-1}.

a Write down the position vector at time t hours.

At the same time, a second ship starts 18 km north of the port and moves with velocity $\mathbf{v}_2 = (3\mathbf{i} - 5\mathbf{j})$ km h^{-1}.

b Write down the position vector of the second ship at time t hours.

c Show that after half an hour, the distance between the two ships is 13.5 km.

d Show that the ships meet, and find the time when this happens.

e How long after the meeting are the ships 18 km apart?

14 A particle is initially at the point A, which has position vector $13.6\mathbf{i}$ m, with respect to an origin O. At the point A, the particle has velocity $(6\mathbf{i} + 2.4\mathbf{j})$ m s^{-1}, and in its subsequent motion, it has a constant acceleration of $(-0.8\mathbf{i} + 0.1\mathbf{j})$ m s^{-2}. The unit vectors $\mathbf{i}$ and $\mathbf{j}$ are directed east and north respectively.

a Find an expression for the velocity of the particle t seconds after it leaves A.

b Find an expression for the position vector of the particle, with respect to the origin O, t seconds after it leaves A

c Find the distance of the particle from the origin O when it is travelling in a north-westerly direction.

[© AQA 2013]

15 A particle has velocity vector $\mathbf{v} = \begin{pmatrix} \sin t \\ e^{\frac{t}{2}} \end{pmatrix}$ and is initially at the origin.

Find the particle's position vector at time t.

Choose from these options.

A $\begin{pmatrix} 1-\cos t \\ 2e^{\frac{t}{2}}-2 \end{pmatrix}$ **B** $\begin{pmatrix} -\cos t \\ 2e^{\frac{t}{2}} \end{pmatrix}$ **C** $\begin{pmatrix} \cos t \\ \frac{1}{2}e^{\frac{t}{2}} \end{pmatrix}$ **D** $\begin{pmatrix} \cos t-1 \\ \frac{1}{2}e^{\frac{t}{2}}-\frac{1}{2} \end{pmatrix}$

16 At time $t = 0$ two aircraft have position vectors $5\mathbf{j}$ and $7\mathbf{k}$. The first moves with constant velocity $3\mathbf{i}-4\mathbf{j}+\mathbf{k}$ and the second with constant velocity $5\mathbf{i}+2\mathbf{j}-\mathbf{k}$.

a Write down the position vector of the first aircraft at time t.

Let d be the distance between the two aircraft at time t.

b Find an expression for d^2 in terms of t. Hence show that the two aircraft will not collide.

c Find the minimum distance between the two aircraft.

17 A position vector of a particle at time t seconds is given by $\mathbf{r} = (5\cos t\,\mathbf{i}+2\sin t\,\mathbf{j})$ m.

a Find the Cartesian equation of the particle's trajectory.

b Find the maximum speed of the particle, and its position vector at the times when it has this maximum speed.

18 A particle of mass 3 kg moves on a horizontal surface under the action of the net force $\mathbf{F} = (36e^{-t}\mathbf{i}-96e^{-2t}\mathbf{j})$ N. The particle is initially at the origin and has velocity $(-6\mathbf{i}+20\mathbf{j})$ m s^{-1}. The unit vectors $\mathbf{i}$ and $\mathbf{j}$ are directed east and north, respectively.

Find the distance of the particle from the origin at the time when it is travelling in the northerly direction.

See Extension Sheet 16 for questions on modelling rotation with vectors.

FOCUS ON... PROOF 2

Deriving the compound angle identities

You are to going to demonstrate that

- $\sin(A+B) = \sin A \cos B + \cos A \sin B$
- $\cos(A-B) = \cos A \cos B + \sin A \sin B.$

You have already seen trigonometric proofs that use right-angled triangles to prove results about more complicated figures. The same approach works here.

Rewind

In Student Book 1, Focus on ... Proof 2, you used this strategy to prove the sine and cosine rules.

PROOF 9

The sine compound angle formula will be proved first: $\sin(A+B) = \sin A \cos B + \cos A \sin B$.

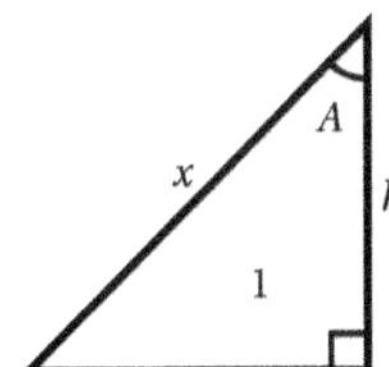

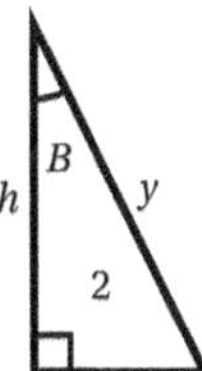

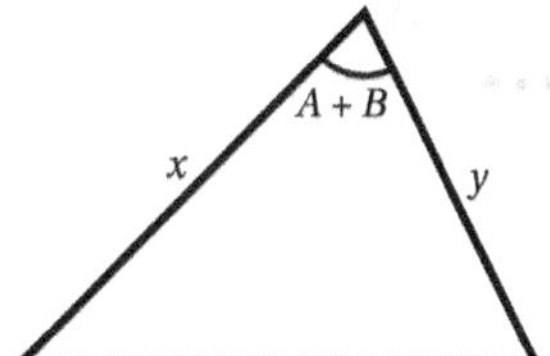

Create a triangle with angle $A+B$ by joining two right-angled triangles with angles A and B.

From triangle 1:
$h = x\cos A$

Now express all the other lengths in terms of x, y, A and B.

Similarly, from triangle 2:
$h = y\cos B$

You can write h in two different ways.

The area of triangle 1 is
$\frac{1}{2}x \times (y\cos B) \times \sin A$

The area of triangle 2 is $\frac{1}{2}y \times (x\cos A) \times \sin B$

Use area $= \frac{1}{2}ab\sin C$ on each of the triangles individually.

The area of the whole triangle is $\frac{1}{2}xy\sin(A+B)$

And then on the triangle as a whole.

Therefore:

$$\frac{1}{2}xy\sin(A+B) = \frac{1}{2}xy\cos B\sin A + \frac{1}{2}xy\cos A\sin B$$

$$\Rightarrow \sin(A+B) = \sin A\cos B + \cos A\sin B$$

Dividing by $\frac{1}{2}xy$.

QUESTIONS

1. Is it possible to draw two right-angled triangles with the same height for any pair of acute angles A and B?

2. Does the identity still hold when the angles A and B are not acute? Can you prove it?

3. To derive the compound angle identities for $\cos(A+B)$ and $\cos(A-B)$, you could use the same triangles and the cosine rule.

   However, there is a simpler proof which uses the relationship between sin and cos:

   $$\sin\left(\frac{\pi}{2}-\theta\right)=\cos\theta \text{ and } \cos\left(\frac{\pi}{2}-\theta\right)=\sin\theta$$

   Write $\cos(A-B)=\sin\left(\frac{\pi}{2}-(A-B)\right)=\sin\left(\left(\frac{\pi}{2}-A\right)+B\right)$

   and use the compound angle identity for $\sin(A+B)$ to prove that

   $\cos(A-B)=\cos A\cos B+\sin A\sin B.$

FOCUS ON... PROBLEM-SOLVING 2

Choosing between analytical and numerical methods

Many problems you have encountered in this course can be solved in more than one way. In real-life situations you are free to choose whatever method and approach you prefer. In making your decision you should think about these questions.

- How difficult is the method?
- How efficient is it? Does it require lots of detailed or repeated calculations?
- How accurate is it? What level of accuracy do you actually need?
- Can it be easily adapted to solve other similar problems?

For example, when solving an equation you have essentially two options: You can try to rearrange the equation, using rules of algebra (this is called an analytical solution), or you can use one of the iterative methods from Chapter 14 (this is a numerical solution). The former may not always be possible but when it works, it gives you an exact solution (for example, $e^{2\pi}$). However, in many applications you only need an answer correct to a couple of decimal places, so you should consider whether the effort required to rearrange the equation is justified.

On the other hand, your equation may have a parameter in it (for example, $x^2 + 3ax + 1 = 0$) and you may want to know how changing the value of the parameter affects the solution. In this case, finding the analytical solution once (in terms of the parameter) may be more efficient than repeating the numerical calculation lots of times.

Here you will look at a problem that can be investigated both analytically and numerically. You can try various approaches and decide for yourself which one suits you best.

The fishing lake problem

The number of fish in a lake can be modelled by the equation

$$x_{n+1} = \alpha x_n - \beta x_n^2 - k$$

In this model, x_n is the number of fish in year n. Each year, due to natural birth and death rates, the population increases by a factor of α, and $\beta x_n{}^2$ die out due to lack of resources or natural causes, and a constant number, k, fish are caught and removed.

The question to ask is: what is the maximum number of fish that can be removed each year without causing the population to die out?

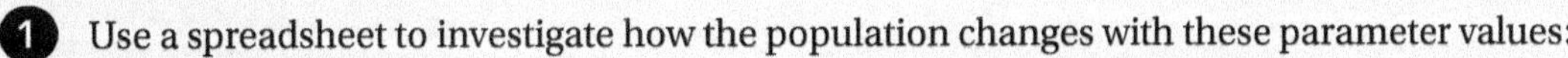

QUESTIONS

1 Use a spreadsheet to investigate how the population changes with these parameter values:

a $\alpha = 1.4, \beta = 0.0002, x_1 = 1000$ and:

i $k = 150$ **ii** $k = 200$ **iii** $k = 205$

b $\alpha = 1.4, \beta = 0.0002, x_1 = 2000$ and:

i $k = 150$ **ii** $k = 200$ **iii** $k = 205$.

What is the largest number of fish that can be caught each year without causing the population to die out? Does this depend on the initial size of the population?

2 Vary the values of α and β slightly. How does this affect the maximum possible value of k?

You need to try lots of different values of the parameters to find the relationship between k, α and β. Is it possible to solve this problem analytically instead, to find an equation linking the three quantities?

The sequence $x_{n+1} = \alpha x_n - \beta x_n^2 - k$ is not one of the types you are familiar with; in fact, it is not possible to find a general formula for x_n. However, you are only interested in the long-term behaviour of the sequence – does it eventually decrease to zero or not? You know from Chapter 14 that, if a sequence $x_{n+1} = g(x_n)$ has a limit, then this limit is a solution of the equation $x = g(x)$.

3 Consider the case when $\alpha = 1.4$ and $\beta = 0.0002$.

a Solve the equation $x = \alpha x - \beta x^2 - k$ when:

i $k = 150$ **ii** $k = 200$ **iii** $k = 205$.

Compare the solution to what you observed in question **1**.

b Find the discriminant of the equation $x = 1.4x - 0.0002x^2 - k$ in terms of k. Hence show that the equation only has a solution when $k \leqslant 200$.

c Use the quadratic formula to write the two solutions in terms of k. Hence show that, when $0 < k < 200$, both solutions are positive. Could you tell, without doing the spreadsheet investigation, which of the two solutions the sequence will converge to?

Rewind

In Chapter 14, Section 5, you learnt that the iteration $x_{n+1} = g(x_n)$ converges to a root of $x = g(x)$ if $|g'(x)| < 1$ near the root.

4 **a** For the general case of the equation $x = \alpha x - \beta x^2 - k$, find the discriminant in terms of α, β and k.

b Hence show that the largest number of fish which can be taken out without causing the population to die out is $k = \dfrac{(\alpha - 1)^2}{4\beta}$.

5 Did you find the theoretical analysis or the spreadsheet investigation easier to follow? Which one do you think gives a more reliable answer? Which helps you understand the problem better?

6 When $\alpha = 1.6$ and $\beta = 0.0007$, the formula you found in question **4 b** says that the maximum number of fish that can be take out is 128. However, with $k = 130$ and $x_1 = 500$, the spreadsheet shows that there are still 400 fish in the lake after 50 years. So is the additional accuracy you get from using the formula always required?

FOCUS ON... MODELLING 2

Translating information into equations

The aim of a mathematical model is to describe a real-life situation using equations, which can then be solved and used to make predictions.

In this section you will look at writing differential equations - these are equations involving the rate of change of a quantity. You need to remember:

- the rate of change of y with respect to x is $\frac{dy}{dx}$
- y is proportional to x means that $y = kx$ for some constant k.

When writing differential equations, you usually have some information about particular values of the quantities involved. Sometimes you can use these to find constants in the equation (such as the k in $y = kx$), but sometimes you need to wait until you have solved the equation.

Tip

In many examples, rate of change means change in time; however, look out for examples where this is not the case!

Rewind

Solving differential equations is covered in Chapter 13. In this section you will just look at setting up equations rather than solving them.

WORKED EXAMPLE 1

The speed of an object decreases at the rate proportional to the square root of its current speed. When the speed is 12 m s^{-1} it is decreasing at the rate of 1.5 m s^{-2}. Using v for speed and t for time, write an equation to represent this information.

$\frac{dv}{dt} = -k\sqrt{v}$

Rate of change means the derivative with respect to time.

The speed is decreasing, so write $-k$ to emphasise this.

When: $v = 12$, $\frac{dv}{dt} = -1.5$

$-1.5 = -k\sqrt{12}$

$\Rightarrow k = \frac{\sqrt{12}}{1.5} = \frac{4\sqrt{3}}{3}$

You can use the given information to find k.

Remember that the rate of change is negative.

So the equation is $\frac{dv}{dt} = -\frac{4\sqrt{3}}{3}\sqrt{v}$

WORKED EXAMPLE 2

In one possible model of population growth, the rate of growth depends on two factors: it is proportional to the current size of the population, and also proportional to cos $30t$. (The second factor represents seasonal breeding patterns.) When the measurements began the population size was 160. Using N for the size of the population and t for time measured in months, write a differential equation to represent this information.

$$\frac{dN}{dt} = kN\cos 30t$$

Both factors need to be multiplied together. Don't forget to include the constant of proportionality.

You only have information about N when $t = 0$, so you can't find k until you have solved the equation.

Rewind

You know from Student Book 1 that the rate of change of velocity is acceleration (Chapter 16) and that the acceleration is proportional to the force acting on an object (Chapter 18). You could therefore use the model in Worked example 1 when there is a resistance force proportional to the square root of the speed, such as air resistance or drag when an object is moving through liquid.

QUESTIONS

Write a differential equation to represent each situation. Where possible, find the values of any constants.

1. A population of a new town (N thousand) increases at a rate proportional to its size. Initially, the size of the population is 3500 and it is increasing at the rate of 120 people per year. Write a differential equation for the rate of change of population.

2. During the decay of a radioactive substance, the rate at which mass is lost is proportional to the mass present at that instant. Use m for the mass of the substance in grams and t for the time in seconds. Initially there is 24 g of the substance and the mass is decreasing at the rate of 1.2 g s^{-1}. Write a differential equation for the rate of change of mass.

3. In an electrical circuit, the voltage is decreasing at a rate proportional to the square of the present voltage. When the voltage is 25 volts it is decreasing at a rate of 2 volts per second.

4 Newton's law of cooling states that the rate at which a body cools is proportional to the difference between its temperature and the temperature of its surroundings. A cup of tea is initially at 100 °C and is cooling at the rate of 2 °C min^{-1} in a room of temperature 24 °C. Use T for the temperature and t for time (in minutes). Write a differential equation for the rate of change of temperature.

5 One of the ends of a metal rod, of length 48 cm, is heated. After a while the temperature remains constant. However, the temperature changes along the length of the rod, decreasing at a rate proportional to the distance from the hot end. The temperatures at the two ends are 230 °C and 52 °C.

Use T for temperature and x for the distance from the hot end (measured in cm). Write a differential equation for how the temperature changes along the rod.

6 The water pressure in the sea increases with depth. The pressure (p) at depth h is proportional to the density (ρ) of the sea water. The density also varies with depth, and is modelled by the equation. $\rho = 1000(1+0.001h)$. Write a differential equation for the rate of increase of pressure with depth.

7 A rumour spreads at a rate proportional to the square root of the number of people who have already heard it, and inversely proportional to the time it has been spreading. After 5 minutes, 25 people have heard the rumour and it is spreading at the rate of three people per minute. Write N for the number of people who have heard the rumour and t for the time, in minutes, since the rumour started. Write a differential equation to model this situation, and explain why the model needs to be modified for small values of t.

8 A cylindrical tank has base radius 1.2 m. Water leaks out of the tank so that the rate at which the volume is decreasing is proportional to the height of the water remaining in the tank. Initially the height of water is 2.5 m and it is decreasing at the rate of 0.05 m per minute. Find an equation for the rate of change of volume of water in the tank.

CROSS-TOPIC REVIEW EXERCISE 2

1 Show that, for small θ, $\theta \sin 2\theta - \cos 2\theta \approx a + b\theta^2$ where a and b are integers to be found.

2 Given that $y = \dfrac{2x^2 - 8x + 5}{(x-2)^2}$, $x \neq 2$, show that $\dfrac{dy}{dx} = \dfrac{k}{(x-2)^3}$ where k is a constant to be found.

3 Find the equation of the curve which has the gradient $\dfrac{dy}{dx} = 3 \tan x$ and passes through the point $(0, 4)$.

4 The graph of $y = 4x^3 - ax^2 + b$ has a point of inflection at $(-1, 4)$. Find the values of a and b.

5 a Use the trapezium rule with five ordinates to find an approximation to $\displaystyle\int_0^2 \frac{x^2}{x^3+4}\,dx$, giving your answer to four significant figures.

b Show that $\displaystyle\int_0^2 \frac{x^2}{x^3+4}\,dx = a \ln b$ where a and b are constants to be found.

6 Find $\displaystyle\int \frac{\sin x + \cos x}{2 \cos x}\,dx$

7 A curve C has parametric equations $x = \sin^2 t,\ y = \cos t,\ -\pi \leqslant t \leqslant \pi$.

a Find $\dfrac{dy}{dx}$ in terms of t.

b The tangent to C at the point where $t = \dfrac{\pi}{6}$ cuts the x-axis at the point P.

Find the x-coordinate of P.

8 The curve $y = x \cos 2x,\ 0 \leqslant x \leqslant \dfrac{\pi}{4}$, is shown in the diagram.

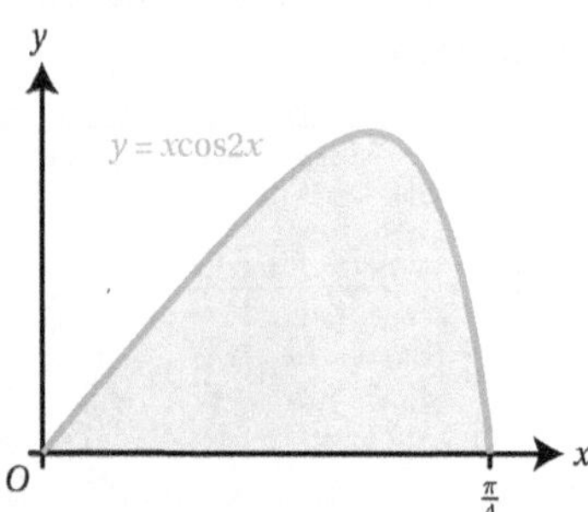

Find the exact value of the shaded area, showing all of your working clearly.

9 a i Solve the equation $\operatorname{cosec}\theta = -4$ for $0° < \theta < 360°$, giving your answers to the nearest 0.1°.

ii Solve the equation

$$2\cot^2(2x + 30°) = 2 - 7\operatorname{cosec}(2x + 30°)$$

for $0° < x < 180°$, giving your answers to the nearest 0.1°.

b Describe a sequence of two geometrical transformations that maps the graph of $y = \operatorname{cosec} x$ onto the graph of $y = \operatorname{cosec}(2x + 30°)$.

[© AQA 2011]

10 The diagram shows a part of the curve with parametric equations $x = t^2$, $y = \sin t$.

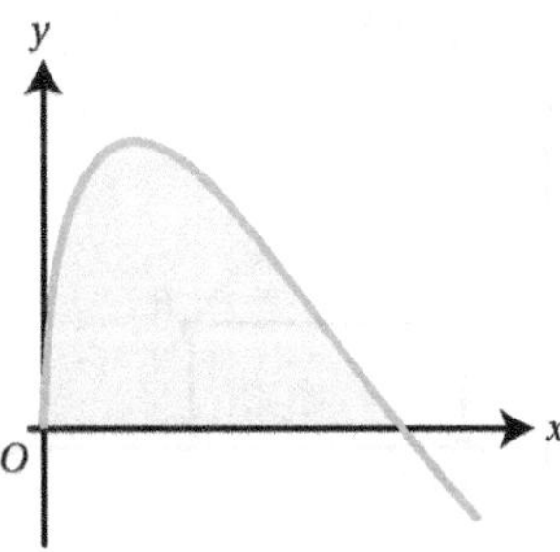

a Find the values of t at the points where the graph crosses the x-axis.

b Find the exact value of the shaded area.

c Find the Cartesian equation of the curve.

11 By taking natural logarithms of both sides, or otherwise, find $\frac{dy}{dx}$, given that $y = x^{\sin x}$.

12 **a** Show that $\frac{d}{dx}(\operatorname{cosec} x) = -\operatorname{cosec} x \cot x$.

b Find the coordinates of the points on the curve $y = \operatorname{cosec} x$, $x \in [0, \pi]$, where the gradient is equal to $2\sqrt{3}$.

13 **a** Simplify $\sin[(A+B)x] - \sin[(A-B)x]$.

b Hence or otherwise find $\int \sin 3x \cos 5x \, dx$.

14 Triangle ABC is made out of a piece of elastic string. Vertices A and B are being pulled apart so that the length of the base AB is increasing at the rate of 3 cm s^{-1} and the height, h, is decreasing at the rate of 2 cm s^{-1}. Initially, $AB = 20$ cm and $h = 30$ cm.

a Show that $AB = 20 + 3t$.

b Find an expression for h in terms of t.

c Find an expression for the rate of change of the area of the triangle in terms of t.

d Find the rate at which the area of the triangle is changing when $AB = 26$ and $h = 26$.

15 A curve is defined by $y = 4\sin x + x^2$, $0 \leqslant x \leqslant 2\pi$.

Find the set of values of x for which the curve is convex.

16 Consider the infinite geometric series $1 + \cos x + \cos^2 x + \cos^3 x + \ldots$ for $0 < x < \pi$.

a Explain why the series converges.

b Show that the sum of the series is $\frac{1}{2}\operatorname{cosec}^2\left(\frac{x}{2}\right)$.

c Find the exact value of $\int_{\frac{\pi}{3}}^{\frac{\pi}{2}} (1 + \cos x + \cos^2 x + \cos^3 x + \ldots)\, dx$.

17 A rectangle is drawn inside the region bounded by the curve $y = \sin x$ and the x-axis, as shown in the diagram. The vertex A has coordinates $(x, 0)$.

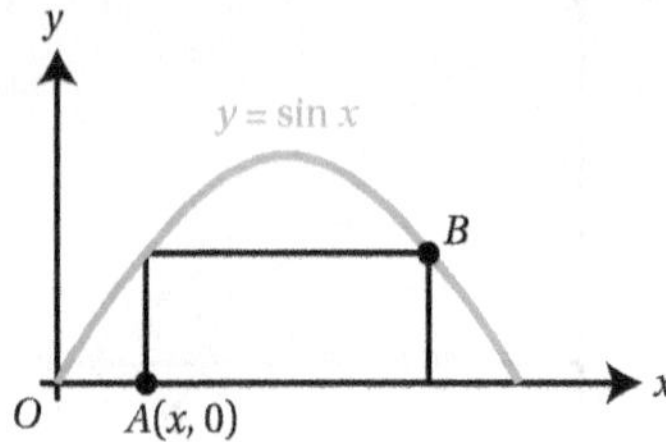

a **i** Write down the coordinates of point B.

ii Find an expression for the area of the rectangle in terms of x.

b **i** Show that the stationary point of the area satisfies the equation $2\tan x = \pi - 2x$.

ii By sketching graphs, show that this equation has one root for $0 < x < \frac{\pi}{2}$.

iii Use the second derivative to show that the stationary point is a maximum.

c **i** The equation for the stationary point can be written as $x = \tan^{-1}\left(\frac{\pi}{2} - x\right)$. Use a suitable iterative formula, with $x_1 = 0.5$, to find the root of the equation $2\tan x = \pi - 2x$ correct to three decimal places.

ii Hence find the maximum possible area of the rectangle.

18 **a** Express $\frac{1}{(3-2x)(1-x)^2}$ in the form $\frac{A}{3-2x} + \frac{B}{1-x} + \frac{C}{(1-x)^2}$.

b Solve the differential equation

$$\frac{dy}{dx} = \frac{2\sqrt{y}}{(3-2x)(1-x)^2}$$

where $y = 0$ when $x = 0$, expressing your answer in the form

$$y^p = q\ln[f(x)] + \frac{x}{1-x}$$

where p and q are constants.

[© AQA 2011]

19 The diagram shows the curves $y = e^{2x} - 1$ and $y = 4e^{-2x} + 2$.

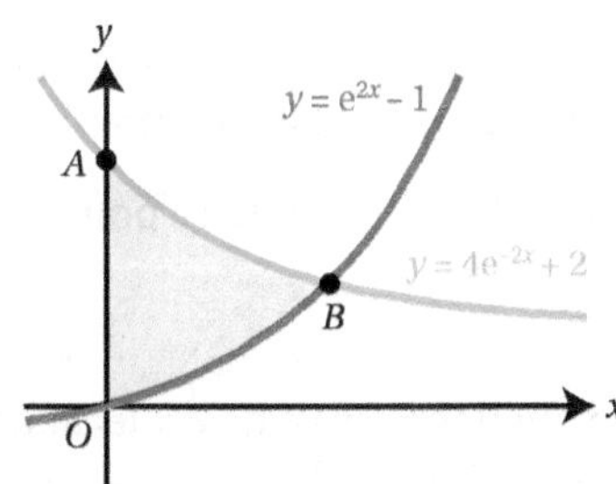

The curve $y = 4e^{-2x} + 2$ crosses the y-axis at the point A and the curves intersect at the point B.

a Describe a sequence of two geometrical transformations that maps the graph of $y = e^x$ onto the graph of $y = e^{2x} - 1$.

b Write down the coordinates of the point A.

c **i** Show that the x-coordinate of the point B satisfies the equation $(e^{2x})^2 - 3e^{2x} - 4 = 0$

ii Hence find the exact value of the x-coordinate of the point B.

d Find the exact value of the area of the shaded region bounded by the curves $y = e^{2x} - 1$ and $y = 4e^{-2x} + 2$ and the y-axis.

[© AQA 2010]

20 Evaluate $\sum_{r=0}^{r=n} \binom{n}{r} \tan^{2r}\left(\frac{\pi}{3}\right)$.

21 **a** State an expression for $\sum_{0}^{n} x^n$.

b Hence or otherwise show that $1 + 2x + 3x^2 + 4x^3 + \ldots + nx^{n-1} = \dfrac{1-(n+1)x^n + nx^{n+1}}{(1-x)^2}$.

22 **a** Use the identity $\cos^2 x + \sin^2 x = 1$ to show that $\cos(\arcsin x) = \sqrt{1-x^2}$.

b The diagram shows part of the curve $y = \sin x$.

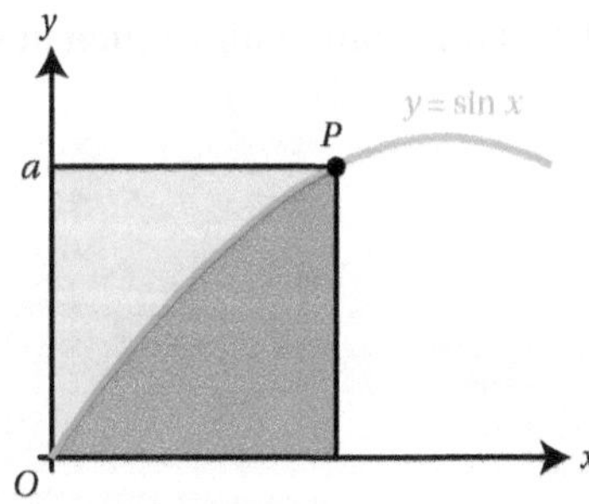

Write down the x coordinate of the point P.

c Find the area of the region shaded red in terms of a, writing your answer in a form without trigonometric functions.

d By considering the area of the region shaded blue, find $\int_0^a \arcsin x \, dx$ for $0 < a < 1$.

23 A function is defined by $f(x) = 2x + \frac{1}{2}\sin 2x - 3\tan x$ for $x \in \left(-\frac{\pi}{2}, \frac{\pi}{2}\right)$.

a Find $f'(x)$.

b Show that the stationary points of $f(x)$ satisfy the equation $2\cos^4 x + \cos^2 x - 3 = 0$.

c Hence show that the function has only one stationary point.

24 A curve has equation $y = (x^2 - a)e^x$.

a Find the range of values of a for which the curve has at least one point of inflection.

b Given that one of the points of inflection is a stationary point, find the value of a.

c For this value of a, find the range of values of x for which the curve is concave.

25 **a** Sketch the graph $y = \ln x$.

b The tangent to this graph at the point $(p, \ln p)$ passes through the origin. Find the value of p.

c For what range of values of k does $\ln x = kx$ have two solutions?

26 **a** Given that $x = \frac{\sin y}{\cos y}$, use the quotient rule to show that

$$\frac{dx}{dy} = \sec^2 y$$

b Given that $\tan y = x - 1$, use a trigonometrical identity to show that

$$\sec^2 y = x^2 - 2x + 2$$

c Show that, if $y = \tan^{-1}(x - 1)$, then

$$\frac{dy}{dx} = \frac{1}{x^2 - 2x + 2}$$

d A curve has equation $y = \tan^{-1}(x - 1) - \ln x$.

i Find the value of the x-coordinate of each of the stationary points of the curve.

ii Find $\frac{d^2y}{dx^2}$.

iii Hence show that the curve has a minimum point which lies on the x-axis.

[© AQA 2012]

17 Projectiles

In this chapter you will learn how to:

- model projectile motion in two dimensions
- find the Cartesian equation of the trajectory of a projectile.

Before you start...

Student Book 1, Chapter 15	You should be able to find the magnitude and direction of a vector.	1 Find the magnitude and direction of the vector $\begin{pmatrix} -3 \\ 2 \end{pmatrix}$.
Student Book 1, Chapter 17	You should be able to use constant acceleration formulae in one dimension.	2 A particle accelerates uniformly from 3 m s^{-1} to 7 m s^{-1} while covering a distance of 60 m in a straight line. Work out the acceleration.
Chapter 16	You should be able to use constant acceleration formulae in two dimensions.	3 A particle initially has velocity $(3\mathbf{i}-4\mathbf{j})\text{ m s}^{-1}$ and accelerates at $(\mathbf{i}+2\mathbf{j})\text{ m s}^{-2}$. Find its velocity after 3 seconds.
Chapter 8	You should know the sine double angle identity.	4 Express $\sin x \cos x$ in terms of $\sin 2x$.
Chapter 8	You should know the definition of $\sec x$.	5 Express $\dfrac{1}{\cos^2 x}$ in terms of $\sec x$.
Chapter 8	You should be able to solve equations involving $\tan x$ and $\sec^2 x$.	6 Solve the equation $2\sec^2 x+\tan x-5=0$ for $0° < x < 180°$.

Motion in two dimensions

In Student Book 1, you used the constant acceleration formulae to analyse the motion of a particle that was projected in one dimension - either horizontally or vertically. In this chapter this idea is extended to look at projectiles moving in two dimensions - vertically and horizontally.

This more realistic model provides a basis for analysing the motion of, for example, a bullet, a golf ball or water from a fountain.

Section 1: Modelling projectile motion

When an object is projected upwards at an angle, it moves in a vertical plane along a symmetrical path.

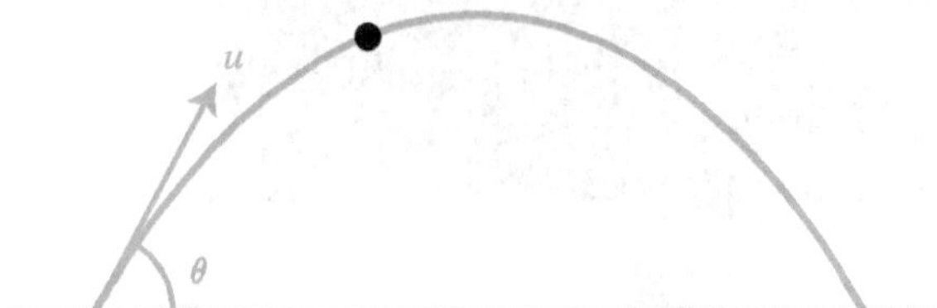

The only force acting on it is gravity so the acceleration is constant with magnitude g and is directed downwards. Note that there is no acceleration horizontally as there is no force acting in this direction.

Key point 17.1

In projectile motion the acceleration of the particle is $\mathbf{a} = \begin{pmatrix} 0 \\ -g \end{pmatrix}$ m s^{-2}.

This means that the motion can be described using the constant acceleration equations in two dimensions.

Fast forward

You will see in Section 2 how to prove that this is a parabola.

Tip

Remember that a particle has no size and, importantly in the context of projectiles, it does not spin.

Rewind

You met the constant acceleration formulae with two-dimensional vectors in Chapter 16, Section 2.

WORKED EXAMPLE 17.1

A particle is projected from point O with the velocity $\begin{pmatrix} 7.2 \\ 4.8 \end{pmatrix}$ m s^{-1}. Use $g = 9.8$ m s^{-2}, giving your final answers to an appropriate degree of accuracy.

a Find the speed and direction of motion of the particle after 0.8 seconds.

b Find the distance of the particle from O at this time.

a $\underline{u} = \begin{pmatrix} 7.2 \\ 4.8 \end{pmatrix}$

First find the velocity vector after 0.8 seconds. Use $\mathbf{v} = \mathbf{u} + \mathbf{a}t$.

$\underline{a} = \begin{pmatrix} 0 \\ -9.8 \end{pmatrix}$

$t = 0.8$

$\underline{v} = \underline{u} + \underline{a}t$

$= \begin{pmatrix} 7.2 \\ 4.8 \end{pmatrix} + 0.8\begin{pmatrix} 0 \\ -9.8 \end{pmatrix}$

$= \begin{pmatrix} 7.2 \\ -3.04 \end{pmatrix}$

Speed $= \sqrt{7.2^2 + 3.04^2} = 7.8$ m s^{-1} (2 s.f.)

Speed is the magnitude of velocity.

Continues on next page

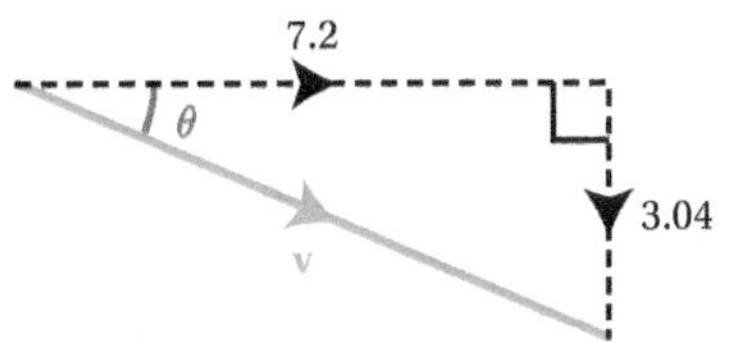

The direction of motion is the direction of the velocity vector.

Draw a diagram to make sure you are finding the correct angle.

$\tan\theta = \dfrac{3.04}{7.2}$

$\Rightarrow \theta = 22^\circ$ (2 s.f.)

The direction of motion is 22° below the horizontal.

b $\mathbf{s} = \mathbf{u}t + \frac{1}{2}\mathbf{a}t^2$

$= 0.8\begin{pmatrix} 7.2 \\ 4.8 \end{pmatrix} + \frac{0.8^2}{2}\begin{pmatrix} 0 \\ -9.8 \end{pmatrix}$

$= \begin{pmatrix} 5.76 \\ 0.704 \end{pmatrix}$

Use $\mathbf{s} = \mathbf{u}t + \frac{1}{2}\mathbf{a}t^2$.

Distance from the origin:

$\sqrt{5.76^2 + 0.704^2} = 5.8$ m (2 s.f.)

Distance is the magnitude of displacement.

The initial velocity may be specified by giving the speed and the angle of projection.

Key point 17.2

If a particle is projected with speed u at an angle θ above horizontal, then the components of the velocity are:

- horizontally: $u_x = u\cos\theta$
- vertically: $u_y = u\sin\theta$

$u\sin\theta$, u, θ, $u\cos\theta$

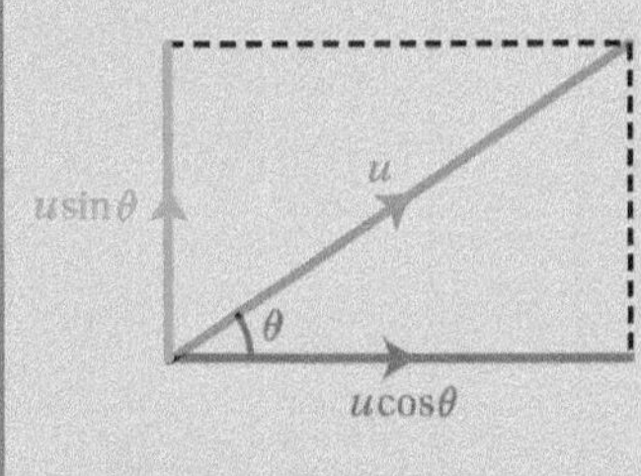

or as a vector: $\mathbf{u} = \begin{pmatrix} u\cos\theta \\ u\sin\theta \end{pmatrix}$

When solving problems involving projectiles it is often useful to consider the horizontal and vertical motion separately, rather than using vectors.

This means using the one-dimensional constant acceleration formulae in each direction separately.

WORKED EXAMPLE 17.2

A particle is projected from ground level with speed 14 m s^{-1} at an angle of 45° above the horizontal.

Find the height of the particle above ground level at the time when its horizontal displacement from the starting point is 11 m. Use $g = 9.81$ m s^{-2}, giving your final answers to an appropriate degree of accuracy.

Horizontally:

$s_x = (u\cos\theta)t$

$11 = (14\cos 45°)t$

$\Rightarrow t = \frac{11}{14\cos 45°} = 1.11$

Find the time first, using the horizontal displacement equation.

Vertically:

$u = 14\sin 45°$

$a = -9.81$

$t = 1.11$

Use $s = ut + \frac{1}{2}at^2$

$s_y = ut + \frac{1}{2}at^2$

$= (14\sin 45°)(1.11) + \frac{1}{2}(-9.81)(1.11)^2$

$= 4.9$ m (2 s.f.)

Tip

Since there is no acceleration horizontally, the horizontal component of velocity is constant throughout the motion, and so the only equation that you can use is $s_x = (u\cos\theta)t$.

If the particle is projected from a point above ground level then the vertical displacement can be negative, corresponding to the particle falling below the starting point.

WORKED EXAMPLE 17.3

A stone is thrown from the top of the cliff which is 30 m above sea level. The initial velocity has magnitude 8 m s^{-1} and is directed at 20° above horizontal.

Find the speed with which the stone hits the sea. Use $g = 10$ m s^{-2}, giving your final answer to an appropriate degree of accuracy.

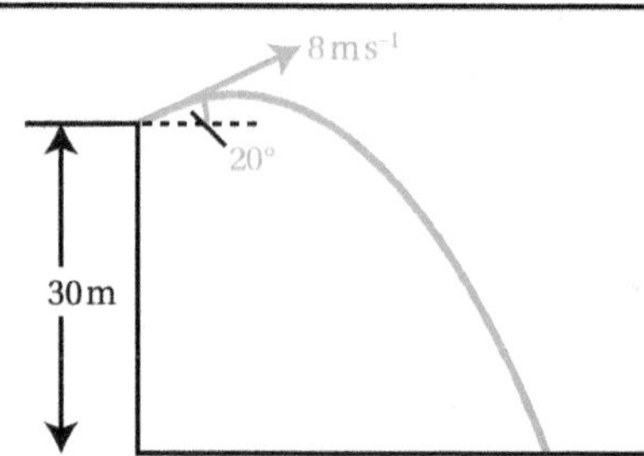

Horizontally:

$v_x = u\cos\theta$

$= 8\cos 20°$

$= 7.518$

The horizontal component of the velocity is constant.

Vertically:

$u = 8\sin 20°$

$a = -10$

$s = -30$

Note that $s = -30$ as the stone finishes 30 m below its starting point.

Continues on next page

$v_y^2 = u^2 + 2as$
$= (8\sin 20°)^2 + 2(-10)(-30)$
$= 607.49$

Use $v^2 = u^2 + 2as$ to find the vertical component of the velocity.

$|\underline{v}|^2 = v_x^2 + v_y^2$
$= 7.518^2 + 607.49$
$= 664$
$\therefore |\underline{v}| = 30\text{ m s}^{-1}$ (1 s.f.)

Now find the magnitude of the velocity.

Questions often ask for the maximum height a projectile reaches, or how far horizontally from the starting point it lands.

Key point 17.3

- A projectile is at its maximum height when $v_y = 0$.
- For a particle projected from ground level, to find the **range** (the maximum horizontal distance travelled) set $y = 0$.

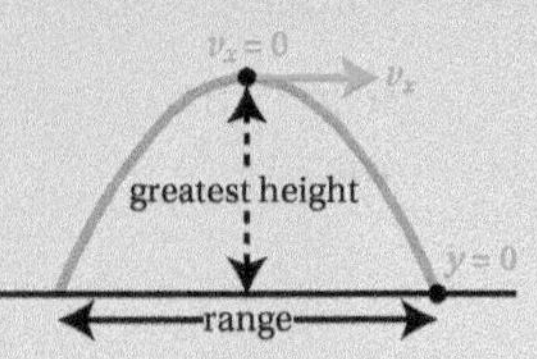

WORKED EXAMPLE 17.4

A particle is projected from ground level with speed u in the direction $\theta°$ above horizontal.

a Show that the maximum height the particle reaches is $\frac{u^2 \sin^2 \theta}{2g}$.

b Given that $u = 10.5\text{ m s}^{-1}$ and $\theta = 30°$, find the range of the projectile. Use $g = 9.81\text{ ms}^{-2}$, giving your final answer to an appropriate degree of accuracy.

a Vertically:
$u = u\sin\theta$
$v = 0$
$a = -g$

The maximum height is reached when $v_y = 0$.

Use $v^2 = u^2 + 2as$ to find the vertical displacement at this point.

$v^2 = u^2 + 2as$
$0 = (u\sin\theta)^2 + 2(-g)s$
$2gs = u^2 \sin^2\theta$
$s = \frac{u^2 \sin^2 \theta}{2g}$

b Vertically:
$u = 10.5\sin 30°$
$a = -9.81$
$s = 0$

The projectile lands when $y = 0$. First find the time the particle lands, using $s = ut + \frac{1}{2}at^2$.

$s = ut + \frac{1}{2}at^2$
$0 = (10.5\sin 30°)t - 4.905t^2$
$0 = 5.25t - 4.905t^2$
$t(5.25 - 4.905t) = 0$
$\therefore t = \frac{5.25}{4.905} = 1.0703$

$t = 0$ corresponds to the starting position, so you want the other value of t.

Continues on next page

$x = (u\cos\theta)t$

$= 10.5\cos 30° \times 1.0703$

$= 9.73\text{ m (3 s.f.)}$

You can now find the horizontal distance at this time.

In all of the worked examples so far the particle was projected upwards at an angle. But the same equations still apply if the particle is projected downwards. The only difference is that the vertical component of the initial velocity is negative.

WORKED EXAMPLE 17.5

A small ball is thrown from a window 12 m above ground with the initial velocity 6 m s^{-1} directed at 30° below the horizontal.

Find how long it takes to reach the ground. Use $g = 9.8$ m s^{-2}, giving your final answer to an appropriate degree of accuracy.

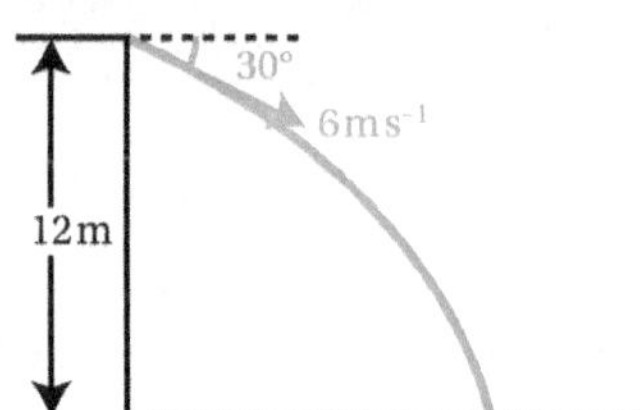

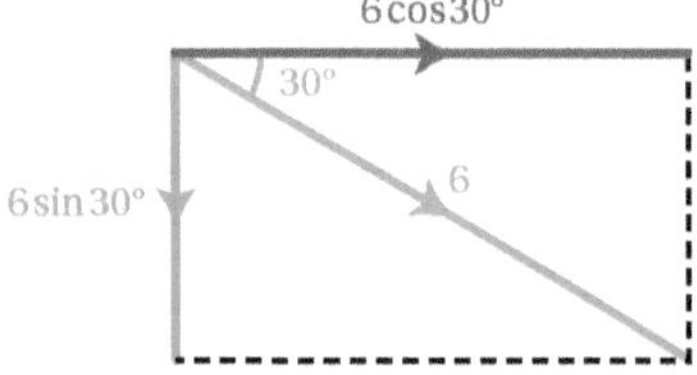

Note that the vertical component of the initial velocity is negative as the particle is going downwards from the start.

Vertically:

$u = -6\sin 30° = -3$

$a = -9.8$

$s = -12$

$$s = ut + \frac{1}{2}at^2$$

$$-12 = -3t + \frac{1}{2}(-9.8)t^2$$

$$4.9t^2 + 3t - 12 = 0$$

$$t = \frac{-3 \pm \sqrt{9 - 4\times(-4.9)\times 12}}{9.8}$$

$$= -1.90 \text{ or } 1.29$$

Solve the quadratic for t.

It takes 1.3 seconds (2 s.f.).

t must be positive.

EXERCISE 17A

In this exercise, unless instructed otherwise, use $g = 9.8\text{ m s}^{-2}$, giving your final answers to an appropriate degree of accuracy.

1 A particle is projected from a point P. Find its velocity vector after 2 seconds if its initial velocity is:

a i $(5\mathbf{i}+21\mathbf{j})\text{ m s}^{-1}$ ii $(7\mathbf{i}+6\mathbf{j})\text{ m s}^{-1}$ b i $(8\mathbf{i}-3\mathbf{j})\text{ m s}^{-1}$ ii $(4\mathbf{i}-9\mathbf{j})\text{ m s}^{-1}$

c i 14 m s^{-1} at $40°$ above the horizontal ii 25 m s^{-1} at $65°$ above the horizontal

d i 8 m s^{-1} at $35°$ below the horizontal ii 20 m s^{-1} at $10°$ below the horizontal.

2 A particle is projected from the origin. Find its position vector after 2 seconds if its initial velocity is as in question 1.

3 A small stone is projected from ground level with speed 10 m s^{-1} at an angle of $25°$ above the horizontal.

a Find its height above the ground after 0.6 seconds.

b What is its speed at this time?

4 A particle is projected with initial velocity $(5\mathbf{i}+2.4\mathbf{j})\text{ m s}^{-1}$. The unit vectors $\mathbf{i}$ and $\mathbf{j}$ are directed to the right and vertically upwards. Find:

a the time it takes the particle to reach its maximum height

b the magnitude and direction of its velocity 1.5 seconds after projection.

5 In this question, use $g = 10\text{ m s}^{-2}$, giving your final answers to an appropriate degree of accuracy

A ball is projected from a point A on a horizontal plane with speed 20 m s^{-1} at an angle of elevation of $50°$. The ball returns to the plane at point B. Find:

a the greatest height above the plane reached by the ball

b the distance AB.

6 A particle is projected with speed 14 m s^{-1} at an angle θ above the horizontal. The greatest height reached above the point of projection is 8 m.

Find, to the nearest degree, the value of θ.

7 A particle is projected from a point O with speed 35 m s^{-1} at an angle of elevation of $30°$.

Find the length of time the particle is more than 15 m above the horizontal level of O.

8 A ball is hit from a point P, 1.5 m above the ground, with speed 28 m s^{-1} at an angle of elevation of $45°$. The ball lands at the point Q, as shown in the diagram.

Find:

a the time taken for the ball to travel from P to Q

b the distance OQ

c the speed with which the particle hits the ground.

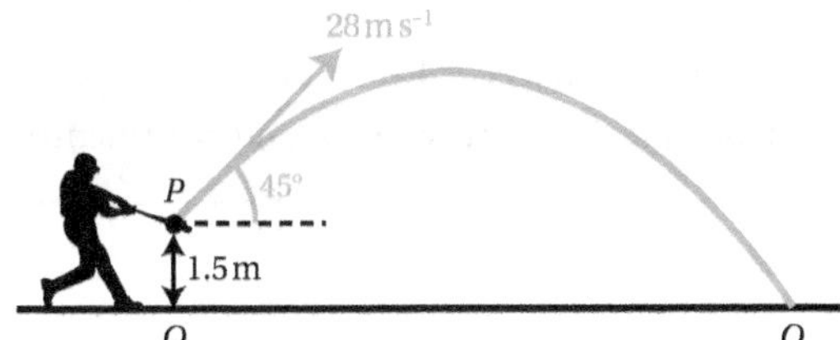

9 A ball is projected with speed 12 $m s^{-1}$ at an angle of 30° above horizontal. A 1.6 m high wall is located 6 m from the point of projection.

Determine whether the ball will clear the wall.

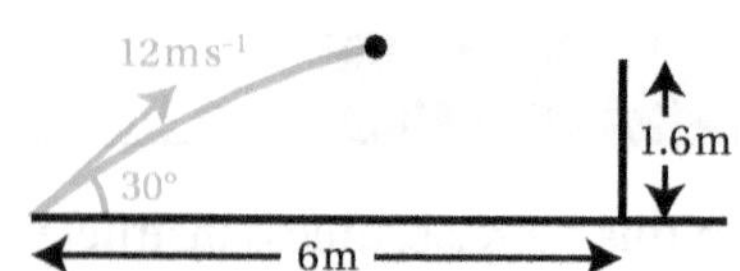

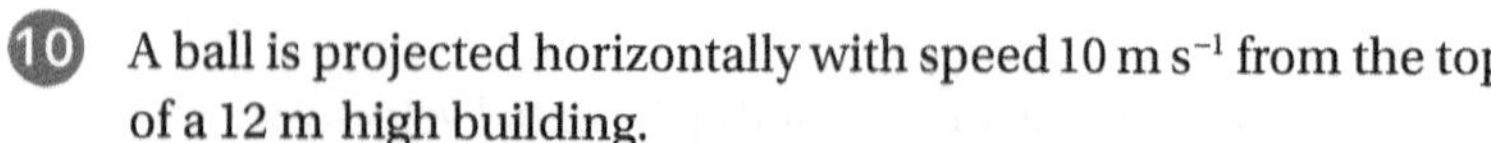

10 A ball is projected horizontally with speed 10 $m s^{-1}$ from the top of a 12 m high building.

a Find the distance from the foot of the building of the point where the ball hits the ground.

A second ball is projected with speed 14 $m s^{-1}$ from the foot of the building.

b Find the possible angles of projection so that the second ball hits the ground at the same place as the first ball.

Elevate

See Support Sheet 17 for an example of finding the angle and speed of projection and for more practice questions.

11 A golfer is aiming to land the ball on the green. The front of the green is 190 m from his position and the green is 10 m long, as shown in the diagram. The ground is horizontal.

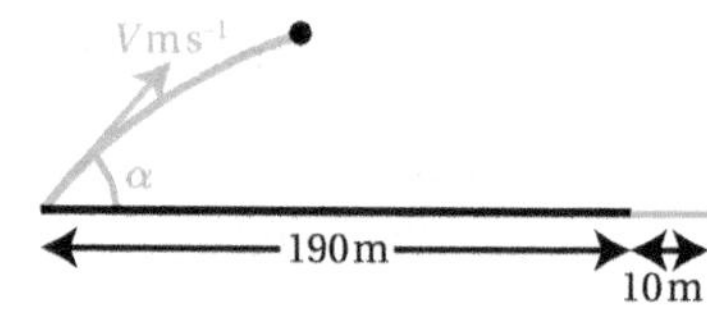

a If the golfer strikes the ball with speed V $m s^{-1}$ at an angle α above the horizontal, show that the horizontal distance travelled by the ball when it lands is $\frac{V^2 \sin 2\alpha}{g}$.

b If $V = 50$, find the range of values of α that will result in the golfer landing the ball on the green.

12 A film director wants to include the 'man fired out of a cannon' stunt in his film. For ethical reasons he uses a scale model of a cannon, which is one tenth of the real size, with a toy fired out of it.

a What is the magnitude of the force of gravity that would be observed by someone watching the film?

b Does the film need to be speeded up or be slowed down to correct this?

Section 2: The trajectory of a projectile

So far you have only looked at how the displacement and velocity of a projectile change with time. But you can also find a relationship between the horizontal and vertical displacements. This leads to an equation describing the path (or **trajectory**) of the projectile.

Rewind

This is an example of **parametric equations** which you met in Chapter 12, Section 2. You can apply the method of eliminating t to find the Cartesian equation of the trajectory to any particle moving in two dimensions.

Key point 17.4

To find an equation for the trajectory of a projectile:

- make t the subject of the $x = (u \cos \theta)t$ equation
- substitute this expression for t into $y = (u \sin\theta)t - \frac{1}{2}gt^2$.

WORKED EXAMPLE 17.6

A particle is projected from ground level with speed 6.5 m s^{-1} at an angle θ above horizontal, such that $\tan\theta = \frac{5}{12}$. Let x and y be the horizontal and vertical displacements from the point of projection, with y measured upwards.

Find an expression for y in terms of x and g.

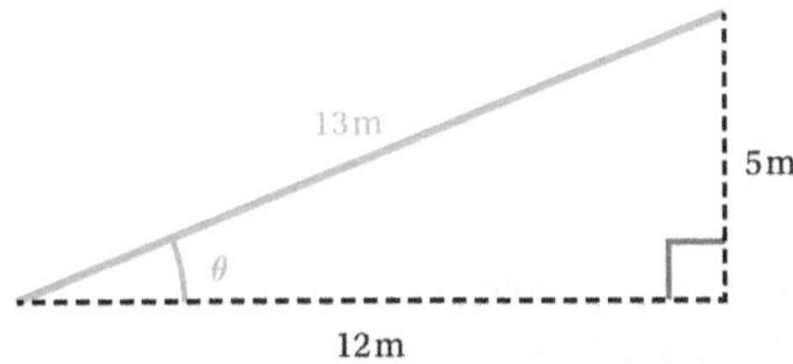

$\cos\theta = \frac{12}{13}$

$\sin\theta = \frac{5}{13}$

First find the components of the velocity.

Horizontally:

$x = (u\cos\theta)t$

$x = 6.5 \times \frac{12}{13}t$

$\Rightarrow t = \frac{x}{6}$

Make t the subject of the horizontal equation.

Vertically:

$y = (u\sin\theta)t - \frac{1}{2}gt^2$

$= \left(6.5 \times \frac{5}{13}\right)t - \frac{g}{2}t^2$

$= \frac{5}{2} \times \frac{x}{6} - \frac{g}{2}\left(\frac{x}{6}\right)^2$

And substitute t into the vertical equation.

$\therefore y = \frac{5}{12}x - \frac{g}{72}x^2$

As you can see, the trajectory of a projectile is a parabola. It is important to remember that the equations that you have been using include gravitational acceleration, but no other force. In particular, this model of a projectile assumes no air resistance. If the air resistance is included the trajectory is no longer a parabola.

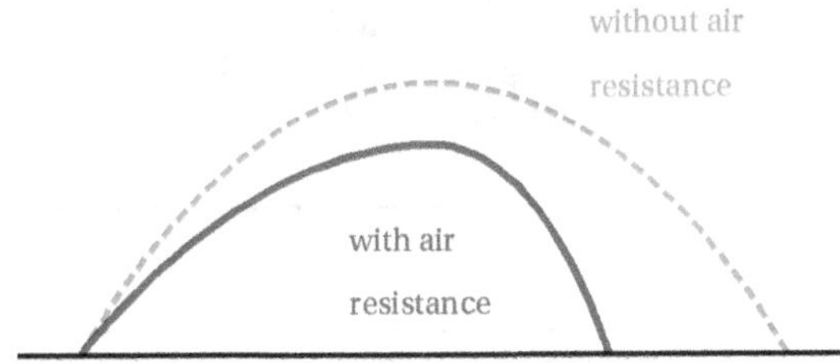

You may need to find the angle of projection in order for the particle to pass through a specific point. There will often be two possible values.

WORKED EXAMPLE 17.7

A particle is projected from a point P on horizontal ground, with speed 21 m s^{-1}, at an angle α above the horizontal. The particle passes through a point Q, which is at a horizontal distance of 12 m from P and a height of 4 m above the ground. Use $g = 9.8$ m s^{-2}, giving your final answer to an appropriate degree of accuracy.

a Show that $2\tan^2\alpha - 15\tan\alpha + 7 = 0$.

b Find the two possible values of α.

a Horizontally:

$$x = (u\cos\alpha)t$$

$$12 = (21\cos\alpha)t$$

$$t = \frac{4}{7\cos\alpha}$$

Make t the subject of the horizontal component equation...

Vertically:

$$y = (u\sin\alpha)t - \frac{1}{2}gt^2$$

$$4 = (21\sin\alpha)t - \frac{1}{2}gt^2$$

$$4 = (21\sin\alpha)\left(\frac{4}{7\cos\alpha}\right) - \frac{1}{2}g\left(\frac{4}{7\cos\alpha}\right)^2$$

... and substitute t into the vertical component equation.

$$4 = 12\frac{\sin\alpha}{\cos\alpha} - \frac{1}{2}\times 9.8 \times \frac{16}{49\cos^2\alpha}$$

$$4 = 12\tan\alpha - \frac{8}{5}\sec^2\alpha$$

$$5 = 15\tan\alpha - 2\sec^2\alpha$$

$\frac{\sin\alpha}{\cos\alpha} = \tan\alpha$ and $\frac{1}{\cos^2\alpha} = \sec^2\alpha$

$$5 = 15\tan\alpha - 2(1 + \tan^2\alpha)$$

$$5 = 15\tan\alpha - 2 - 2\tan^2\alpha$$

$$2\tan^2\alpha - 15\tan\alpha + 7 = 0$$

$\sec^2\alpha = 1 + \tan^2\alpha$

b $(2\tan\alpha - 1)(\tan\alpha - 7) = 0$

$$\tan\alpha = \frac{1}{2} \text{ or } 7$$

$$\therefore \alpha = 27^\circ \text{ or } 82^\circ \text{ (2 s.f.)}$$

Factorise and solve for α.

Note that there are two possible trajectories that pass through Q.

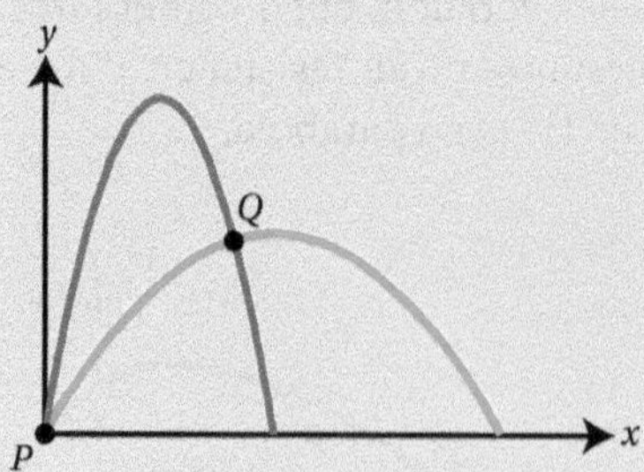

WORK IT OUT 17.1

A ball is thrown from a point P, at ground level, with speed 15 m s^{-1}, at an angle $\theta°$ above the horizontal. It lands at a point 12 m horizontally from P on a platform that is 4 m above the ground. The ledge starts 10 m from P and is 4 m long as shown.

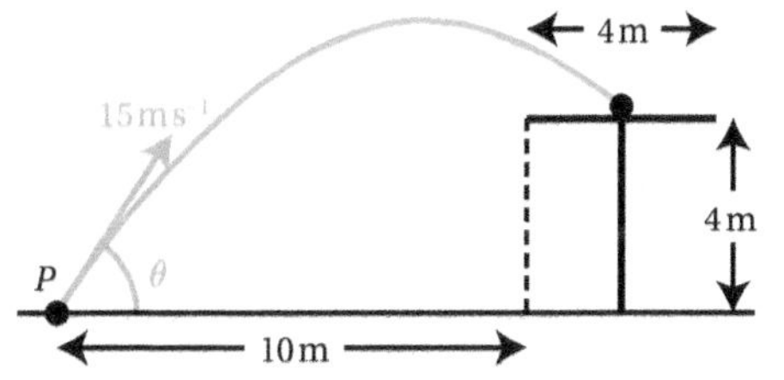

The trajectory of the ball is given by $y = x\tan\theta - \dfrac{gx^2}{2u^2}(1+\tan^2\theta)$.

Find the angle θ. Use $g = 10\text{ m s}^{-2}$, giving your final answer to an appropriate degree of accuracy.

Which is the correct solution? Identify the errors made in the incorrect solutions.

Solution 1	Solution 2	Solution 3
$x=12, y=4, u=15$ and $g=10$	$x=12, y=4, u=15$ and $g=10$	$x=10, y=4, u=15$ and $g=10$
$4 = 12\tan\theta - \dfrac{10(12)^2}{2(15)^2}(1+\tan^2\theta)$	$4 = 12\tan\theta - \dfrac{10(12)^2}{2(15)^2}(1+\tan^2\theta)$	$4 = 10\tan\theta - \dfrac{10(10)^2}{2(15)^2}(1+\tan^2\theta)$
$4 = 12\tan\theta - \dfrac{16}{5}(1+\tan^2\theta)$	$4 = 12\tan\theta - \dfrac{16}{5}(1+\tan^2\theta)$	$4 = 10\tan\theta - \dfrac{20}{9}(1+\tan^2\theta)$
$4 = 15\tan\theta - 4 - 4\tan^2\theta$	$5 = 15\tan\theta - 4 - 4\tan^2\theta$	$18 = 45\tan\theta - 10 - 10\tan^2\theta$
$4\tan^2\theta - 15\tan\theta + 9 = 0$	$4\tan^2\theta - 15\tan\theta + 9 = 0$	$10\tan^2\theta - 45\tan\theta + 28 = 0$
$(4\tan\theta - 3)(\tan\theta - 3) = 0$	$(4\tan\theta - 3)(\tan\theta - 3) = 0$	$\tan\theta = 0.746$ or 3.75
$\tan\theta = \dfrac{3}{4}$ or 3	$\tan\theta = \dfrac{3}{4}$ or 3	$\theta = 40°$ (1 s.f.) or $80°$ (1 s.f.)
$\theta = 37°$ or $72°$	$\theta = 37°$ or $72°$	
$\therefore \theta = 40°$ (1 s.f.) as this is the first point of contact with the platform	$\therefore \theta = 70°$ (1 s.f.) as the particle needs to be on the way down	

EXERCISE 17B

In this exercise, unless otherwise instructed, use $g = 9.8\text{ m s}^{-2}$, giving your final answers to an appropriate degree of accuracy.

1. A particle is projected horizontally with speed 8 m s^{-1} from the top of a 25 m high cliff. Let the origin of an x–y coordinate system be located at the top of the cliff, with the y-axis vertical and the x-axis horizontal in the direction of projection.

   Find, in terms of g, the Cartesian equation of the trajectory of the particle.

2. A rugby ball is kicked at an angle of elevation of 45° towards a cross bar that is y m high and x m away horizontally from the kicker.

   a If the ball hits the crossbar, show that $y = x - \dfrac{g\,x^2}{u^2}$.

   b If the cross bar is 3 m high and the kick is to be taken 35 m from the posts, find the minimum velocity with which the ball must be kicked to clear the bar.

   c State two modelling assumptions you needed to make.

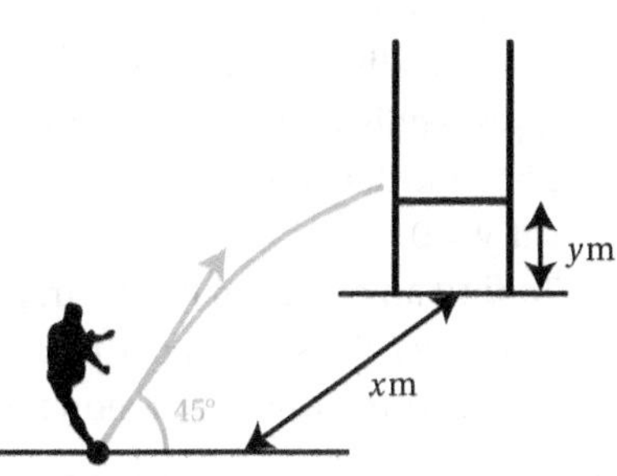

3 In this question, use $g = 9.81\text{ m s}^{-2}$, giving your final answer to an appropriate degree of accuracy.

a Show that the equation of the trajectory of a particle projected at speed $u\text{ m s}^{-1}$ at an angle of elevation θ is $y = x\tan\theta - \dfrac{gx^2}{2u^2\cos^2\theta}$.

b An archer fires an arrow from a point 1.5 m above the ground with speed 44 m s^{-1} at an angle of elevation of 7°. The target is 50 m away and the centre of the target is 1.3 m above the ground. The diameter of the target is 15 cm.

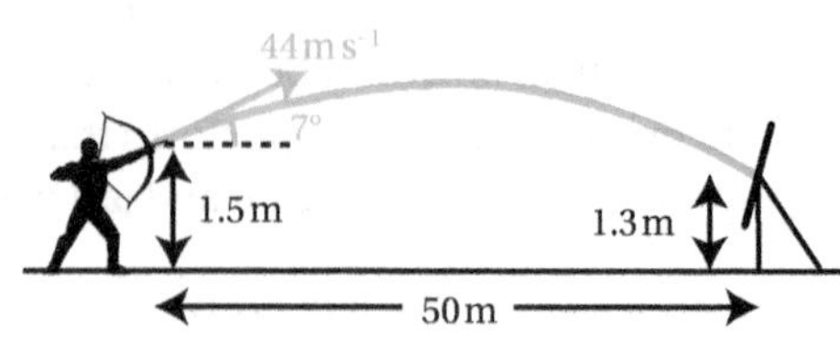

Determine whether the arrow hits the target.

4 A particle is projected from the origin with speed $V\text{ m s}^{-1}$ at an angle of elevation α. It passes through the point with position vector $4\mathbf{i} + 8\mathbf{j}$.

a Show that $\dfrac{2g}{V^2\cos^2\alpha} = \tan\alpha - 2$.

b The particle subsequently passes through the point with position vector $6\mathbf{i} + 5\mathbf{j}$.

i Show that $\tan\alpha = \dfrac{13}{3}$. **ii** Find V.

5 A particle is projected from a point on horizontal ground with speed u at an angle of elevation θ.

The maximum height reached by the particle is 28 m and the particle hits the ground 84 m from the point of projection.

Find u and θ.

6 In this question use $g = 10\text{ m s}^{-2}$, giving your final answer to an appropriate degree of accuracy.

A basketball player shoots at the hoop. The hoop is 3 m from the ground and the basketball player stands 6 m horizontally from the hoop. The ball is released from 2 m above the ground at an angle of α above the horizontal.

The initial speed of the ball is 10 m s^{-1}.

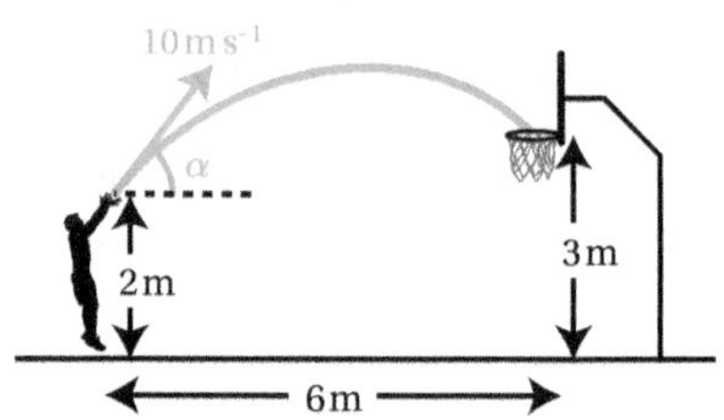

a Given that the ball passes through the hoop, show that $9g\tan^2\alpha - 300\tan\alpha + 50 + 9g = 0$.

b Hence find the angle at which the ball was released, justifying your answer.

Checklist of learning and understanding

- In projectile motion the acceleration of q particle is $\mathbf{a} = \begin{pmatrix} 0 \\ -g \end{pmatrix}\text{ m s}^{-2}$.
- If a particle is projected with speed u at an angle $\theta°$ above horizontal, then the components of the velocity are:
  - horizontally: $u_x = u\cos\theta$
  - vertically: $u_y = u\sin\theta$
- A projectile is at its maximum height when $v_y = 0$.
- For a particle projected from ground level, to find the range (the maximum horizontal distance travelled) set $y = 0$.
- To find an equation for the trajectory of a projectile:
  - make t the subject of the $x = (u\cos\theta)t$ equation
  - substitute this expression for t into $y = (u\sin\theta)t - \frac{1}{2}gt^2$.

Mixed practice 17

In this exercise, unless otherwise instructed, use $g = 9.8\text{ m s}^{-2}$, giving your final answers to an appropriate degree of accuracy.

1 A particle is projected at a velocity of $\begin{pmatrix} 7 \\ 24 \end{pmatrix}\text{m s}^{-1}$. Find the minimum speed during its flight.

Choose from these options.

A 0 m s^{-1} **B** 7 m s^{-1} **C** 24 m s^{-1} **D** 25 m s^{-1}

2 A particle of mass m is projected horizontally off the edge of a cliff and lands in the sea below at a distance x from the base of the cliff.

A second particle of mass $2m$ is projected horizontally from the same point at the same speed and lands in the sea at a distance y from the base of the cliff.

Which one of these statements is true?

A $x > y$ **B** $y > x$

C $x = y$ **D** It depends on the height of the cliff.

3 A ball is thrown horizontally at 12 m s^{-1} from the top of a 30 m high building. Find:

a the time it takes to hit the ground below

b the distance from the bottom of the building to the point where the ball hits the ground.

4 In this question, use $g = 9.81\text{ m s}^{-2}$, giving your final answers to an appropriate degree of accuracy.

A projectile is launched from the point O with velocity $(10\mathbf{i} + 15\mathbf{j})\text{ m s}^{-1}$. After t seconds it is at the point A and is travelling with velocity $(10\mathbf{i} - 9.5\mathbf{j})\text{ m s}^{-1}$. Find:

a the value of t **b** the distance OA.

5 In this question, use $g = 10\text{ m s}^{-2}$, giving your final answers to an appropriate degree of accuracy.

A ball is thrown from a window 5 m above the ground with velocity $(9\mathbf{i} + 8\mathbf{j})\text{ m s}^{-1}$.

It is caught by a child 1 m above the ground. The child stands x m from the point vertically below the window at ground level. Find:

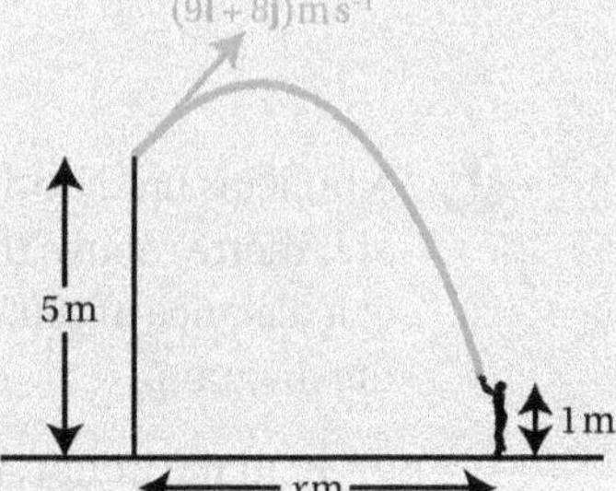

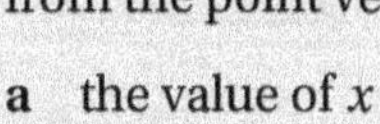

a the value of x

b the speed of the ball as it is caught.

6 A particle is projected from a point A on a horizontal plane with speed $u\text{ ms}^{-1}$ at an angle $\theta°$ above the horizontal. It lands on the plane at the point B.

a **i** Show that $AB = \dfrac{u^2 \sin 2\theta}{g}$.

ii Hence deduce that for fixed u, the maximum range is achieved when $\theta = 45°$.

b If $AB = \dfrac{2u^2}{3g}$ find, to the nearest degree, the two possible angles at which the particle was projected.

7 A child stands on a bridge that passes over a river which is 6 m below. He throws a stone with speed 15 m s^{-1} directly at a small stationary rock which is 8 m from the base of the bridge as shown in the diagram.

How far away from the rock does the stone land in the river?

8 A particle is projected from O with velocity $(8\mathbf{i} + 11\mathbf{j})$ m s^{-1}. It passes through the point P at time t seconds.

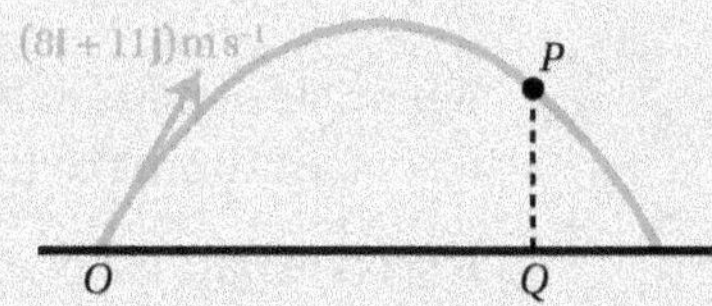

a Given that $OQ = 2PQ$, find the value of t.

b At point R the ball has the same speed as at P. Find the time taken for the ball to travel from O to R.

9 A tennis ball is projected from a point O with a velocity of $\left(4\sqrt{3}\mathbf{i} + 4\mathbf{j}\right)$ m s^{-1}, where $\mathbf{i}$ and $\mathbf{j}$ are horizontal and vertical unit vectors respectively. The ball travels in a vertical plane through O, which is 30 cm above the horizontal surface of a tennis court. During its flight, the horizontal and upward vertical distances of the ball from O are x metres and y metres respectively.

Model the ball as a particle.

a Show that, during the flight, the equation of the trajectory of the ball is given by

$$y = \frac{x}{\sqrt{3}} - \frac{49x^2}{480}$$

b The ball hits a vertical net at a point A. The net is at a horizontal distance of 4 m from O.

Determine the height of the point A, above the surface of the tennis court. Give your answer to the nearest centimetre.

c State a modelling assumption, other than the ball being a particle, that you need to make to answer this question.

[© AQA 2014]

10 A bullet is fired horizontally from the top of a vertical cliff, at a height of h metres above the sea. It hits the sea 4 seconds after being fired, at a distance of 1000 metres from the base of the cliff, as shown in the diagram.

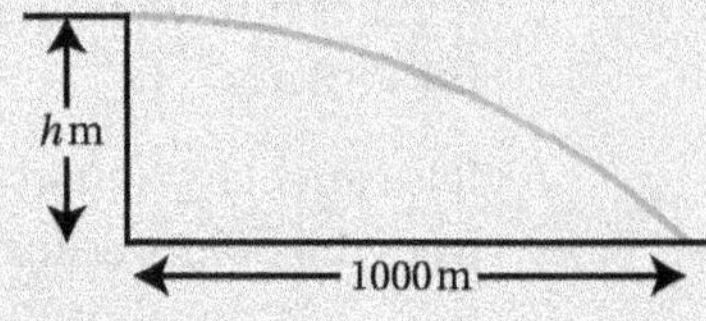

a Find the initial speed of the bullet.

b Find h.

c Find the speed of the bullet when it hits the sea.

d Find the angle between the velocity of the bullet and the horizontal when it hits the sea.

[© AQA 2011]

11 In this question, use $g = 10\text{ m s}^{-2}$, giving your final answers to an appropriate degree of accuracy.

A particle is projected from a point 30 m above ground level with speed $u = (3\mathbf{i} + 4\mathbf{j})\text{ m s}^{-1}$ as shown in the diagram.

Find its distance from the point of projection at the instant the particle is moving in a direction perpendicular to its initial velocity.

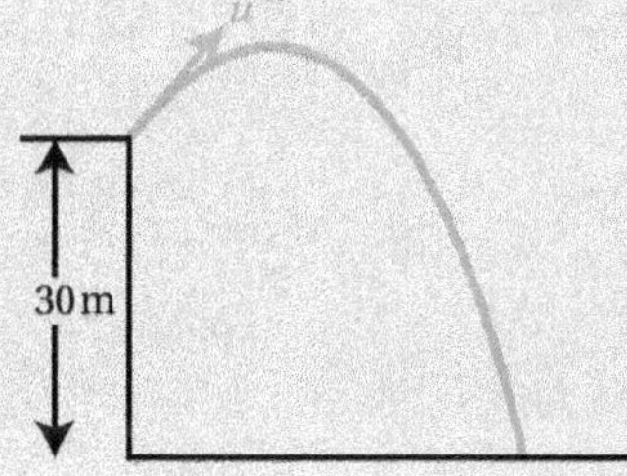

12 A ball is projected with speed u m s^{-1} at an angle of elevation α above the horizontal so as to hit a point P on a wall. The ball travels in a vertical plane through the point of projection. During the motion, the horizontal and upward vertical displacements of the ball from the point of projection are x metres and y metres respectively.

a Show that, during the flight, the equation of the trajectory of the ball is given by

$$y = x \tan\alpha - \frac{gx^2}{2u^2}(1 + \tan^2\alpha)$$

b The ball is projected from a point 1 metre vertically below and R metres horizontally from the point P.

i By taking $g = 10\text{ m s}^{-2}$, show that R satisfies the equation

$$5R^2\tan^2\alpha - u^2R\tan\alpha + 5R^2 + u^2 = 0$$

ii Hence, given that u and R are constants, show that, for $\tan\alpha$ to have real values, R must satisfy the inequality

$$R^2 \leqslant \frac{u^2(u^2 - 20)}{100}$$

iii Given that $R = 5$, determine the minimum possible speed of projection.

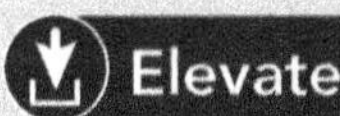

See Extension Sheet 17 for a selection of more challenging problems.

18 Forces in context

In this chapter you will learn how to:

- resolve forces in a given direction in order to calculate the resultant force
- use a model for friction
- determine the acceleration of a particle moving on an inclined plane.

Before you start...

Student Book 1, Chapter 18	You should be able to add vectors and find magnitudes.	1 Three horizontal forces act on a particle. In newtons, the forces are: $\mathbf{F}_1 = 2\mathbf{i}$, $\mathbf{F}_2 = 3\mathbf{j}$ and $\mathbf{F}_3 = \mathbf{i} - 2\mathbf{j}$. Calculate the magnitude of the resultant force and its angle from the direction $\mathbf{i}$.
Student Book 1, Chapter 17	You should be able to solve problems involving motion with constant acceleration.	2 A force of 5 N acts upon a particle with mass 2 kg. If the particle is initially at rest, after how many seconds will its displacement be equal to 5 m?

Improving the model

In Student Book 1, Chapter 18, you learnt to add and subtract forces in vector form and to determine the angle and magnitude of the resultant force. This chapter deals with more complex situations involving strings and planes in different orientations, where the forces are not perpendicular. To solve these problems you will need to combine your knowledge of forces, vectors and trigonometry.

In Student Book 1 you met problems where there is a constant resistance force acting on an object. Here the model of friction will be improved to enable more real-world situations to be modelled.

Section 1: Resolving forces

When considering forces acting on a moving body, it can be useful to split each force into horizontal and vertical components. This process is called **resolving** the force.

WORKED EXAMPLE 18.1

Two forces act on a particle in the horizontal $x-y$ plane. Force $\mathbf{F}_1$ has magnitude 10 N and acts at angle 30° to the positive x-axis. Force $\mathbf{F}_2$ has magnitude 9 N and acts in the direction of the negative x-axis. Find the magnitude and direction of the resultant force.

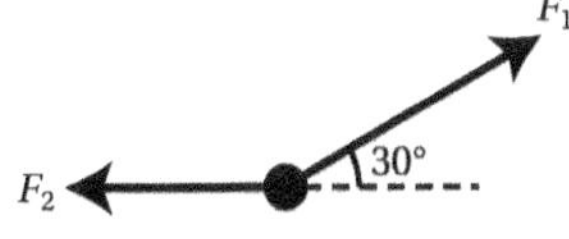

Always draw a diagram.

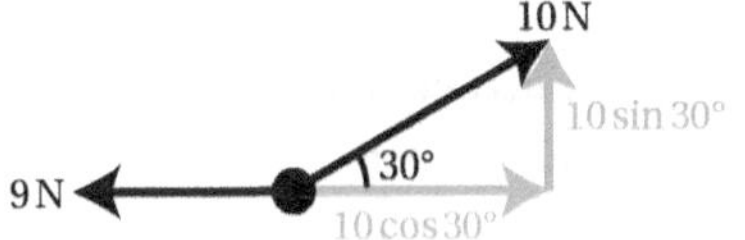

Form a right-angled triangle with the force as hypotenuse and the sides horizontal and vertical. In this case, since the force $\mathbf{F}_2$ is in the x-direction, you only need to resolve force $\mathbf{F}_1$.

$R(\rightarrow): X = 10\cos 30° - 9$

$= 5\sqrt{3} - 9 = -0.3397$

$R(\uparrow): Y = 10\sin 30° = 5$

Calculate horizontal and vertical components of $\mathbf{F}_1$ and add $\mathbf{F}_2$ to the horizontal direction
The notation R (→) is a standard shorthand for 'resolve horizontally' and R (↑) stands for 'resolve vertically'.

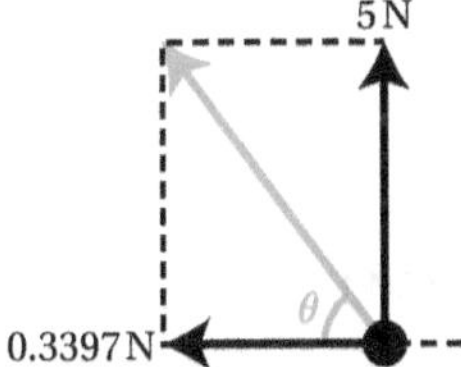

Draw another triangle showing the resultant force. This helps find its magnitude and direction.
Note that −0.3397 means that the force is acting in the negative x direction.

Magnitude: $\sqrt{0.3397^2 + 5^2} = 5.01\text{ N}$

Direction: $\theta = \arctan\left(\frac{5}{0.3397}\right) = 86.1°$

$180° - 86.1° = 93.9°$, so the direction is 93.9° from the positive x-axis.

Once you have found forces, you may have to work with them, using Newton's second law ($F = ma$).

Rewind

You met Newton's second law in Student Book 1, Chapter 18.

WORKED EXAMPLE 18.2

A particle with mass 5 kg moves on a horizontal (xy) plane with constant acceleration $a\mathbf{i}$, where $a > 0$, acted upon by two forces in the plane, $\mathbf{F}_1$ and $\mathbf{F}_2$.

$\mathbf{F}_1$ has magnitude 3 N and acts at angle 30° anticlockwise from the positive x-axis.
$\mathbf{F}_2$ has magnitude 5 N and acts at acute angle θ clockwise from the positive x-axis.

a Show that $\cos\theta = \frac{\sqrt{91}}{10}$.
b Calculate the exact value of a.

a

3 N, a, 30°, θ, 5 N

Always draw a diagram.

$$R(\uparrow): 3\sin 30° = 5\sin\theta$$

$$\sin\theta = \frac{3}{5}\sin 30°$$

$$= \frac{3}{10}$$

Resolve vertically: no acceleration means the forces are equal.

Rearrange to find $\sin\theta$.

$$\sin^2\theta + \cos^2\theta = 1$$

$$\cos\theta = \sqrt{1-\sin^2\theta}$$

$$= \sqrt{1-\frac{9}{100}}$$

$$= \frac{\sqrt{91}}{10}$$

Use $\sin^2\theta + \cos^2\theta = 1$ to convert $\sin\theta$ into $\cos\theta$. Since the angle is acute you only need to take the positive root as $\cos\theta > 0$ for acute angles.

b $R(\rightarrow): F = ma$

$$3\cos 30° + 5\cos\theta = 5a$$

$$a = \frac{1}{5}\left(\frac{3\sqrt{3}}{2} + \frac{5\sqrt{91}}{10}\right)$$

$$= \frac{3\sqrt{3}+\sqrt{91}}{10}\ \text{m s}^{-2}$$

Resolve horizontally and use Newton's second law.

You need to be able to apply previous techniques, in particular equations of constant acceleration, to problems involving objects moving under a combination of forces.

WORKED EXAMPLE 18.3

A toy helicopter of mass 0.2 kg has its rotors set so that the force provided by the engine is given by $\mathbf{F} = a\mathbf{i} + b\mathbf{j}$ where a and b can be varied using a remote control, but the magnitude of $\mathbf{F}$ is always 3 N. The unit vector $\mathbf{i}$ is horizontal, $\mathbf{j}$ is vertical. Use $g = 9.8\text{ m s}^{-2}$, giving your final answers to an appropriate degree of accuracy.

a What is the acceleration if the helicopter is set to fly horizontally?

b The helicopter accelerates at 45° above the horizontal. How long would it take for the helicopter to travel 5 metres?

Always draw a diagram. You could define θ to be the angle the force $\mathbf{F}$ makes with the vertical or the horizontal. Here, it is the vertical.

a $R(\uparrow): 3\cos\theta = 0.2g$

$$\cos\theta = \frac{1.96}{3}$$

$$\theta = 49.2°$$

There is no vertical acceleration.

$R(\rightarrow): 3\sin\theta = 0.2a$

$$a = 15\sin\theta$$

$$= 11\text{ m s}^{-2}\ (2\text{ s.f.})$$

Next resolve horizontally and use Newton's second law.

b The 3 N force acts at angle θ above 45°:

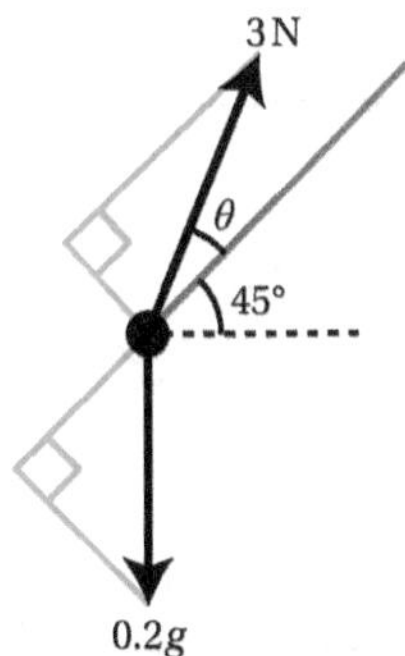

The direction of the resultant force is 45° above the horizontal, so choose this as one of the directions to resolve in. The net force in the perpendicular direction is zero.

(You could solve this problem by resolving horizontally and vertically. However, resolving in the direction of motion gives simpler equations.)

Perpendicular to movement:

$R(\nwarrow): 3\sin\theta = 0.2g\cos 45°$

$$\sin\theta = \frac{0.2 \times 9.8}{3\sqrt{2}} = 0.462$$

$$\theta = 27.5°$$

Continues on next page

In direction of movement:

$R(\nearrow): F = ma$

$3\cos\theta - 0.2g\sin 45° = 0.2a$

$a = 6.37\text{ m s}^{-2}$

Constant acceleration problem:

$u = 0, s = 5, a = 6.37$. Find t.

$s = ut + \frac{1}{2}at^2$

$5 = \frac{1}{2} \times 6.37t^2$

$t = \sqrt{\frac{10}{6.37}} = 1.3$ seconds (2 s.f.)

Once you have found the acceleration you can use the constant acceleration formulae. Notice that, although the forces act in two dimensions, the helicopter moves in a straight line. Hence you can use the one-dimensional version of the constant acceleration formulae.

EXERCISE 18A

In this exercise, unless instructed otherwise, use $g = 9.8\text{ m s}^{-2}$, giving your final answers to an appropriate degree of accuracy.

1 In each system, a particle on a horizontal surface is affected by two forces $\mathbf{F}_1$ and $\mathbf{F}_2$. Resolve the resultant force into $\mathbf{i}$ and $\mathbf{j}$ components where $\mathbf{j}$ represents due north, $\mathbf{i}$ represents due east.

a **i** $\mathbf{F}_1 = 3\mathbf{i} + 2\mathbf{j}$, $\mathbf{F}_2$ has magnitude 5 N and acts at a bearing of 060°.

ii $\mathbf{F}_1 = 5\mathbf{i} + 7\mathbf{j}$, $\mathbf{F}_2$ has magnitude 12 N and acts at a bearing of 105°.

b **i** $\mathbf{F}_1$ has magnitude 7 N and acts at a bearing of 015°, $\mathbf{F}_2$ has magnitude 8 N and acts at a bearing of 210°.

ii $\mathbf{F}_1$ has magnitude 7 N and acts at a bearing of 075°, $\mathbf{F}_2$ has magnitude 7 N and acts at a bearing of 285°.

2 A particle of mass m kg hangs at rest attached to two light, inextensible strings, at angle θ_1 and θ_2 from the horizontal.

For each system, find the tension in each string.

a **i** $m = 5\text{ kg}, \theta_1 = 0°, \theta_2 = 10°$ **ii** $m = 8\text{ kg}, \theta_1 = 30°, \theta_2 = 30°$

b **i** $m = 15\text{ kg}, \theta_1 = 50°, \theta_2 = 70°$ **ii** $m = 10\text{ kg}, \theta_1 = 20°, \theta_2 = 30°$

3 Find the resultant force acting on the object in the diagram, giving your answer in the form $a\mathbf{i} + b\mathbf{j}$.

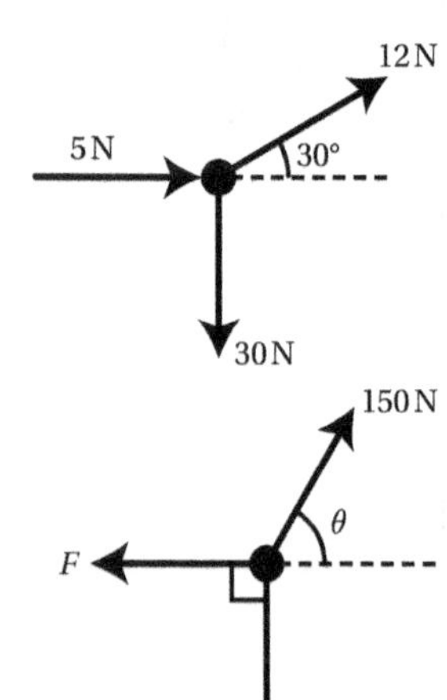

4 In the diagram the particle is in equilibrium.

a Find the angle θ. **b** Find the force F.

5 In this question, use $g = 10 \text{ m s}^{-2}$, giving your final answers to an appropriate degree of accuracy. A mass of 200 g is being held by a string at an angle θ to the vertical with tension T newtons and a horizontal force of 5 N.

a Show that $T \cos \theta = 2$ **b** Find the value of θ. **c** Find the value of T.

6 Three friends are pulling a 100 kg load across a smooth horizontal surface. Alf pulls with a force of 200 N. Bashir pulls with a force 150 N and is positioned at an angle 15° clockwise from Alf. Charlie pulls with force k N and is positioned at an angle 40° clockwise from Bashir.

Modelling the load as a particle, and given the load begins to move exactly towards Bashir, find k and determine the initial acceleration of the load.

7 Two forces of magnitudes 8 N and 15 N act at a point P.

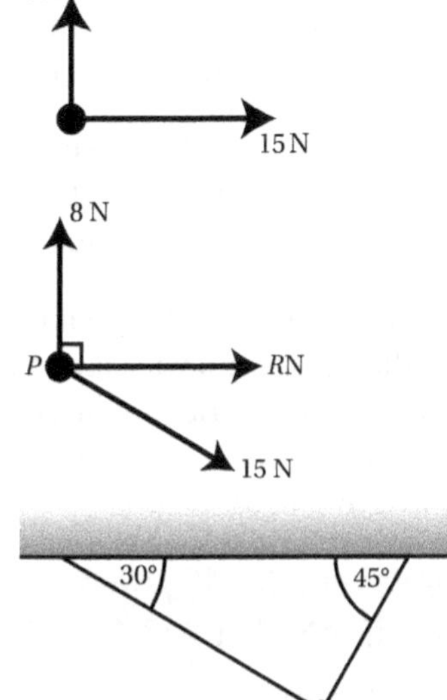

a Given that the two forces are perpendicular to each other, find:

i the angle between the resultant and the 15 N force

ii the magnitude of the resultant force.

b It is given instead that the resultant of the two forces acts in a direction perpendicular to the 8 N force.

i Find the angle between the resultant and the 15 N force

ii Find the magnitude of the resultant.

8 A particle with mass 1 kg hangs at rest suspended by two light inextensible strings. One string is at an angle of 30° to the horizontal and the other is at an angle of 45° to the horizontal.

Find the tension in the two strings.

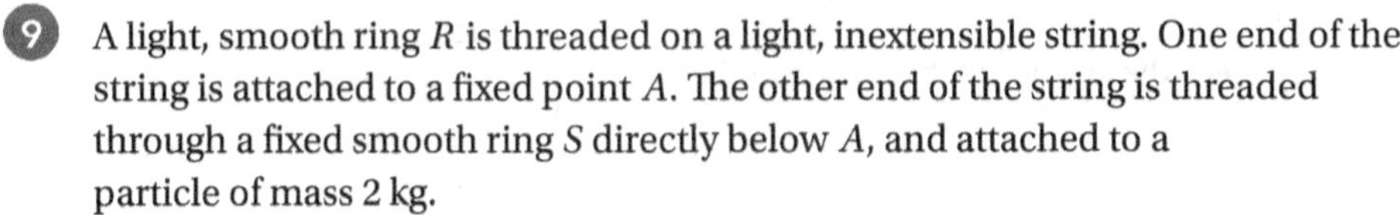

9 A light, smooth ring R is threaded on a light, inextensible string. One end of the string is attached to a fixed point A. The other end of the string is threaded through a fixed smooth ring S directly below A, and attached to a particle of mass 2 kg.

A force of magnitude F N is applied to ring R at an angle θ° with the horizontal.

a Given the string is taut, find, in terms of g, the exact value of F required to maintain the system in equilibrium with angle $ARS = 90°$ and $\theta = 0°$.

b If angle $ARS = 90°$ and the system is in equilibrium, find the relationship between θ and angle SAR.

c If ring R actually had mass 0.1 kg, find the new force F required to hold the system in equilibrium with angle $ARS = 90°$ and $\theta = 0°$, and determine angle SAR.

10 A small, smooth ring R of weight 5 N is threaded on a light, inextensible, taut string. The ends of the string are attached to fixed points A and B at the same horizontal level. A horizontal force of magnitude 4 N is applied to R. In the equilibrium position the angle ARB is a right angle, and the portion of the string attached to B makes an angle θ with the horizontal

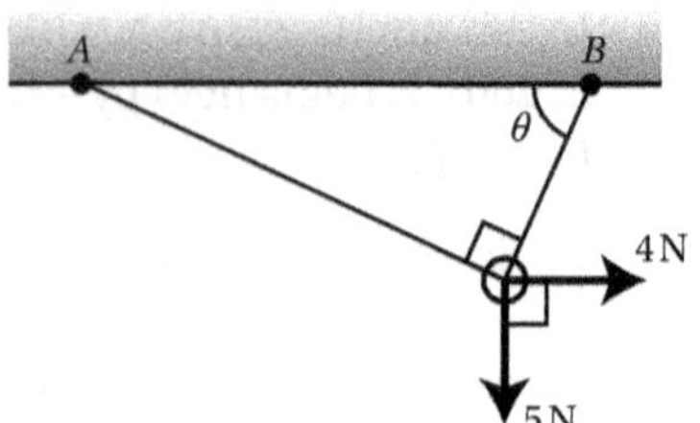

a Explain why the tension T is the same in each part of the string.

b Find T and θ.

Section 2: Coefficient of friction

In Student Book 1 you met friction as a constant force opposing motion. However, friction is not always constant; it depends on:

- the force pushing the object into the surface
- how rough the object and the surface are
- the driving force applied to the object.

If the driving force is strong enough, friction will be overcome and an object that was at rest will start to move. There must therefore be a maximum value of friction; this is called **limiting friction**. When friction is at its limiting value, the object is on the point of moving and is said to be in **limiting equilibrium**.

Key point 18.1

The maximum or limiting value of friction, F_{max}, is given by

$$F_{max} = \mu R$$

where R is the normal reaction force between the object and the surface and μ is the **coefficient of friction**.

Focus on...

Focus on... Problem solving 3 asks you to decide whether values of certain quantities (including frictional force and µ) are reasonable in the given contexts.

This coefficient deals with how rough the two surfaces are – for example, for an ice skate on an ice rink it is about 0.05, while for a car tyre on a road it is about 1. If the coefficient of friction is zero the surface is described as **smooth** and there is no friction force.

If the driving force is not sufficiently strong to overcome F_{max}, the frictional force will just match it so that the object remains at rest. In this case $F < \mu R$.

Once the object is moving, the frictional force is constant at F_{max}.

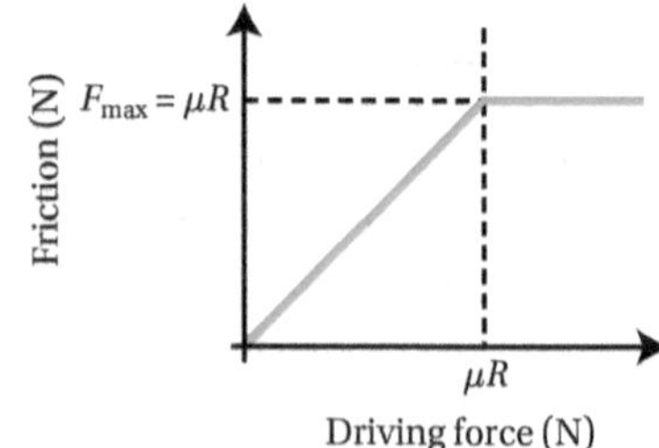

Key point 18.2

- If the resultant of the driving forces is smaller than F_{max}, the object will remain at rest and the friction force, F, will equal the resultant of the driving forces.
- If the resultant of the driving forces is larger than F_{max}, the object will move and $F = \mu R$.

WORKED EXAMPLE 18.4

A block of mass 9 kg lies at rest on a rough horizontal surface, with coefficient of friction $\mu = 0.2$. Attached to opposite ends of the block are two light strings, each of which is under 5 N tension. Use $g = 10\text{ m s}^{-2}$, giving your final answers to an appropriate degree of accuracy.

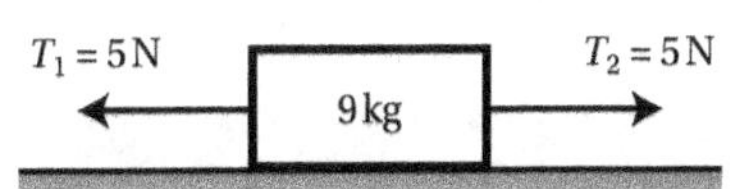

a The tension in the left string is raised to 10 N.

- **i** Calculate the direction and magnitude of the frictional force.
- **ii** If the block moves, determine its acceleration.

b The tension in the right string is then raised to 30 N.

- **i** Calculate the direction and magnitude of the frictional force.
- **ii** If the block moves, determine its acceleration.

a

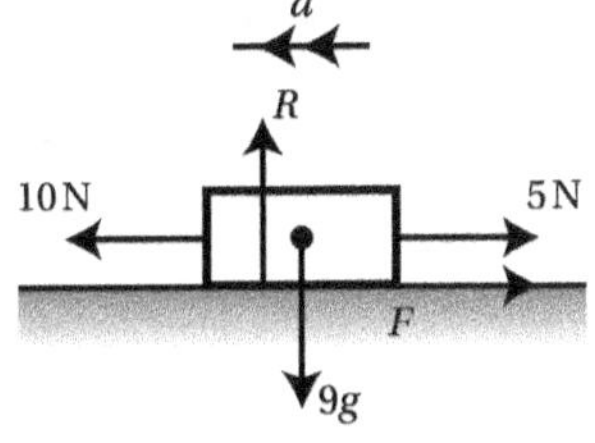

Always draw a diagram.
If the block moves it will be to the left so friction acts to the right.

$R(\uparrow): R = 9g$

$R = 90\text{ N}$

Since any movement will be horizontal, there is no vertical acceleration. Resolving vertically gives R.

Maximum friction is $\mu R = 18\text{ N}$

Calculate limiting friction μR.

$R(\leftarrow): X = 10 - 5 - F$

$= 5 - F$

Resolve horizontally.

i If $X = 0, 5 - F = 0$

$F = 5$

$\therefore F = 5\text{ N}$ to the right.

ii The block will not move.

Friction will oppose the resultant up to a maximum of limiting friction. Since the resultant driving force is less than limiting friction, the object will not move. Friction will just balance the driving force.

b As before $|F| \leq 18$

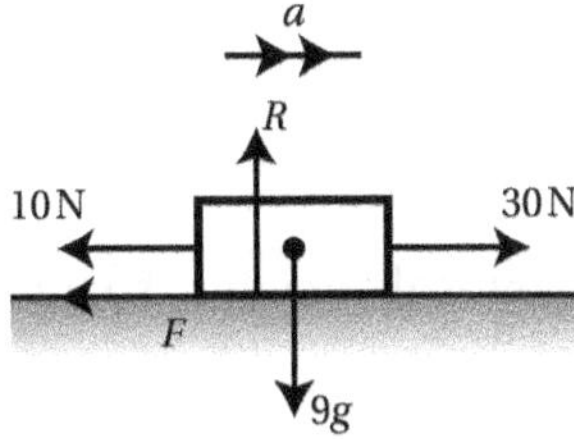

If the block moves it will now be to the right so friction acts to the left.

$R(\rightarrow): X = 30 - 10 - F$

$= 20 - F$

i If $X = 0, 20 - F = 0$

$F = 20$

But $F_{max} = 18\text{ N}$

$\therefore F = 18\text{ N}$ to the left.

The resultant driving force exceeds limiting friction, so friction takes its limiting value.

ii $F = ma$

$20 - 18 = 9a$

$a = \dfrac{(20-18)}{9} = 0.2\text{ m s}^{-2}$ (1 s.f.)

The resultant force is 20 − 18 N. Use $F = ma$.

Notice that the frictional force acts in different directions in parts **a** and **b** of Worked example 18.4 because it must always oppose the resultant force which would drive motion.

Tip

In some questions you may be uncertain initially which direction the friction acts. If you calculate a negative friction force in your answer, you have probably drawn the friction in the wrong direction. Change your diagram and adjust your equations.

Focus on...

See Focus on... Proof 3 for proofs of formulae for the minimum force required to move a particle on a rough surface in various situations.

WORKED EXAMPLE 18.5

Two blocks, A and B, lie at rest on a rough surface, with block A on top of block B. The coefficient of friction between the blocks is 0.15, the mass of block A is 5 kg and the mass of block B is 8 kg. The coefficient of friction between block B and the surface is 0.3. Use $g = 9.8\text{ m s}^{-2}$, giving your final answers to an appropriate degree of accuracy.

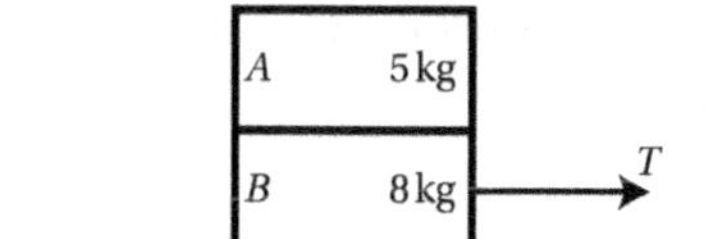

A light inextensible string is attached to block B, and the tension in the string is $T = 5$ N.

a Show that the system remains at rest.

The tension in the string is increased to 100 N and block A begins to slide on block B.

b What is the acceleration of each block?

a Treating the blocks as a single object of mass 13 kg, with frictional force between B and the surface being F_{BS}:

Since you are interested in the motion of the whole system, treat the two blocks as a single object.

Consider horizontal and vertical components separately. Remember to include the normal reaction force in your diagram.

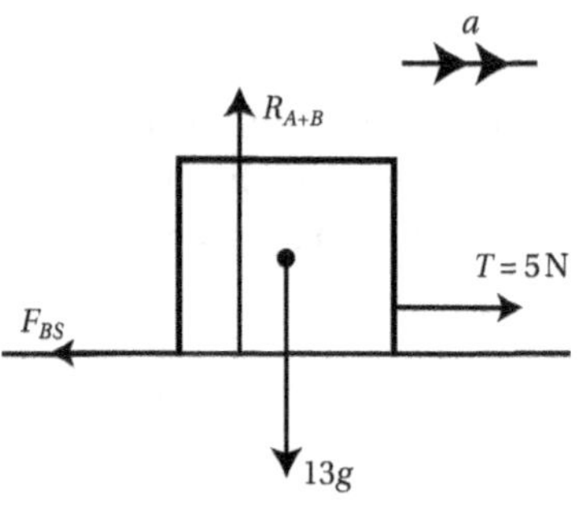

$R(\uparrow): R_{A+B} = 13g$

There is no movement in the vertical direction, so the net force is zero.

Limiting friction: $\mu R_{A+B} = 3.9g = 38\text{ N (2 s.f.)}$

$R(\rightarrow): X = 5 - F_{BS}$

Friction will oppose the resultant up to a maximum of limiting friction.

The tension is less than the maximal frictional force, so $F_{BS} = 5$ N and the system remains at rest.

b 100 N exceeds μR so the blocks will move, with a net driving force $100 - 38.22 = 61.78$ N.

Once the blocks are moving, the friction force equals μR.

Continues on next page

Let the frictional force between the blocks be F_{AB}.

For the upper block, A:

Since the two blocks will move with different accelerations, consider forces for each block separately, starting with block A.

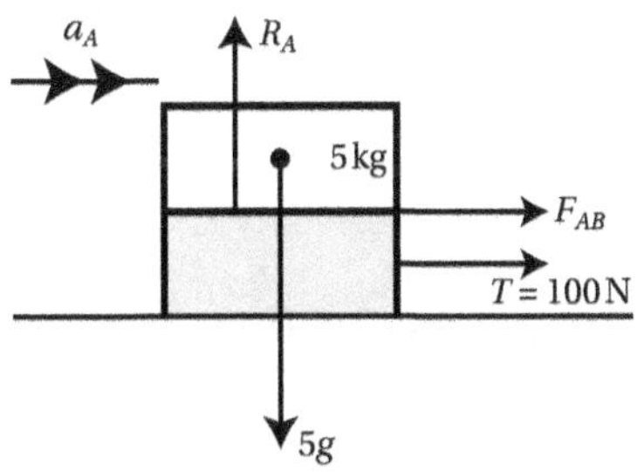

$\text{R}(\uparrow): R_A = 5g \qquad (1)$

Limiting friction: $\mu R_A = 0.75g = 7.35$ N

You can now find the limiting friction acting on A.

$\text{R}(\rightarrow): F = ma \qquad (2)$

$F_{AB} = 5a_A$

$a_A = \frac{7.35}{5} = 1.5 \text{ m s}^{-2}$ (2 s.f.)

Since block A is sliding on block B, the friction is maximal and $F_{AB} = 7.35$ N.

For block B:

Now look at the lower block, considering vertical components first. According to Newton's third law, the normal reaction force R_A acts downwards on block B.

$\text{R}(\uparrow): R_{A+B} = 8g + R_A$

Substituting for R_A from (1):

$R_{A+B} = 13g$

Limiting friction: $\mu R_{A+B} = 3.9g = 38.22$ N

$\text{R}(\rightarrow): F = ma$

$100 - F_{AB} - F_{BS} = 8a_B$

Since the system is moving, friction with the surface is also limiting, so $F_{BS} = 38.22$ N

Substituting F_{AB} from (2):

You can again use the fact that the block is moving to infer that friction is at the limiting level.

$a_B = \frac{100 - 38.22 - 7.35}{8} = 6.8 \text{ m s}^{-2}$ (2 s.f.)

It may be necessary to resolve all driving forces before you can calculate the direction of the frictional force.

WORKED EXAMPLE 18.6

A particle of mass 2.3 kg rests on a rough horizontal surface, with coefficient of friction between the particle and the surface 0.25. The unit vectors **i** and **j** are both in the horizontal plane.

Two horizontal forces act on the particle: $\mathbf{F}_1 = (3\mathbf{j} - 2\mathbf{i})$ N and $\mathbf{F}_2 = 6\mathbf{i}$ N.

a The frictional force $\mathbf{F} = (a\mathbf{i} + b\mathbf{j})$ N. Find a and b.

b A third force $\mathbf{F}_3 = -13\mathbf{i}$ N is applied. Find the new frictional force in the form $\mathbf{F} = (p\mathbf{i} + q\mathbf{j})$ N.

a $\mathbf{F}_1 + \mathbf{F}_2 = (4\mathbf{i} + 3\mathbf{j})$ N — Find the resultant driving force.

Magnitude of the resultant driving force is

$\sqrt{4^2 + 3^2} = 5$ N

Normal reaction equals weight since there is no vertical movement: — Calculate the magnitude of maximal friction and compare to the driving force.

$R = 2.3g = 22.54$ N

Limiting friction $\mu R = 5.635 > 5$

The driving force is less than maximal friction so the particle will not move and friction will exactly counter the driving force. — The driving force is less than maximal friction, so the forces will be in equilibrium.

$\therefore \mathbf{F} = -(4\mathbf{i} + 3\mathbf{j})$

b $\mathbf{F}_1 + \mathbf{F}_2 + \mathbf{F}_3 = (3\mathbf{j} - 9\mathbf{i})$ N — Find the overall driving force.

Magnitude of the overall driving force is

$\sqrt{(-9)^2 + 3^2} = 3\sqrt{10}$ N $= 9.49$ N

Limiting friction $5.635 < 9.49$ — Calculate the magnitude of maximal friction (as found in part **a**) and compare to the driving force.

Friction is limiting.

The friction force has magnitude 5.635 N in the direction of $-(3\mathbf{j} - 9\mathbf{i})$. Hence the friction force is: — The magnitude of the friction force is 5.635 N and its direction is opposite to the direction of the overall driving force (which has magnitude 9.49). You need to change the magnitude of the force without changing its direction, so divide by 9.49 and multiply by 5.635.

$$\frac{5.635}{9.49}(-3\mathbf{j} + 9\mathbf{i}) = (3.44\mathbf{i} - 1.78\mathbf{j})\text{ N}$$

EXERCISE 18B

In this exercise, unless instructed otherwise, use $g = 9.8\,\text{m s}^{-2}$, giving your final answers to an appropriate degree of accuracy.

1 In this question use $g = 10\,\text{m s}^{-2}$, giving your final answers to an appropriate degree of accuracy.

In each problem a block of mass m kg is pulled along a rough horizontal surface by a light horizontal rope with tension T newtons. The acceleration is $a\,\text{m s}^{-2}$. The coefficient of friction is μ.

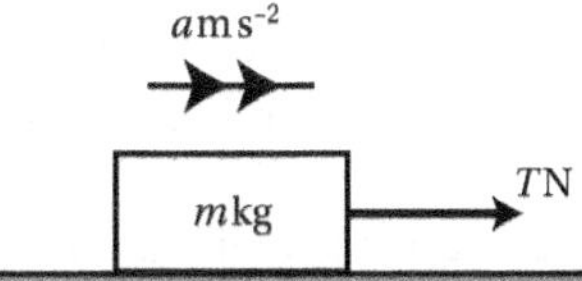

a Find μ when:

i $m = 2, T = 5, a = 0.1$ **ii** $m = 8, T = 2, a = 0.2$

b Find a when:

i $m = 3, T = 30, \mu = 0.5$ **ii** $m = 5, T = 1, \mu = 0.01$

2 In this question use $g = 10\,\text{m s}^{-2}$, giving your final answers to an appropriate degree of accuracy.

A car of mass 1000 kg has a coefficient of friction of 0.9 with the road. What horizontal force is required to move the car?

3 A block of mass 1.5 kg lies in limiting equilibrium on a horizontal surface, with a horizontal force of 6 N applied to it. Find the coefficient of friction between the block and the surface.

4 In this question use $g = 9.81\,\text{m s}^{-2}$, giving your final answers to an appropriate degree of accuracy.

A child pulls a toybox, of mass 3.5 kg, across a rough floor, using a light string tied to the box at one end.

The tension in the string is 28 N and the string remains at an angle of 25° to the horizontal.

If the coefficient of friction between the box and the floor is 0.7, what is the acceleration of the box?

5 In this question use $g = 10\,\text{m s}^{-2}$, giving your final answers to an appropriate degree of accuracy.

Two particles, A and B, are connected by a light inextensible string.

Particle A, which weighs 8 N, is placed on a rough horizontal surface. The connecting string runs horizontally to a smooth pulley P at the end of the surface, then vertically downward to B, which weighs 5 N.

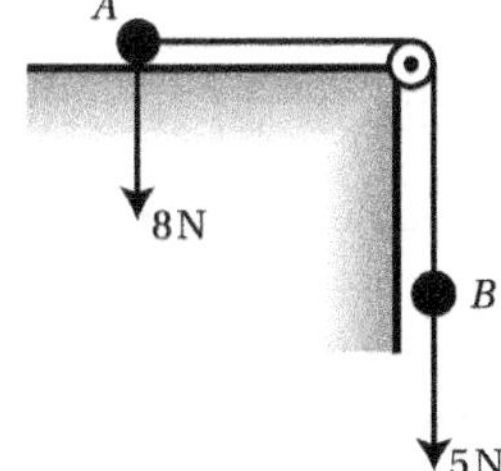

a The system is in limiting equilibrium. Calculate the coefficient of friction between A and the surface.

b A smooth ring with weight 1 N is threaded on to the string so that it lies on particle B. Calculate the downward acceleration of B.

6 Two particles, P and Q, are connected by a taut, light, inextensible string and lie at rest on a plane, with one particle at point A and the other at point B. P weighs 5 N and Q weighs 8 N.

The coefficient of friction between either particle and point A is μ_A and the coefficient of friction between either particle and point B is μ_B.

When P is at A, a horizontal force of 7 N is applied to P acting in direction $\overrightarrow{QP}$, and the system is in limiting equilibrium.

When P is at B, a horizontal force of 6 N is applied to P acting in direction $\overrightarrow{QP}$, and the system is once again in limiting equilibrium.

Calculate μ_A and μ_B.

7 A car of weight 12 000 N, travelling at 25 m s^{-1}, skids to a halt in 50 m taking 4 seconds.

a Assuming a constant braking force, find the deceleration of the car.

b Assume that friction is the only force acting on the car and that $g = 10\text{ m s}^{-2}$. Find the coefficient of friction between the car and the road.

8 A particle of mass 1 kg lies on a rough horizontal surface, with the coefficient of friction between surface and particle equal to 0.75.

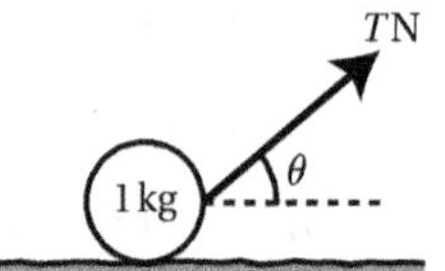

A light inextensible string is attached to the particle, and tension applied with the string at an angle θ above the horizontal where $0° < \theta < 90°$.

a Given the system is in limiting equilibrium, show that the tension T in the string satisfies the equation $T = \frac{5.88}{\sin(\theta + \alpha)}$ for some value α and find α.

b The string will break if $T > 6$ N. Find the range of values for θ for which the particle will be caused to move by tension in the intact string.

c Find the maximum possible acceleration for the particle.

9 In the model for friction used in this section, which of these factors affect the frictional force of a surface acting on an object?

A The contact surface area between the surface and object

B The speed of the object

C The acceleration of the object

D Lubrication between the surface and the object

Section 3: Motion on a slope

If an object is on a slope (sometimes called an inclined plane) it is often convenient to resolve forces parallel or perpendicular to the slope. You use the same technique as for all resolving problems – drawing forces as the hypotenuse of a right-angled triangle with sides parallel and perpendicular to the slope.

R ($\|$) is a standard notation for 'resolve parallel to the slope'.

R ($\perp$) is a standard notation for 'resolve perpendicular to the slope'.

WORKED EXAMPLE 18.7

A block of mass 2 kg lies on a smooth slope inclined at 20° to the horizontal and slides down under gravity. Calculate the component of the gravitational force acting in the direction of the slope. Use $g = 10\ \mathrm{m\,s^{-2}}$, giving your final answers to an appropriate degree of accuracy.

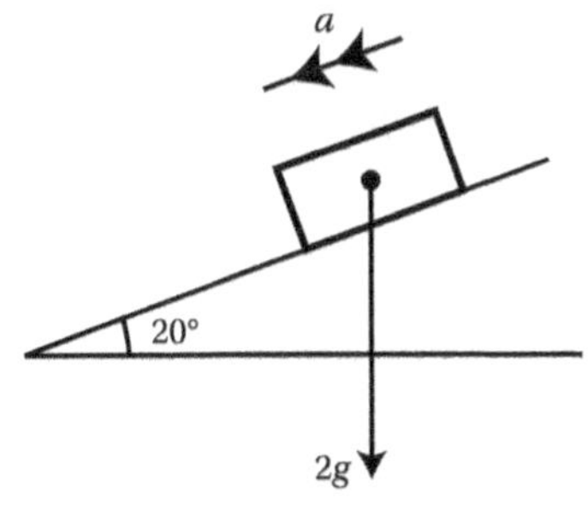

Always draw a diagram.

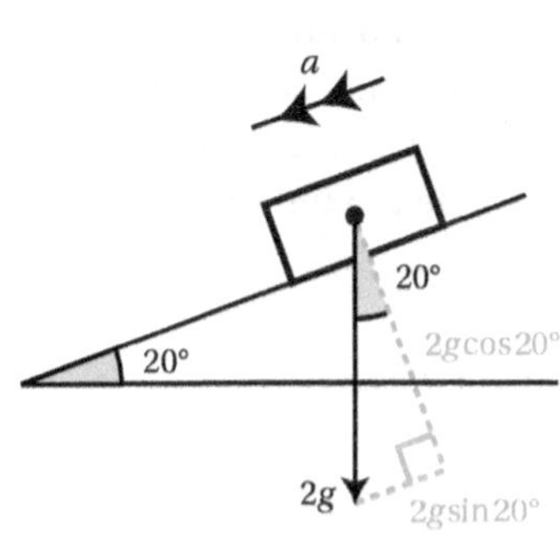

Form a right-angled triangle with the force as hypotenuse and the sides parallel and perpendicular to the slope. Label relevant angles.

$R(||): 2g \sin 20° = 7\ \mathrm{N}$ (1 s.f.)

Gravity comes up so often in inclined plane problems that it is useful to remember these general results.

Key point 18.3

The components of gravity acting on a slope inclined at an angle θ to the horizontal are:

- $R(||): mg \sin\theta$
- $R(\perp): mg \cos\theta$

Tip

A good way of remembering this is to think about what happens when θ is zero. Then there is no component of gravity parallel to the plane, and a component mg perpendicular to the plane.

WORKED EXAMPLE 18.8

A block with mass 3.5 kg lies on a smooth slope inclined at 15° to the horizontal, and is held in equilibrium by a string parallel to the slope, with tension T.
Calculate T and the normal reaction force R. Use $g = 9.8\ \text{m s}^{-2}$, giving your final answers to an appropriate degree of accuracy.

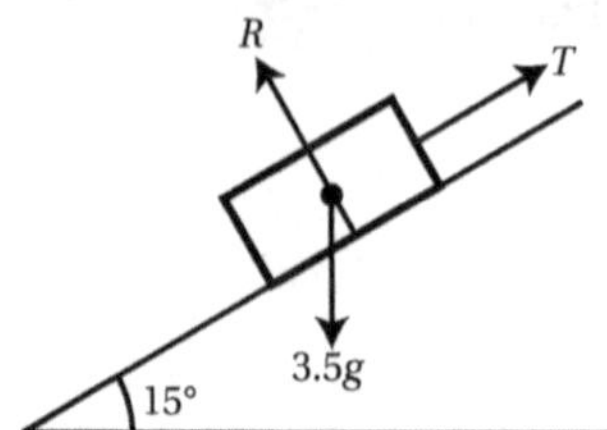

Always draw a diagram.

Resolve along and perpendicular to the slope.

The system is in equilibrium, so resolving in any direction should give a zero result. Since two of the three forces (R and T) are aligned to the slope, this is the preferred option.

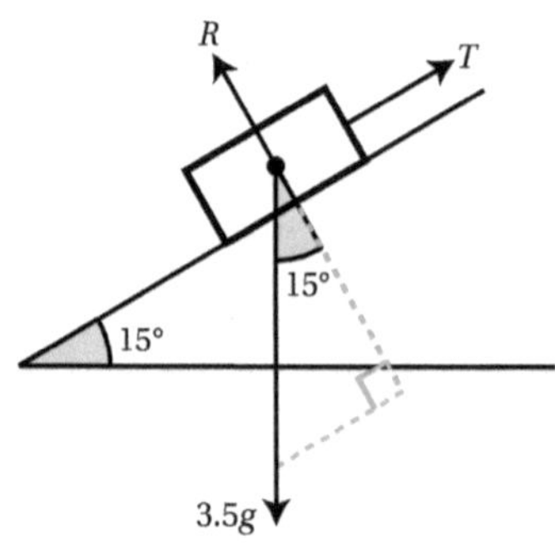

For each force not parallel or perpendicular to the slope, draw a right-angled triangle to take components in these directions.

The block is in equilibrium.

R($\perp$): $R = 3.5g \cos 15° = 33$ N (2 s.f.)

Resolving perpendicular to the slope gives an equation for R.

Think carefully where the 15° angle is.

R($\parallel$): $T = 3.5g \sin 15° = 8.9$ N (2 s.f.)

Resolving parallel to the slope gives an equation for T.

Some problems about motion on a slope also involve friction.
As previously, identify the direction of movement in the absence of friction, and assign friction to act in the opposite direction.

WORKED EXAMPLE 18.9

A block B with mass 5 kg lies on a rough surface inclined at 15° to the horizontal. The coefficient of friction between block and surface is 0.3.

A light inextensible string attached to the block passes over a smooth peg at the highest point of the surface (so that the length between block and peg is parallel to the slope) and is attached at its other end to a particle of mass 3 kg which hangs freely.

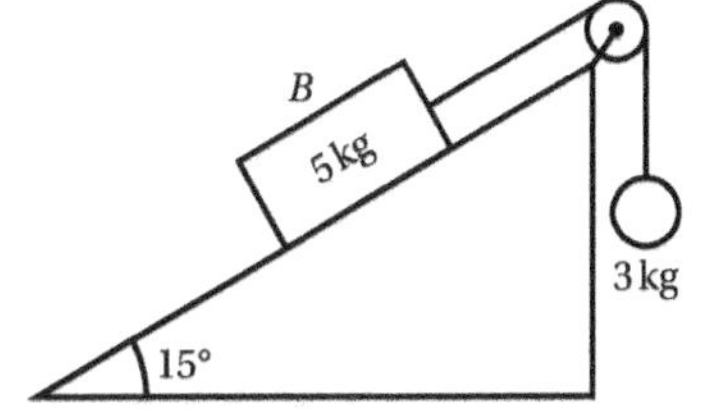

The system is initially held at rest and then released. Use $g = 9.8\text{ m s}^{-2}$, giving your final answers to an appropriate degree of accuracy.

a Calculate the direction and magnitude of the frictional force between B and the surface.

b After the block has travelled 1 metre, the string breaks. Calculate the total distance travelled in the subsequent 10 seconds.

a

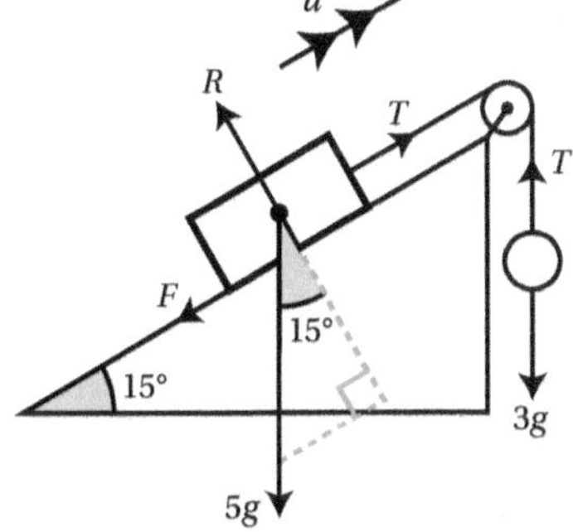

Always start by drawing a force diagram.

For the particle:

R(↓): $F = ma$

$3g - T = 3a$ (1)

For the block:

R(∥): $F = ma$

$T - F - 5g\sin 15° = 5a$ (2)

Resolve separately for each component of the system, in their respective directions of movement.

Adding the two calculations will eliminate the tension term.

R(⊥): $R = 5g\cos 15°$

$R = 47.3\text{ N}$

Resolve perpendicular to the direction of movement of the object on the slope to find the normal reaction force.

Limiting friction: $\mu R = 14.2\text{ N}$

Calculate maximum possible friction μR.

Continues on next page

(1) + (2): $3g - F - 5g\sin 15° = 8a$

Driving force for motion: $3g - 5g\sin 15° = 16.7\text{ N} > \mu R$

The driving force is greater than maximum frictional force so the system will move.

The system will be at limiting friction with $F = 14\text{ N}$ (2 s.f.) acting downslope.

> Determine whether the force driving movement in the system is sufficient to overcome friction.

b $a = \dfrac{3g - F - 5g\sin 15°}{8} = 0.315\text{ m s}^{-2}$

> Separate the movement into three phases:
>
> 1 Moving up the slope with string attached

Phase 1: Moving up the slope with string attached.

Constant acceleration problem:

$u = 0, s = 1, a = 0.315$. Find v.

$v = \sqrt{u^2 + 2as} = 0.794\text{ m s}^{-1}$

Phase 2: Moving up the slope without string.

> 2 Moving up the slope after the string breaks.

In equation (2), T is now zero.

$a = -\dfrac{F + 5g\sin 15°}{5} = -5.38\text{ ms}^{-2}$

Constant acceleration problem:

$u = 0.794, v = 0, a = -5.38$. Find s.

$s = \dfrac{v^2 - u^2}{2a} = 0.0586\text{ m}$

Phase 3: Movement after coming to rest (friction now acting up the slope to prevent sliding)

> 3 Subsequent movement
>
> Note that the frictional force changes direction.

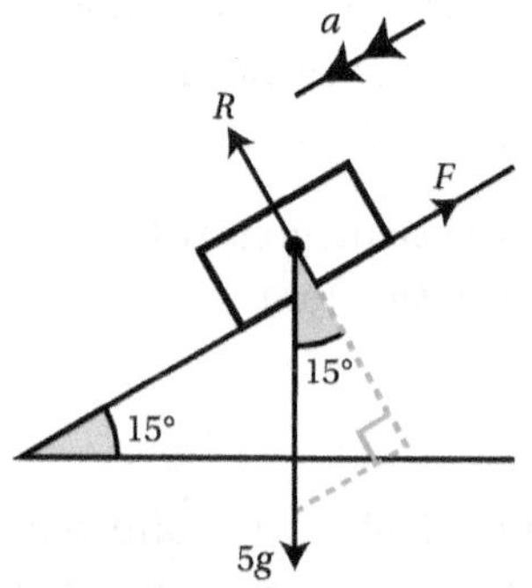

Total force acting down the slope is $5g\sin 15° = 12.7\text{ N} < 14.2\text{ N}$

The driving force is less than maximal friction so the block will remain at rest.

Total distance travelled after the string breaks is therefore equal to 0.059 m (2 s.f.).

EXERCISE 18C

In this exercise, unless instructed otherwise, use $g = 9.8\ \text{m s}^{-2}$, giving your final answers to an appropriate degree of accuracy.

1 A particle of mass m kg is released from rest on a slope inclined at an angle θ to the horizontal, with coefficient of friction between the particle and the slope being μ.

For each case determine the force of friction acting on the particle.

a **i** $m = 5, \theta = 30°, \mu = 0$ **ii** $m = 8, \theta = 45°, \mu = 0$

b **i** $m = 1, \theta = 30°, \mu = 0.8$ **ii** $m = 3, \theta = 20°, \mu = 1$

c **i** $m = 2, \theta = 60°, \mu = 0.2$ **ii** $m = 5, \theta = 45°, \mu = 0.1$

2 A particle P of mass m kg and a particle Q of mass M kg are connected by a light, inextensible string.

Particle P lies on a slope inclined at an angle θ to the horizontal. The string passes from P parallel to the line of greatest slope, and runs over a smooth peg at the top of the slope, then descends vertically to Q.

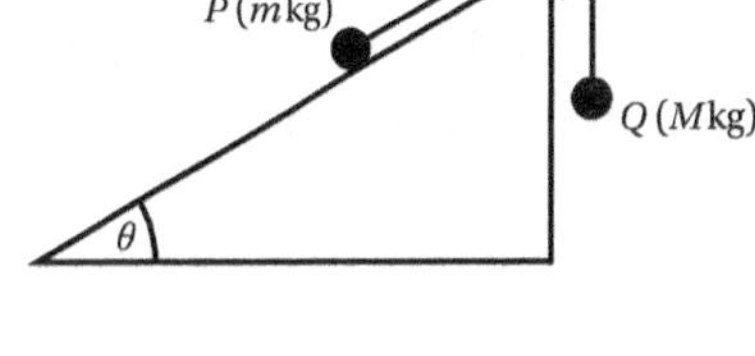

The coefficient of friction between P and the slope is μ.

For each system, find:

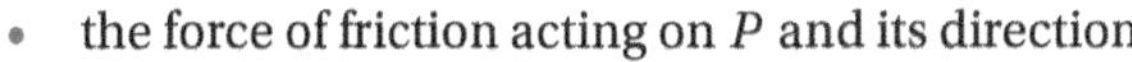

- the force of friction acting on P and its direction
- the acceleration of P when the system is released from rest.

a **i** $m = 5, M = 3, \theta = 30°, \mu = 0$ **ii** $m = 5, M = 3, \theta = 60°, \mu = 0$

b **i** $m = 8, M = 8, \theta = 45°, \mu = 0.2$ **ii** $m = 8, M = 4, \theta = 15°, \mu = 0.3$

c **i** $m = 5, M = 1, \theta = 30°, \mu = 0.7$ **ii** $m = 5, M = 1, \theta = 60°, \mu = 0.7$

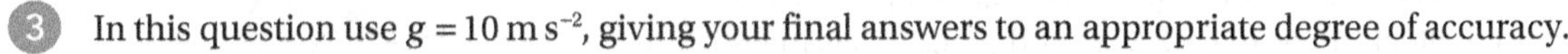

3 In this question use $g = 10\ \text{m s}^{-2}$, giving your final answers to an appropriate degree of accuracy.

A particle of mass 5 kg is projected with velocity $20\ \text{m s}^{-1}$ up the line of steepest slope of a long smooth plane inclined at 30° to the horizontal.

a Calculate the reaction force from the plane acting on the particle.

b Calculate the total distance travelled in the first 2 seconds.

4 A child of mass 30 kg is sliding down a slide at an angle of 40° to the horizontal.

a Find the normal reaction of the slide on the child.

b Find the acceleration if:

i the slide is smooth

ii there is a coefficient of friction of 0.1 between child and slide.

Elevate

See Support Sheet 18 for a further example of limiting equilibrium on a rough inclined plane and for more practice questions.

5 A block of mass 3 kg lies at rest on a rough board, with $\mu = 0.45$.

Find, to the nearest degree, the minimum angle to which the board can be raised before the block begins to slide.

6 A particle P is projected upwards along a line of greatest slope from the foot of a surface inclined at 45° to the horizontal. The initial speed of P is 8 m s^{-1} and the coefficient of friction is 0.3. The particle P comes to instantaneous rest before it reaches the top of the inclined surface.

a Calculate the distance P moves before coming to rest.

b Calculate the time P takes before coming to rest.

c Find the time taken for P to return to its initial position from its highest point.

7 A block B of mass 1 kg lies on a smooth plane inclined at 35° to the horizontal. A light, inextensible string is attached at one end to B, and runs from B up the slope parallel to the line of greatest slope on the plane to a smooth peg, and is attached at the other end to a particle P of mass 2 kg, which hangs vertically below the peg, exactly 1 m from the floor.

The system is released from rest.

a Find the acceleration of B up the slope.

When P hits the floor, the string breaks.

b Assuming the initial distance between the block and the peg is sufficiently great that the block will not reach the peg, find the total distance travelled by the block when it instantaneously passes through its initial position.

8 A block of mass m lies on a flat, rough surface. The coefficient of friction between the block and the surface is μ.

The surface is initially horizontal, and its inclination is gradually increased. When the inclination of the surface exceeds 42° to the horizontal, the block begins to move. Find μ.

9 In this question use $g = 9.81\text{ m s}^{-2}$, giving your final answers to an appropriate degree of accuracy.

Two blocks A and B, connected by a light inextensible string, lie with B above A on the line of greatest slope of a rough plane inclined at 30° to the horizontal. Block A has mass 2.5 kg and block B has mass 1.5 kg. The connecting string is 1 m long, and block A begins 2 m above the foot of the slope.

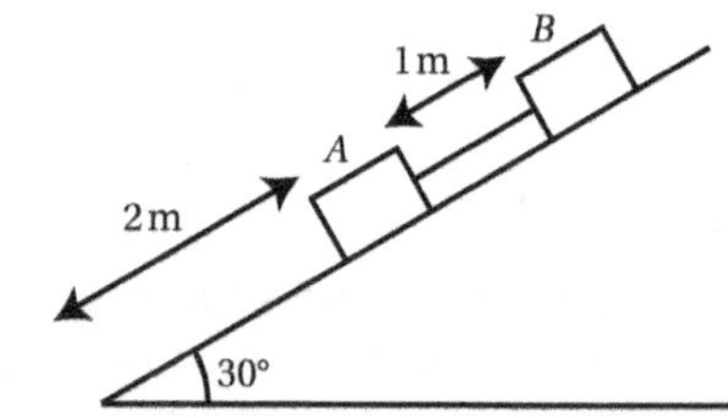

The coefficient of friction between each block and the surface is 0.15.

The system is released from rest at time $t = 0$; when block A reaches the foot of the slope, its motion stops immediately. Find the time at which B impacts with A.

10 A particle P with mass 1.8 kg lies on a rough plane inclined at 30° to the horizontal. A light, inextensible string connects to P then runs parallel with the line of greatest slope of the plane to a smooth peg, then vertically downwards, through a smooth, free ring R with mass 2 kg and then vertically upward to attach to a fixed point S.

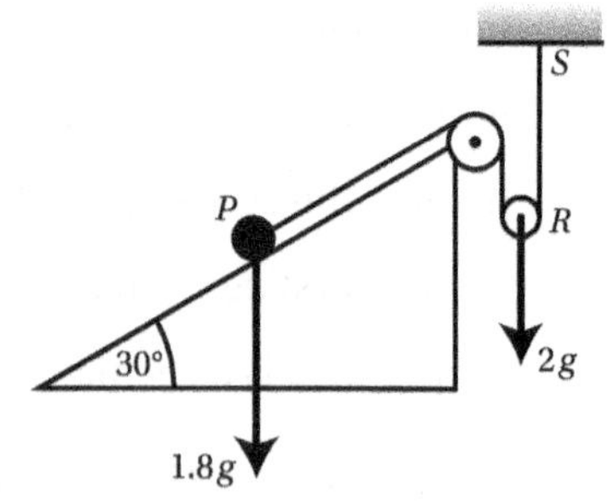

The coefficient of friction between P and the plane is 0.15.

a By resolving forces vertically at R, show that the acceleration of R when the system is released from rest is related to the tension T in the string by $a + T = g$.

b By resolving forces at P, find an equation linking a and T with friction F.

c Find the direction and magnitude of the frictional force.

d Determine whether P will remain stationary, move upslope or move downslope when the system is released from rest

11 A toy hovercraft with mass 0.5 kg is placed on a rough surface inclined at 30° to the horizontal.

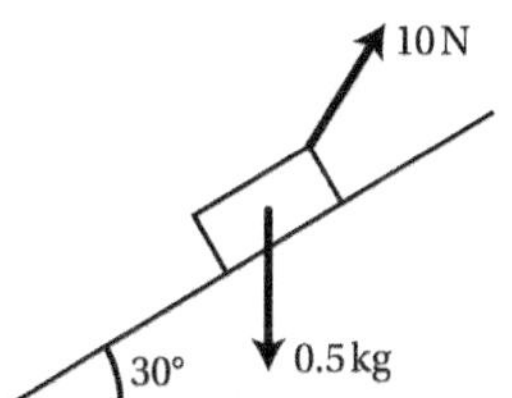

Propulsion of the toy is achieved by directing two small fans, one either side of the toy, set at an angle to provide a driving force for movement. The force produced by the fans is 10 N.

The coefficient of friction between the toy and the surface is 0.1.

a Find the acceleration of the toy up the slope if the fans are set to blow horizontally.

b Find the acceleration of the toy up the slope if the fans are set to blow parallel to the slope.

c The fans are set so that the direction of the driving force is $\theta°$ greater than the slope (i.e. at $(30+\theta)°$ above the horizontal). Find the value of θ that maximises the acceleration.

12 A block B of mass 3 kg lies on a surface inclined at 45° to the horizontal. A light, inextensible string is attached at one end to B, and runs from B up the slope parallel to line of greatest slope on the plane to a smooth peg P. The string passes over the peg, through a smooth ring R of mass 4 kg, and is attached to a wall at W.

Given the angle PRW equals 120° and the system is in equilibrium, determine the possible values for μ, the coefficient of friction between block B and the surface.

13 A particle is projected upwards on a rough slope inclined at an angle θ to the horizontal. There is a coefficient of friction μ between the particle and the slope. The acceleration on the way up is twice the acceleration on the way down. Prove that $\tan\theta = 3\mu$.

Checklist of learning and understanding

- To resolve a force in a given direction, draw a right-angled triangle with the force as the hypotenuse and the other sides of the triangle parallel and perpendicular to the direction of interest.
- When calculating motion on a slope, resolve forces parallel and perpendicular to the slope rather than vertically and horizontally. In particular the components of gravity acting on a slope inclined at an angle θ to the horizontal are:
  - $\text{R}(\parallel): mg\sin\theta$
  - $\text{R}(\perp): mg\cos 6.$
- The maximum or limiting value of friction, F_{max}, between an object and a surface is given by:

$$F_{max} = \mu R$$

  where R is the normal reaction force between the object and the surface and μ is the coefficient of friction.
  - If the resultant of the driving forces is smaller than F_{max}, the object will remain at rest and the friction force, F, will equal the resultant of the driving forces.
  - If the resultant of the driving forces is larger than F_{max}, the object will move and $F = \mu R$.

Mixed practice 18

In this exercise, unless instructed otherwise, use $g = 9.8\text{ m s}^{-2}$, giving your final answers to an appropriate degree of accuracy.

1 A canal boat of mass 1500 kg is being pulled along a straight, smooth canal by a horse. The horse has a rope with tension 1200 N acting at an angle 15° to the canal. Assuming the canal boat can only go in the direction of the canal and no other forces are acting, find the acceleration of the canal boat to three significant figures. Choose from these options.

A 0.207 m s^{-2} **B** 0.773 m s^{-2} **C** 0.800 m s^{-2} **D** 0.828 m s^{-2}

2 The diagram shows two forces acting on a particle.

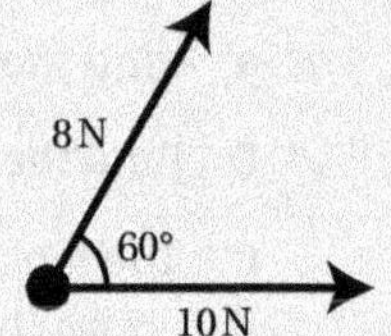

a Find the component of the resultant force in the direction of the 10 N force.

b The direction of the 8 N force is allowed to vary. Find the maximum and minimum value of the magnitude of the resultant force.

3 A block of mass 5 kg slides freely from rest down a smooth slope inclined at 25° to the horizontal. What is the acceleration of the block down the slope?

4 In this question use $g = 10\text{ m s}^{-2}$, giving your final answers to an appropriate degree of accuracy.

A particle is projected with speed 8 m s^{-1} across a rough horizontal surface and comes to rest 10 m from its starting point.

a Calculate the coefficient of friction between the particle and the surface.

b A second, identical particle is projected across the same surface and comes to rest 20 m from its starting point. Determine its initial speed.

5 A particle of mass 4 kg is suspended in equilibrium by two light strings, *AP* and *BP*. The string *AP* makes an angle of 30° to horizontal and the other string, *BP*, is horizontal, as shown in the diagram.

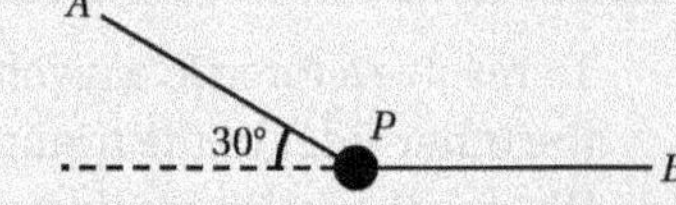

a Draw and label a diagram to show the forces acting on the particle.

b Show that the tension in the string *AP* is 78.4 N.

c Find the tension in the horizontal string *BP*.

[© AQA 2008]

6 A box, of mass 3 kg, is placed on a rough slope inclined at an angle of 40° to the horizontal. It is released from rest and slides down the slope.

a Draw a diagram to show the forces acting on the box.

b Find the magnitude of the normal reaction force acting on the box.

c The coefficient of friction between the box and the slope is 0.2.
Find the magnitude of the friction force acting on the box.

d Find the acceleration of the box.

e State an assumption that you have made about the forces acting on the box

[© AQA 2013]

7 Three forces are in equilibrium in a vertical plane, as shown in the diagram. There is a vertical force of magnitude 40 N and a horizontal force of magnitude 60 N. The third force has magnitude F newtons and acts at an angle θ above the horizontal.

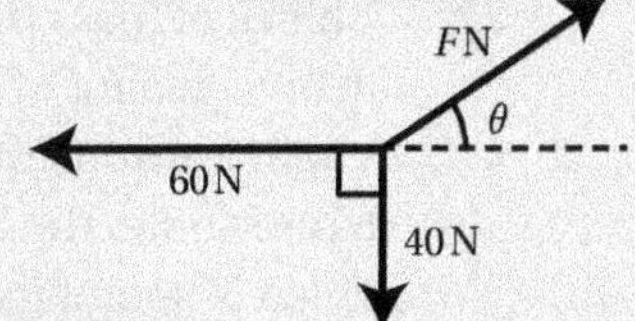

a Find F.

b Find θ.

[© AQA 2014]

8 A particle of mass 3 kg is on a smooth slope inclined at 60° to the horizontal. The particle is held at rest by a force of T newtons parallel to the slope, as shown in the diagram.

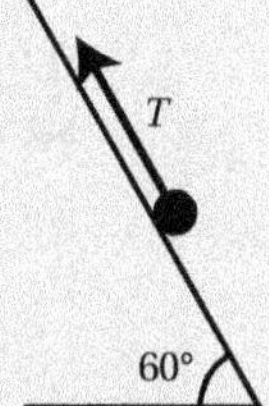

a Draw a diagram to show all the forces acting on the particle.

b Show that the magnitude of the normal reaction acting on the particle is 14.7 newtons.

c Find T.

[© AQA 2010]

9 A box of mass 4 kg is held at rest on a plane inclined at an angle of 40° to the horizontal. The box is then released and slides down the plane.

a A simple model assumes that the only forces acting on the box are its weight and the normal reaction from the plane. Show that, according to this simple model, the acceleration of the box would be $6.30\ \text{m s}^{-2}$, correct to three significant figures.

b In fact, the box moves down the plane with constant acceleration and travels 0.9 metres in 0.6 seconds By using this information, find the acceleration of the box.

c Explain why the answer to part **b** is less than the answer to part **a**.

[© AQA 2009]

10 A block, of mass 4 kg, is made to move in a straight line on a rough horizontal surface by a horizontal force of 50 newtons, as shown in the diagram.

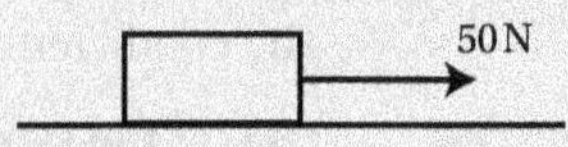

Assume that there is no air resistance acting on the block.

a Draw a diagram to show all the forces acting on the block.

b Find the magnitude of the normal reaction force acting on the block.

c The acceleration of the block is $3\ \text{m s}^{-2}$. Find the magnitude of the friction force acting on the block.

d Find the coefficient of friction between the block and the surface.

e Explain how and why your answer to **d** would change if you assumed that air resistance did act on the block.

[© AQA 2012]

11 A block is projected up a rough slope that makes an angle 30° with the horizontal. The coefficient of friction is $\frac{\sqrt{3}}{2}$. Find the magnitude of the acceleration of the block during its motion.

Choose from these options.

A $\frac{g}{4}$ **B** $\frac{\sqrt{3}g}{4}$ **C** $\frac{5g}{4}$ **D** $\frac{3\sqrt{3}g}{4}$

12 A block of mass 8 kg is held at rest on a rough horizontal surface. The coefficient of friction between the block and the surface is 0.3. A light inextensible string which passes over a smooth peg is attached at one and to the block and at the other end to a particle of mass 5 kg, which is hanging at rest. The system is released from rest. Find:

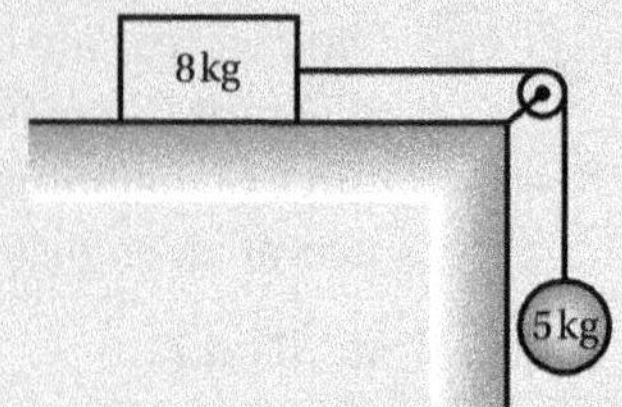

a the magnitude of the frictional force acting on the block

b the acceleration of the block and the tension in the string.

13 In this question use $g = 10\text{ m s}^{-2}$, giving your final answers to an appropriate degree of accuracy.

A block with mass 2 kg is pulled from rest up a smooth slope inclined at 20° to the horizontal by a string with tension T, maintained at an angle of 30° to the horizontal.

a After 5 seconds, the block has moved 1 m along the slope. Calculate T.

b The block is allowed to come to rest again, then the tension is increased so that the block is about to lift off the slope. Calculate the minimum tension needed to achieve this.

14 A particle P, mass 1 kg, is projected up a line of greatest slope of a rough plane inclined at 30° to the horizontal with initial velocity 8 m s^{-1}. The particle comes to instantaneous rest on the plane after travelling 4 m.

a Calculate the frictional force acting on P, and the coefficient of friction between P and the plane.

b Determine the acceleration of P down the plane subsequent to the instant of rest.

15 A particle of mass 2.2 kg is projected with velocity 5 m s^{-1} up the line of steepest slope of a plane inclined at 20° to the horizontal.

a If the plane is smooth, determine the velocity of the particle after 10 seconds, assuming it does not reach the end of the plane.

b If the plane is rough, with coefficient of friction 0.3 between particle and plane, calculate the maximum vertical height above its starting point that the particle will reach.

16 A particle is projected with speed 5 m s^{-1} down the line of steepest slope of an inclined rough plane. The coefficient of friction between the particle and the plane is 1.

It takes the same time T seconds for a particle of mass 2 kg to travel 1 m from its start position if the plane is inclined at angle 2θ to the horizontal as it takes for a particle of mass 1 kg to travel 1 m from its start position if the plane is inclined at angle θ to the horizontal.

Show that $(4\cos\theta - 1)(\cos\theta - \sin\theta) = 2$

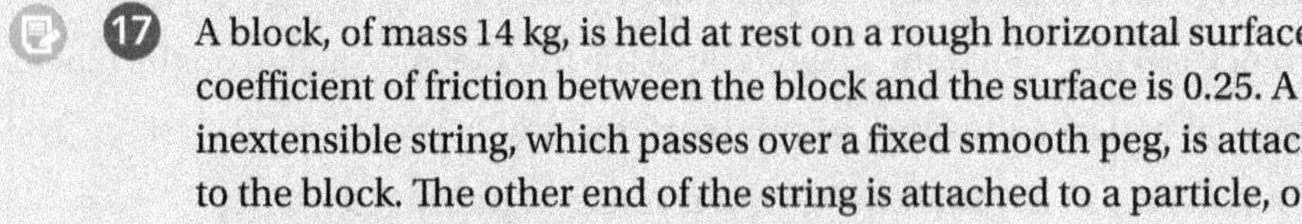

17 A block, of mass 14 kg, is held at rest on a rough horizontal surface. The coefficient of friction between the block and the surface is 0.25. A light inextensible string, which passes over a fixed smooth peg, is attached to the block. The other end of the string is attached to a particle, of mass 6 kg, which is hanging at rest.

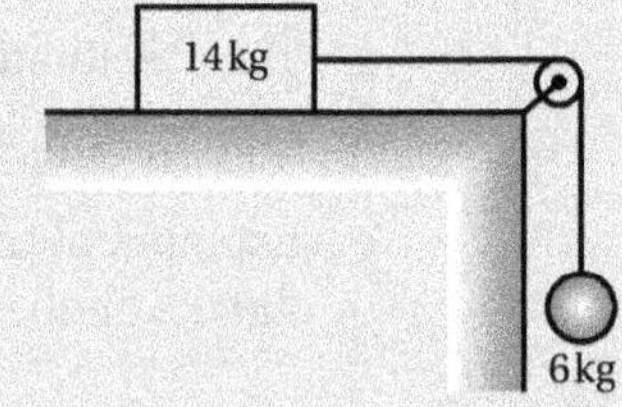

The block is released and begins to accelerate.

a Find the magnitude of the friction force acting on the block.

b By forming two equations of motion, one for the block and one for the particle, show that the magnitude of the acceleration of the block and the particle is 1.225 m s^{-2}.

c Find the tension in the string.

d When the block is released, it is 0.8 metres from the peg. Find the speed of the block when it hits the peg.

e When the block reaches the peg, the string breaks and the particle falls a further 0.5 metres to the ground. Find the speed of the particle when it hits the ground.

[© AQA 2009]

18 A cyclist freewheels, with constant acceleration, in a straight line down a slope. As the cyclist moves 50 metres, his speed increases from 4 m s^{-1} to 10 m s^{-1}.

a **i** Find the acceleration of the cyclist.

ii Find the time that it takes the cyclist to travel this distance.

b The cyclist has a mass of 70 kg. Calculate the magnitude of the resultant force acting on the cyclist.

c The slope is inclined at an angle α to the horizontal.

i Find α if it is assumed that there is no resistance force acting on the cyclist.

ii Find α if it is assumed that there is a constant resistance force of magnitude 30 newtons acting on the cyclist.

d Make a criticism of the assumption described in part **c ii**.

[© AQA 2012]

19 A child pulls a sledge, of mass 8 kg, along a rough horizontal surface, using a light rope. The coefficient of friction between the sledge and the surface is 0.3. The tension in the rope is T newtons. The rope is kept at an angle of 30° to the horizontal, as shown in the diagram.

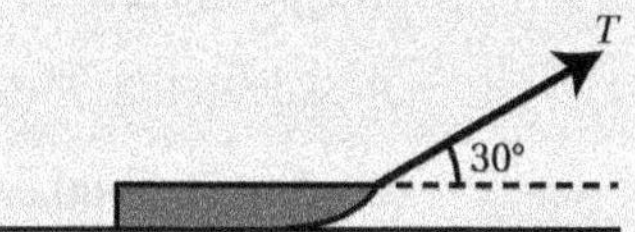

Model the sledge as a particle.

a Draw a diagram to show all the forces acting on the sledge.

b Find the magnitude of the normal reaction force acting on the sledge, in terms of T.

c Given that the sledge accelerates at 0.05 m s^{-2}, find T.

[© AQA 2012]

20 Two right-angled triangular prisms of equal height, with angle of greatest slope 30° and 45° are positioned as shown, with a smooth peg P between the two highest points.

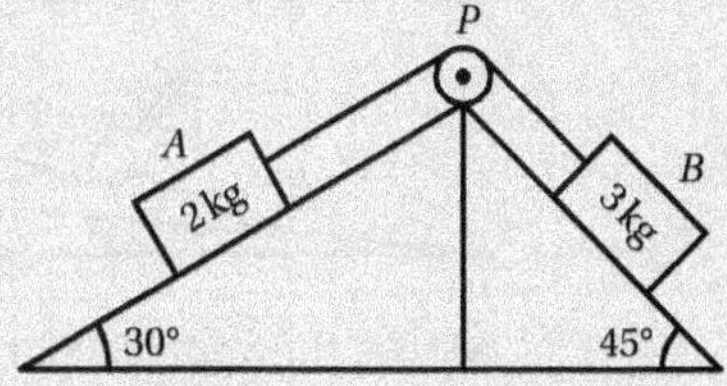

Block A, with mass 2 kg, is placed on the 30° slope and block B, with mass 3 kg, is placed on the 45° slope. The two blocks are connected by a light, inextensible string which runs parallel to the line of greatest slope of each prism and passes over the smooth peg.

The coefficient of friction between block A and the 30° slope surface is μ and the coefficient of friction between block B and the 45° slope surface is 2μ.

At time $t = 0$, block A is projected down the 30° slope with speed 9 m s^{-1}.

a Calculate the acceleration of block A in terms of μ.

It is determined that $\mu = 0.1$.

b Calculate whether the blocks will return to their original positions, and if so, the time at which this will occur.

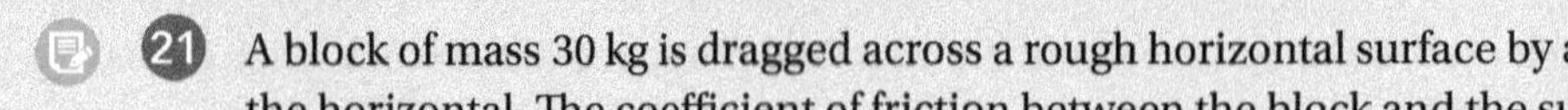

21 A block of mass 30 kg is dragged across a rough horizontal surface by a rope that is at an angle of 20° to the horizontal. The coefficient of friction between the block and the surface is 0.4.

a The tension in the rope is 150 newtons.

i Draw a diagram to show the forces acting on the block as it moves.

ii Show that the magnitude of the normal reaction force on the block is 243 newtons, correct to three significant figures.

iii Find the magnitude of the friction force acting on the block.

iv Find the acceleration of the block.

b When the block is moving, the tension is reduced so that the block moves at a constant speed, with the angle between the rope and the horizontal unchanged. Find the tension in the rope when the block is moving at this constant speed.

c If the block were made to move at a greater **constant** speed, again with the angle between the rope and the horizontal unchanged, how would the tension in this case compare to the tension found in part **b**?

[© AQA 2013]

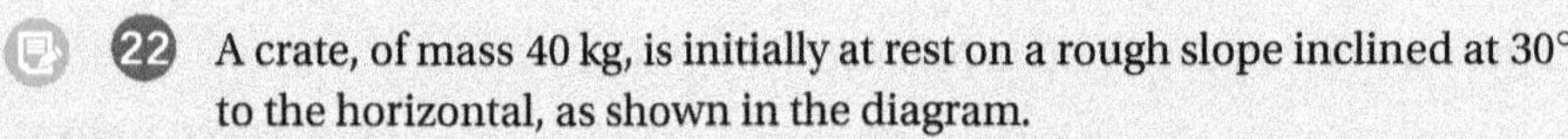

22 A crate, of mass 40 kg, is initially at rest on a rough slope inclined at 30° to the horizontal, as shown in the diagram.

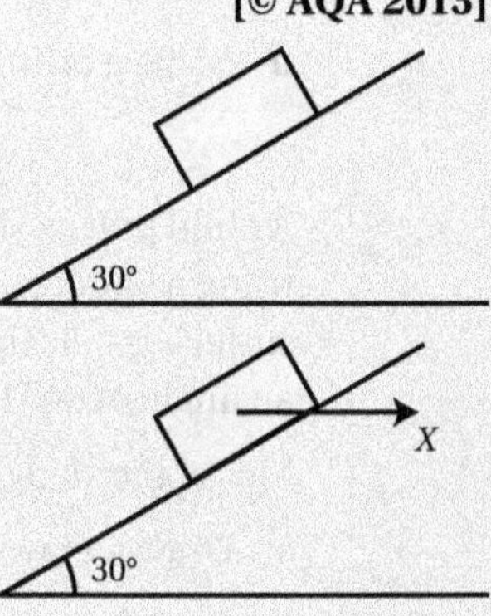

The coefficient of friction between the crate and the slope is μ.

a Given that the crate is on the point of slipping down the slope, find μ.

b A horizontal force of magnitude X newtons is now applied to the crate, as shown in the diagram.

i Find the normal reaction on the crate in terms of X.

ii Given that the crate accelerates up the slope at $0.2\ \text{m s}^{-2}$, find X.

[© AQA 2014]

23 A van, of mass 2000 kg, is towed up a slope inclined at 5° to the horizontal. The tow rope is at an angle of 12° to the slope. The motion of the van is opposed by a resistance force of magnitude 500 newtons. The van is accelerating up the slope at $0.6\ \text{m s}^{-2}$.

Model the van as a particle.

a Draw a diagram to show the forces acting on the van.

b Show that the tension in the tow rope is 3480 newtons, correct to three significant figures.

[© AQA 2011]

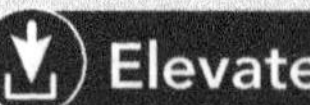

See Extension Sheet 18 for a selection of more challenging problems.

19 Moments

In this chapter you will learn how to:

- find the turning effect of a force
- work with uniform rods and laminas
- understand and use rotational equilibrium.

Before you start...

Student Book 1, Chapter 18	You should be able to recognise types of force acting on a particle.	1 A particle is pulled across a smooth horizontal table by a string that is parallel to the table. Draw a diagram and label all the forces acting on the particle.
Student Book 1, Chapter 18	You should understand when a particle is in equilibrium.	2 Three forces act on a particle as shown. 5N F 12N The particle is in equilibrium. Find the magnitude of F.

Modelling rotating systems

Until now you have used the particle model to analyse forces and motion. However, there are situations in which this is not appropriate.

Consider, for example, closing a door.

A force is applied to push the door closed, but when the door moves it does not do so in a straight line. Instead it rotates about the hinge.

In this chapter you will find out how situations like this can be modelled.

Section 1: The turning effect of a force

From your experience of closing doors you probably know that it is easier to push if your hand is further away from the hinge. This is because the **moment** – the turning effect of a force – depends upon both the force applied and the distance away from the pivot point.

Key point 19.1

The moment of a force F about a point P is:

$$\text{moment} = Fd$$

where d is the perpendicular distance of the line of action of the force from P.

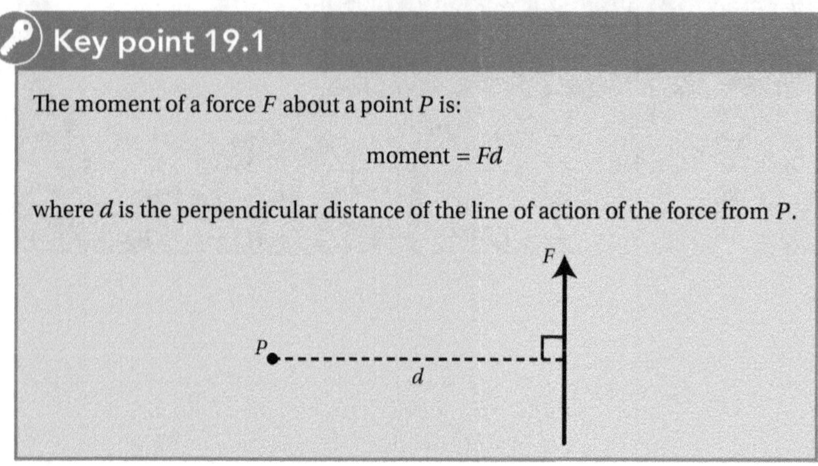

The units of a moment are newton metres (N m).

The moment will cause either clockwise or anticlockwise rotation about a point.

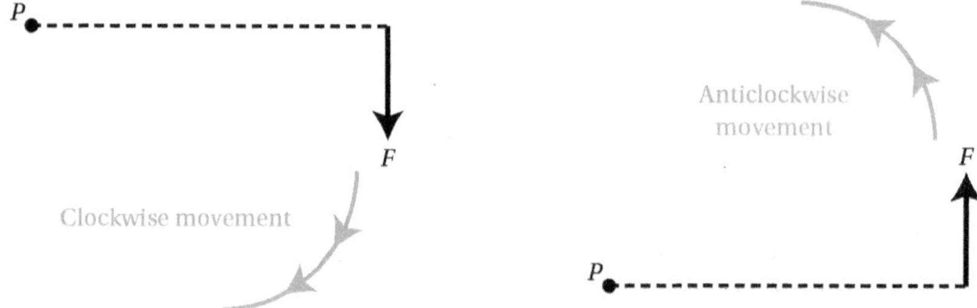

If the line of action of the force acts through P, then the moment about P will be zero (as the perpendicular distance from P is 0 m). So in this case there is no rotational effect from the force.

Many situations involving forces that cause rotation can be modelled using two basic shapes.

- A **uniform lamina**, which is a two-dimensional object usually in the shape of a rectangle. You might use this to model objects such as a door or a book.
- A **uniform rod**, which has just one dimension. You might use this to model a see-saw, snooker cue or a plank.

In both cases **uniform** means that the object has the same density throughout. The key fact you need to know is where the **centre of mass** is for both of these shapes.

Key point 19.2

The centre of mass is the point at which the object's weight acts.

- For a uniform rod, this is at the midpoint.
- For a uniform lamina, this is at the intersection of the diagonals.

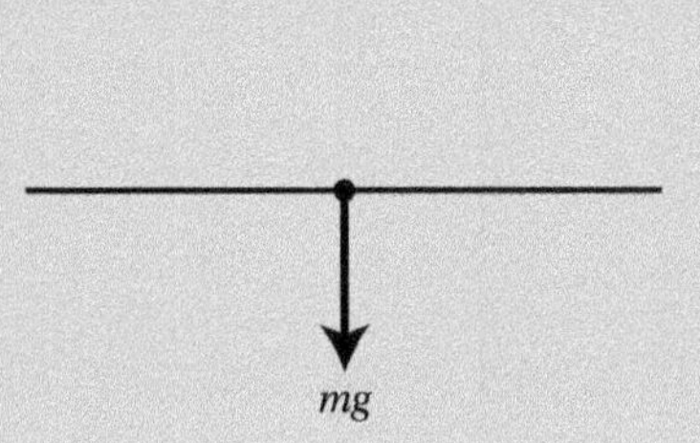

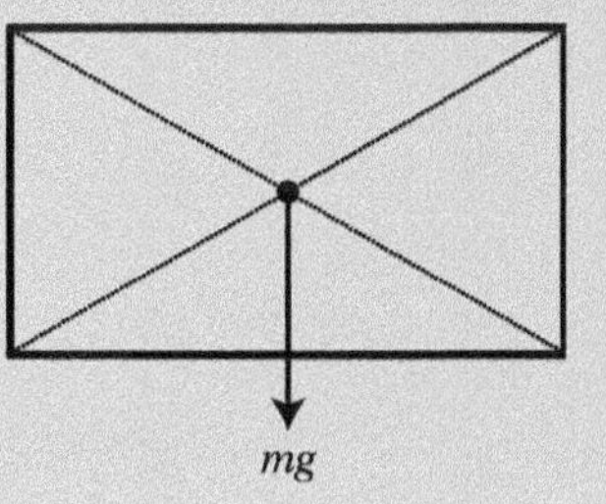

Fast forward

Although this might sound obvious, determining the centre of mass of more complex shapes can be quite difficult. You will see how to do this if you study the Mechanics option of Further Mathematics.

WORKED EXAMPLE 19.1

A uniform rod of length 6 m has weight 5 N.

Find the moment of its weight about the point:

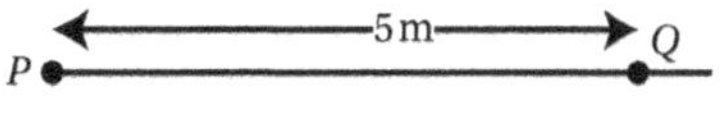

a P **b** Q.

a

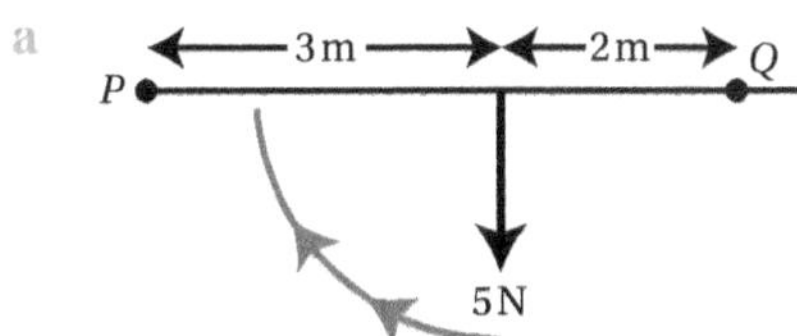

Moment $= Fd = 5 \times 3$

$= 15$ N m clockwise

The perpendicular distance from P to the weight is 3 m. The weight acts at the midpoint. It causes the rod to rotate clockwise about P.

b

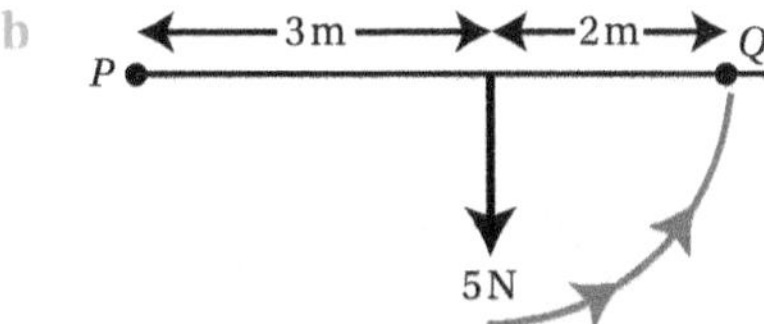

Moment $= Fd = 5 \times 2$

The perpendicular distance from Q to the weight is 2 m.

$= 10$ N m anticlockwise

The weight will cause the rod to rotate anticlockwise about Q.

WORKED EXAMPLE 19.2

A uniform rectangular lamina 0.4 m by 0.6 m has weight 8 N.

The lamina is free to rotate in a vertical plane about P.

Find the moment about P of the:

a weight **b** 12 N force.

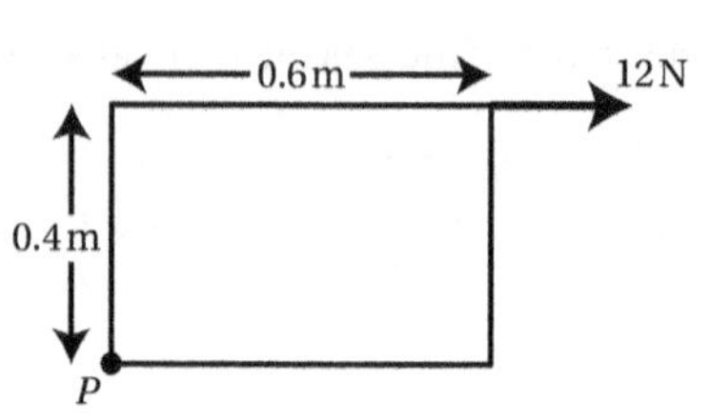

a

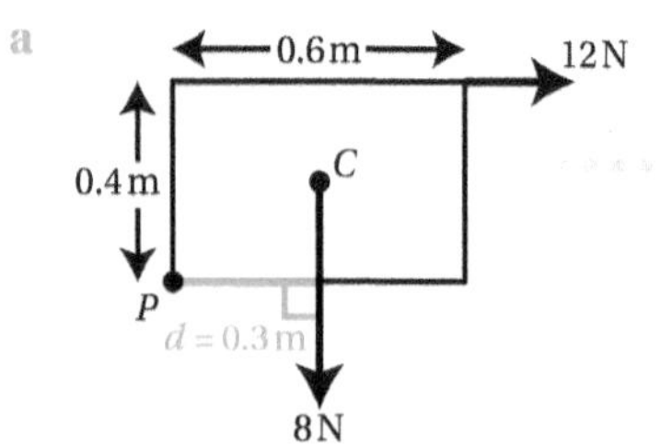

The weight acts at the centre of the lamina.

Moment $= Fd$
$= 8 \times 0.3$
$= 2.4$ N m clockwise

The perpendicular distance from P to the line of action of the weight is $d = 0.3$ m.

The rotation will be clockwise.

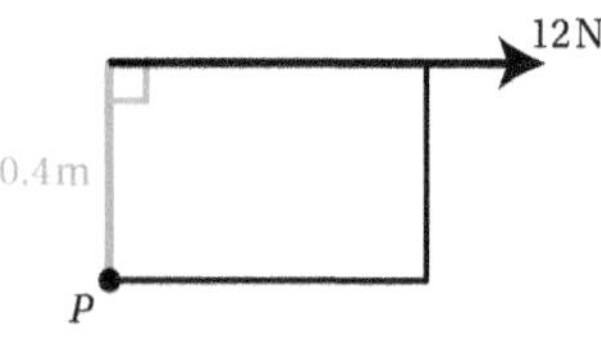

b Moment $= Fd$
$= 12 \times 0.4$
$= 4.8$ N m clockwise

The perpendicular distance from P to the line of action of the 12 N force is $d = 0.4$ m.

The rotation will be clockwise.

In just the same way that when several forces act on a body they combine to give an overall resultant force, the same is true of moments.

Key point 19.3

To find the resultant moment about a point, find the sum of the clockwise and anticlockwise moments separately.

The resultant moment will be the difference between the two sums, in the direction of the larger moment.

Focus on...

Focus on ... Modelling 3 applies moment to the context of levers.

WORKED EXAMPLE 19.3

A uniform rod of length 8 m and weight 10 N is acted on by a 7 N and 3 N force as shown.

Find the resultant moment about the point P.

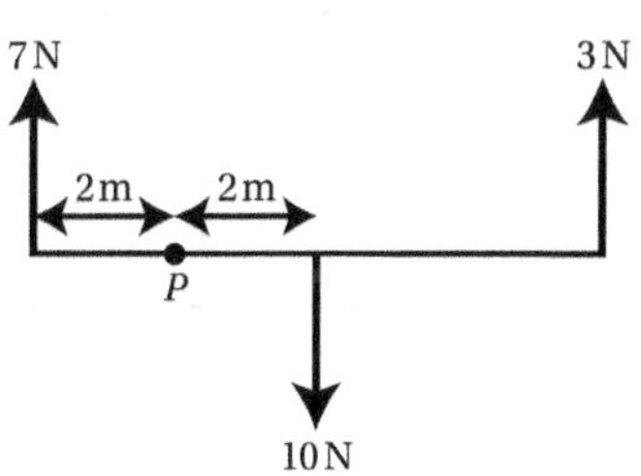

Moment of 7 N force $= 7 \times 2$
$= 14$ N m clockwise

Find the moment of each force in turn.

Moment of weight $= 10 \times 2$
$= 20$ N m clockwise

Moment of 3 N force $= 3 \times 6$
$= 18$ N m anticlockwise

The 3 N force is $2 + 4 = 6$ m from P.

Total clockwise moments $= 14 + 20 = 34$ N m
Total anticlockwise moments $= 18$ N m

Find the sum of the moments that act in the same direction.

Resultant moment about $P = 34 - 18 = 16$ N m clockwise

The sum of the clockwise moments is greater.

EXERCISE 19A

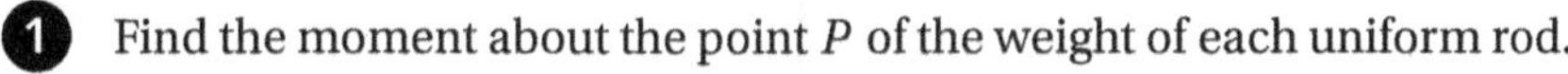

1 Find the moment about the point P of the weight of each uniform rod.

a **i**

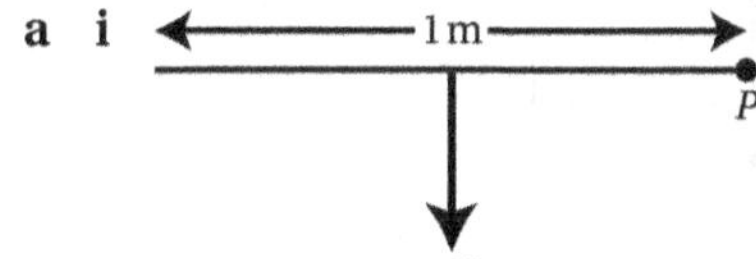

ii

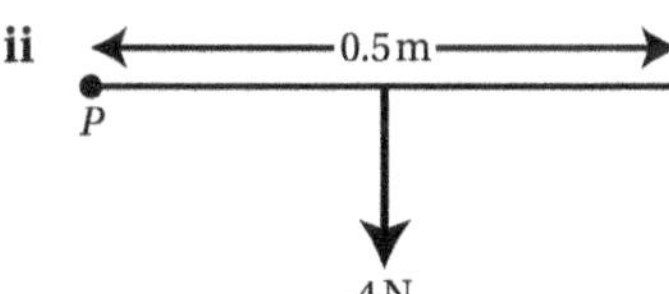

b **i**

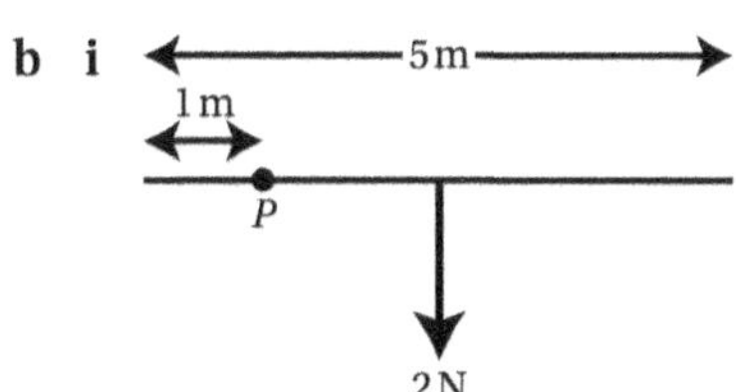

ii

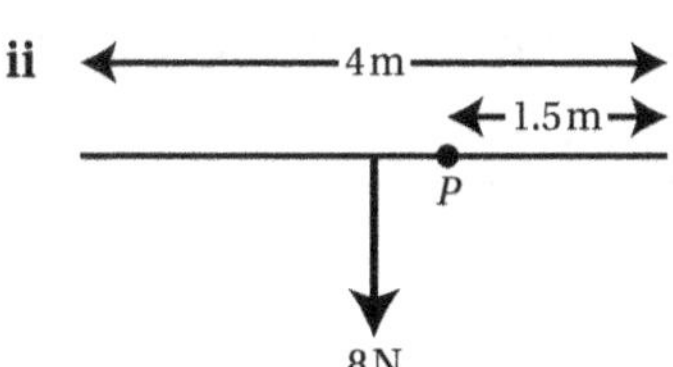

c **i**

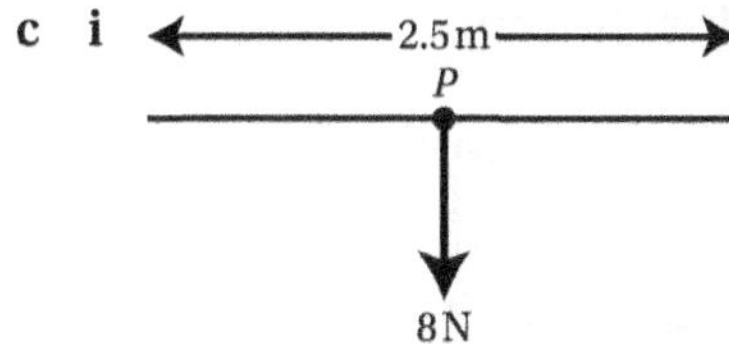

ii

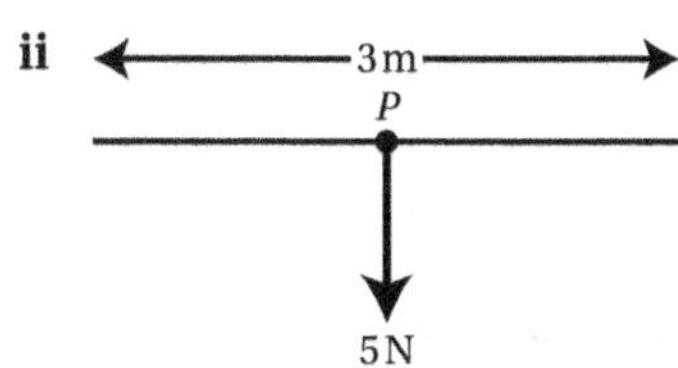

2 Find the moment about the point P of the weight of each uniform lamina.

a **i**

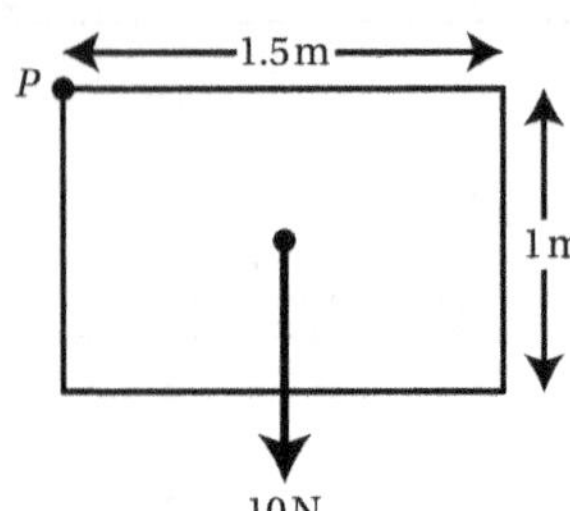

ii

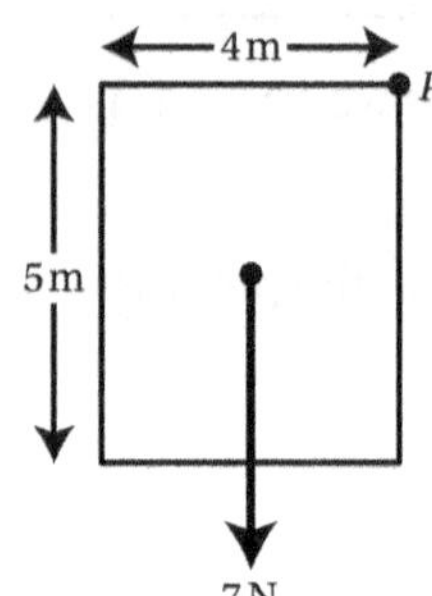

b **i**

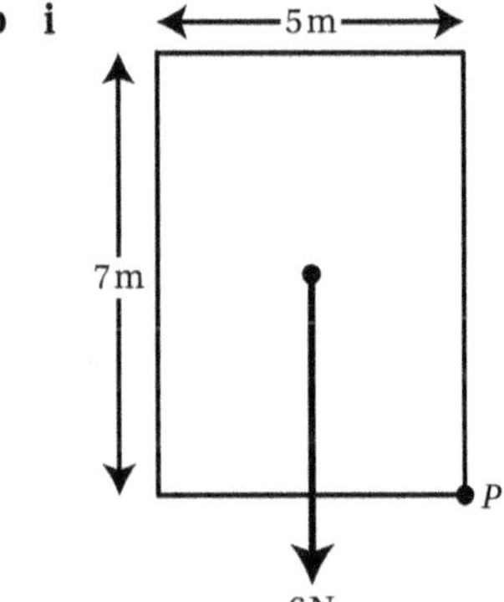

ii

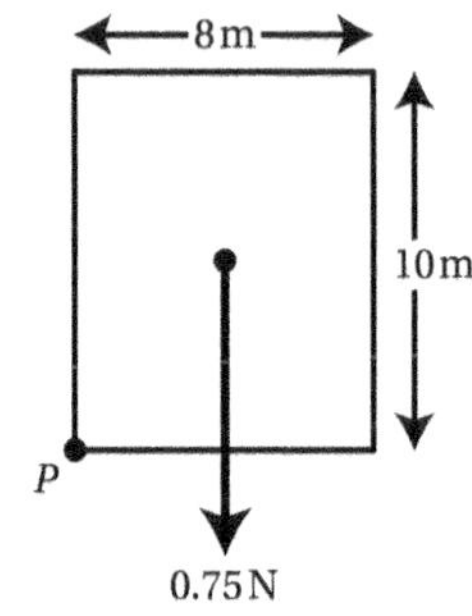

c **i**

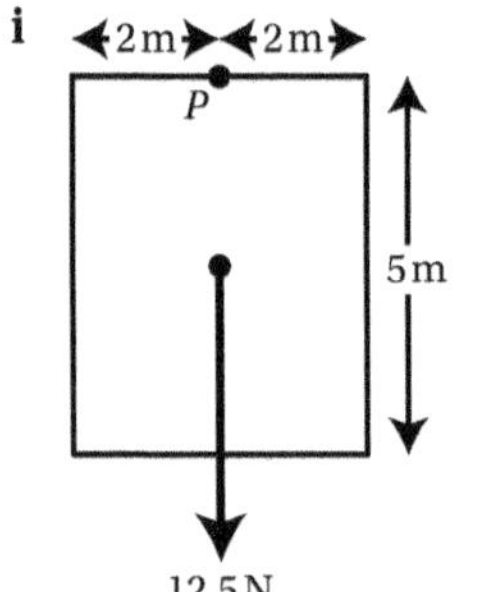

ii

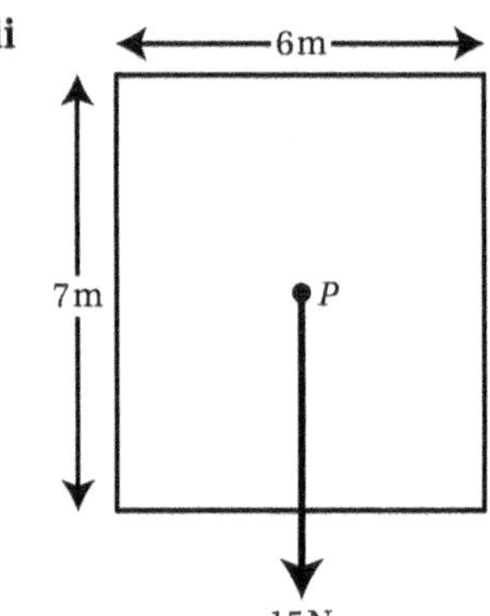

3 A 4 m long uniform rod weighs 50 N.

Find the net moment about the point P in each situation.

a **i** 4m
P

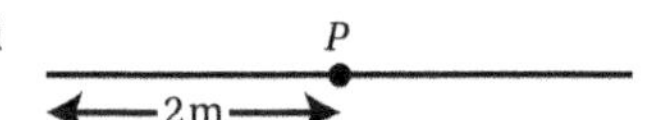

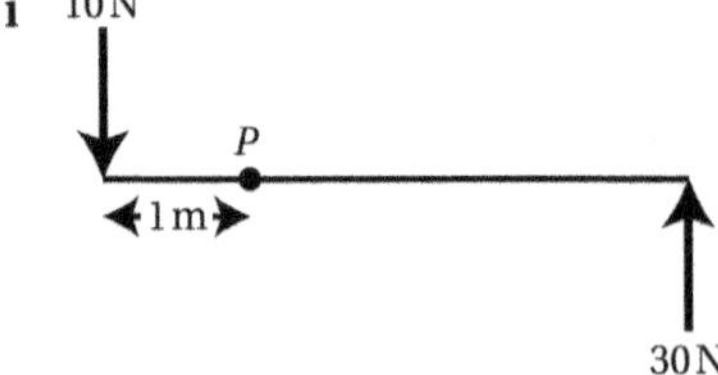

ii

b **i** P
2m

ii

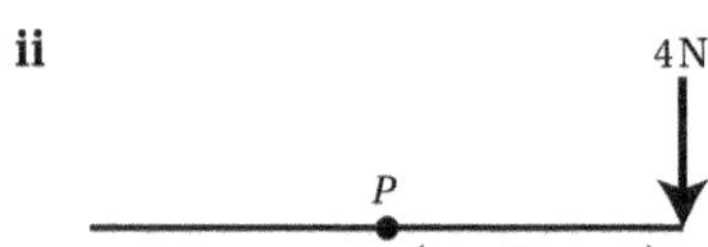

c **i** 10N
P
1m
30N

ii

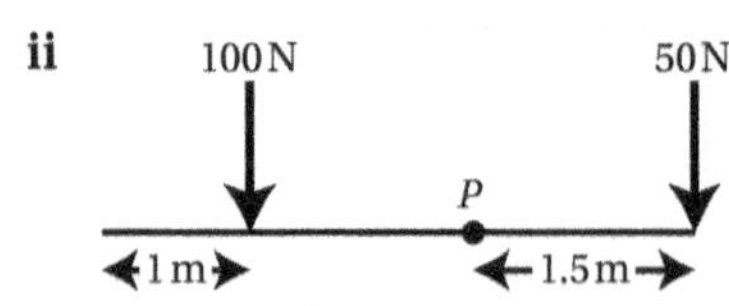

4 A 4 m by 6 m uniform rectangular lamina weighs 150 N.

Find the net moment about the point P in each situation, given that the lamina is in a vertical plane.

a i

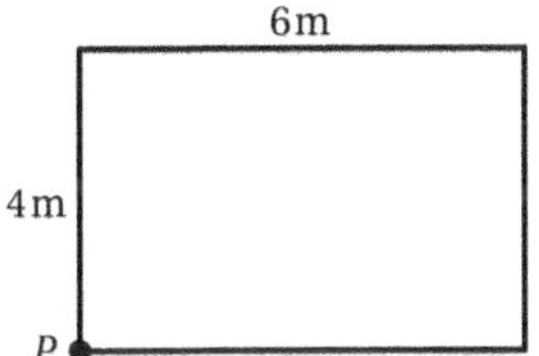

ii

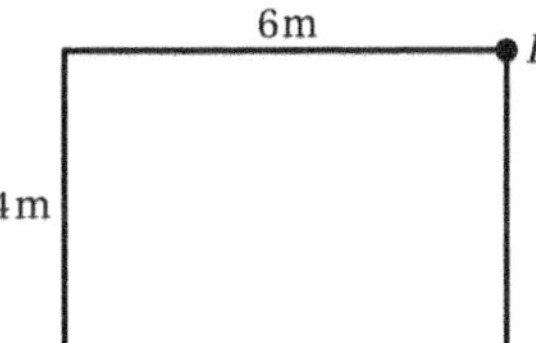

b i

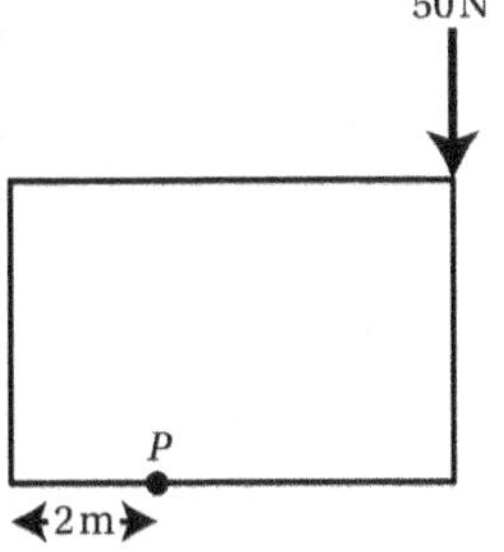

ii

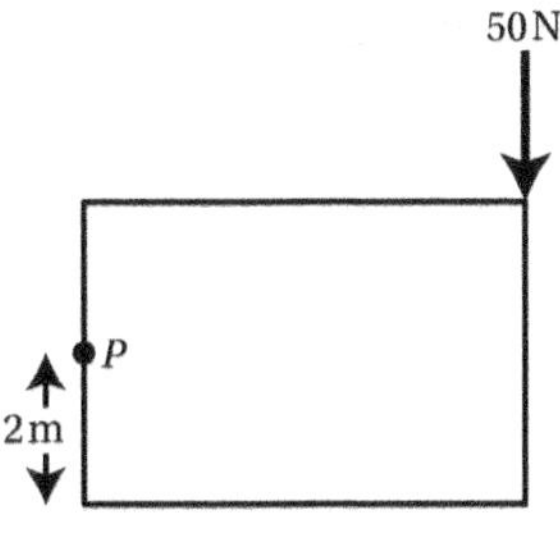

c i

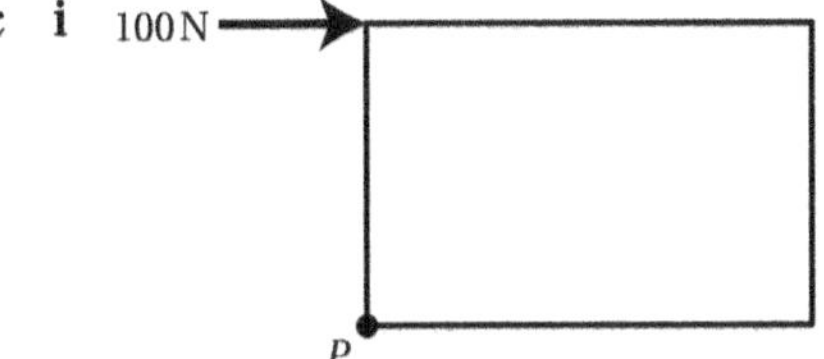

ii

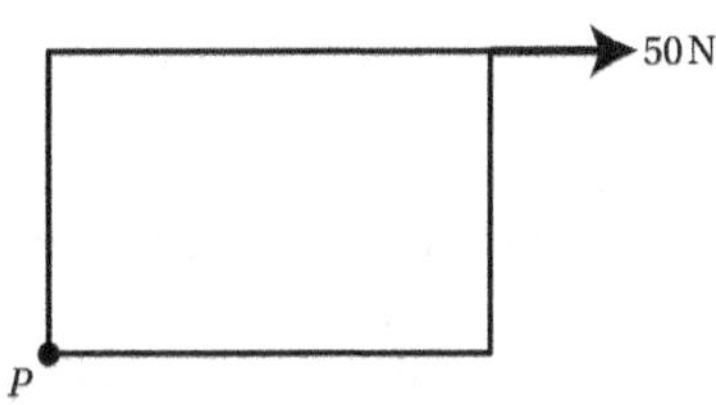

d i

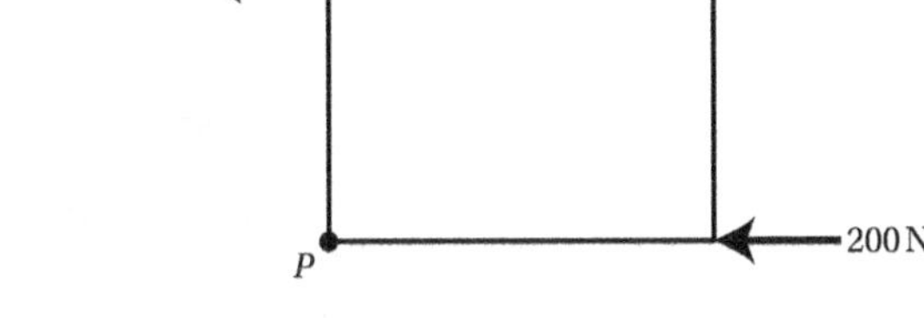

ii

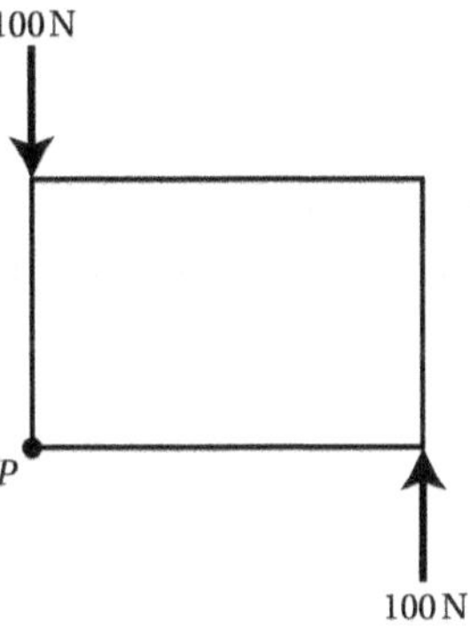

e i

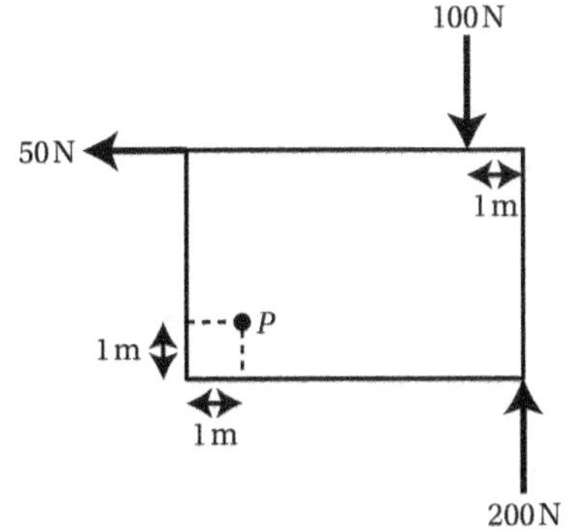

ii

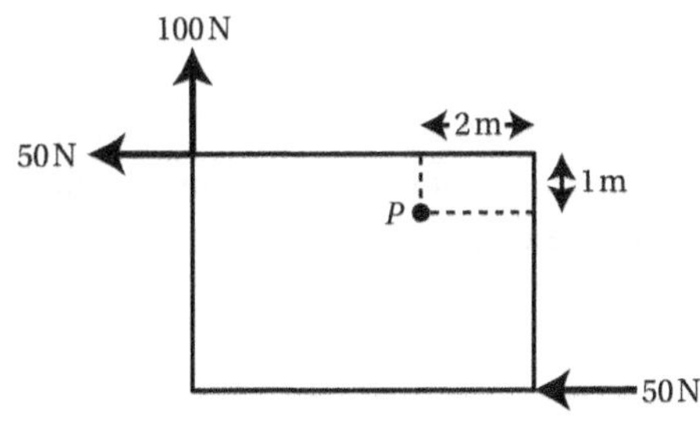

5 A steering wheel is modelled as a ring of diameter 30 cm. The driver applies two clockwise forces of 3 N tangentially at diametrically opposite sides of the steering wheel. Find the net moment about the centre of the wheel.

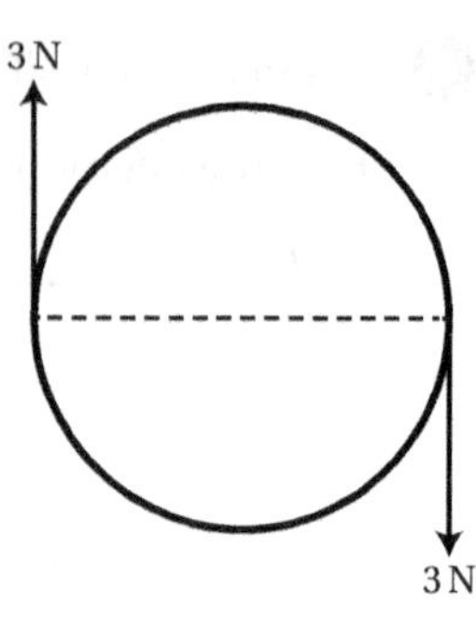

6 Find the resultant moment, including the direction, about the point P in the diagram.

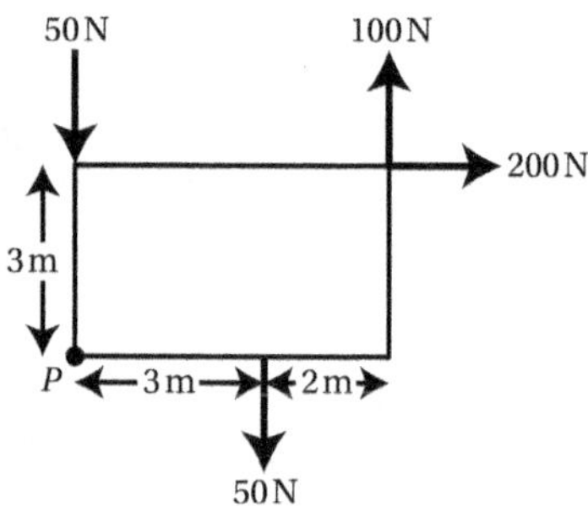

7 In this question use $g = 10$ m s^{-2}, giving your final answer to an appropriate degree of accuracy.

a A diving board is modelled by a uniform plank of length 3 m and mass 100 kg. A diver of mass 70 kg stands at one end of the plank. Find the total moment around the opposite end of the board.

b Explain how you used the fact that the plank is rigid in your calculation in part **a**.

8 The diagram shows a regular hexagon. Find the resultant moment, including the direction, about the centre of the hexagon.

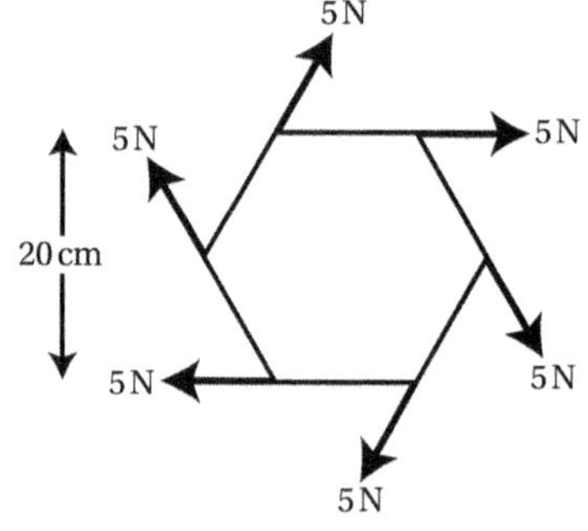

9 The diagram shows a square of side 1 m with four equal forces acting at the corners. Prove that the net moment is the same about any point on the interior of the square.

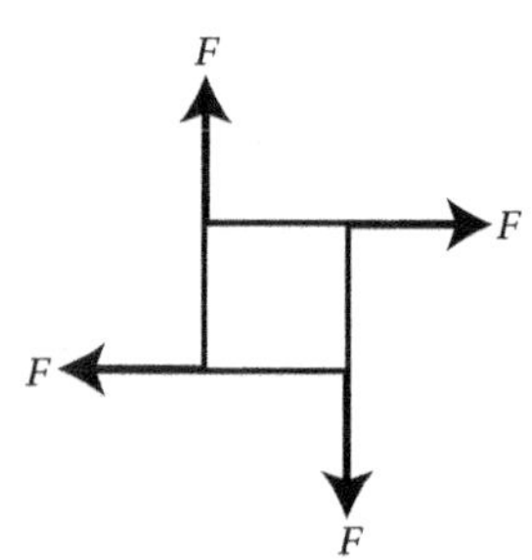

10 In this question use $g = 9.8$ m s^{-2}, giving your final answer to an appropriate degree of accuracy.

A badminton racquet is modelled as a uniform rod of length 30 cm and mass 20 g connected to a square lamina of side length 15 cm and mass 50 g. The racquet is held horizontally at a point, P, 5 cm from the end of the rod. Find the net moment of the racquet about the point P.

Section 2: Equilibrium

To counterbalance an object's weight and maintain it in equilibrium there will either be:

- smooth supports providing a normal reaction force, or
- light strings providing a tension.

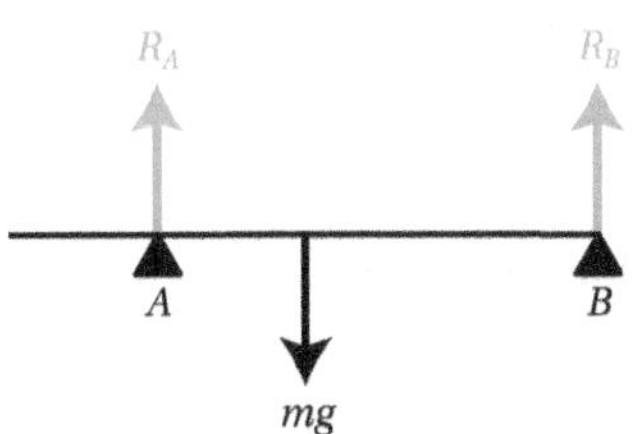

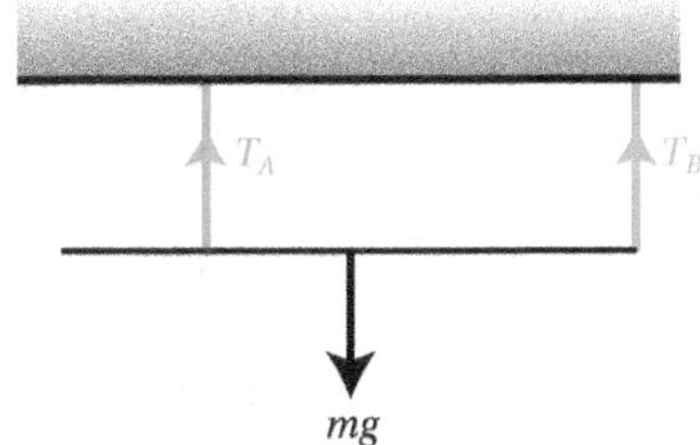

For an object to be at rest in equilibrium, the condition that there is no rotation needs to be added.

Key point 19.4

If an object is in equilibrium there is zero resultant force and zero resultant moment about any point.

Rewind

Recall from Chapter 18 that 'smooth' means there is no friction (so here the force at the support will be normal to the surface of the object), and 'light' means the string has no mass.

Notice that whereas the resultant moment will in general be different about different points, if the object is in equilibrium, the resultant moment will be zero about **any point**.

Therefore, since you have a choice of which point to take moments about, in general it is a good idea to choose a point where at least one force acts. Since the moment of that force will be zero, it eliminates it and makes the calculation simpler.

WORKED EXAMPLE 19.4

A plank of length 6 m and mass 20 kg rests in equilibrium on two identical chairs, as shown in the diagram.

The chairs can provide a reaction force of 10g N before breaking. Assuming that the plank can be modelled as a uniform rod, determine if either chair will break.

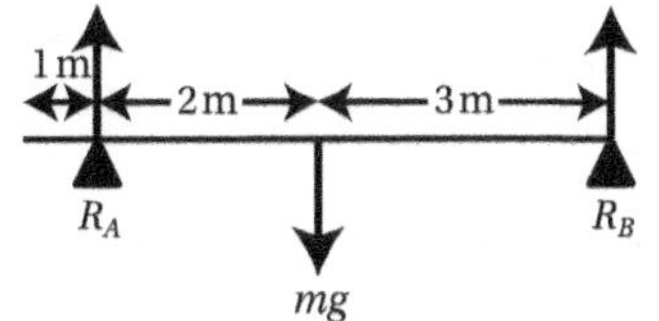

Since the plank is a uniform rod the centre of mass is 3 metres from either end so 2 metres from A.

Continues on next page

Taking moments around A:

$2\times 20g = 5\times R_B$

$R_B = \frac{40g}{5}$

$= 8g$ N

As the plank is in equilibrium, sum of clockwise moments = sum of anticlockwise moments.

Notice that taking moments about either A or B allows you to ignore one of the unknown forces. You could choose either point to start with.

So the chair will not break at B.

$8g < 10g$ so the chair does not break.

Vertical forces:

$R_A + R_B = mg$

$R_A = mg - R_B$

$= 12g$ N

Equilibrium also means that forces up = forces down.

This exceeds the breaking force so the chair will break at A.

$12g > 10g$ so the chair does break.

WORKED EXAMPLE 19.5

A shop sign is formed from a rectangular plastic sheet $ABCD$ of weight 40 N. It is held in equilibrium by a vertical wire at A and horizontal wires at A and C, as shown in the diagram below.

If the sign can be modelled as a uniform lamina with $AB = 0.5$ m and $AD = 0.4$ m, find T_1, T_2, and T_3.

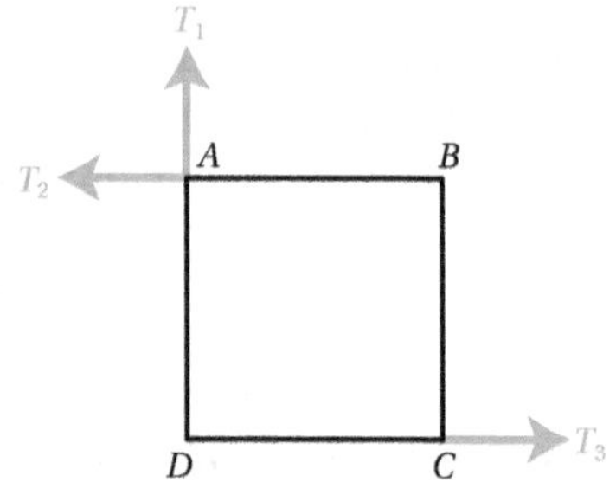

Take moments about A:

$0.25\times 40 = 0.4\times T_3$

$T_3 = \frac{10}{0.4}$

$= 25$ N

Moments about A eliminates two of the unknown forces.

Since it is in equilibrium, sum of clockwise moments = sum of anticlockwise moments.
The centre of mass is 0.25 m to the right of A.

Horizontal forces:

$T_3 = T_2 = 25$ N

Forces right = forces left

Vertical forces:

$T_1 = 40$ N

Forces up = forces down

If a rod is supported at two points, it can be in limiting equilibrium on the point of turning (or tilting) about one of these points. In this situation the force at the other support point will be zero.

WORKED EXAMPLE 19.6

A wooden plank AB of mass 18 kg and length 4 m is suspended horizontally from the ceiling by two ropes 0.6 m from each end of the plank at C and D.

When a mass m is placed on the plank at B, the plank is on the point of turning about D.

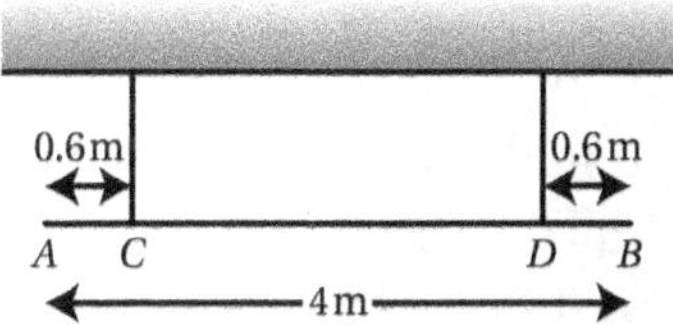

a Find m.

b State any modelling assumptions you have made.

a

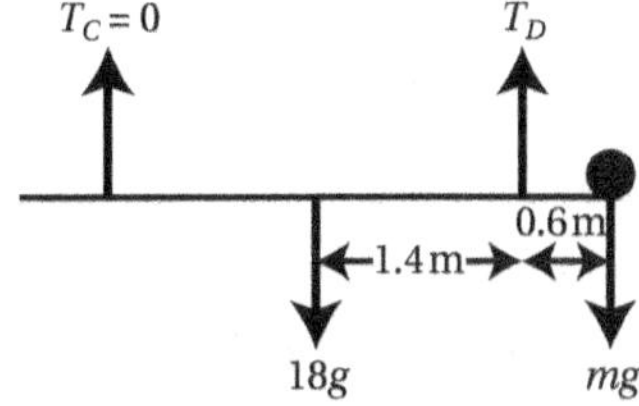

When the plank is on the point of turning about D, the rope will be on the point of going slack at C so $T_C = 0$.

Taking moments about D:

$$1.4 \times 18g = 0.6 \times mg$$
$$25.2 = 0.6m$$
$$m = 42$$

Taking moment about D eliminates the unknown tension.

b The plank is uniform.

The mass is a particle.

The ropes are light.

WORK IT OUT 19.1

A uniform rectangular lamina $ABCD$ of width $CD = 30$ cm and height $AD = 20$ cm has a weight of 80 N and is attached to a fixed bolt at point D, about which it can rotate freely. A horizontal string is attached midway along CB under a tension of 130 N. A downward force of 20 N is applied at a point P which lies on AB with $AP = x$ so that the lamina hangs in equilibrium with DC horizontal. Find the value of x.

Which is the correct solution? Identify the errors made in the incorrect solutions.

Solution A	Taking moments about D: Clockwise moments: $20x$ Anticlockwise moments: $130 \times 10 = 1300$ So $20x = 1300$ $x = 65$ cm

Continues on next page

Solution B	Taking moments about the centre of mass: Clockwise moments: $20(x-15)+100\times 10 = 20x+700$ Anticlockwise moments: $130\times 15 = 1950$ So $20x+700 = 1950$ $x = 62.5$ cm
Solution C	Taking moments about D: Clockwise moments: $20x+80\times 15 = 20x+1200$ Anticlockwise moments: $130\times 10 = 1300$ So $20x+1200 = 1300$ $x = 5$ cm

EXERCISE 19B

1 In each diagram a uniform rod of length 10 m and weight 50 N is being held in equilibrium by two wires. Find the unknown values.

a i

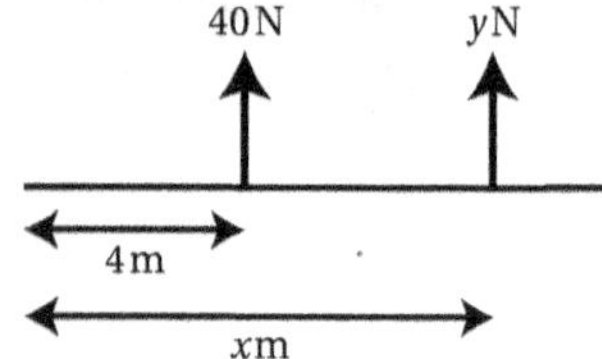

ii

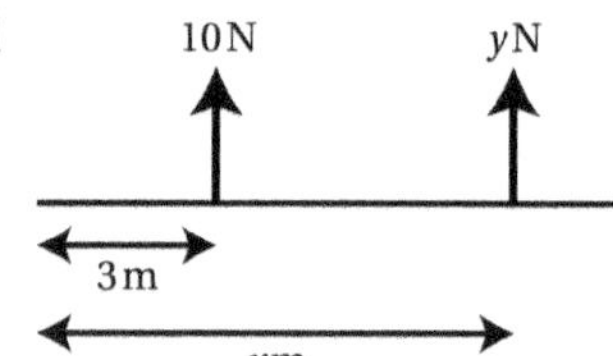

b i

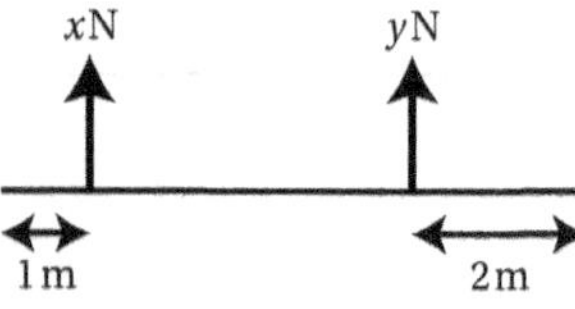

ii

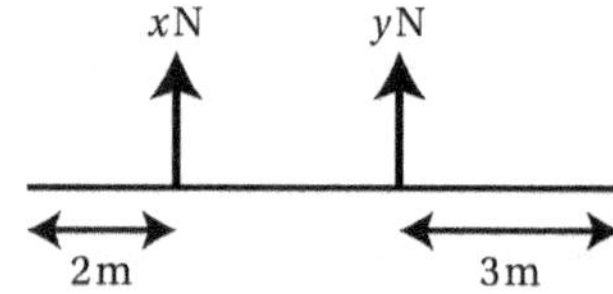

2 Each uniform rectangular lamina has a weight of 100 N and is hanging in equilibrium in a vertical plane. Find the unknown values.

a i

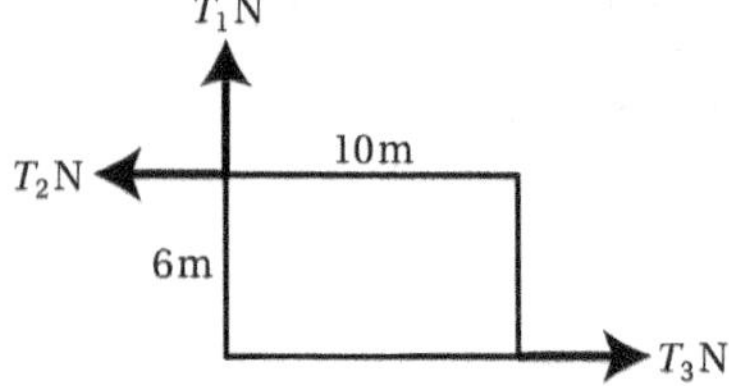

ii

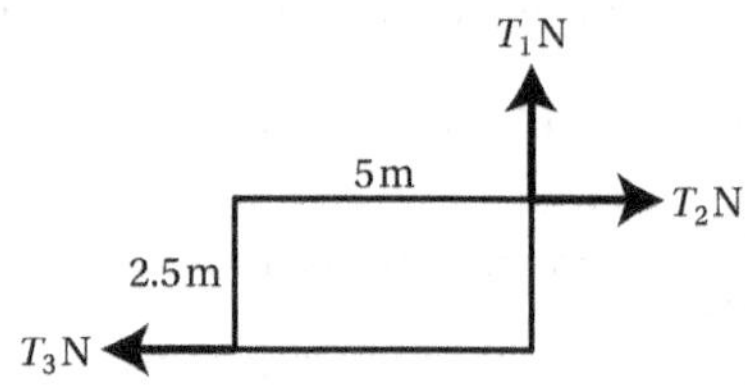

b **i**

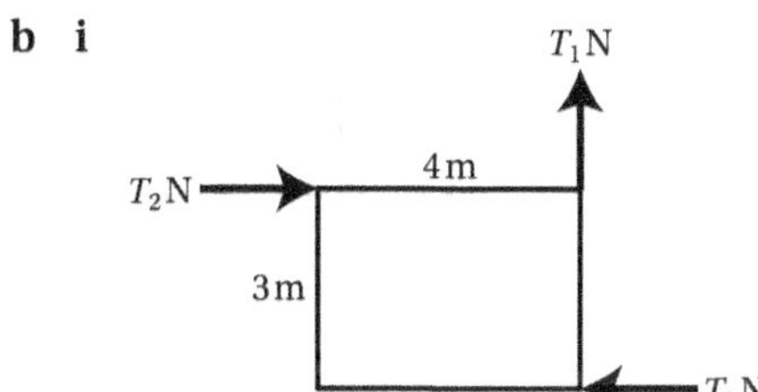

ii

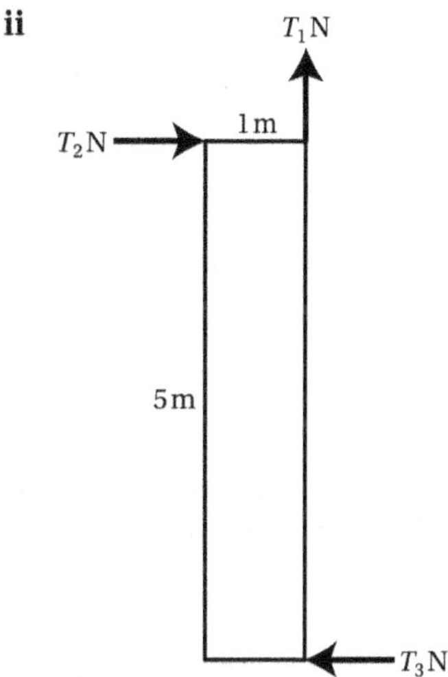

c **i**

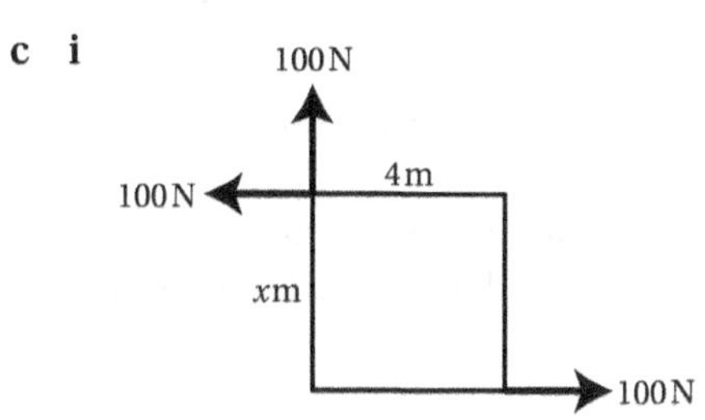

ii

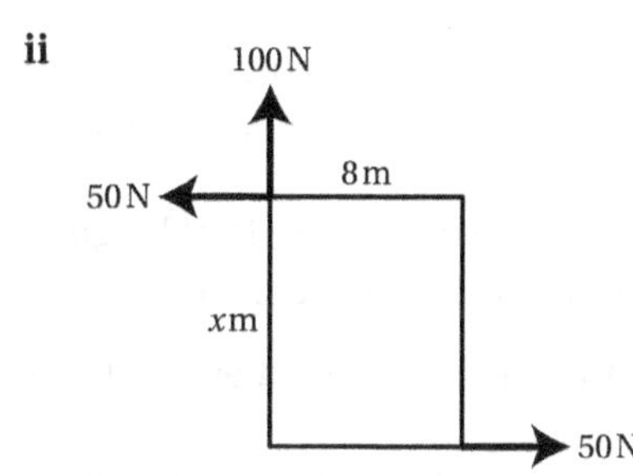

3 Two children sit on a seesaw formed from a uniform rod of length 4 m, balanced in the middle. One child, of mass 30 kg, sits on one end. How far from the other end should the other child, of mass 40 kg, sit so that the seesaw is balanced in a horizontal position?

4 A door is 120 cm wide. A perpendicular force of 80 N is applied 90 cm from the hinge but a wedge at the end of the door opposite the hinge is keeping it shut. Find the frictional force acting though the wedge.

5 **a** In a simplified model of a crane the arm PQ is modelled as a uniform rod of length 20 m and mass 1000 kg.

The arm is attached to the main body of the crane 5 m from P. A mass of 5000 kg is suspended 2 m from Q. A counter mass of 15 000 kg can be moved along the arm to keep the crane in equilibrium. Find the distance from P it should be attached.

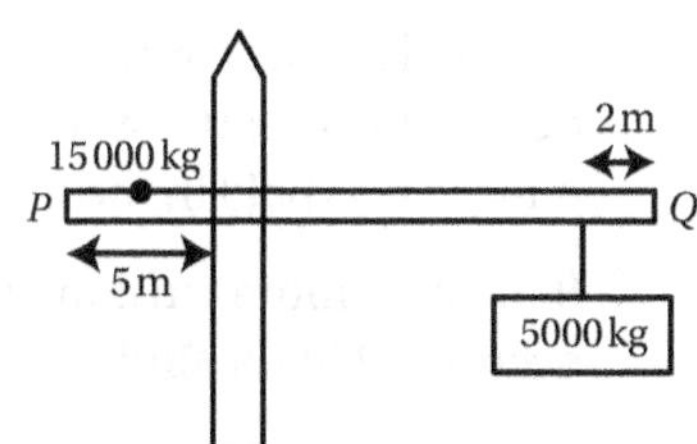

b Explain why the counter mass would not need to be placed precisely at the position found in **a**.

6 A plank AB of mass 6 kg and length 5 m rests on supports at points C and D where $AC = 1.5$ m and $BD = 2$ m. When a 15 kg mass is placed on the plank at E the plank is on the point of tilting about D.

Modelling the plank as a uniform rod and the mass as a particle, find the distance AE.

7 A uniform rod AB of weight 40 N and length 8 m is suspended horizontally from the ceiling by two vertical wires at C and D, as shown.

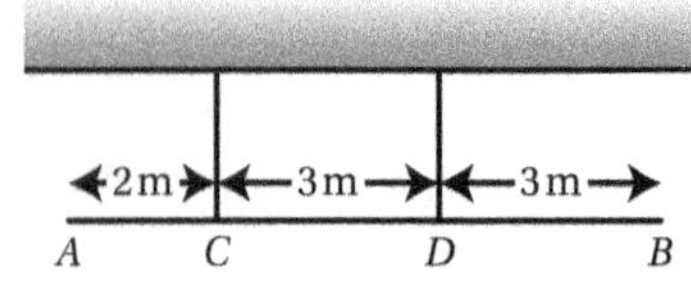

A weight of 25 N is placed on the rod at B.

a Show that the rod will turn about D.

A weight of W N is now placed on the rod at A.

b Find the minimum value of W so that the rod is restored to equilibrium.

8 A uniform plank of length 3 m and mass 5 kg rests in a horizontal equilibrium on two supports - one at the end of the plank and the other 1 m from the other end. Find, in terms of g, the reaction force supplied by each support.

9 In this question use $g = 10 \text{ m s}^{-2}$, giving your final answers to an appropriate degree of accuracy.

A uniform plank of length 2 m and mass 20 kg is suspended horizontally by wires at either end. A painter of weight 80 kg is standing 0.5 m from one end of the beam. Find the tension in each of the wires.

10 a A pole vaulter holds a uniform pole of length 4 m and weight 40 N with one hand pushing down on the end and the other pushing up x cm away. Find the vertical forces exerted by his hands in terms of x.

b State one additional assumption have you made in part **a**.

11 In this question use $g = 10 \text{ m s}^{-2}$, giving your final answers to an appropriate degree of accuracy.

A spade is modelled as a uniform rod of mass 2 kg and length 90 cm attached to a uniform square lamina of side 20 cm and mass 0.5 kg. A gardener holds the spade horizontally with hands 30 cm and 60 cm from the end of the rod. Find the vertical forces exerted by the gardener's hands.

12 In this question use $g = 9.8 \text{ m s}^{-2}$, giving your final answers to an appropriate degree of accuracy.

A model for the elbow joint models the bicep muscle connecting to the horizontal forearm by a vertical tendon 4 cm from the elbow joint. A mass m is held in the hand 30 cm from the elbow joint.

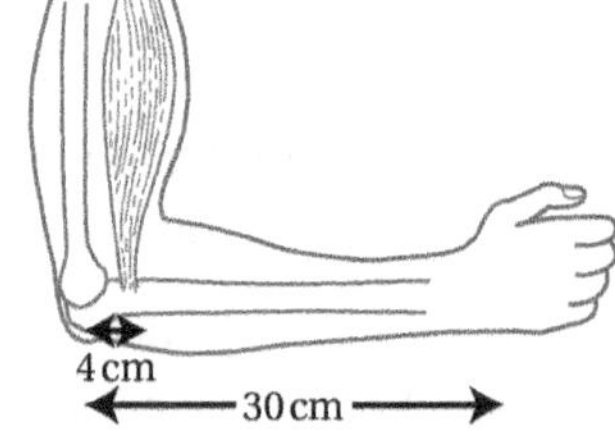

If the maximum tension that can be exerted by the tendon before injury occurs is 2300 N, find the maximum mass that can be held in this way.

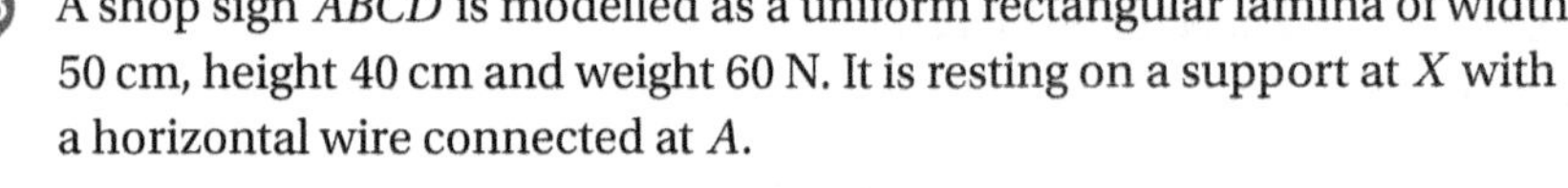

13 A shop sign $ABCD$ is modelled as a uniform rectangular lamina of width 50 cm, height 40 cm and weight 60 N. It is resting on a support at X with a horizontal wire connected at A.

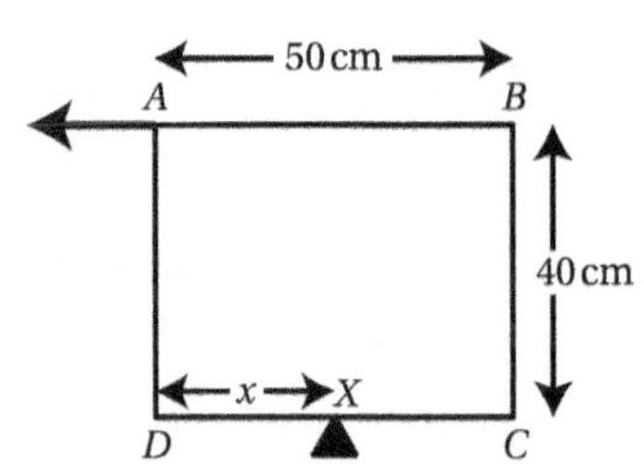

a Find, in terms of x where necessary:

i the tension in the wire

ii the normal reaction at X

iii the friction force at X.

b How would your answer to part **a i** be different if a light rod rather than a wire were attached at A?

14 A door is modelled as a rectangular lamina of weight 150 N with height 2 m and width 1.2 m, supported in equilibrium by two hinges at X and Y.

If the force at X is entirely horizontal, find the magnitude of the forces at X and Y.

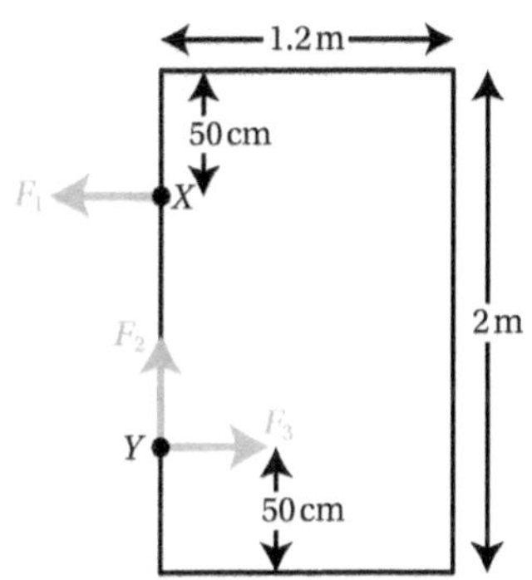

15 A snooker cue is formed by connecting end-to-end two uniform rods of length 80 cm. One has mass 1.5 kg and the other has mass 2 kg. It balances above a point a distance x from the exterior end of the 2 kg section. Find the value of x.

16 In the film 'The Italian Job' a coach is balancing on the edge of a cliff with gold bullion at one end and a group of people at the other. Model the coach as a uniform rod of length 10 m and mass 15 000 kg with 1000 kg of gold at the end overhanging the cliff and 500 kg of people at the other end. How much of the coach can overhang the cliff before it falls?

Checklist of learning and understanding

- The moment of a force F about a point P is:

  moment $= Fd$

  where d is the perpendicular distance of the line of action of the force from P.
- The centre of mass is the point at which the object's weight acts.
  - For a uniform rod, this is at the midpoint.
  - For a uniform lamina, this is at the intersection of the diagonals.
- To find the resultant moment about a point, find the sum of the clockwise and anticlockwise moments separately.
- The resultant moment will be the difference between the two sums (in the direction of the larger).
- If an object is in equilibrium there is zero resultant force and zero resultant moment about any point.

Mixed practice 19

1 What is the resultant moment about point O on the uniform rectangular lamina of weight 10 N shown in the diagram?

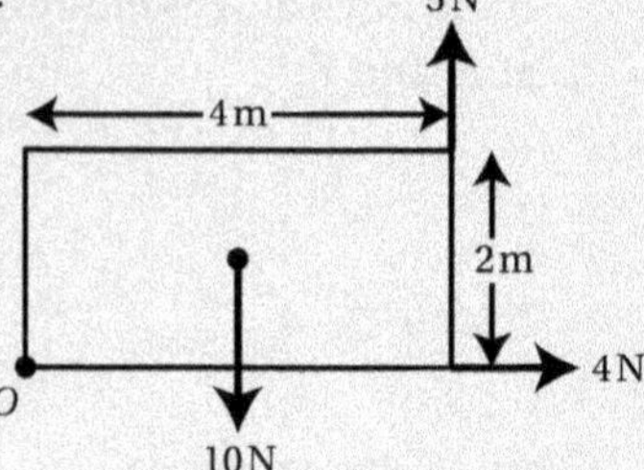

Choose from these options.

A 0 N m **B** 5 N m clockwise

C 10 N m clockwise **D** 12 N m anticlockwise

2 A rigid uniform rod of length 8 m and weight 40 N rests on two supports as shown. A 30 N weight sits 2 m from support A.

Find the reaction force of the support at A.

Choose from these options.

A 14 N **B** 26 N

C 44 N **D** 70 N

3 The diagram shows a uniform rod of length 2 m and weight 50 N lying in equilibrium, suspended from two wires, under tension 20 N and 30 N, and set at distances x m and $2x$ m respectively. Find the value of x.

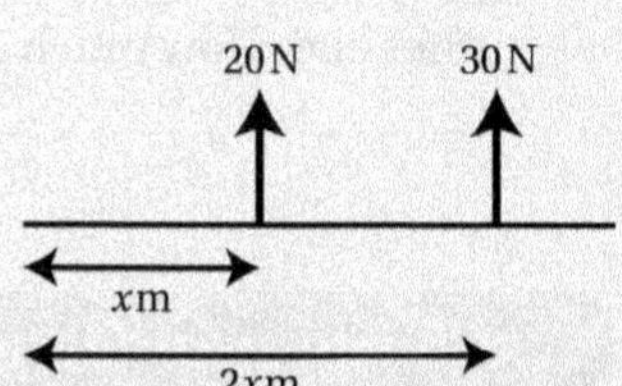

4 A uniform rod AB of weight 30 N and length 2 m is freely hinged at A with a vertical string attached at B. The tension, T newtons, in the string is sufficient to maintain the rod in a horizontal equilibrium.

a Find the value of T.

b Find the magnitude and direction of the force provided by the hinge.

5 In this question use $g = 10\text{ m s}^{-2}$, giving your final answers to an appropriate degree of accuracy.

A uniform plank, of length 8 metres, has mass 30 kg. The plank is supported in equilibrium in a horizontal position by two smooth supports at the points A and B, as shown in the diagram. A block, of mass 20 kg, is placed on the plank at point A.

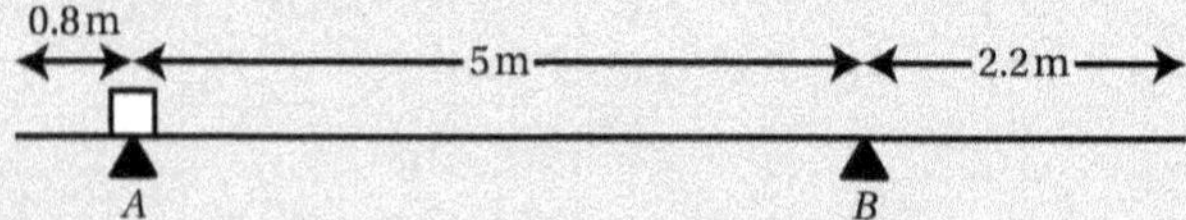

a Draw a diagram to show the forces acting on the plank.

b Show that the magnitude of the force exerted on the plank by the support at B is $19.2g$ newtons.

c Find the magnitude of the force exerted on the plank by the support at A.

d Explain how you have used the fact that the plank is uniform in your solution.

6 In this question use $g = 9.8 \text{ m s}^{-2}$, giving your final answers to an appropriate degree of accuracy.

A uniform plank AB, of length 6 m, has mass 25 kg. It is supported in equilibrium in a horizontal position by two vertical inextensible ropes. One of the ropes is attached to the plank at the point P and the other rope is attached to the plank at the point Q, where $AP = 1$ m and $QB = 0.8$ m, as shown in the diagram.

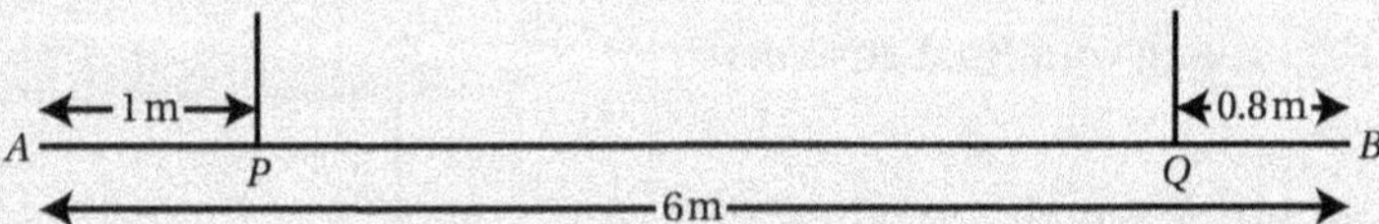

a **i** Find the tension in each rope.

ii State how you have used the fact that the plank is uniform in your solution.

b A particle of mass m kg is attached to the plank at point B, and the tension in each rope is now the same.

Find m.

[© AQA 2013]

7 Ken is trying to cross a river of width 4 m. He has a uniform plank, AB, of length 8 m and mass 17 kg. The ground on both edges of the river bank is horizontal. The plank rests at two points, C and D, on fixed supports which are on opposite sides of the river. The plank is at right angles to both river banks and is horizontal. The distance AC is 1 m, and the point C is at a horizontal distance of 0.6 m from the river bank. Ken, who has mass 65 kg, stands on the plank directly above the middle of the river, as shown in the diagram.

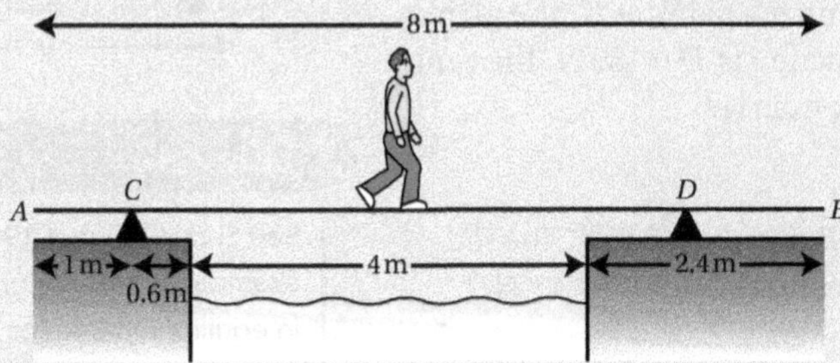

a Draw a diagram to show the forces acting on the plank.

b Given that the reaction on the plank at the point D is $44g$ N, find the horizontal distance of the point D from the nearest river bank.

c State how you have used the fact that the plank is uniform in your solution.

[© AQA 2011]

8 An oar is modelled as a uniform rod PQ of length 3 m and mass 5 kg with an additional mass of 1 kg attached to end P. The oar is hung by a single wire. How far from Q must this be attached if the oar is to hang horizontally?

9 In this question use $g = 10\text{ m s}^{-2}$, giving your final answers to an appropriate degree of accuracy.

A uniform plank PQ of mass 10 kg and length 5 m rests on supports 1 m from P and 2 m from Q. A mass 60 kg is placed in a position to equalise the reaction forces on the two supports. Find the distance of the mass from P.

10 In this question use $g = 9.8\text{ m s}^{-2}$, giving your final answers to an appropriate degree of accuracy.

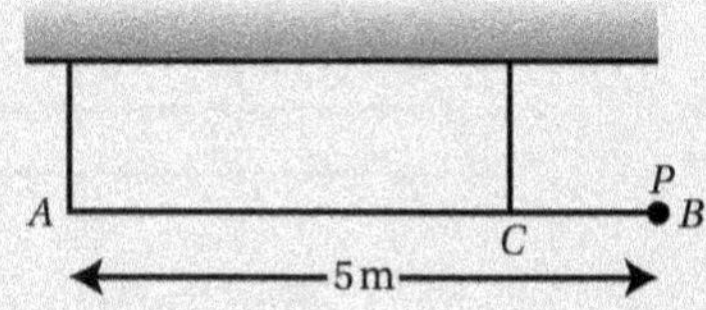

A uniform plank AB of mass 40 kg and length 5 m hangs from two vertical ropes attached to A and C.

When a particle, P, of weight 28 N is attached to B the plank rests horizontally in equilibrium.

If the tension in the rope at C is three times the tension at A, find:

a the tension at C **b** the distance CB.

11 A uniform plank AB of length 6 m and weight 100 N rests horizontally in equilibrium on two smooth supports as shown.

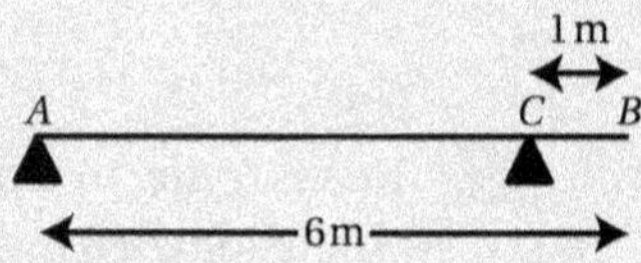

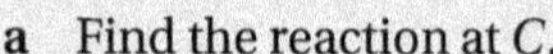

a Find the reaction at C.

A child of weight 400 N stands on the plank at D. The plank remains in equilibrium. The reactions on the plank at A and C are now equal.

b Find the distance AD.

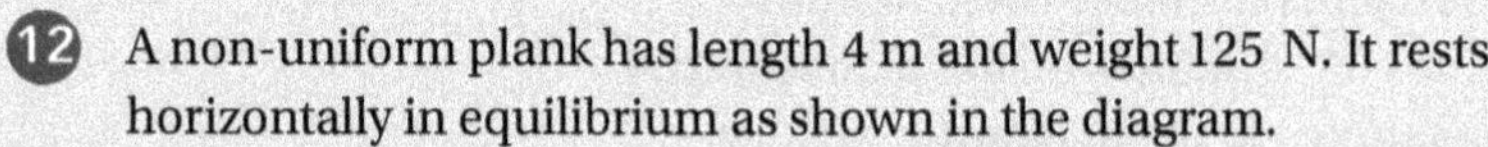

12 A non-uniform plank has length 4 m and weight 125 N. It rests horizontally in equilibrium as shown in the diagram.

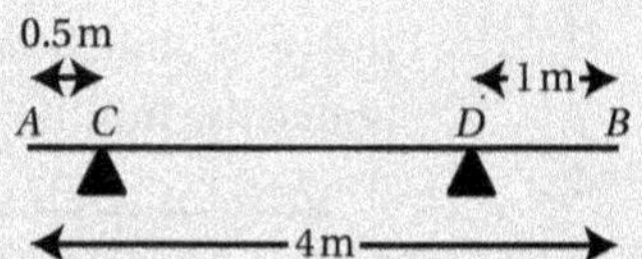

A particle of weight W N is placed on the plank at B. The plank remains in equilibrium and the reaction at D is 139 N. The centre of mass of the plank is a distance x from A.

a Show that $7W + 250x = 820$.

The particle is now removed from B and placed at A. The rod remains in equilibrium and the reaction at D is now 43 N.

b Find W and x.

Elevate

See Support Sheet 19 for an example of non-uniform rods in equilibrium and for more practice questions.

13 A uniform plank AB of length 6 m and weight 200 N rests horizontally in equilibrium on two smooth supports C and D as shown.

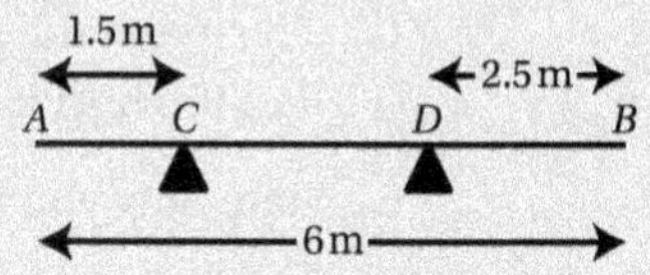

A particle of weight W N is attached to a point on the plank x m from A. The plank remains in equilibrium and the reactions at C and D are now equal.

a Show that $W = \dfrac{200}{11 - 4x}$.

b Hence find the range of possible values of x.

14 A rectangular lamina is formed by connecting two square uniform laminas of side 2 metres and masses M and m where $M > m$.

The rectangular lamina balances on a point a distance x from the join line, as shown in the diagram. Find an expression for x in terms of M and m.

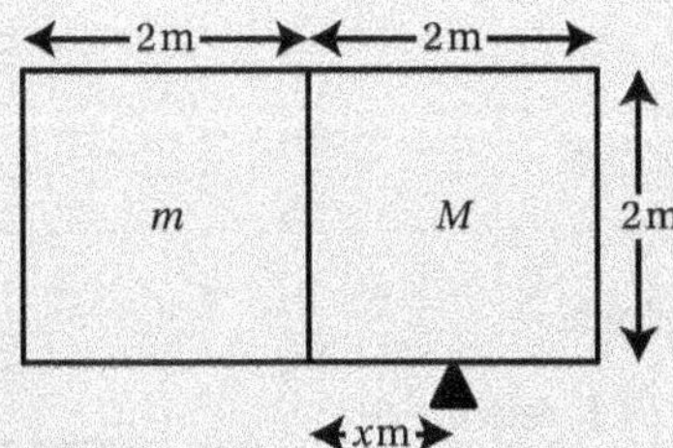

Elevate

See Extension Sheet 19 for some questions involving laminas formed from a rectangle and a triangle.

15 A uniform rod of length 6 m and weight 100 N is being pushed over a roller (R) with negligible friction and a support (S) with a coefficient of friction 0.6. R and S are 3 m apart. x is the length of the rod overhanging S with $0 < x < 3$.

Find, as a function of x, the force required for the rod to move at a constant speed.

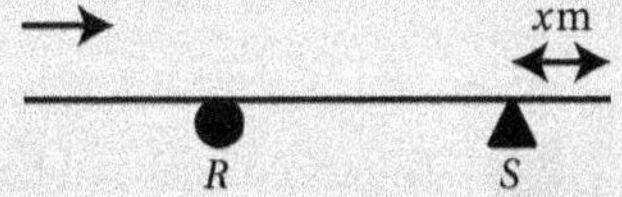

16 A car and contents is modelled by a uniform rectangular laminar (since depth will not be relevant) of mass 1200 kg and width 4 m. The wheels are located 1.5 m from the front of the car and 1 m from the rear of the car.

a By taking moments about the rear wheel, show that the normal reaction of the ground on the front wheel is 800 g N.

b The car has front wheel drive. Assuming that the car has sufficient power and the coefficient of friction between the car and the road is 0.5, find, in terms of g, the maximum acceleration of the car.

Rewind

The coefficient of friction was covered in Chapter 18, Section 2.

FOCUS ON... PROOF 3

Overcoming friction

A block of weight W rests on rough horizontal ground. The coefficient of friction between the block and the ground is μ. You are going to derive the formulae for the minimum force required to move the block in various situations.

QUESTIONS

1 A horizontal force of magnitude F acts on the block.

Find, in terms of W and μ, the minimum value of F required to move the block.

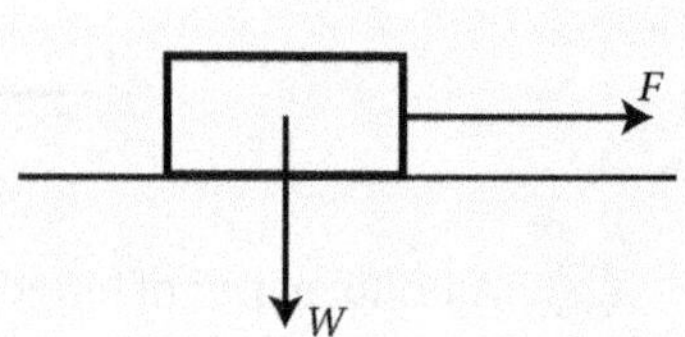

2 The force acts at a fixed angle θ above the horizontal.

Prove that the minimum value of F is $\frac{\mu W \sec\theta}{1+\mu\tan\theta}$.

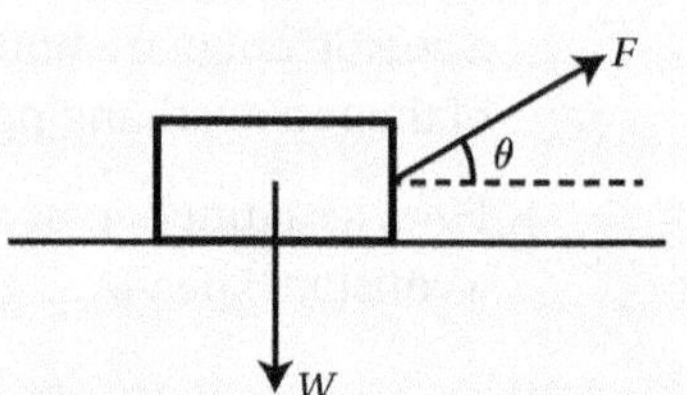

3 Both F and θ can vary.

a **i** Find the maximum value of $\cos\theta+\mu\sin\theta$ for $0° \leqslant \theta \leqslant 90°$.

ii Hence find the minimum magnitude of the force required to move the box.

iii Prove that this minimum magnitude is always less than W. (In other words, however large μ is, it is always possible to move the box using a force smaller than its weight.)

b Prove that the angle at which the minimal force needs to act is $\theta = \arctan\mu$.

4 The force pushes the block at an angle θ to the horizontal, as shown.

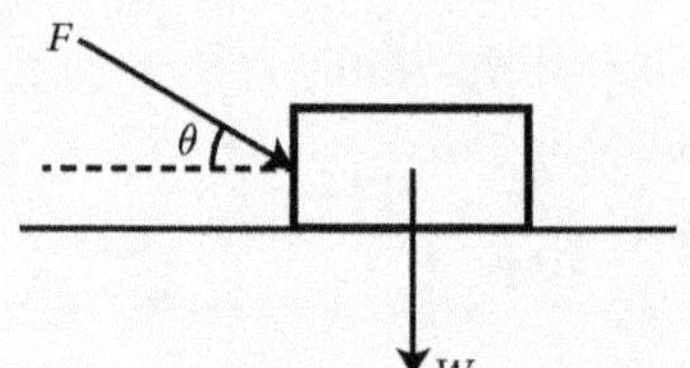

a Prove that $F(\cos\theta-\mu\sin\theta) \geqslant \mu W$.

b By considering the graph of $y=\cos\theta-\mu\sin\theta$ for $0° \leqslant \theta \leqslant 90°$, or otherwise, prove that the required magnitude of the force is minimal when the force is horizontal, and state this magnitude.

FOCUS ON... PROBLEM SOLVING 3

Checking for reasonableness

When solving problems in real contexts answers are rarely easy numbers to work with. It can therefore be difficult to tell whether your answer is correct. However, there are still some checks you can do, such as confirming that your calculation gives correct units and making sure that the answer is not completely unreasonable.

QUESTIONS

Here are proposed answers to some mechanics problems. Decide which ones are obviously wrong.

1 A car's deceleration is $260\ \text{m s}^{-2}$.

2 The tension in the cable supporting a lift is 0.36 N.

3 The frictional force acting on a box is 30 N.

4 The stone was dropped from a height of 2.8×10^{-3} m.

5 The road is inclined at 72° to the horizontal.

6 The mass of a car is 2600 kg.

7 The coefficient of friction between a box and the floor is 36.2.

8 An athlete runs 100 m at an average speed of $32\ \text{km h}^{-1}$.

9 The maximum height reached by the projectile is 3600 km.

10 The two trains will meet after 0.45 seconds.

11 The coefficient of friction between the skates and the ice is 0.6.

12 The car accelerates at $47\,000\ \text{km h}^{-2}$.

13 The coefficient of friction between the box and the ice is 0.035.

14 The ball takes 12.3 seconds to fall from the tenth floor.

15 The journey from London to Manchester takes 18 hours.

Modelling with moments

It is alleged that Archimedes once said, 'Give me a place to stand and I will move the Earth.'

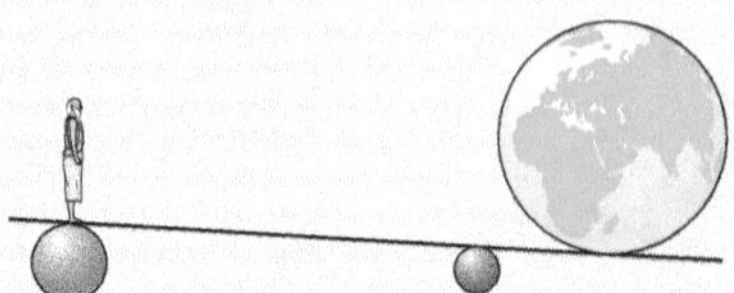

It is theoretically possible to move any weight if you use a sufficiently long lever. Archimedes would need to be standing on a different planet and he would also need a support for the lever.

QUESTIONS

1. Suppose the Earth is modelled as a particle of mass 6×10^{24} kg, resting on one end of the lever, and that Archimedes is a particle of mass 60 kg, sitting on the other end. The lever rests on a support 1 m from the Earth. How long would the other side of the lever need to be?

2. Other than modelling the Earth and Archimedes as particles, what other assumption did you make in your calculation in question **1**?

3. Is modelling the Earth as a particle realistic? What could you change in your model to make the particle assumption reasonable? (The radius of the Earth is 6.4×10^6 m.) How does this change your answer in question **1**?

4. Using the modelling assumptions from question **1**, suppose Archimedes moves his end of the lever for a year, at the average speed of 6 km h^{-1}. By how much would he move the Earth?

Did you know?

The Ancient Greek mathematician Archimedes is perhaps best known for exclaiming 'Eureka!' on realising that a body submerged in water displaces its own volume of water, an idea now commonly known as Archimedes' principle. However, his work extended well beyond this discovery and his work on levers. As well as designing the screw pump for raising water, compound pulleys and siege machines, he also developed the fundamentals of integral calculus 2000 years before Newton and Leibniz eventually formalised it.

CROSS-TOPIC REVIEW EXERCISE 3

In this exercise, unless instructed otherwise, use $g = 9.8 \text{ m s}^{-2}$, giving your final answers to an appropriate degree of accuracy.

1 Points A and B have position vectors $\mathbf{a} = 3\mathbf{i} - \mathbf{j} + 5\mathbf{k}$ and $\mathbf{b} = 2\mathbf{j} - \mathbf{k}$.

a Find the vector $\overrightarrow{AB}$.

b Find the exact distance between the two points.

2 Points M, N and P have coordinates $(2, 1, -5), (5, -3, 2)$ and $(-2, -3, 7)$, respectively.

a Find the coordinates of the point Q so that $MNPQ$ is a parallelogram.

b Show that $MNPQ$ is a rhombus.

c Find the coordinates of the point S on the line MN such that $NS = MN$.

3 A particle moves on a horizontal plane, in which the unit vectors $\mathbf{i}$ and $\mathbf{j}$ directed east and north respectively.

At time t seconds, the position vector of the particle is $\mathbf{r}$ metres, where

$$\mathbf{r} = \left(2e^{\frac{1}{2}t} - 8t + 5\right)\mathbf{i} + (t^2 - 6t)\mathbf{j}$$

a Find an expression for the velocity of the particle at time t.

b i Find the speed of the particle when $t = 3$.

ii State the direction in which the particle is travelling when $t = 3$.

c Find the acceleration of the particle when $t = 3$.

d The mass of the particle is 7 kg.

Find the magnitude of the resultant force on the particle when $t = 3$.

[© AQA 2009]

4 A uniform plank of mass 12 kg and length 4.6 m rests on two supports, A and B. The support A is 0.8 m from one end of the plank and support B is 0.3 m from the other end. The plank is horizontal. A box of mass 7 kg is placed on the plank, 1.2 m from A.

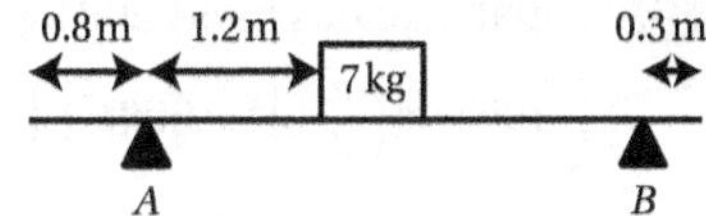

Find, in terms of g, the force acting on the plank at each support.

5 A non-uniform plank of mass 8 kg and length 4 m rests on a support at its mid-point. A particle of mass 3 kg rests on the plank, 0.6 m to the left of the support. Given that the plank is in equilibrium, find, in terms of g:

a the position of its centre of mass

b the magnitude of the force acting on the plank at the support.

 **6** A particle, of mass m kg, remains in equilibrium under the action of three forces, which act in a vertical plane, as shown in the diagram. The force with magnitude 60 N acts at 48° above the horizontal and the force with magnitude 50 N acts at an angle θ above the horizontal.

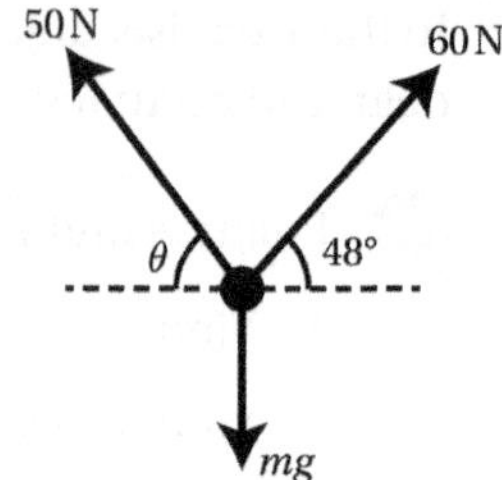

a By resolving horizontally, find θ.

b Find m.

[© AQA 2010]

7 Three points have coordinates $T(-3, 1, 4)$, $U(0, 2, 3)$ and $V(5, a, b)$. Find the values of a and b so that TUV is a straight line.

8 The position vectors of two particles, A and B, at time t, are given by $\mathbf{r}_A = (4\cos(\pi t) - 2)\mathbf{i} - 5\mathbf{j}$ and $\mathbf{r}_B = 2\mathbf{i} + (5 - 4\sin(\pi t))\mathbf{j}$. Prove that the distance between the two particles is constant.

 **9** The constant forces $\mathbf{F}_1 = (8\mathbf{i} + 12\mathbf{j})$ newtons and $\mathbf{F}_2 = (4\mathbf{i} - 4\mathbf{j})$ newtons act on a particle. No other forces act on the particle.

a Find the resultant force on the particle.

b Given that the mass of the particle is 4 kg, show that the acceleration of the particle is $(3\mathbf{i} + 2\mathbf{j})$ m s^{-2}.

c At time t seconds, the velocity of the particle is $\mathbf{v}$ m s^{-1}.

i When $t = 20$, $\mathbf{v} = 40\mathbf{i} + 32\mathbf{j}$. Show that $\mathbf{v} = -20\mathbf{i} - 8\mathbf{j}$ when $t = 0$.

ii Write down an expression for $\mathbf{v}$ at time t.

iii Find the times when the speed of the particle is 8 m s^{-1}.

[© AQA 2010]

 **10** A particle has mass 200 kg and moves on a smooth horizontal plane. A single horizontal force, $\left(400\cos\left(\frac{\pi}{2}t\right)\mathbf{i} + 600t^2\mathbf{j}\right)$ newtons, acts on the particle at time t seconds.

The unit vectors $\mathbf{i}$ and $\mathbf{j}$ are directed east and north respectively.

a Find the acceleration of the particle at time t.

b When $t = 4$, the velocity of the particle is $(-3\mathbf{i} + 56\mathbf{j})$ m s^{-1}. Find the velocity of the particle at time t.

c Find t when the particle is moving due west.

d Find the speed of the particle when it is moving due west.

[© AQA 2010]

11 In this question use $g = 9.81 \text{ m s}^{-2}$, giving your final answers to an appropriate degree of accuracy.

A uniform rectangular lamina has sides of length 68 cm and 42 cm and mass 620 grams. The lamina is freely hinged at point A and is held in equilibrium, with the longer side horizontal, by means of a light inextensible string attached to point B.

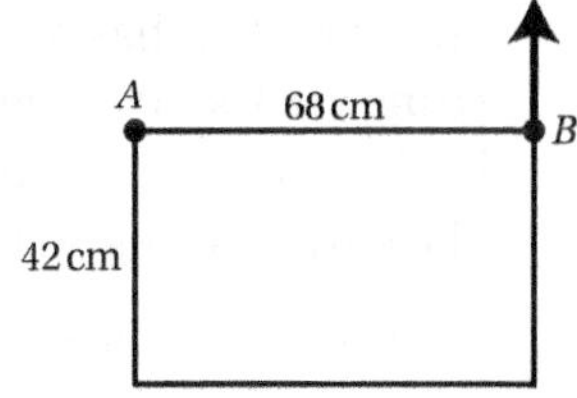

a Find the tension in the string.

b Find the magnitude and direction of the force acting at the lamina at A.

12 A block, of mass 5 kg, slides down a rough plane inclined at 40° to the horizontal. When modelling the motion of the block, assume that there is no air resistance acting on it.

a Draw and label a diagram to show the forces acting on the block.

b Show that the magnitude of the normal reaction force acting on the block is 37.5 N, correct to three significant figures.

c Given that the acceleration of the block is 0.8 m s^{-2}, find the coefficient of friction between the block and the plane.

d In reality, air resistance does act on the block. State how this would change your value for the coefficient of friction, and explain why.

[© AQA 2008]

13 In this question use $g = 10 \text{ m s}^{-2}$, giving your final answers to an appropriate degree of accuracy.

Two small blocks, each of mass 3 kg, are connected by a light inextensible string passing over a smooth pulley. One block rests on a rough horizontal table and it other hangs freely with the string vertical. The coefficient of friction between the block and the table is 0.4.

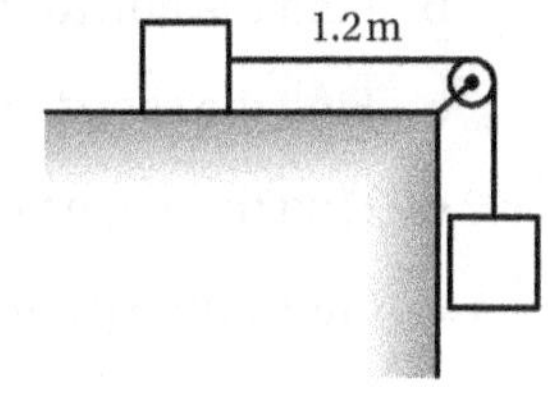

The first block is initially 1.2 m from the edge of the table when the system is released from rest.

a Find, in terms of g, the acceleration of the system.

b How long does it take for the first block to reach the edge of the table?

c Find the magnitude and direction of the force acting on the pulley.

d How did you use the modelling assumption that:

i the pulley is smooth **ii** the string in inextensible?

14 Two particles, A and B, are connected by a light inextensible string which passes over a smooth peg. Particle A has mass 2 kg and particle B has mass 4 kg. Particle A hangs freely with the string vertical. Particle B is at rest in equilibrium on a rough horizontal surface with the string at an angle 30° to the vertical. The particles, peg and string are shown in the diagram.

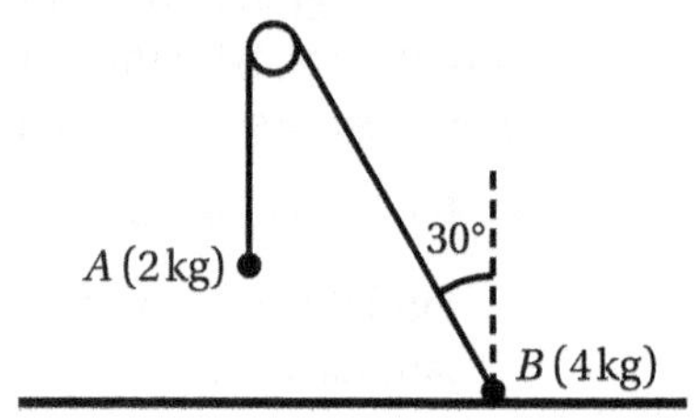

a By considering particle A, find the tension in the string.

b Draw a diagram to show the forces acting on particle B.

c Show that the magnitude of the normal reaction force acting on particle B is 22.2 newtons correct to three significant figures.

d Find the least possible value of the coefficient of friction between B and the surface.

[© AQA 2011]

15 An arrow is fired from a point at a height of 1.5 metres above horizontal ground. It has an initial velocity of 12 m s^{-1} at an angle of 30° above the horizontal. The arrow hits a target at a height of 1 metre above horizontal ground. The path of the arrow is shown in the diagram.

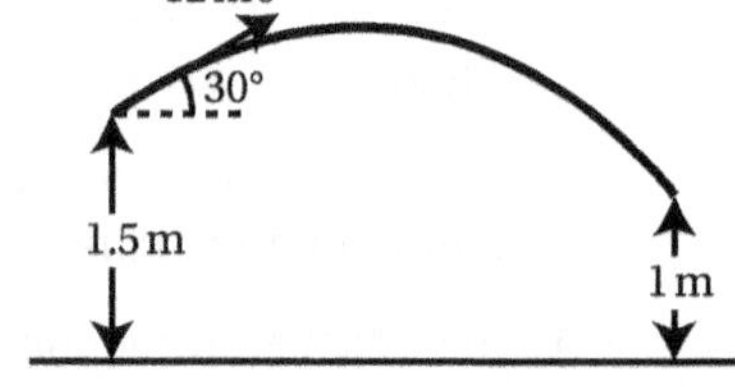

Model the arrow as a particle.

a Show that the time taken for the arrow to travel to the target is 1.3 seconds.

b Find the horizontal distance between the point where the arrow is fired and the target.

c Find the speed of the arrow when it hits the target.

d Find the angle between the velocity of the arrow and the horizontal when the arrow hits the target.

e State one assumption that you have made about the forces acting on the arrow.

[© AQA 2011]

16 A car, of mass m kg, is moving along a straight horizontal road. At time t seconds, the car has speed v ms^{-1}. As the car moves, it experiences a resistance force of magnitude $2mv^{\frac{5}{4}}$ newtons. No other horizontal force acts on the car.

a Show that:

$$\frac{dv}{dt} = -2v^{\frac{5}{4}}$$

b The initial speed of the car is 16 m s^{-1}. Show that:

$$v = \left(\frac{2}{t+1}\right)^4$$

[© AQA 2011]

17 A particle moves in a straight line with acceleration $\mathbf{a} = (-3e^{t}\,\mathbf{i} + 5e^{-t}\,\mathbf{j})$ m s^{-2}. The particle is initially at the origin and has velocity $\mathbf{v} = (3\mathbf{i} - 5\mathbf{j})$ m s^{-1}. Prove that the particle moves in a straight line.

18 A golfer hits a ball which is on horizontal ground. The ball initially moves with speed V ms^{-1} at an angle of 40° above the horizontal. There is a pond further along the horizontal ground. The diagram below shows the initial position of the ball and the position of the pond.

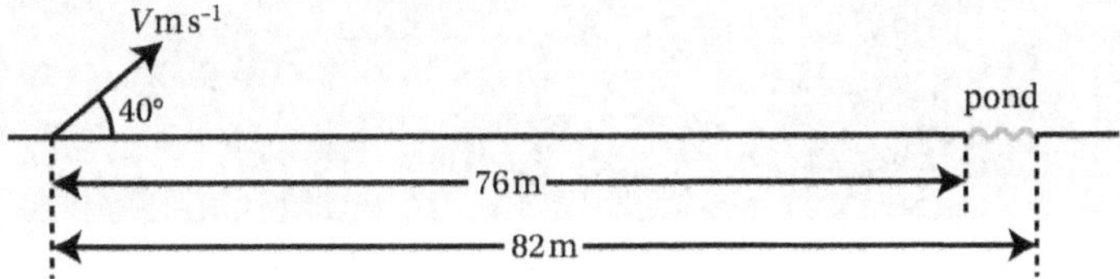

a State **two** assumptions that you should make in order to model the motion of the ball.

b Show that the horizontal distance, in metres, travelled by the ball when it returns to ground level is:

$$\frac{V^2 \sin 40° \cos 40°}{4.9}$$

c Find the range of values of V for which the ball lands in the pond.

[© AQA 2008]

19 In this question use $g = 10$ m s^{-2}, giving your final answers to an appropriate degree of accuracy.

A netball player takes a shot at the hoop, 3 m above the ground. The player stands 5 m from the foot of the post and releases a ball at the height of 1.5 m, as shown in the diagram. The ball is released with speed V at an angle $\alpha°$ above the horizontal.

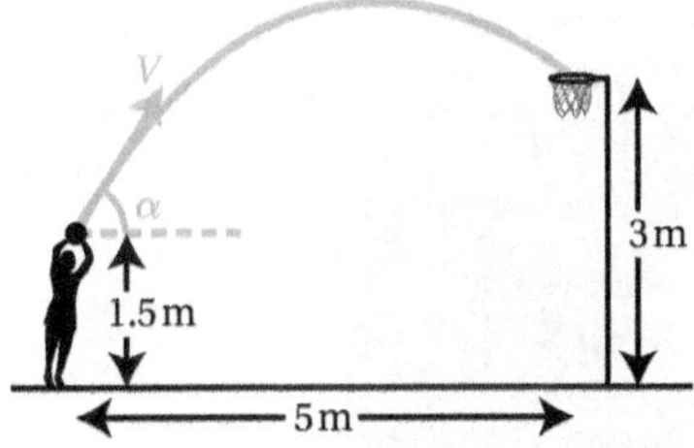

a Given that the ball passes through the hoop, show that

$$\frac{50g}{V^2} = 3 + 10\sin 2\alpha + 3\cos 2\alpha.$$

b i Find the maximum value of $10\sin 2\alpha + 3\cos 2\alpha$.

ii Hence find the minimum speed with which the ball can be released and still pass through the hoop.

20 Conditional probability

In this chapter you will learn how to:

- use set notation to describe probabilities
- work with conditional probabilities in the context of Venn diagrams, two-way tables and tree diagrams
- use a formula for conditional probability.

Before you start...

GCSE	You should be able to use tree diagrams to solve problems.	1 A mother has two children, who are not twins. What is the probability that they are both boys or both girls?
Student Book 1, Chapter 1	You should be able to use set notation.	2 Write out this set. {prime numbers} $\cap$ {even numbers}
Student Book 1, Chapter 21	You should understand the basic laws of probability including the terms mutually exclusive and independent.	3 In a family having a car is mutually exclusive of having a motorbike. The probability of having a car is $\frac{1}{2}$. The probability of having a motorbike is $\frac{1}{3}$. What is the probability of having neither a car nor a motorbike?
Student Book 1, Chapter 21	You should understand probability distributions, including the binomial distribution.	4 What is the probability of getting 4 heads when 6 fair coins are tossed?

What is conditional probability?

What is the probability that you will become a millionaire? You could just look at data for how many millionaires there are in the world, but this is not likely to give you a very accurate answer because it depends on lots of other factors. Where you were born, what your parents do and your attitude towards risk all change the probability. You might be glad to know that the fact that you are doing Mathematics A Level immediately increases your probability of becoming a millionaire!

Information often changes probabilities. A probability that takes into account information is called a **conditional probability**. In reality nearly all probabilities are conditional – the probability of a patient having heart disease might change depending upon their age, the probability of a defendant being guilty might change depending upon their prior convictions, the probability of a football team winning might depend upon the team they are playing.

Most people have a very poor intuition for conditional probabilities. In this chapter you will see various ways to visualise conditional probabilities and how you can use them solve problems.

Focus on...

See Focus on... Problem solving 4 for the approach of using extreme values to suggest possible solutions.

Section 1: Set notation and Venn diagrams

What is more likely when you roll a dice once:

- getting a prime number *and* an odd number
- getting a prime number *or* an odd number?

The first possibility is restrictive - both conditions have to be satisfied. The second opens up many more possibilities - either condition can be satisfied. So the second must be more likely.

These are examples of two of the most common way of combining events: intersection (in normal language 'and') and union (in normal language 'or'). You can use set notation to describe the different ways of combining events.

Key point 20.1

- $A \cap B$ is the **intersection** of A and B, meaning when both A and B happen.
- $A \cup B$ is the **union** of A and B, meaning when either A happens, or B happens, or both happen.
- A' is the **complement** of A, meaning everything that could happen other than A.

Did you know?

Why do mathematicians not just use simple words? One of the reasons for this is the ambiguity of everyday language. If I say that I play rugby or hockey some people may think this means I do not play both.

You can use Venn diagrams to illustrate the concepts of union and intersection.

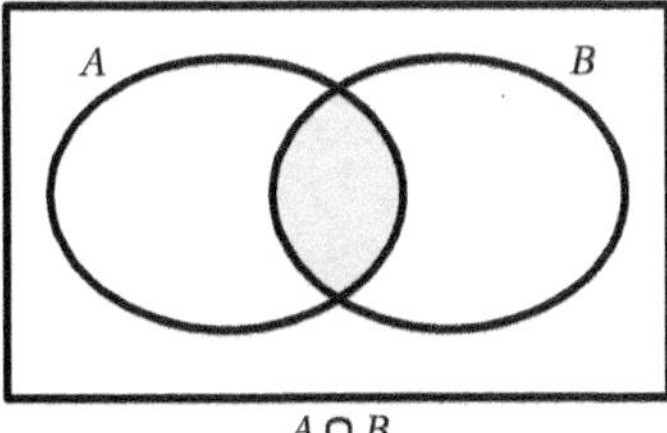

$A \cap B$

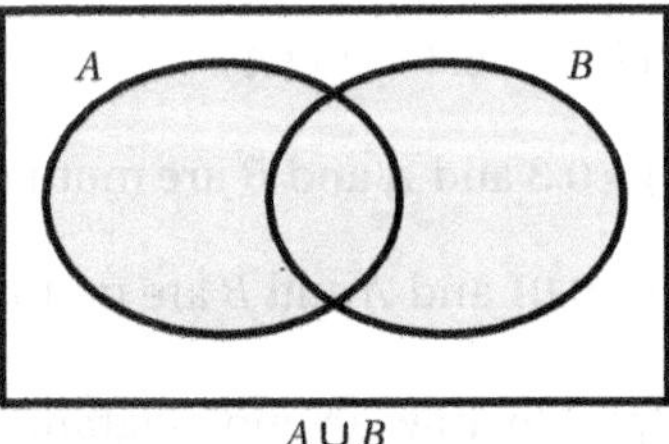

$A \cup B$

There is an important result that comes from looking at the Venn diagrams.

Key point 20.2

$$P(A \cup B) = P(A) + P(B) - P(A \cap B)$$

This will be given in your formula book.

You can interpret this formula as saying, 'If you want the number of ways of getting A or B, take the number of ways of getting A and add to that the number of ways of getting B. However, you have counted the number of ways of getting A and B twice, so compensate by subtracting it'.

If there is no possibility of A and B occurring at the same time, then $\mathrm{P}(A \cap B) = 0$. These events are **mutually exclusive**, and the formula reduces to $\mathrm{P}(A \cup B) = \mathrm{P}(A) + \mathrm{P}(B)$.

WORKED EXAMPLE 20.1

A chocolate is selected randomly from a box. The probability of its containing nuts is $\frac{1}{4}$. The probability of its containing caramel is $\frac{1}{3}$. The probability of its containing both nuts and caramel is $\frac{1}{6}$.

What is the probability of a randomly chosen chocolate containing either nuts or caramel or both?

$$\begin{aligned}\mathrm{P}(\text{nuts} \cup \text{caramel}) &= \mathrm{P}(\text{nuts}) + \mathrm{P}(\text{caramel}) - \mathrm{P}(\text{nuts} \cap \text{caramel})\\ &= \frac{1}{4} + \frac{1}{3} - \frac{1}{6}\\ &= \frac{5}{12}\end{aligned}$$

Use the formula: $\mathrm{P}(A \cup B) = \mathrm{P}(A) + \mathrm{P}(B) - \mathrm{P}(A \cap B)$

EXERCISE 20A

1 a i $\mathrm{P}(A) = 0.4, \mathrm{P}(B) = 0.3$ and $\mathrm{P}(A \cap B) = 0.2$. Find $\mathrm{P}(A \cup B)$.

ii $\mathrm{P}(A) = \frac{3}{10}, \mathrm{P}(B) = \frac{4}{5}$ and $\mathrm{P}(A \cap B) = \frac{1}{10}$. Find $\mathrm{P}(A \cup B)$.

b i $\mathrm{P}(A) = \frac{2}{3}, \mathrm{P}(B) = \frac{1}{8}$ and $\mathrm{P}(A \cup B) = \frac{5}{8}$. Find $\mathrm{P}(A \cap B)$.

ii $\mathrm{P}(A) = 0.2, \mathrm{P}(B) = 0.1$ and $\mathrm{P}(A \cup B) = 0.25$. Find $\mathrm{P}(A \cap B)$.

c i $\mathrm{P}(A \cap B) = 20\%, \mathrm{P}(A \cup B) = 0.4$ and $\mathrm{P}(A) = \frac{1}{3}$. Find $\mathrm{P}(B)$.

ii $\mathrm{P}(A \cup B) = 1, \mathrm{P}(A \cap B) = 0$ and $\mathrm{P}(B) = 0.8$. Find $\mathrm{P}(A)$.

d i Find $\mathrm{P}(A \cup B)$ if $\mathrm{P}(A) = 0.4, \mathrm{P}(B) = 0.3$ and A and B are mutually exclusive.

ii Find $\mathrm{P}(A \cup B)$ if $\mathrm{P}(A) = 0.1, \mathrm{P}(B) = 0.01$ and A and B are mutually exclusive.

2 In each question, you might find it helpful to draw a Venn diagram.

a i When a fruit pie is selected at random, P(it contains pears) = $\frac{1}{5}$ and P(it contains apples) = $\frac{1}{4}$. 10% contain both apples and pears.

Find P(apples $\cup$ pears).

ii In a library 80% of books are classed as fiction and 70% are classed as 20th century. Half of the books are 20th century fiction.

What proportion of the books are either fiction or from the 20th century?

b **i** 95% of students in a school play either football or tennis. The probability of a randomly chosen student playing football is $\frac{6}{10}$ and the probability of their playing tennis is $\frac{5}{8}$.

What percentage of students play both football and tennis?

ii 2 in 5 students in a school study Spanish and 1 in 3 study French. Half of the school study either French or Spanish.

What fraction study both French and Spanish?

3 $P(A) = 0.6, P(B) = 0.5$ and $P(A \cup B) = 0.9$.

a By drawing a Venn diagram or otherwise find $P(A \cap B)$.

b Find $P(A \cap B')$.

4 90% of students in a class have social media account A and 3 out of 5 have social media account B. One twentieth of students have neither of these social media accounts.

What percentage have both of these social media accounts?

5 Events A and B satisfy $P((A \cup B)') = 0.2, P(A) = P(B) = 0.5$.

Find $P(A' \cap B)$.

6 25% of teams in a football league have French players and 30% have Italian players. 60% have neither French nor Italian players.

What percentage have French but not Italian players?

7 $P(A \cap B') = 0.3$ and $P(B) = 0.4$. Events A and B are independent.

Find $P(A \cap B)$.

8 The probability of having brown eyes and wearing glasses is 0.15. The probability of neither having brown eyes nor wearing glasses is 0.3. Given that eye colour and wearing glasses are independent, find the probability of:

a having brown eyes **b** wearing glasses.

Conditional probability

The probability of A given that B has happened can be written as $P(A \mid B)$.

Venn diagrams provide a good way of thinking about conditional probability.

If B has happened then you must be in the hatched region. The probability of A now happening depends upon the relative probability of the shaded region compared to the hatched region. This leads to an extremely useful formula for conditional probability.

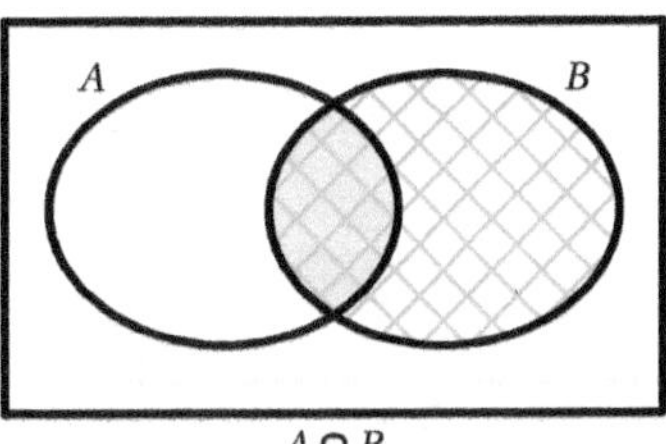

$A \cap B$

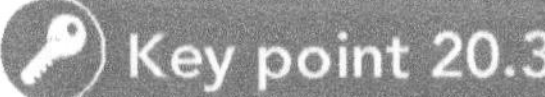

Key point 20.3

The probability of A given B has happened is:

$$P(A \mid B) = \frac{P(A \cap B)}{P(B)}$$

WORKED EXAMPLE 20.2

The probability that a randomly chosen resident of a city in Japan is a millionaire is $\frac{1}{10\,000}$. The probability that a randomly chosen resident lives in a mansion is $\frac{1}{30\,000}$. Only 1 in 40 000 are millionaires who live in mansions. What is the probability of a randomly chosen individual being a millionaire given that they live in a mansion?

$$\text{P(millionaire}\mid\text{mansion)} = \frac{\text{P(millionaire} \cap \text{mansion)}}{\text{P(mansion)}}$$

Write the required probability in the 'given' notation and apply the formula.

$$= \frac{\left(\frac{1}{40\,000}\right)}{\left(\frac{1}{30\,000}\right)}$$

$$= \frac{3}{4}$$

WORKED EXAMPLE 20.3

$\text{P}(A) = \frac{1}{2}$, $\text{P}(B) = \frac{1}{4}$ and $\text{P}((A \cup B)') = \frac{5}{12}$.

a Find $\text{P}(A \cap B)$.

b Hence find $\text{P}(A \mid B)$.

a

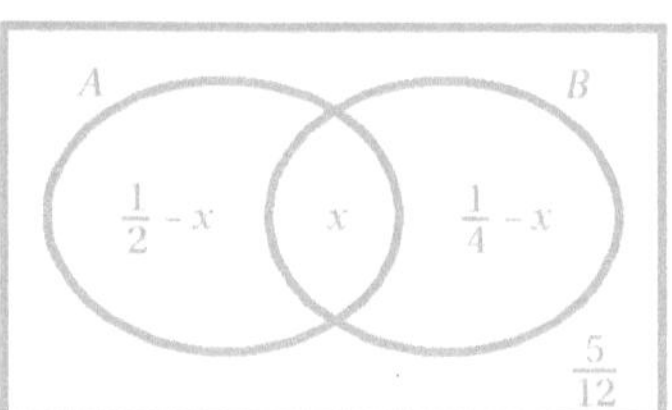

It is often a good idea to start with labelling the intersection with an unknown and writing in all the remaining information in terms of the unknown.

$$\text{So} \left(\frac{1}{2} - x\right) + x + \left(\frac{1}{4} - x\right) + \frac{5}{12} = 1$$

The probabilities sum to 1.

$$\frac{7}{6} - x = 1$$

$$x = \frac{1}{6}$$

$$\therefore \text{P}(A \cap B) = \frac{1}{6}$$

Make it clear that you know x is $\text{P}(A \cap B)$.

b $$\text{P}(A \mid B) = \frac{\text{P}(A \cap B)}{\text{P}(B)} = \frac{\left(\frac{1}{6}\right)}{\left(\frac{1}{4}\right)}$$

Use Key point 20.3.

$$= \frac{2}{3}$$

You can use Venn diagrams for more than two groups. They can represent the number in the group as well as the probability.

WORKED EXAMPLE 20.4

In a class of 32 students, 19 students have a bicycle, 21 students have a mobile phone and 16 students have a laptop computer. 11 students have both a bike and a phone, 12 students have both a phone and a laptop and 6 students have both a bike and a laptop. 2 students have none of these objects.

a How many students have a bike, a phone and a laptop?

b What is the probability that they have all three of the items, given that they have at least two of them?

a

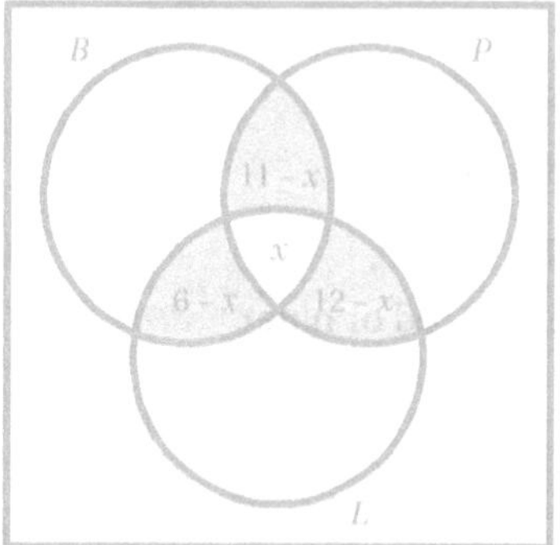

Draw a Venn diagram showing three overlapping groups, and label the size of the central region as x. Then work outwards. For example, the number who have a bicycle and a phone but not a laptop will be $11 - x$.

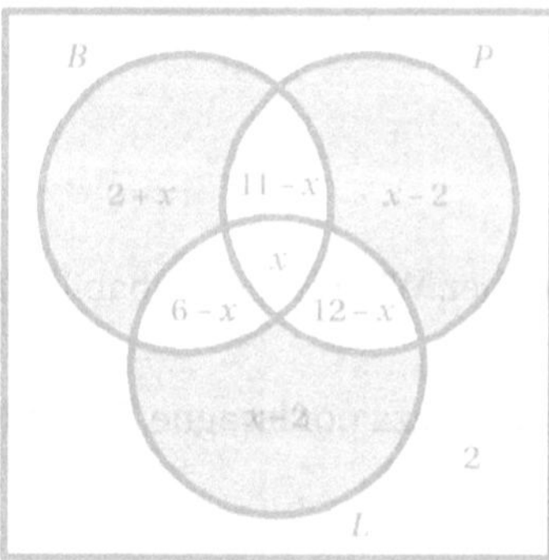

Continue working outwards. For example, the total of all the bicycle regions must be 19, so the remaining section is $19 - (11 - x) - (6 - x) - x$ which is $2 + x$.

You know there are 2 students outside B, L and P.

$$(2 + x) + (11 - x) + (6 - x) + x + (x - 2) + (12 - x) + (x - 2) + 2 = 32$$

$$29 + x = 32$$

$$x = 3$$

Therefore three students have a bicycle, a phone and a laptop.

Use the fact that there are 32 students in the class to form an equation.

b There are 23 students who have at least two items and three with three items.

So $\text{P}(3 \text{ items} \mid \text{at least 2 items}) = \frac{3}{23}$.

Use the Venn diagram to find the total number of students in overlapping regions.

EXERCISE 20B

1 For each question, write in mathematical notation the probability required. You should write an expression rather than a number.

a Find the probability that the outcome on a dice is prime and odd.

b Find the probability that a person is from either Senegal or Taiwan.

c A student is studying A Levels. Find the probability that the student is also French.

d If a playing card is a red card find the probability that it is a heart.

e What proportion of German people live in Munich?

f What is the probability that someone is wearing neither black nor white socks?

g What is the probability that a vegetable is a potato if it is not a cabbage?

h What is the probability that a ball drawn is red, given that the ball is either red or blue?

2 a i If $\mathrm{P}(X) = 0.3$ and $\mathrm{P}(X \cap Y) = \frac{1}{5}$ find $\mathrm{P}(Y \mid X)$.

ii If $\mathrm{P}(Y) = 0.8$ and $\mathrm{P}(X \cap Y) = \frac{3}{7}$ find $\mathrm{P}(X \mid Y)$.

b i If $\mathrm{P}(X) = 0.4$, $\mathrm{P}(Y) = 0.7$ and $\mathrm{P}(X \cap Y) = \frac{1}{4}$ find $\mathrm{P}(X \mid Y)$.

ii If $\mathrm{P}(X) = 0.6$, $\mathrm{P}(Y) = 0.9$ and $\mathrm{P}(X \cap Y) = \frac{1}{2}$ find $\mathrm{P}(Y \mid X)$.

3 In these questions, you might find it helpful to draw a Venn diagram.

a In a class of 30 students 20 take French, 12 take German and 4 take neither. What is the probability that someone who takes German also takes French?

b In a survey 60% of people like pizza and 50% like lasagne. 10% like nether pizza nor lasagne. Find the probability that someone likes lasagne given that they like pizza.

4 Simplify each expression where possible.

a $\mathrm{P}(x > 2 \cap x > 4)$ b $\mathrm{P}(y \leqslant 3 \cup y < 2)$ c $\mathrm{P}(a < 3 \cap a > 4)$ d $\mathrm{P}(a < 5 \cup a \geqslant 0)$

e P(apple $\cup$ fruit) f P(apple $\cap$ fruit) g P(multiple of 4 $\cap$ multiple of 2) h P(square $\cup$ rectangle)

i P(cat $\cap$ dog′) j P(cat $\cap$ pet′) k P(blue $\cap$ (blue $\cup$ red)) l P(blue $\cap$ (blue $\cap$ red))

m P(rectangle | square) n $\mathrm{P}(x^2 = 9 \mid x = 2)$

5 Out of 145 students in a college, 34 play football, 18 play badminton, and 5 play both sports.

a Draw a Venn diagram showing this information.

b How many students play neither sport?

c What is the probability that a randomly chosen student plays badminton?

d Given that the chosen student plays football, what is the probability that they also play badminton?

6 Out of 145 students in a college, 58 study Mathematics, 47 study Economics and 72 study neither of the two subjects.

a Draw a Venn diagram to show this information.

b How many students study both subjects?

c A student tells you that he studies Mathematics. What is the probability that he studies both Mathematics and Economics?

7 **a** In a survey, 60% of people are in favour of a new primary school and 85% are in favour of a new library. Half of all those surveyed would like both a new primary school and a new library. What percentage supported neither a new library nor a new primary school?

b What proportion of those wanting a new primary school also wanted a new library?

8 If $P(A) = 0.2$, $P(A \cap B) = 0.1$ and $P(A \cup B) = 0.7$:

a find $P(B)$ **b** find $P(A \mid B)$.

9 An integer is chosen at random from the first one thousand positive integers. Find the probability that the integer chosen is:

a a multiple of 6 **b** a multiple of **both** 6 and 8 **c** a multiple of 8 given that it is a multiple of 6.

10 Denise conducts a survey about food preferences in the college. She asks students which of the three meals (spaghetti bolognese, chilli con carne, and vegetable curry) they would prefer to eat. She finds out that, out of the 145 students:

- 43 would eat spaghetti bolognese
- 80 would eat vegetable curry
- 20 would eat both spaghetti and curry
- 24 would eat both curry and chilli
- 35 would eat both chilli and spaghetti
- 12 would eat all three meals
- 10 would prefer not to eat any of the three meals.

a Draw a Venn diagram showing this information.

b How many students would eat only spaghetti?

c How many students would eat chilli?

d What is the probability that a randomly selected student would eat only one of the three meals?

e Given that a student would eat only one of the three meals, what is the probability that they would eat curry?

f Find the probability that a randomly selected student would eat at least two of the three meals.

11 The probability that a person has dark hair is 0.7, the probability that they have blue eyes is 0.4 and the probability that they have both dark hair and blue eyes is 0.2.

a Draw a Venn diagram showing this information.

b Find the probability that a person has neither dark hair nor blue eyes.

c Given that a person has dark hair, find the probability that they also have blue eyes.

d Given that a person does not have dark hair, find the probability that they have blue eyes.

e Are the characteristics of having dark hair and having blue eyes independent? Explain your answer.

12 The probability that it rains on any given day is 0.45 and the probability that it is cold is 0.6. The probability that it is neither cold nor raining is 0.25.

a Find the probability that it is both cold and raining.

b Draw a Venn diagram showing this information.

c Given that it is raining, find the probability that it is not cold.

d Given that it is not cold, find the probability that it is raining.

e Are the events 'it's raining' and 'it's cold' independent? Explain your answer and show any supporting calculations.

13 If $P(A|B) = \frac{1}{2}$ and $P(B|A) = \frac{1}{3}$ find $\frac{P(A \cup B)}{P(A \cap B)}$.

14 If $P(A) = P(B) = \frac{4}{5}P(A \cup B)$ find $P(A|B)$.

15 If $P(A) = 0.8$ and $P(B) = 0.4$ find the maximum and minimum values of $P(A|B)$.

16 **a** If $P(X)$ represents a probability, state the possible values that $P(X)$ can take.

b Express $P(A) - P(A \cap B)$ in terms of $P(A)$ and $P(B|A)$.

c By considering an expression for $P(A \cup B) - P(A \cap B)$ show that $P(A \cup B) \geqslant P(A \cap B)$.

Section 2: Two-way tables

Another useful way of looking at conditional probability is to use two-way tables. This lists all the possible outcomes varying along two factors.

Fast forward

If you study the Statistics Option in Further Mathematics you will see a method called the chi-squared test to see if the two factors are independent.

WORKED EXAMPLE 20.5

The table shows the arrival time of a random sample of 100 letters, as well as whether they were posted first class or second class.

	Next day	Later
1st class	64	16
2nd class	12	8

Find:

a P(1st class and next day)

b P(next day | 1st class)

c P(1st class | next day)

a $\frac{64}{100} = 0.64$ — There are 64 letters in the 1st class and next day categories, out of 100 letters.

b $\frac{64}{80} = 0.8$ — There are 64 next day letters out of 80 1st class letters.

c $\frac{64}{76} = 0.842$ — There are 64 1st class letters out of 76 next day letters.

EXERCISE 20C

1 In each two-way table, find $P(A \mid X)$.

a **i**

	A	B
X	12	18
Y	14	16

ii

	A	B
X	15	35
Y	12	16

b **i**

	X	Y
A	3	7
B	6	4

ii

	A	B
X	10	16
Y	15	13

c **i**

	X	Y	Z
A	6	5	4
B	8	7	2
C	10	9	0

ii

	A	B	C
X	3	8	9
Y	5	6	2
Z	7	4	1

2 The two-way table lists the numbers of students in different year groups in a school.

	Year 9	Year 10	Year 11	Total
Girls		85		
Boys	88			240
Total	186		163	509

a Copy and complete the table.

b Find the probability that a randomly selected student is a girl from year 11.

c Find the probability that a randomly selected girl is from year 11.

3 The two-way table describes the additions made to coffee in a drinks machine in one day.

	Milk	No milk
Sugar	28	14
No sugar	32	16

a Find the probability of sugar being added.

b Find the probability that sugar is added if milk is added.

c Show that whether milk is added is independent of whether sugar is added.

4 The table shows the numbers of returned shoes to three stores in one day.

	Store A	Store B	Store C
Returned	7	14	6
Unreturned	18	20	16

a Find P(returned).

b Find P(returned | Store A).

c Find P(returned | Store A′).

5 The table shows a general two-way table.

	Q	R
S	a	b
T	c	d

Find, in terms of a, b, c and d:

a $P(Q \cap S)$ **b** $P(Q \cup S)$ **c** $P(Q)$

d $P(S \mid Q)$ **e** $P(S \mid Q')$.

6 The table shows the results of the top three countries in the 2016 Olympics.

	Gold	Silver	Bronze
USA	46	37	38
GB	27	23	17
China	26	18	26

A random result is chosen from amongst these 258 results. Find:

a P(gold | GB) **b** P(gold) **c** P(gold $\cap$ GB) **d** P(gold $\cup$ GB).

7 A company makes three different sizes of T-shirt in three different colours. This table shows the sales in a week.

	S	M	L
White	14	22	18
Black	24	36	33
Green	23	30	23

Find:

a P(S $\cap$ white) **b** P(S $\cup$ white) **c** P(S)

d P(S | white) **e** P(S′ | white′).

8 This table gives the probabilities of different midday temperatures in different air pressures in the UK, based on long-term observations.

	> 1000 hPa	< 1000 hPa
< 10 °C	0.08	0.12
10 °C to 20 °C	0.27	0.29
> 20 °C	0.14	0.10

a For a randomly chosen day find:

i P(> 1000 hPa) **ii** P(< 1000 hPa $\cap$ > 20 °C) **iii** P(> 1000 hPa | > 10 °C).

9 James investigates two identities:

1 $P(A \mid B) + P(A \mid B') \equiv 1$

2 $P(A' \mid B) + P(A \mid B) \equiv 1$

Use a counterexample to show that one of these identities is incorrect.

Section 3: Tree diagrams

A tree diagram is a useful way of illustrating situations where one outcome depends upon another. For example, this diagram shows the experience of a restaurant trying to predict how many portions of chips it serves.

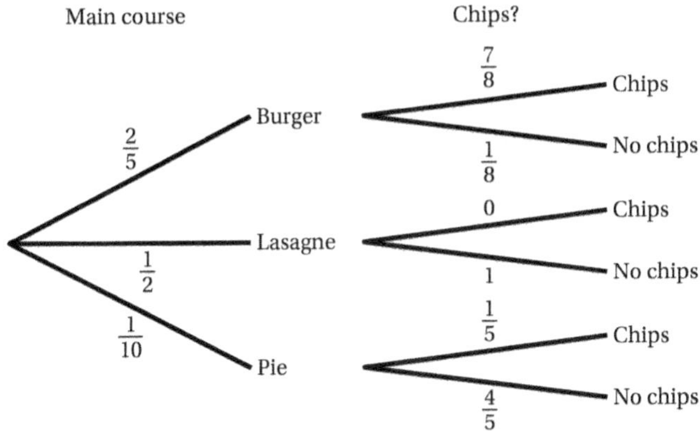

The second probabilities all depend on the first, so they are conditional probabilities. For example:

$$\text{P(chips} \mid \text{burger)} = \frac{7}{8}$$

You may have learnt that the probability of being on a branch is found by multiplying along the branch. For example, this means:

$$\begin{aligned}\text{P(chips} \cap \text{burger)} &= \text{P(burger)} \times \text{P(chips} \mid \text{burger)} \\ &= \frac{2}{5} \times \frac{7}{8} \\ &= \frac{7}{20}\end{aligned}$$

Key point 20.4

For a tree diagram, use the formula:

$$\mathrm{P}(A \cap B) = \mathrm{P}(B) \times \mathrm{P}(A \mid B)$$

This will be given in your formula book.

Tip

This is actually just a rearrangement of Key point 20.3:

$$\mathrm{P}(A \mid B) = \frac{\mathrm{P}(A \cap B)}{\mathrm{P}(B)}$$

$$\Leftrightarrow \mathrm{P}(A \cap B) = \mathrm{P}(B) \times \mathrm{P}(A \mid B)$$

If you wanted to find the overall probability of chips, you need to add different branches together:

$$\begin{aligned}\text{P(chips)} &= \text{P(chips} \cap \text{burger)} + \text{P(chips} \cap \text{lasagne)} + \text{P(chips} \cap \text{pie)} \\ &= \left(\frac{2}{5} \times \frac{7}{8}\right) + \left(\frac{1}{2} \times 0\right) + \left(\frac{1}{10} \times \frac{1}{5}\right) \\ &= \frac{37}{100}\end{aligned}$$

WORKED EXAMPLE 20.6

If Lucas revises there is an 80% chance he will pass the test, but if he doesn't revise there is only a 30% chance of his passing. He revises for $\frac{3}{4}$ of tests. What proportion of tests does he pass?

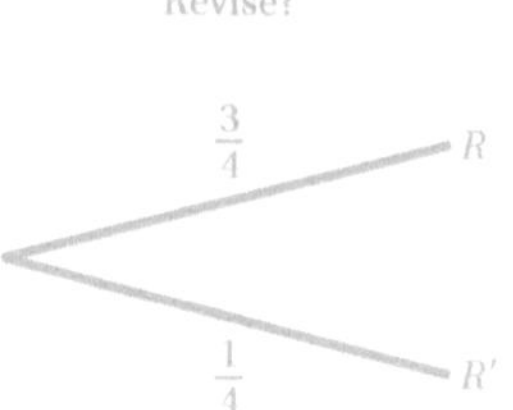

Decide which probability is not conditional. Start the tree diagram with this event. The probability of passing the test is conditional on revision, so the revision branches have to come first.

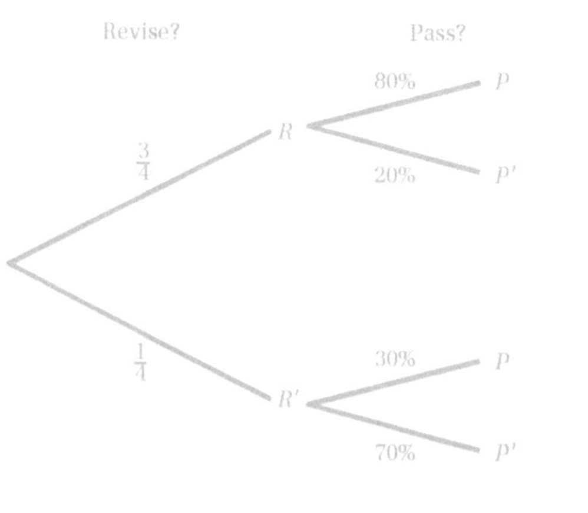

Add the conditional event.

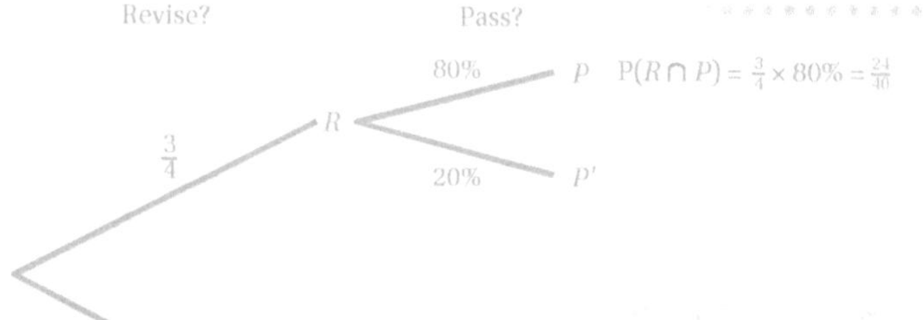

Identify which branches result in passing the test.

Multiply to find the probability at the end of each branch.

$$\text{P(passing)} = \text{P(revising} \cap \text{passing)} + \text{P(not revising} \cap \text{passing)}$$

$$= \frac{24}{40} + \frac{3}{40}$$

$$= \frac{27}{40}$$

Sometimes, you can use the information found from a tree diagram to find another conditional probability.

Tip

Worked examples 20.6 and 20.7 use the terms 'chance' and 'proportion'. These are just other words for probability.

WORKED EXAMPLE 20.7

If it is raining in the morning there is a 90% chance that Shivani will take her umbrella. If it is not raining in the morning there is only a $\frac{1}{5}$ chance of Shivani taking her umbrella. On any given morning the probability of rain is 0.1.

a What is the probability that Shivani takes an umbrella?

b If you see Shivani with an umbrella, what is the probability that it was raining that morning?

a

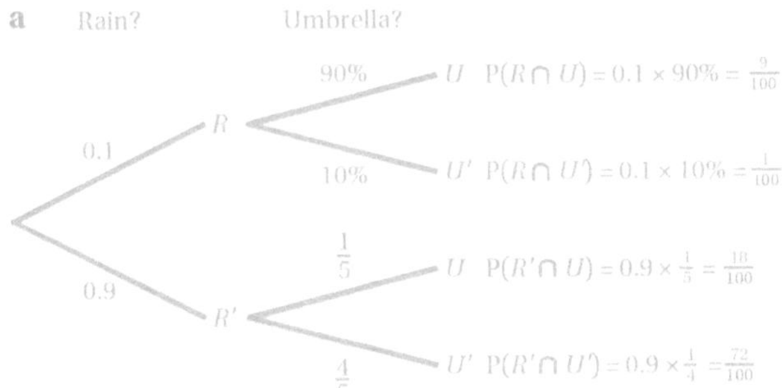

First draw a tree diagram.

$$\text{P(rain} \cap \text{umbrella)} = \frac{9}{100}$$

$$\text{P(rain}' \cap \text{umbrella)} = \frac{18}{100}$$

$$\text{P(umbrella)} = \frac{9}{100} + \frac{18}{100} = \frac{27}{100}$$

Use the tree diagram to find relevant probabilities.

b

$$\text{P(rain | umbrella)} = \frac{\text{P(rain} \cap \text{umbrella)}}{\text{P(umbrella)}}$$

$$= \frac{\left(\frac{9}{100}\right)}{\left(\frac{27}{100}\right)} = \frac{1}{3}$$

Since you need to find a conditional probability, use Key point 20.3.

In Worked example 20.7 you were given P(umbrella | rain) and found P(rain | umbrella). You can use Key point 20.3 to derive a formula for this.

Key point 20.3 says that $P(A \cap B) = P(B) \times P(A \mid B)$ but you could equally well have swapped A and B and written $P(B \cap A) = P(A) \times P(B \mid A)$.

However, $P(A \cap B)$ and $P(B \cap A)$ are the same thing (they are both the probability of A and B). So the right-hand sides of the two expressions are equal. This leads to a formula that could be used to solve Worked example 20.7.

Tip

Quite often in questions like this you use the answer to the first part in the second part.

Key point 20.5

$P(A \mid B)$ and $P(B \mid A)$ can be related by using the formula:

$$P(B) \times P(A \mid B) = P(A) \times P(B \mid A)$$

Focus on...

See Focus on... Proof 4 for the use of conditional probability and tree diagrams in analysing a claim of guilt in a legal case.

WORK IT OUT 20.1

The probability of an acorn landing more than 2 m from the original tree is 80%. Of those that land more than 2 m from the tree, 20% germinate. Of those that land less than 2 m from the tree 5% germinate.

An acorn germinates. What is the probability that it landed more than 2 m from the tree?

Which is the correct solution? Identify the errors made in the incorrect solutions.

Solution A

Work out P(germinate) from a tree diagram.

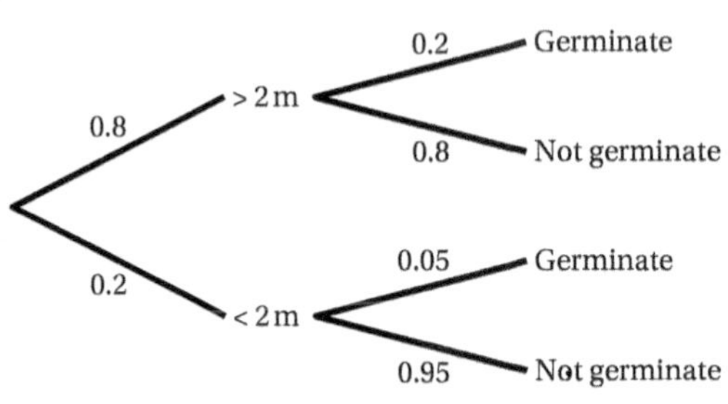

$$P(>2\text{ m})\times P(\text{germinate}\,|>2\text{ m}) = P(\text{germinate})\times P(>2\text{ m}\,|\,\text{germinate})$$

$$0.8\times 0.2 = [0.8\times 0.2+0.2\times 0.05]\times P(>2\text{ m}\,|\,\text{germinate})$$

$$\text{So } P(>2\text{ m}\,|\,\text{germinate}) = 0.16\div 0.17 \approx 0.941$$

Solution B

$$P(>2\text{ m}\,|\,\text{germinate}) = \frac{P(>2\text{ m}\cap\text{germinate})}{P(\text{germinate})}$$

$$= \frac{P(>2\text{ m})\,P(\text{germinate})}{P(\text{germinate})} = P(>2\text{ m}) = 0.8$$

Solution C

Using a Venn diagram:

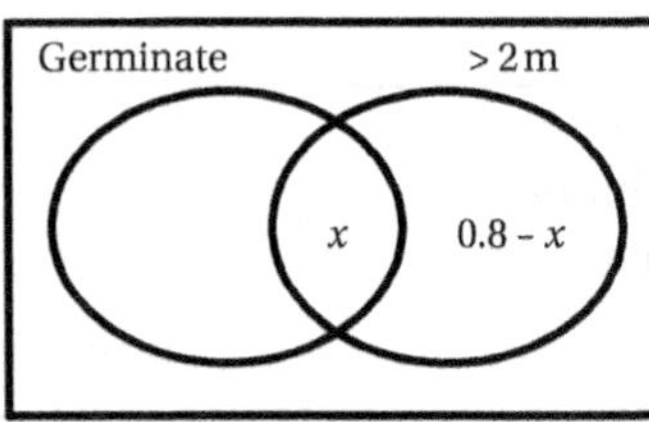

Since $P(\text{germinate}\,|>2\text{ m}) = 0.2$:

$$\frac{x}{0.8} = 0.2$$

So $P(\text{germinate AND} > 2\text{ m}) = 0.16$

EXERCISE 20D

1 In these questions you might find it helpful to write the information in a suitable tree diagram.

a **i** $P(A) = 0.4$ and $P(B \mid A) = 0.3$. Find $P(A \cap B)$.

ii $P(X) = \frac{3}{5}$ and $P(Y \mid X) = 0$. Find $P(X \cap Y)$.

b **i** $P(A) = 0.3, P(B) = 0.2$ and $P(B \mid A) = 0.8$. Find $P(A \cap B)$.

ii $P(A) = 0.4, P(B) = 0.8$ and $P(A \mid B) = 0.3$. Find $P(A \cap B)$.

c **i** $P(A) = \frac{2}{5}, P(B) = \frac{1}{3}$ and $P(A \mid B) = \frac{1}{4}$. Find $P(A \cup B)$.

ii $P(A) = \frac{3}{4}$, $P(B) = \frac{1}{4}$ and $P(B \mid A) = \frac{1}{3}$. Find $P(A \cup B)$.

2 A class contains 6 boys and 8 girls. Two students are picked at random. What is the probability that they are both boys?

3 A bag contains 4 red balls, 3 blue balls and 2 green balls. A ball is chosen at random from the bag and is not replaced A second ball is chosen. Find the probability of choosing one green ball and one blue ball in any order.

4 $P(A) = 0.3; P(B \mid A) = 0.6; P(B \mid A') = 0.8$.

a Illustrate this information on a tree diagram.

b Find $P(A \cap B)$.

c Find $P(A \cup B)$.

d Find $P(B)$.

Elevate

See Support Sheet 20 for a further example on conditional probability and tree diagrams and for more practice questions.

5 Given that $P(X) = \frac{1}{3}, P(Y \mid X) = \frac{2}{9}$ and $P(Y \mid X') = \frac{1}{3}$, find:

a $P(Y')$ **b** $P(X' \cup Y')$.

6 A factory has two machines for making widgets. The older machine has a larger capacity, so it makes 60% of the widgets, but 6% are rejected by quality control. The newer machine has only a 3% rejection rate. Find the probability that a randomly selected widget is rejected.

7 The school tennis league consists of 12 players. Daniel has a 30% chance of winning any game against a higher-ranked player, and a 70% chance of winning any game against a lower-ranked player. If Daniel is currently in third place, find the probability that he wins his next game against a random opponent.

8 Box *A* contains 6 red balls and 4 green balls. Box *B* contains 5 red balls and 3 green balls. A standard fair cubical dice is thrown. If a six is obtained, a ball is selected from box *A*; otherwise a ball is selected from box *B*.

a Calculate the probability that the ball selected was red.

b Given that the ball selected was red, calculate the probability that it came from box *B*.

9 Robert travels to work by train every weekday from Monday to Friday. The probability that he catches the 7.30 am train on Monday is $\frac{2}{3}$. The probability that he catches the 7.30 am train on any other weekday is 90%. A weekday is chosen at random.

a Find the probability that he catches the 7.30 am train on that day.

b Given that he catches the 7.30 am train on that day, find the probability that the chosen day is Monday.

10 Bag 1 contains 6 red cubes and 10 blue cubes. Bag 2 contains 7 red cubes and 3 blue cubes.

Two cubes are drawn at random, the first from Bag 1 and the second from Bag 2.

a Find the probability that the cubes are of the same colour.

b Given that the cubes selected are of different colours, find the probability that the red cube was selected from Bag 1.

11 On any day in April there is a $\frac{2}{3}$ chance of rain in the morning. If it is raining there is a $\frac{4}{5}$ chance David will remember his umbrella, but if it is not raining there is only a $\frac{2}{5}$ chance he will remember his umbrella.

a On a random day in April, what is the probability David has his umbrella?

b Given that David has his umbrella on a day in April, what is the probability that it was raining?

12 A new blood test has been devised for early detection of a disease. Studies show that the probability that the blood test correctly identifies someone with this disease is 0.95, and the probability that the blood test correctly identifies someone without that disease is 0.99. The incidence of this disease in the general population is 0.0003.

The result of the blood test on one patient indicates that he has the disease. What is the probability that this patient has the disease?

13 Louise has two coins: one is a normal fair coin with heads on one side and tails on the other. The second coin has heads on both sides. Louise randomly picks a coin and flips it. The result comes up heads. What is the probability that Louise chooses the fair coin?

14 There are 36 discs in a bag. Some of them are black and the rest are white. Two are simultaneously selected at random. Given that the probability of selecting two discs of the same colour is equal to the probability of selecting two discs of different colour, how many black discs are there in the bag?

15 Prove that if A and B are independent then so are A' and B'.

Checklist of learning and understanding

- You can use set notation when describing probabilities:
  - $A \cap B$ is the **intersection** of A and B, meaning both A and B happen
  - $A \cup B$ is the **union** of A and B, meaning A happens or B happens or both happen
  - A' is the **complement** of A, meaning everything that could happen other than A.
- From a Venn diagram: if $P(A \cup B) = P(A) + P(B) - P(A \cap B)$. If $P(A \cap B) = 0$, then A and B are **mutually exclusive**, and the formula reduces to $P(A \cup B) = P(A) + P(B)$.
- $P(A \mid B)$ is the probability of A happening if B has happened. This can be visualised in Venn diagrams, two-way tables or tree diagrams.
  - You can also use the formula $P(A \mid B) = \frac{P(A \cap B)}{P(B)}$.
  - In a tree diagram this formula is often rearranged to get:

    $P(A \cap B) = P(A) \times P(B \mid A)$
  - You can relate $P(A \mid B)$ and $P(B \mid A)$ by using the formula:

    $P(B) \times P(A \mid B) = P(A) \times P(B \mid A)$

Mixed practice 20

1. A box of 10 chocolates contains 6 milk and 4 dark. Three are chosen at random, without replacement. Given that the first chocolate chosen is milk, find the probability that all three are milk.

   Choose from these options.

   A $\frac{1}{6}$ B $\frac{27}{125}$ C $\frac{5}{18}$ D $\frac{2}{9}$

2. $P(A \mid B) = \frac{1}{4}, P(A' \cap B') = \frac{3}{10}, P(B) = \frac{3}{5}$.

   Which statement is true? Choose from these options.

   A A and B are mutually exclusive.

   B A and B are independent.

   C A and B are neither mutually exclusive nor independent.

   D More information is required to decide.

3. A drawer contains 6 red socks, 4 black socks and 8 white socks. Two socks are picked at random without replacement.

   What is the probability that two socks of the same colour are drawn?

4. Out of 100 flies studied in an experiment 24 have the gene ig9 and 52 have the gene xar3. 28 have neither gene.

   a Illustrate this information on a Venn diagram.

   b Find P(ig9 $\cap$ xar3).

   c Find P(ig9 | xar3).

5. The table shows the colour of hair and the colour of eyes of a sample of 750 people from a particular population.

		Colour of hair					
		Black	Dark	Medium	Fair	Auburn	Total
Colour of eyes	Blue	6	51	68	66	24	215
	Brown	14	92	97	90	47	340
	Green	0	37	55	64	39	195
	Total	20	180	220	220	110	750

   Calculate, **to three decimal places**, the probability that a person, selected at random from this sample, has:

   i fair hair

   ii auburn hair and blue eyes

   iii either auburn hair or blue eyes but not both

   iv green eyes, given that the person has fair hair

   v fair hair, given that the person has green eyes.

[© AQA 2014]

6 $P(X) = 0.2; P(A \mid X) = 0.4; P(A \mid X') = 0.5$.

a Illustrate this information on a tree diagram.

b Find $P(A \cap X)$.

c Find $P(A \cup X)$.

d Find $P(A)$.

7 The number of hours spent practising before a music examination was recorded by a teacher for each of her students.

Hours practised	Male	Female
10 or fewer	4	4
11 to 20	8	12
21 or more	7	5

a Find P(male | 10 or fewer hours practised).

b Two different students are randomly selected for further interviews. Find the probability that both are male.

8 Alison is a member of a tenpin bowling club which meets at a bowling alley on Wednesday and Thursday evenings.

The probability that she bowls on a Wednesday evening is 0.90. Independently, the probability that she bowls on a Thursday evening is 0.95.

a Calculate the probability that, during a particular week, Alison bowls on:

  i two evenings

  ii exactly one evening.

David, a friend of Alison, is a member of the same club.

The probability that he bowls on a Wednesday evening, given that Alison bowls on that evening, is 0.80. The probability that he bowls on a Wednesday evening, given that Alison does not bowl on that evening, is 0.15.

The probability that he bowls on a Thursday evening, given that Alison bowls on that evening, is 1.

The probability that he bowls on a Thursday evening, given that Alison does not bowl on that evening, is 0.

b Calculate the probability that, during a particular week:

  i Alison and David bowl on a Wednesday evening

  ii Alison and David bowl on both evenings

  iii Alison, but not David, bowls on a Thursday evening

  iv neither bowls on either evening.

[© AQA 2013]

9 The probability that a student plays badminton is 0.3. The probability that a student plays neither football nor badminton is 0.5 and the probability that a student plays both sports is x.

a Draw a Venn diagram showing this information.

b Find the probability that a student plays badminton, but not football.

c Given that a student plays football, the probability that they also play badminton is 0.5.

Find the probability that a student plays both badminton and football.

d Hence complete your Venn diagram. What is the probability that a student plays only badminton?

e Given that a student plays only one sport, what it the probability that they play badminton?

10 Only two international airlines fly daily into an airport. Pi Air has 40 flights a day and Lambda Air has 25 flights a day. Passengers flying with Pi Air have a $\frac{1}{10}$ probability of losing their luggage and passengers flying with Lambda Air have a $\frac{1}{4}$ probability of losing their luggage. Someone complains that their luggage has been lost. Find the probability that they travelled with Pi Air.

11 A girl walks to school every day. If it is not raining, the probability that she is late is $\frac{1}{5}$. If it is raining, the probability that she is late is $\frac{2}{3}$. The probability that it rains on a particular day is $\frac{1}{4}$. On one particular day the girl is late. Find the probability that it was raining on that day.

12 The probability that a man leaves his umbrella in any shop he visits is $\frac{1}{5}$. After visiting two shops in succession, he finds he has left his umbrella in one of them. What is the probability that he left his umbrella in the second shop?

13 a A large bag of sweets contains 8 red and 12 yellow sweets. Two sweets are chosen at random from the bag without replacement. Find the probability that two red sweets are chosen.

b A small bag contains 4 red and n yellow sweets. Two sweets are chosen without replacement from this bag. If the probability that two red sweets are chosen is $\frac{2}{15}$, show that $n = 6$.

Ayesha has one large bag and two small bags of sweets. She selects a bag at random and then picks two sweets without replacement.

c Calculate the probability that two red sweets are chosen.

d Given that two red sweets are chosen, find the probability that Ayesha had selected the large bag.

14 The probability that it rains during a summer's day in a certain town is 0.2. In this town, the probability that the daily maximum temperature exceeds 25 °C is 0.3 when it rains and 0.6 when it does not rain. Given that the maximum daily temperature exceeded 25 °C on a particular summer's day, find the probability that it rained on that day.

15 Given that $P((A \cup B)') = 0$, $P(A' \mid B) = \frac{1}{5}$ and $P(A) = \frac{14}{15}$ find $P(B)$.

16 Two events, A and B satisfy $P(A) = P(B) = k\,P(A \cup B)$

a Find the possible values k can take.

b Find $P(A \mid B)$ in terms of k.

See Extension Sheet 20 for a selection of more challenging problems.

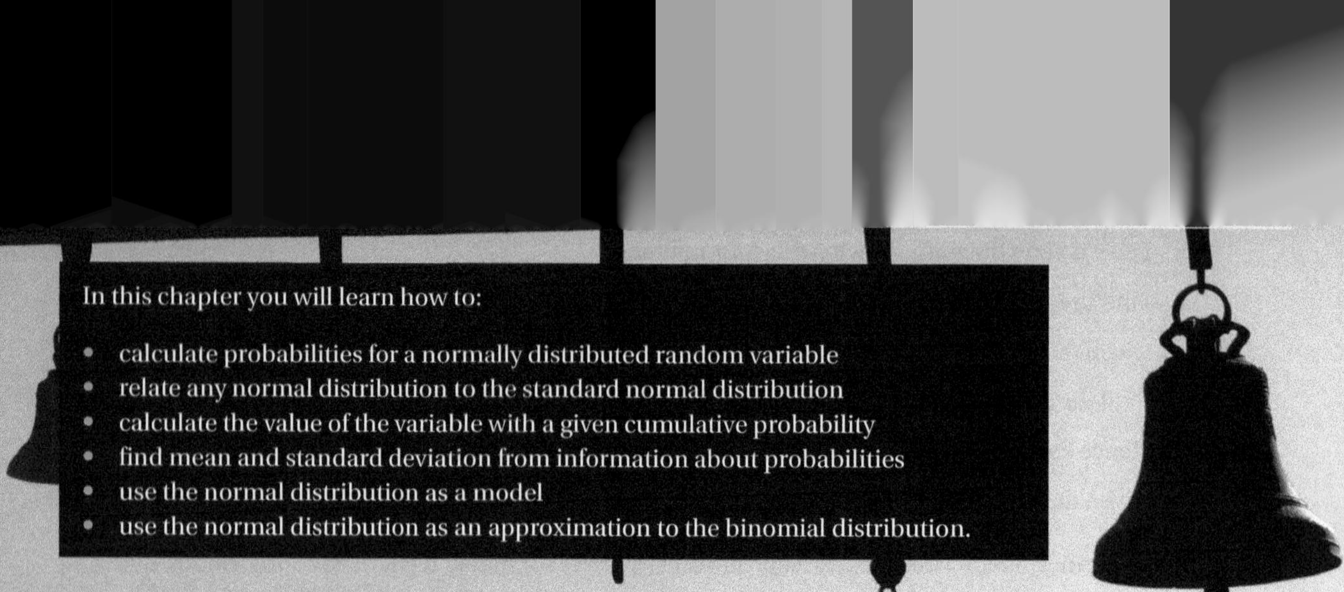

Before you start...

GCSE	You should be able to solve simultaneous equations.	1 Solve these simultaneous equations. $x+4.72y=7.32$ $x-1.28y=0.435$
Student Book 1, Chapter 20	You should be able to interpret histograms.	2 What is the frequency density of a group of 60 people with masses strictly between 40 kg and 50 kg?
Student Book 1, Chapter 21	You should be able to use the binomial distribution.	3 The probability of rolling a six on a biased dice is $\frac{1}{5}$. The dice is rolled four times. Find the probability of getting exactly 1 six.
Chapter 20	You should be able to work with tree diagrams.	4 There is a 20% chance of it raining. If it rains there is a 60% chance I am late. If it does not rain there is a 25% chance I am late. What is the probability that I am late?
Chapter 20	You should be able to calculate conditional probabilities.	5 Two dice are rolled – one red and one blue. If the total score is 5, what is the probability that the score on the red dice is 3?

What is the normal distribution?

You have already met discrete random variables, which you can describe by listing all possible values and their probabilities. With a continuous variable, such as height or time, it is impossible to list all values so you need a different way to describe how the probability is distributed across the possible values.

To do this you can use a similar idea to a histogram: draw a curve such that the area under the curve represents probability. This curve is called the **probability density function**.

 Rewind

For a reminder of discrete probability distributions, see Student Book 1, Chapter 21.

 Rewind

You met histograms in Student Book 1, Chapter 20.

There are many situations where a variable is very likely to be close to its average value, with values further away from the average becoming increasingly unlikely. You can model many such situations by using the **normal distribution**, which is symmetrical about the average value. Natural measurements, such as heights of people or masses of animals, tend to follow a normal distribution.

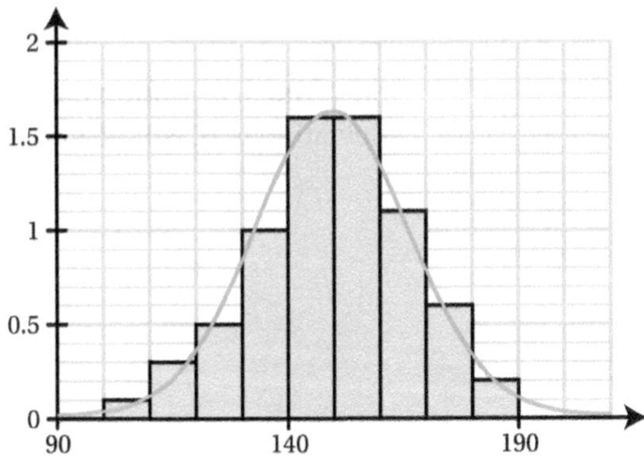

> **Elevate**
>
> See Extension Sheet 21 for an investigation that explores why the normal distribution occurs so often.

Section 1: Introduction to normal probabilities

To specify a normal distribution fully you need to know its mean (μ) and variance (σ^2). If a variable follows this distribution you use the notation $X \sim \mathrm{N}(\mu, \sigma^2)$.

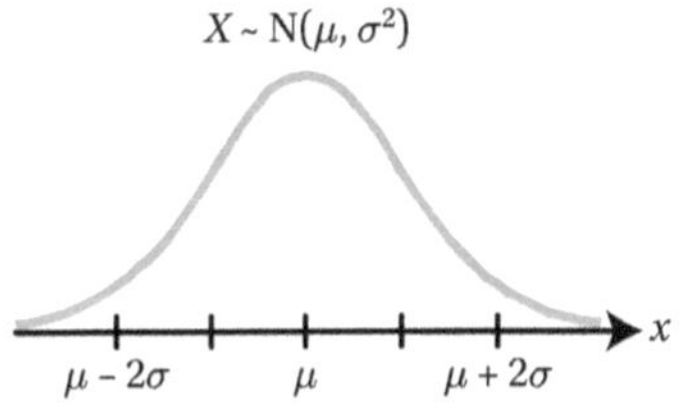

The probability density function is symmetrical about the mean. The standard deviation affects the width of the curve - the larger σ is, the wider the curve is. Remember that the probability corresponds to the **area** under the graph.

You know from Student Book 1 that you can find the area under a curve by integration. However, the equation for the normal distribution curve cannot be integrated exactly, so most calculators have a built-in function to find approximate probabilities.

> **Common error**
>
> Be careful with the notation: σ^2 is the variance, so $X \sim \mathrm{N}(10, 9)$ has standard deviation $\sigma = 3$.

You may find it helpful to sketch a diagram to get a visual representation of the probability you are trying to find.

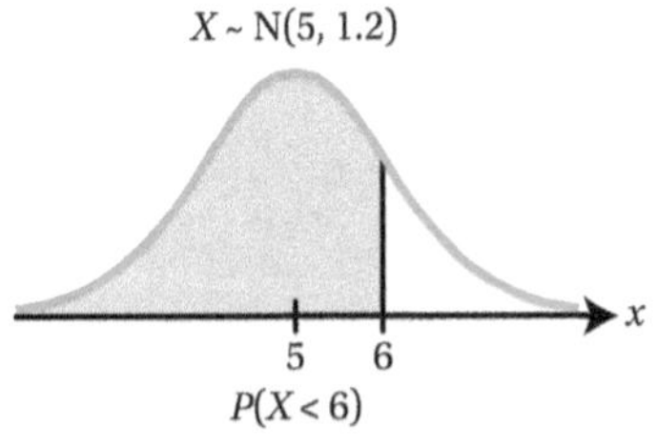

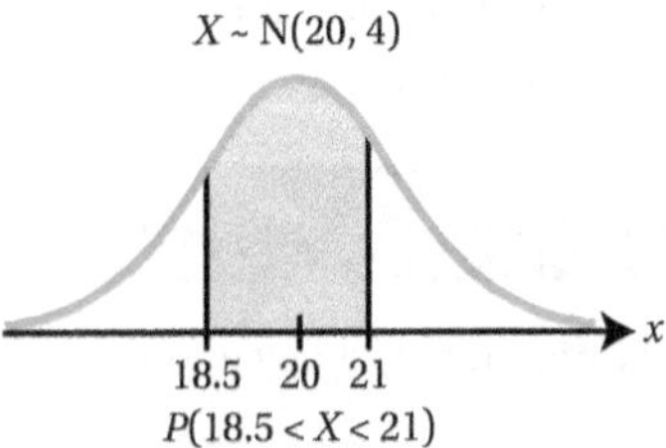

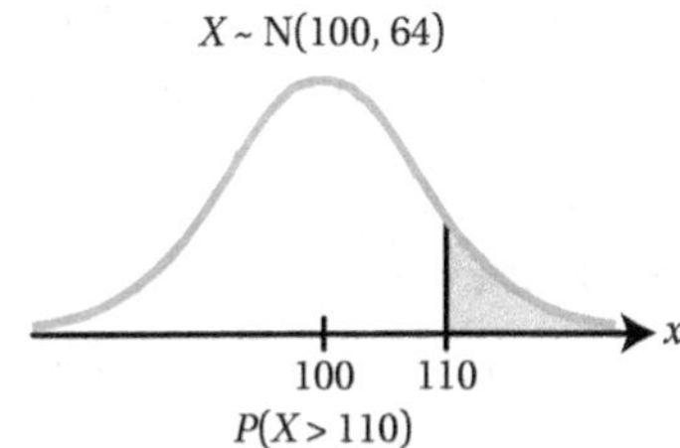

The diagrams can also provide a useful check, as you can see whether to expect the probability to be smaller or greater than 0.5.

> **Tip**
>
> With a continuous variable, $\mathrm{P}(X \leqslant k)$ and $\mathrm{P}(X < k)$ mean exactly the same thing, so you don't have to worry about whether end-points should be included. This is not the case for discrete variables!

WORKED EXAMPLE 21.1

The average height of people in a town is 170 cm with standard deviation 10 cm. What is the probability that a randomly selected resident:

a is less than 165 cm tall

b is between 180 cm and 190 cm tall?

X is the height of a town resident so

$X \sim \mathrm{N}(170, 100)$

State the distribution used. The standard deviation is 10, so the variance is 100.

a

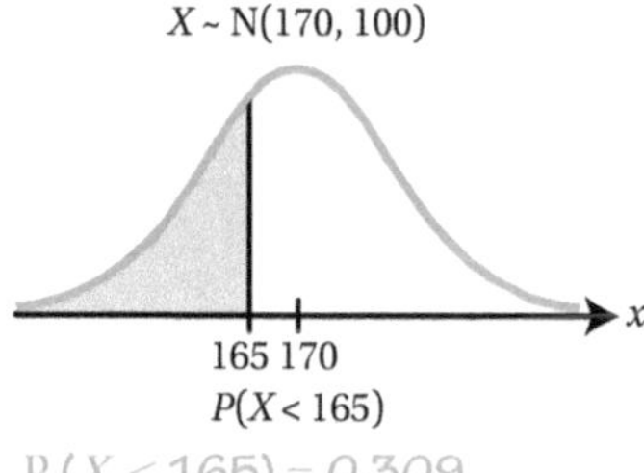

$\mathrm{P}(X < 165) = 0.309$

State the probability to be found and use your calculator.

b

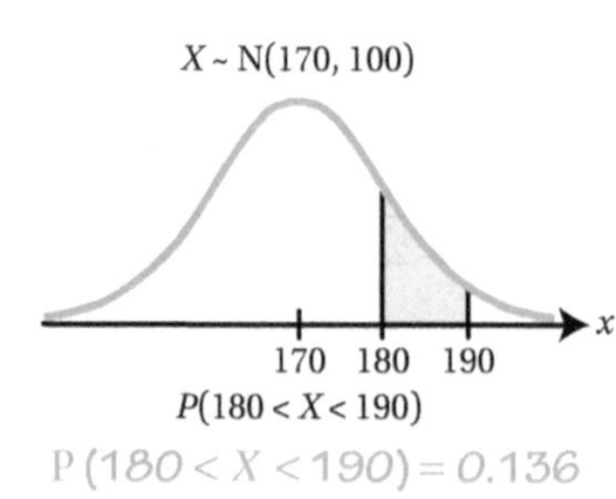

$\mathrm{P}(180 < X < 190) = 0.136$

State the probability to be found and use your calculator.

Note that the domain of the normal probability density function is all real numbers – the x-axis is an asymptote to the graph. This means that, in theory, a normal variable could take any real value. However, most of the data lies within three standard deviations of the mean. It is useful to remember the percentages summarised in Key point 21.1.

Key point 21.1

- Approximately 99.7% of the data lie within three standard deviations of the mean.
- Approximately 95% of data lie within two standard deviations of the mean.
- Approximately two-thirds of the data lie within one standard deviation of the mean.

It turns out that the points of inflection of the normal distribution curve lie one standard deviation from the mean. You can use this to estimate the standard deviation.

PROOF 10

The curve corresponding to the normal distribution with mean zero and standard deviation σ is $y = ke^{-\left(\frac{x^2}{2\sigma^2}\right)}$.

Prove that the points of inflection of $y = ke^{-\left(\frac{x^2}{2\sigma^2}\right)}$ occur at $x = \pm\sigma$.

$$\frac{dy}{dx} = -\frac{2kx}{2\sigma^2}e^{-\left(\frac{x^2}{2\sigma^2}\right)}$$

$$= -\frac{kx}{\sigma^2}e^{-\left(\frac{x^2}{2\sigma^2}\right)}$$

To find a point of inflection, use the fact that $\frac{d^2y}{dx^2} = 0$.

First, use the chain rule to differentiate.

$$\frac{d^2y}{dx^2} = -\frac{k}{\sigma^2}e^{-\left(\frac{x^2}{2\sigma^2}\right)} + \left(-\frac{2x}{2\sigma^2}\right)\left(-\frac{kx}{\sigma^2}\right)e^{-\left(\frac{x^2}{2\sigma^2}\right)}$$

$$= -\frac{k}{\sigma^2}e^{-\left(\frac{x^2}{2\sigma^2}\right)} + \frac{kx^2}{\sigma^4}e^{-\left(\frac{x^2}{2\sigma^2}\right)}$$

Use both the chain rule and the product rule to find the second derivative.

At a point of inflection, $\frac{d^2y}{dx^2} = 0$:

$$-\frac{k}{\sigma^2}e^{-\left(\frac{x^2}{2\sigma^2}\right)} + \frac{kx^2}{\sigma^4}e^{-\left(\frac{x^2}{2\sigma^2}\right)} = 0$$

$$\frac{k}{\sigma^2}e^{-\left(\frac{x^2}{2\sigma^2}\right)}\left(-1 + \frac{x^2}{\sigma^2}\right) = 0$$

Factorise and solve, noting that $\frac{k}{\sigma^2}e^{-\left(\frac{x^2}{2\sigma^2}\right)} \neq 0$.

Since $\frac{k}{\sigma^2}e^{-\left(\frac{x^2}{2\sigma^2}\right)} \neq 0$,

$$-1 + \frac{x^2}{\sigma^2} = 0$$

$$x^2 = \sigma^2$$

$$x = \pm\sigma$$

If a normally distributed random variable has mean 120, should a value of 150 be considered unusually large? The answer depends on how spread out the variable is - and this is measured by its standard deviation. If the standard deviation were 30 then a value around 150 would be quite common; however, if the standard deviation were 5 then 150 would indeed be very unusual.

The probability of a normally distributed random variable being less than a given value ($P(X \leqslant x)$, called the **cumulative probability**) depends only on the number of standard deviations x is away from the mean. This is called the **z-score**.

Key point 21.2

For $X \sim N(\mu, \sigma^2)$ the z-score measures the number of standard deviations away from the mean:

$$z = \frac{x - \mu}{\sigma}$$

WORKED EXAMPLE 21.2

Given that $X \sim \mathrm{N}(15, 6.25)$:

a work out how many standard deviations $x = 16.1$ is away from the mean
b find the value of X which is 1.2 standard deviations below the mean.

a $z = \frac{x-\mu}{\sigma}$ — The number of standard deviations away from the mean is measured by the z-score.

$\sigma = \sqrt{6.25} = 2.5$ — 6.25 is the variance so take the square root to get the standard deviation.

$\therefore z = \frac{16.1-15}{2.5} = 0.44$

16.1 is 0.44 standard deviations away from the mean.

b $z = -1.2$ — Values below the mean have a negative z-score.

$\therefore -1.2 = \frac{x-15}{2.5}$

$x - 15 = -3$

$x = 12$

If you are given a random variable $X \sim \mathrm{N}(\mu, \sigma^2)$ you can create a new random variable Z which takes the values equal to the z-scores of the values of X.

Whatever the original mean and standard deviation of X, this new random variable Z always has normal distribution with mean 0 and variance 1, called the **standard normal distribution**: $Z \sim \mathrm{N}(0, 1)$.

Fast forward

This is an extremely important property of the normal distribution, which needs to be used in situations when the mean and standard deviation of X are not known (see Section 3).

Key point 21.3

The probabilities of X and Z are related by:

$$\mathrm{P}(X \leqslant x) = \mathrm{P}\left(Z \leqslant \frac{x-\mu}{\sigma}\right)$$

WORKED EXAMPLE 21.3

Let $X \sim \mathrm{N}(6, 0.5^2)$. Write, in terms of probabilities of Z:

a $\mathrm{P}(X \leqslant 6.1)$ **b** $\mathrm{P}(5 < X < 7)$

a $\mathrm{P}(x \leqslant 6.1) = \mathrm{P}\left(Z \leqslant \frac{6.1-6}{0.5}\right)$ — You are given that $x = 6.1$ so you can calculate Z.

$= \mathrm{P}(Z \leqslant 0.2)$

b $\mathrm{P}(5 < X < 7) = \mathrm{P}\left(\frac{5-6}{0.5} < Z < \frac{7-6}{0.5}\right)$ — You now have two x-values, so find the corresponding Z value for each of them.

$= \mathrm{P}(-2 < Z < 2)$

Questions on the normal distribution can be combined with other probability facts – in particular, watch out for questions that bring in conditional probability or the binomial distribution.

Rewind

Conditional probability was covered in Chapter 20.

The binomial distribution was covered in Student Book 1, Chapter 21.

Did you know?

Before graphical calculators (which was not so long ago!) people used tables showing cumulative probabilities of the standard normal distribution. Because of their importance they were given special notation: $\Phi(z) = P(X \leqslant z)$. Although you do not have to use this notation, you should understand what it means.

WORKED EXAMPLE 21.4

The mass of fish caught by a trawler follows a normal distribution with mean 4 kg and standard deviation 0.5 kg. A juvenile fish is classified as one with a mass less than 3 kg.

a What is the probability that a randomly chosen fish is a juvenile?

b If 300 fish are caught, what is the probability that there are more than ten juveniles? State any assumptions you have to make, and comment on their validity.

c If a fish is a juvenile, what is the probability that its mass is more than 2.5 kg?

a Let X be mass of fish so $X \sim N(4, 0.25)$.

State the distribution. Variance $= 0.5^2 = 0.25$.

$P(X < 3) = 0.0228$

Use your calculator to find the required probability.

b Let Y be number of juveniles so $Y \sim B(300, 0.0228)$.

Write down the names of all the random variables to make it obvious when a distribution has changed.

$P(Y > 10) = 1 - P(Y \leqslant 9)$
$= 0.150$

Even though you write down parameters to three significant figures, use exact values in your calculations.

c $P(X > 2.5 \mid X < 3) = \dfrac{P(X > 2.5 \cap X < 3)}{P(X < 3)}$

Use the formula for conditional probability.

$= \dfrac{P(2.5 < X < 3)}{P(X < 3)}$

If X is bigger than 2.5 and less than 3 it is between 2.5 and 3.

$= \dfrac{P(X < 3) - P(X < 2.5)}{P(X < 3)}$

$= \dfrac{0.0228 - 0.00135}{0.0228}$

$= 94.1\%$

It is very common in this type of question that you can use some of the probabilities already calculated in an earlier part.

EXERCISE 21A

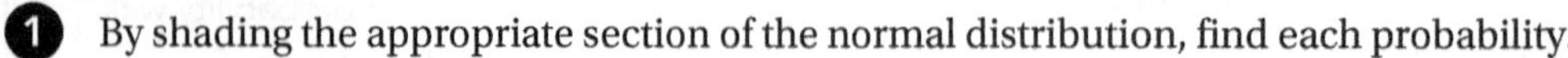

1 By shading the appropriate section of the normal distribution, find each probability.

a $X \sim \mathrm{N}(20, 100)$

i $\mathrm{P}(X \leqslant 32)$ ii $\mathrm{P}(X < 12)$

b $Y \sim \mathrm{N}(4.8, 1.44)$

i $\mathrm{P}(Y > 5.1)$ ii $\mathrm{P}(Y \geqslant 3.4)$

c $R \sim \mathrm{N}(17, 2)$

i $\mathrm{P}(16 < R < 20)$ ii $\mathrm{P}(17.4 < R < 18.2)$

d Q has a normal distribution with mean 12 and standard deviation 3.

i $\mathrm{P}(Q > 9.4)$ ii $\mathrm{P}(Q < 14)$

e If F has a normal distribution with mean 100 and standard deviation 25.

i $\mathrm{P}(|F - 100| < 15)$ ii $\mathrm{P}(|F - 100| > 10)$

> **Tip**
>
> Remember that for a continuous variable, $\mathrm{P}(X \leq k)$ is the same as $\mathrm{P}(X < k)$.

2 Find the z-score corresponding to each given value of X.

a i $X \sim \mathrm{N}(12, 2^2), x = 13$ ii $X \sim \mathrm{N}(38, 7^2), x = 45$

b i $X \sim \mathrm{N}(20, 9), x = 15$ ii $X \sim \mathrm{N}(162, 25), x = 160$

3 Given that $X \sim \mathrm{N}(16, 2.5^2)$, write each probability in terms of probabilities of the standard normal variable.

a i $\mathrm{P}(X < 20)$ ii $\mathrm{P}(X < 19.2)$

b i $\mathrm{P}(X \geqslant 14.3)$ ii $\mathrm{P}(X \geqslant 8.6)$

c i $\mathrm{P}(12.5 < X < 16.5)$ ii $\mathrm{P}(10.1 \leqslant X \leqslant 15.5)$

4 It is found that the lifespan of a certain brand of laptop battery follows a normal distribution with mean 16 hours and standard deviation 5 hours. A particular battery has a lifespan of 10.2 hours.

a How many standard deviations below the mean is this?

b What is the probability that a randomly chosen laptop battery has a lifespan shorter than this?

5 When Ali competes in long jump competitions, the lengths of his jumps are normally distributed with mean 5.2 m and standard deviation 0.7 m.

a What is the probability that Ali will record a jump between 5 m and 5.5 m?

b Ali needs to jump 6 m to qualify for the school team.

i What is the probability that he will qualify with a single jump?

ii If he is allowed three jumps, what is the probability that he will qualify for the school team?

c What assumptions did you have to make in your answer to part **b ii**? Are these likely to be met in this situation?

6 Masses of a species of cat have a normal distribution with mean 16 kg and variance 16 kg^2. Given a sample of 2000 such cats, estimate the number that will have a mass above 13 kg.

7 A normal curve has points of inflexion with x coordinates 5 and 11. Find the mean and standard deviation of this distribution.

8 Copy the diagram and mark on the approximate position of the points of inflexion. Hence estimate the mean and standard deviation of the normal curve.

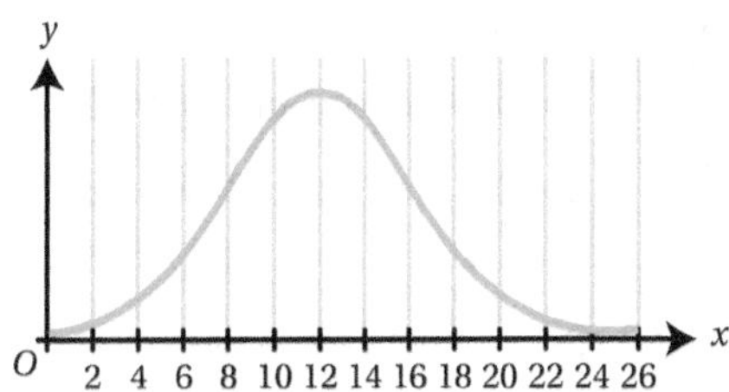

9 Estimate the mean and standard deviation of the normal curve.

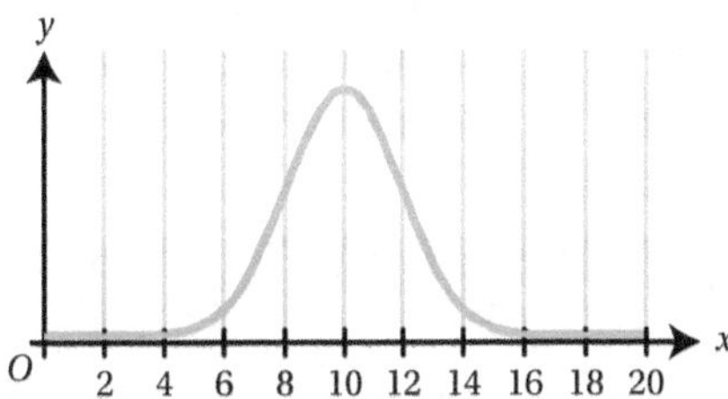

10 The 400 m time of a group of athletes can be modelled by a normal distribution with mean 60 seconds and standard deviation 2 seconds.

- **a** Find the probability that a randomly chosen athlete will run the 400 m in under 59 seconds.
- **b** Show that, if the binomial distribution can be used, the probability that all four athletes in a 4×400 m team run under 59 seconds is 0.9%.
- **c** Miguel says that this means that the probability of the team breaking the school record of 3 minutes and 56 seconds is only 0.9%. Give three reasons why this is likely to be incorrect. Do you think the real value will be greater or less than 0.9%?

11 If $D \sim \mathrm{N}(250, 400)$, find:

a $\mathrm{P}(D > 265 \cap D < 280)$ **b** $\mathrm{P}(D > 265 \mid D < 280)$ **c** $\mathrm{P}(D < 242 \cup D > 256)$.

12 If $Q \sim \mathrm{N}(4, 160)$, find:

a $\mathrm{P}(|Q| > 5)$ **b** $\mathrm{P}(Q > 5 \mid |Q| > 5)$.

13 The masses of apples are normally distributed with mean mass 150 g and standard deviation 25 g. Supermarkets classify apples as medium if their masses are between 120 g and 170 g.

- **a** What proportion of apples are medium?
- **b** In a bag of 10 apples what is the probability that there are at least eight medium apples?

14 The wingspans of a species of pigeon are normally distributed with mean length 60 cm and standard deviation 6 cm. A pigeon is chosen at random.

- **a** Find the probability that the length of its wingspan is greater than 50 cm.
- **b** Given that its length is greater than 50 cm, find the probability that the length of its wingspan is greater than 55 cm.

15 Grains of sand are believed to have a normal distribution with mean size 2 mm and variance 0.25 mm^2.

- **a** Find the probability that a randomly chosen grain of sand is larger than 1.5 mm.
- **b** The sand is passed through a filter that blocks grains wider than 2.5 mm. The sand that passes through is examined. What is the probability that a randomly chosen grain of filtered sand is larger than 1.5 mm?

16 The amount of paracetamol per tablet is believed to be normally distributed with mean 500 mg and standard deviation 160 mg. A dose of less than 300 mg is ineffective in dealing with toothache. In a trial of 20 people suffering toothache, what is the probability that two or more of them have less than the effective dose?

17 A variable has a normal distribution with a mean that is 7 times its standard deviation. What is the probability of the variable taking a value less than 5 times the standard deviation?

18 If $X \sim N(\mu, \sigma^2)$ and $P(X \leqslant x) = k$ find $P(X \leqslant 2\mu - x)$ in terms of k.

Section 2: Inverse normal distribution

You now know how to find probabilities, given information about the variable. In real life it is often useful to work backwards from probabilities to estimate information about the data. This requires the **inverse normal distribution.**

Tip

Note that many textbooks use the $\Phi(z)$ notation mentioned in Section 1 to write inverse normal distribution:

If $P(X \leqslant x) = p$, then

$\Phi^{-1}(p) = z = \frac{x-\mu}{\sigma}$.

Key point 21.4

For a given value of probability p, the inverse normal distribution gives the value of x such that $P(X \leqslant x) = p$.

WORKED EXAMPLE 21.5

The length of men's feet is thought to be normally distributed with mean 22 cm and variance 25 cm^2.
A shoe manufacturer wants only 5% of men to be unable to find shoes large enough for them.
How big should their largest shoe be?

If X is length of a man's foot then $X \sim N(22, 25)$. — Convert the information into mathematical terms.

$P(X > x) = 0.05$ — You want to find the value of x such that $P(X > x) = 0.05$.

$P(X \leqslant x) = 1 - P(X > x) = 0.95$ — You may have to convert into a probability of the form $P(X \leqslant x)$.

$\Rightarrow x = 30.2$ cm — Use the inverse normal distribution on your calculator.

So their largest shoe must fit a foot 30.2 cm long.

WORK IT OUT 21.1

Given that $X \sim N(10, 25)$ and $P(X > x) = 0.75$ find the value of x.

Which is the correct solution? Identify the errors made in the incorrect solutions.

Solution 1	Solution 2	Solution 3
$P(X < x) = 0.25$ From your calculator the corresponding z-score is -0.674. Therefore $\frac{x-10}{5} = -0.674$ $x - 10 = -3.37$ $x = 6.63$ (3 s.f.)	$\Phi^{-1}(0.75) = 0.674$ Therefore $\frac{x-10}{25} = 0.674$ $x - 10 = 16.9$ $x = 26.9$ (3 s.f.)	$\frac{x-10}{25} = 0.75$ $x - 10 = 18.75$ $x = 28.75$ (4 s.f.)

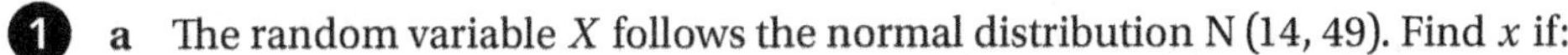

EXERCISE 21B

1 a The random variable X follows the normal distribution $N(14, 49)$. Find x if:

i $P(X < x) = 0.8$ ii $P(X < x) = 0.46$

b The random variable X follows the normal distribution $N(36.5, 10)$. Find x if:

i $P(X > x) = 0.9$ ii $P(X > x) = 0.4$

c The random variable X follows the normal distribution $N(0, 12)$. Find x if:

i $P(|X| < x) = 0.5$ ii $P(|X| < x) = 0.8$

2 IQ tests are designed to have a mean of 100 and a standard deviation of 20. What IQ score is needed to be in the top 2% of IQ scores?

3 Rabbits' masses are normally distributed with an average mass of 2.6 kg and a variance of 1.44 kg^2. A vet decides that the top 20% of rabbits are obese. What is the minimum mass for an obese rabbit?

4 The amount of coffee dispensed by a machine follows a normal distribution with mean 150 ml and standard deviation 5 ml.

a Calculate the probability that the machine dispenses less than 142 ml of coffee.

b Find the value of a if 20% of cups contain more than a ml of coffee. Give your answer to 1 decimal place.

5 The times taken for students to complete a test are normally distributed with a mean of 32 minutes and standard deviation of 6 minutes.

a Find the probability that a randomly chosen student completes the test in less than 35 minutes.

b 90% of students complete the test in less than t minutes. Find the value of t.

c A random sample of eight students had their times for the test recorded. Find the probability that exactly two of these students complete the test in less than 30 minutes.

6 An old textbook says that the range of data can be estimated as 6 times the standard deviation. If the data is normally distributed, what percentage of the data is within this range?

7 The time taken to do a maths question can be modelled by a normal distribution with mean 80 seconds and standard deviation 15 seconds. The probability of two randomly chosen questions both taking longer than a seconds is 0.063 752. Find the value of a.

8 The concentration of salt in a cell, X, can be modelled by a normal distribution with mean μ and standard deviation 2%. Find the value of α such that: $P(\mu - \alpha < X < \mu + \alpha) = 0.9$.

9 For a normal distribution find the ratio:

a $\dfrac{\text{median}}{\text{mean}}$ b $\dfrac{\text{standard deviation}}{\text{inter quartile range}}$

10 Evaluate $\Phi^{-1}(x) + \Phi^{-1}(1 - x)$, where $\Phi^{-1}(x)$ is the inverse normal distribution function for the standard normal distribution.

11 If $Z \sim N(0, 1)$ and $P(Z < k) = \Phi(k)$ find $P(|Z| < k)$ in terms of $\Phi(k)$.

12 Most calculators have a random number generator which generates random numbers from 0 to 1. These random numbers are uniformly distributed, which means that the probability is evenly spread over all possible values. How can you use these to form random numbers drawn from a normal distribution?

Section 3: Finding unknown μ or σ

One of the main applications of statistics is to determine parameters of the population, given information about the data. But how can you use the normal distribution calculations if the mean or the standard deviation is unknown? This is where the standard normal distribution comes in useful: replace all the X values by their z-scores as they follow a known distribution, $\mathrm{N}(0,1)$.

Tip

This involves solving equations, and sometimes simultaneous equations. As the numbers usually have many decimal places you might want to use your calculator.

WORKED EXAMPLE 21.6

The random variable X follows a normal distribution with standard deviation $\sigma = 1.2$. An experiment estimated that $\mathrm{P}(X > 3.4) = 0.2$. Estimate the mean of X correct to two significant figures.

$\mathrm{P}(X > 3.4) = 1 - \mathrm{P}(X \leqslant 3.4)$
$= 1 - 0.2$
$= 0.8$

Get the probability in the form $\mathrm{P}(X \leqslant k)$.

If $Z = \frac{X-\mu}{\sigma}$, $Z \sim \mathrm{N}(0,1)$:

Since you do not know μ, convert the probability into information about Z.

$\mathrm{P}(Z \leqslant z) = 0.8$
$\Rightarrow z = 0.8416$

Find z from your calculator.

$\therefore 0.8416 = \frac{3.4-\mu}{1.2}$
$3.4 - \mu = 1.01$
$\mu = 2.4$

Relate z to the given x-value using $z = \frac{x-\mu}{\sigma}$.

WORKED EXAMPLE 21.7

The masses of gerbils are thought to be normally distributed. If 30% of gerbils have a mass more than 65 g and 20% have a mass less than 40 g, estimate the mean and the variance of the mass of a gerbil.

Let X be mass of a gerbil. Then $X \sim \mathrm{N}(\mu, \sigma^2)$

Convert the information into mathematical terms.

$\mathrm{P}(X < 40) = 0.2$ (1)
$\mathrm{P}(X > 65) = 0.3$
$\Rightarrow \mathrm{P}(X \leqslant 65) = 0.7$ (2)

Get the second statement in the form $\mathrm{P}(X \leqslant k)$.

From (1):
$\mathrm{P}(Z < z) = 0.2$
$\Rightarrow z = -0.842$
$\frac{40-\mu}{\sigma} = -0.842$
$40 - \mu = -0.842\sigma$ (3)

Use the inverse normal distribution for Z and relate it to the given X values.

Continues on next page

From (2):

$P(Z \leqslant z) = 0.7$

$\Rightarrow z = 0.524$ .... Use your calculator again to find z.

$\frac{65-\mu}{\sigma} = 0.524$

$65 - \mu = 0.524\sigma \quad (4)$

$(4)-(3)$: .... Solve the simultaneous equations.

$25 = 1.366\sigma$

$\sigma = 18.3$

$\sigma^2 = 335g^2$

$\therefore \mu = 55.4g$

EXERCISE 21C

1 a Given that $X \sim N(\mu, 4)$ find μ if:

i $P(X > 4) = 0.8$ ii $P(X > 9) = 0.2$.

b Given that $X \sim N(8, \sigma^2)$ find σ if:

i $P(X \leqslant 19) = 0.6$ ii $P(X \leqslant 0) = 0.3$.

2 Given that $X \sim N(\mu, \sigma^2)$ find μ and σ if:

Elevate

See Support Sheet 21 for a further example of finding unknown mean and standard deviation and for more practice questions.

a i $P(X > 7) = 0.8$ and $P(X < 6) = 0.1$

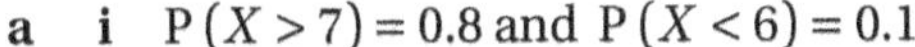

ii $P(X > 150) = 0.3$ and $P(X < 120) = 0.4$

b i $P(X > 0.1) = 0.4$ and $P(X \geqslant 0.6) = 0.25$

ii $P(X > 700) = 0.8$ and $P(X \geqslant 400) = 0.99$.

3 A manufacturer knows that their machines produce bolts with diameters that follow a normal distribution with standard deviation 0.02 cm. The manager takes a random sample of bolts and finds that 6% of them have diameter greater than 2 cm. Find the mean diameter of the bolts.

4 The energy of an electron can be modelled by a normal distribution with mean 12 eV and standard deviation σ eV. Given that 20% of electrons have an energy above 15 eV find the value of σ.

5 It is known that the heights of a certain plant follow a normal distribution. In a sample of 200 plants, 32 are less than 45 cm tall and 50 are more than 88 cm tall. Estimate the mean and the standard deviation of the heights.

6 The time taken for a computer to start is modelled by a normal distribution. It is tested 100 times and on 40 times it takes longer than 30 seconds. On 25 times it takes less than 15 seconds. Estimate the mean and standard deviation of the start-up times.

7 The actual voltage of a brand of 9 V battery is thought to be normally distributed with standard deviation 0.8 V and mean $(9.2 - t)$ V, where t is the time, in hours, for which the battery has been used. The batteries can no longer power a lamp when they drop below 7 V. A batch of batteries is tested and it is found that only 10% can power the lamp. If the model is correct, estimate how long the batteries have been used, assuming that they were all used for the same amount of time.

8 A scientist noticed that 36% of temperature measurements were 4 Celsius degrees lower than the average. Assuming that the measurements follow a normal distribution, estimate the standard deviation.

9 The waiting time for a train is normally distributed with mean 10 minutes. 80% of the time the waiting time is over 8 minutes. Find the probability that a person waits over 15 minutes on exactly two out of three times they wait for the train.

10 The random variable X models the temperature in an oven in °C. It follows a normal distribution where the mean value is the temperature set on the oven. The probability of being within five Celsius degrees of the temperature set is 40%. Find the probability of being within 10 Celsius degrees of the temperature set.

Section 4: Modelling with the normal distribution

Key point 21.5

For a normal distribution you would expect:

- a histogram to show an approximately symmetrical distribution with only one mode and no sharp cut off
- summary statistics to show that nearly all of the data fall within three standard deviations of the mean.

Focus on...

Many real-world situations, such as the heights of people, masses of gerbils or error in experimental measurements, can be modelled by using the normal distribution. However, you should not just assume that a variable follows a normal distribution. There are various useful checks that you can use. You can explore this in more detail in Focus on... Modelling 4.

WORKED EXAMPLE 21.8

For each histogram, explain why the normal distribution is not a good model for the data.

a
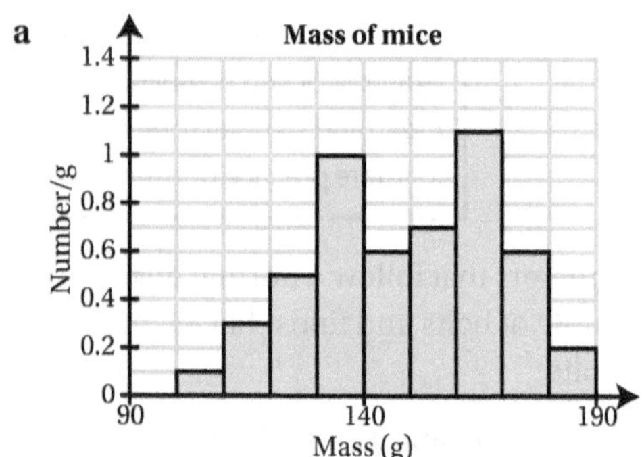

b
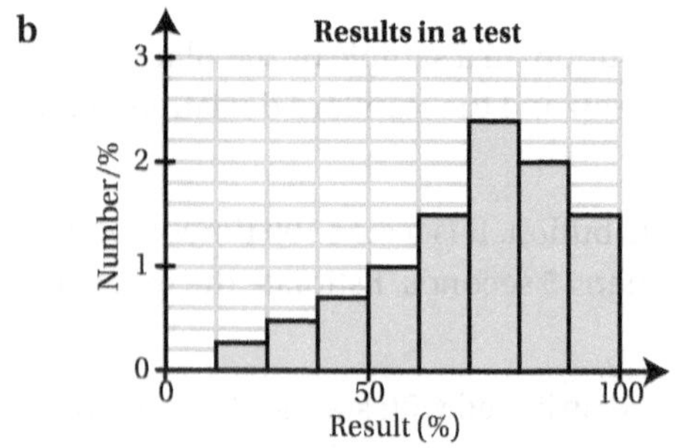

a There appear to be two modes.

b These data do not seem to be symmetrical and there is a sharp cut off at the top end because it is not possible to score over 100%.

Fast forward

If you study the Statistics option of Further Mathematics you can find out about a statistical test, called the chi-squared test, which allows you to decide more precisely whether data follow a normal distribution.

WORKED EXAMPLE 21.9

In a sample it is found that the mean time taken to complete a puzzle is 42 seconds with a standard deviation of 28 seconds. Explain why the normal distribution would not be an appropriate model to predict the time taken to complete this puzzle.

There is a cut-off at zero seconds which is 1.5 standard deviations below the mean. The normal distribution would therefore predict that a significant number of people complete the puzzle in a negative amount of time, which is not possible.

With only the mean and standard deviation provided (no graph of the distribution), consider whether virtually all the data fall within three standard deviations of the mean.

Many other statistical distributions can be approximated by the normal distribution. Before people had computers and graphical calculators this was very important, because calculations with the normal distribution were often much easier.

For example, consider the bar charts for the binomial distribution.

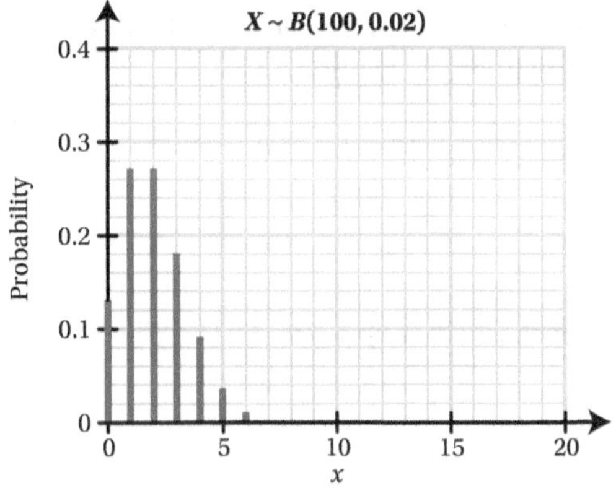

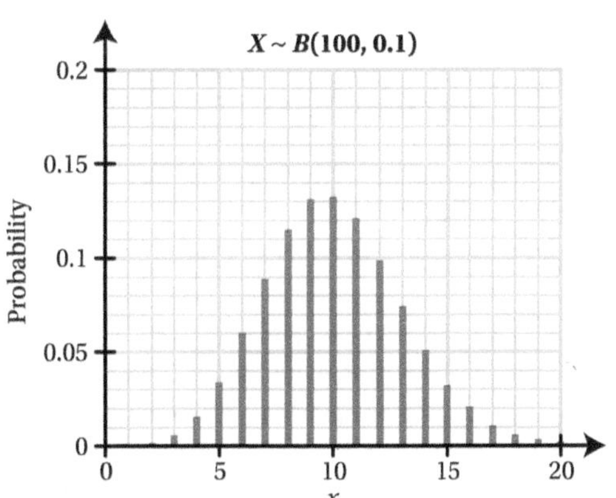

The distribution on the left has a sharp cut off at 0 and is not symmetrical, but the distribution on the right appears to have roughly the shape of a normal distribution. It turns out that it can indeed be approximated by a normal distribution, although the proof of this is beyond what you need.

Key point 21.6

If $X \sim \text{B}(n, p)$ with $np > 5$ and $n(1-p) > 5$ then X can be approximated by the normal distribution $\text{N}(np, np(1-p))$.

Did you know?

Because the binomial distribution deals with discrete variables and the normal distribution deals with continuous variables, technically you should use something called a continuity correction – this means saying that in the binomial distribution $\text{P}(X = 5)$ is equivalent to $\text{P}(4.5 < X < 5.5)$ in the normal distribution. However, you do not need to use this in the A Level course.

WORKED EXAMPLE 21.10

A dice is rolled 300 times. The random variable X models the number of sixes thrown.

a Explain why X follows a binomial distribution.

b The random variable X can be approximated by a normal distribution.
  i Find the mean and variance of this normal distribution.
  ii What properties of X make this approximation valid?

c Let Y be a random variable with the normal distribution from part **b**. Find $P(Y \leqslant 40)$.

d **i** Find $P(X \leqslant 40)$.
  ii Explain why this is not the same as your answer to part **c**.

You met the conditions for a binomial distribution to be appropriate in Student Book 1, Chapter 21.

a The events are independent with constant probability. Classify the outcomes into six or not six.

b i $n = 300$ and $p = \frac{1}{6}$

$np = 50$ and $np(1-p) = \frac{250}{6}$

So the approximate normal distribution is $N\left(50, \frac{250}{6}\right)$.

Use the fact that $N(np, np(1-p))$ is the approximate distribution.

ii $np = 50 > 5$ and $n(1-p) = 250 > 5$
This means that the binomial distribution will be reasonably symmetrical, making a normal approximation valid.

Check whether $np > 5$ and $n(1-p) > 5$.

c For $Y \sim N\left(50, \frac{250}{6}\right)$,
$(Y \leqslant 40) = 0.0607$

State the distribution of Y and use your calculator to find the required probability.

d i For $X \sim B\left(300, \frac{1}{6}\right)$,
$P(X \leqslant 40) = 0.0675$

Use your calculator to find the probability for the original binomial distribution.

ii X and Y do not have exactly the same distribution – X is discrete and Y is continuous.

The normal distribution is only an approximation to the binomial, so the probabilities won't be exactly the same.

The normal approximation is useful in binomial hypothesis tests when the value of n is large, because the exact binomial probabilities can be difficult to find.

WORKED EXAMPLE 21.11

A survey of a random sample of 5000 people in a large city was conducted to test a hypothesis about the proportion of people who cycle to work. In this sample, c people cycled to work. The test concluded that there is evidence, at the 5% significance level, that this proportion is less than 0.06.

Use an appropriate normal distribution to estimate the maximum possible value of c.

Let X be the number of people who cycle to work.

Then $X \sim B(5000, p)$

The hypotheses are:

$H_0: p = 0.06$

$H_1: p < 0.06$

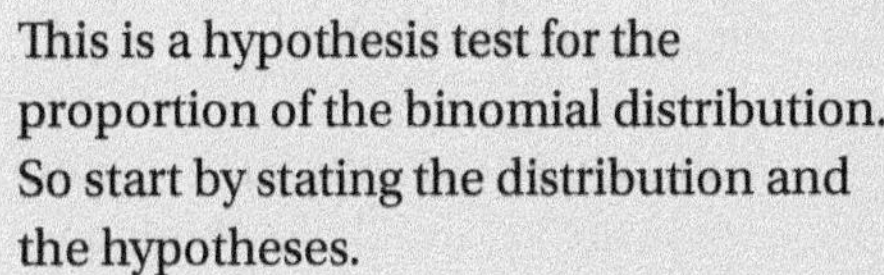

This is a hypothesis test for the proportion of the binomial distribution. So start by stating the distribution and the hypotheses.

Need c such that:

$P(X \leqslant c) < 0.05$, where $X \sim B(5000, 0.06)$

c is the critical value for this test. The significance level is 5%.

Approximate normal distribution:

$np = 300, np(1-p) = 282$

So $X \approx N(300, 282)$

You can use a normal approximation because $np = 300 > 5$ and $n(1-p) = 4700 > 5$.

Using this normal distribution,

Use the inverse normal distribution to find the value of c.

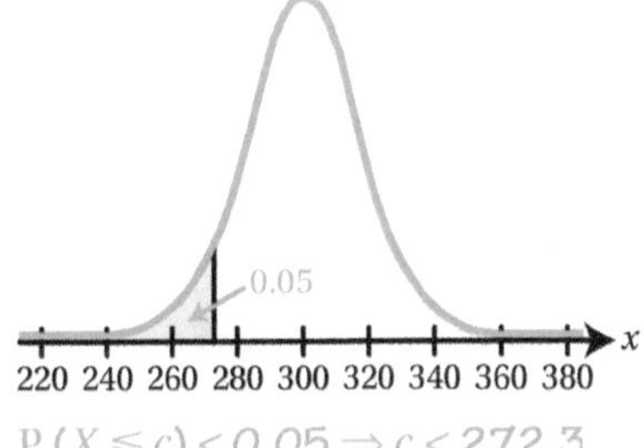

$P(X \leqslant c) < 0.05 \Rightarrow c < 272.3$

So $c \approx 272$

c is the number of people who cycle to work, so needs to be a whole number.

EXERCISE 21D

1 For each histogram, decide if the data could be modelled by a normal distribution. If they cannot, give a reason.

a

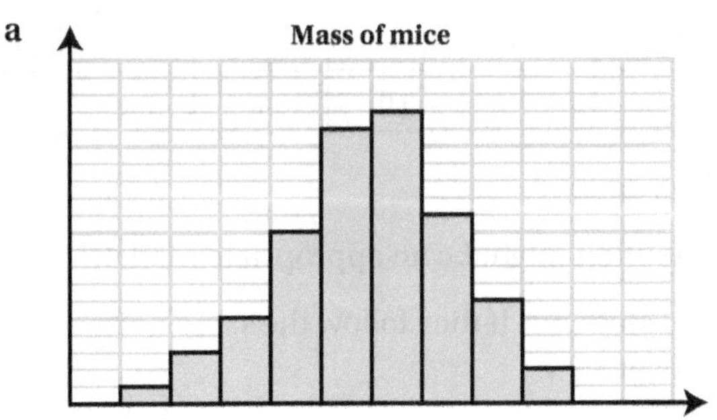

b

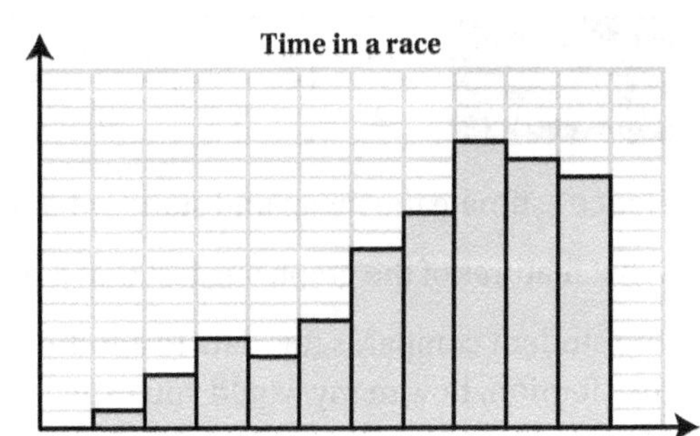

c

GDP of a country

d

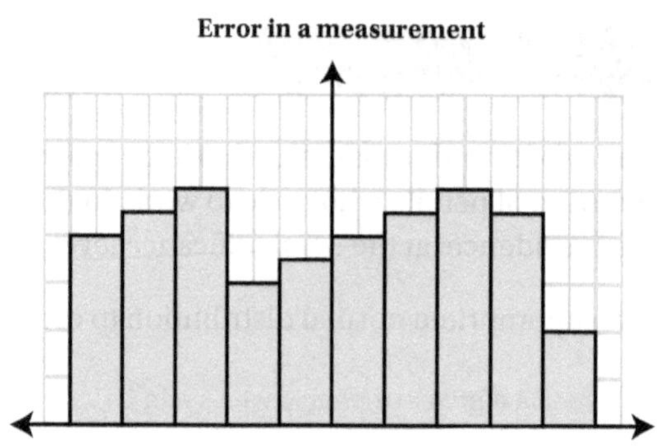

2 The mean number of children in a family in the UK is 2.4 with a standard deviation of 1.1. Use these figures to explain why a normal distribution would not be a good model of the number of children in a family in the UK.

3 The quantities of bread purchased by a random sample of 150 people in a one-week period were recorded.

The smallest quantity purchased was 296 g and the largest was 828 g.

The summary statistics for the sample are:

$\sum x = 84\,345$ and $\sum x^2 = 48\,626\,834$

a Find the sample mean and standard deviation.

b Explain whether the normal distribution would be a suitable model for the weekly quantity of bread purchased.

4 A psychology student asks people to estimate the value of an angle. Her results are summarised in the histogram.

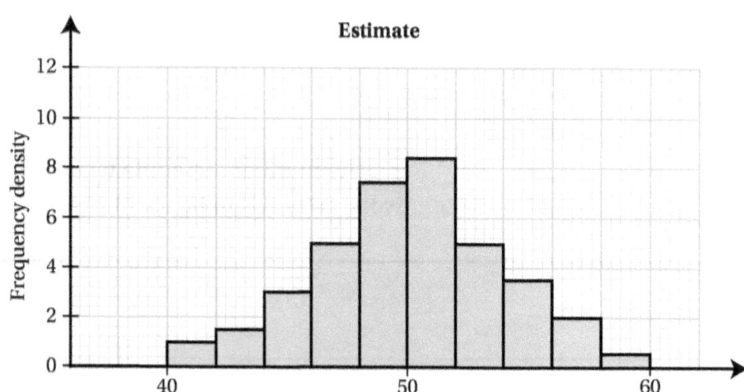

a Copy and complete the frequency table.

Angle, °	40–42	42–44	44–46	46–48	48–50	50–52	52–54	54–56	56–58	58–60
Frequency	2	3	6	10			10	7	4	1

b Hence estimate the mean and standard deviation of the data.

c What features of the graph suggest that a normal distribution might be an appropriate model?

d The student compares this data to another group of 100 students. If they follow the same normal distribution, how many would you expect to estimate over 55°?

5 A fair coin is tossed 100 times.

a Find the probability that there are more than 60 heads, using:

i the binomial distribution.

ii the normal approximation.

b Find the percentage error in using the normal distribution in this situation.

6 The number of people voting for a particular party in an election can be modelled by a binomial distribution. There are n voters and the probability of each voting for this party is 0.48. Use a normal approximation to the binomial to find the probability that this party wins a majority if:

a $n = 100$

b $n = 10\,000$.

7 The random variable X has the binomial distribution $\mathrm{B}(n, p)$.

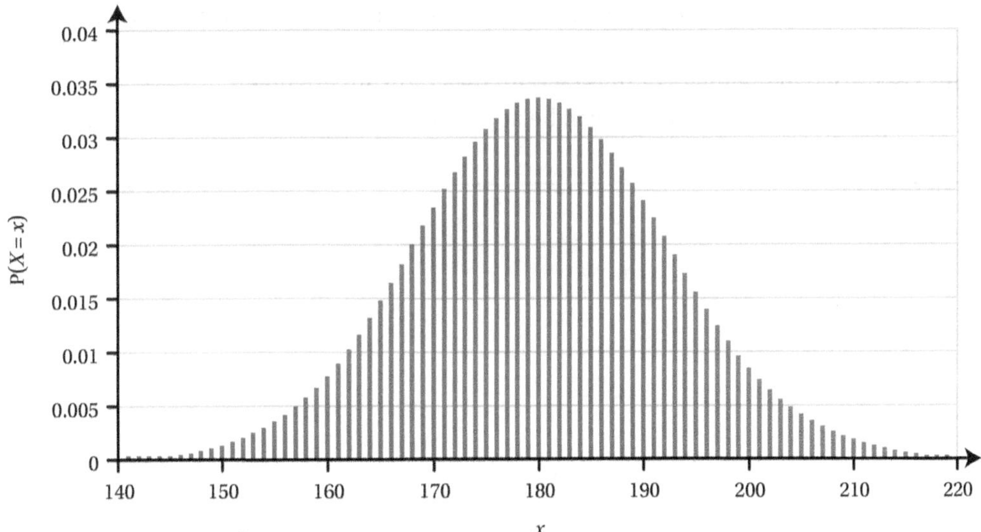

a The distribution of X can be approximated by a normal distribution. Estimate the mean and standard deviation of this normal distribution, showing your method clearly.

b Hence estimate the values of n and p.

8 Data collected over a long period of time indicate that 23% of children contract a certain disease. Following a public awareness campaign, a doctor conducts a survey to find out whether this proportion has decreased. The doctor uses a random sample of 3000 children and conducts a hypothesis test at the 2.5% significance level.

Use an appropriate normal distribution to find the approximate critical region for this test.

9 a Prove that if $np > 5$ and $n(1-p) > 5$ then $n > 10$.

Use a counterexample to show that the reverse is not true.

b A binomial distribution $\mathrm{B}(n, 0.8)$ has a probability of 0.181 of being above 100. Find the value of n.

10 The random variable $X \sim \mathrm{B}(n, p)$ is approximated by the normal distribution $\mathrm{N}(np, np(1-p))$.

Using this approximation, prove that if $np > 9$ then all values of X within three standard deviations of the mean are positive.

Checklist of learning and understanding

- The **normal distribution** models many physical situations. It is completely described once you know its mean (μ) and its variance (σ^2). Calculators can provide the probabilities of being in any given range.
- These values are useful to know:
  - approximately 99.7% of the data lie within three standard deviations of the mean
  - approximately 95% of the data lie within two standard deviations of the mean.
  - approximately two-thirds of the data lie within one standard deviation of the mean.
- The **z-score** is the number of standard deviations above the mean that have a given cumulative probability. It is related to the original variable through the equation

  $$z = \frac{x - \mu}{\sigma}$$

- If you know probabilities relating to a variable with a normal distribution you can use the **inverse normal distribution** to deduce information about the variable.
- You need to use the z-score when the mean or the standard deviation are unknown.
- If $X \sim \text{B}(n, p)$ with $np > 5$ and $n(1-p) > 5$ then X can be approximated by the normal distribution $\text{N}(np, np(1-p))$

Mixed practice 21

1. $X \sim N(7, 2^2)$. What is $P(X = 7)$?

   Choose from these options.

   A 0 **B** 0.1 **C** 0.5 **D** 1

2. The test scores of a group of students are normally distributed with mean 62 and variance 144.

   a Find the percentage of students with scores above 80.

   b What is the lowest score achieved by the top 50% of the students?

3. The masses of kittens are normally distributed with mean 1.2 kg and standard deviation 0.3 kg.

   a Out of a group of 20 kittens, how many would be expected to have a mass of less than 1 kg?

   b 30% of kittens have a mass of more than m kg. Determine the value of m.

4. The random variable X is normally distributed with a mean of 3 and standard deviation of 1.5. By sketching a normal curve or otherwise, find the value of k such that $P(2.6 < X < k) = 0.32$.

5. The masses, M kg, of babies born at a certain hospital satisfy $M \sim N(3.2, 0.72)$.

   Find the value of m such that 35% of the babies have masses between m kg and 3.2 kg where $m < 3.2$.

6. The volume, V litres, of Cleanall washing-up liquid in a 5-litre container may be modelled by a normal distribution with a mean, μ, of 5.028 and a standard deviation of 0.015.

   a Determine the probability that the volume of Cleanall in a randomly selected 5-litre container is:

   i less than 5.04 litres

   ii more than 5 litres.

   b Determine the value of v such that $P(\mu - v < V < \mu + v) = 0.95$.

   [© AQA 2013]

7. $X \sim N(\mu, \sigma^2)$ and the interquartile range is 10. Find the value of σ.
   Choose from these options.

   A 0.25 **B** 7.41 **C** 14.8 **D** 25

8. The adult female of a breed of dog has average height 0.7 m with variance 0.05 m^2.

   a If the height follows a normal distribution, find the probability that a randomly selected adult female dog is more than 0.75 m tall.

   b Find the probability that in six independently selected adult female dogs of this breed exactly four are above 0.75 m tall.

9. Heights of trees in a forest are distributed normally with mean 26.2 m and standard deviation 5.6 m.

   a Find the probability that a tree is more than 30 m tall.

   b What is the probability that among 16 randomly selected trees at least two are more than 30 m tall?

10 It is known that the scores on a test follow a normal distribution $N(\mu, \sigma^2)$. 20% of the scores are above 82 and 10% of the scores are below 47.

a Show that $\mu + 0.8416\sigma = 82$.

b By writing another similar equation, find the mean and the standard deviation of the scores.

11 200 people were asked to estimate the size of an angle. 16 gave an estimate that was less than 25° and 42 gave an estimate that was more than 35°. Assuming that the data follows a normal distribution, estimate the mean and the standard deviation of the results.

12 The volume, X millilitres, of energy drink in a bottle can be modelled by a normal random variable with mean 507.5 and standard deviation 4.0.

a Find:

i $P(X < 515)$

ii $P(500 < X < 515)$

iii $P(X \neq 507.5)$.

b Determine the value of x such that $P(X < x) = 0.96$.

c The energy drink is sold in packs of 6 bottles. The bottles in each pack maybe regarded as a random sample. Calculate the probability that the volume of energy drink in at least 5 of the 6 bottles in a pack is between 500 ml and 515 ml.

[© AQA 2013]

13 50% of students in a university are female. The discrete random variable X, which is the number of female students in a group of size n, is assumed to follow a binomial distribution.

a Explain why, if n is large, the binomial distribution can be approximated by the normal distribution and state its parameters.

b A dancing club contains 200 students. Assuming the binomial distribution is valid, use the normal approximation to find the probability that more than 60% are female.

c Are the conditions for the binomial distribution met in this situation? Explain your answer.

14 The results of an examination have a mean of 54%, a median of 55% and a standard deviation of 12%.

a Explain why a normal distribution is a plausible model for this data.

Grades are awarded in the following way:

- The top 20% get an distinction.
- The next 30% get a merit.
- The next 40% get a pass.
- The remaining people get a fail.

b Assuming a normal model, find the grade boundaries for this examination.

15 A company makes a large number of steel links for chains. They know that the force required to break any individual link is modelled by a normal distribution with mean 20 kN. The company tests chains consisting of 4 links. If any link breaks, the chain will break. A force of 18 kN is applied to all of the chains and 30% break.

a Estimate the probability of a single link breaking.

b Hence estimate the standard deviation in the breaking strength of the links.

16 a 30% of sand from Playa Gauss falls through a sieve with gaps of 1 mm, but 90% passes through a sieve with gaps of 2 mm. Assuming that the sand's diameter is normally distributed, estimate the mean and standard deviation of the size of the grains of sand.

b 80% of sand from Playa Fermat falls through a sieve with gaps of 2 mm. 40% of this filtered sand passes through a sieve with gaps of 1 mm. Assuming that the sand's diameter is normally distributed, estimate the mean and standard deviation of the sand.

22 Further hypothesis testing

In this chapter you will learn how to:

- treat the sample mean as a random variable and see how it is distributed
- test whether the mean of a normally distributed population is different from a predicted value
- test whether a set of bivariate data provides evidence for significant correlation.

Before you start...

Student Book 1, Chapter 20	You should be able to interpret correlation coefficients.	1 Information on height, mass, waist size and average time spent exercising per week was recorded from a random sample of adult males. Match the values of the product moment correlation coefficient with each of the sets of variables: A height and mass B height and time spent exercising C waist measurement and time spent exercising. 1 $r = -0.82$ 2 $r = 0.13$ 3 $r = 0.71$
Student Book 1, Chapter 22	You should be able to conduct hypothesis tests using the binomial distribution.	2 A dice is rolled ten times and four sixes are obtained. It is claimed that the dice is biased in favour of getting a six. Test this claim at the 10% level.
Chapter 21	You should be able to conduct calculations using the normal distribution.	3 $X \sim \text{N}(175, 10^2)$. Find: a $\text{P}(X < 190)$ b $\text{P}(150 < X < 185)$ c a such that $\text{P}(X > a) = 0.01$.

Testing means and correlation coefficients

A note on a cereal packet claims that it has an average mass of 500 g. A sample of 10 packets contains a mean of 499 g of cereal. Is this evidence that the company is systematically underfilling the packets?

Intuition suggests probably not – you would not expect the mean of every sample to be exactly 500 g and the result seems to be reasonably close. But how far below 500 g would the mean have to be before there was significant evidence? To answer questions like this you can use a hypothesis test.

 **Rewind**

You met hypothesis testing with the binomial distribution in Student Book 1, Chapter 22.

In this chapter you will look at two different types of hypothesis test. The first is a test to see if the mean of a sample is very different from a predicted value - such as in the cereal packet example. To do this you first need to establish some theory about the **distribution of sample means**.

In Student Book 1, Chapter 20, you saw that correlation coefficients could be used to describe the strength of correlation, but it was not clear how big the coefficient needed to be to have significant evidence of correlation. In Section 3 you will see how hypothesis tests can be used to decide if the sample correlation coefficient provides evidence for correlation in the population.

Section 1: Distribution of the sample mean

If data follow a normal distribution then every observation is a random variable, meaning that each observation can take a different value. The histogram shows samples taken from an $N(10,16)$ distribution.

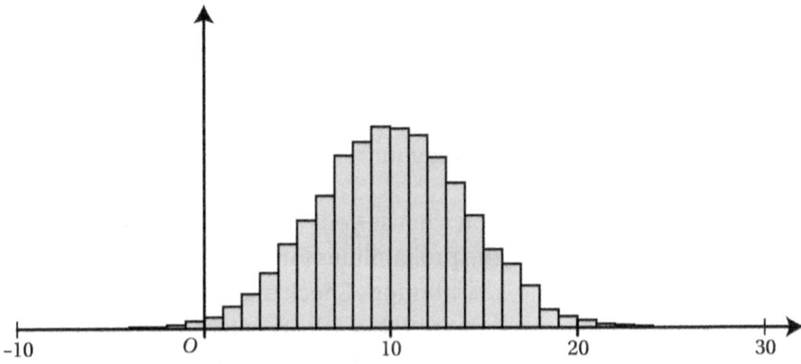

Instead of looking at a single value, you can look at a sample of n observations and take a mean. This might take a different value every time you do it, so it is also a random variable and given the symbol $\bar{X}_n$. For the distribution shown, take lots of samples of size 20 and create a histogram of their mean, shown in dark blue.

Tip

You might like to use technology to see if you can create a similar histogram.

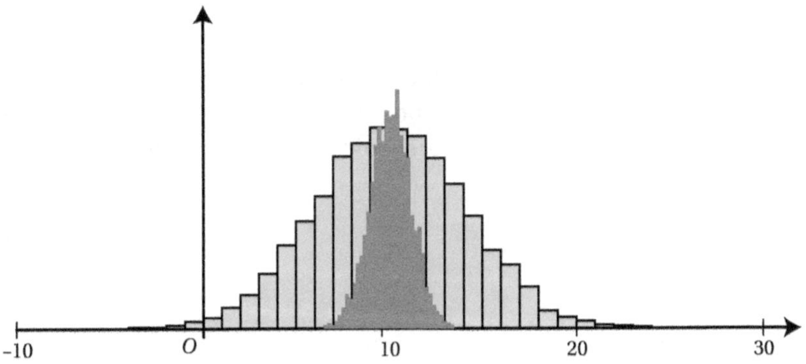

There are two important things to note about the sample means.

- They are clustered around the same mean as the original data.
- They are less spread out.

It can be shown that the sample mean of n observations of a normal distribution also follows a normal distribution with parameters relating to the original distribution and n.

Key point 22.1

If the original distribution was $N(\mu, \sigma^2)$ then:

$$\bar{X}_n \sim N\left(\mu, \frac{\sigma^2}{n}\right)$$

Fast forward

You will meet a proof of this if you study the Statistics option of Further Mathematics.

You know from Chapter 21, Section 1, that if you take a normally distributed random variable, subtract its mean and divide by its standard deviation, the new random variable is $N(0, 1)$. Doing that here to $\bar{X}$ means that the random variable $Z = \frac{\bar{X} - \mu}{\frac{\sigma}{\sqrt{n}}}$ is $N(0,1)$. You can use this fact in hypothesis testing (see Section 2).

WORKED EXAMPLE 22.1

A distribution is $X \sim N(10, 16)$.

a Find $P(9 < X < 11)$. **b** Find $P(9 < \bar{X}_{20} < 11)$. **c** Comment on your results in parts **a** and **b**.

a $P(9 < X < 11) = 0.197$ (from calculator)

You can use your calculator to find the probabilities for a given normal distribution. Check that your answer is reasonable – the required region is within less than one standard deviation away from the mean, so the result must be less than two-thirds.

b $\bar{X}_{20} \sim N\left(10, \frac{16}{20}\right)$

Use $\bar{X}_n \sim N\left(\mu, \frac{\sigma^2}{n}\right)$ with $n = 20$

So the standard deviation is $\frac{4}{\sqrt{20}} \approx 0.89$

$P(9 < \bar{X}_{20} < 11) = 0.736$ (from calculator)

Here the required region is within just over one standard deviation of the mean, so the answer should be just over two-thirds.

c The mean of 20 observations is much more likely to be within 1 unit of the true mean (10) than a single observation is.

Common error

Make sure you know whether you are working with an observation, X, or the mean of several observations, $\bar{X}$. If you are working with $\bar{X}$, don't forget to divide the variance by n.

EXERCISE 22A

1 Write down the distribution of the sample mean, given the original distribution.

a i If $X \sim N(4,100)$ find $\bar{X}_4$. ii If $X \sim N(20,125)$ find $\bar{X}_5$.

b i If $X \sim N(0,1)$ find $\bar{X}_{10}$. ii If $X \sim N(0,10)$ find $\bar{X}_4$.

2 Find each probability.

a i If $X \sim N(4,100)$ find $P(\bar{X}_4 < 6)$. ii If $X \sim N(20,125)$ find $P(\bar{X}_5 > 16)$.

b i If $X \sim N(0,1)$ find $P(-0.5 < \bar{X}_{10} < 1)$. ii If $X \sim N(0,10)$ find $P(0 < \bar{X}_4 < 3)$.

3 X is the energy (in eV) of beta particles emitted from a radioactive isotope. It is known that $X \sim N(40, 25)$. $\bar{X}$ is the average energy of 100 beta particles.

a Stating one necessary assumption, write down the distribution of $\bar{X}$ along with its parameters.

b Find $P(39 < X < 41)$. c Find $P(39 < \bar{X} < 41)$.

4 The mass of a breed of dog is known to follow a normal distribution with a mean of 10 kg and a standard deviation of 2.5 kg. A random sample of four dogs is measured. What is the probability that their mean is more than 9 kg?

5 The volume of apple juice in a carton follows a normal distribution with a mean of 152 ml and a standard deviation of 4 ml. A quality control process rejects a batch if a random sample of 16 cartons has a mean of less than 150 ml. Find the probability that a batch gets rejected.

6 Eggs are sold in boxes of 6. The masses of eggs have a normal distribution with mean μ g and variance 50 g^2. What is the minimum value of μ that must be chosen if the average mass of an egg must be more than 75 g in at least 90% of boxes?

7 The lifetime of a bulb, X hours, is modelled by $N(10\,000, \sigma^2)$. 5% of samples of 100 bulbs have a mean lifetime of less than 9900. Find the value of σ.

8 The length of a species of fly follows a normal distribution with mean 8 mm and standard deviation σ mm. 10% of samples of 50 flies have a mean of more than 9.2 mm. Find the value of σ.

9 The diameter of an apple has mean 8 cm and standard deviation 1 cm. A sample of n apples is chosen and their mean diameter measured.

a What is the probability that the mean diameter is between 7.9 cm and 8.1 cm if $n = 3$?

b The probability that the mean diameter is between 7.9 cm and 8.1 cm must be at least 0.3. What is the smallest value of n that must be chosen?

c The probability of the mean diameter being between 7.9 and 8.1 is required to double to 0.6. How many times bigger than in part **b** must n now be?

10 The mass of a student in a group is X kg where X follows the $N(70, 9)$ distribution. In a sample of four observations find the probability that:

a the total mass is less than 300 kg

b the heaviest student has a mass less than 75 kg.

Section 2: Hypothesis tests for a mean

One very common decision you have to make is if a mean is different from a predicted value - for example, you may be told that the average IQ is 100 and want to see if students in a school have above average IQ. If you only have a sample from the school it is possible that the mean is above 100 just through chance. You can conduct a hypothesis test to see if the difference above 100 is big enough to be significant.

Tip

Make sure you distinguish between the mean of the sample ($\bar{X}$) and the true mean of the population (μ). You use $\bar{X}$ to make a decision about μ.

In hypothesis testing, you assume that the conservative position - called the null hypothesis - is true and then see how likely you are to see something like the observed data. If the probability of seeing the observed data is very low, you reject the null hypothesis.

Rewind

The terminology associated with hypothesis tests was introduced in Student Book 1, Chapter 22.

To conduct a hypothesis test you need to make some assumptions about the underlying distribution. The test you will study requires the underlying distribution to be normal with known variance.

Key point 22.2

To test the value of a population mean, μ, against a suggested value, μ_0, at significance level α:

1 Set up appropriate hypotheses, depending on the context, using one of:

- $H_0: \mu = \mu_0$ $H_1: \mu \neq \mu_0$
- $H_0: \mu = \mu_0$ $H_1: \mu > \mu_0$
- $H_0: \mu = \mu_0$ $H_1: \mu < \mu_0$

2 Conduct the test with one of these methods.

- See if your observed mean falls into the **acceptance region** or the **critical region**. To do this write down the distribution of $\bar{X}$ (using Key point 22.1). Then find the regions at the ends of the distribution which have a total probability of α (the critical region). Where these regions are depends on the alternative hypothesis.

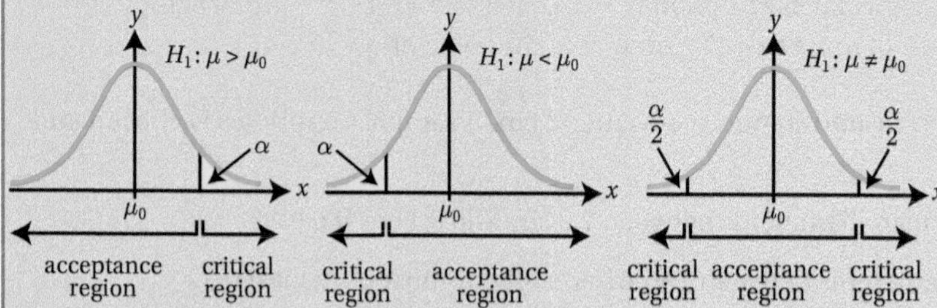

- Use your calculator to find the **p-value**, p, of the observed mean. This is the probability of getting the observed value or more extreme (according to the null hypothesis).

3 Reject the null hypothesis if the mean falls into the critical region or if $p \leqslant \alpha$.
It is important that you put your conclusion in context and that it is not overly certain - you must show an appreciation that you have only found evidence rather than stating a certain conclusion.

Testing using the critical region

The method using the critical region (or acceptance region) will be looked at first.

WORKED EXAMPLE 22.2

The level of testosterone in blood is normally distributed with mean 24 nmol l^{-1} and standard deviation 6 nmol l^{-1}. After completing a race, a sprinter gives two samples with an average of 34 nmol l^{-1}. Is this sufficiently different (at 1% significance) to suggest that the sprinter's testosterone level is above average?

X = level of testosterone in the sprinter's blood

$X \sim \mathrm{N}(\mu, 36)$

Define the variables.

H_0: $\mu = 24$

H_1: $\mu > 24$

State the hypotheses. This is a one-tailed test because the question is only looking for evidence of high testosterone.

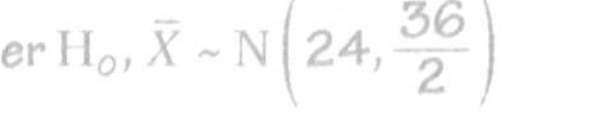

State the distribution of $\bar{X}$ under H_0. The sprinter gave two samples.

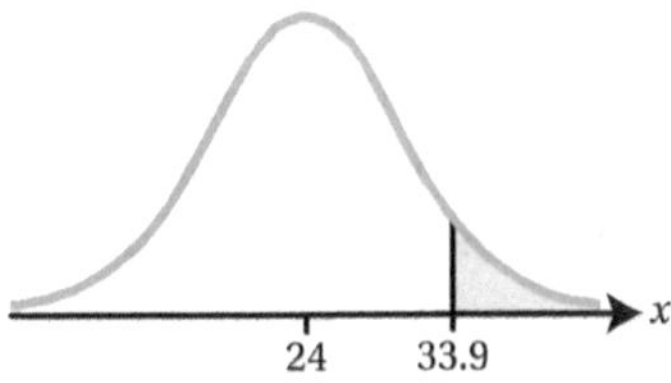

Find the critical region by sketching the normal curve and using the inverse normal distribution.

$\mathrm{P}(X < a) = 0.99$ so $a = 33.9$ (from calculator).

$\therefore$ critical region : $\bar{X} > 33.9$

$34 > 33.9$

So, 34 nmol l^{-1} falls in the rejection region; reject the null hypothesis.

There is evidence that the sprinter's testosterone level is above average.

Draw a conclusion in context.

Sometimes questions only ask you to find the critical region without actually performing a test.

WORKED EXAMPLE 22.3

The temperature of a water bath is normally distributed with a mean of 60 °C and a standard deviation of 1 °C. After the equipment is serviced it is assumed that the standard deviation is the same. The temperature is measured on five independent occasions and a test is performed at the 5% significance level to see if the temperature has changed from 60 °C. What range of mean temperatures would result in accepting that the temperature has changed?

X = temperature of water bath

$X \sim \mathrm{N}(\mu, 1)$ ········ Define the variables.

$\mathrm{H}_0: \mu = 60$

$\mathrm{H}_1: \mu \neq 60$ ········ State the hypotheses. This is a two-tailed test because the question does not specify that you are looking for evidence that the bath is too warm or too cold.

$\bar{X} \sim \mathrm{N}\left(60, \frac{1}{5}\right)$ ········ State the distribution of $\bar{X}$ under H_0. Five measurements were taken.

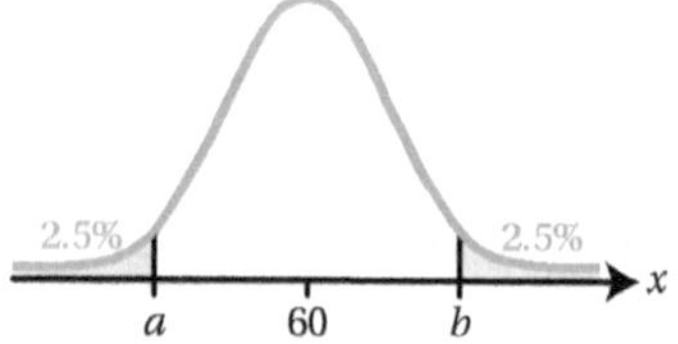

Use the inverse normal distribution to find the critical values of $\bar{X}$ for the two-tailed region. The probability of being in each tail is half of the significance level, so it is 2.5%.

$\mathrm{P}(\bar{X} < a) = 0.025 \Rightarrow a = 59.1$ (from calculator)

by symmetry about the mean, $b = 60.9$

$\therefore\ \bar{X} < 59.1$ or $\bar{X} > 60.9$ ········ Write down the critical region.

You might be asked to work with the standard normal variable Z, rather than with $\bar{X}$.

WORKED EXAMPLE 22.4

Jennifer believes that her average discus throw is above 30 m. She knows that her discus throws are normally distributed with standard deviation 3 m. She decides to take an average of 10 throws and work out the test statistic:

$$Z = \frac{\overline{X} - \mu}{\frac{3}{\sqrt{10}}}$$

She then conducts a hypothesis test at the 5% significance level.

a Write down Jennifer's null and alternative hypotheses.
b Write down the distribution of Z, including the value of any parameters.
c Find the critical region in terms of Z.
d The average of Jennifer's ten throws was 32 m.
Find the value of Z and hence decide if the null hypothesis can be rejected.

a H_0: $\mu = 30$
H_1: $\mu > 30$

This a one-tailed test, as you are looking for an increase in average distance.

b $Z \sim N(0, 1)$

You are given that $\overline{X} \sim N\left(\mu, \frac{3^2}{10}\right)$ so $\frac{\overline{X} - \mu}{\frac{3}{\sqrt{10}}} \sim N(0, 1)$.

c $P(Z > a) = 0.05$
$\Rightarrow a = 1.64$ (from calculator)
$\therefore Z > 1.64$

Use the inverse normal distribution to find the critical value of Z for the one-tailed region at 5%.

d $\overline{X} = 32$
$\therefore Z = \frac{32 - 30}{\frac{3}{\sqrt{10}}} = 2.11$

You are given that $\overline{X} = 32$. Use this to work out Z.

$2.11 > 1.64$ so reject H_0

2.11 is in the critical region.

Testing using the p-value

You might find the p-value method more straightforward.

Tip

Many calculators can calculate the p-value. You should check your manual to see how to do this with yours.

WORKED EXAMPLE 22.5

Traditional light bulbs have an average lifetime of 800 hours and a standard deviation of 100 hours. A manufacturer claims that the lifetimes of their bulbs have the same standard deviations but that they last longer. A sample of 50 of the manufacturer's light bulbs has an average lifetime of 829.4 hours. Test the manufacturer's claim at the 5% significance level.

X = lifetime of a bulb
$X \sim \mathrm{N}(\mu, 100^2)$

Define the variables.

$\mathrm{H_0}$: $\mu = 800$
$\mathrm{H_1}$: $\mu > 800$

State the hypotheses.

$\bar{X} \sim \mathrm{N}\left(800, \frac{100^2}{50}\right)$

State the test statistic and its distribution.

p-value $= \mathrm{P}(\bar{X} \geqslant 829.4)$
$= 0.0188$ (from calculator)

Use your calculator to find the p-value.

$0.0188 < 0.05$

Therefore reject $\mathrm{H_0}$ – there is evidence to support the manufacturer's claim that their light bulbs last longer than 800 hours.

Compare to the significance level and conclude.

WORK IT OUT 22.1

The wingspan of a species of butterflies is known to be normally distributed with mean 10 cm and standard deviation 1 cm. A scientist thinks he may have found a new species. The mean of a sample of six of these butterflies is 11.2 cm. Test the scientist's claim at the 5% significance level, assuming the wingspans are still normally distributed with standard deviation 1 cm.

Which is the correct solution? Identify the errors made in the incorrect solutions.

Solution A	$\mathrm{H_0}: \mu = 10; \mathrm{H_1}: \mu \neq 10$ If $X \sim \mathrm{N}(10, 1)$ then $\mathrm{P}(\bar{X} > 11.2) = 0.115$ (from calculator) $0.115 > 0.05$ so accept $\mathrm{H_0}$ – a new species has not been found.
Solution B	$\mathrm{H_0}: \mu = 10; \mathrm{H_1}: \mu \neq 10$ If $\bar{X} \sim \mathrm{N}\left(10, \frac{1}{\sqrt{6}}\right)$ then $\mathrm{P}(\bar{X} > 11.2) = 0.001\,64$ (from calculator) The p-value is twice this which is $0.003\,29$. $0.003\,29 < 0.05$ so reject $\mathrm{H_0}$ – there is evidence for a new species.
Solution C	$\mathrm{H_0}: \mu = 10; \mathrm{H_1}: \mu > 10$ If the mean is 10 and the standard deviation is $\frac{1}{6}$ then the critical region is $\bar{X} > 10.3$. $11.2 > 10.3$ so the observed value lies in the critical region. Reject $\mathrm{H_0}$ – there is a new species.

EXERCISE 22B

1 Write null and alternative hypotheses for each situation.

a i The average IQ, μ, in a school over a long period of time has been 102. It is thought that changing the menu in the cafeteria might have an effect upon the average IQ.

ii It is claimed that the average size, μ, of photos created by a camera is 1.2 Mb. A computer scientist believes that this figure is inaccurate.

b i A consumer believes that steaks sold in portions of 250 g are on average underweight.

ii A careers adviser believes that the average extra amount earned over a lifetime by people with a degree is more than the \$150 000 figure he has been told at a seminar.

c i The mean breaking tension, μ_T, of a brake cable does not normally exceed 3000 N. A new brand claims that it regularly does exceed this value.

ii The average time, μ_t, taken to match a fingerprint is normally more than 28 minutes. A new computer program claims to be able to do better.

2 In each situation, it is believed that $X \sim \text{N}(\mu, 100)$. Find the critical region in each case.

a i $H_0: \mu = 60$; $H_1: \mu \neq 60$; 5% significance; $n = 16$

ii $H_0: \mu = 120$; $H_1: \mu \neq 120$; 10% significance; $n = 30$

b i $H_0: \mu = 80$; $H_1: \mu > 80$; 1% significance; $n = 18$

ii $H_0: \mu = 750$; $H_1: \mu > 750$; 2% significance; $n = 45$

c i $H_0: \mu = 80.4$; $H_1: \mu < 80.4$; 10% significance; $n = 120$

ii $H_0: \mu = 93$; $H_1: \mu < 93$; 5% significance; $n = 400$

3 In each situation, it is believed that $X \sim \text{N}(\mu, 400)$. Find the p-value of the observed sample mean. Hence decide the result of the test if it is conducted at the 5% significance level

a i $H_0: \mu = 85$; $H_1: \mu \neq 85$; $n = 16$; $\bar{x} = 95$

ii $H_0: \mu = 144$; $H_1: \mu \neq 144$; $n = 40$; $\bar{x} = 150$

b i $H_0: \mu = 85$; $H_1: \mu > 85$; $n = 16$; $\bar{x} = 95$

ii $H_0: \mu = 144$; $H_1: \mu > 144$; $n = 40$; $\bar{x} = 150$

c i $H_0: \mu = 265$; $H_1: \mu < 265$; $n = 14$; $\bar{x} = 256.8$

ii $H_0: \mu = 377$; $H_1: \mu < 377$; $n = 100$; $\bar{x} = 374.9$

d i $H_0: \mu = 95$; $H_1: \mu < 95$; $n = 12$; $\bar{x} = 96.4$

ii $H_0: \mu = 184$; $H_1: \mu > 184$; $n = 50$; $\bar{x} = 183.2$

4 The average height of 18-year-olds in England is 168.8 cm and the standard deviation is 12 cm. Caroline believes that the students in her class are taller than average. To test her belief she measures the heights of 16 students in her class.

a State the hypotheses for Caroline's test.

You can assume that the heights follow a normal distribution and that the standard deviation of heights in Caroline's class is the same as the standard deviation for the whole population. The students in Caroline's class have average height of 171.4 cm.

b Test Caroline's belief at the 5% level of significance.

5 All students in a large school are given a typing test and it was found that the times taken to type one page of text are normally distributed with mean 10.3 minutes and standard deviation 3.7 minutes. The students were given a month-long typing course and then a random sample of 20 students was asked to take the typing test again. The mean time was 9.2 minutes. Test at 10% significance level whether there is evidence that the time the students take to type a page of text has decreased.

6 The national mean score in GCSE Mathematics is 4.73 with a standard deviation of 1.21. In a particular school the average of 50 students is 4.81.

a State two assumptions that are needed to perform a hypothesis test to see if the mean is better in this school than the background population.

b Assuming that these assumptions are met, test at the 5% significance level whether the school is producing better results than the national average.

7 A farmer knows from experience that the average height of apple trees is 2.7 m with standard deviation 0.7 m. The farmer buys a new orchard and wants to test whether the average height of apple trees is different. She assumes that the standard deviation of heights is still 0.7 m.

a State the hypotheses she should use for her test.

b The farmer measures the heights of 45 trees and finds their average.

Find the critical region for the test at the 10% level of significance.

c Given that the average height of the 45 trees is 2.3 m, state the conclusion of the hypothesis test.

Elevate

See Support Sheet 22 for a further example of testing for the mean of a normal distribution and for more practice questions.

8 A doctor has a large number of patients starting a new diet in order to lose mass. Before the diet the mass of the patients was normally distributed with mean 82.4 kg and standard deviation 7.9 kg. The doctor assumes that the diet does not change the standard deviation of the masses. After the patients have been on the diet for a while, the doctor takes a sample of 40 patients and finds their mean mass.

a The doctor believes that the average mass of the patients has decreased following the diet. He wishes to test his belief at the 5% level of significance. Find the critical region for this test.

b State an additional assumption required in your answer to part **a**.

c The average mass of the 40 patients after the diet was 78.4 kg. State the conclusion of the test.

9 A geologist is measuring the volume of bubbles in an underwater lava flow. He believes that the bubbles are normally distributed with variance 25 cm^6, but that they are smaller than in exposed lava flow where the mean is 20 cm^3.

a Write down the geologist's null and alternative hypotheses.

b The geologist calculates the test statistic

$$Z = \frac{\bar{X} - \mu}{\frac{\sigma}{\sqrt{n}}}$$

and rejects the null hypothesis if $Z < -2$.

What significance level does this correspond to?

c The geologist finds a mean of 17 cm^3. For what values of n will the geologist reject the null hypothesis?

10 The school canteen sells coffee in cups claiming to contain 250 ml. It is known that the amount of coffee in a cup is normally distributed with standard deviation 6 ml. Adam believes that on average the cups contain less coffee than claimed. He wishes to test his belief at the 5% significance level.

a Adam measures the amount of coffee in 10 randomly chosen cups and finds the average to be 248 ml. Can he conclude that the average amount of coffee in a cup is less than 250 ml?

b Adam decides to collect a larger sample. He finds the average to be 248 ml again, but this time there is sufficient evidence to conclude that the average amount of coffee in a cup is less than 250 ml. What is the minimum sample size he must have used?

11 The null hypothesis $\mu = 30$ is tested and a value $X = 35$ is observed. Will it have a higher p-value if the alternative hypothesis is $\mu \neq 30$ or $\mu > 30$?

Section 3: Hypothesis tests for correlation coefficients

You already know that values of the correlation coefficient close to 1 or −1 represent strong positive or negative correlation respectively.

However, values in between can be difficult to interpret. For example, is $r = 0.6$ evidence of significant correlation? The answer depends on the number of data points and how certain you want to be.

You need to distinguish between two related values. The correlation coefficient of the sample has the symbol r and the correlation coefficient of the underlying population has the symbol ρ (the Greek letter rho). You can conduct a hypothesis test using r to decide if there is evidence that ρ is not zero. In a two-tailed test you are looking to see if there is correlation in either direction - positive or negative - so the alternative hypothesis is $\rho \neq 0$. In a one-tailed test you are looking for correlation in just one direction, so the alternative hypothesis would be either $\rho > 0$ or $\rho < 0$.

Finding the distribution of r goes beyond what you need to know, but you can find the critical values in tables that will be provided. They are calculated assuming that both variables follow a normal distribution. If the modulus of r is larger than the appropriate critical value then you reject the null hypothesis.

One tail **Two tail**	**10%** **20%**	**5%** **10%**	**2.5%** **5%**	**1%** **2%**
n				
4	0.8000	0.9000	0.9500	0.9800
5	0.6870	0.8054	0.8783	0.9343
6	0.6084	0.7293	0.8114	0.8822
7	0.5509	0.6694	0.7545	0.8329
8	0.5067	0.6215	0.7067	0.7887

Rewind

See Student Book 1, Chapter 20, for a reminder of how to interpret the correlation coefficient.

Did you know?

The correlation coefficient used in this course is called the Pearson product-moment correlation coefficient. This is just one type of correlation coefficient – you could also use Spearman's rank correlation coefficient or Kendall's tau. They all have advantages and disadvantages.

WORKED EXAMPLE 22.6

The correlation coefficient between the mass and height of students in a sample of six students is 0.85. Test at the 5% significance level whether height and mass of students from this school are positively correlated.

$H_0: \rho = 0$
$H_1: \rho > 0$

Write down the hypotheses. You are only looking for positive correlation, so it is one-tailed.

One tail	10%	5%	2.5%
Two tail	20%	10%	5%
n			
4	0.8000	0.9000	0.9500
5	0.6870	0.8054	0.8783
6	0.6084	0.7293	0.8114
7	0.5509	0.6694	0.7545

Look for the critical value at the intersection of the 5% one-tail column and the $n = 6$ row.

The critical value from the table is 0.729.

$0.85 > 0.729$

so reject H_0. There is evidence that mass and height are positively correlated.

State the conclusion in context.

Did you know?

There are many examples of spurious correlations. For example, there is a very strong negative correlation between the number of pirates and global warming!
Can you explain this?

WORKED EXAMPLE 22.7

Twenty students were asked about the number of hours they spent watching television each week and their results in a reading test. The correlation coefficient for their results was –0.31. Test for evidence of correlation at the 10% significance level.

$H_0: \rho = 0$
$H_1: \rho \neq 0$

Write down the hypotheses. You are not told to look for correlation in any particular direction, so it is two-tailed.

One tail Two tail	10% 20%	5% 10%	2.5% 5%
n			
19	0.3077	0.3877	0.4555
20	0.2992	0.3783	0.4438
21	0.2914	0.3687	0.4329

Write down the critical values from the table. Remember that it is a two-tailed test so the critical region is $X > 0.3783$ or $X < -0.3783$.

The critical values are ± 0.3783.

$|-0.31| < 0.3783$ so do not reject H_0. There is not significant evidence for correlation between hours watching television and results in a reading test.

State the conclusion in context.

EXERCISE 22C

1 Test each sample correlation coefficient for positive correlation at 5% significance. n is the sample size.

a i $r = 0.4; n = 20$ ii $r = 0.3; n = 100$ b i $r = 0.4; n = 15$ ii $r = 0.5; n = 9$

2 Test each sample correlation coefficient for correlation at 5% significance. n is the sample size.

a i $r = -0.5; n = 15$ ii $r = -0.3; n = 30$ b i $r = 0.25; n = 100$ ii $r = 0.65; n = 10$

3 Information for 20 students is used to investigate the hypothesis that there is a correlation between IQ and results in a mathematics test.

a Write down the null and alternative hypotheses for this investigation.

b Data are collected and the p-value for the correlation coefficient is 0.002 18. What is the conclusion of the hypothesis test at the 5% significance level?

4 The average speed of cars is measured at six different checkpoints at varying distances from a junction. There is a belief that in general cars get faster as they are further from the junction.

a Write down the null and alternative hypotheses for this investigation.

b The p-value of the observed data is found to be 0.084. Test the data at the 5% significance level.

5 The amount spent by a government on unemployment support is expected to be negatively correlated with the amount spent on education. Data were collected across 50 countries in 2013.

a What is the population associated with this sample?

b Write down appropriate null and alternative hypotheses.

c The sample correlation coefficient was found to be –0.36. What is the conclusion of the hypothesis test at 5% significance?

6 The correlation coefficient between the amount of water used in a town on 30 summer days and the temperature is 0.85.

a Jane thinks that there is a correlation between water usage and temperature on a summer day. Write down the null and alternative hypotheses that Jane should use to test her suspicion.

b Conduct the test at the 5% significance level.

c Karl says that if people use more water the days will be warmer. Give two reasons why your hypothesis test does not support Karl's statement.

7 The level of antibodies in blood is thought to go down as the dose of a medical drug is increased.

a State appropriate null and alternative hypotheses to test this statement.

b In a sample of 65 patients the correlation coefficient between these two variables is found to be –0.34. Conduct an appropriate hypothesis test at the 1% significance level.

c What are the advantages of using a 1% significance level rather than a 5% significance level for medical tests?

8 Data are collected on the height of a cake and the temperature at which it is baked.

a If a hypothesis test is conducted to test for positive correlation the p-value is 0.032. Is this evidence of positive correlation at the 5% significance level?

b If the same data were instead used to test for correlation in either direction, what would be the p-value? Is there evidence of correlation at the 5% significance level?

c If the correlation coefficient increases, does this increase or decrease the p-value found in part **a**? Justify your answer.

9 It is suspected that there is correlation between the height of a tree and the total surface area of its leaves. A random sample of n trees is measured and the correlation coefficient is found to be 0.402.

What is the smallest value of n that makes this significant at 5% significance?

10 Why do critical value tables for correlation coefficients start at $n = 3$?

Checklist of learning and understanding

- The sample mean, $\bar{X}$, is a random variable.
  If $X \sim \text{N}(\mu, \sigma^2)$, then $\bar{X}_n \sim \text{N}\left(\mu, \frac{\sigma^2}{n}\right)$
- To test the value, μ, of a population mean against a suggested value, μ_0:
  - set up appropriate hypotheses depending on the context, using one of:
    - $\text{H}_0: \mu = \mu_0$; $\text{H}_1: \mu \neq \mu_0$
    - $\text{H}_0: \mu = \mu_0$; $\text{H}_1: \mu > \mu_0$
    - $\text{H}_0: \mu = \mu_0$; $\text{H}_1: \mu < \mu_0$
  - then, use the distribution of $\bar{X}$ and your calculator to either find the p-value or set up the critical region for the given significance level, α
  - if $p \leqslant \alpha$ (or if $\bar{X}$ is in the critical region), reject H_0.
- To test whether there is correlation between two variables:
  - set up appropriate hypotheses depending on the context, using one of:
    - $\text{H}_0: \rho = 0$; $\text{H}_1: \rho \neq 0$
    - $\text{H}_0: \rho = 0$; $\text{H}_1: \rho > 0$
    - $\text{H}_0: \rho = 0$; $\text{H}_1: \rho < 0$
  - then look up in the tables the critical value for the given significance level and sample size
  - if the modulus of the sample correlation coefficient, r, is greater than the critical value, then reject H_0.

Mixed practice 22

1 A random sample of 20 people have their heights and masses measured. Which correlation coefficient is the smallest that would suggest evidence of positive correlation between height and mass, using a 5% significance level? Choose from these options.

A −0.6 **B** 0.24 **C** 0.41 **D** 0.52

2 The breaking load of steel wire is known to be normally distributed with mean 80 N and standard deviation 4 N. Find the probability that the mean breaking load of a sample of 10 such wires is between 80 N and 81 N.

3 Data are collected on HIV rates and literacy rates of 40 countries in 2015.

a What is the population from which this sample is drawn?

b A hypothesis test is conducted to see if there is correlation between HIV rates and literacy rates. Write down appropriate null and alternative hypotheses.

c The p-value for the observed correlation coefficient is 0.12. What is the conclusion of the hypothesis test at the 10% significance level?

4 The mass of cakes produced by a bakery is known to be normally distributed with mean 300 g and standard deviation 40 g. A new baker is employed.

a State appropriate null and alternative hypotheses to test if the mean mass of cakes has changed.

b The mean of 12 cakes is found to be 292 g.

i What is the p-value of these data?

ii What is the conclusion of the hypothesis test at the 10% significance level?

5 As a special promotion, a supermarket offers cartons of orange juice containing '25% extra' with no increase in price.

A random sample of cartons of orange juice was checked. The percentages by which the contents exceeded the nominal quantity were recorded, with the following results.

23.3 27.5 25.7 20.9 24.3 22.6 21.5 22.1

Examine whether the mean percentage by which the contents exceed the nominal quantity is less than 25. Use the 5% significance level. Assume that the data are from a normal distribution with standard deviation 2.3.

[© AQA 2011]

6 The results of a group of students in a test is thought to follow a N $(\mu, 25)$ distribution. The mean of a random sample of 20 students is used to test the hypothesis $\mu = 100$ against $\mu < 100$.

What is the critical region at the 10% significance level (given to 3 s.f.)? Choose from these options.

A $\bar{X} < 91.8$ **B** $\bar{X} < 92.8$ **C** $\bar{X} < 93.6$ **D** $\bar{X} < 98.6$

7 A test is conducted to see if the time spent revising correlates with the results in a test.

a State a distributional assumption required to use tables of critical values for the correlation coefficient.

b A random sample of 25 students is surveyed. The correlation coefficient for their responses is 0.55. Conduct a hypothesis test at the 5% significance level.

8 The time taken for a full kettle to boil is known to follow a normal distribution with mean 40 seconds and standard deviation 5 seconds. After cleaning, the kettle is boiled 10 times to test if the time taken for it to boil has decreased.

a Stating two necessary assumptions, find the critical region for this hypothesis test at the 5% significance level.

b The mean time is found to be 37 seconds. State the outcome of the hypothesis test.

c Why should there be a long delay between the ten observations for this test to be valid?

9 Paul knows the time it takes him to complete the crossword in a particular newspaper is normally distributed with mean 45 minutes and standard deviation 10 minutes.

He changes his strategy to attempt all the down clues first, and wishes to test whether this has affected his average time.

To do this he finds the mean of his next 20 attempts and calculates the test statistic

$$Z = \frac{\bar{X} - \mu}{\frac{\sigma}{\sqrt{n}}}$$

a State the null and alternative hypothesis for the test.

b If the test is conducted at the 5% significance level, find the critical region in terms of Z.

c The mean of his next 20 attempts is 41 minutes.

Find the value of Z and hence state the conclusion of the test.

d What would be the conclusion if Paul was testing at the 5% level whether his average time had improved? Explain your answer fully.

10 A company produces low-energy light bulbs. The bulbs are described as using 9 watts of power. Amir, the production manager, asked Jenny to measure the power used by each of a sample of 120 bulbs from the latest batch produced and to test the hypothesis that the mean for the batch is 9.0 watts. Jenny is to carry out the test at the 5% significance level.

a What assumption must be made about the sample of 120 bulbs if the result of this test is to be valid?

b Jenny found that the mean for the sample was 9.2 watts and that the standard deviation was 1.3 watts. Carry out the test asked for by Amir.

c When Jenny reported the conclusion of the test to Amir, he said that he had intended to ask Jenny to test whether the bulbs in the batch use more than 9.0 watts on average.

For the test in part **b**, state the effect that this new information would have on:

i the alternative hypothesis

ii the critical value(s)

iii the conclusion.

11 Yukun wants to test the null hypothesis $H_0: \mu = 10$ against the alternative hypothesis $H_1: \mu > 10$. He finds the mean sample of the data, $\bar{X}$, is 11.2. His calculator tells him that the p-value of his data is 0.04. Which of these expressions defines the p-value of his hypothesis test?

A $P(H_0 \mid \bar{X} = 11.2)$ **B** $P(H_0 \mid \bar{X} \geqslant 11.2)$ **C** $P(\bar{X} = 11.2 \mid H_0)$ **D** $P(\bar{X} \geqslant 11.2 \mid H_0)$

12 The distance an athlete jumps in a long jump is known to be normally distributed with mean 5.84 m and standard deviation 0.31 m. After a change to her technique she looks at the average of n jumps to see if her average distance has changed. She uses a 5% significance level.

a Write down appropriate null and alternative hypotheses for this test.

b If the jumps are still normally distributed with standard deviation 0.31 m, find the acceptance region in terms of n.

c If the mean is found to be 5.6 m, find the smallest value of n that would result in the null hypothesis being rejected.

13 The viewing figures of a long-running television series are given as X million. In the past it was known to follow a normal distribution with a standard deviation of 0.3 million. A producer wants to know if a new presenter has changed the viewing figures across 12 episodes. He conducts a hypothesis test, assuming the viewing figures are still normally distributed with standard deviation 0.3 million. The acceptance region is found to be $6.258 < \bar{X} < 6.542$.

a Deduce the null and alternative hypotheses.

b Find the significance level of this hypothesis test, giving your answer as a percentage to the nearest whole number.

Elevate

See Extension Sheet 22 for some questions to make you think about appropriate significance levels in various situations.

FOCUS ON... PROOF 4

The prosecutor's fallacy

A man is accused of robbing a jeweller's shop and stealing diamonds. The only evidence against him is a bag of diamonds found in his car during the police investigation.

The prosecution argues that the probability of the bag being in the car if the man is innocent is 1 in 10 000, and hence the probability of him being guilty is 9999 in 10 000 (or 99.99%). You are going to investigate whether this is a valid argument.

Investigating the argument

QUESTIONS

1 Let p denote the probability that the man is guilty of stealing the diamonds. Without taking into account any information (such as any findings of the police investigations), estimate the value of p.

2 Denote by G the event that the man is guilty, so $P(G) = p$, and let E be the event that the evidence (in this case, the bag of diamonds) is found in his car. Then the prosecution's statement says that $P(E \mid G') = 0.0001$. You also need to estimate the probability of finding the evidence if he is guilty; let's assume this is quite likely, so set $P(E \mid G) = 0.99$.

a Complete the tree diagram.

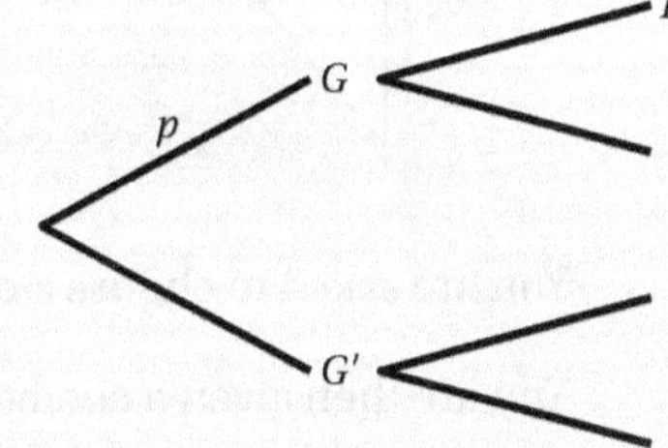

b Using the tree diagram:

i find $P(E)$

ii use the conditional probability formula to find $P(G \mid E)$ and $P(G' \mid E)$ in terms of p.

c Prove that, when $p < \frac{1}{10\,000}$, $P(G \mid E) < P(G' \mid E)$. What does this tell you about the prosecutor's argument?

3 Why was it reasonable to assume that p is very small? Think about what p represents.

4 How do $P(G \mid E)$ and $P(G' \mid E)$ compare to each other if p is larger (say, 0.1)?

5 Would it make sense to swap the branches on the tree diagram, so that the first 'level' has E and E'?

6 The prosecutor stated that the probability that the bag is found in the car if the man is innocent is 1 in 10 000. Write this event using the conditional probability notation. Hence write down and interpret the event which has the probability of 9999 in 10 000.

7 In law, the phrase 'proof beyond reasonable doubt' is used. What do you think this means? How does this compare to mathematical proof?

FOCUS ON... PROBLEM SOLVING 4

Using extreme values

Probability can often be counter-intuitive, and people find it difficult to evaluate their solutions. One useful strategy can be to use extreme values to make the answer more obvious.

The Monty Hall problem

This game featured in a US television show.

> You are shown three closed doors and told that one door hides a car while the other two hide goats. You will win whatever is behind the door that you choose. The show host knows which door hides the car.

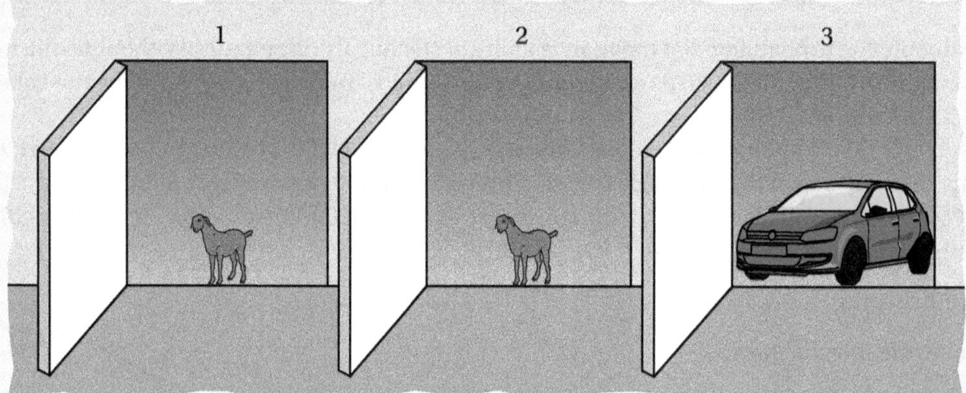

> You are asked to choose a door. The host then opens one of the other two doors to reveal a goat.
>
> You are then given a choice: stick with the original door, or switch to the third one. What should you do to maximise your probability of winning the car?

Most people's intuition is that it doesn't matter – there are two closed doors and the car is equally likely to be behind either of them. But this does not take into account the fact the host **knows** where the car is, so would never open that door.

QUESTIONS

1. Imagine that instead there are 100 doors and you initially choose door 1. The host opens 98 of the other doors, leaving out door 43. What would you do?

   The fact that the host knew which door to leave closed gives you additional information, so the probability has changed from the original equally likely for each door.

2. Suppose you play the game 300 times. Call the door hiding the car Door 1. Your initial choice of the door is random. If you picked Door 1 then the host can choose whether to open Door 2 or Door 3; assume the host chooses randomly. If you initially picked Door 2 or 3, then the host has no choice about which one to open.

Copy and complete this table showing the possible outcomes, assuming the car is behind Door 1.

		You choose			
		1	2	3	
Host opens	1				
	2				
	3				
		100	100	100	300

Use the table to find the probability that you win the car if you switch.

Two children in the garden

There are two children in the garden. One of them is a girl. What is the probability that both of them are girls?

You may assume that both genders are equally likely and that the genders of the two children are independent of each other – so, for example, they are not twins.

3 a Which of these arguments do you find most convincing?

i Both genders are equally likely, and the genders of the two children are independent, so the probability that the second one is also a girl is $\frac{1}{2}$.

ii The two genders are independent, so the probability of two girls is $\frac{1}{2} \times \frac{1}{2} = \frac{1}{4}$.

iii The options for the two genders are GG, BB and GB. Hence the probability that both are girls is $\frac{1}{3}$.

iv The options for the two genders are GG, BB, GB and BG. Hence the probability of two girls is $\frac{1}{4}$.

v You already know that one child is a girl, so there are fewer options to choose from. Hence the probability of two girls is more than $\frac{1}{4}$.

b Now suppose there are 10 children in the garden and 9 of them are girls. Do you think that the probability that all 10 are girls is bigger or smaller than $\left(\frac{1}{2}\right)^{10}$?

c The table shows possible outcomes for two children. Copy and complete the probabilities and hence find the probability that both children are girls, given that at least one of them is a girl.

		First child	
		girl	boy
Second child	girl		
	boy		

FOCUS ON... MODELLING 4

When can you use the normal distribution?

The normal distribution is commonly used as a model in many applications. However, it is important to be aware that this model is not always suitable. This section focuses on the properties of the normal distribution, which you should consider when deciding whether to use it as a model.

WORKED EXAMPLE 1

The marks for a group of 1000 students on a Statistics exam are summarised in the histogram. The mean mark is 50 and the standard deviation of the marks is 20.6.

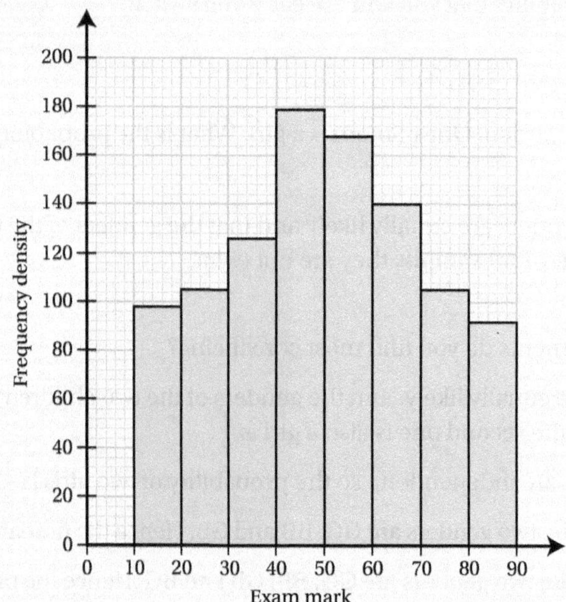

a Estimate the number of students whose marks are more than two standard deviations away from the mean.

b Hence state, with a reason, whether a normal distribution could be used to model the distribution of the marks.

a $50 + 2 \times 20.6 = 91.2$
$50 - 2 \times 20.6 = 8.8$

No students had marks more than two standard deviations from the mean.

On the histogram, there are no data values below 10 or above 90.

b For a normal distribution, the number of students with marks more than two standard deviations from the mean should be around 5%.
Hence the normal distribution does not seem to be a good model for these marks.

For a normal distribution, 95% of the data should be within two standard deviations of the mean.

QUESTIONS

1 The table summarises heights of a group of 80 schoolchildren.

Height (cm)	Frequency
120–130	9
130–140	32
140–150	26
150–160	11
160–170	4

a Draw a histogram to represent these data (you may want to use technology to do this).

b Hence explain whether a normal distribution would be a suitable model for these heights.

2 a An old textbook says that the range of the data is about six times the standard deviation. For a normal distribution, what percentage of the values is contained in this range?

The box plot summarises results of a discus throw competition (length, measured in metres).

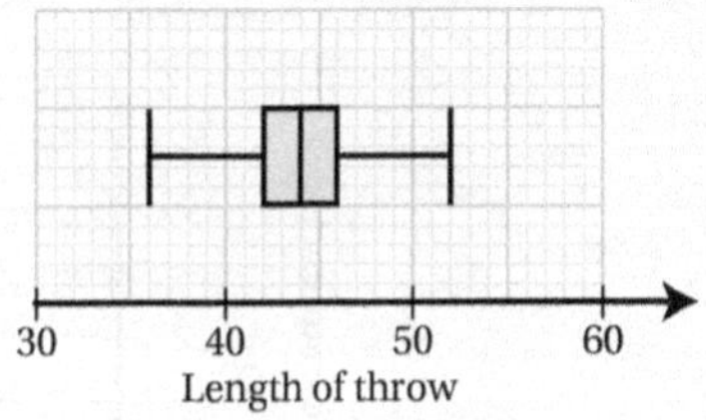

b Assuming the data follow a normal distribution, use the box plot to estimate its mean and standard deviation. Find the inter-quartile range for this normal distribution.

c Hence state, with a reason, whether a normal distribution is a suitable model for the lengths of the throws.

3 The mean, median and standard deviation for two sets of data are given. For each set of data decide, based on this information, whether a normal distribution would be a suitable model.

a mean = 231; median = 252; SD = 168

b mean = 165; median = 153; SD = 2.7

4 In social science research, subjects are often asked to rank their opinions on a five-point scale, called the Likert scale (for example: strongly disagree, disagree, no opinion, agree, strongly agree). For the purpose of statistical analysis, these responses are sometimes translated into numbers (e.g. strongly disagree = 1, strongly agree = 5). Give a reason why data measured on a Likert scale should not be modelled using a normal distribution.

5 a A law firm has found that the average length of phone calls made by the employees is 5.8 minutes. Give a reason why a normal distribution might not be a suitable model for the length of phone calls.

b It is often said that many naturally occurring measurements approximately follow a normal distribution. Discuss whether a normal distribution would be an appropriate model in each situation.

i The mass of red squirrels in a forest.

ii The heights of all parents and children at a nursery school open day.

iii The number of children in a family.

6 The diagram shows three cumulative frequency curves. Which curves show:

a a symmetrical distribution

b a normal distribution?

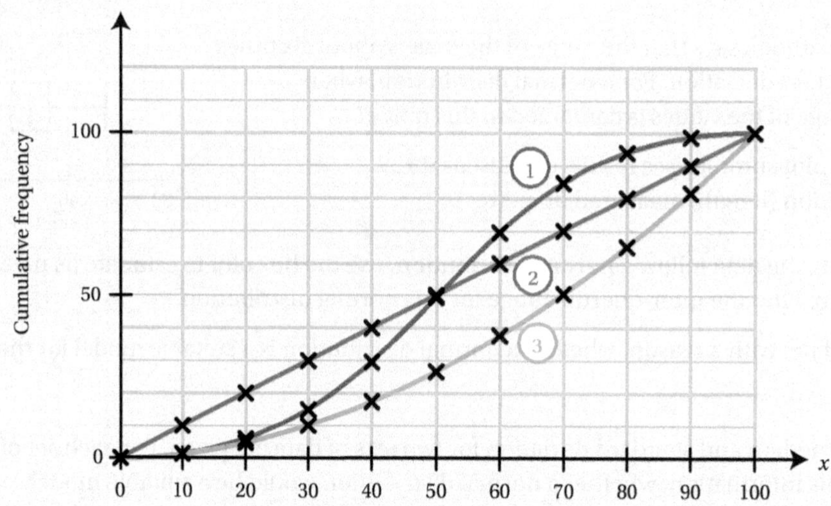

CROSS-TOPIC REVIEW EXERCISE 4

1. Asher has a large bag of sweets, half of which are red. He rolls a fair six-sided dice once. If the dice shows 1 or 2, he randomly picks two sweets from the bag. If the dice shows any other number, he randomly picks three sweets from the bag.

   Find the probability that Asher picks at least one red sweet.

2. Elsa is investigating whether there is any correlation between the average daily temperature and the daily amount of rainfall.

   a State suitable null and alternative hypotheses for her test.

   Elsa collects the data for a random sample of 12 days and calculates that the correlation coefficient between the average temperature and the amount of rainfall is –0.52.

   b Conduct a hypothesis test at the 5% level of significance. State your conclusion in context.

3. It is known that the heights of a certain type of rose bush follow a normal distribution with mean 86 cm and standard deviation 11 cm. Larkin thinks that the roses in her garden have the same standard deviation of heights, but are taller on average. She measures the heights of 12 rose bushes in her garden and finds that their average height is 92 cm.

   a State suitable hypotheses to test Larkin's belief.

   b Showing your method clearly, test at the 5% level of significance whether there is evidence that Larkin's roses are taller than average.

4. The lifetime of a certain type of lightbulb, T hours, is modelled by the distribution $N(620, \sigma^2)$. It is given that $P(T > 670) = 0.15$.

   a Find the value of σ.

   b Find the probability that, in a sample of 40 randomly selected lightbulbs, at least 10 last more than 670 hours.

   c By considering the range of values that contains nearly all values of T, discuss whether the normal distribution is a reasonable model for the lifetime of the lightbulbs.

5. During June 2011, the volume, X litres, of unleaded petrol purchased per visit at a supermarket's filling station by private-car customers could be modelled by a normal distribution with a mean of 32 and a standard deviation of 10.

   a Determine:

   i $P(X < 40)$

   ii $P(X > 25)$

   iii $P(25 < X < 40)$.

   b Given that during June 2011 unleaded petrol cost £1.34 per litre, calculate the probability that the unleaded petrol bill for a visit during June 2011 by a private-car customer exceeded £65.

   c Give **two** reasons, in context, why the model $N(32, 10^2)$ is unlikely to be valid for a visit by **any** customer purchasing fuel at this filling station during June 2011.

[© AQA 2012]

6 A data set consists of four numbers: 1, 4, 5 and x. Find the value of x for which the standard deviation of the data is the minimum possible.

7 Two events A and B are such that $P(A) = \frac{3}{4}$, $P(B \mid A) = \frac{1}{5}$ and $P(B' \mid A') = \frac{4}{7}$. By use of a tree diagram, or otherwise, find:

a $P(A \cap B)$

b $P(B)$

c $P(A \mid B)$.

8 Theo repeatedly rolls a fair dice until he gets a six.

a Show that the probability of him getting this six on the third roll is $\frac{25}{216}$.

b p_r is the probability of getting his first six on the rth roll. Find an expression for p_r in terms of r.

c Prove algebraically that $\sum_{r=1}^{r=\infty} p_r = 1$

9 Roger is an active retired lecturer. Each day after breakfast, he decides whether the weather for that day is going to be fine (F), dull (D) or wet (W). He then decides on only one of four activities for the day: cycling (C), gardening (G), shopping (S) or relaxing (R). His decisions from day to day may be assumed to be independent.

The table shows Roger's probabilities for each combination of weather and activity.

		Weather		
		Fine (F)	**Dull (D)**	**(W)**
Activity	**Cycling (C)**	0.30	0.10	0
	Gardening (G)	0.25	0.05	0
	Shopping (S)	0	0.10	0.05
	Relaxing (R)	0	0.05	0.10

a Find the probability that, on a particular day, Roger decided:

i that it was going to be fine and that he would go cycling;

ii on either gardening or shopping;

iii to go cycling, given that he had decided that it was going to be fine;

iv **not** to relax, given that he had decided that it was going to be dull;

v that it was going to be fine, given that he did **not** go cycling.

b Calculate the probability that, on a particular Saturday and Sunday, Roger decided that it was going to be fine and decided on the same activity for both days.

[© AQA 2013]

10 The masses of bags of sugar are normally distributed with mean 150 g and standard deviation 12 g.

a Find the probability that a randomly chosen bag of sugar weighs more than 160 g.

b Find the probability that in a box of 20 bags there are at least two that weigh more than 160 g.

c Darien picks out bags of sugar from a large crate at random. What is the probability that he has to pick up exactly 4 bags before he finds one that weighs more than 160 g?

11 A machine, which cuts bread dough for loaves, can be adjusted to cut dough to any specified set weight. For any set weight, μ grams, the actual weights of cut dough are known to be approximately normally distributed with a mean of μ grams and a fixed standard deviation of σ grams.

It is also known that the machine cuts dough to within 10 grams of any set weight.

a Estimate, with justification, a value for σ.

b The machine is set to cut dough to a weight of 415 grams.

As a training exercise, Sunita, the quality control manager, asked Dev, a recently employed trainee, to record the weight of each of a random sample of 15 such pieces of dough selected from the machine's output. She then asked him to calculate the mean and the standard deviation of his 15 recorded weights.

Dev subsequently reported to Sunita that, for his sample, the mean was 391 grams and the standard deviation was 95.5 grams.

Advise Sunita on whether or not **each** of Dev's values is likely to be correct. Give numerical support for your answers.

c Maria, an experienced quality control officer, recorded the weight, y grams, of each of a random sample of 10 pieces of dough selected from the machine's output when it was set to cut dough to a weight of 820 grams. Her summarised results were as follows.

$$\sum y = 8210.0 \text{ and } \sum (y - \bar{y})^2 = 110.00$$

Explain, with numerical justifications, why **both** of these values are likely to be correct.

12 A random variable has a normal distribution with mean 0 and standard deviation 1.

a Use your calculator to find $P(0 < X < 1)$ correct to five decimal places.

The exact value of the probability is given by:

$$\frac{1}{\sqrt{2\pi}} \int_0^1 e^{-\frac{x^2}{2}} \, dx$$

b Use the trapezium rule with six strips to estimate the value of this integral. Give your answer correct to five decimal places.

c Find the percentage error in using the trapezium rule to estimate this probability.

13 The random variable X has binomial distribution $B(n, p)$. The line graph shows the probability distribution of X.

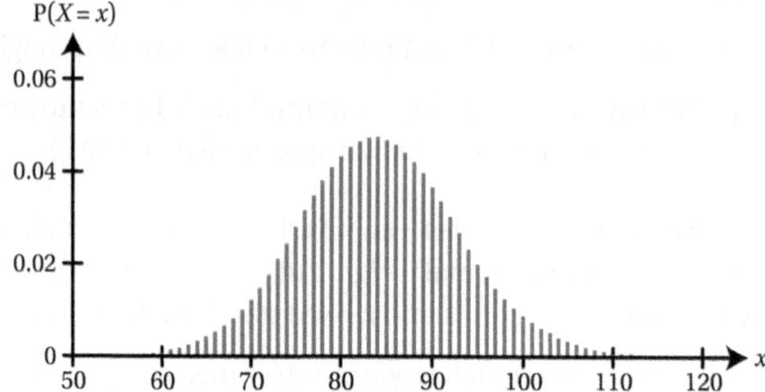

The distribution of X can be approximated by a normal distribution.

a Use the graph to estimate the mean and standard deviation of the normal distribution. Explain clearly how you arrived at your answer.

b Hence estimate the values of n and p.

14 A student is investigating whether there is any correlation between the amount of time spent revising and marks gained on a test. She uses information from a sample of six tests. In order to have a larger sample, she collects data from two of her friends.

She finds that the correlation coefficient between hours spent revising and the percentage mark on the test is -0.511. She therefore suggests that there is negative correlation between the amount of time spent revising and the test marks.

a State the null and alternative hypotheses for her test.

b Test the hypotheses at the 5% level of significance and interpret the conclusion in context.

The scatter graph of her data is shown here.

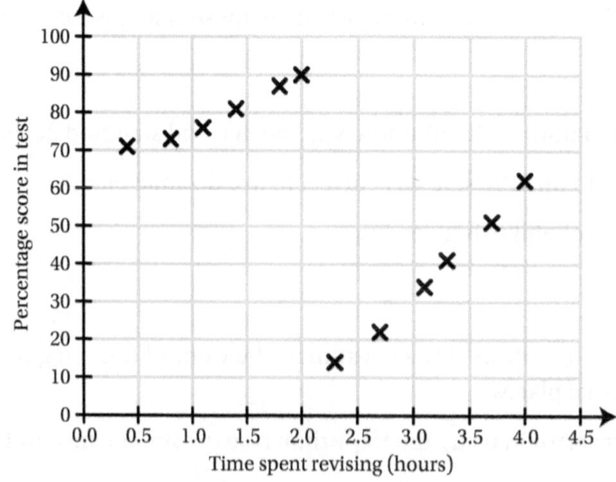

c Suggest one possible interpretation of the data and the result of the hypothesis test.

15 A company manufactures bath panels. The bath panels should be 700 mm deep, but a small amount of variability is acceptable. The depths are known to be normally distributed with standard deviation 2.1 mm.

a In order to check that the mean depth is 700 mm, Amir takes a random sample of 6 bath panels from the current production and measures their depths, in millimetres, with the following results.

701.2 698.2 704.4 699.4 695.5 698.9

Test whether the current mean is 700 mm, using the 5% significance level.

b Isabella, a manager, tells Amir that, in order to check whether the current mean is 700 mm, it is necessary to take a larger sample. Amir therefore takes a random sample of size 40 from the current production and finds that the mean depth is 701.34 mm.

Test whether the current mean is 700 mm, using the data from this second sample and the 5% significance level.

[© AQA 2011]

16 Daniel and Theo play a game with a biased coin. There is a probability of $\frac{1}{5}$ of the coin showing a head and a probability of $\frac{4}{5}$ of it showing a tail. They take it in turns to toss the coin. If the coin shows a head the player who tossed the coin wins the game. If the coin shows a tail, the other player has the next toss. Daniel plays first and the game continues until there is a winner.

a Write down the probability that Daniel wins on his first toss.

b Calculate the probability that Theo wins on his first toss.

c Calculate the probability that Daniel wins on his second toss.

d Show that the probability of Daniel winning is $\frac{5}{9}$.

e State the probability of Theo winning.

f They play the game with a different coin and find that the probability of Daniel winning is twice the probability of Theo winning. Find the probability of this coin showing a head.

17 In a large typesetting company the time taken for a typesetter to set one page is normally distributed with mean 7.80 minutes and standard deviation 1.22 minutes. A new training scheme is introduced and, after all the typesetters have completed the scheme, the times taken by a random sample of 35 typesetters are recorded. The mean time for the sample is 7.24 minutes. You may assume that the population standard deviation is unchanged.

a Test, at the 1% significance level, whether the mean time to set a page has decreased.

b It is required to redesign the test so that the probability of incorrectly rejecting the null hypothesis is less than 0.01 when the sample mean is 7.50. Find the smallest sample size needed.

18 A normal distribution can be represented by a curve, as shown in the diagram.

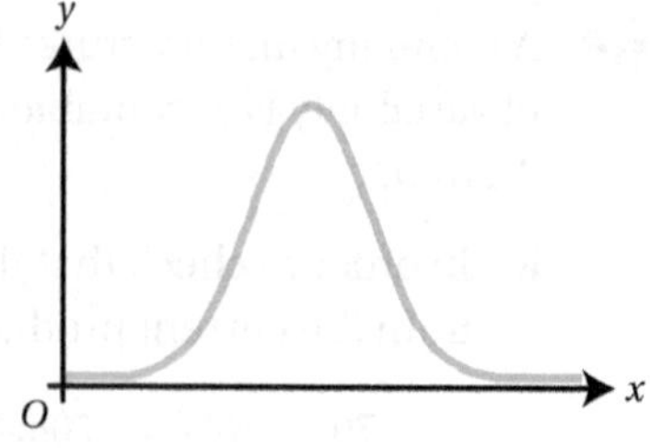

The equation of the curve is $y = \frac{1}{\sqrt{8\pi}} e^{-\frac{1}{8}(x-7)^2}$

a Find the x-coordinates of the points of inflection of the curve.

The random variable X follows this normal distribution.

b Use the trapezium rule with four strips to estimate $P(4.6 < X < 5.4)$. Give your answer to three significant figures.

c Is it possible to tell, without doing any further calculations, whether the answer in part **b** is an over-estimate or an under-estimate? Explain your answer.

PRACTICE PAPER 1

A Level Mathematics: Paper 1

2 hours, 100 marks

1 If $y = x^2$ then $y^{-\frac{2}{3}}$ is equivalent to which one of these?

A $x^{\frac{4}{3}}$ B $-x^{\frac{3}{4}}$ C $\frac{1}{x^{\frac{4}{3}}}$ D $-\frac{1}{x^{\frac{3}{4}}}$ [1 mark]

2 The graph of $y = x^2 - x + 1$ is translated two units to the right. What is the equation of the resulting graph?

Choose from these options.

A $y = x^2 - 3x + 7$ B $y = x^2 - x + 3$ C $y = x^2 - 5x + 7$ D $y = x^2 + 3x + 3$ [1 mark]

3 Find the coordinates of the centre, C, and the radius, r, of the circle with equation $x^2 - 2x + y^2 + 10y = 30$.

Choose from these options.

A $C(1, -5), r = 2\sqrt{14}$ B $C(-1, 5), r = 2\sqrt{14}$ C $C(1, -5), r = 2$ D $C(-1, 5), r = 2$ [1 mark]

4 Find the derivative of $\sin^2 3x$.

Choose from these options.

A $2 \sin 3x \cos 3x$ B $6 \cos 3x$ C $\frac{1}{6} \cos^3 3x$ D $3 \sin 6x$ [1 mark]

5 Solve the equation $\ln(x + 2) = 3 + \ln(x - 1)$.

Give your answer in terms of e. [4 marks]

6 Find an approximate expression for $\cos 2\theta(1 - 4 \sin \theta)$ when θ is small enough to neglect the terms in θ^3 and above. [3 marks]

7 The triangle in the diagram has area 22 cm^2. The angle θ is obtuse.

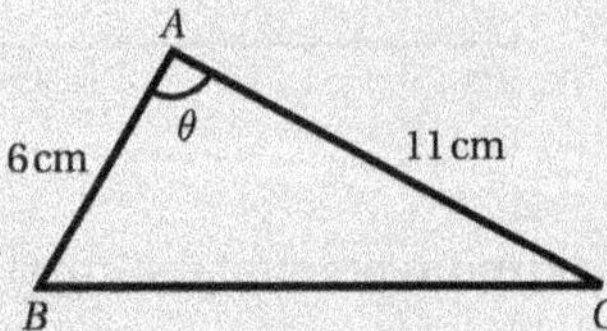

Find the length of BC correct to one decimal place. [5 marks]

8 Find the equation of the normal to the curve $y = 2e^{-x}$ at the point where $x = \ln 3$.

Give your answer in the form $ax + by = p \ln q + c$, where a, b, c, p and q are integers. [6 marks]

9 **a** Solve the inequality $3x^2 + 3x - 2 > 0$. [3 marks]

b A sequence is given by $u_n = n^3 - 3n + 4$ for $n \geqslant 1$.

Prove that the sequence is increasing. [4 marks]

10 The sector shown in the diagram has perimeter P.

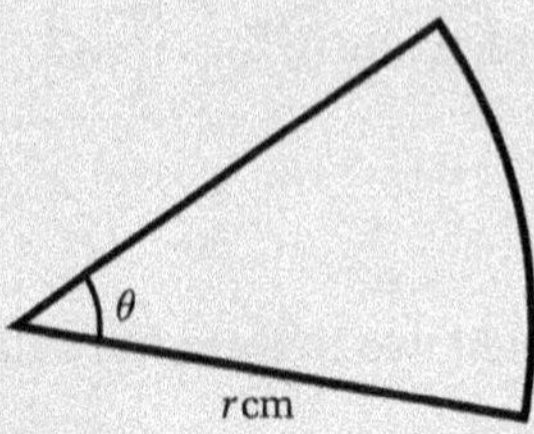

Find, in terms of P, the largest possible area of the sector. [6 marks]

11 **a** Express $\frac{3x+4}{(1-x)(2+5x)}$ in the form $\frac{A}{1-x}+\frac{B}{2+5x}$, where A and B are integers. [3 marks]

b Hence, or otherwise, find the binomial expansion of $\frac{3x+4}{(1-x)(2+5x)}$ up to and including the term in x^2. [6 marks]

c Find the range of values of x for which the binomial expansion of $\frac{3x+4}{(1-x)(2+5x)}$ is valid. [1 mark]

12 **a** Three consecutive terms in an arithmetic sequence are $3e^a, 8, 5\,e^{-a}$.
Find the possible values of a. [5 marks]

b Prove that there is no value of b for which $3e^b, 8, 5\,e^{-b}$ are consecutive terms of a geometric sequence. [3 marks]

13 Use a substitution to find $\int_2^3 \frac{3x^5}{(x^3-3)^2}\,dx$ in the form $\ln p + q$ where p and q are rational numbers. [7 marks]

14 The diagram shows a part of the graph with equation $y = \ln(x)\sin(x)$

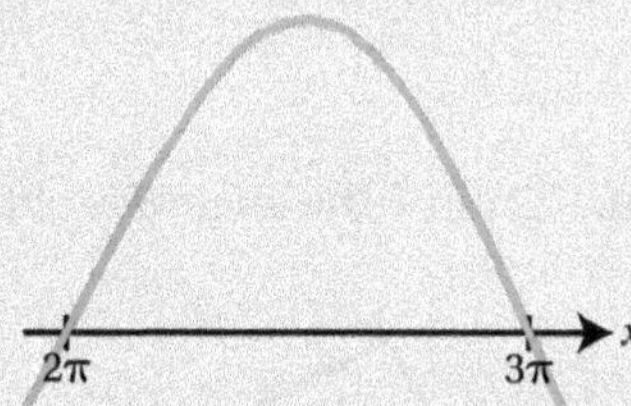

a Show that the x-value of the maximum point of the curve lies between $\frac{5\pi}{2}$ and $\frac{8\pi}{3}$. [4 marks]

b Use three rectangles of equal width to find values of L and U such that:

$$L < \int_{2\pi}^{\frac{5\pi}{2}} \ln(x)\sin(x)\,dx < U$$ [3 marks]

c Use the trapezium rule with six strips to find an approximate value of $\int_{2\pi}^{3\pi} \ln(x)\sin(x)\,dx$.

State, with a reason, whether your answer is an overestimate or an underestimate. [3 marks]

15 A curve is defined by the implicit equation $x^2 - \frac{1}{2}y^2 + 2xy + 5 = 0$.

a Find an expression for $\frac{dy}{dx}$ in terms of x and y. [5 marks]

b Hence prove that the curve has no tangents parallel to the y-axis. [3 marks]

16 Consider the equation $\ln(x-2) = \frac{1}{2}\sin x$, where x is measured in radians.

a By means of a sketch, show that this equation has only one solution. [3 marks]

b Show that this solution lies between 3 and 4. [2 marks]

c Show that the equation can be rearranged into the form $x = e^{a\sin x} + b$, where a and b are constants to be found. [2 marks]

d Hence use a suitable iterative formula to find an approximate solution to the equation $\ln(x-2) = \frac{1}{2}\sin x$ correct to three decimal places. [3 marks]

17 The polynomial f(x) is defined by $f(x) = 9x^3 - 7x - 2$.

a Use the factor theorem to show that $(3x+1)$ is a factor of f(x). [2 marks]

b Hence express f(x) as a product of three linear factors. [2 marks]

c **i** Show that the equation $9\cos 2\theta \sin\theta + 5\sin\theta + 4 = 0$ can be written as $9x^3 - 7x - 2 = 0$, where $x = \sin\theta$.

ii Hence find all solutions of the equation $9\cos 2\theta \sin\theta + 5\sin\theta + 4 = 0$ in the interval $0° < \theta < 360°$, giving your solutions to the nearest degree. [8 marks]

PRACTICE PAPER 2

A Level Mathematics: Paper 2

2 hours, 100 marks

Section A

1 What is a possible equation of this graph?

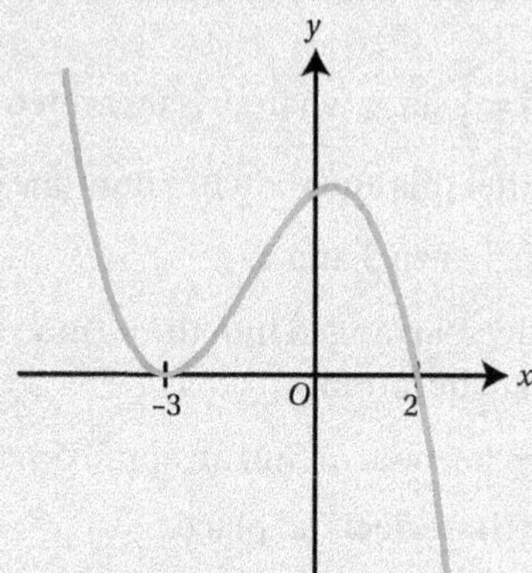

Choose from these options.

A $y = (x+3)^2(x-2)$ **B** $y = -(x+3)(x-2)$

C $y = (x+3)^2(2-x)$ **D** $y = (x+3)(x-2)$ [1 mark]

2 Let a be the smallest positive solution of the equation $\cos 2x = k$, where x is in radians.

Find the next smallest positive solution in terms of a.

Choose from these options.

A $2\pi + a$ **B** $2\pi - a$ **C** $\pi - a$ **D** $\pi - \frac{a}{2}$ [1 mark]

3 The diagram shows the graphs of $y = x^2$ and $y = 6 - x$.

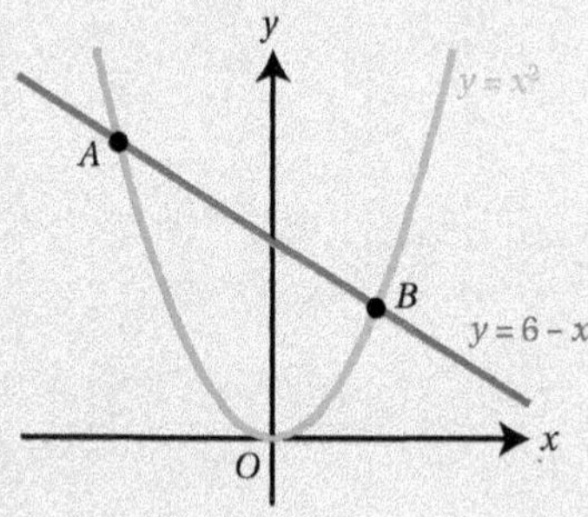

a Find the coordinates of points A and B. [2 marks]

b Shade the region determined by the inequalities $y \geqslant x^2$ and $y \leqslant 6 - x$. [1 mark]

c State the smallest integer value of x that satisfies both inequalities. [1 mark]

4 Points A and B have position vectors $\mathbf{a} = 3\mathbf{i} - 7\mathbf{j} + 3\mathbf{k}$ and $\mathbf{b} = \mathbf{i} - \mathbf{j} + 5\mathbf{k}$.

a Find the position vector of the midpoint of AB. [2 marks]

b Point C has position vector $2t\mathbf{i} + t\mathbf{j} - 3\mathbf{k}$. It is given that $CA = CB$.

Find the value of t. [3 marks]

5 Find the exact coordinates of the point of inflection on the graph of $y = xe^{-3x}$. [5 marks]

6 Prove by contradiction that $\log_3 5$ is an irrational number. [4 marks]

7 **a** Show that the equation $3\operatorname{cosec}^2\theta + 5\cot\theta = 5$ can be written in the form $a\cot^2\theta + b\cot\theta + c = 0$. State the values of the constants a, b and c. [3 marks]

b Hence solve the equation

$$3\operatorname{cosec}^2\theta + 5\cot\theta = 5 \text{ for } 0° \leqslant \theta \leqslant 360°$$

giving your answers to the nearest degree. [4 marks]

8 Find $\int_1^2 x^3 \ln x \, dx$, giving your answer in the form $\ln p - q$. [5 marks]

9 The circle shown in the diagram has equation $x^2 + y^2 - 10x - 10y + 25 = 0$.

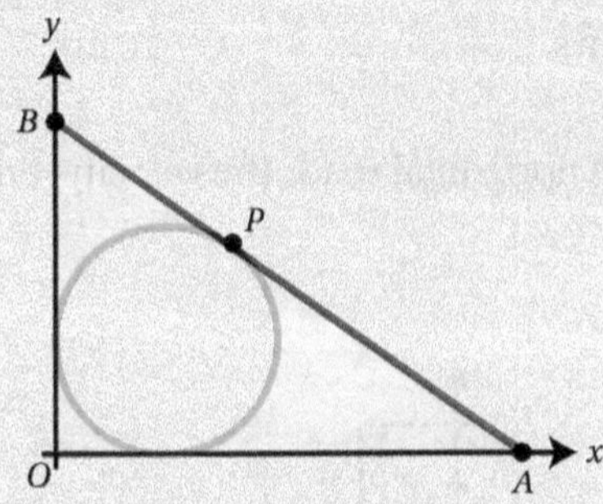

a Show that the point $P(8, 9)$ lies on the circle. [2 marks]

b The line AB is tangent to the circle at P.

Find the exact value of the area shaded in the diagram. [5 marks]

10 The diagram shows the curves with equations $y = 3e^x - 3$ and $y = 13 - 5e^{-x}$.

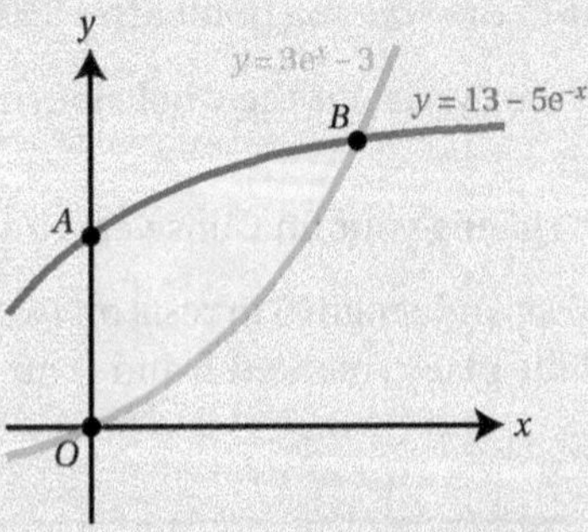

a Write down the coordinates of point A. [1 mark]

b The curves intersect at the point B.

i Show that the x-coordinate of B satisfies the equation $3e^{2x} - 16e^x + 5 = 0$.

ii Hence find the exact coordinates of B. [4 marks]

c Find the exact value of the area of the shaded region. [6 marks]

Section B

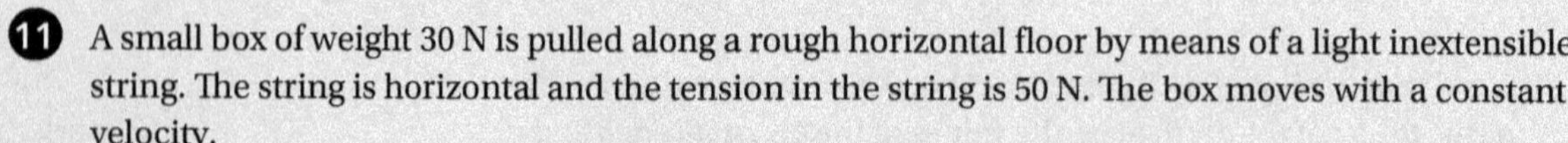

11 A small box of weight 30 N is pulled along a rough horizontal floor by means of a light inextensible string. The string is horizontal and the tension in the string is 50 N. The box moves with a constant velocity.

Find the coefficient of friction between the box and the floor, correct to one decimal place.

Choose from these options.

A 0.6 N   **B** 0.6   **C** 1.7 N   **D** 1.7   [1 mark]

12 A particle moves in the plane with constant acceleration. Its velocity changes from $(23\mathbf{i} - 7\mathbf{j})\ \text{m s}^{-1}$ to $(17\mathbf{i} + \mathbf{j})\ \text{m s}^{-1}$ during 5 seconds.

Find the magnitude of the acceleration.

Choose from these options.

A $4\ \text{m s}^{-2}$ **B** $2\ \text{m s}^{-2}$ **C** $20\ \text{m s}^{-2}$ **D** $0.4\ \text{m s}^{-2}$ [1 mark]

13 A cyclist moves along a straight horizontal road. The velocity–time graph of her journey is shown in the diagram.

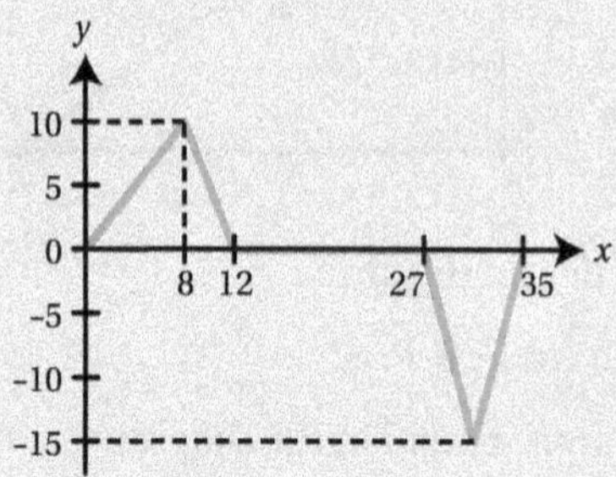

a Calculate the acceleration of the cyclist during the first 8 seconds. [2 marks]

b Show that the cyclist returns to the starting point after 35 seconds. [3 marks]

c State the average **velocity** of the cyclist for the whole journey. [1 mark]

14 In this question use $g = 9.8\ \text{m s}^{-2}$, giving your final answers to an appropriate degree of accuracy.

A non-uniform plank of mass 16 kg and length 5 m rests on two supports, A and B. The supports are located 0.50 m from the ends of the plank, marked L and R on the diagram.

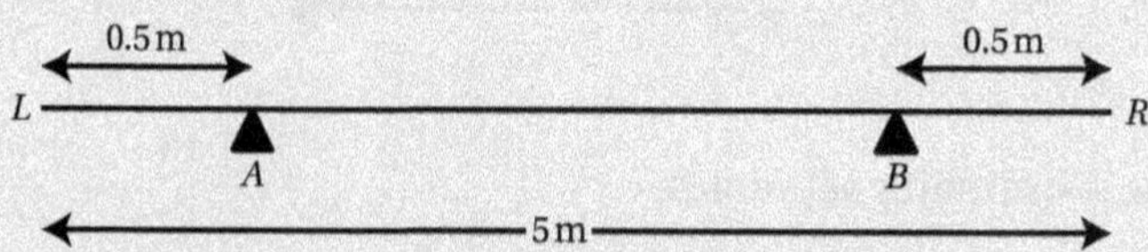

Given that the reaction force acting on the plank at the support A is 95 N:

a find the reaction force acting on the plank at B [1 mark]

b find the distance of the centre of mass of the plank from the end marked L [3 marks]

15 Four forces act on a particle in a horizontal plane, as shown in the diagram. The unit vectors **i** and **j** are directed east and north, respectively.

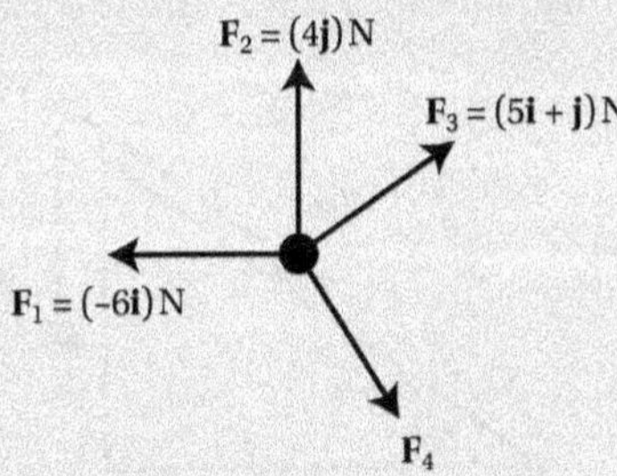

a The particle is in equilibrium. Find the force $\mathbf{F}_4$ in the form $(a\mathbf{i} + b\mathbf{j})$ N. [2 marks]

b The force $\mathbf{F}_4$ is suddenly removed. Given that the mass of the particle is 600 grams, find:

i the magnitude of the acceleration of the particle

ii the angle the acceleration vector makes with the direction of the vector **i**. [5 marks]

16 **a** A particle moves in a straight line with constant acceleration a m s^{-2}. The initial velocity of the particle is U m s^{-1}. At time T seconds, the velocity of the particle is V m s^{-1} and its displacement from the initial position is S m.

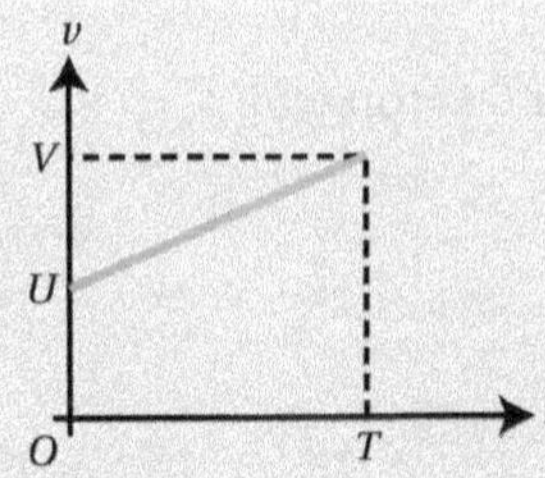

i By considering the velocity–time graph, write down an equation relating a, T, U and V.

ii Hence show that $S = VT - \frac{1}{2}aT^2$. [4 marks]

b A car is moving along a straight horizontal road when the driver applies the brakes. The car decelerates at a constant rate of 2.3 m s^{-2}. In the next 6 seconds is travels 145 m.

Find its speed at the end of the 6 seconds. [3 marks]

17 In this question use $g = 10$ m s^{-2}, giving your final answers to an appropriate degree of accuracy.

A child kicks a ball from ground level at an angle θ above the horizontal. The speed of projection is V m^{-2}. The ball is modelled as a particle and air resistance can be ignored.

a At time t seconds the ball's displacement from the point of projection is $(x\mathbf{i} + y\mathbf{j})$ m.

Show that $y = (\tan\theta)x - \left(\frac{g\sec^2\theta}{2V^2}\right)x^2$. [4 marks]

b The angle of projection is $\theta = 50°$. The ball passes over a 2.4 m tall fence which is 8.6 m from the point of projection. Find the minimum possible value of V. [3 marks]

c How would the answer change if your model included air resistance? [1 mark]

18 Two particles, P and Q, are connecting by a light inextensible string which passes over a smooth pulley. Particle P has mass m kg and lies on a rough plane. The plane is inclined at 35° to the horizontal, and the coefficient of friction between P and the plane is μ. Particle Q has mass 5.2 kg and hangs freely below the pulley.

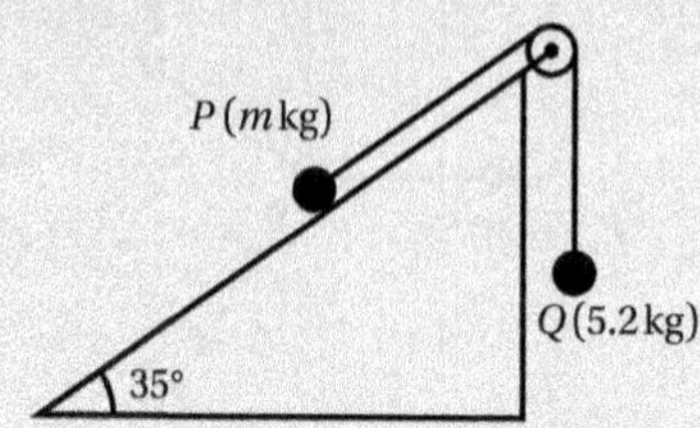

The system is in equilibrium.

a Show that $m \geqslant \dfrac{5.2}{\sin 35° + \mu\cos 35°}$ [5 marks]

b Obtain another similar inequality for m. [2 marks]

c In the case when $\mu = 0.48$, find the range of values of m for which the system remains in equilibrium. [2 marks]

19 A particle of mass 3 kg moves in a straight line. At time t seconds the velocity of the particle is v m s^{-1} and the force acting on the particle is F newtons, where $F = 36\sqrt{v}\cos(2t)$. When $t = 0$ the velocity of the particle is 16 m s^{-1}.

Find the velocity of the particle after 5 seconds. [7 marks]

PRACTICE PAPER 3

A Level Mathematics: Paper 3

2 hours, 100 marks

Section A

1 The diagram shows the graph of the function $y = f(x)$. The graph crosses the x-axis at the points $(1, 0)$, $(3, 0)$ and $(6, 0)$. The area labelled R equals 15 and the area labelled S equals 26.

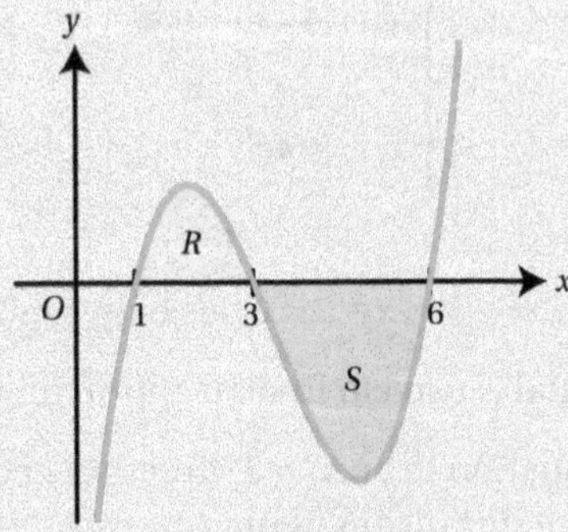

What is the value of $\int_1^6 f(x)\,dx$?

Choose from these options.

A 11   **B** -11 **C** 41   **D** -41 [1 mark]

2 Find the largest possible domain of the function $f(x) = \ln(x^2 + x - 6)$.

Choose from these options.

A $-3 < x < 2$ **B** $x > 0$ **C** $x < -3$ or $x > 2$ **D** $x \neq -3$ and $x \neq 2$ [1 mark]

3 The first term of an arithmetic sequence is -4 and the 15th term is 31.

Find the sum of the first 15 terms. [3 marks]

4 The polynomial $2x^4 - x^3 + ax^2 - 26x + b$ has factors $(2x + 1)$ and $(x - 3)$.

Find the values of a and b. [5 marks]

5 **a** Write $6x^2 - 12x + 11$ in the form $a(x - p)^2 + q$. [3 marks]

b A function is defined by $f(x) = 2x^3 - 6x^2 + 11x + 2$ for $x \in \mathbb{R}$.

i Prove that $f(x)$ is an increasing function. [2 marks]

ii Does $f(x)$ have an inverse function? Explain your answer. [2 marks]

iii Find the range of values of x for which $f(x)$ is convex. [3 marks]

6 **a** Express $\sqrt{5}\sin x + \sqrt{7}\cos x$ in the form $R\sin(x + \alpha)$, where $0 < \alpha < \frac{\pi}{2}$.
Give the value of α correct to three decimal places. [3 marks]

b Hence find the minimum value of $\dfrac{48}{2 - \left(\sqrt{5}\sin x + \sqrt{7}\cos x\right)}$.

Give your answer in the form $a + b\sqrt{3}$. [3 marks]

7 The diagram shows a part of the graph with equation $y = x^2 - 3\ln x - 3$.

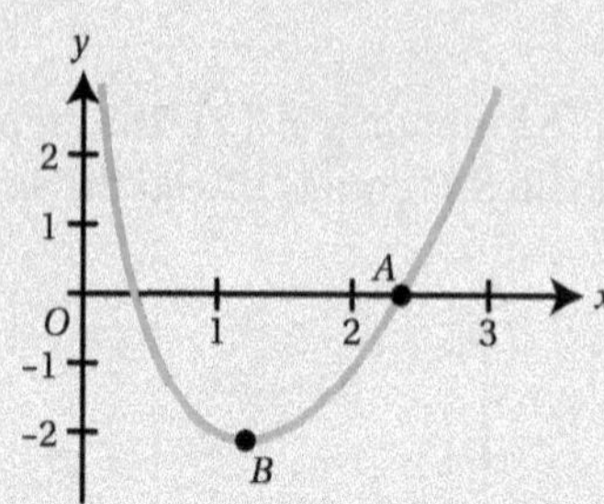

One of the roots of the equation $x^2 - 3\ln x - 3 = 0$ lies between 2 and 3.

a The Newton–Raphson method is used to find an approximation to the root at A.

i Taking the first approximation to be $x_1 = 3$, find the second approximation, x_2. Give your answer to three significant figures.

ii Illustrate the relationship between x_1 and x_2 on a copy of the diagram. [5 marks]

b The point B is the minimum point on the curve.

i Find the exact x-coordinate of B.

ii Explain why an iteration starting at B would not converge to root of the equation $x^2 - 3\ln x - 3 = 0$. [4 marks]

8 A student is investigating the number of friends people have on a large social networking site. He collects some data on the percentage (P) of people who have n friends and plots this graph.

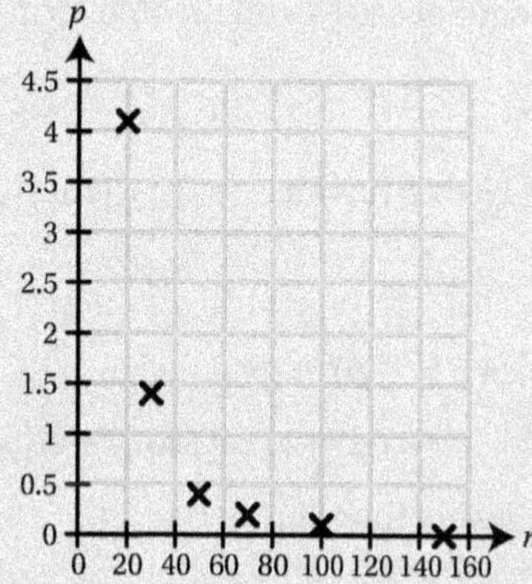

The student proposes two possible models for the relationship between n and P.

Model 1: $P = ak^{-n}$ Model 2: $P = an^{-k}$

To check which model is a better fit, he plots the graph of y against x, where $y = \log P$ and $x = \log n$.

The graph is approximately a straight line with equation $y = 1.2 - 2.6x$.

a Is model 1 or model 2 a better fit for the data? Explain your answer. [2 marks]

b Find the values of a and k. [3 marks]

9 A curve is given by parametric equations $x = \cos t$, $y = \sin 2t$, for $t \in [0, 2\pi)$. The curve crosses the x-axis at point A, and B is a maximum point on the curve.

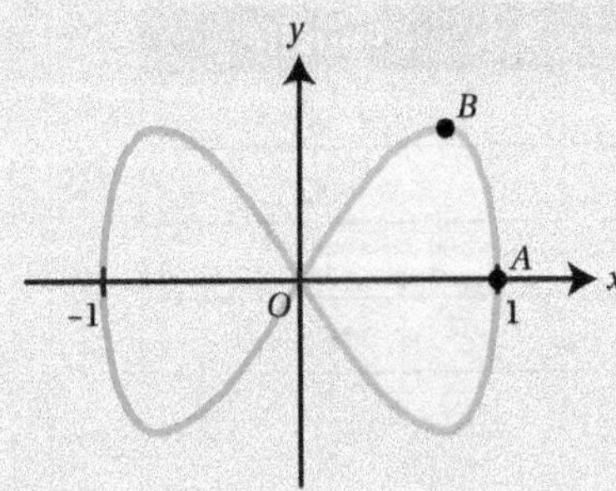

a Find the exact coordinates of B. [4 marks]

b **i** Find the values of t at the points O and A.

ii Find the area of the shaded region. [6 marks]

Section B

10 **a** A university wants to investigate ways to encourage more students to apply for its courses. It takes a list of all schools in the UK and randomly selects ten to visit. The university asks 100 students at each of the ten schools to complete questionnaires. What type of sampling is this?

Choose from these options.

A Cluster sampling **B** Stratified sampling

C Simple random sampling **D** Quota sampling [1 mark]

b In the questionnaire the number of students who said they were considering applying to the university in each school, x, is summarised as:

$$\Sigma x = 253,\ \Sigma x^2 = 7094$$

Which of these is the best estimate of the standard deviation between the schools of the percentage applying to the university?

Choose from these options.

A 8.33 **B** 8.78 **C** 69.3 **D** 77.0 [1 mark]

11 In a survey of 40 families the total number of takeaways a family buys each week and their total food budget has a correlation coefficient of −0.25.

a A test is carried out using the following hypotheses:

$H_0: \rho = 0;\ H_1: \rho \neq 0$

Show that at the 5% significance level H_0 is not rejected. [2 marks]

b Decide if each of these statements are supported by the test, explaining you reasoning.

i The number of takeaways is independent of the total food budget. [2 marks]

ii There is no correlation between the total number of takeaways and their total food budget. [2 marks]

12 The two-way table shows the number of people in a survey with a special food requirement along with their gender.

	Male	Female
Vegan	6	8
Vegetarian	23	32
Pescetarian	14	16
No requirement	56	62

a Find P(male $\cap$ vegan). [1 mark]

b Find P(male $\cup$ vegan). [2 marks]

c Find P(male | vegan). [2 marks]

d Two vegans are chosen without replacement. Find the probability that they are both male. [2 marks]

e Use appropriate figures from this table to provide a counterexample to the identity

$P(A\mid B)+P(A\mid B')\equiv 1$ [2 marks]

13 There are 655 girls at a school. 312 of them describe themselves as having blonde hair.

a Use the binomial distribution to find the probability that a randomly selected group of 14 girls contains exactly 10 blonde girls. [2 marks]

b A club contains 14 girls of whom 10 have blonde hair. Use the binomial distribution to test at 5% significance whether this is more than the expected number of girls with blonde hair. State your null and alternative hypotheses and your p-value. [4 marks]

c Give two reasons why the binomial distribution may not be an appropriate model for the number of girls with blonde hair in this club. [2 marks]

14 The distribution of the number of books, B, borrowed by members of a library follows this distribution.

b	0	1	2	3
$P(B=b)$	0.4	0.35	0.15	a

a Find the value of a. [1 mark]

b Find the probability that a randomly chosen member has borrowed at least one book. [1 mark]

c The librarian proposes this model for the probability of a book begin overdue:
If a member has borrowed b books, the probability that he has an overdue book is $P(\text{overdue}) = 1-(0.5)^b$.

Find the probability that a randomly chosen member has an overdue book. [3 marks]

15 Two events, A and B, satisfy: $P(B)=\frac{7}{10}$, $P(A\mid B)=\frac{2}{7}$ and $P(A\cup B)=\frac{69}{70}$.

a Find $P(A)$. [3 marks]

b Show that A and B are not independent. [2 marks]

16 Alex keeps a record of the number of calories consumed each day over the course of several years. The mean is 2502 kilocalories and standard deviation 208 kilocalories. The lowest value is 1955 and the highest value is 3197.

a Give two reasons why the numerical information given supports a normal model for the data. [2 marks]

b Assuming that a normal model holds, find the probability that Alex eats over 2600 kilocalories in a day. [1 mark]

c Stating a necessary assumption, find the probability that Alex eats over 2600 kilocalories on at least 5 days in a week. [4 marks]

d After attending a healthy eating course Alex finds that he eats 16 849 kilocalories in a week. Test at the 5% significance level if his average daily calorie intake is different after the healthy eating course, assuming that it still comes from a normal distribution with standard deviation 208 kcal. State your null and alternative hypotheses. [4 marks]

e Jane eats more than 3000 kilocalories on 5% of days and less than 2000 kilocalories on 10% of days. Assuming a normal distribution, estimate the value of the mean and variance. [4 marks]

FORMULAE

Pure mathematics

Binomial series

$$(a+b)^n = a^n + \binom{n}{1} a^{n-1}b + \binom{n}{2} a^{n-2}b^2 + \ldots + \binom{n}{r} a^{n-r}b^r + \ldots + b^n \quad (n \in \mathbb{Z}^+)$$

where $\binom{n}{r} = {}^nC_r = \dfrac{n!}{r!(n-r)!}$

$$(1+x)^n = 1 + nx + \frac{n(n-1)}{1.2}x^2 + \ldots + \frac{n(n-1)\ldots(n-r+1)}{1.2\ldots r}x^r + \ldots \quad (|x| < 1, n \in \mathbb{R})$$

Arithmetic series

$$S_n = \frac{1}{2}n(a+l) = \frac{1}{2}n[2a+(n-1)d]$$

Geometic series

$$S_n = \frac{a(1-r^n)}{1-r}$$

$$S_\infty = \frac{a}{1-r} \quad \text{for} \quad |r| < 1$$

Trigonometry: small angles

For small angle θ,

$\sin\theta \approx \theta$

$\cos\theta \approx 1 - \dfrac{\theta^2}{2}$

$\tan\theta \approx \theta$

Trigonometric identities

$\sin(A \pm B) \equiv \sin A\cos B \pm \cos A\sin B$

$\cos(A \pm B) \equiv \cos A\cos B \mp \sin A\sin B$

$\tan(A \pm B) \equiv \dfrac{\tan A \pm \tan B}{1 \mp \tan A\tan B} \quad \left(A \pm B \neq \left(k+\frac{1}{2}\right)\pi\right)$

$\sin A + \sin B \equiv 2\sin\dfrac{A+B}{2}\cos\dfrac{A-B}{2}$

$\sin A - \sin B \equiv 2\cos\dfrac{A+B}{2}\sin\dfrac{A-B}{2}$

$\cos A + \cos B \equiv 2\cos\dfrac{A+B}{2}\cos\dfrac{A-B}{2}$

$\cos A - \cos B \equiv 2\sin\dfrac{A+B}{2}\sin\dfrac{A-B}{2}$

FORMULAE

A Differentiation

$\mathbf{f}(x)$	$\mathbf{f}'(x)$
$\tan kx$	$k\sec^2 kx$
$\operatorname{cosec} x$	$-\operatorname{cosec} x \cot x$
$\sec x$	$\sec x \tan x$
$\cot x$	$-\operatorname{cosec}^2 x$
$\dfrac{f(x)}{g(x)}$	$\dfrac{f'(x)g(x)-f(x)g'(x)}{(g(x))^2}$

Integration

$$\int u\frac{dv}{dx}\,dx = uv - \int v\frac{du}{dx}\,dx$$

(+ constant; $a > 0$ where relevant)

$\mathbf{f}(x)$	$\int \mathbf{f}(x)\,\mathbf{d}x$
$\tan x$	$\ln\lvert\sec x\rvert$
$\cot x$	$\ln\lvert\sin x\rvert$
$\operatorname{cosec} x$	$-\ln\lvert\operatorname{cosec} x + \cot x\rvert = \ln\lvert\tan\frac{1}{2}x\rvert$
$\sec x$	$\ln\lvert\sec x + \tan x\rvert = \ln\lvert\tan(\frac{1}{2}x + \frac{1}{4}\pi)\rvert$
$\sec^2 kx$	$\frac{1}{k}\tan kx$

Numerical solution of equations

The Newton-Raphson iteration for solving $f(x) = 0$: $x_{n+1} = x_n - \dfrac{f(x_n)}{f'(x_n)}$

Numerical integration

The trapezium rule: $\int_a^b y\,dx \approx \frac{1}{2}h\{(y_0 + y_n) + 2(y_1 + y_2 + \ldots + y_{n-1})\}$, where $h = \dfrac{b-a}{n}$

Mechanics

Constant acceleration

$s = ut + \frac{1}{2}at^2$

$s = vt - \frac{1}{2}at^2$

$v = u + at$

$s = \frac{1}{2}(u+v)t$

$v^2 = u^2 + 2as$

A

$\mathbf{s} = \mathbf{u}t + \frac{1}{2}\mathbf{a}t^2$

$\mathbf{s} = \mathbf{v}t - \frac{1}{2}\mathbf{a}t^2$

$\mathbf{v} = \mathbf{u} + \mathbf{a}t$

$\mathbf{s} = \frac{1}{2}(\mathbf{u} + \mathbf{v})t$

FORMULAE

Probability and statistics

Probability

A $\mathrm{P}(A \cup B) = \mathrm{P}(A) + \mathrm{P}(B) - \mathrm{P}(A \cap B)$

$\mathrm{P}(A \cap B) = \mathrm{P}(A) \times \mathrm{P}(B \mid A)$

Standard discrete distributions

Distribution of X	$\mathrm{P}(X = x)$	Mean	Variance
Binomial $\mathrm{B}(n, p)$	$\binom{n}{x} p^x (1-p)^{n-x}$	np	$np(1-np)$

Sampling distributions

For a random sample $X_1, X_2, \ldots, X_n$ of n independent observations from a distribution having mean μ and variance σ^2:

S^2 is an unbiased estimator of σ^2, where $S^2 = \dfrac{\Sigma(X_i - \bar{X})^2}{n-1}$

A For a random sample of n observations from $\mathrm{N}(\mu, \sigma^2)$

$$\frac{\bar{X} - \mu}{\frac{\sigma}{\sqrt{n}}} \sim \mathrm{N}(0,1)$$

FORMULAE

A Critical values of the product moment correlation coefficient

The table gives the critical values, for different significance levels, of the product moment correlation coefficient, r, for varying sample sizes, n.

One tail	10%	5%	2.5%	1%	0.5%	One tail
Two tail	**20%**	**10%**	**5%**	**2%**	**1%**	**Two tail**
n						n
4	0.8000	0.9000	0.9500	0.9800	0.9900	4
5	0.6870	0.8054	0.8783	0.9343	0.9587	5
6	0.6084	0.7293	0.8114	0.8822	0.9172	6
7	0.5509	0.6694	0.7545	0.8329	0.8745	7
8	0.5067	0.6215	0.7067	0.7887	0.8343	8
9	0.4716	0.5822	0.6664	0.7498	0.7977	9
10	0.4428	0.5494	0.6319	0.7155	0.7646	10
11	0.4187	0.5214	0.6021	0.6851	0.7348	11
12	0.3981	0.4973	0.5760	0.6581	0.7079	12
13	0.3802	0.4762	0.5529	0.6339	0.6835	13
14	0.3646	0.4575	0.5324	0.6120	0.6614	14
15	0.3507	0.4409	0.5140	0.5923	0.6411	15
16	0.3383	0.4259	0.4973	0.5742	0.6226	16
17	0.3271	0.4124	0.4821	0.5577	0.6055	17
18	0.3170	0.4000	0.4683	0.5425	0.5897	18
19	0.3077	0.3887	0.4555	0.5285	0.5751	19
20	0.2992	0.3783	0.4438	0.5155	0.5614	20
21	0.2914	0.3687	0.4329	0.5034	0.5487	21
22	0.2841	0.3598	0.4227	0.4921	0.5368	22
23	0.2774	0.3515	0.4132	0.4815	0.5256	23
24	0.2711	0.3438	0.4044	0.4716	0.5151	24
25	0.2653	0.3365	0.3961	0.4622	0.5052	25
26	0.2598	0.3297	0.3882	0.4534	0.4958	26
27	0.2546	0.3233	0.3809	0.4451	0.4869	27
28	0.2497	0.3172	0.3739	0.4372	0.4785	28
29	0.2451	0.3115	0.3673	0.4297	0.4705	29
30	0.2407	0.3061	0.3610	0.4226	0.4629	30
31	0.2366	0.3009	0.3550	0.4158	0.4556	31
32	0.2327	0.2960	0.3494	0.4093	0.4487	32
33	0.2289	0.2913	0.3440	0.4032	0.4421	33
34	0.2254	0.2869	0.3388	0.3972	0.4357	34
35	0.2220	0.2826	0.3338	0.3916	0.4296	35
36	0.2187	0.2785	0.3291	0.3862	0.4238	36
37	0.2156	0.2746	0.3246	0.3810	0.4182	37
38	0.2126	0.2709	0.3202	0.3760	0.4128	38
39	0.2097	0.2673	0.3160	0.3712	0.4076	39
40	0.2070	0.2638	0.3120	0.3665	0.4026	40
41	0.2043	0.2605	0.3081	0.3621	0.3978	41
42	0.2018	0.2573	0.3044	0.3578	0.3932	42
43	0.1993	0.2542	0.3008	0.3536	0.3887	43
44	0.1970	0.2512	0.2973	0.3496	0.3843	44
45	0.1947	0.2483	0.2940	0.3457	0.3801	45
46	0.1925	0.2455	0.2907	0.3420	0.3761	46
47	0.1903	0.2429	0.2876	0.3384	0.3721	47
48	0.1883	0.2403	0.2845	0.3348	0.3683	48
49	0.1863	0.2377	0.2816	0.3314	0.3646	49
50	0.1843	0.2353	0.2787	0.3281	0.3610	50
60	0.1678	0.2144	0.2542	0.2997	0.3301	60
70	0.1550	0.1982	0.2352	0.2776	0.3060	70
80	0.1448	0.1852	0.2199	0.2597	0.2864	80
90	0.1364	0.1745	0.2072	0.2449	0.2702	90
100	0.1292	0.1654	0.1966	0.2324	0.2565	100

Answers

1 Proof and mathematical communication

BEFORE YOU START

1 a $\Leftrightarrow$ b $\Leftarrow$

2–4 Proof.

EXERCISE 1A

1 For example, $x = 225°$

2–5 Proof.

6 For example, 10, 10, 20, 400, 401, 403, 404, 405, 406, 407.

7 a–c Proof.

8 a–b Proof.

c Its diagonals are perpendicular (so it must be a kite).

9 a–c Proof.

10 a $k = -1$

b Proof.

11 Proof.

EXERCISE 1B

1–12 Proof.

EXERCISE 1C

1 a 4 b Line 2.

2 a Proof. b $\Leftarrow$

3 a $x = -1$ b Line 3.

4 Line 2.

5 Line 3: $x = \pm 1$

Line 5: $\frac{dy}{dx} = 0$ only for $x = \pm 1$. It should be $\frac{d}{dx}(3x^2 - 3)$.

Line 7: $\frac{d^2y}{dx^2} = 0$ doesn't imply a minimum.

6 If $q = 0$ the suggested factorisation is not necessarily true.

7 Line 8 does not follow from line 7.

MIXED PRACTICE 1

1 C

2–4 Proof.

5 Line 2 is not equivalent to line 1.
Line 3 does not necessarily follow from line 2.
Line 5 is not equivalent to line 4.

6 C

7–10 Proof.

11 a Line 4 has an incorrect sign in the bracket.
Line 5 has a missing $\pm$.
Line 6 is equivalent to line 5 as given, but earlier errors mean that the final solution may not be a solution to the original equation, and/or there may be other solutions.

b Proof.

12 Line 3.

13–15 Proof.

2 Functions

BEFORE YOU START

1 a -1 b 6

2 a $x \in (3, 6]$ b $x \in (-\infty, 3) \cup [6, \infty)$

3 $\left(x + \frac{5}{2}\right)^2 - \frac{13}{4}$; $\left(-\frac{5}{2}, -\frac{13}{4}\right)$

4 $x < -1$ or $x > 5$

5 a $x = \frac{1}{2}(\ln y + 1)$ b $x = \frac{1}{3}(e^y - 4)$

6 $x > \frac{16}{9}$

EXERCISE 2A

1 a Function; many-to-one.

b Mapping.

c Mapping.

d Function; one-to-one.

e Function; one-to-one.

f Function; one-to-one.

2 a i Yes ii Yes

b i Yes ii Yes

c i No ii No

3 a i One-to-one. ii Many-to-one.

b i Many-to-one. ii Many-to-one.

c i Many-to-one. ii One-to-one.

4 a i Yes ii Yes

b i No ii No

c i No ii Yes

5 A is correct. Yes, as $-\sqrt{x}$.

EXERCISE 2B

1 a Domain: $\mathbb{R}$; range: $(0, \infty)$

b Domain: $\mathbb{R}$; range: $(0, \infty)$

c Domain: $(0, \infty)$; range: $\mathbb{R}$

d Domain: $(0, \infty)$; range: $\mathbb{R}$

2 a i $x \neq -2$ ii $x \neq 7$

b i $x \neq 2$ or -4 ii $x \neq \pm 3$

c i $x \geqslant 1$ ii $x \geqslant -3$

d i $a > 1$ ii $a < \frac{2}{5}$

e i $x \neq 0$ or -1 ii $x \geqslant -1$

f i $x \geqslant 0$ ii $x \geqslant -\frac{3}{2}$

3 a i $f(x) \leqslant 7$ ii $f(x) \geqslant 3$

b i $g(x) \geqslant 12$ ii $g(x) > 8$

c i $h(x) < 3, h(x) \in \mathbb{Z}$

ii $h(x) > 4, h(x) \in \mathbb{Z}$

d i $q(x) \leqslant -1$ or $q(x) > 0$

ii $q(x) > 0$

4 a i $x \in \mathbb{R}; f(x) \geqslant -5$

ii $x \in \mathbb{R}; f(x) \geqslant 4$

b i $x \in \mathbb{R}; g(x) \leqslant 5$

ii $x \in \mathbb{R}; g(x) \leqslant 3$

c i $x \leqslant -\sqrt{5}$ or $x \geqslant \sqrt{5}; f(x) \geqslant 0$

ii $-3 \leqslant x \leqslant 3; f(x) \geqslant 0$

d i $x \leqslant 2$ or $x \geqslant 4; f(x) \geqslant 0$

ii $x \leqslant -3$ or $x \geqslant 1; f(x) \geqslant 0$

5 a $2\left(x+\frac{3}{2}\right)^2 - \frac{15}{2}$ b $f(x) \geqslant -\frac{15}{2}$

6 $x > -\frac{3}{2}; g(x) \in \mathbb{R}$

7 $x \geqslant 5$

8 $x \geqslant 1, x \neq 2, x \neq 3$

9 a

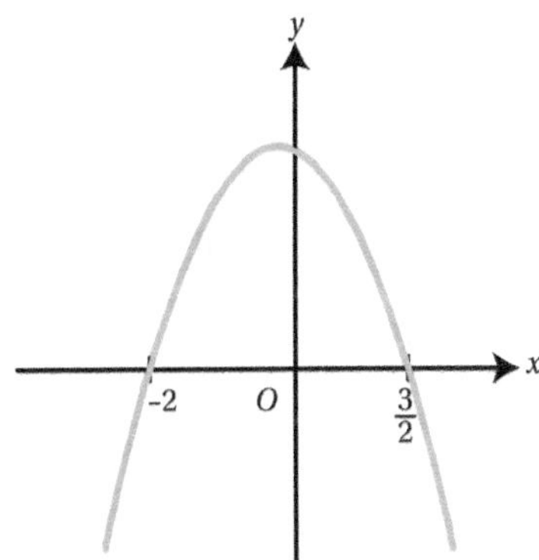

b $-2 \leqslant x \leqslant \frac{3}{2}$

10 $x < -2$ or $x > -1$

11 $x \leqslant \frac{1}{2}$ or $x > 12$

12 a i $a \leqslant x < b$ ii $\varnothing$

b $f(a) = \begin{cases} \ln(b-a) & \text{for } a < b \\ \text{undefined} & \text{for } a \geqslant b \end{cases}$

WORK IT OUT 2.1

Solution 2 is correct.

EXERCISE 2C

1 a i 5 ii 26

b i 17 ii 32

2 a i $3x^2 + 5$ ii $x^4 + 2x^2 + 2$

b i $9x + 8$ ii $9x^2 + 12x + 5$

3 a i $9\sqrt{a} + 17$ ii $y^4 - 4y^3 + 8y^2 - 8y + 5$

b i $4x - x^2$ ii $1 + 2x + x^2$

4 a $9y^2 + 17$ b $27z^2 + 36z + 17$

5 a i x^2 ii x^3

b i $3x - 5$ ii $x^2 + 5x + 6$

c i $x + 4$ ii $x^{\frac{2}{3}}$

d i $\ln(\ln x)$ ii $\ln\left(\frac{x+1}{3}\right)$

6 $x = 0, -2$

7 $x = -\frac{1}{3}$

8 a $\sqrt[3]{2x+3}$ b $2\sqrt[3]{x} + 3$

9 a $a = -\frac{4}{3}, b = -\frac{2}{3}$ b $y \geqslant 0$

10 a x^2 is not always > 3.

b $x \in (-\infty, -\sqrt{3}) \cup (\sqrt{3}, \infty)$

11 $\frac{x}{6} - \frac{1}{3}$

WORK IT OUT 2.2

Solution 1 is correct.

EXERCISE 2D

1 **a** **i** $\frac{x-1}{3}$ **ii** $\frac{x+3}{7}$

b **i** $\frac{2x}{3x-2}, x \neq \frac{2}{3}$ **ii** $\frac{x}{1-2x}, x \neq \frac{1}{2}$

c **i** $\frac{xb-a}{x-1}, x \neq 1$ **ii** $\frac{x-1}{bx-a}, x \neq \frac{a}{b}$

d **i** $1-x$ **ii** $\frac{x-2}{3}$

e **i** $\frac{x^2+2}{3}, x \geqslant 0$ **ii** $\frac{(2-x^2)}{5}, x \geqslant 0$

f **i** $\frac{1-e^x}{5}$ **ii** $\frac{e^x-2}{2}$

g **i** $2\ln\left(\frac{x}{7}\right), x > 0$ **ii** $\frac{1}{10}\ln\left(\frac{x}{9}\right), x > 0$

h **i** $5-\sqrt{x+19}$ **ii** $\sqrt{x+10}-3$

2 **a**

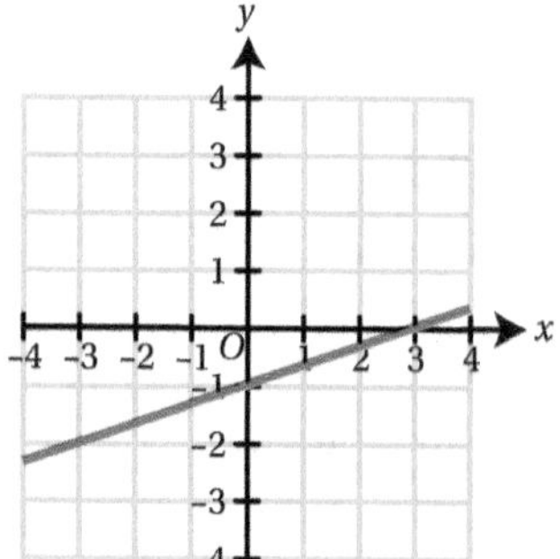

b

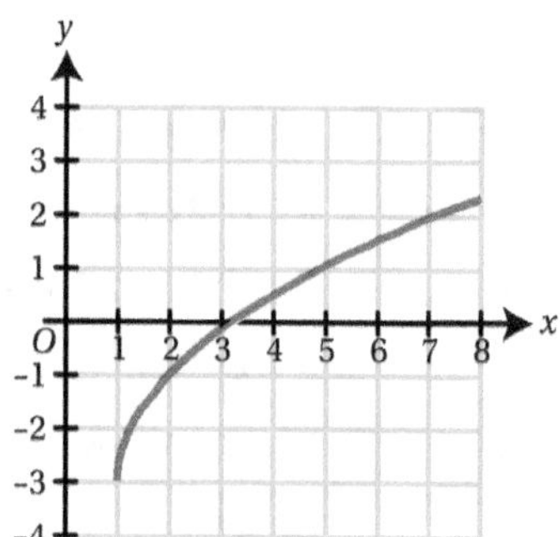

c

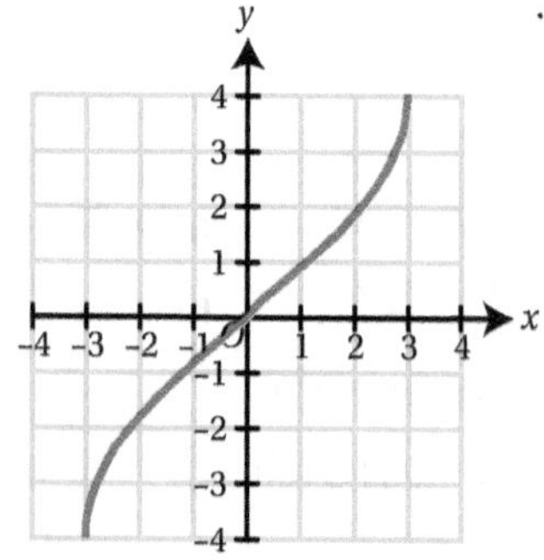

d

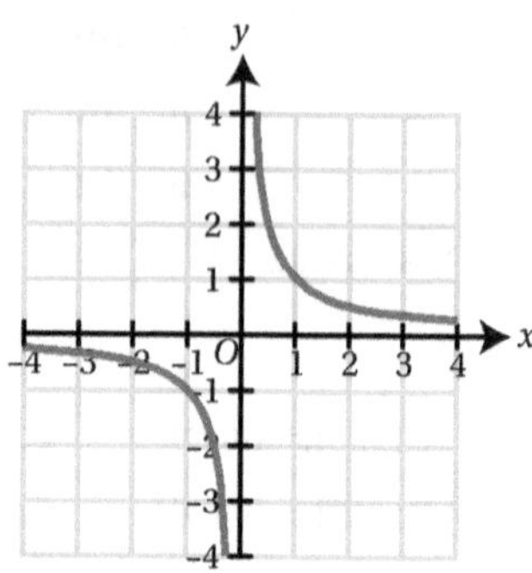

3 Each function is one-to-one so domain of f = range of f^{-1} and range of f = domain of f^{-1}.

a **i** $f^{-1}(x) = \frac{2x+1}{3-x}$; domain of f: $x \neq -2$; range of f: $f(x) \neq 3$.

ii $f^{-1}(x) = \frac{2x+3}{x-2}$; domain of f: $x \neq 2$; range of f: $f(x) \neq 2$.

b **i** $f^{-1}(x) = \frac{x^2-4x+13}{27}$; domain of f: $x \geqslant \frac{1}{3}$; range of f: $f(x) \leqslant 2$.

ii $f^{-1}(x) = 8x - 4x^2$; domain of f: $x \leqslant 4$; range of f: $f(x) \geqslant 1$.

c **i** $f^{-1}(x) = \frac{e^{x-3}+3}{4}$; domain of f: $x > \frac{3}{4}$; range of f: $f(x) \in \mathbb{R}$.

ii $f^{-1}(x) = e^{\frac{x+1}{2}} - 3$; domain of f: $x > -3$; range of f: $f(x) \in \mathbb{R}$.

d **i** $f^{-1}(x) = \ln\left(\frac{3-x}{2}\right)+2$; domain of f: $x \in \mathbb{R}$; range of f: $f(x) < 3$.

ii $f^{-1}(x) = \frac{1}{2}\left(5-\ln\left(\frac{x-1}{3}\right)\right)$; domain of f: $x \in \mathbb{R}$; range of f: $f(x) > 1$.

4 **a** -1 **b** 1

5 -23

6 **a** $f^{-1}(x) = \frac{1}{2}\ln\left(\frac{x}{3}\right)$

b Domain: $x > 0$; range: $f^{-1}(x) \in \mathbb{R}$.

7 $(f \circ g)^{-1}(x) = \sqrt[3]{\frac{x-3}{2}}$

8 Proof.

9 **a**

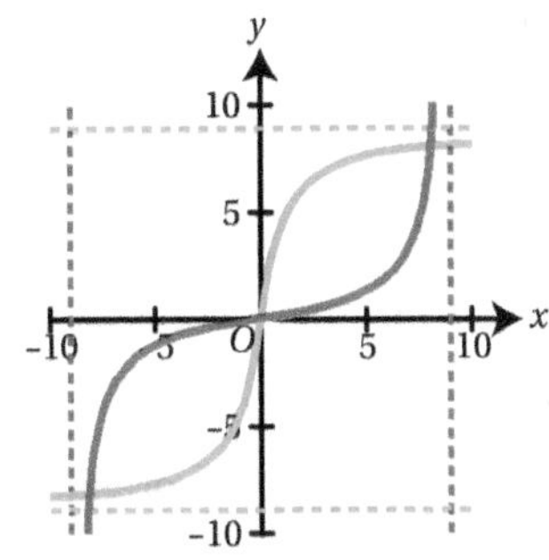

b Domain: $-9 < x < 9$; range: $f^{-1}(x) \in \mathbb{R}$.

c $x = -8, 0, 8$

10 **a** $\ln 3$ **b** Proof.

11 $f^{-1}(x) = -\sqrt{\frac{9x+4}{1-x}}$ for $-\frac{4}{9} \leqslant x < 1$

12 **a** $f^{-1}(x) = \frac{e^x}{3} + 1$ **b** $(g \circ f)(x) = 3x - 3$

13 **a**

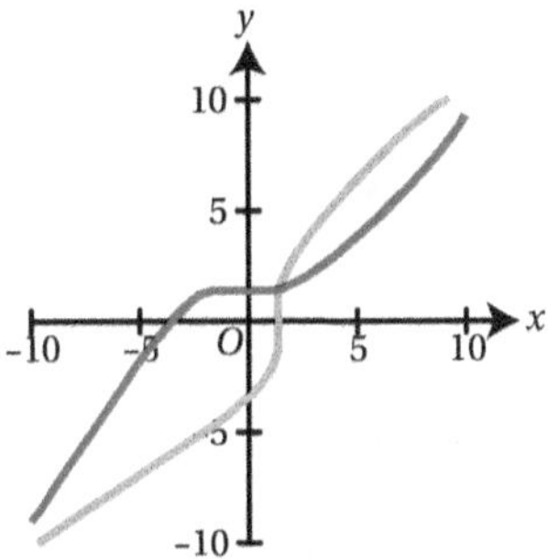

b $x = \frac{3}{2}$

WORK IT OUT 2.3

Solution 3 is correct.

EXERCISE 2E

1 **a** $k = 0; f^{-1}(x) = -\sqrt{x}, x \geqslant 0$

b $k = -1; f^{-1}(x) = \sqrt{x-2} - 1, x > 2$

c $k = 1; f^{-1}(x) = 1 - \sqrt{6-x}, x \leqslant 6$

d $k = -2; f^{-1}(x) = \sqrt{x+1} - 2, x > -1$

2 **a** $x \leqslant -5$

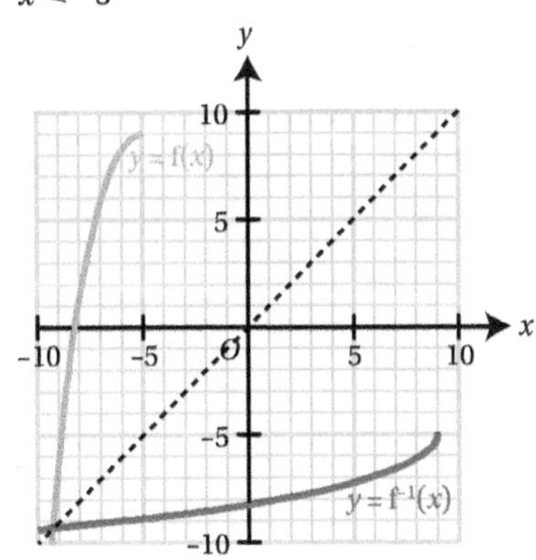

b $x \in [-5, 2]$

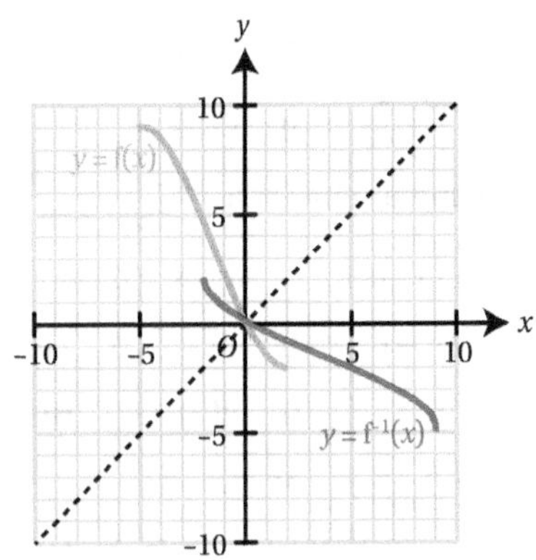

c $1 \leqslant x < \infty$

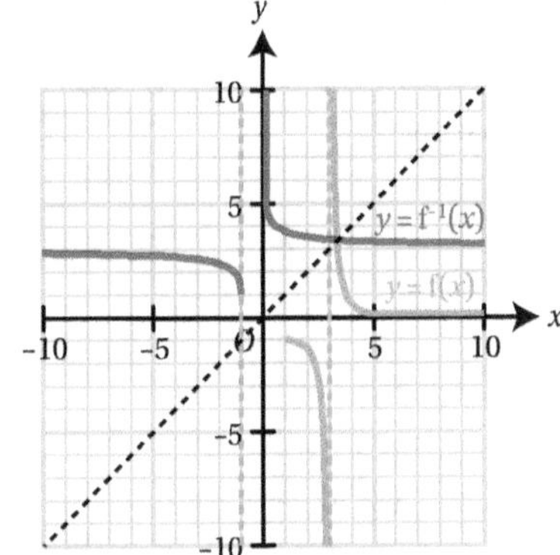

3 **a** $e^x - 3$

b $f(x)$ has a turning point, so is not one-to-one.

c $\ln(3)$

4 $k = -3$

5 $a \in [0, 1]$

MIXED PRACTICE 2

1 B

2 **a** $3^x - 3$ **b** $\sqrt[3]{\ln\left(\frac{x}{3}\right) + 1}, x > 0$

3 **a** $y = \log_2 x$ **b** $(1, 0)$

4 $x = \frac{1}{4}$

5 **a** 10 **b** $4 - x^2$

c Reflection in the line $y = x$.

d **i** $\sqrt{x-1}$ **ii** $f^{-1}(x) \geqslant 3$ **iii** $x \geqslant 10$

e No solutions.

6 a $f(x) \geqslant 0$

b i $\sqrt{\frac{20}{x} - 5}$ ii $x = \frac{2}{3}$

c i $\frac{x^2+5}{2}$ ii $x = 3$

7 a Not one-to-one. b $\frac{1-x}{2x}$

c $g^{-1}(x) \neq -0.5$ d $x = 0$

8 a $(x-3)^2 + 1$ b $\sqrt{x-1} + 3$ c $x \geqslant 1$

9 a i 15 ii $f(x) \in \mathbb{R}$ iii $2z + 1$

iv $\frac{3x+5}{x-1}$ v $4x + 3$

b $f(x)$ can be 1, which is not in the domain of g.

c i $\frac{x+3}{x-1}$

ii Reflection in the line $y = x$.

iii $x \neq 1$ iv $g^{-1}(x) \neq 1$

10 a $(x+2)^2 + 5$

b

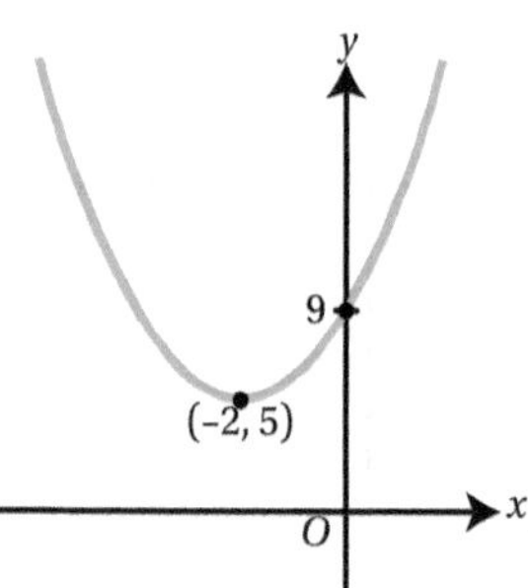

c Range of f is $f(x) \geqslant 5$; range of g is $g(x) > 0$.

d $h(x) > 9$

11 a i $\frac{e^x + 3}{2}$ ii $f^{-1}(x) > \frac{3}{2}$

iii

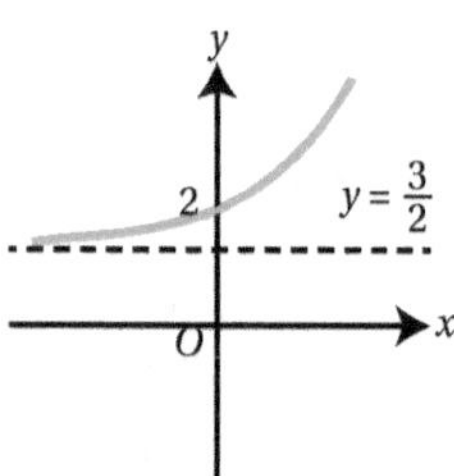

b i $(2x-3)^2 - 4$ ii $x = \frac{1}{2} \ln 8 = \ln \sqrt{8}$

12 a $(x-3)^2 - 7$ b $h(x) \geqslant -7$ c $\sqrt{x+7} + 3$

13 a Proof.

b $f\left(\frac{1}{x}\right) + 2\,f(x) = \frac{2}{x} + 1$

c $\frac{4 + x - 2x^2}{3x}$

14 a $a = -2, b = 1$ b $f \circ g\,(x) \geqslant 0$

15 a Proof.

b Rotation 180° about the origin.

c–d Proof.

e Reflection in the y-axis.

f–g Proof.

3 Further transformations of graphs

BEFORE YOU START

1 a

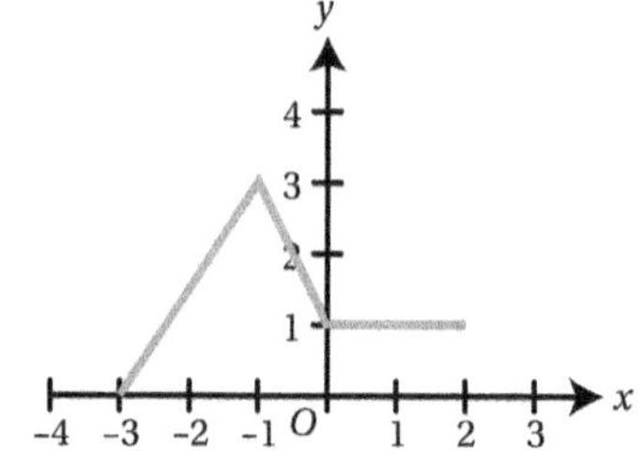

b

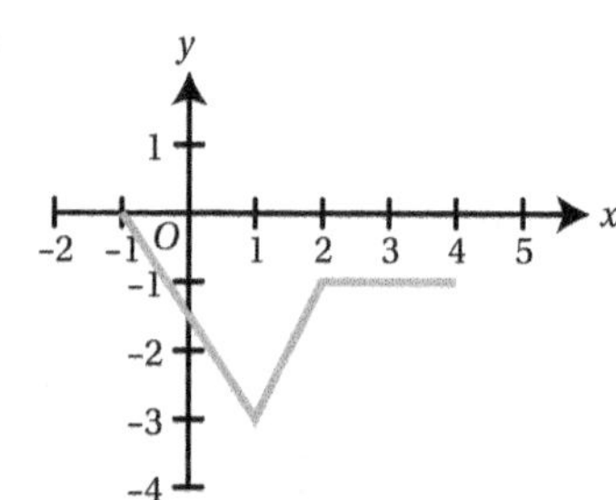

2 a $y = x^2 - 3x + 5$ b $y = \frac{1}{4}x^2 - \frac{3}{2}x + 5$

3 a $x \in [3, \infty)$ b $x \in (-3, -2)$

WORK IT OUT 3.1

Solution 1 is correct.

EXERCISE 3A

1 a i

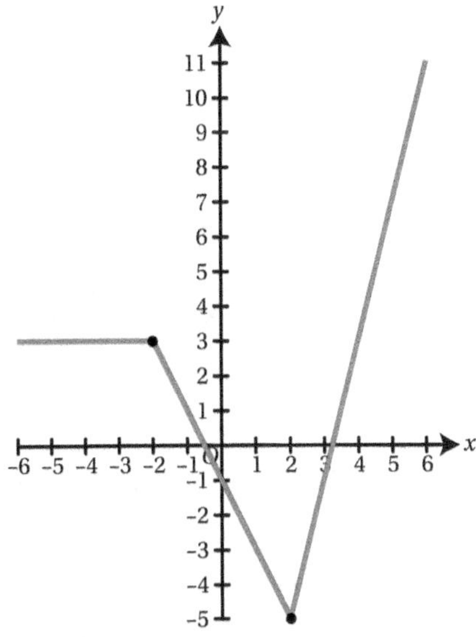

ii

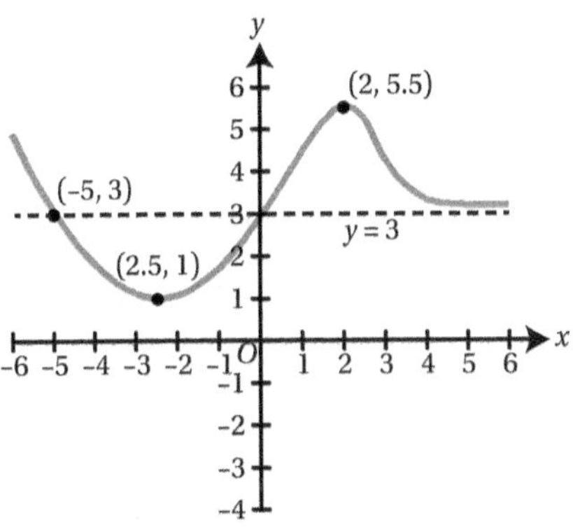

b i

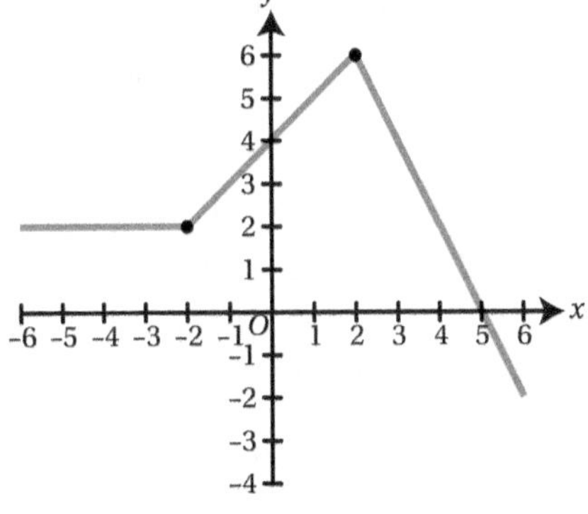

ii

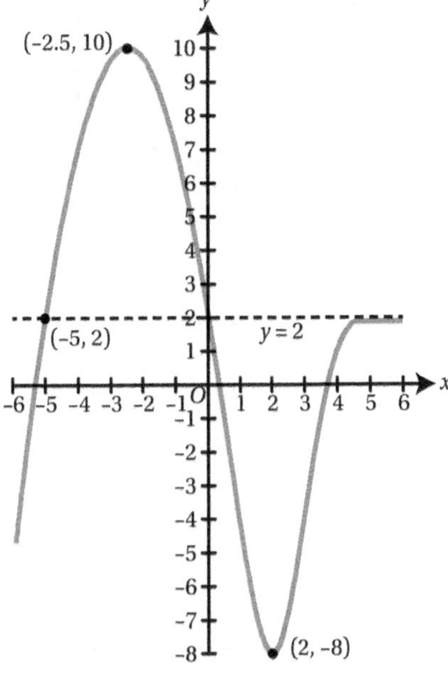

c i

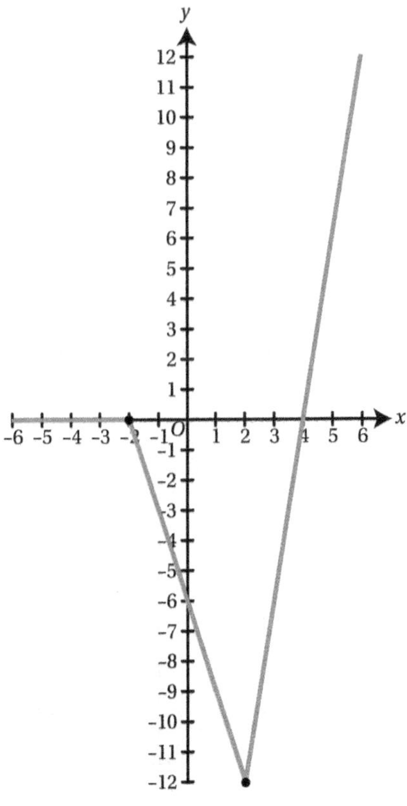

ii

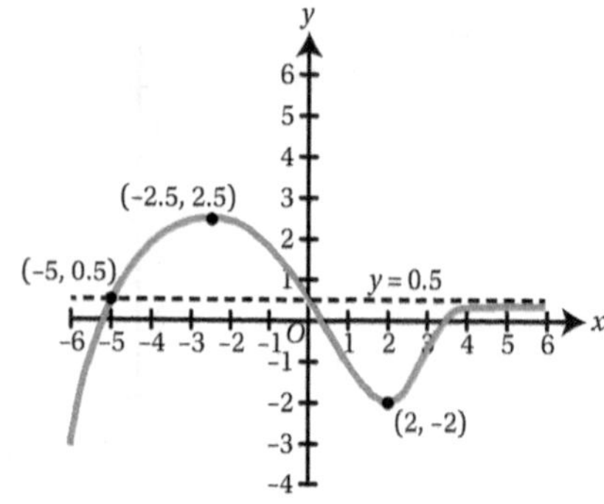

d i

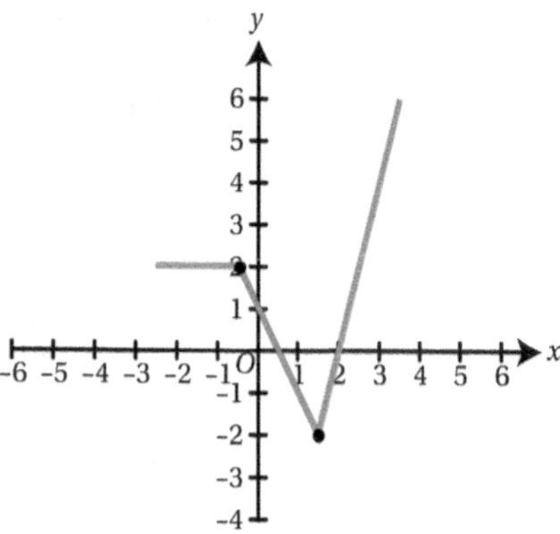

ii

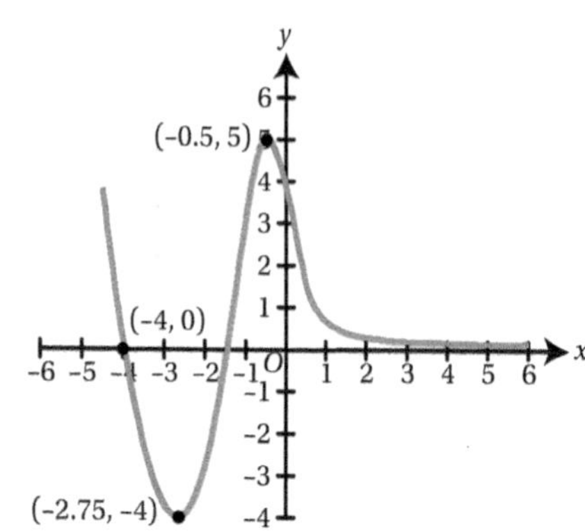

e i

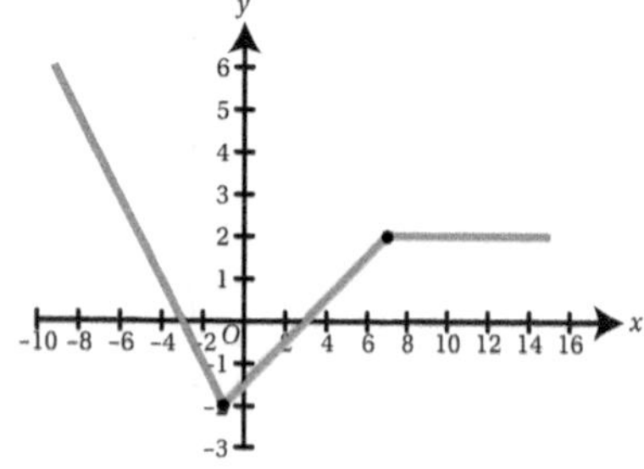

ii

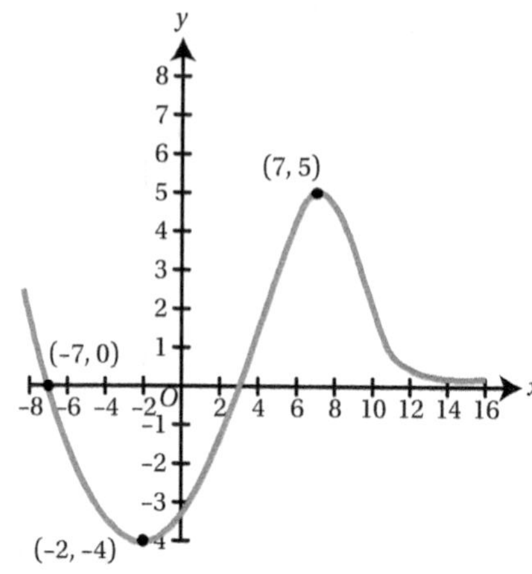

2 **a** i $k(x) = 2\,f(x) - 6$; vertical stretch of scale factor 2 and translation $\begin{pmatrix} 0 \\ -6 \end{pmatrix}$ (in either order).

ii $k(x) = 5\,f(x) + 4$; stretch of scale factor 5 relative to $y = 0$ and translation $\begin{pmatrix} 0 \\ 4 \end{pmatrix}$.

b i $h(x) = 5 - 3\,f(x)$; stretch of scale factor 3 relative to $y = 0$, reflection in $y = 0$ and translation $\begin{pmatrix} 0 \\ 5 \end{pmatrix}$.

ii $h(x) = 4 - 8\,f(x)$; stretch of scale factor 8 relative to $y = 0$, reflection in $y = 0$ and translation $\begin{pmatrix} 0 \\ 4 \end{pmatrix}$.

3 **a** i $g(x) = 6x^2 - 6$ ii $g(x) = x^2 + 1$

b i $g(x) = x^2 + 4$ ii $g(x) = 7x^2 - 4$

c i $g(x) = 3 - 2x^2$ ii $g(x) = 6 - 2x^2$

d i $g(x) = 5 - x^2$ ii $g(x) = -3 - 3x^2$

4 **a** i $g(x) = f(x+1) = f(-x-1)$; translation $\begin{pmatrix} 1 \\ 0 \end{pmatrix}$ then reflection in y-axis; or reflection in y-axis then translation $\begin{pmatrix} -1 \\ 0 \end{pmatrix}$.

ii $g(x) = f(x-3) = f(3-x)$; translation $\begin{pmatrix} -3 \\ 0 \end{pmatrix}$ then reflection in y-axis; or reflection in y-axis then translation $\begin{pmatrix} 3 \\ 0 \end{pmatrix}$.

b **i** $k(x) = f(2x+2)$; translation $\begin{pmatrix} -2 \\ 0 \end{pmatrix}$ then stretch of scale factor $\frac{1}{2}$ relative to $x = 0$; or stretch of scale factor $\frac{1}{2}$ relative to $x = 0$ then translation $\begin{pmatrix} -1 \\ 0 \end{pmatrix}$.

ii $k(x) = f(3x-1)$; translation $\begin{pmatrix} 1 \\ 0 \end{pmatrix}$ then stretch of scale factor $\frac{1}{3}$ relative to $x = 0$; or stretch of scale factor $\frac{1}{3}$ relative to $x = 0$ then translation $\begin{pmatrix} \frac{1}{3} \\ 0 \end{pmatrix}$.

5 **a** **i** $g(x) = 32x^2 - 16x - 2$

ii $g(x) = 8x^2 + 16x + 4$

b **i** $g(x) = 8x^2 + 64x + 124$

ii $g(x) = \frac{9x^2}{2} - 9x + \frac{1}{2}$

c **i** $g(x) = 2x^2 - 12x + 14$

ii $g(x) = 2x^2 + 12x + 14$

6 **a** $y = p(\sin x + c)$

b $y = p\sin x + c$

c $y = \sin\left(\frac{x+d}{q}\right)$

d $y = \sin\left(\frac{x}{q} + d\right)$

7 **a** $y = x^2$ **b** $q = e^{-2}$

c Horizontal stretch with scale factor e^2.

8 **a**

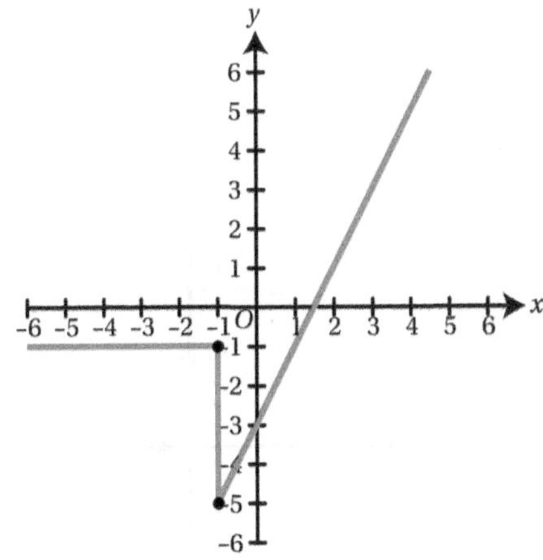

b

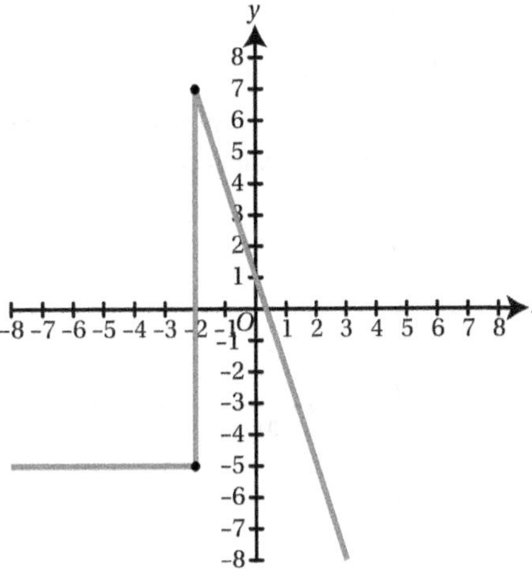

9 **a**

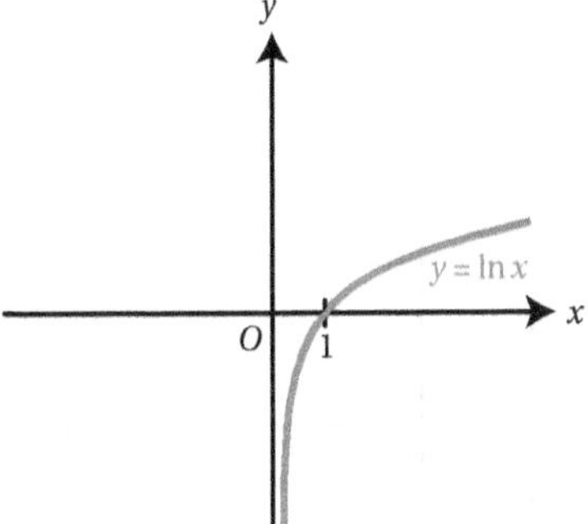

b

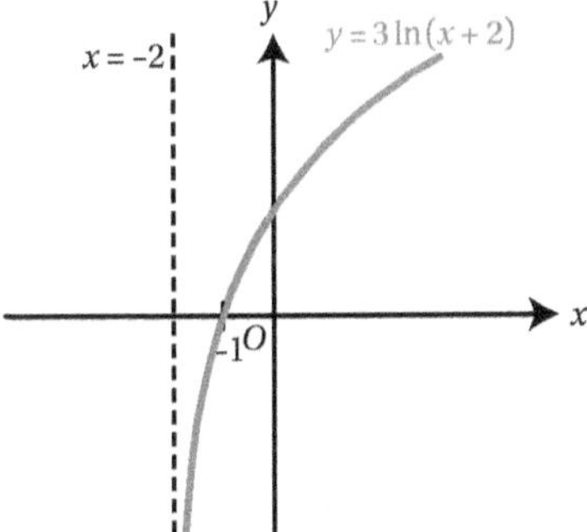

c

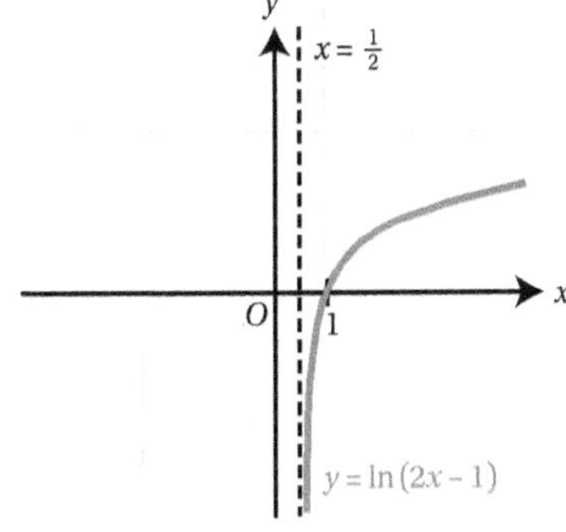

10 $y = -x^2 + 7x - 10$

11 $a = 5, b = 7$

12 $a = 16, b = 0, c = -25$

13 $h(x) = 4^{x+1} + 16x - 4$

14 **a** Vertical stretch with scale factor 3; horizontal stretch with scale factor 2.

b

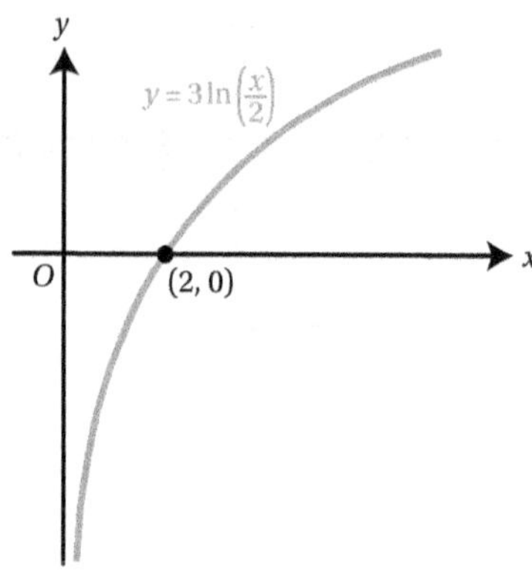

c

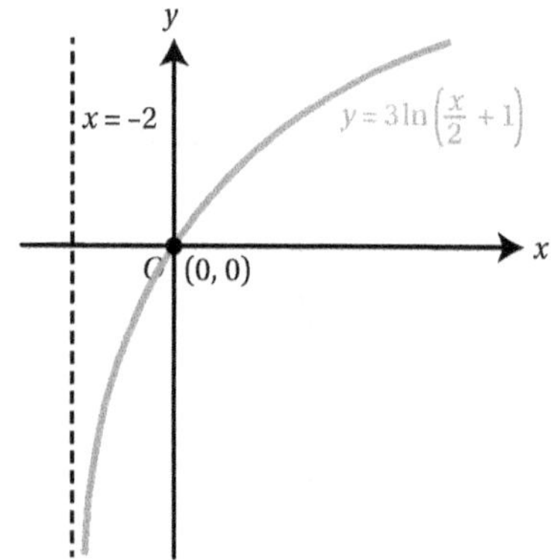

EXERCISE 3B

1 **a** **i**

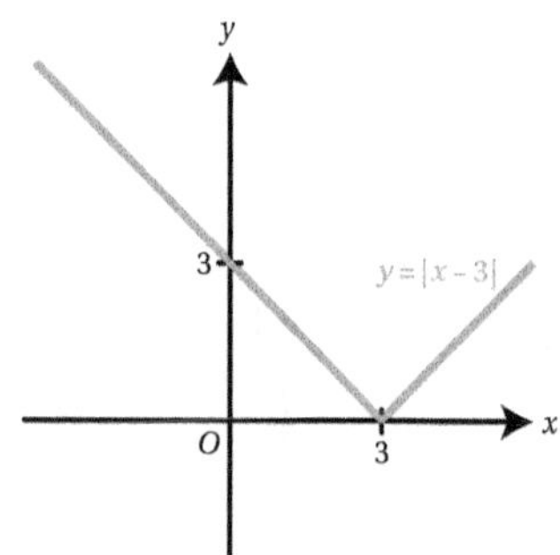

ii

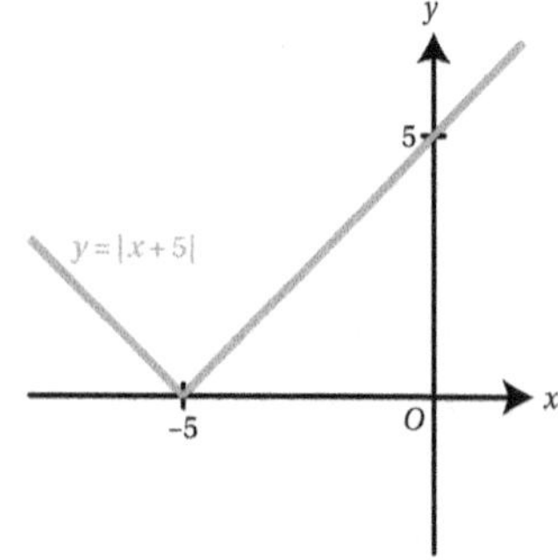

b **i**

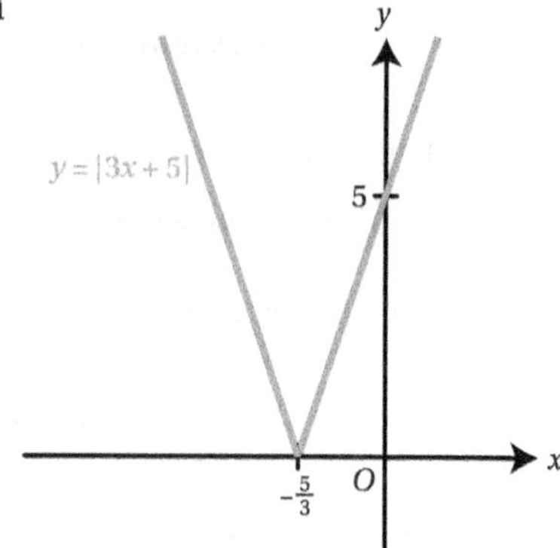

ii

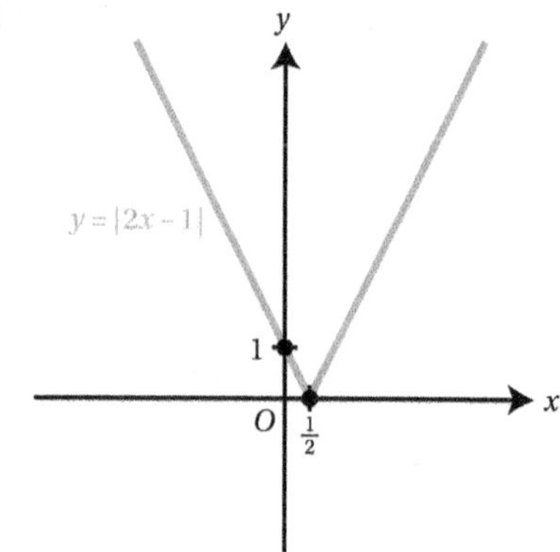

c **i**

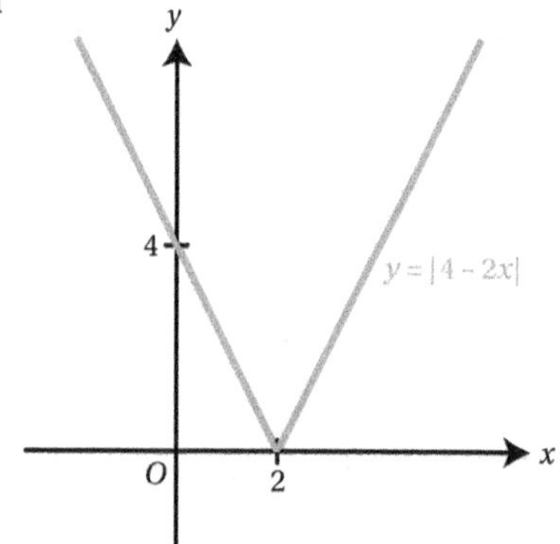

ii

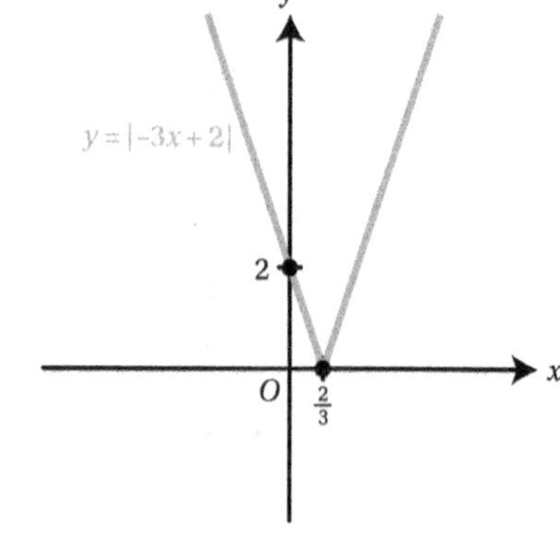

d i

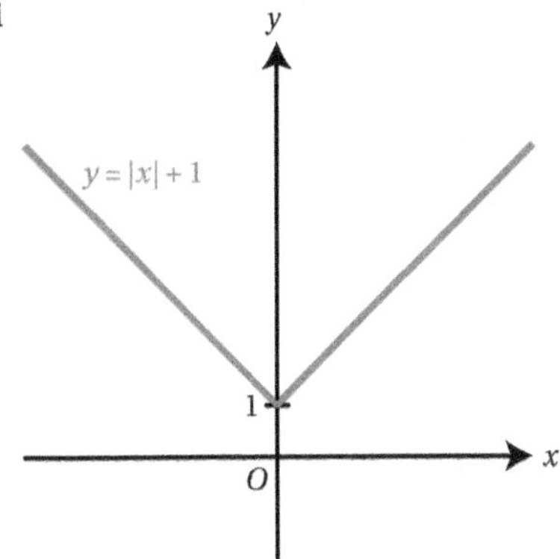

ii

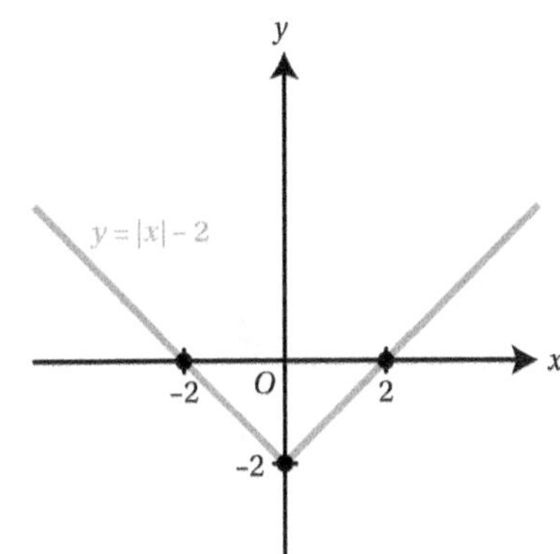

2 a i $y=|x-4|$ ii $y=|x+2|$

b i $y=|2x+4|$ ii $y=|3x+6|$

c i $y=|3x-3|$ ii $y=|2x-5|$

3 a i $-3<x<13$ ii $-7<x<11$

b i $-6<x<4$ ii $-9<x<-3$

4 a i $x>6$ or $x<-2$ ii $x>8$ or $x<-6$

b i $x>6$ or $x<-16$ ii $x>-2$ or $x<-6$

5 a i $|x-9|<4$ ii $|x-18|<7$

b i $|x+2|<12$ ii $|x+7|<9$

6 a i $|x-7|>4$ ii $|x-12|>7$

b i $|x+6|>3$ ii $|x+4|>10$

7 a

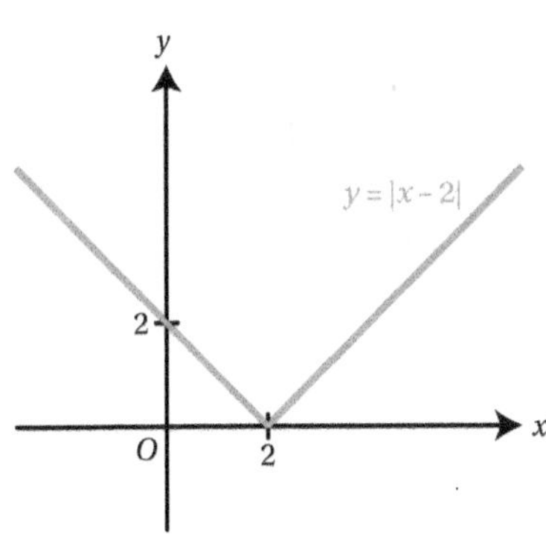

b

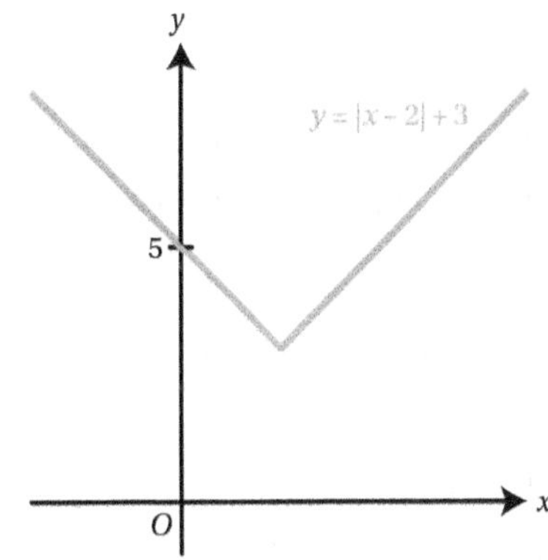

8

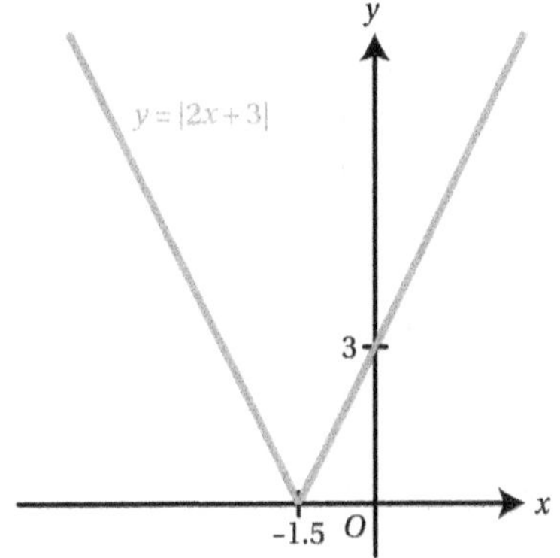

9

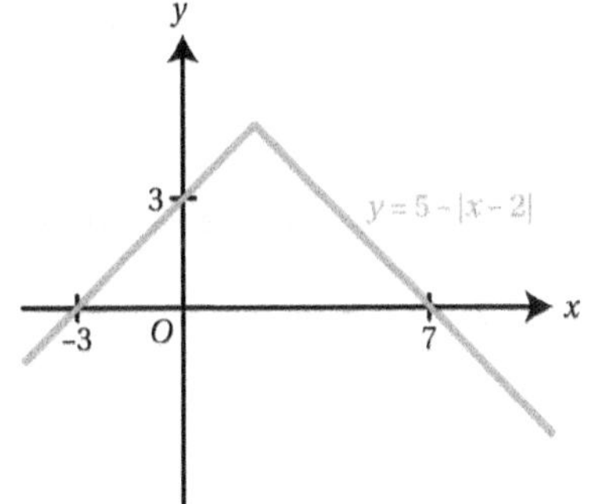

10

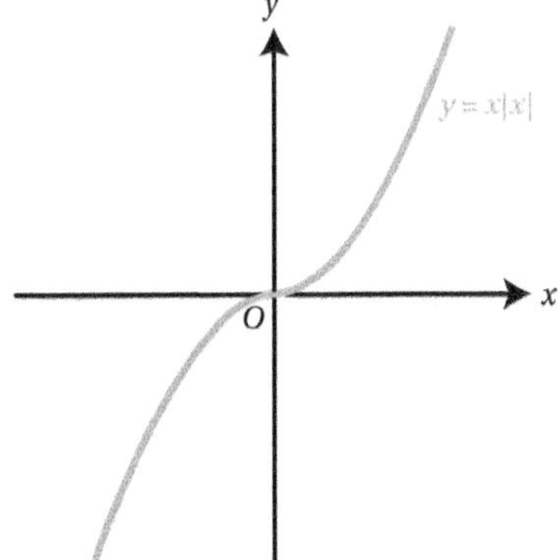

WORK IT OUT 3.2

Solution 1 is correct.

EXERCISE 3C

1 a i $x = \pm 4$ ii $x = \pm 18$

b i $x = 0, 4$ ii $x = -1, \frac{1}{3}$

c i $x = -8, 0$ ii $x = \frac{2}{3}, 8$

d i $x = -2, 3$ ii $x = -\frac{1}{2}$

e i $x = \frac{2}{3}, 4$ ii $x = \frac{1}{4}, \frac{9}{2}$

f i $x = \frac{1}{2}$ ii $x = -1, -2$

2 a i $x \in (-\infty, -5) \cup (5, \infty)$

ii $x \in (-\infty, -2) \cup (2, \infty)$

b i $-3 < x < 3$

ii $-10 < x < 10$

c i $x \in \left(-\infty, -\frac{5}{2}\right] \cup \left[\frac{3}{2}, \infty\right)$

ii $x \in \left[-\frac{1}{3}, \frac{5}{3}\right]$

d i $\frac{4}{3} < x < 6$

ii $x < 1$ or $x > 5$

e i $x \in \left(-\infty, -\frac{5}{3}\right) \cup (3, \infty)$

ii $x \in \left(-\infty, \frac{3}{5}\right) \cup (5, \infty)$

f i $-\frac{4}{3} \leqslant x \leqslant 4$ ii $-1 \leqslant x \leqslant 1$

3 a $-\frac{4}{5}, \frac{2}{3}$ b $-\frac{4}{5} < x < \frac{2}{3}$

4 a $x = 1, 7$ b $x \in (-\infty, 1] \cup [7, \infty)$

5 a

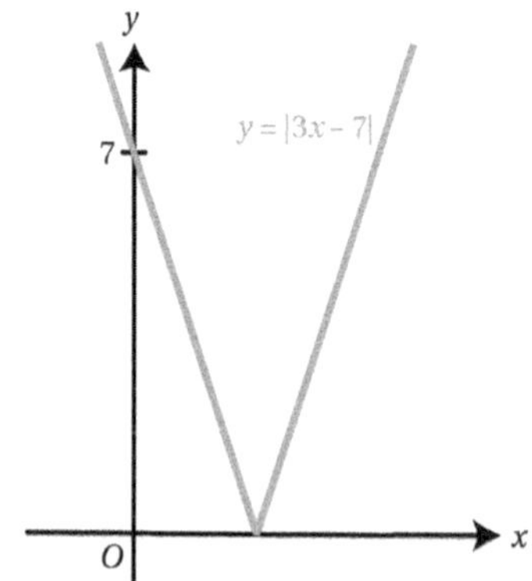

b No solutions.

6 $x \in \mathbb{R}$

7 $0 \leqslant x \leqslant \frac{2k}{3}$

8 $x \geqslant 0$

MIXED PRACTICE 3

1 a

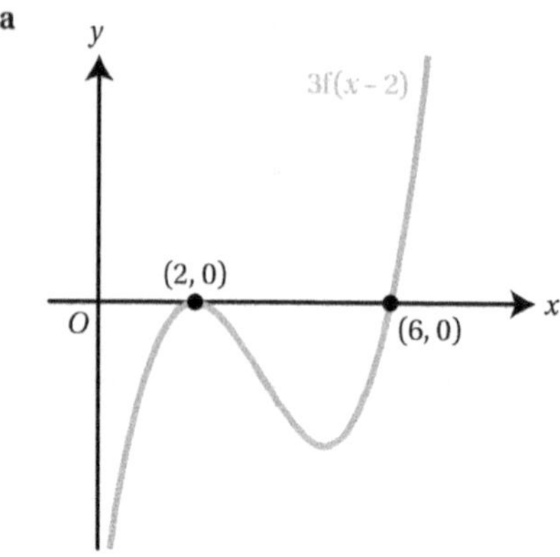

b

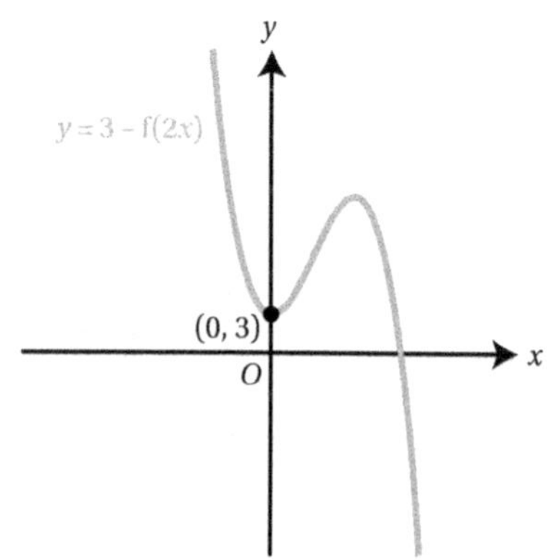

2 B

3 $y = 2x^3 - 12x^2 + 24x - 18$

4 Translation by $\begin{pmatrix} -3 \\ 0 \end{pmatrix}$ and vertical stretch with scale factor 3.

5 a

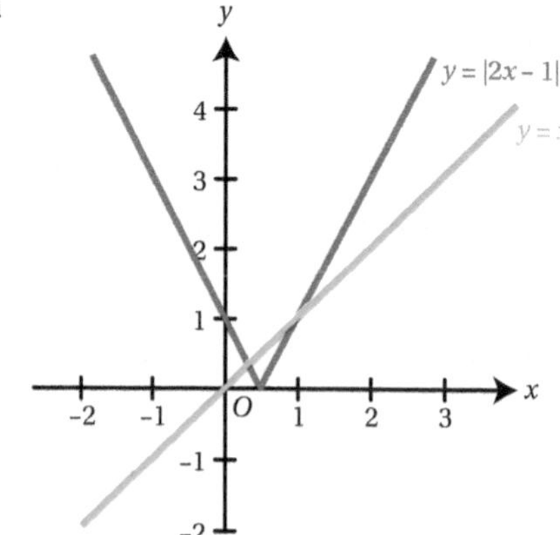

b $\frac{1}{3} < x < 1$

6 a $x = 1, 3$ b $x \leqslant 1, x \geqslant 3$

7 D

8 $x < -5$ or $x > \frac{7}{3}$

9 a Translation by $\begin{pmatrix} 2 \\ 0 \end{pmatrix}$ and vertical stretch with scale factor 3.

b Translation by $\begin{pmatrix} 3 \\ 0 \end{pmatrix}$ and translation by $\begin{pmatrix} 0 \\ 10 \end{pmatrix}$.

c Translation by $\begin{pmatrix} 5 \\ 10 \end{pmatrix}$ and vertical stretch with scale factor 3.

10 a Vertical stretch with scale factor 3; reflection in the x-axis; translation 5 units up.

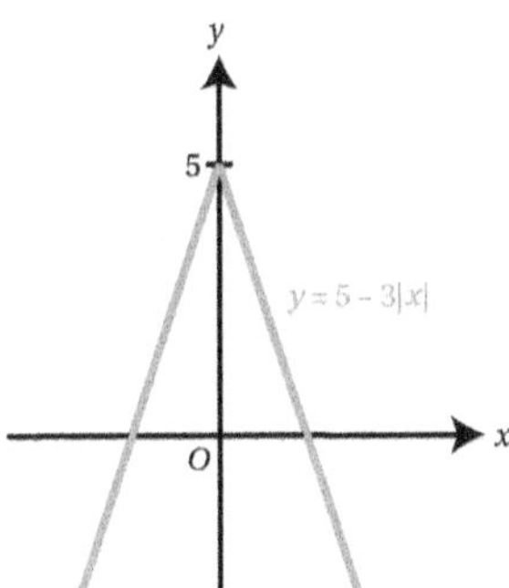

b $1.2, -0.8$

c $-\frac{4}{5} \leqslant x \leqslant \frac{6}{5}$

11 a Translation by $\begin{pmatrix} -2 \\ 0 \end{pmatrix}$.

b

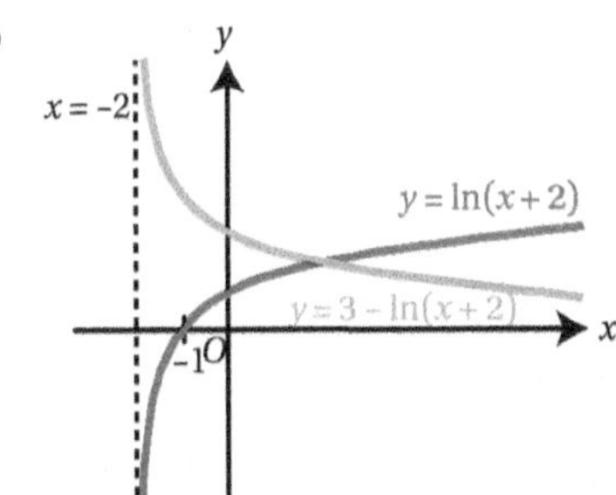

c i $\begin{pmatrix} 2 \\ 0 \end{pmatrix}$

ii $a = -1, b = 6, c = -9, d = -3$

12 a Translation $\begin{pmatrix} -1 \\ 0 \end{pmatrix}$; vertical stretch, scale factor 4; translation $\begin{pmatrix} 0 \\ -2 \end{pmatrix}$.

b $(0, -2)$, $\left(e^{0.5} - 1, 0\right)$

13 $x \geqslant 0$

14

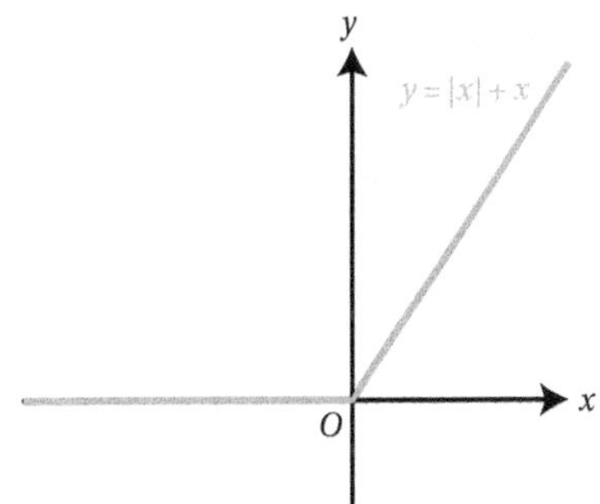

4 Sequences and series

BEFORE YOU START

1 a $3n - 1$ b $-4n + 19$

2 $u_2 = 10, u_3 = 28$

3 $a = -4, b = 3$

4 7

5 17

6 $-3, -2, -1, 0, 1, 2, 3$

EXERCISE 4A

1 a i $3.1, 8.1, 13.1, 18.1, 23.1$

ii $10, 6.2, 2.4, -1.4, -5.2$

b i $0, 1, 4, 13, 40$

ii $1, -1, -19, -181, -1639$

c i $1, 3, \frac{5}{3}, \frac{11}{5}, \frac{21}{11}$

ii $3, -1, 3, -1, 3$

d i $0, 4, 8, 12, 16$

ii $13, 11, 9, 7, 5$

2 a i $5, 8, 11, 14, 17$

ii $-4.5, -3, -1.5, 0, 1.5$

b i $0, 7, 26, 63, 124$

ii $5, 20, 45, 80, 125$

c i $3, 9, 27, 81, 243$

ii $4, 2, 1, \frac{1}{2}, \frac{1}{4}$

d i $1, 4, 27, 256, 3125$

ii $1, 0, -1, 0, 1$

3 a i Increases, converges to 10.

ii Decreases, diverges.

b i Periodic, period 2.

ii Periodic, period 2.

4 a $u_1 = 2, u_2 = 8$ b $a = 2$

5 a $\frac{47}{17}$

b Converges to 3 (and oscillates).

6 $\frac{2}{5}$

7 a Proof. b $\frac{1}{2}$

8 a $\frac{1}{2}, -2, -\frac{1}{3}, 0, \frac{1}{2}$ b $\frac{1}{2}$

9 a 1 b $-1 \leqslant u_1 \leqslant 1$

10 Proof.

EXERCISE 4B

1 a i 27 ii 39

b i 116 ii $\frac{665}{48}$

c i $14b$ ii $19p$

2 a i $\sum_{2}^{43} r$ ii $\sum_{3}^{30} 2r$

b i $\sum_{1}^{6} \frac{1}{2^{r+1}}$ ii $\sum_{0}^{5} \frac{2}{3^r}$

c i $\sum_{2}^{10} 7ar$ ii $\sum_{0}^{19} r^b$

3 39

4 a Proof. b $\frac{3}{2}$

5 $15 \ln 3$

6 a 0 b $\sqrt{3}$

EXERCISE 4C

1 a i $u_n = 9 + 3(n-1)$

ii $u_n = 57 + 0.2(n-1)$

b i $u_n = 12 - (n-1)$

ii $u_n = 18 - \frac{1}{2}(n-1)$

c i $u_n = 1 + 3(n-1)$

ii $u_n = 9 + 10(n-1)$

d i $u_n = 4 - 4(n-1)$

ii $u_n = 27 - 7(n-1)$

e i $u_n = -17 + 11(n-1)$

ii $u_n = -32 + 10(n-1)$

2 a i 33 ii 29

b i 100 ii 226

3 a $u_n = 5 + 8(n-1) = 8n - 3$ b 50

4 121

5 25

6 17 rungs.

7 $a = 2, b = -3$

8 a Proof. b 456 pages.

EXERCISE 4D

1 a i 3060 ii 1495

b i 9009 ii 23 798

c i −204 ii 1470

d i 667.5 ii 14.25

2 a i 13 ii 32

b i $\frac{x}{2}$ ii $5\sqrt{x}$

3 30

4 a 1, 5, 9 b $u_n = 4n - 3$

5 $a = 15, d = -8$

6 $a = 2, d = 5$

7 $a = -7, d = 3$

8 Proof.

9 559

10 55

11 $u_n = 6n - 5$

12 $\theta = 20°$

13 a 71 071 b 429 429

14 10 300

15 a $590 \ln x$ b e^5

16 23 926

EXERCISE 4E

1 a i $u_n = 6 \times 2^{n-1}$ ii $u_n = 12 \times \left(\frac{3}{2}\right)^{n-1}$

b i $u_n = 20 \times \left(\frac{1}{4}\right)^{n-1}$ ii $u_n = \left(\frac{1}{2}\right)^{n-1}$

c i $u_n = (-2)^{n-1}$ ii $u_n = 5 \times (-1)^{n-1}$

d i $2(\sqrt{3})^{n-1}$ ii $\frac{4}{(\sqrt{2})^{n-1}}$

e i $u_n = ax^{n-1}$ ii $u_n = 3 \times (2x)^{n-1}$

2 a i 13 ii 7

b i 10 ii 10

c i 10 ii 8

3 a $u_n = 2 \times 3^{n-1}$ b 39 366

4 a $u_1 = 3, r = \pm 2$ b ± 384

5 a $a = -3.5$ or 7 b 137 781 or 68 890.5

6 10

7 a Proof. b 8

8 32

9 23

10 $\pm 0.5, 1$

11 $a = -2, b = 4$

12 $m = 7$

EXERCISE 4F

1 a i 17 089 842 ii 2303.4375

b i 514.75 ii 9.487 171

c i 39 368 ii 9840

d i 191.953 125 or 63.984 375

ii 24 414 062.5 or 16 276 041.67

2 a i $r = 0.15$ ii $r = 2.7$

b i $r = -1.3, 0.3$ ii $r = -1.7, 0.7$

3 a 5 b $S_n = \frac{375(5^n - 1)}{4}$

4 a $a = \frac{10}{3}, r = \frac{2}{3}$ b 9.83

5 1.52 or −2.52

6 $a = 5, r = \frac{3}{2}$

7 a $\frac{3}{2}$ b 160

8 a $a = \frac{1}{2}, r = 4$ b 11

9 a Proof. b 2 446 675

10 a Proof. b 1 : 1024

EXERCISE 4G

1 a i $\frac{27}{2}$ ii $\frac{196}{3}$

b i $\frac{1}{3}$ ii $\frac{26}{33}$

c i Divergent. ii Divergent.

d i $\frac{25}{3}$ ii $\frac{18}{5}$

e i Divergent. ii $\frac{7}{3}$

2 a i $|x| < 1$ ii $|x| < 1$

b i $|x| < \frac{1}{3}$ ii $|x| < \frac{1}{10}$

c i $|x| < \frac{1}{5}$ ii $|x| < \frac{1}{3}$

d i $|x| < 4$ ii $|x| < 12$

e i $|x| < 3$ ii $|x| < \frac{4}{5}$

f i $|x| > 2$ ii $|x| > \frac{1}{2}$

g i $1 < x < 2$ ii $0 < x < 4$

h i $\frac{1}{2} < x < 1$ ii $x < -\frac{1}{2}$

i i $|x| < 1$ ii $|x| < \frac{1}{\sqrt[3]{4}}$

3 $-\frac{54}{5}$

4 $\frac{27}{2}$

5 128 or 384

6 a $a = 4\sqrt{2}, r = \frac{1}{\sqrt{2}}$ b $m = 8, n = 2$

7 a $\frac{2}{3}$ b 9

8 $\frac{1}{8}$

9 a $|x| < \frac{3}{2}$ b 5

10 9

11 a $1 < x < \frac{5}{3}$ b 7

12 a 3 b ∞

13 a $x < 0$ b $x = -3$

14 a Proof. b ln 3

15 Proof.

WORK IT OUT 4.1

Solution C is correct.

EXERCISE 4H

1 a £34.78 b £1194.05

2 a £60 500 b 22 years

3 a 5000×1.063^n b £6786.35

c i $5000 \times 1.063^n > 10\,000$

ii 12 years

4 a 10 b 23.7%

5 a $V = £265.33$ b 235 months

6 a 12 days b Day 102

7 Proof.

8 a 0.8192 m b 15.3 m

c Physical interactions haven't been considered.

9 a Proof.

b $25\,000(1.04^n - 1)$ c Year 29

MIXED PRACTICE 4

1 B

2 $a = 11, d = -\frac{7}{2}$

3 97.2

4 a $a = 3072, r = \frac{1}{4}$ b 4096

5 2

6 a $u_1 = 12$ b 4

c i $k = 13$ ii 67 108 860

7 A

8 A

9 13th term

10 a $2n - 1$ b 6 c 64

11 $n < 19$ years

12 a $-3, 7$ b 54

13 4.5

14 19 264

15 a $-\frac{1}{2} \leqslant k \leqslant 2$ b $-\frac{12}{5}$

16 a Proof. b 56

17 a Proof. b -4 c 1953

18 Proof.

19 $d = 0, -\frac{1}{4}$

20 $\frac{-1+\sqrt{5}}{2}$

21 $\ln\left(\frac{a^{69}}{b^{138}}\right)$

22 a n b $\frac{n(n+1)}{2}$

c $\frac{n(n-1)}{2} + 1$ d Proof.

e 32

23 a Proof.

b $150\,000 \times 1.03^n - \frac{1000\,000(1.03^n - 1)}{3}$

c 21 years

5 Rational functions and partial fractions

BEFORE YOU START

1 $\frac{2+2x}{x(2+x)}$

2 $(2x+1)(3x+2)$

3 $x^2 - 3x + 3$

4 $x - 1$

WORK IT OUT 5.1

Solution 2 is correct.

EXERCISE 5A

1 a i $(2x+1)(x-2)(x-3)$

ii $(3x-1)(x-1)^2$

b i $(2x-3)(x+2)(x+3)$

ii $(3x-5)(x+4)(x-4)$

2 $(2x+5)(2x+1)(x-3)$

3 $a = -2$

4 a $a = -8$ b $2(3x+1)(x-1)(2x+1)$

5 a Proof. b $(2x-a)(x-a)^2$

6 Proof; $x = a, 2a, 3a$

7 Proof.

8 $b = \pm\sqrt{\frac{a^3}{a-1}}$

9 a $a = -23, b = -6$

b $x = -\frac{1}{2}, -\frac{1}{3}, -2, \frac{3}{5}$

10 $x = \frac{1}{2}, 2, -3$

WORK IT OUT 5.2

Solution 2 is correct.

EXERCISE 5B

1 a i $2x+3$ ii $2x+4$

b i $\frac{1}{2}$ ii $\frac{1}{5}$

c i $3x+4$ ii $5x-7$

d i -1 ii -1

e i $\frac{2}{x-2}$ ii $\frac{4}{x+3}$

f i $\frac{x+1}{x+4}$ ii $\frac{x+2}{x+4}$

g i $\frac{3x+1}{4x+1}$ ii $\frac{4x-5}{3x-2}$

2 a i $2x$ ii $3x$

b i $\frac{x}{2}$ ii $5x^2$

c i $\frac{1}{6}$ ii $\frac{5}{2}$

d i $4x$ ii $2x$

e i $\frac{x+1}{x-1}$ ii $\frac{x-3}{x-1}$

3 a i 2 ii $\frac{5}{3}$

b i $\frac{2}{x+1}$ ii $\frac{3x}{5}$

c i $\frac{x}{3}$ ii $\frac{x+5}{x+1}$

d i $\frac{x}{3x+2}$ ii $\frac{5}{7x-2}$

4 a i $\frac{1}{2}x-\frac{1}{4}+\frac{\frac{1}{4}}{2x+1}$

ii $\frac{1}{2}x-\frac{3}{4}+\frac{\frac{9}{4}}{2x+3}$

b i $x+1+\frac{3}{2x+1}$ ii $x+\frac{4}{5x+3}$

5 $\frac{x+5}{x+4}$

6 $\frac{x+3}{2x^2}$

7 a $x+2$ b $x=-2$ or -1

8 $x=\pm 6$

9 Quotient: 2; remainder: 25.

10 Quotient: $x-1$; remainder: 3.

11 $2+\frac{3}{x+2}$

12 $x+1$

13 a Proof. b $x-a$

14 $a=-1$; quotient: $x+1$.

15 Proof.

16 a $\frac{3ab}{2a+b}$ b Proof.

EXERCISE 5C

1 a i $\frac{1}{x}+\frac{1}{x+2}$ ii $\frac{1}{x-3}-\frac{1}{x}$

b i $\frac{1}{x+1}+\frac{2}{x+2}$ ii $\frac{2}{x-1}+\frac{3}{x+2}$

c i $\frac{1}{x-3}-\frac{2}{x+4}$ ii $\frac{2}{x-5}-\frac{3}{x+6}$

d i $\frac{1}{2x-1}+\frac{1}{x+3}$ ii $\frac{2}{3x-2}-\frac{1}{2x+5}$

2 a i $\frac{1}{x}+\frac{1}{x-2}-\frac{2}{x-3}$

ii $\frac{2}{x}-\frac{1}{x+2}-\frac{1}{x+1}$

b i $\frac{5}{x-1}-\frac{2}{x+1}-\frac{3}{x+2}$

ii $\frac{4}{x-3}-\frac{1}{x+4}-\frac{3}{x+1}$

c i $\frac{8}{2x+1}+\frac{1}{x+3}-\frac{5}{x+1}$

ii $\frac{3}{3x-1}-\frac{4}{2x+3}+\frac{1}{x+4}$

3 $A=-\frac{2}{21}, B=\frac{5}{21}$

4 $\frac{1}{2(2x-1)}-\frac{1}{2(2x+1)}$

5 $\frac{1}{x}+\frac{1}{x+1}-\frac{2}{x+2}$

6 $\frac{1}{3x+1}+\frac{1}{3x-1}-\frac{2}{3x}$

7 $\frac{1}{6}\left(\frac{1}{x+1}-\frac{1}{x-1}-\frac{1}{2x+4}+\frac{1}{2x-4}\right)$

8 a Proof. b $1-\frac{3}{x-1}+\frac{3}{x-2}$

9 $\frac{1}{x-a}-\frac{1}{x}$

10 $\frac{1}{x-a}+\frac{1}{x-2a}$

11 a $R=1$ b $1+\frac{1}{5(x+1)}-\frac{1}{5(x+6)}$

12 Discussion.

EXERCISE 5D

1 a i $\frac{1}{x+1}-\frac{1}{(x+1)^2}$ ii $\frac{1}{x-2}+\frac{2}{(x-2)^2}$

b i $\frac{4}{x^2}-\frac{1}{x}+\frac{1}{x+4}$ ii $\frac{1}{x-1}-\frac{1}{x^2}-\frac{1}{x}$

c i $\frac{2}{x-1}+\frac{3}{(x+2)^2}-\frac{2}{x+2}$

ii $\frac{1}{x-2}+\frac{6}{(x-2)^2}-\frac{1}{x+1}$

d i $\frac{1}{1-x}+\frac{2}{2x+3}+\frac{10}{(2x+3)^2}$

ii $\frac{2}{2x-5}-\frac{1}{x+1}-\frac{2}{(x+1)^2}$

2 $\frac{1}{x-2}-\frac{1}{x}-\frac{2}{x^2}$

3 $\frac{1}{x^2}-\frac{4}{x+1}+\frac{4}{x}$

4 $\frac{1}{16(x+2)}-\frac{1}{16(x-2)}+\frac{1}{4(x-2)^2}$

5 a Proof. b $(2x-1)(x+1)^2$

c $\frac{2}{2x-1}-\frac{1}{x+1}+\frac{3}{(x+1)^2}$

6 a Proof.

b $\frac{2}{2x-3}-\frac{4}{(2x-3)^2}-\frac{3}{3x+2}$

7 a $x+2+\frac{2}{x^3+3x^2}$

b $x+2+\frac{2}{3x^2}-\frac{2}{9x}+\frac{2}{9(x+3)}$

8 a Proof. b $\frac{1}{x-1}+\frac{3-x}{(x-2)^2}$

9 $\frac{1}{x-a}+\frac{1}{(x-a)^2}-\frac{1}{x}$

MIXED PRACTICE 5

1 C

2 a x^2-x+1 b $7, 11, 13$

3 a Proof.

b $3(2x+3)(x-1)(x+4)$

c $x=-\frac{3}{2}, 1, -4$ d $\frac{3(x-1)}{2}$

4 $\frac{1}{x-3}-\frac{1}{x+2}$

5 $\frac{1}{x}-\frac{2}{x+3}+\frac{1}{x-3}$

6 $\frac{1}{12(3x+2)}+\frac{1}{12(3x-2)}$

7 $\frac{1}{4x}-\frac{1}{4(x+2)}-\frac{1}{2(x+2)^2}$

8 $\frac{\sqrt{10}}{x-\sqrt{10}}-\frac{\sqrt{10}}{x+\sqrt{10}}$

9 $\frac{3}{1-3x}+\frac{1}{1+x}-\frac{4}{(1+x)^2}$

10 a 4

b i Proof.

ii $(3x+2)(3x-1)^2$ iii $3x-1$

11 A

12 a $\frac{2}{2x-1}-\frac{1}{2x+1}$ b $3x-\frac{2(2x+3)}{4x^2-1}$

c $3x-\frac{4}{2x-1}+\frac{2}{2x+1}$

13 a $\frac{3}{(2x+3)(x+2)}$ b $\frac{6}{2x+3}-\frac{3}{x+2}$

14 a Proof. b $(x-1)(3x+1)^2$

c $x=1, -\frac{1}{3}$

d $\frac{1}{2(x-1)}-\frac{3}{2(3x+1)}-\frac{2}{(3x+1)^2}$

15 a Proof. b $x=\pm0.5, -1$

c $\frac{1}{2x-1}+\frac{1}{2x+1}-\frac{1}{x+1}$

16 a $2x+1+\frac{3x+7}{x^2+5x+6}$

b $2x+1+\frac{1}{x+2}+\frac{2}{x+3}$

17 a $a=35, b=6$

b $(2x+1)(3x+1)(x+2)(x+3)$

c $x=-\frac{1}{2}, -\frac{1}{3}, -2, -3$

18 a $\frac{1}{x+1}-\frac{1}{x+5}$

b $\frac{1}{(x+1)^2}+\frac{1}{(x+5)^2}+\frac{1}{2(x+5)}-\frac{1}{2(x+1)}$

19 a $1+\frac{8}{u-4}$ b $1+\frac{2}{x-2}-\frac{2}{x+2}$

20 $\frac{1}{x}+\frac{a}{x^2}+\frac{2}{x-a}$

21 $A=1, B=2, k=3$

22 $b=-2a^2$

23 0

24 Proof.

25 a $R_T=\frac{R_1R_2}{R_1+R_2}$ b Proof.

6 General binomial expansion

BEFORE YOU START

1 $2x^2$

2 $81-216x+216x^2-96x^3+16x^4$

3 $-1<x<5$

4 $\frac{1}{x-2}-\frac{2}{x}$

EXERCISE 6A

1 a i $1-2x+3x^2$; $|x|<1$

ii $1-3x+6x^2$; $|x|<1$

b i $1+\frac{x}{3}-\frac{x^2}{9}$; $|x|<1$

ii $1+\frac{x}{4}-\frac{3x^2}{32}$; $|x|<1$

c i $1-x-\frac{x^2}{2}$; $|x|<\frac{1}{2}$

ii $1-\frac{3x}{2}-\frac{9x^2}{8}$; $|x|<\frac{1}{3}$

d i $\frac{1}{4}+\frac{x}{16}+\frac{x^2}{64}$; $|x|<4$

ii $\frac{1}{5}+\frac{x}{25}+\frac{x^2}{125}$; $|x|<5$

2 $1+x+\frac{2x^2}{3}+\dots$

3 $\frac{1}{2}-\frac{x}{48}+\frac{x^2}{576}-\frac{7x^3}{41\,472}+\dots$

4 a $1-4x+12x^2+\dots$

b $|x|<\frac{1}{2}$

5 a Proof. b $|x| < 9$

c $3 + \frac{x}{6} - \frac{x^2}{216} + \dots$ d 3.162 0

6 a $1 - 2x - 2x^2 - 4x^3 + \dots$

b $|x| < \frac{1}{4}$

c i 9.797 96 ii 0.077 6

7 4

8 –540

9 $-\frac{1}{2}$

10 $a_0 = 1,\ a_1 = \frac{1}{2},\ a_2 = -\frac{1}{8}$; consistent

WORK IT OUT 6.1

Solution B is correct.

EXERCISE 6B

1 $x + 2x^2 + 3x^3 + \dots$

2 $x + x^2 - \frac{x^3}{2}$; $|x| < \frac{1}{2}$

3 $1 + 2x + 2x^2 + \dots$

4 $1 - 2x + 5x^2 + \dots$

5 a $\frac{1}{1+x} + \frac{1}{1+2x}$ b $2 - 3x + 5x^2 + \dots$

c $|x| < \frac{1}{2}$

6 a $\frac{1}{(x+1)^2} + \frac{2}{2-x} + \frac{2}{x+1}$ b $4 - \frac{7x}{2} + \frac{21x^2}{4} + \dots$

c $|x| < 1$

7 $1 - 3x + 7x^2 + \dots$

8 a $1 - \frac{7}{2}x + \dots$ b $|x| < \frac{1}{12}$ c 3.86

9 $\frac{8}{3}$

10 $\sqrt{(x-1)}$ is not real for values $|x| < 1$ so even some cunning rearrangement does not allow convergence.

MIXED PRACTICE 6

1 D

2 $1 + 4x + 12x^2 + 32x^3 + \dots$; $|x| < \frac{1}{2}$

3 $\frac{1}{2} - \frac{x}{16} + \frac{3x^2}{256} + \dots$; $|x| < 4$

4 a $\frac{1}{(1+x)^2}$ b $1 - 2x + 3x^2 - 4x^3 + \dots$

5 $1 - \frac{x^2}{2} - \frac{x^4}{8} + \dots$

6 B

7 B

8 a $2 + \frac{x}{12} - \frac{x^2}{288} + \dots$ b 20.08

9 a $\frac{1}{3-x} + \frac{1}{x+2} + \frac{2}{(x+2)^2}$

b $\frac{4}{3} - \frac{23}{36}x + \frac{29}{54}x^2 + \dots$

c $|x| < 2$

10 $\frac{1}{6} - \frac{x}{36} - \frac{11x^2}{216} + \dots$; $|x| < 2$

11 $a = 1, b = 3$ or $a = 3, b = 1$

12 a i $1 - \frac{1}{3}x + \frac{2}{9}x^2 + \dots$ ii $1 - \frac{x}{4} + \frac{x^2}{8} + \dots$

b $a = 4, b = -1, c = \frac{1}{2}$

13 a $1 + 4x - 4x^2 + \dots$ b $4 + 2x - \frac{1}{4}x^2 + \dots$

c $\frac{167}{36}$

14 a $1 + x + \frac{2x^3}{3} + \dots$ b 2.080 09

15 –270

16 $1 - x + x^3 + \dots$

17 a $A = 1, B = -1, C = 1$; $|x| < 1$

b Proof.

c $P = 1, Q = -1, R = 1$; $|x| > 1$

d 0.990 1

e 0.019 608

18 a $mc^2 + \frac{1}{2}mv^2 + \frac{3m}{8c^2}v^4 + \dots$

b i 0.003 73% ii 17.7%

c Proof.

Focus on ... Proof 1

Arithmetic series proof

$[a + (n-1)d] + [a + (n-2)d]$

$[a + (n-1)d] + \dots + [a + (n-1)d]$

n

$\frac{n}{2}$

Geometric series proof

Multiply through by r.

$= ar^n - a$

$S_n(r-1) = a(r^n - 1)$

1 $r - 1 = 0$ so can't divide by it on last line; $S_n = an$.

2 Yes. $S_n = \frac{a(1-(-1)^n)}{2}$

3 a $\frac{1+\sqrt{5}}{2}$ b $\frac{1}{2}$

Focus on ... Problem solving 1

1 45

2 n^2

3 $2n+5$

4 89

Focus on ... Modelling 1

1 a i 0.333 ii 8000

iii

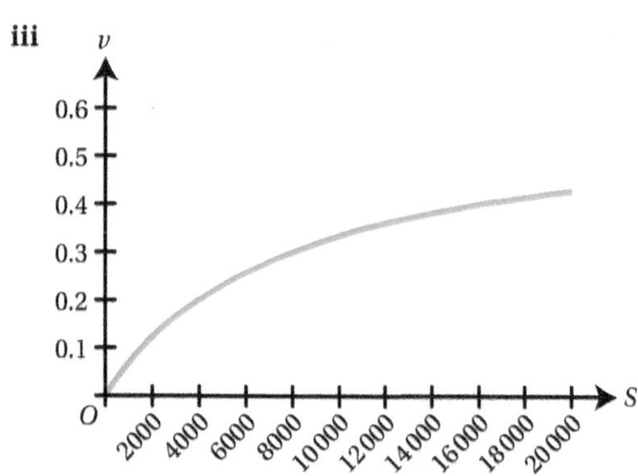

b The maximum possible rate (asymptote, can't be reached).

2 $b=2, c=1$

3 a Rational function.

b Exponential function.

Cross-topic review exercise 1

1 Proof.

2 4

3 $\frac{a(r^n-1)}{n(r-1)}$

4 $a=4, b=-1$ or $a=-4, b=3$

5 a $x \geqslant -3$; $f(x) \leqslant 5$

b i Decreasing so one-to-one.

ii $f^{-1}(x) = (5-x)^2 - 3; x \leqslant 5$

c $-3 \leqslant x \leqslant 22$

6

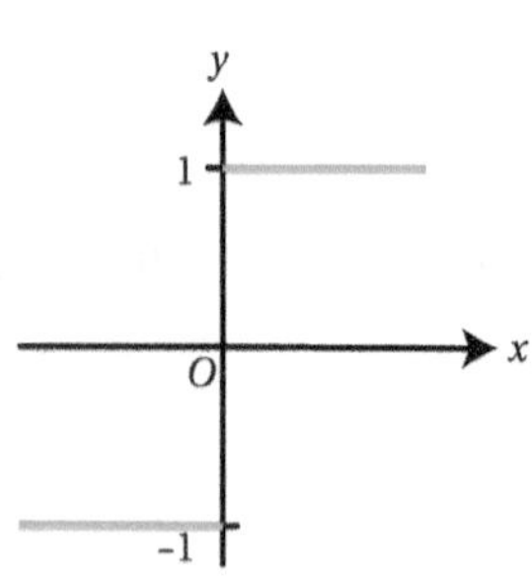

7 a

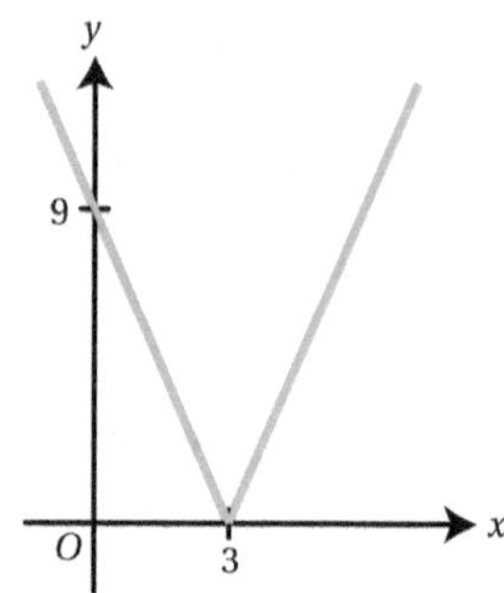

b $x=1$ or 5

c $x<1$ or $x>5$

8 a i 0

ii Proof.

b $\frac{3x}{(x+1)(3x+2)}$

9 a $f(x) \geqslant 7$

b Proof.

10 a $\frac{2047}{1024}$ b $-55 \ln 2$

11 $-0.618 < x < 1.62$

12 a $A=2,\ B=-5$

b

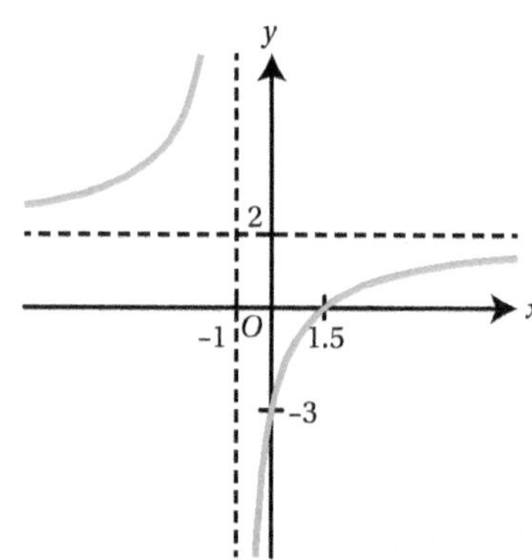

$y \neq 2$

c Domain: $x<-1$ or $x \geqslant 1.5$; range: $g \circ h(x) \geqslant 0$ and $g \circ h(x) \neq \sqrt{2}$.

13 $\frac{3}{64}$

14 a–b Proof.

c 18

d Underestimate, as the kitten will grow more slowly as it gets older.

15 a $A=2, B=-1$

b i $-\frac{1}{3}+\frac{29}{9}x-\frac{241}{27}x^2+\ldots$

ii Expansion is only valid for $|x|<\frac{1}{3}$ and 0.4 is not this interval.

16 e^8-1

17 a No b 6 c −7

18 a $(-y, x)$ b $y = f^{-1}(-x)$

19 a $\frac{f(x)}{(x-a)^2} = g(x) + \frac{mx+c}{(x-a)^2}$

b $2(x-a)\,g(x) + (x-a)^2\,g'(x) + m$

c Proof.

d $f(a) = 0, f'(a) = 0$

7 Radian measure

BEFORE YOU START

1 $\frac{\sqrt{3}}{2}$

2 $x = 30, 150, 210, 330$

3 $26.4°$

4 $y = 2\cos(x - 30°)$

5 $2 + 6x^2 + 18x^4$

WORK IT OUT 7.1

Solution 2 is correct.

EXERCISE 7A

1 a

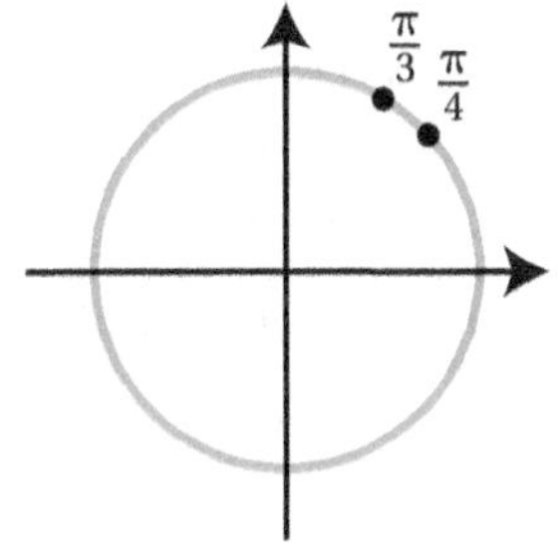

b

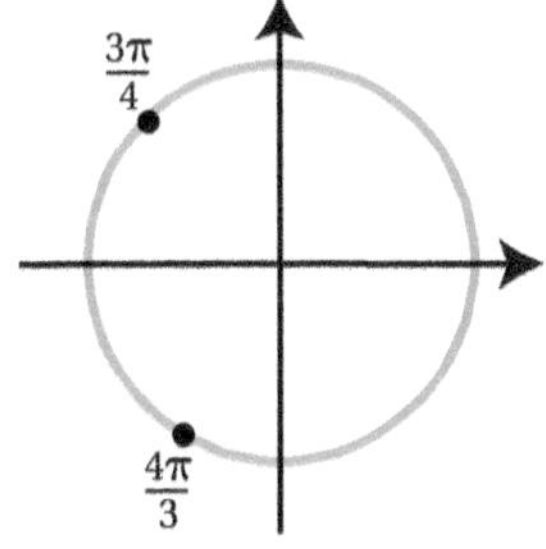

c

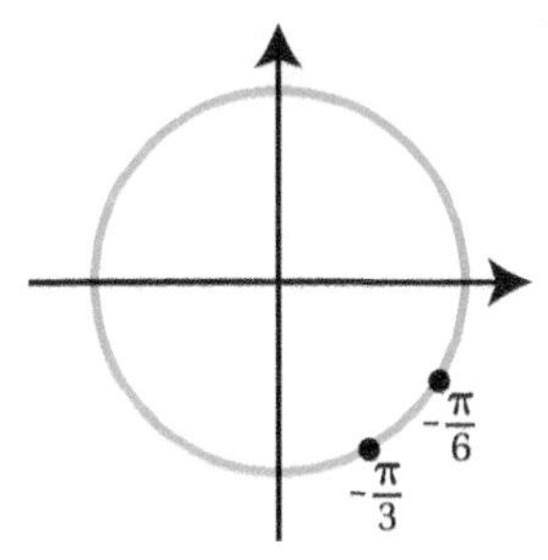

d

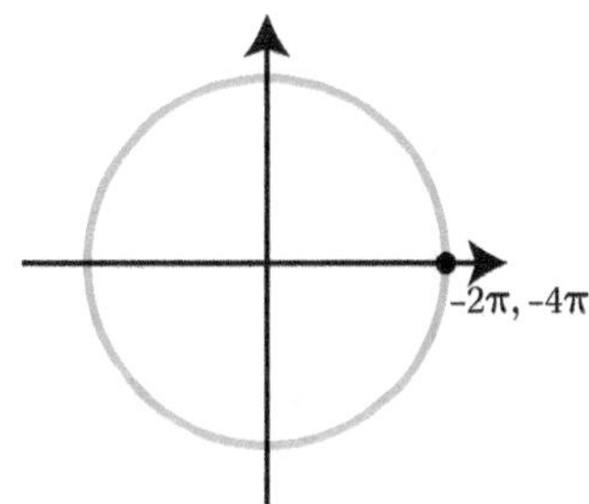

2 a i $\frac{3\pi}{4}$ ii $\frac{\pi}{4}$ b i $\frac{\pi}{2}$ ii $\frac{3\pi}{2}$

c i $\frac{2\pi}{3}$ ii $\frac{5\pi}{6}$ d i $\frac{5\pi}{18}$ ii $\frac{4\pi}{9}$

3 a i 5.585 ii 0.349

b i 4.712 ii 1.571

c i 1.134 ii 2.531

d i 1.745 ii 1.449

4 a i 60° ii 45° b i 150° ii 120°

c i 270° ii 300° d i 69.9° ii 265°

5 a i

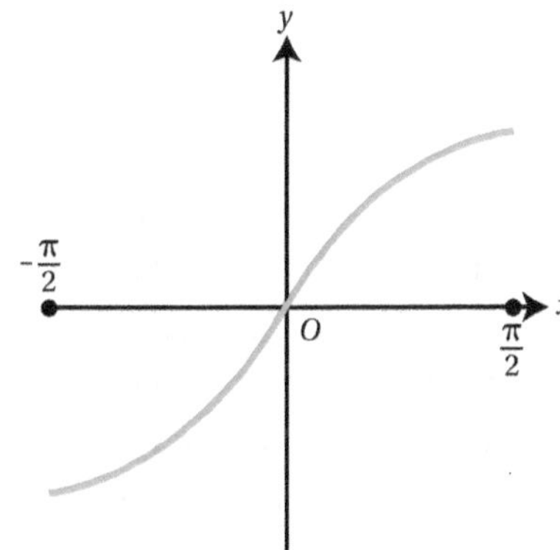

ii

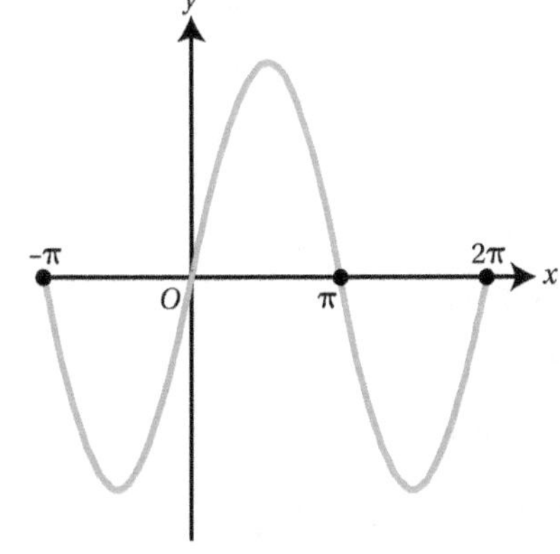

b **i**

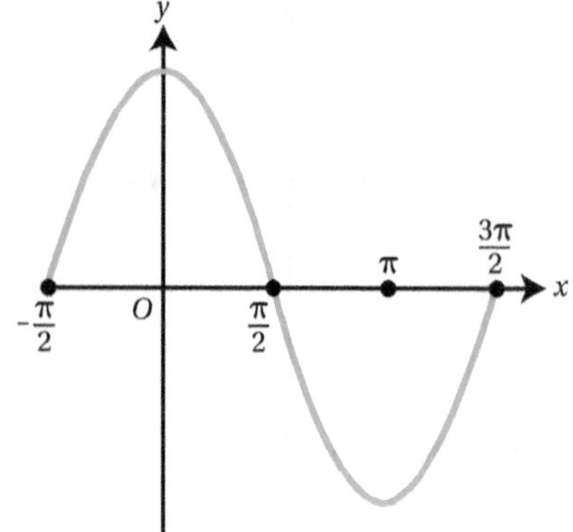

ii

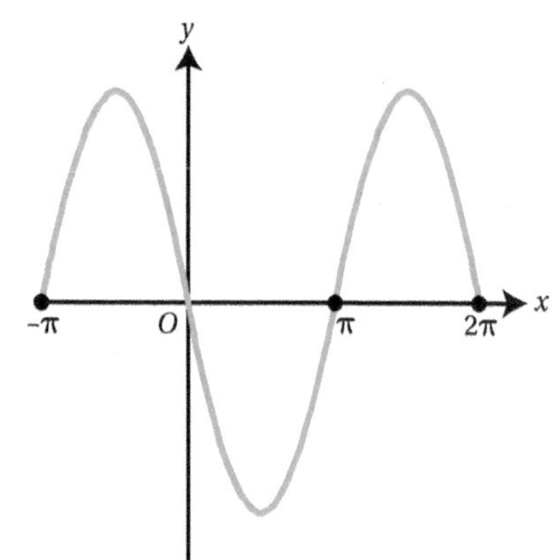

6 **a** 0.434 **b** 0.434
c −0.434 **d** −0.434

7 **a** −0.809 **b** 0.809
c 0.809 **d** −0.809

8 **a** 0.414 **b** −0.414
c 0.414 **d** −0.414

9 **a** **i** $-\frac{\sqrt{2}}{2}$ **ii** $-\frac{\sqrt{2}}{2}$
b **i** $-\frac{1}{2}$ **ii** $\frac{-\sqrt{3}}{2}$
c **i** −1 **ii** −1

10 **a** $\frac{3}{4}$ **b** $\frac{\sqrt{2}+\sqrt{3}}{2}$
c $\frac{1-\sqrt{3}}{2}$

11–12 Proof.

13 $-2\cos x$

14 $\sin x$

WORK IT OUT 7.2

Solution 1 is correct.

WORK IT OUT 7.3

Solution 3 is correct.

EXERCISE 7B

1 **a** **i** 0.927 **ii** 0.201
b **i** −1.25 **ii** −0.927

2 **a** **i** $\frac{\pi}{6}$ **ii** $\frac{\pi}{6}$
b **i** $-\frac{\pi}{3}$ **ii** $\frac{3\pi}{4}$
c **i** $-\frac{\pi}{2}$ **ii** $\frac{\pi}{4}$

3 **a** **i** $\frac{\pi}{3}$ **ii** $\frac{5\pi}{6}$
b **i** $\frac{\pi}{3}$ **ii** π
c **i** $\frac{\pi}{3}$ **ii** $-\frac{\pi}{4}$
d **i** $-\frac{\pi}{4}$ **ii** $-\frac{\pi}{6}$

4 **a** $\frac{\sqrt{3}}{2}$ **b** $\frac{\sqrt{3}}{4}$ **c** $\frac{3-\sqrt{3}}{9}$

5 **a** **i** $\frac{\pi}{6}, \frac{11\pi}{6}$ **ii** $\frac{\pi}{4}, \frac{7\pi}{4}$
b **i** $\frac{2\pi}{3}, \frac{4\pi}{3}$ **ii** $\frac{5\pi}{6}, \frac{7\pi}{6}$
c **i** $\frac{\pi}{4}, \frac{3\pi}{4}$ **ii** $\frac{\pi}{3}, \frac{2\pi}{3}$
d **i** $\frac{\pi}{6}, \frac{7\pi}{6}$ **ii** $\frac{3\pi}{4}, \frac{7\pi}{4}$

6 **a** **i** 0.644, 5.64, 6.93, 11.9
ii 0.841, 5.44, 7.12, 11.7
b **i** −2.21, − 0.927, 4.07, 5.36
ii −2.78, − 0.36, 3.50, 5.93
c **i** −0.588, 2.55 **ii** −1.25, 1.89
d **i** $0, 2\pi, 4\pi$ **ii** $\frac{\pi}{2}, \frac{3\pi}{2}, \frac{5\pi}{2}, \frac{7\pi}{2}$

7 $-\frac{\pi}{6}, -\frac{5\pi}{6}$

8

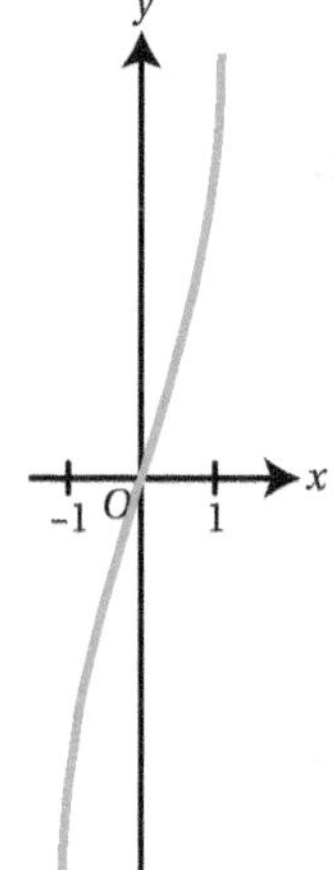

9 $\frac{\pi}{12}, \frac{\pi}{3}, \frac{7\pi}{12}, \frac{5\pi}{6}$

10 **a** Proof.

b 1.01, 2.13

11 $0, \frac{\pi}{3}, \pi$

12 Proof.

13 $\pm\sqrt{\frac{\pi}{6}}, \pm\sqrt{\frac{5\pi}{6}}, \pm\sqrt{\frac{13\pi}{6}}, \pm\sqrt{\frac{17\pi}{6}}$

14 **a** Proof.

b $\arcsin x = \frac{\pi}{2} - \arccos x$

c $x = 1$

EXERCISE 7C

1 **a** Amplitude: 3; period: $\frac{\pi}{2}$

b Amplitude: 1; period: 4π

c Amplitude: 1; period: $\frac{2\pi}{3}$

d Amplitude: 2; period: 2

2 **a** **i**

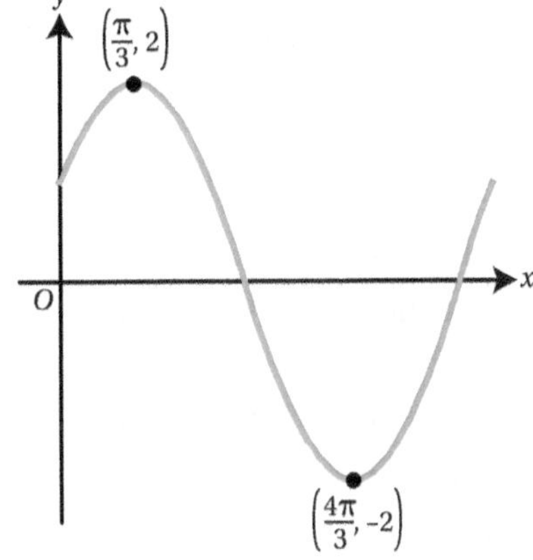

ii

b **i**

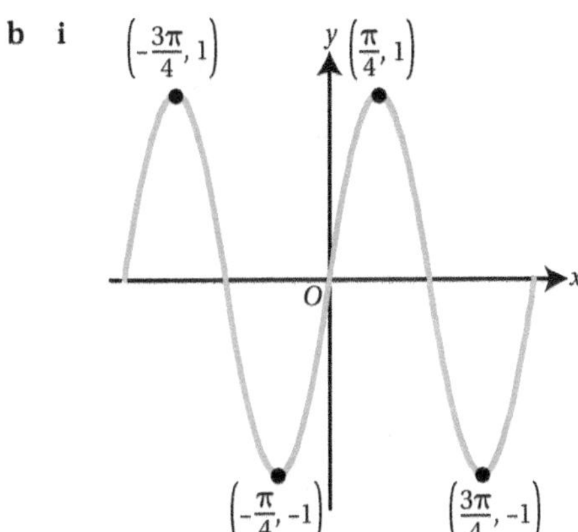

ii

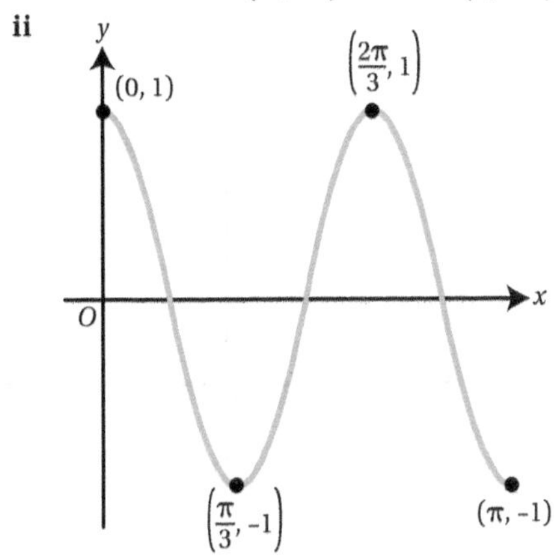

c **i**

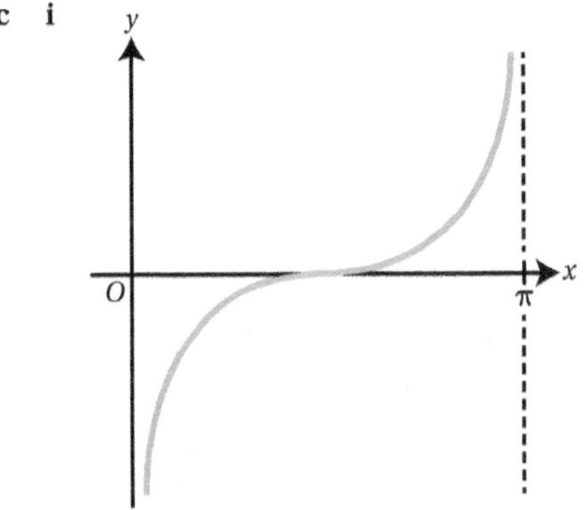

ii

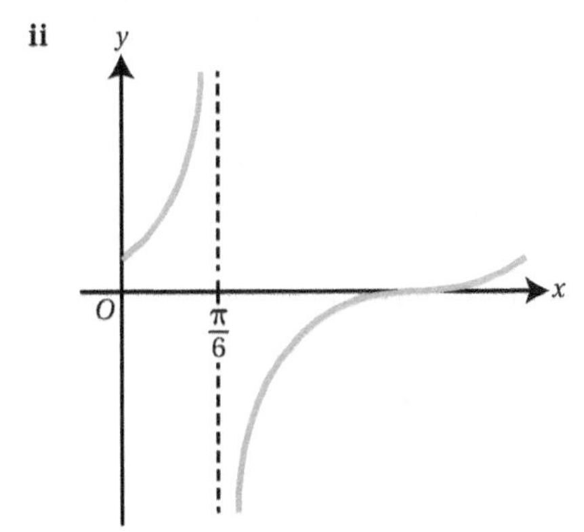

d **i**

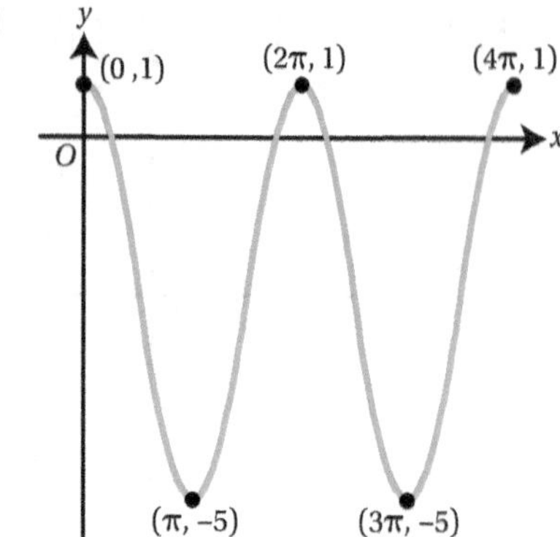

ii

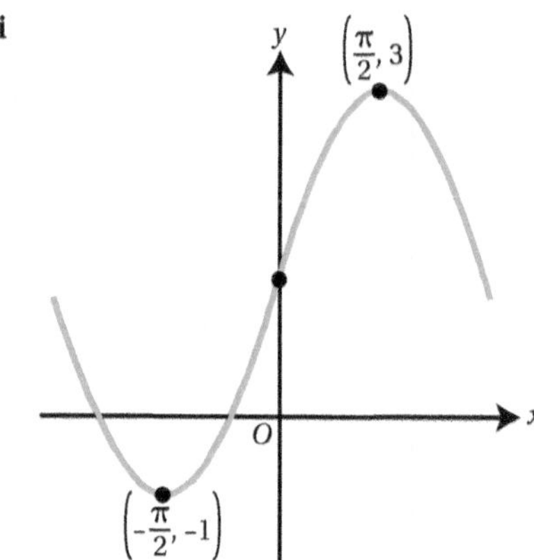

3 **a** 9 m; 23 m **b** 3 a.m.; 3 p.m.

4 **a** 4 **b** 2 seconds

5 $p = 5, q = 2$

6 $a = 2, b = 20°$

7 **a**

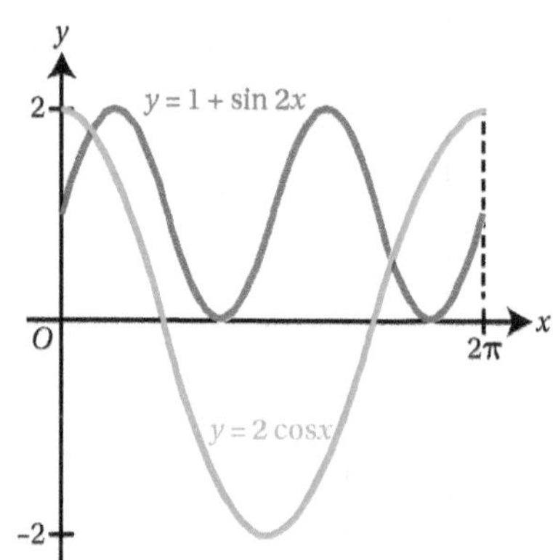

b Two intersections.

c 8 solutions.

8 $a = 1.5, b = \frac{\pi}{6}, m = 4.5$

9 **a**

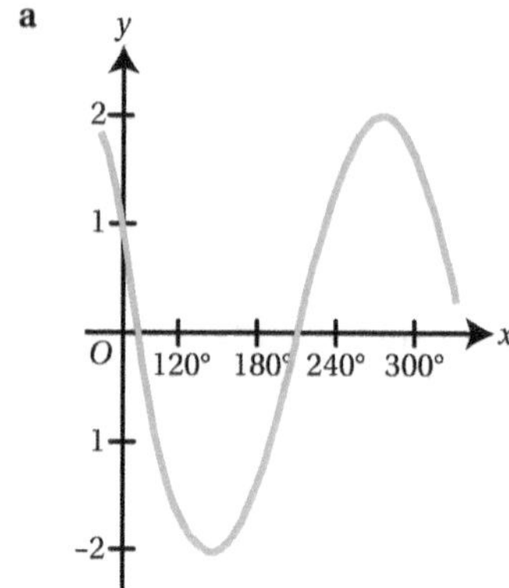

b maximum: (300°, 2), minimum: (120°, −2)

c maximum: (300°, 1), minimum: (120°, −3)

10 **a** $a = 5, k = \frac{\pi}{5}$ **b** 7.5 s

11 **a** 110 cm; 130 cm

b $\frac{\pi}{5}$ s

c $\frac{\pi}{10}$ s

12 **a** $h = 14 - 12\cos\theta$

b $\theta = \frac{\pi t}{2}$

c $h = 14 - 12\cos\left(\frac{\pi t}{2}\right)$; 1 minute 20 seconds

EXERCISE 7D

1 **a** 7.8 cm **b** 1.8 cm

2 **a** 82.2 cm **b** 6.84 cm

3 **a** 16.25 cm^2 **b** 0.072 cm^2

4 **a** 463 cm^2 **b** 4.79 cm^2

5 25 cm

6 **a** 0.9375 **b** 53.7°

7 2.53 radians

8 7.5 cm

9 0.8 radians

10 167°

11 6.69 cm

12 9.49 cm

13 48.4 cm^2

14 15.7 cm

15 31.6 cm

16 11.3 cm

17 $\left(10 + \frac{25\pi}{6}\right)$ cm

18 5 cm

19 5.14 cm^2

20 2 cm or 1.5 cm

21 2.54 radians

22 1.2π radians

EXERCISE 7E

1 a i 0.935 cm ii 3.39 cm

b i 21.7 cm ii 15.8 cm

2 a i 1.89 cm ii 6.99 cm

b i 52.5 cm ii 37.1 cm

3 a i 0.06 cm^2 ii 1.21 cm^2

b i 149 cm^2 ii 70.1 cm^2

4 Proof.

5 a Proof.

b 70.1° c 3.67 cm^2

EXERCISE 7F

1 a i 0.2 ii −0.14

b i 0.955 ii 0.98

c i 0.12 ii −0.2

2 a i 2θ ii $-3x$

b i $1-4.5x^2$ ii $1-12.5\theta^2$

c i x^2 ii $\frac{\theta^2}{2}$

3 a i $1-\frac{13}{2}\theta^2$ ii $1-\frac{65}{8}\theta^2$

b i $1+2\theta-2\theta^2$ ii $1-2\theta-\frac{1}{2}\theta^2$

c i $2-3\theta-2\theta^2$ ii $-3+8\theta+3\theta^2$

4 a $1+\theta-6\theta^2$ b 0.96

5 a $1-\frac{1}{2}\theta^2$ b $6\sqrt{2-\sqrt{3}}$

6 a $1+6x+12x^2$

b i 0.0838% ii 41.6%

7 $1+3\theta+\frac{17}{2}\theta^2+\ldots$

8 $1-3\theta+7\theta^2+\ldots$

9 $\sin\theta\approx\frac{\pi\theta}{180}$; $\cos\theta\approx 1-\frac{\pi^2\theta^2}{64\,800}$

10 a $1+\frac{1}{2}\theta-\frac{1}{8}\theta^2+\ldots$

b 0.338

11 $\frac{1}{3}$

12 $2+\frac{1}{2}\theta+\frac{\theta^2}{8}+\ldots$

13 a sin x: 0.244; cos x: 0.662; tan x: 0.173

b i Proof.

b ii Because for small (positive or negative) θ, $\cos\theta>0$.

MIXED PRACTICE 7

1 C

2 a 1.4 m b 2.09 m

3 $\pm 2.41, \pm 0.730$

4 1.564

5 a $\frac{\pi}{3}$

b 28.9 cm^2

c 23.8 cm

6 80 cm^2

7 a 10.2 cm^2 b 18.8 cm

8 a 9 cm^2

b i 3 cm ii $k=5$

9 B

10 $a=5, b=\frac{\pi}{4}$

11 a 3π m b 5.05 m c 1.50 m

12 a 78.5 s

b 377 m

c 4.8 m s^{-1}

13 a π

b $\left(\frac{\pi}{3},0\right),\left(\frac{5\pi}{6},0\right),\left(\frac{4\pi}{3},0\right)$ and $\left(\frac{11\pi}{6},0\right)$

c

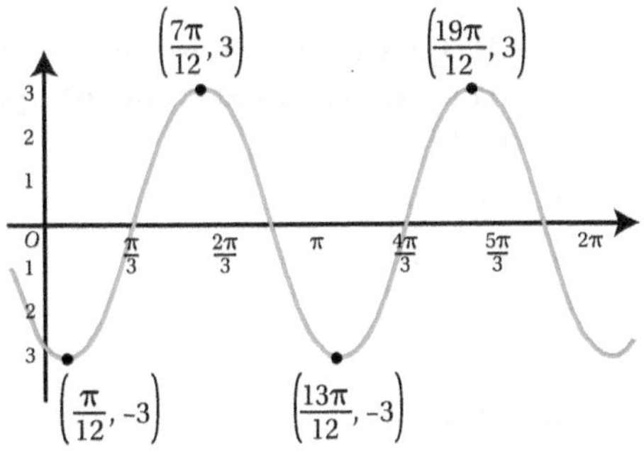

$\left(\frac{7\pi}{12},3\right)$ $\left(\frac{19\pi}{12},3\right)$

$\left(\frac{\pi}{12},-3\right)$ $\left(\frac{13\pi}{12},-3\right)$

14 $1-\theta+\frac{1}{2}\theta^2+\ldots$

15 a $ASBT$ is a square.

b $r\sqrt{2}$

c $\frac{\pi r^2}{4}$

d $\left(\frac{\pi}{2}-1\right)r^2$

16 5.48 cm^2

17 a

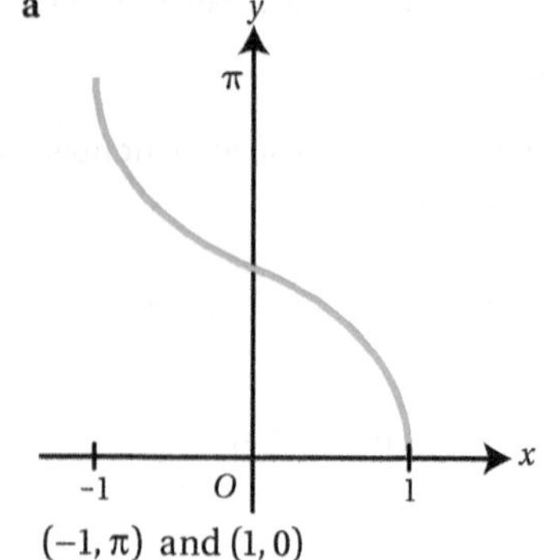

$(-1, \pi)$ and $(1, 0)$

b

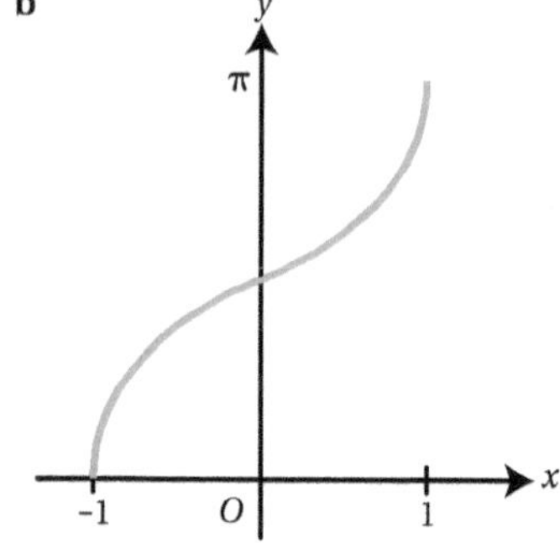

$(-1, 0)$ and $(1, \pi)$

18 $-\frac{7\pi}{24}, -\frac{\pi}{24}, \frac{17\pi}{24}, \frac{23\pi}{24}$

19 $\frac{\pi}{2}, \frac{3\pi}{2}, \frac{2\pi}{3}, \frac{4\pi}{3}$

20 a Proof.

b 0.361, 2.78, 3.50, 5.92

21 a $2 + \frac{3}{4}x - \frac{9}{64}x^2 + \ldots$

b 0.840

22 a $\frac{\pi}{2}$, right angle between a tangent and a radius.

b ABO_2P is a rectangle, because there are right angles at A and B, and AB is parallel to PO_2.

c 24.5 cm

d 1.369 radians

e 85.6 cm

23 a x

b Proof.

c $\frac{\sqrt{2}}{2}$

8 Further trigonometry

BEFORE YOU START

1 a $\frac{2\sqrt{2}}{3}$ b $2\sqrt{2}$

2 $\left(\frac{\pi}{4}, -2\right)$

3 a 21.1, 81.1 b $\frac{-\pi}{2}, \frac{\pi}{6}, \frac{5\pi}{6}$

WORK IT OUT 8.1

Solution 3 is correct.

EXERCISE 8A

1 a $\frac{1}{2}\sin x + \frac{\sqrt{3}}{2}\cos x$

b $\frac{\sqrt{2}}{2}\sin x - \frac{\sqrt{2}}{2}\cos x$

c $-\frac{\sqrt{2}}{2}\sin x - \frac{\sqrt{2}}{2}\cos x$

d $-\sin x$

2 a $\frac{\sqrt{3}-1}{2\sqrt{2}}$ b $\frac{1+\sqrt{3}}{2\sqrt{2}}$

c $\frac{\sqrt{3}+1}{1-\sqrt{3}}$ d $\frac{\sqrt{3}-1}{2\sqrt{2}}$

3 a $\frac{56}{65}$ b $\frac{8+3\sqrt{5}}{15}$

4 a Proof.

b $\sqrt{2}\cos x$

5 a $\frac{\tan\theta - 1}{\tan\theta + 1}$

b $-\frac{1}{2}, -\frac{1}{3}$

c 2.68, 2.82

6 a $\sin\left(x + \frac{\pi}{4}\right)$; 1; $x = \frac{\pi}{4}$

b $2\cos(x - 25°)$; 2; $x = 25°$

7 $\frac{\sqrt{3}}{2} + \frac{1}{2}x - \frac{\sqrt{3}}{4}x^2$

8 a Proof.

b $\frac{\pi}{3}$

9 a Proof.

b $\frac{\pi}{4}, \frac{3\pi}{4}, \frac{5\pi}{4}, \frac{7\pi}{4}$

EXERCISE 8B

1 a i $-\frac{7}{8}$ ii $\frac{1}{9}$

b i $\frac{4\sqrt{2}}{9}$ ii $\frac{24}{25}$

c i $-\frac{3\sqrt{7}}{8}$ ii $-\frac{4\sqrt{5}}{9}$

2 a i $\frac{2-\sqrt{2}}{4}$ ii $\frac{2+\sqrt{2}}{4}$

b i $\frac{2-\sqrt{3}}{4}$ ii $\frac{2+\sqrt{3}}{4}$

3 a $\cos(6A)$ b $2\sin 10x$

c $3\cos b$ d $\frac{5}{2}\sin\left(\frac{2x}{3}\right)$

4 a $0, \pi, 2\pi$ b $90°$

c $-\frac{\pi}{2}, \frac{\pi}{2}, 0.305, 2.84$ d $0°, 180°, 360°$

5 a-d Proof.

6 0.955, −0.955, 2.19, −2.19

7–8 Proof.

9 a $\pm\frac{\sqrt{3}}{2}$ b $\frac{\pi}{6}, \frac{5\pi}{6}$

10 a $4\cos^3 A - 3\cos A$ b $\frac{3\tan A - \tan^3 A}{1 - 3\tan^2 A}$

11 a $8\cos^4\theta - 8\cos^2\theta + 1$

b $8\sin^4\theta - 8\sin^2\theta + 1$

12 a i-ii Proof. b $\frac{1-\cos x}{1+\cos x}$

13 $\frac{2a-b}{4a}$

14–15 Proof.

EXERCISE 8C

1 i $2\sqrt{13}\sin(x + 0.983)$

ii $\sqrt{10}\sin(x + 0.322)$

2 i $2\sqrt{2}\sin(x - 45°)$

ii $2\sin(\theta - 60°)$

3 i $2\sqrt{2}\cos\left(x + \frac{\pi}{6}\right)$

ii $5\sqrt{2}\cos\left(x + \frac{\pi}{4}\right)$

4 i $9.22\cos(x - 40.6°)$ ii $13\cos(x - 22.6°)$

5 a $13\sin(x + 1.18)$

b Vertical stretch with scale factor 13; translation 1.18 units to the left.

6 a $\sqrt{58}\sin(x - 1.17)$ b $f(x) \in [-\sqrt{58}, \sqrt{58}]$

7 a $\sqrt{41}\cos(x + 0.896)$ b 0.675

8 a $2\cos\left(x - \frac{\pi}{3}\right)$

b Minimum: $\left(\frac{4\pi}{3}, -2\right)$; maximum: $\left(\frac{\pi}{3}, 2\right)$

9 1.57, 2.50

10 $-\pi, -\frac{3\pi}{4}, 0, \frac{\pi}{4}, \pi$

EXERCISE 8D

1 a i 2.760 ii 1.480

b i −2.670 ii 1.212

c i 1.051 ii 0.5774

2 a i $\frac{2\sqrt{3}}{3}$ ii $\sqrt{2}$

b i $-\sqrt{2}$ ii $-\frac{2\sqrt{3}}{3}$

c i −1 ii $\frac{\sqrt{3}}{3}$

d i −1 ii 0

3 $\operatorname{cosec} A = \frac{5}{4}$, $\sec B = \frac{3}{\sqrt{5}}$

4 a i 1.05, 5.24 ii 1.23, 5.05

b i 0.730, 2.41 ii 0.379, 2.76

c i 0.197, 3.34 ii 1.11, 4.25

d i 0.615, 2.53, 3.76, 5.67

ii 0.126, 1.44, 3.27, 4.59

5 a i $-\frac{\pi}{6}, -\frac{5\pi}{6}$ ii $-\frac{\pi}{2}$

b i $\frac{\pi}{6}, -\frac{5\pi}{6}$ ii $\frac{\pi}{4}, -\frac{3\pi}{4}$

c i 0 ii $\frac{5\pi}{6}, -\frac{5\pi}{6}$

d i $\frac{\pi}{2}, -\frac{\pi}{2}$ ii $-\frac{\pi}{4}, \frac{3\pi}{4}$

6 a i $\frac{5}{3}$ ii $\frac{\sqrt{29}}{5}$

b i $2\sqrt{6}$ ii $2\sqrt{2}$

c i $\frac{1}{\sqrt{10}}$ ii $\frac{2}{\sqrt{5}}$

d i $\pm\frac{3}{\sqrt{7}}$ ii $\pm\frac{2}{\sqrt{3}}$

7 Proof.

8 $60°, 120°, 240°, 300°$

9 $\pm\frac{\pi}{2}, 0.340, 2.80$

10–12 Proof.

13 a Proof.

b $\frac{\pi}{4}, \frac{5\pi}{4}, 1.11, 4.25$

14 a Proof.

b $\pm\frac{\pi}{2}, -2.98, 0.165$

15 a Proof.

b $16.9°, 107°$

16 $1 + \frac{9}{2}\theta^2 + \ldots$

17 $-\frac{11\pi}{12}, -\frac{7\pi}{12}, -\frac{\pi}{4}, \frac{\pi}{12}, \frac{5\pi}{12}, \frac{3\pi}{4}$

18 $\arccos\left(\frac{1}{x}\right)$

MIXED PRACTICE 8

1 C

2 a Proof.

b $\frac{\pi}{6}, \frac{5\pi}{6}, \frac{3\pi}{2}$

3 **a** $\frac{1}{2}\cos x-\frac{\sqrt{3}}{2}\sin x$

b $-2\pi,-\pi,0,\pi,2\pi$

4 **a** $AB=2r\sin\theta$; $BC=2r\cos\theta$

b $2r^2\sin\theta\cos\theta$

c $r^2\sin\theta\cos\theta$

d $\frac{1}{2}$

5 **a** Proof.

b -1

c $1+\sqrt{2}$

6 **a** $a=1.2, p=\frac{2\pi}{3}$

b Amplitude = 0.9 m, period = 3 s

c $\frac{3}{2}\sin\left(\frac{2\pi}{3}x+0.927\right)$

d Amplitude = $\frac{3}{2}$ m, period = 3 s

e 1.06 m

f 0.0579 m, 0.557 m

7 **a** $\sqrt{10}\sin(x-71.6°)$

b 32°, 291°

8 B

9 **a** Proof.

b $\pi, \frac{2\pi}{3}, \frac{4\pi}{3}$

10 **a** $(t+1)(t^2-4t+1)$

b Proof.

c 1

d $\tan 15°=2-\sqrt{3}$, $\tan 75°=2+\sqrt{3}$

11 1.23, 5.05

12 **a** Proof.

b 0.17, 0.43, 1.74, 2.00

13 **a** $2\sqrt{5}\sin\left(2x+\frac{\pi}{6}\right)$

b **i** $2+\frac{4}{5}\sqrt{5}$ **ii** $x=\frac{2\pi}{3}$

14 **a** x

b Proof.

c $2x\sqrt{1-x^2}$

15 **a** **i** $x^2-2\sqrt{10}x+10$

ii Proof.

iii $x=\sqrt{10}$

b–c Proof.

d $\sqrt{10}$

e $d=\sqrt{10}, \theta=0.443$ radians

9 Calculus of exponential and trigonometric functions

BEFORE YOU START

1 $\ln 3+4\ln x$

2 $\frac{dy}{dx}=6x+\frac{4}{3x^3}$

3 $\frac{3}{2}$

4 $\frac{1}{2}$

EXERCISE 9A

1 **a** **i** $3e^x$ **ii** $\frac{2e^x}{5}$

b **i** $\frac{-2}{x}$ **ii** $\frac{1}{3x}$

c **i** $\frac{1}{5x}-3+4e^x$ **ii** $-\frac{e^x}{2}+\frac{3}{x}$

d **i** $3\cos x$ **ii** $-2\sin x$

e **i** $2+5\sin x$ **ii** $\sec^2 x$

f **i** $\frac{\cos x-2\sin x}{5}$ **ii** $\frac{1}{2}\sec^2 x-\frac{1}{3}\cos x$

2 **a** **i** $\frac{3}{x}$ **ii** $\frac{10}{x}$

b **i** $\frac{3}{x}$ **ii** $\frac{1}{x}$

c **i** e^3e^x **ii** $\frac{e^x}{e^3}$

d **i** $2x$ **ii** $3e^2x^2$

e **i** $3\sec^2 x$ **ii** $4\sec^2 x$

f **i** e^x **ii** $-\frac{1}{2}e^x$

3 $2-\frac{7}{\ln 4}$

4 $3-\frac{1}{2\ln 3}$

5 $x=\ln 3$

6 $x=3$

7 π

8 $\frac{11}{6}-\frac{\pi^2}{12}$

9 $y=3x-\ln 4+2$

10 $y=2x+2.26$

11 $x=\frac{\pi}{4}, \frac{5\pi}{4}$

12 Tangent: $4x-y+1-\pi=0$;
normal: $4x+16y-\pi-16=0$

13 $x=\frac{\pi}{3}, \frac{2\pi}{3}, \frac{4\pi}{3}, \frac{5\pi}{3}$

14 (0.245, 4.12) local maximum; (3.39, −4.12) local minimum.

15 Proof.

16 $y \geqslant 6 - 4\ln 4$

17 a (4, ln 4 − 2) local maximum.

b (ln 2.5, 5 − 5 ln 2.5) local minimum.

18 a 40 million litres.

b $t = \frac{\pi}{2}$ (1.6 days), $\frac{3\pi}{2}$ (4.7 days)

WORK IT OUT 9.1

Statement 4 is correct.

WORK IT OUT 9.2

Solution 2 is the correct method, but the value of the shaded area is ∞.

EXERCISE 9B

1 a i $5e^x + c$ ii $9e^x + c$

b i $\frac{2e^x}{5} + c$ ii $\frac{7e^x}{11} + c$

c i $\frac{e^x}{2} + \frac{3x^2}{4} + c$

ii $\frac{e^x}{5} + \frac{x^4}{20} + c$

d i $3\sin x + c$ ii $-4\cos x + c$

e i $-\frac{1}{2}\cos x - \sin x + c$

ii $\frac{2}{3}\sin x + \frac{1}{3}\cos x + c$

f i $\frac{2}{3}x^{\frac{3}{2}} - \cos x + c$ ii $\sin x - 2\sqrt{x} + c$

2 a i $2\ln x + c$ ii $3\ln x + c$

b i $\frac{1}{2}\ln x + c$ ii $\frac{1}{3}\ln x + c$

c i $\frac{5}{2}\ln x + c$ ii $\frac{2}{3}\ln x + c$

d i $\frac{x^2}{2} - \ln x + c$ ii $\frac{x^3}{3} + 5\ln x + c$

e i $3\ln x - \frac{2}{x} + c$ ii $-\frac{3}{x} - 5\ln x + c$

f i $2\ln x + 6\sqrt{x} + c$ ii $\frac{2}{3}x^{\frac{3}{2}} - 4\ln x + c$

3 a i $3(e^2 - 1)$ ii $2(e^3 - e^1)$

b i 2 ii 8

c i $4 - 3e + 2\ln 2$ ii $4\ln 3 - 13 + 3e$

d i $1.5\ln 3$ ii $\frac{4}{3}\ln\left(\frac{5}{2}\right)$

e i 2 ii $\frac{9}{2}$

f i 1 ii −2

g i 0 ii 0

4 $\frac{2}{3}\ln 3$

5 $\frac{3}{2}$

6 $\frac{1}{2} + \sqrt{3}$

7 15

8 a $-2\ln 3$ b $2\ln 3$

9 $y = \sin x - \cos x$

10 $-\frac{1}{2}\sin x + c$

11 a $f(x) = \frac{1}{2}\ln x + c$

b $y = \frac{1}{2}\ln x - \frac{1}{2}\ln 2 + 7$

12 0.838

13 a Proof.

b $\frac{15}{2} - 4\ln\left(\frac{8}{3}\right)$

14 Proof.

15 $y = \ln\left|\frac{e^5}{x}\right|$

MIXED PRACTICE 9

1 D

2 $y = e^{\frac{\pi}{2}}x - \frac{\pi}{2}e^{\frac{\pi}{2}} + e^{\frac{\pi}{2}} + 2$

3 $f(x) = \frac{1}{2} - \cos x$

4 i $2 - \frac{7}{\ln 4}$ ii $3 - \frac{1}{2\ln 3}$

5 $e^{\pi} + \pi + 1$

6 $\frac{21}{2} + 10\ln\left(\frac{2}{5}\right)$

7 $\frac{dy}{dx} = 3e^{3x} + \frac{1}{x}$

8 i $\frac{dy}{dx} = 2e^{2x} - 10e^x + 12$

ii $\frac{d^2y}{dx^2} = 4e^{2x} - 10e^x$

9 C

10 $\ln|x| + \frac{2}{5}x^{\frac{5}{2}} + c$

11 (0.644, 5) local maximum; (3.79, −5) local minimum.

12 $\frac{\pi}{6}, -\frac{5\pi}{6}$ local minima; $-\frac{\pi}{6}, \frac{5\pi}{6}$ local maxima.

13 $x + 6y = 36 + \ln 2$

14 a i 11 000 ii $t = 9.55$ hours

b i $\frac{dP}{dt} = e^t - 3$ ii 8.70 hours

c i $\frac{d^2P}{dt^2} = e^t$ ii 9704

15 $\sqrt{3}-1$

16 $\sin x - \cos x + c$

10 Further differentiation

BEFORE YOU START

1 a $6x^2 - \frac{3}{2\sqrt{x}}$

b $\frac{5}{x} - \frac{1}{x^4}$

c $5e^x$

d $4\cos x + 3\sin x + 2\sec^2 x$

2 a $y = -x + 2; y = x$

b $(2, 2 - \ln 4)$

3 a 3 b $\frac{3}{4}\tan x$

4 a $\frac{\sin x}{\cos^2 x}$ b $\frac{1}{\cos x}$

5 a $\frac{x^2 + x - 1}{-3x - 1}$ b $\frac{x}{(x-1)^{\frac{3}{2}}}$

6 a $y = \frac{1}{2}(1 + \ln x)$ b $y = -\frac{x}{x+1}$

EXERCISE 10A

1 a i $15(3x+4)^4$ ii $35(5x+4)^6$

b i $4(5-x)^{-5}$ ii $7(1-x)^{-8}$

c i $\frac{3}{2\sqrt{3x-2}}$ ii $\frac{1}{2\sqrt{x+1}}$

d i $\frac{1}{(3-x)^2}$ ii $-\frac{4}{(2x+3)^3}$

e i $5e^{5x-3}$ ii $10e^{10x+1}$

f i $-2e^{1-2x}$ ii $-3e^{4-3x}$

g i $4\cos 4x$ ii $\pi\cos\pi x$

h i $-2\pi\sin 2\pi x$

ii $-3\sin 3x$

i i $5\sec^2 5x$ ii $\frac{\pi}{4}\sec^2\left(\frac{\pi}{4}x\right)$

j i $4\sin(1-4x)$

ii $-\cos(2-x)$

k i $4\sec 4x\tan 4x$

ii $2\sec(2x+1)\tan(2x+1)$

l i $-3\text{cosec}^2(3x)$ ii $-5\text{cosec}(5x)\cot(5x)$

m i $\frac{5}{5x+2}$ ii $\frac{1}{x-4}$

n i $-\frac{1}{5-x}$ ii $-\frac{2}{3-2x}$

2 a i $7(2x-3)(x^2-3x+1)^6$

ii $15x^2(x^3+1)^4$

b i $(2x-2)e^{x^2-2x}$ ii $-3x^2e^{4-x^3}$

c i $-6e^x(2e^x+1)^{-4}$ ii $20e^x(2-5e^x)^{-5}$

d i $6x\cos(3x^2+1)$

ii $-(2x+2)\sin(x^2+2x)$

e i $-3\sin x\cos^2 x$

ii $4\cos x\sin^3 x$

f i $\frac{2-15x^2}{2x-5x^3}$ ii $\frac{8x}{4x^2-1}$

g i $\frac{16}{x}(4\ln x - 1)^3$ ii $-\frac{5}{x}(\ln x + 3)^{-6}$

h i $\frac{3x}{\sqrt{3x^2+1}}$ ii $-\frac{2x}{\sqrt{5-2x^2}}$

3 a i $6\sec^2(3x)\tan(3x)$

ii $4\tan(2x)\sec^2(2x)$

b i $6\sin(3x)\cos(3x)e^{\sin^2(3x)}$

ii $\frac{2\ln(2x)}{x}e^{(\ln 2x)^2}$

c i $-16\sin(2x)\cos(2x)(1-2\sin^2(2x))$

ii $-24\sin 3x(4\cos 3x+1)$

d i $\frac{6\sin 2x}{1-3\cos 2x}$

ii $\frac{5\sin 5x}{2-\cos 5x}$

4 $y = 66x - 11$

5 $y = \frac{27\sqrt{2}}{8}x - \frac{77}{12}$

6 7

7 $(0, -216), (\sqrt{2}, 0), (-\sqrt{2}, 0)$

8 $(0, -1), (1, 0)$

9 a $-\sqrt{3}\text{ m s}^{-1}$

b $\frac{1}{3}\text{ m s}^{-2}$

10 $\left(\frac{\pi}{2}, e\right), \left(\frac{3\pi}{2}, e^{-1}\right)$

11 $\left(6, -\frac{1}{9}\right)$

12 495

13 a $-2\text{cosec}^2 x\cot x$

b $-\frac{\pi}{4}, \frac{3\pi}{4}$

14 a Left post.

b Proof.

c $\frac{3\sqrt[3]{2}}{2}$

15 a $0, \frac{\pi}{3}, \pi, \frac{5\pi}{3}, 2\pi$

b $(0.568, 0.369), (2.21, -1.76), (4.08, 1.76), (5.72, -0.369)$

c

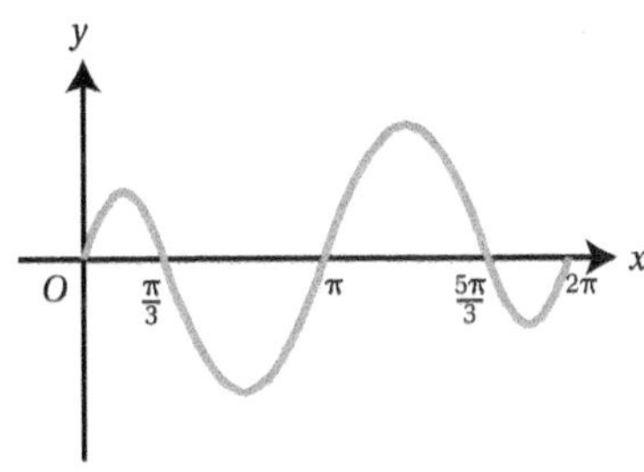

WORK IT OUT 10.1

Solution 2 is correct.

EXERCISE 10B

1 **a** **i** $2x\cos x - x^2\sin x$

ii $-x^{-2}\sin x + x^{-1}\cos x$

b **i** $-2x^{-3}\ln x + x^{-3}$ **ii** $\ln x + 1$

c **i** $3x^2\sqrt{2x+1} + x^3(2x+1)^{-\frac{1}{2}}$

ii $-x^{-2}\sqrt{4x} + 2x^{-1}(4x)^{-\frac{1}{2}}$

d **i** $2\,e^{2x}\tan x + e^{2x}\sec^2 x$

ii $e^{x+1}\sec 3x + 3\,e^{x+1}\sec 3x\tan 3x$

2 **a** **i** $3(x+1)^3(x-2)^4(3x-1)$

ii $(x-3)^6(x+5)^3(11x+23)$

b **i** $(2x-1)^3(1-3x)^2(-42x+17)$

ii $(1-x)^4(4x+1)(-28x+3)$

3 $e^{2x}(6x^2+4x+3)$

4 $(9x^2+12x+2)\,e^{3x}$

5 $x = -\frac{1}{2}, 2$

6 $x = -\frac{1}{3}, 3, \frac{7}{4}$

7 $e^x(1+x)\cos(x\,e^x)$

8 **a** $\ln x + 1$

b $x\ln x - x + c$

9 $\left(\frac{3\pi}{4}, -\frac{\sqrt{2}}{2}e^{-\frac{3\pi}{4}}\right)$

10 $a = 4, b = 5$

11 **a** $y = e^{x\ln x}$

b $(1+\ln x)x^x$

c $(e^{-1}, e^{-e^{-1}})$

12 **a** $\dfrac{pb+qa}{p+q}$

b

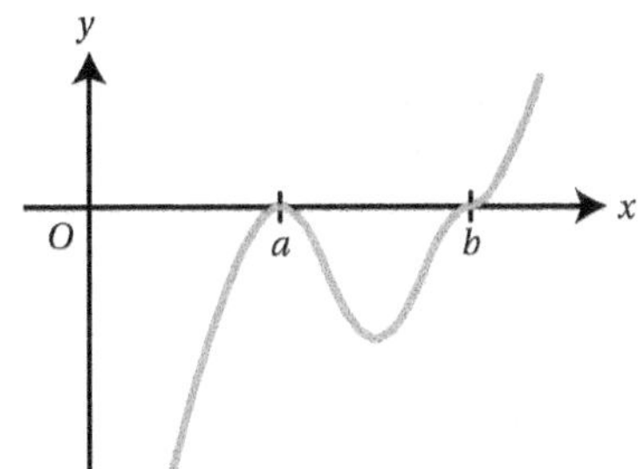

c q is even.

WORK IT OUT 10.2

All solutions are correct.

EXERCISE 10C

1 **a** **i** $\dfrac{2}{(x+1)^2}$ **ii** $\dfrac{-5}{(x-3)^2}$

b **i** $\dfrac{x(2x+1)^{-\frac{1}{2}} - (2x+1)^{\frac{1}{2}}}{x^2}$

ii $\dfrac{2x(x-1)^{\frac{1}{2}} - \frac{1}{2}x^2(x-1)^{-\frac{1}{2}}}{x-1}$

c **i** $\dfrac{2(x^2-x-2)}{(x^2+2)^2}$

ii $-\dfrac{x^2+2x+4}{(1+x)^2}$

d **i** $\dfrac{1-\ln 3x}{x^2}$ **ii** $\dfrac{x-2x\ln 2x}{x^4}$

2 $y = \frac{\pi^2}{4}x + \frac{16-\pi^4}{8\pi}$

3 $(0, 0), (1, 1)$

4 $a = -1$

5 $\left(e, \frac{1}{e}\right)$ local maximum.

6 $x \in (0, 1) \cup (1, 2)$

7 $a = 3, b = 4, p = \frac{3}{2}$

8 Proof.

EXERCISE 10D

1 **a** **i** $\frac{2}{3}$ **ii** $\frac{1}{2}$

b **i** 0 **ii** -1

c **i** -1 **ii** $\frac{5}{3}$

d **i** -1 **ii** $-\frac{1}{2}$

2 **a** **i** $\dfrac{2x}{y^2}$

ii $-\dfrac{2x^3}{3y}$

b i $\frac{y(8x-y)}{2x(y-2x)}$

ii $\frac{y}{2y-x}$

c i $\frac{1-2y}{2x-4y-1}$

ii $\frac{y}{2y-x}$

d i $\frac{y(2x-e^y)}{xy\,e^y-4}$

ii $\frac{\cos x-3\sin y}{3x\cos y-2\sin y}$

3 a $(3,2),(-3,-2)$ b $(\sqrt{2},4\sqrt{2}),(-\sqrt{2},-4\sqrt{2})$

4 a i $3\ln 3$ ii $25\ln 5$

b i $4\ln\left(\frac{1}{2}\right)$ ii $3\ln\left(\frac{1}{3}\right)$

c i $\frac{3\ln 2}{8}$ ii $4\ln 4$

d i $-3\ln 3$ ii $-\frac{\ln 5}{5}$

5 2

6 a Proof.

b $6x+5y-13=0$

7 a Proof.

b $3y-x-8=0$

8 4

9 $17x-8y+6=0$

10 $x=2$

11 $\frac{y2^y}{1-xy2^y\ln 2}$

12 $(2,e^4)$

13 a $y=3x-4$

b Proof.

c $(1,-1)$

EXERCISE 10E

1 a $x=y^2$ b $\frac{dy}{dx}=\frac{1}{2y}=\frac{1}{2\sqrt{x}}$

2 $\frac{dy}{dx}=\frac{1}{4}e^x$

3 $a=9$

4 a Proof.

b $\frac{1}{3}$

5 a $f^{-1}(x)=a^x$

b Proof.

6 a-b Proof.

c $2+\sqrt{2}$

7 a $\cos y$ b $\frac{1}{\sqrt{1-x^2}}$

8 Proof.

MIXED PRACTICE 10

1 D

2 A

3 a i $5e^{5x}$ ii $\frac{3}{2\sqrt{3x+2}}$

b $\frac{e^{5x}(30x+23)}{2\sqrt{3x+2}}$

4 $\frac{16}{225}$

5 a Proof.

b $\frac{1}{4}$

6 $\frac{729}{80}$

7 $9\ln 3$

8 a $3x^2\ln x+x^2$

b i $y=4e^2x-3e^3$

ii $x=\frac{3}{4}e$

9 $\frac{5}{2}$

10 $(3,3e^{-3})$

11 a $k=\ln a$

b Proof.

12 $(15\ln 5-20)\text{ m s}^{-2}$

13 a 3 seconds b 30 m s^{-1}

14 a $x=\frac{1}{2}$

b $(0,0),(1,-1)$

c $(0,0)$ local minimum; $(1,-1)$ local maximum.

d

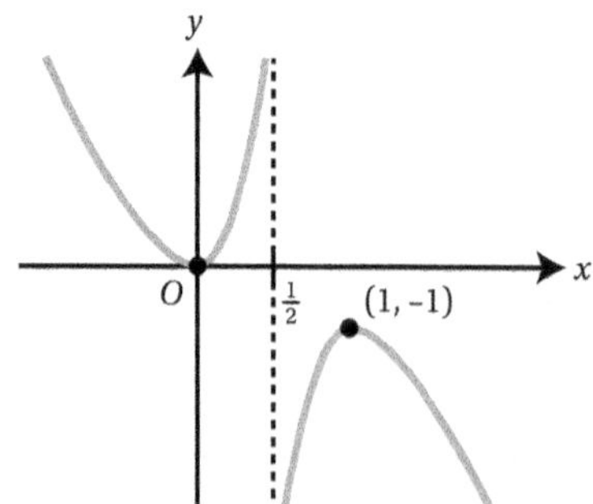

15 a Proof.

b $y-2=-2\sqrt{3}\left(x-\frac{\pi}{6}\right)$

16 a $g'(x) = 3 + \frac{1}{x} > 0$ b $\frac{1}{4}$

17 $\left(\frac{1}{3}, 1\right)$ and $\left(-\frac{1}{3}, -1\right)$

18 a $(2, 4), (-2, -4)$

b Proof.

c $(2, 4)$ local maximum; $(-2, -4)$ local minimum.

19 a Proof.

b $y - \frac{\pi}{6} = -\frac{4}{3}\left(x - \frac{1}{\sqrt{3}}\right)$

20 a $A = 1, B = 3$ b $(4, 0.272)$ c $\frac{4}{3}$

11 Further integration techniques

BEFORE YOU START

1 a $\frac{4}{3}x^3 + 3\ln x + c$ b $-5\cos x + c$

2 a $4e^x$ b $4e - 4$

3 a $4\cos 4x$ b $\frac{2x}{x^2+1}$

4 $0.8, -0.8$

5 $\frac{4}{x-1} - \frac{4}{x+2} - \frac{12}{(x+2)^2}$

WORK IT OUT 11.1

Solutions B and C are correct.

EXERCISE 11A

1 a i $(x+3)^5 + c$ ii $\frac{1}{6}(x-2)^6 + c$

b i $\frac{1}{32}(4x-5)^8 + c$ ii $2\left(\frac{1}{8}x+1\right)^4 + c$

c i $-\frac{8}{7}\left(3 - \frac{1}{2}x\right)^7 + c$ ii $-\frac{1}{9}(4-x)^9 + c$

d i $\frac{1}{3}(2x-1)^{\frac{3}{2}} + c$ ii $-\frac{4}{5}(2-5x)^{\frac{7}{4}} + c$

e i $4\left(2 + \frac{x}{3}\right)^{\frac{3}{4}} + c$ ii $2(4-3x)^{-1} + c$

2 a i $e^{3x} + c$ ii $\frac{1}{2}e^{2x+5} + c$

b i $6e^{\frac{2x-1}{3}} + c$ ii $2e^{\frac{1}{2}x} + c$

c i $2e^{-3x} + c$ ii $-\frac{1}{4}e^{-4x} + c$

d i $8e^{-\frac{x}{4}} + c$ ii $-\frac{3}{2}e^{-\frac{2}{3}x} + c$

3 a i $\ln|x+4| + c$ ii $\ln|5x-2| + c$

b i $\frac{2}{3}\ln(3x+4) + c$ ii $-4\ln|2x-5| + c$

c i $\frac{3}{4}\ln|1-4x| + c$ ii $-\frac{1}{2}\ln|7-2x| + c$

d i $x + 3\ln|5-x| + c$ ii $3x - \ln|3-x| + c$

4 a $\operatorname{cosec} x + c$

b $\tan 3x + c$

c $\frac{1}{3}\cos(2-3x) + c$

d $-4\cot\left(\frac{1}{4}x\right) + c$

e $\frac{1}{2}\sin 4x + c$

f $2\sec\frac{x}{2} + c$

5 $\frac{1}{5}$

6 $\frac{3}{2}(e^{-2} - e^{-8})$

7 18

8 0.492

EXERCISE 11B

1 a i $\frac{1}{4}(x^2+3)^4 + c$

ii $\frac{1}{6}(x^2-1)^6 + c$

b i $\frac{1}{15}(3x^2-15x+4)^5 + c$

ii $\frac{1}{12}(x^3+3x^2-5)^4 + c$

c i $\ln|x^2+3| + c$

ii $\ln|x^3-4x+5| + c$

d i $\frac{1}{2}\ln|x^2+8x-3| + c$

ii $\frac{1}{3}\ln|x^3+3x^2-15x+1| + c$

e i $\sqrt{x^2+2} + c$

ii $-\frac{1}{3(x^3-4)} + c$

f i $-\frac{2}{3}\cos^6 x + c$

ii $\frac{1}{8}\sin^4 2x + c$

g i $\frac{1}{4}\tan^4 x + c$

ii $-\frac{1}{5}\cot^5 x + c$

h i $\frac{1}{2}e^{3x^2-1} + c$

ii $\frac{3}{2}e^{x^2} + c$

i i $\frac{1}{2}\ln(e^{2x+3}+4) + c$

ii $\frac{1}{4}\ln|3+4\sin x| + c$

2 $-e^{\cos x} + c$

3 $e^5 - e^{-1}$

4 $k = 8$

5 $-\frac{1}{12\sin^4 3x} + c$

6 $-\frac{1}{10}\operatorname{cosec}^5 2x + c$

EXERCISE 11C

1 **a** **i** $\frac{2}{5}(x+1)^{\frac{5}{2}}-\frac{2}{3}(x+1)^{\frac{3}{2}}+c$

ii $\frac{2}{7}(x-2)^{\frac{7}{2}}+\frac{8}{5}(x-2)^{\frac{5}{2}}+\frac{8}{3}(x-2)^{\frac{3}{2}}+c$

b **i** $\frac{2}{9}(x-5)^9+\frac{5}{4}(x-5)^8+c$

ii $\frac{1}{7}(x+3)^7-\frac{1}{2}(x+3)^6+c$

2 **a** **i** $2\ln(\sqrt{x}+1)+c$

ii $\frac{1}{2}\ln(3+4\sqrt{x})+c$

b **i** $\ln(\ln x)+c$

ii $-\frac{1}{2(\ln x)^2}+c$

3 **a** **i** $\frac{1}{24}(2x-1)^6+\frac{1}{20}(2x-1)^5+c$

ii $\frac{1}{7}(3x+2)^7-\frac{1}{3}(3x+2)^6+c$

b **i** $\frac{2}{5}(x-3)^{\frac{5}{2}}+2(x-3)^{\frac{3}{2}}+c$

ii $\frac{2}{125}(5x-6)^{\frac{5}{2}}+\frac{22}{75}(5x-6)^{\frac{3}{2}}+$

c **i** $\frac{2}{5}(x-5)^{\frac{5}{2}}+\frac{20}{3}(x-5)^{\frac{3}{2}}+50(x-5)^{\frac{1}{2}}+c$

ii $-\frac{1}{(2x-3)}-\frac{13}{2(2x-3)^2}+c$

4 **a** **i** 2732.8 **ii** 1.8

b **i** $\frac{1}{6}$ **ii** $\frac{1}{3}$

c **i** $9-8\ln 2$ **ii** $-5\frac{1}{15}+12\ln\frac{5}{3}$

5 $\frac{2}{3}(x-2)^{\frac{3}{2}}+4(x-2)^{\frac{1}{2}}+c$

6 **a** Proof. **b** $\ln|x^2+x+1|+c$

7 $\frac{1}{4}\tan(\ln(x^2))+c$

8 $2\sqrt{3}-2;\ a=2, b=3, c=-2$

9 $\frac{\pi}{12}$

EXERCISE 11D

1 **a** **i** $\frac{1}{2}x\sin 2x+\frac{1}{4}\cos 2x+c$

ii $-2x\cos\left(\frac{x}{2}\right)+4\sin\left(\frac{x}{2}\right)+c$

b **i** $-2xe^{-2x}-e^{-2x}+c$

ii $\frac{1}{4}xe^{4x}-\frac{1}{16}e^{4x}+c$

c **i** $x^2\ln 5x-\frac{1}{2}x^2+c$

ii $\frac{1}{2}x^2\ln x-\frac{1}{4}x^2+c$

d **i** $\frac{1}{8}x^4\ln x-\frac{1}{32}x^4+c$

ii $\frac{1}{2}x^6\ln 2x-\frac{1}{12}x^6+c$

2 1

3 $(e^3-1)\ln 2+\frac{2}{3}e^3+\frac{1}{3}$

EXERCISE 11E

1 **a** $\frac{1}{3}x^2\sin 3x+\frac{2}{9}x\cos 3x-\frac{2}{27}\sin 3x+c$

b $-\frac{x^2}{2}\cos 2x+\frac{x}{2}\sin 2x+\frac{1}{4}\cos 2x+c$

c $x^2e^{\frac{x}{4}}-8xe^{\frac{x}{4}}+32e^{\frac{x}{4}}+c$

d $\frac{x^2}{6}(x+2)^6-\frac{x}{21}(x+2)^7+\frac{1}{168}(x+2)^8+c$

2 **a** $2x\ln(3x)-2x+c$

b $-x\ln x+x+c$

3 **a** $\frac{\pi}{2}-1$ **b** $\frac{1}{2}(1-\ln 2)$

4 Discussion.

5 $-\frac{2}{3}xe^{-3x}-\frac{2}{9}e^{-3x}+c$

6 $\frac{5e^6+1}{36}$

7 $e-2$

8 **a** Proof.

b $x\tan x-\ln|\sec x|+c$

9 e^2+1

10 **a** Proof.

b $\frac{e^x}{2}(\sin x+\cos x)+k$

WORK IT OUT 11.2

They are all right, with different $+c$:

$\frac{1}{2}\sin^2 x=-\frac{1}{2}\cos^2 x+\frac{1}{2}=-\frac{1}{4}\cos 2x+\frac{1}{4}$

EXERCISE 11F

1 **a** $\frac{1}{3}\sec 3x+c$

b $-\cot x+c$

c $-\frac{1}{4}\cos 4x+c$

d $\frac{1}{2}(-3\cot 2x+\operatorname{cosec} 2x)+c$

e $\sin x+\cos x+c$

2 **a** $\frac{1}{3}\sin^3 x-\frac{1}{5}\sin^5 x+c$

b $-\frac{1}{\sin x}-\sin x+c$

c $-\frac{1}{4}e^{\cos 2x}+c$

d $\frac{1}{15}\tan^5 3x+c$

e $-\frac{1}{4}\sqrt{1+\cos 4x}+c$

3 **a** **i** $x+\frac{1}{2}\sin 2x+c$

ii $\frac{1}{2}\left(\frac{1}{6}\sin 6x+x\right)+c$

b **i** $4\tan\left(\frac{x}{2}\right)-2x+c$

ii $\frac{1}{3}\tan 3x-x+c$

4 a i $\frac{\pi}{2}$

ii $\frac{9\pi}{8}-1$

b i $1-\ln 2$

ii $6\sqrt{3}-2\pi$

5 a i $\ln 2$

ii $2-2\ln 2$

b i $\frac{\pi}{9}+\frac{\sqrt{3}}{6}$

ii $\frac{\pi}{12}-\frac{\sqrt{3}}{16}$

c i $\frac{1}{2}\left(x\sqrt{1-x^2}-\arccos x\right)+c$

ii $18\arcsin x+3x\sqrt{1-x^2}+c$

6 $\frac{1}{2}x-\frac{3}{4}\sin\left(\frac{2x}{3}\right)+c$

7 a Proof.

b $\frac{1}{2}\tan^2 x+\ln|\cos x|+k$, or, alternatively, $\frac{1}{2}\tan^2 x-\ln|\sec x|+k$

8 3

9 a Proof.

b $-\frac{2}{3}\cos^3 x+\cos x+c$

10 $2\arcsin\left(\sqrt{x}\right)+c$

11 $\arctan x+c$

12 a Proof. b 4

13 a $y=\sqrt{25-x^2}$ b Proof.

EXERCISE 11G

1 a $\frac{3}{2}\ln|x^2-4|+c$

b $2\ln|x^2-3x+1|+c$

c No

d $\frac{5}{2}\ln|x^2+1|+c$

e No

f $\frac{1}{3}\ln|x^3-9x|+c$

g $4\ln|x^3+2|+c$

2 a $\frac{1}{6}(2x-3)^3+c$

b $-\frac{1}{5}\ln|2-5x|+c$

c $\frac{1}{3}\ln|x-1|+c$

d $x-\ln|x|+c$

3 a i $3\ln|x-10|+2\ln|x-3|+c$

ii $2\ln|x+1|-\ln|x-3|+c$

b i $\frac{1}{2}\ln|x-1|-\frac{1}{2}\ln|x+1|+c$

ii $\frac{1}{2}\ln|x-1|+\frac{1}{2}\ln|x+1|+c$

c i $3\ln|x-2|-\ln|1-x|+c$

ii $-\ln|1-x|-2\ln|1+x|+c$

d i $5\ln|x+3|+\ln|x|-\frac{3}{x}+c$

ii $2\ln|x-2|+2\ln|x|+\frac{1}{x}+c$

e i $\ln|x-1|-\frac{2}{x-1}-\ln|x+3|+c$

ii $\ln|x+1|-\frac{1}{x+1}-\ln|x-2|+c$

4 a i $x-\ln|x+2|+c$

ii $2x+5\ln|x-1|+c$

b i $\frac{1}{2}x^2+3x+11\ln|x-3|+c$

ii $\frac{1}{2}x^2-3x+14\ln|x+5|+c$

5 c $\ln|x-2|-\ln|x+2|+c$

e $2\ln|x|+3\ln|x-3|-\ln|x+3|+c$

6 a $\frac{1}{x-2}-\frac{1}{x+3}$

b $\ln\left|\frac{x-2}{x+3}\right|+c$

7 $-\ln 3=\ln\left(\frac{1}{3}\right)$

8 $3\left(1+\ln\frac{4}{7}\right)$

9 a $\frac{1}{2-x}+\frac{2}{x+1}$

b 8

10 $\ln\left(\frac{4}{9}\right)+\frac{1}{2}$

MIXED PRACTICE 11

1 C

2 $\frac{\pi}{2}$

3 $\frac{1}{2}x\sin 2x+\frac{1}{4}\cos 2x+c$

4 6.36

5 $\frac{\sqrt{3}}{4}$

6 a $-\frac{1}{3}\ln|1-3x|+c$

b $-\frac{1}{2}(2x+3)^{-1}+c$

7 $\frac{1}{6}xe^{6x}-\frac{1}{36}e^{6x}+c$

8 a $\frac{dy}{dx}=12x^2-6$ b $\frac{1}{6}\ln\left(\frac{13}{3}\right)$

9 A

10 $x\ln x-x+c$

11 **a** $\frac{e^{-2x}}{e^{-2x}-3}$ or $\frac{1}{1-3e^{2x}}$

b $\frac{1}{2}\ln\left|\frac{1}{e^{-2x}-3}\right|+c$

12 $\arcsin\left(\frac{x}{2}\right)+c$

13 **a** $\frac{2}{x-1}-\frac{1}{x+2}$ **b** $\ln\left(\frac{7}{4}\right)$

14 $\ln|\ln x|+c$

15 $\frac{8}{3}\left(\frac{1}{2}x-1\right)^{\frac{3}{2}}+8\left(\frac{1}{2}x-1\right)^{\frac{1}{2}}+c$

16 **a** $A=-6$, $B=4$

b $3x-2\ln|3x-1|+4\ln|x-1|+c$

17 $\ln\left(\frac{3}{2}\right)-\frac{1}{3}$

18 $(6+4\ln 4)$ m

19 $\ln|x-2|-\frac{5}{x-2}+c$

20 $\frac{e-1}{e+1}$

21 **a** $\frac{9\pi}{2}$ **b** Proof.

22 **a** $x+c$ **b** $\ln|\sin x+\cos x|+d$

c $\frac{1}{2}(x-\ln|\sin x+\cos x|)+k$

12 Further applications of calculus

BEFORE YOU START

1 **a** $2e^{2x}\sin 3x+3e^{2x}\cos 3x$

b $\frac{2x-2x\ln(x^2+1)}{(x^2+1)^2}$

2 $x=0.285$ or 2.86

3 $\frac{\sqrt{3}}{4}$

4 **a** $\frac{1}{2}x-\frac{1}{4}\sin 2x+c$

b $\frac{1}{2}\ln|x^2-2|+c$

c $\frac{1}{3}x\sin 3x+\frac{1}{9}\cos 3x+c$

WORK IT OUT 12.1

Solutions 2 and 3 are correct.

EXERCISE 12A

1 **a** **i** Decreasing, convex.

ii Concave.

iii Increasing, convex.

b **i** Increasing.

ii Increasing, concave.

iii Increasing, convex.

2 **a**

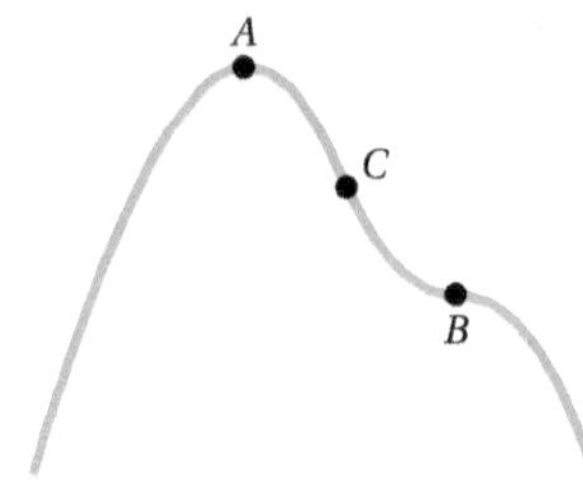

b

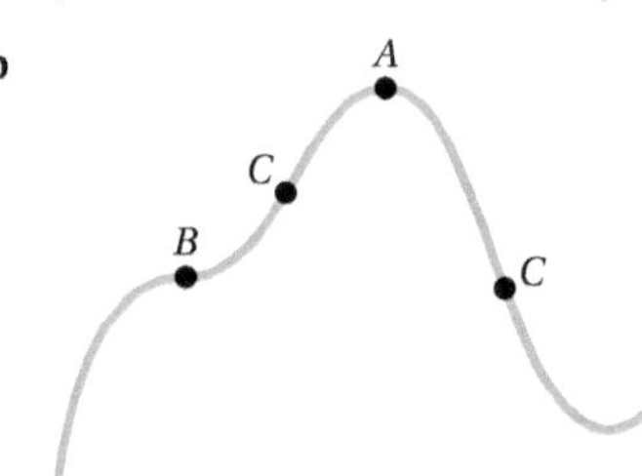

3

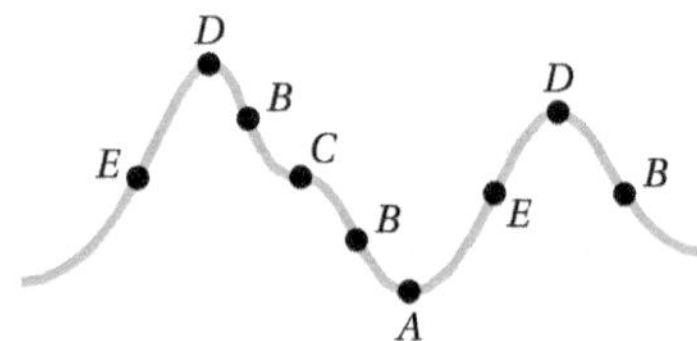

4 $(\ln 2, 2-(\ln 2)^2)$

5 $(1, 4)$, $(-1, -10)$

6 Proof.

7 $x<-3$ or $x>2$

8 $\left(\frac{\pi}{2}, \frac{\pi}{2}\right)$, $\left(\frac{3\pi}{2}, \frac{3\pi}{2}\right)$

9 $(0, 0)$

10 $x=0$ maximum; $x=4$ minimum.

11 Proof.

12 $\frac{1}{9}$

13 $2-\sqrt{2}<x<2+\sqrt{2}$

14

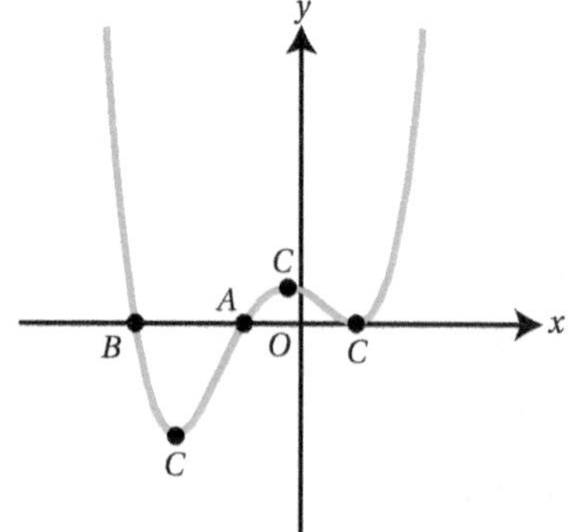

EXERCISE 12B

1 a i

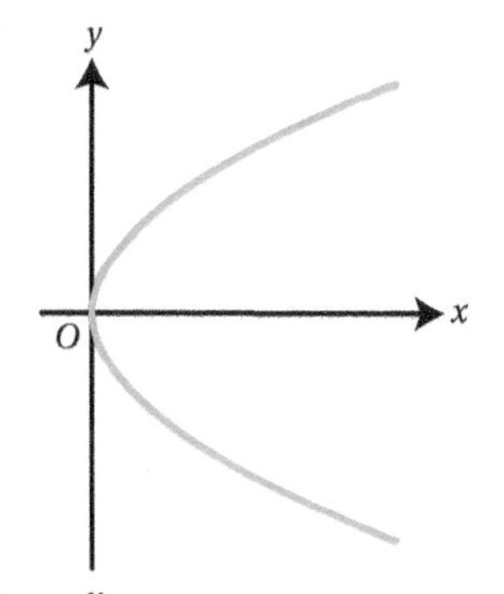

ii

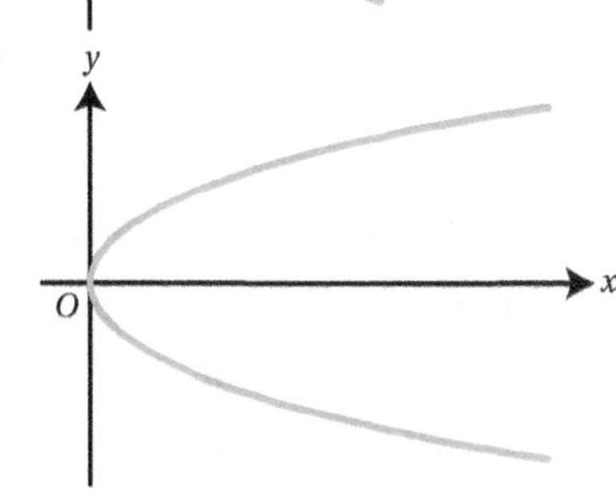

b i

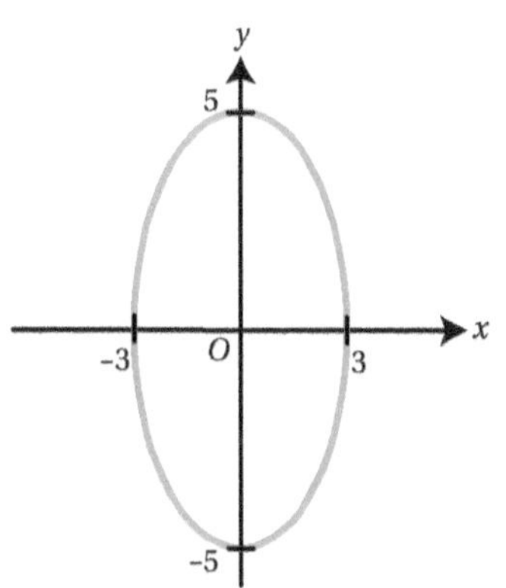

ii

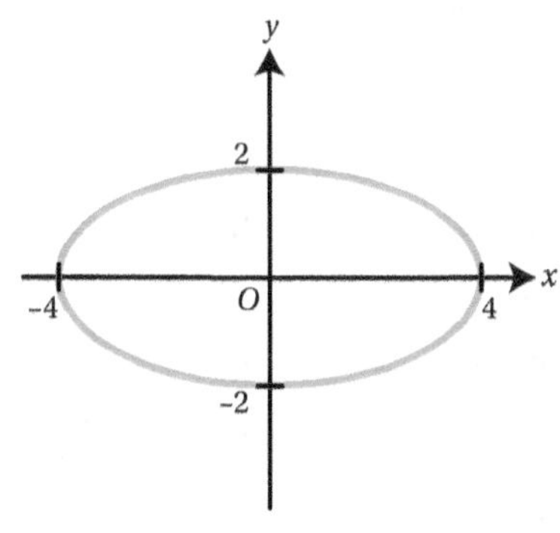

c i

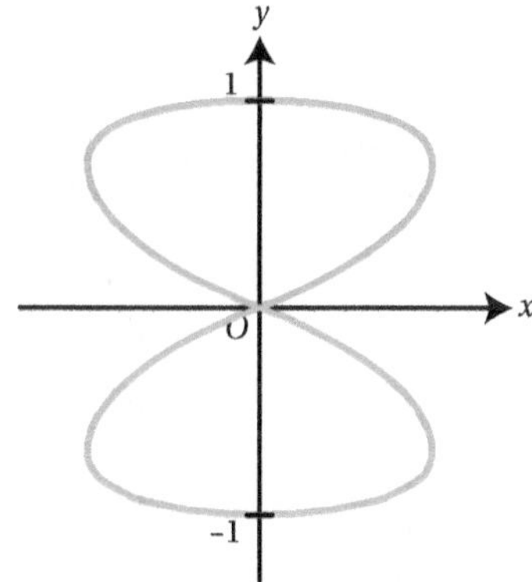

ii

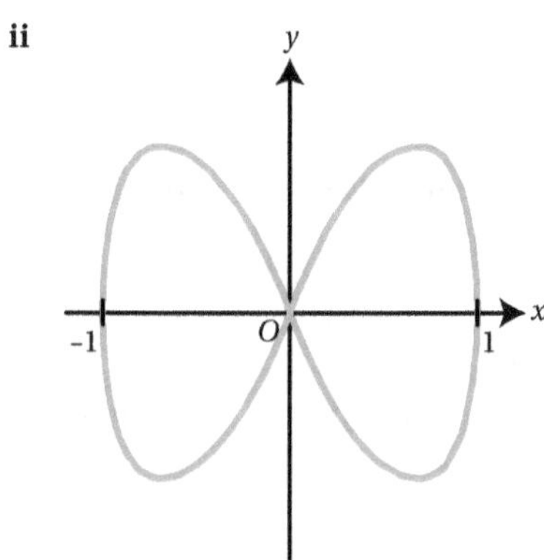

2 a **i** −1 **ii** 2

b **i** −2 **ii** −3

c **i** 2.50 **ii** $\frac{\pi}{3}$

3 a **i** $4x = 3y^2$ **ii** $x = 5(2-y)^2$

b **i** $x^3 = 8y^2$ **ii** $125x^2 = 4y^3$

c **i** $x^2 + y^2 = 25$ **ii** $9x^2 + y^2 = 36$

d **i** $x = 4y^2 - 2$ **ii** $9x = 9 - 2y^2$

e **i** $y^2 - x^2 = 1$ **ii** $4x^2 - 9y^2 = 36$

4 a 4.4 m **b** 8.57 m

5 a $2 - \frac{1}{\sqrt{3}}$

b 2; the distance along the line.

c $x - \sqrt{3}y = 1 - 2\sqrt{3}$

WORK IT OUT 12.2

Solution 2 is correct.

EXERCISE 12C

1 a i $\frac{1}{3t}$ ii $-\frac{1}{10t}$

b i $\frac{\sin\theta}{4\sin 2\theta}$ ii $-\frac{3\cos\theta}{2\sin 2\theta}$

c i $\sin\theta$ ii $\frac{2}{3}\operatorname{cosec}\theta$

2 a i $-\frac{1}{3}$ ii $-\frac{1}{20}$

b i $-\frac{3}{2}$ ii $-\frac{5}{9}$

c i $\frac{3}{4}$ ii $-\sqrt{3}$

3 $y=-6x+17$

4 $y=-1$

5 $\frac{9}{4\mathrm{e}}$

6 a $y=16x-255$ b $\left(-\frac{1}{16}, -256\right)$

7 (27, 18)

8 Proof.

9 Proof.

10 a $(2a+aq^2, 0)$ b Proof.

EXERCISE 12D

1 a i 45 ii $\frac{112}{3}$

b i $4-\ln 5$ ii $6-4\ln 2$

c i 6 ii $4\pi-8$

2 a (1, 0), (16, 0) b 22.5

3 a 0 and ln 2 b $\ln 2-\frac{5}{8}$

4 a Proof. b 3π

5 a 3 b $\frac{80}{9}-2\ln 3$

c $x^2-y^2=4$

EXERCISE 12E

1 a i $144x^3$ ii $6x^2(x^3+1)$

b i $-6x\sin(3x^2)$ ii $2x\sec^2(x^2+1)$

2 a i 50 ii -12

b i -6 ii 1

c i $\pm\frac{1}{3}$ ii -2

3 a i 22 ii 38

b i 45 ii 176

c i 0.24 ii 0.006 67

4 $113\text{ cm}^2\text{ s}^{-1}$

5 2 cm s^{-1}

6 $768\pi\text{ cm}^3\text{ s}^{-1}$

7 $75\text{ cm}^2\text{ s}^{-1}$

8 2 cm s^{-1}

9 $160\pi\text{ cm}^3\text{ s}^{-1}$

10 19.1 units per second

WORK IT OUT 12.3

Solution 3 is correct.

EXERCISE 12F

1 a i $\frac{11}{4}$ ii $\frac{79}{6}$

b i $\frac{32}{3}$ ii $\frac{1}{6}$

c i 9 ii $\frac{1}{3}$

d i $\frac{9}{8}$ ii $\frac{1}{3}$

e i $\frac{15}{4}-4\ln 2$ ii $12-5\ln 5$

2 a i 9.13 ii 2.50

b i 0.828 ii 41.3

c i 2.35 ii 5.38

3 $\frac{28}{3}$

4 $\frac{32}{3}$

5 $\mathrm{e}^2-\frac{11}{3}$

6 Proof.

7 6

8 e^2-3

9 $12-5\ln 5$

10 $\frac{71}{6}$

11 $m=4$

12 $a=1+\sqrt{3}$

MIXED PRACTICE 12

1 D

2 $\left(2, -\frac{2}{3}\right)$

3 a $t=1$ b $y-1=-\frac{1}{6}(x-3)$

4 36

5 a 6 b $\frac{22}{3}$

6 a $-\frac{1}{2}$ b $x=\frac{8}{(y+1)^2}+1$

7 D

8 25

9 $\frac{1024}{5}$

10 a $0, \frac{\pi}{3}, \pi$ b $\frac{5}{4}$

11 a Proof. b 4

12 a (0, 0) local minimum; (0.8, 0.082) local maximum.
   b Proof.

13 Proof.

14 3

15 a $100h-\frac{1}{2}h^2$ b Proof c 6

16 a $-\frac{1}{4}e^{4t}$
   b i −4 ii (−2, 12) iii (−50, 0)
   c $xy+4y-4x=32$

17 a $\frac{\pi}{4}$ and $\frac{5\pi}{4}$ seconds b 1.93 m

18 Proof.

19 a 0, $\frac{\pi}{2}$ b $\frac{5\pi}{2}, 10\pi$

20

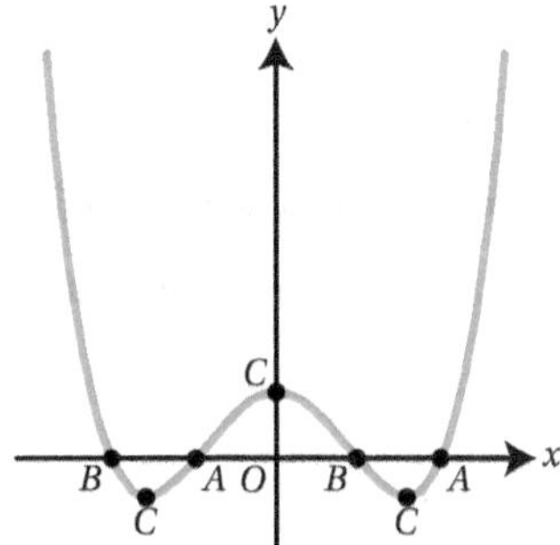

21 a Proof. b $(-a, 0)$ and $(3a, 8a^2)$
   c $\frac{64}{3}a^3$ d $\frac{15}{16}$

13 Differential equations

BEFORE YOU START

1 $v=3+5e^t$

2 27 N downwards.

3 $\frac{dV}{dt}=20\pi r^2\sqrt{r}$

4 $2\ln\left(\frac{2+x}{2-x}\right)+c$

5 a $2\ln(x^2+3)+c$
  b $\frac{1}{3}x^3\ln x-\frac{1}{9}x^3+c$

6 $\frac{1}{2}\tan 2x-x+c$

EXERCISE 13A

1 a i $y=-\frac{3}{2}\cos 2x+c$
    ii $y=12\sin\left(\frac{x}{3}\right)+c$
  b i $y=\frac{1}{3}e^{2x}+c$
    ii $y=8e^{\frac{x}{2}}+c$
  c i $y=3\tan x+c$
    ii $y=\tan x-x+c$
  d i $y=-\frac{1}{2}x^{-2}\ln x-\frac{1}{4}x^{-2}+c$
    ii $y=\sec x+c$

2 a i $y=\frac{4}{3}\sqrt{3x+9}-2$
    ii $y=-2\sqrt{4-x}+3$
  b i $y=\ln(x^2+1)-\ln 2$
    ii $y=\frac{1}{4}x^2+\frac{1}{2}\ln|x|+\frac{3}{4}$
  c i $y=-2e^{-3x}+2$ ii $y=-2e^{1-2x}+2$
  d i $y=\frac{1}{4}\sin^4 x+\frac{1}{16}$ ii $y=\frac{1}{2}\sec^2 x+4$

WORK IT OUT 13.1

Solution 2 is correct.

EXERCISE 13B

1 a i $y=\frac{2}{3}x^{\frac{3}{2}}$ ii $y=\frac{1}{1-2x^2}$
  b i $y=2x^4$ ii $y=3e^{-x^3}$

2 a i $\sin y=\frac{1}{2}-\cos x$
    ii $\tan y=\tan x-\sqrt{3}$
  b i $\ln y=\frac{1}{3}x^3$
    ii $-\frac{1}{y}=\ln x-1$
  c i $e^{-2y}=-4e^x+5$
    ii $e^y=e^x+e^2-1$

3 a i $y=\pm\sqrt{x^3+c}$ ii $y=-\frac{1}{x^2+c}$
  b i $y=\arcsin(\ln|x|+c)$
    ii $y=\arctan(\ln|x-2|+c)$
  c i $y=Ae^x(x-1)-3$
    ii $y=\frac{A}{1-x}$

4 $y=e^{-(x-1)^2}$

5 $H=10$

6 Proof.

7 $y=\frac{2+2Ax^4}{1-Ax^4}$

8 $k=3$

EXERCISE 13C

1 a i $\frac{dN}{dt}=5N$ ii $\frac{dM}{dt}=-3M$
  b i $\frac{dv}{dt}=\frac{kv}{\sqrt{t}}$ ii $\frac{dN}{dt}=k\sqrt{N}\sqrt[3]{t}$
  c i $\frac{dr}{dt}=\frac{k}{2\pi\sqrt{r}}$ ii $\frac{dr}{dt}=-\frac{0.2}{\pi r^2}$

2 a Proof.
  b $N=700e^{0.2t}$; after 20 minutes
    $N=700e^4\approx 38\,200$.

3 a Proof. b 3.47 seconds

4 a Proof.

b $v = \frac{8}{1+2t}$; 1.5 s

5 a When $t = 0$, $\frac{dN}{dt} = 1.6 > 0$; decreases from 5.7 years.

b $N = 2e^{0.8t-0.07t^2}$

c 19 665

d It will decay to zero.

6 a $\frac{dQ}{dP}$ is the rate of change. 'Proportional to Q and inversely proportional to P' means that $\frac{dQ}{dP} = \epsilon\frac{Q}{P}$, so $\frac{1}{Q}\frac{dQ}{dP} = \frac{\epsilon}{P}$.
ϵ is negative because demand decreases as price increases, so the rate of change $\frac{dQ}{dP}$ is negative.

b $Q = AP^{\epsilon}$

c i

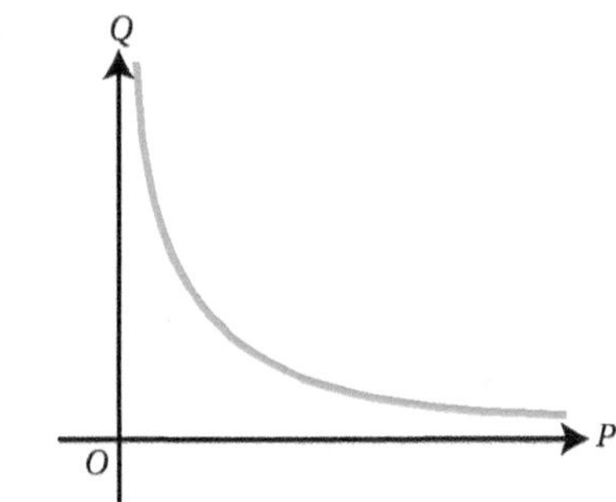

Demand is inversely proportional to price.

ii

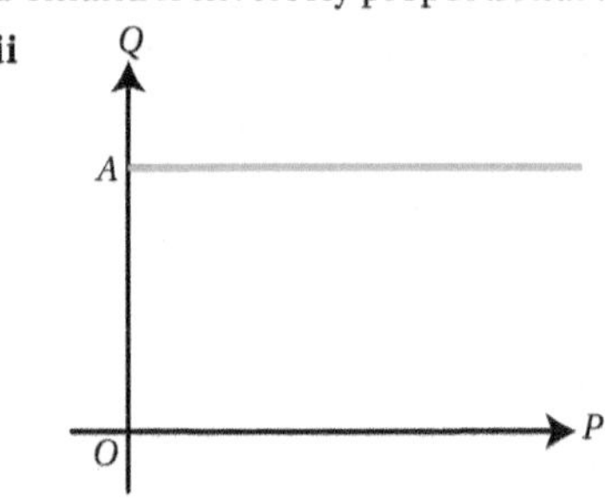

Demand is independent of price.

7 a Proof.

b $\theta = 19 - 14e^{-0.3t}$; 11 minutes

8 a $v = 5 - 5e^{-0.8t}$

b The velocity approaches 5 m s^{-1}.

9 a Proof.

b $\frac{2\pi}{3}$ seconds

10 a Proof.

b $v = 10e^{-0.3t}$

c $x = \frac{100}{3}(1 - e^{-0.3t})$; it approaches $\frac{100}{3}$ m.

MIXED PRACTICE 13

1 C

2 B

3 $y = 3 + e^{x^2}$

4 $y = \arctan\left(\sin x + \frac{1}{2}\right)$

5 $A = 6, B = 4$

6 a $\frac{1}{3}(x^2+3)^{\frac{3}{2}} + c$

b $y = \frac{1}{2}\ln\left(\frac{2}{3}(x^2+3)^{\frac{3}{2}} - \frac{13}{3}\right)$

7 a Proof.

b $N = 250e^{0.04t}$; 34.7 months

c Not suitable, as it predicts indefinite growth.

d $N = 250e^{0.04\left(t + \frac{15}{\pi}\sin\left(\frac{\pi t}{6}\right)\right)}$

8 a $Q = AP^{\varepsilon}$ where $A = e^c$

b The demand decreases as the price increases.

c Luxury goods; the demand increases with price.

9 a $h = 10 - \frac{20}{t+2}$

b It will never fill, as $h < 10$ for all t.

10 a Proof.

b $C = 2$, $k = 4.9$

c The velocity increases towards 2 m s^{-1} as t increases.

11 a $\frac{dA}{dt} = -k$ b i Proof. ii 12

12 a Decrease in size due to, for example, competition for food.

b $N = \frac{3}{1+e^{-1.2t}}$

c Increases with the limit of 3000.

13 Proof.

14 a $\frac{dv}{dx} = -\frac{8e^{-4x}}{v}$ b Proof.

c $x = \frac{1}{2}\ln(4t+1)$, $v = \frac{2}{4t+1}$

14 Numerical solution of equations

BEFORE YOU START

1 1, 3, −3, −33

2 a $x = e^{\frac{x}{3}} - 2$ b $x = \frac{1}{2}\sqrt{x^2+12}$

c $x = \arctan\frac{x}{3}$

3 a $6x \tan x + 3x^2 \sec^2 x$

b $\frac{x - 2x \ln x}{x^4} = \frac{1 - 2\ln x}{x^3}$

c $6xe^{3x^2} - \frac{1}{x}$

WORK IT OUT 14.1

Solution C is correct.

EXERCISE 14A

1 a Exact solution.

b Cannot rearrange.

c Exact solution.

d Exact solution.

e Cannot rearrange.

f Exact solution.

g Cannot rearrange.

h Cannot rearrange.

2 a 2 and 3 b 0 and 1

c 1 and 2 d −2 and −1

3–5 Proof.

6 a Proof. b $k = 2$

7 a Proof.

b i $f(2) = -6$, $f(3) = 11$

ii $f(x)$ is not continuous (has an asymptote) at $x = 2.5$.

8 a i

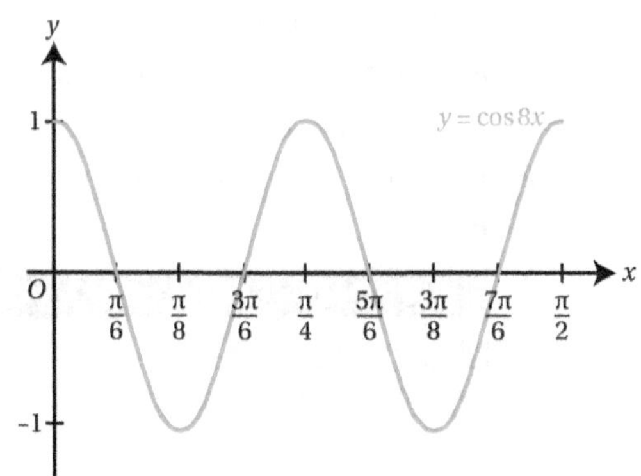

ii 2

b i $g(0) = g\left(\frac{\pi}{4}\right) = 1$

ii $g(x)$ changes sign twice.

iii Use, for example, $x = 0, \frac{\pi}{8}, \frac{\pi}{4}$

EXERCISE 14B

1 a i 3.611 ii 0.995

b i 2.257 ii 0.5

2 a i 1.31 ii −0.347

b i 1.78 ii 11.6

3 −0.7

4 a Proof. b 1.32, 1.29

5 6.5904, 6.5915

6 a Proof. b 3.23 c Proof.

7 a Proof. b 1.327

8 0.407

EXERCISE 14C

1 a i $x_1 = 2$; x_0 close to stationary point.

ii $x_1 = 4$; x_0 close to stationary point.

b i $x_1 = 3.7$; on the other side of an asymptote.

ii $x_1 = -4.52$; outside the domain.

2 a $x = 0$ and $\frac{1}{3}$ b Proof.

c It is close to a stationary point.

3 $x = 3$; the tangent crosses the x-axis further away from the root, so subsequent values of x_n increase.

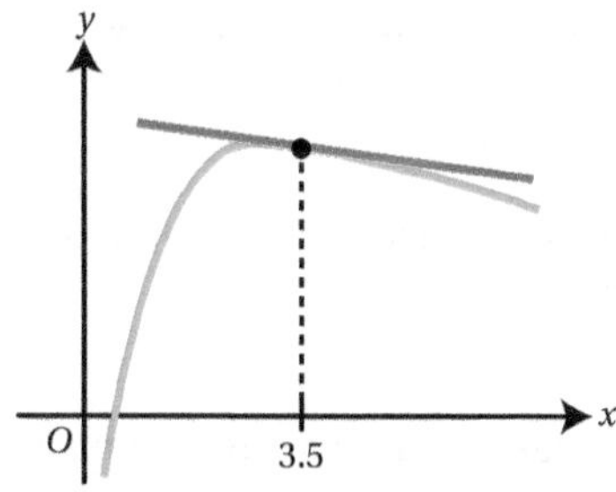

4 a $0, \frac{\pi}{2}, \frac{3\pi}{2}, \frac{5\pi}{2}$ b $x_1 = 3.55$

c $x_1 > \frac{\pi}{2}$, so falls on the far side of a discontinuity.

5 a $(0, -2)$, $(2, 2)$

b i $x = \alpha$ ii Not converge.

c $x_0 > 2$

6 **a** −2 and −1, −1 and 1, 1 and 2.

b (−1, 5), (1, − 3)

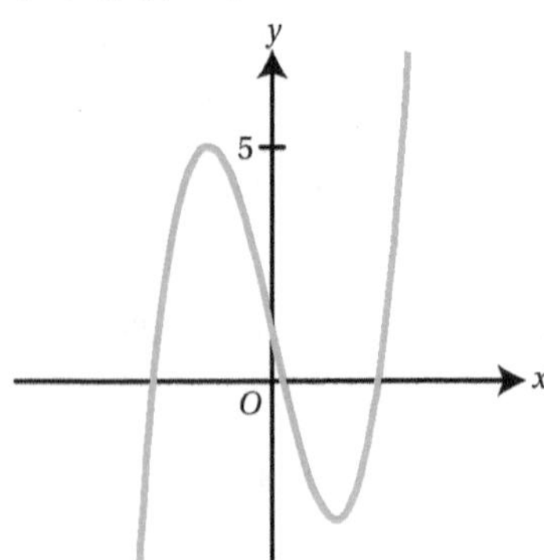

c (−1.68, 0)

d It will converge to a.

e 0.168

7 **a** (1, 1.5)

b

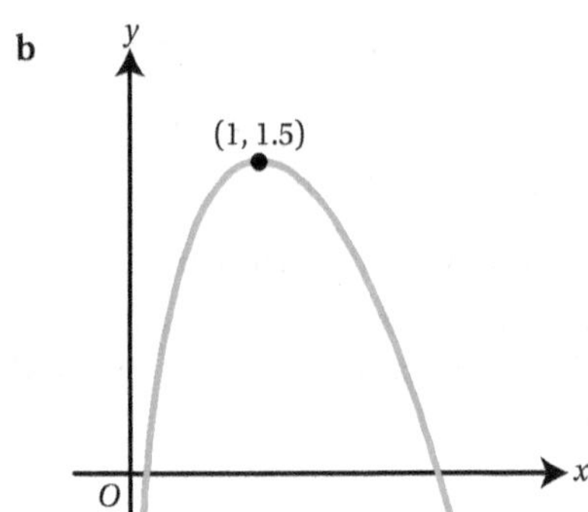

c 2 and 3

d $x_1 = -0.29$ which is outside of the domain of f(x).

e $x_0 > 1$

8 **a** **i** Too close to stationary point at 1.22.

ii $1.31 < x_0 < 1.83$

b **i** Too close to stationary point at 1.10.

ii $0 < x_0 < 0.869$

c **i** $x_1 < 0$, so lies in a different region.

ii $0 < x_0 < 0.978$

d **i** $x_1 > \frac{3\pi}{2}$, so lies in a different region.

ii $4.30 < x_0 < \frac{3\pi}{2}$

9 **a** Proof.

b $1 < x_0 < \alpha$ and $x_0 > \alpha$ where $\alpha = \sqrt{5}$.

c Proof.

10 **a**

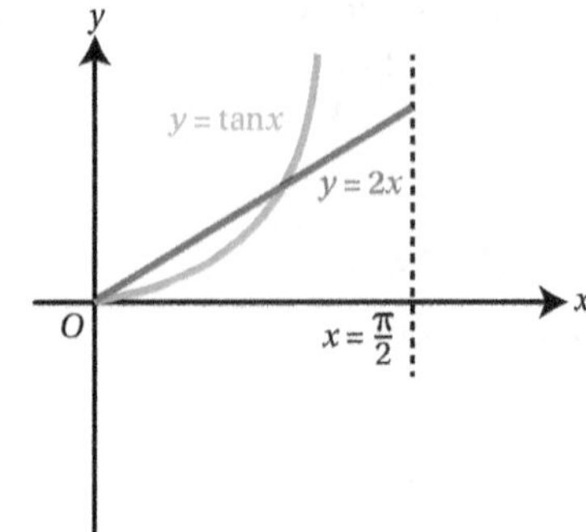

b Proof.

c

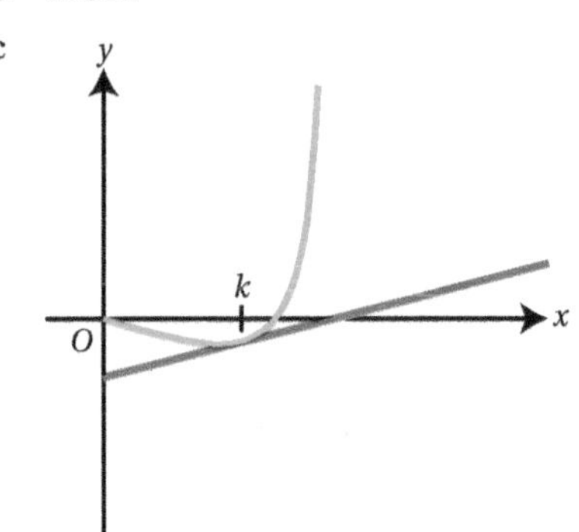

d Does not converge, or converges to another root (in a different interval).

EXERCISE 14D

1 **a** **i** 3, 3.219, 3.305, 3.337, 3.349

ii 4, 2.982, 2.949, 2.948, 2.948

b **i** −1, 0, −0.250, −0.200, −0.211

ii 1, 0.540, 0.997, 0.546, 0.996

c **i** 0.5, 1.057, 0.938, 1.056 0.940

ii 1, 1.364, 1.435, 1.397, 1.420

2 **a** **i** 0.70 **ii** 0.95

b **i** 4.51 **ii** 0.41

3

	i	ii	iii
a	No	No	N/A
b	No	No	N/A
c	Yes	Yes	Convergence to upper root only. Convergence for $x_0 >$ lower root.
d	Yes	Yes	Convergence to lower root only. Convergence for $x_0 <$ upper root.

4 2.58

5 **a** 0.9502 **b** $f(x) = x - \cos\left(\frac{x}{3}\right)$, 0.95

6 $x_1 = 0.775\,40$, $x_2 = 0.855\,78$, $x_3 = 0.880\,65$; $x = 0.892$

7 a $-0.8041, -0.8780$

b $a = 4, b = 3, c = 5$

8 x_3

9 a Proof.

b $x_2 = 1.611, x_3 = 1.572, x_4 = 1.584$

c $x = 1.58$ (2 d.p.)

10 a Proof.

b 5.24 c $3+\sqrt{5}$; 0.0751%

11 a

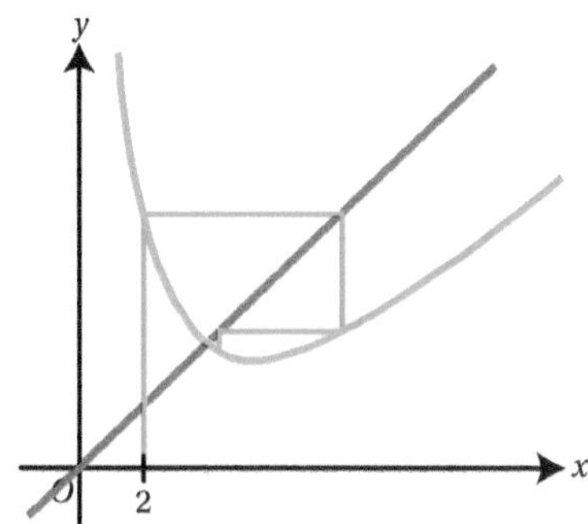

b Proof.

EXERCISE 14E

1 a $0, -1, 2, 11, 362\ldots$ (large, positive); diverges (increases without a limit).

b $1, 4.5, -5.125, -8.133, -28.071\ldots$ (large, negative); diverges (decreases without a limit).

c $0, 2, 1.6, 1.539, 1.545\ldots 1.544, 1.544$; converges to 1.54 (and oscillates).

d $2, 0.641, 0.115, -0.142, -0.271\ldots -0.397, -0.397$; converges to -0.397 (decreases).

e $2, 1, 2, 1, 2\ldots 1, 2$; oscillates between 1 and 2.

f $2, 2.079, 2.196, 2.36, 2.576\ldots 4.524, 4.528$; converges (increases to 4.54).

g $1.5, 1.216, 0.588, -1.595$, not real; undefined after x_4.

2 a Converges.

b Converges.

c Converges to B.

d Converges to A.

e Converges to C.

f Converges to B.

g Converges to B.

3 a i $x = \sin^{-1}\left(\frac{x}{3}\right)$ ii $x = \tan^{-1}\left(\frac{x}{5}\right)$

b i $x = e^x + 3$ ii $x = \ln x + 2$

c i $x = \left(\frac{x}{3}\right)^2 + 1$ ii $x = \left(\frac{x}{2}\right)^3 - 5$

d i $x = \sqrt{\frac{x+1}{3}}$ ii $x = \sqrt{7x-1}$

4 a i $a = 2, b = -0.5$ ii $a = 2, b = -6$

b i $a = 4, b = 1$ ii $a = 1, b = 0.5$

c i $a = \frac{1}{2}, b = -\frac{5}{2}$ ii $c = \frac{1}{3}, d = \frac{2}{3}$

d i $a = 1, b = 2$ ii $a = 3, b = 1$

5 These rearrangements are not unique - there are alternatives which will converge to the various roots.

a lower root: $x_{n+1} = 2 + e^{\frac{(x_n-2)}{3}}$;
upper root: $x_{n+1} = 2 + 3\ln(x_n - 2)$

b lower root: $x_{n+1} = 2 + \tan\left(\frac{x_n - 1}{3}\right)$;
upper root: $x_{n+1} = 1 + 3\arctan(x_n - 2)$

c lower and upper roots:
$x_{n+1} = \sqrt[3]{3x_n^2 + 5x_n - 3}$;
central root: $x_{n+1} = \frac{(x_n^3 - 3x_n^2 + 3)}{5}$

6 a α b γ

7 a Converges to P. b Converges to P.

c Converges to S.

8 a $g(x) = x - \frac{f(x)}{f'(x)}$

b $g'(x) = \frac{f(x)f''(x)}{(f'(x))^2}$, so $g'(a) = 0$

9 a Proof.

b $x < -2\ln 2$; the positive root.

10 a 1 and 2 b β

c $c = 1, k = 15, n = -3; \alpha = 0.067$

11 a 0 or $\frac{k-1}{k}$

b $2 - k$

c $1 < k < 3$

MIXED PRACTICE 14

1 A

2 B

3 a Proof.

b 0.248

4 a-b Proof.

c 2.77

5 a 2 solutions;

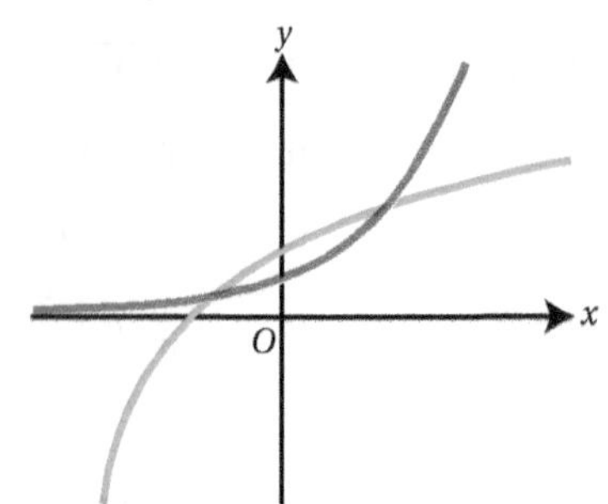

b Proof.

c 1.13

6 a–b Proof.

c 2.065

7 9.859

8 a Proof.

b $x_1 = 1.5$

c x_1 is on the other side of the asymptote.

d −1.365

9 a–b Proof.

c $x = \pi + \arctan\left(\frac{x}{2}\right)$; 4.275

10 a Proof.

b 3.880, 3.918

c Proof.

11 a 6.5 m

b i Proof.

ii Proof.

iii $t = -1.53$ seconds. Not suitable in context, the required solution must be positive.

iv $t_{n+1} = \frac{15 - 6.5e^{-t_n}}{9.8}$

v 1.36 seconds

c 3.08 seconds

15 Numerical integration

BEFORE YOU START

1 7.5

2

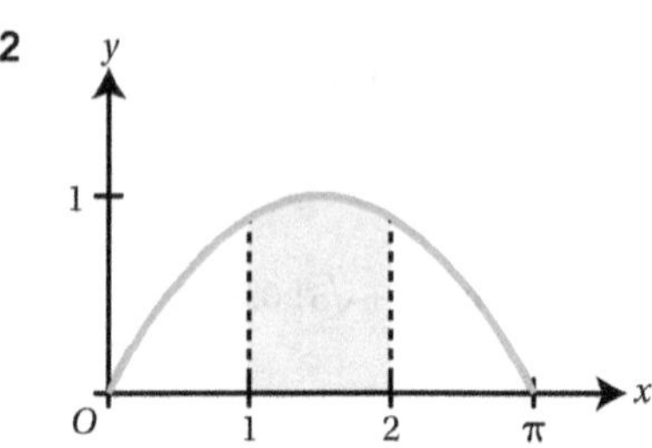

3 63 m

WORK IT OUT 15.1

Solution 2 is correct.

EXERCISE 15A

1 a i

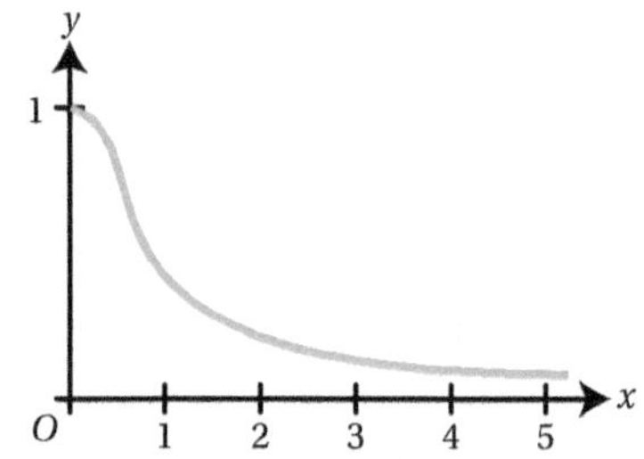

Lower bound: 0.386; upper bound: 1.39.

ii

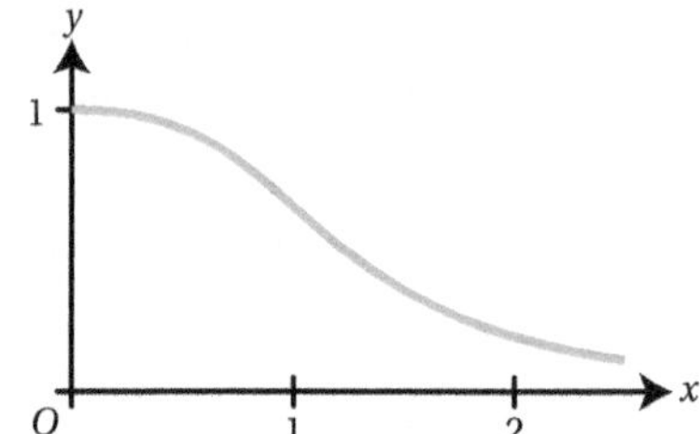

Lower bound: 0.910; upper bound: 1.27.

b i

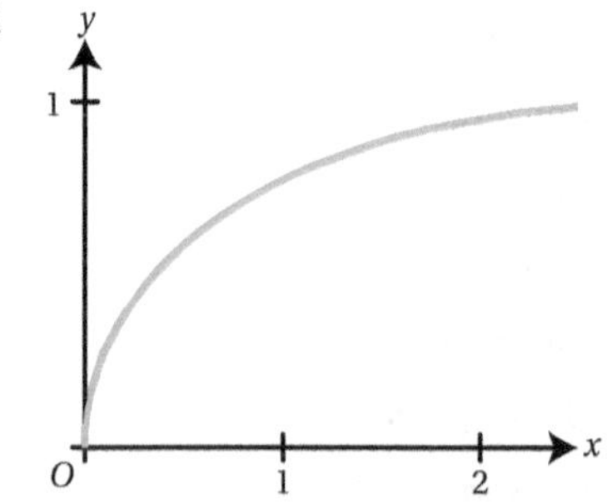

Lower bound: 1.29; upper bound: 1.68.

ii

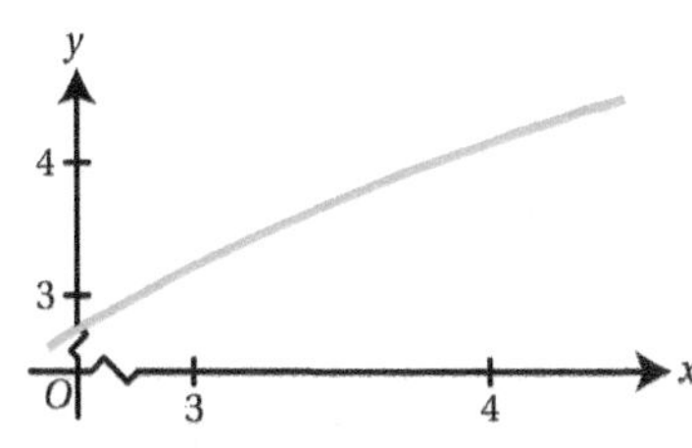

Lower bound: 3.61; upper bound: 3.79.

c i

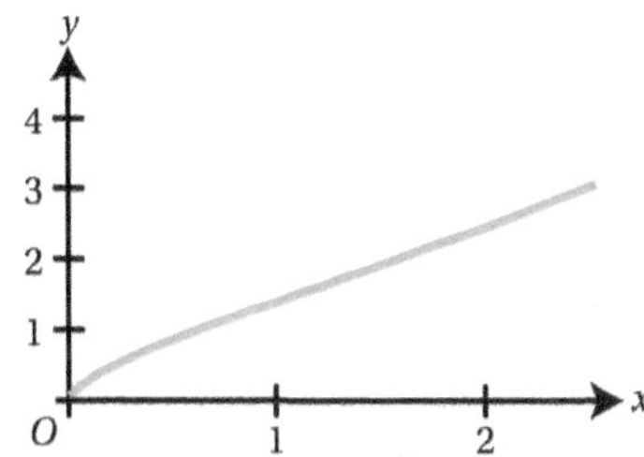

Lower bound: 2.74; upper bound: 3.99.

ii

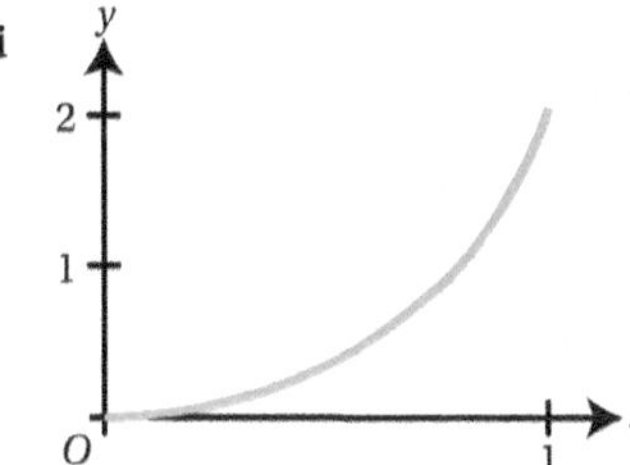

Lower bound: 0.264; upper bound: 0.576.

2 a i Difference 0.072 21.

ii Difference 0.036 10.

iii Difference 0.018 05.

b i Difference 0.186 31.

ii Difference 0.093 15.

iii Difference 0.046 58.

c i Difference 0.2451.

ii Difference 0.1226.

iii Difference 0.0613.

d i Difference 0.031 56.

ii Difference 0.015 78.

iii Difference 0.007 89.

In general, doubling the number of rectangles from 10 to 20 or from 20 to 40 halves the difference between the upper and lower bounds.

3 Lower bound: 1.83; upper bound: 2.33.

4 a Lower bound: 2.89; upper bound: 4.97.

b More rectangles.

5 a

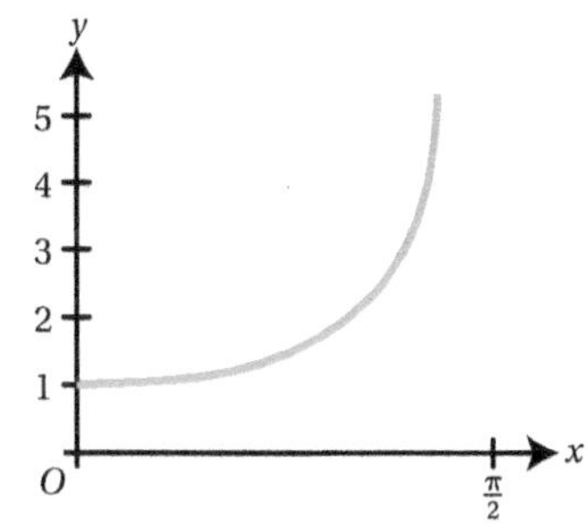

b 1.35 c Decrease.

6 $\frac{\pi}{4}$

7 a $\left(\frac{\pi}{2}, 1\right)$ b 2.01

WORK IT OUT 15.2

Solution 3 is correct.

EXERCISE 15B

1 a i 0.886 ii 1.09

b i 1.48 ii 3.70

c i 3.38 ii 0.455

2 i and ii See table in Worked solutions.

iii Usually error decreases by a factor of 3 to 4.

3 a Approximation 1.26.

b $2\left(\sqrt{5}-\sqrt{2}\right)$

c $\frac{\pi}{2}$

d Approximation 0.609.

e $3\ln(6.75)-3$

f $\ln\left(\frac{\operatorname{cosec}(0.5)+\cot(0.5)}{\operatorname{cosec}(2)+\cot(2)}\right)^2 \approx 3.62$

4 a

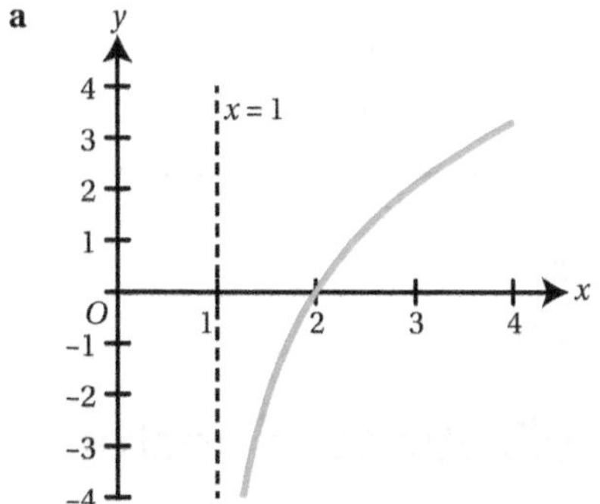

b 3.86

c Concave curve: underestimate.

5 a 1.98

b More trapezia; exact integration is possible.

6 a $a = \sqrt{\frac{\pi}{2}}$

b 0.957

c Concave curve: underestimate.

7 10.2 m

8 a $p = \pi^2$, $q = 4\pi^2$

b 22.3 m

MIXED PRACTICE 15

1 D

2 18.1

3 a 1650

b Concave curve: underestimate.

c More intervals.

4 1.39

5 a 6.43

b More intervals.

6 $L = 4.14$, $U = 5.20$,

7 B

8 2.50

9 a $t = 0, \frac{\pi}{3}, \frac{2\pi}{3}, \frac{4\pi}{3}, \frac{5\pi}{3}, \pi$ and 2π

b 1.05

10 $K = \left(\frac{2542}{17}\right)^5$

11 a

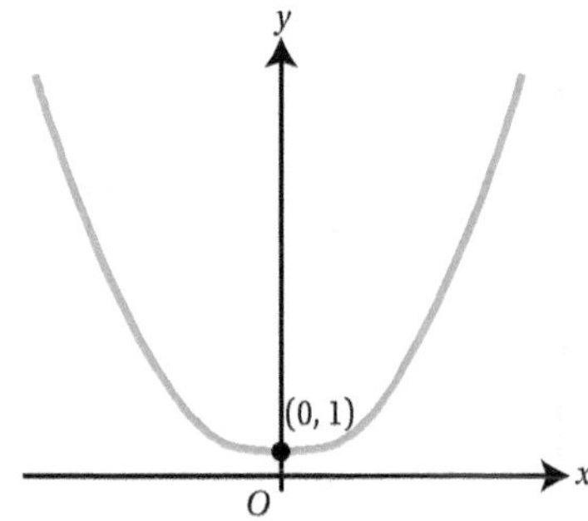

b 2.09

12 4.35 m s^{-1}

16 Applications of vectors

BEFORE YOU START

1 a $\begin{pmatrix} -3 \\ -2 \end{pmatrix}$

b $\begin{pmatrix} 11 \\ -3 \end{pmatrix}$

c $\begin{pmatrix} 32 \\ -20 \end{pmatrix}$

d $\begin{pmatrix} 5 \\ -7 \end{pmatrix}$

2 $\sqrt{13}$; 146° from horizontal.

3 a −12 m s^{-1} b 48 m s^{-1}

4 a 34.2 b $x = 2e^t - \frac{1}{3}t^3 - 2$

5 0.333 m s^{-2}

6 $x = -1 + 4y - 2y^2$

EXERCISE 16A

1 a i $(-\mathbf{i} + 1.67\mathbf{j})$ m s^{-1}

ii $(-2.25\mathbf{i} + \mathbf{j})$ m s^{-1}

b i $(1.2\mathbf{i} + 0.2\mathbf{j})$ m s^{-1}

ii $(0.75\mathbf{i} - 1.25\mathbf{j})$ m s^{-1}

c i $(-1.13\mathbf{i} + 0.5\mathbf{j})$ m s^{-1}

ii $(0.545\mathbf{i} + 0.091\mathbf{j})$ m s^{-1}

2 a i $(0.2\mathbf{i} + 0.5\mathbf{j})$ m s^{-2}; 0.539 m s^{-2}

ii $(0.5\mathbf{i} + 0.625\mathbf{j})$ m s^{-2}; 0.800 m s^{-2}

b i $(0.8\mathbf{i} - 0.4\mathbf{j})$ m s^{-2}; 0.894 m s^{-2}

ii $(-0.3\mathbf{i} + 0.4\mathbf{j})$ m s^{-2}; 0.5 m s^{-2}

3 a $(1.5\mathbf{i} + 0.33\mathbf{j})$ m s^{-1}; 1.54 m s^{-1}

b $(0\mathbf{i} + 0\mathbf{j})$ m s^{-1}; 1.24 m s^{-1}

4 a 6.08 m b $y = x^2 + 6x + 11$

5 a 2.11 m s^{-2} b 128° from horizontal.

6 a 0.950 m s^{-1} b $(6\mathbf{i} - 2\mathbf{j})$ m

c $(0.316\mathbf{i} - 0.105\mathbf{j})$ m s^{-1}

d 0.972 m s^{-1}; the particle changes direction during the motion.

7 $\begin{pmatrix} 8 \\ 2 \end{pmatrix}$ m

8 a 9.22 m

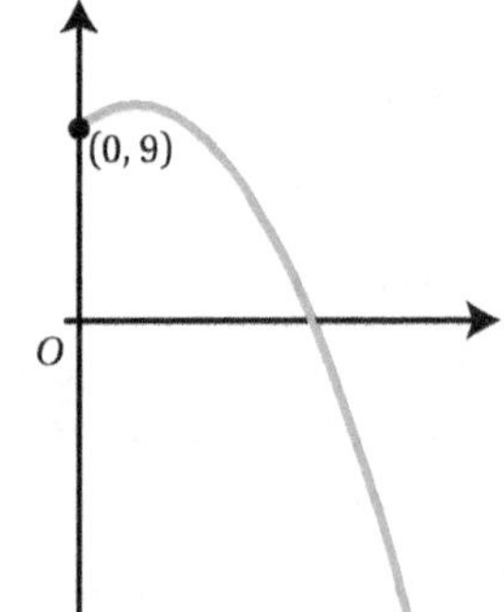

b y (0, 9) O x

9 a $x+2y=-5$

b $\sqrt{5}=2.24$ m

10 4 m

WORK IT OUT 16.1

Solution 3 is correct.

EXERCISE 16B

1 a i $\begin{pmatrix}0.6\\1.8\end{pmatrix}$ m s^{-1}

ii $(12.4\mathbf{i}-2.2\mathbf{j})$ m s^{-1}

b i $(-6.25\mathbf{i}-7.5\mathbf{j})$ m

ii $(6.3\mathbf{i}+0.15\mathbf{j})$ m

c i $(-3\mathbf{i}-4.5\mathbf{j})$ m s^{-1}

ii $(1.67\mathbf{i}-2.33\mathbf{j})$ m s^{-1}

d i 2 s ii 6 s

2 a i 20.6 m s^{-1} ii 58.3 m s^{-1}

b i 221 m s^{-1} ii 94.9 m s^{-1}

3 $\mathbf{v}=(7.7\mathbf{i}-0.4\mathbf{j})$ m s^{-1}, $\mathbf{s}=(39.2\mathbf{i}+7\mathbf{j})$ m

4 a 25.1 m

b 10.3° from horizontal.

5 a $(-2\mathbf{i}+1.5\mathbf{j})$ m s^{-2} b 4 s

6 1.02 m s^{-2}

7 153° from horizontal.

8 1.89 m s^{-2}

9 10 s

10 $(3\mathbf{i}+0.5\mathbf{j})$ m s^{-1}

11 Proof.

12 4.92 m

WORK IT OUT 16.2

Solution 2 is correct.

EXERCISE 16C

1 a i $\mathbf{v}=(3-\cos t)\mathbf{i}+(1-2t)\mathbf{j}$,

$\mathbf{a}=\sin(t)\mathbf{i}-2\mathbf{j}$,

$v(3)=6.40$ m s^{-1}

ii $\mathbf{v}=(2\mathrm{e}^{2t}-1)\mathbf{i}+(2t+2\mathrm{e}^{2t})\mathbf{j}$,

$\mathbf{a}=(4\mathrm{e}^{2t})\mathbf{i}+(2+4\mathrm{e}^{2t})\mathbf{j}$,

$v(3)=1145$ m s^{-1}

b i $\mathbf{v}=-12\sin(3t)\mathbf{i}+6\cos(2t)\mathbf{j}$,

$\mathbf{a}=-36\cos(3t)\mathbf{i}-12\sin(2t)\mathbf{j}$,

$v(3)=7.59$ m s^{-1}

ii $\mathbf{v}=\left(\frac{3}{t+1}-1\right)\mathbf{i}+\left(2t+\frac{1}{t+1}\right)\mathbf{j}$,

$\mathbf{a}=\left(-\frac{3}{(t+1)^2}\right)\mathbf{i}+\left(2-\frac{1}{(t+1)^2}\right)\mathbf{j}$,

$v(3)=6.25$ m s^{-1}

2 a i $\mathbf{v}=\left(3t-\frac{1}{3}t^3+2\right)\mathbf{i}+(t^2+5)\mathbf{j}$;

$\mathbf{r}=\left(\frac{3}{2}t^2-\frac{1}{12}t^4+2t\right)\mathbf{i}+\left(\frac{1}{3}t^3+5t\right)\mathbf{j}$;

distance = 27.2 m

ii $\mathbf{v}=\left(\frac{1}{2}t^2+t+1\right)\mathbf{i}+(3t-2)\mathbf{j}$;

$\mathbf{r}=\left(\frac{1}{6}t^3+\frac{1}{2}t^2+t\right)\mathbf{i}+\left(\frac{3}{2}t^2-2t\right)\mathbf{j}$;

distance = 14.2 m

b i $\mathbf{v}=(3\mathrm{e}^t-4)\mathbf{i}+(-2\mathrm{e}^{-t}+3)\mathbf{j}$;

$\mathbf{r}=(3\mathrm{e}^t-4t-3)\mathbf{i}+(2\mathrm{e}^{-t}+3t-2)\mathbf{j}$;

distance = 45.8 m

ii $\mathbf{v}=\left(-\frac{3}{2}\cos 2t+\frac{9}{2}\right)\mathbf{i}+\left(\frac{3}{2}\sin 2t\right)\mathbf{j}$;

$\mathbf{r}=\left(-\frac{3}{4}\sin 2t+\frac{9}{2}t\right)\mathbf{i}+\left(-\frac{3}{4}\cos 2t+\frac{3}{4}\right)\mathbf{j}$;

distance = 13.7 m

c i $\mathbf{v}=(2\sin t)\mathbf{i}+(-3\cos t+3)\mathbf{j}$;

$\mathbf{r}=(-2\cos t+2)\mathbf{i}+(-3\sin t+3t)\mathbf{j}$;

distance = 9.46 m

ii $\mathbf{v}=\left(\frac{1}{2}\mathrm{e}^{2t}-\frac{1}{2}\right)\mathbf{i}+(3\mathrm{e}^t-3)\mathbf{j}$;

$\mathbf{r}=\left(\frac{1}{4}\mathrm{e}^{2t}-\frac{1}{2}t-\frac{1}{4}\right)\mathbf{i}+(3\mathrm{e}^t-3t-3)\mathbf{j}$;

distance = 110 m

3 a $\mathbf{v}=(2\mathrm{e}^{2t})\mathbf{i}+\mathbf{j}$ b 44 100 m s^{-1} (3 s.f.)

4 a $\mathbf{v}=\begin{pmatrix}3t^2\\\frac{1}{2}\sin 2t\end{pmatrix}$

b $\begin{pmatrix}27\\0.009\,96\end{pmatrix}$ m

5 a 3.61 m s^{-1} b 3.72 m s^{-2}

c $\left(\left(3t+\frac{1}{2}\cos(2t)-\frac{1}{2}\right)\mathbf{i}+\sin(2t)\mathbf{j}\right)$ m

6 2.35 m

7 Proof.

8 a $\frac{5}{7}$ b $\begin{pmatrix}77\\55\end{pmatrix}$

9 a–b Proof.

10 1.06 s; 2.18 m

EXERCISE 16D

1 a i $\begin{pmatrix}4\\0\\0\end{pmatrix}$ ii $\begin{pmatrix}0\\-5\\0\end{pmatrix}$

b i $\begin{pmatrix}3\\0\\1\end{pmatrix}$ ii $\begin{pmatrix}0\\2\\-1\end{pmatrix}$

2 a i $\begin{pmatrix}21\\3\\36\end{pmatrix}$ ii $\begin{pmatrix}20\\-8\\12\end{pmatrix}$

b i $\begin{pmatrix}2\\3\\9\end{pmatrix}$ ii $\begin{pmatrix}6\\-1\\5\end{pmatrix}$

c i $\begin{pmatrix}11\\-3\\8\end{pmatrix}$ ii $\begin{pmatrix}-3\\5\\6\end{pmatrix}$

d i $\begin{pmatrix}10\\-3\\11\end{pmatrix}$ ii $\begin{pmatrix}17\\6\\35\end{pmatrix}$

3 a i $-5\mathbf{i}+5\mathbf{k}$ ii $4\mathbf{i}+8\mathbf{j}$

b i $\mathbf{i}-3\mathbf{j}+3\mathbf{k}$ ii $2\mathbf{j}+\mathbf{k}$

c i $4\mathbf{i}+7\mathbf{k}$ ii $5\mathbf{i}-4\mathbf{j}+15\mathbf{k}$

4 $|\mathbf{a}|=\sqrt{21}, |\mathbf{b}|=\sqrt{2}, |\mathbf{c}|=\sqrt{21}, |\mathbf{d}|=\sqrt{2}$

5 a i $\sqrt{19}$ ii $\sqrt{38}$

b i $\sqrt{74}$ ii $\sqrt{13}$

6 a $\sqrt{53}$ b $\sqrt{94}$

c $\sqrt{53}$ d $\sqrt{2}$

7 a $-4\mathbf{i}+2\mathbf{j}-\mathbf{k}$ b $-\frac{8}{3}\mathbf{i}+\frac{4}{3}\mathbf{j}-\frac{2}{3}\mathbf{k}$

c $4\mathbf{i}-3\mathbf{j}+\mathbf{k}$ d $-\frac{1}{2}\mathbf{i}+\mathbf{j}-\frac{1}{2}\mathbf{k}$

8 $\begin{pmatrix}2\\0\\-\frac{3}{4}\end{pmatrix}$

9 −2

10 $\pm 2\sqrt{6}$

11 $3, -\frac{5}{3}$

12 $-2, -\frac{23}{15}$

13 $t=\frac{1}{3}$; $d=\sqrt{\frac{14}{3}}$

EXERCISE 16E

1 a i $\frac{1}{3}\begin{pmatrix}2\\2\\1\end{pmatrix}$ ii $\frac{1}{3}\begin{pmatrix}2\\2\\-1\end{pmatrix}$

b i $\frac{1}{\sqrt{3}}\begin{pmatrix}1\\1\\1\end{pmatrix}$ ii $\frac{1}{5}\begin{pmatrix}4\\-1\\2\sqrt{2}\end{pmatrix}$

2 a $\begin{pmatrix}1\\-3\\7\end{pmatrix}$ b $\begin{pmatrix}3.5\\-0.5\\1.5\end{pmatrix}$

3 $\begin{pmatrix}0\\-1\\6\end{pmatrix}$

4 $-\frac{4}{3}$

5 −2

6 $\begin{pmatrix}1.6\\0.8\\1.8\end{pmatrix}$

7 a $\frac{3}{2}\mathbf{i}+\frac{3}{2}\mathbf{j}-2\mathbf{k}$ b $\left(\frac{1}{2}, \frac{13}{2}, 0\right)$

8 $p=\frac{3}{8}$, $q=\frac{1}{8}$

9 a $\pm\begin{pmatrix}4\sqrt{2}\\-\sqrt{2}\\\sqrt{2}\end{pmatrix}$ b $\pm\begin{pmatrix}\sqrt{6}\\-\frac{\sqrt{6}}{2}\\\frac{\sqrt{6}}{2}\end{pmatrix}$

10 $\mathbf{m}=\frac{q}{p+q}\mathbf{a}+\frac{p}{p+q}\mathbf{b}$

11 a–b Proof.

MIXED PRACTICE 16

1 B

2 a 9.77 N

b 6.08 m s^{-1}, 9.46° from the horizontal.

3 $\frac{3}{2}$

4 a $-5\mathbf{i}+4\mathbf{j}-2\mathbf{k}$

b Proof.

5 6.25 m s^{-1}

6 a Proof.

b $3\mathbf{i}+6\mathbf{j}+19\mathbf{k}$

7 a $\mathbf{a}=(0.3\mathbf{i}+0.4\mathbf{j})$ m s^{-2}

b 68.0 m

c $(5.5\mathbf{i}+4\mathbf{j})$ m s^{-1}

8 D

9 0.569 s

10 a 014° b 22.0 s

c It stays at a constant height.

11 $\frac{\sqrt{474}}{5}$

12 57.4° below the horizontal.

13 a $\begin{pmatrix} 3t \\ 4t \end{pmatrix}$ b $\begin{pmatrix} 3t \\ 18-5t \end{pmatrix}$

c Proof.

d $t = 2$ e 2 hours

14 a $\mathbf{v} = 6\mathbf{i} + 2.4\mathbf{j} + (-0.8\mathbf{i} + 0.1\mathbf{j})t$

b $\mathbf{r} = 13.6\mathbf{i} + (6\mathbf{i} + 2.4\mathbf{j})t + \frac{1}{2}(-0.8\mathbf{i} + 0.1\mathbf{j})t^2$

c 45.6 m

15 A

16 a $3t\mathbf{i} + (5-4t)\mathbf{j} + t\mathbf{k}$

b $d^2 = 44t^2 - 88t + 74$; Proof.

c 5.48 km

17 a $4x^2 + 25y^2 = 100$

b 5 m s^{-1} when $\mathbf{r} = \pm 2\mathbf{j}$

18 8.96 m

Focus on ... Proof 2

1 Yes, by varying the base (or hypotenuse) accordingly.

2 Yes.

3 Proof.

Focus on ... Problem solving 2

1 a i Tends to 1500.

ii Stays at 1000.

iii Dies out after 48 years.

b Same as in **a**.

$k = 200$ is the largest value without the population dying out. This doesn't depend on the initial size of the population.

2 Increases with α, decreases with β.

3 a i 500, 1500 ii 1000

iii No solutions.

b $0.16 - 0.0008k$

c $x = \frac{0.4 \pm \sqrt{0.16 - 0.0008k}}{0.0004}$

4 a $(\alpha - 1)^2 - 4\beta k$ b Proof.

5–6 Discussion.

Focus on ... Modelling 2

1 $\frac{dN}{dt} = 0.0343N$

2 $\frac{dm}{dt} = -0.005m$

3 $\frac{dV}{dt} = -0.0032V^2$

4 $\frac{dT}{dt} = -0.0263(T - 24)$

5 $\frac{dT}{dx} = -kx$

6 $\frac{dp}{dh} = 1000k(1 + 0.01h)$

7 $\frac{dN}{dt} = \frac{3\sqrt{N}}{t}$; this model predicts infinite initial rate of growth.

8 $\frac{dV}{dt} = -0.0905h$ or $-0.002V$

Cross-topic review exercise 2

1 $a = -1, b = 4$

2 $k = 6$

3 $y = 3\ln|\sec x| + 4$

4 $a = -12, b = -4$

5 a 0.3662 b $a = \frac{1}{3}, b = 3$

6 $\frac{1}{2}\ln|\sec x| + \frac{x}{2} + c$

7 a $-\frac{1}{2}\sec t$ b $\frac{7}{4}$

8 $\frac{1}{8}(\pi - 2)$

9 a i 194.5°, 345.5° ii 82.2°, 157.8°

b Translate $\begin{pmatrix} -30 \\ 0 \end{pmatrix}$ and then stretch scale factor $\frac{1}{2}$ parallel to x-axis.

10 a $0, \pi$ b 2π

c $y = \sin\sqrt{x}$

11 $x^{\sin x}\left(\cos x \ln x + \frac{\sin x}{x}\right)$

12 a Proof.

b $\left(\frac{5\pi}{6}, 2\right)$

13 a $2\sin(Bx)\cos(Ax)$

b $\frac{1}{4}\cos 2x - \frac{1}{16}\cos 8x + c$

14 a Proof. b $h = 30 - 2t$

c $25 - 6t$ d $13\text{ cm}^2\text{ s}^{-1}$

15 $x \in \left[0, \frac{\pi}{6}\right) \cup \left(\frac{5\pi}{6}, 2\pi\right]$

16 a $|\cos x| < 1$ for $0 < x < \pi$

b Proof.

c $\sqrt{3} - 1$

17 a i $(\pi - x, \sin x)$

ii $(\pi - 2x)\sin x$

b i–iii Proof.

c i 0.710 ii 1.12

18 a $A = 4,\ B = -2,\ C = 1$

b $y^{\frac{1}{2}} = 2\ln\left(\frac{3-3x}{3-2x}\right) + \frac{x}{1-x}$

19 a Stretch scale factor $\frac{1}{2}$ parallel to x-axis.

Translation $\begin{pmatrix} 0 \\ -1 \end{pmatrix}$.

b (0, 6)

c i Proof.

ii ln 2

d 3 ln 2

20 4^n

21 a $\frac{1-x^{n+1}}{1-x}$

b Proof.

22 a Proof.

b $\arcsin a$

c $1-\sqrt{1-a^2}$

d $a \arcsin a + \sqrt{1-a^2} - 1$

23 a $2 + \cos 2x - 3\sec^2 x$

b–c Proof.

24 a $a \geqslant -2$ b $a = -1$ c $-3 < x < -1$

25 a

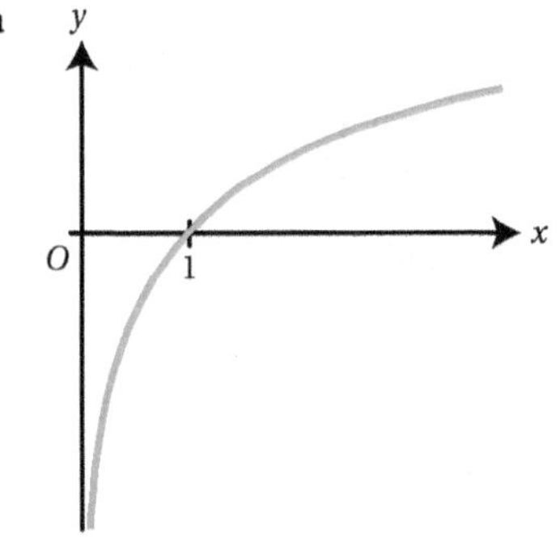

b e

c $0 < k < \frac{1}{e}$

26 a–c Proof.

d i $x = 1, 2$

ii $-(x^2 - 2x + 2)^{-2}(2x - 2) + x^{-2}$

iii Proof.

17 Projectiles

BEFORE YOU START

1 $\sqrt{13}$ at 146° from vector **i**.

2 $0.333\ \text{m s}^{-2}$

3 $(6\mathbf{i} + 2\mathbf{j})\ \text{m s}^{-1}$

4 $\frac{1}{2}\sin 2x$

5 $\sec^2 x$

6 45°, 124°

EXERCISE 17A

1 a i $(5\mathbf{i} + 1.4\mathbf{j})\ \text{m s}^{-1}$ ii $(7\mathbf{i} - 14\mathbf{j})\ \text{m s}^{-1}$

b i $(8\mathbf{i} - 23\mathbf{j})\ \text{m s}^{-1}$ ii $(4\mathbf{i} - 29\mathbf{j})\ \text{m s}^{-1}$

c i $(11\mathbf{i} - 11\mathbf{j})\ \text{m s}^{-1}$ ii $(11\mathbf{i} + 3.1\mathbf{j})\ \text{m s}^{-1}$

d i $(6.6\mathbf{i} - 24\mathbf{j})\ \text{m s}^{-1}$ ii $(20\mathbf{i} - 23\mathbf{j})\ \text{m s}^{-1}$

2 a i $(10\mathbf{i} + 22\mathbf{j})$ m ii $(14\mathbf{i} - 7.6\mathbf{j})$ m

b i $(16\mathbf{i} - 26\mathbf{j})$ m ii $(8\mathbf{i} - 38\mathbf{j})$ m

c i $(21\mathbf{i} - 1.6\mathbf{j})$ m ii $(21\mathbf{i} + 26\mathbf{j})$ m

d i $(13\mathbf{i} - 29\mathbf{j})$ m ii $(39\mathbf{i} - 27\mathbf{j})$ m

3 a 0.77 m (2 s.f.) b $9.2\ \text{m s}^{-1}$ (2 s.f.)

4 a 0.24 s (2 s.f.)

b $13\ \text{m s}^{-1}$; 68° below the horizontal. (2 s.f.)

5 a 10 m (1 s.f.) b 40 m (1 s.f.)

6 63° (2 s.f.)

7 0.71 s (2 s.f.)

8 a 4.1 s (2 s.f.)

b 81 m (2 s.f.)

c $29\ \text{m s}^{-1}$ (2 s.f.)

9 Yes; height is 1.83 m at that point.

10 a 16 m (2 s.f.) b 26° or 64° (2 s.f.)

11 a Proof.

b $24.1° < \alpha < 25.8°$ or $64.2° < \alpha < 65.9°$

12 a Ten times normal.

b Slowed down by a factor of $\sqrt{10} \approx 3.2$.

WORK IT OUT 17.1

Solution 2 is correct.

EXERCISE 17B

1 $y = \frac{-g}{128}x^2$

2 a Proof.

b $19\ \text{m s}^{-1}$ (2 s.f.)

c No air resistance; rugby ball is a particle (has no size and doesn't spin).

3 a Proof.

b Yes

4 a Proof.

b i Proof. ii $13\ \text{m s}^{-1}$ (2 s.f.)

5 $u = 29\ \text{m s}^{-1}$; $\theta = 53°$ (2 s.f.)

6 a Proof.

b 70° (1 s.f.); it must be falling when it goes through the hoop.

MIXED PRACTICE 17

1 B

2 C

3 a 2.5 s (2 s.f.) b 30 m (2 s.f.)

4 a 2.50 s (3 s.f.) b 26.0 m (3 s.f.)

5 a 18 b 15 m s^{-1}

6 a i–ii Proof

b 21°, 69°

7 1.8 m (2 s.f.)

8 a 1.4 s (2 s.f.) b 0.82 s (2 s.f.)

9 a Proof.

b 0.98 m or 98 cm

c No air resistance **or** the ball doesn't spin **or** no loss of energy.

10 a 250 m s^{-1} b 78 m (2 s.f.)

c 250 m s^{-1} (2 s.f.) d 8.9° (2 s.f.)

11 2 m (1 s.f.)

12 a Proof.

b i–ii Proof.

iii 8 m s^{-1} (1 s.f.)

18 Forces in context

BEFORE YOU START

1 $\mathbf{F} = 3\mathbf{i} + \mathbf{j}$; magnitude $\sqrt{10}$ N; angle 18.4° from direction **i**.

2 2 s

EXERCISE 18A

1 a i $7.33\mathbf{i} + 4.5\mathbf{j}$ ii $16.6\mathbf{i} + 3.89\mathbf{j}$

b i $-2.19\mathbf{i} - 0.167\mathbf{j}$ ii $3.62\mathbf{j}$

2 Answers given to 2 s.f.

a i $T_1 = 280$ N, $T_2 = 280$ N

ii $T_1 = T_2 = 78$ N

b i $T_1 = 58$ N, $T_2 = 110$ N

ii $T_1 = 110$ N, $T_2 = 120$ N

3 $15.4\mathbf{i} - 24\mathbf{j}$ N

4 a 41.8° b 112 N

5 a Proof.

b 70° (1 s.f.) c 5 N (1 s.f.)

6 $k = 80.5$ N; $a = 4.05$ m s^{-2}

7 a i 28.1° ii 17 N

b i 32.2° ii 12.7 N

8 7.2 N, 8.8 N (2 s.f.)

9 a $2\sqrt{2}g$ N

b $\theta° = S\hat{A}R - 45°$

c $F = 28$ N (2 s.f.); $S\hat{A}R = 43°$ (2 s.f.)

10 a They are two parts of the same string.

b 4.5 N; 84° (2 s.f.)

EXERCISE 18B

1 a i 0.2 ii 0.005

b i 5 m s^{-2} ii 0.1 m s^{-2}

2 Force > 9000 N

3 0.41 (2 s.f.)

4 2.75 m s^{-2} (3 s.f.)

5 a 0.625 b 0.7 m s^{-2} (1 s.f.)

6 $\mu_A = \frac{1}{3}, \mu_B = \frac{2}{3}$

7 a 6.25 m s^{-2} b 0.625

8 a $\alpha = \tan^{-1}\frac{4}{3} = 53°$ (2 s.f.)

b $25° < \theta < 48°$ (2 s.f.) c 0.15 m s^{-2} (2 s.f.)

9 C and D only.

EXERCISE 18C

1 Answers given to 2 s.f.

a i 0 N ii 0 N

b i 4.9 N ii 10 N

c i 2.0 N ii 3.5 N

2 Answers given to 2 s.f.

a i 0 N; 0.61 m s^{-2} up the slope.

ii 0 N; 1.6 m s^{-2} down the slope.

b i 11 N down the slope; 0.74 m s^{-2} up the slope.

ii 19 N down the slope; 0 m s^{-2}.

c i 15 N up the slope; 0 m s^{-2}.

ii 17 N up the slope; 2.6 m s^{-2} down the slope.

3 a 40 N (1 s.f.)

b 30 m

4 a 230 N (2 s.f.)

b i 6.3 m s^{-2} (2 s.f.)

ii 5.5 m s^{-2} (2 s.f.)

5 24° (2 s.f.)

6 a 3.6 m (2 s.f.) b 0.89 s (2 s.f.) c 1.2 s (2 s.f.)

7 a 4.7 m s^{-2} (2 s.f.)

b 3.7 m (2 s.f.)

8 0.90 (2 s.f.)

9 0.78 s (2 s.f.)

10 a Proof.

b $T - 0.9g\sqrt{2} - F = 2a_R$

c Up the slope with magnitude 1.9 N (2 s.f.).

d Down the slope, with acceleration 0.27 m s^{-2} (2 s.f.) .

11 a 11 m s^{-2} (2 s.f.)

b 14 m s^{-2} (2 s.f.)

c 5.7° (2 s.f.)

12 $\mu > 0.47$ (2 s.f.)

13 Proof.

MIXED PRACTICE 18

1 B

2 a 14 N

b Maximum 18 N, minimum 2 N.

3 4.1 m s^{-2} (2 s.f.)

4 a 0.3 (1 s.f.)

b 10 m s^{-1} (1 s.f.)

5 a

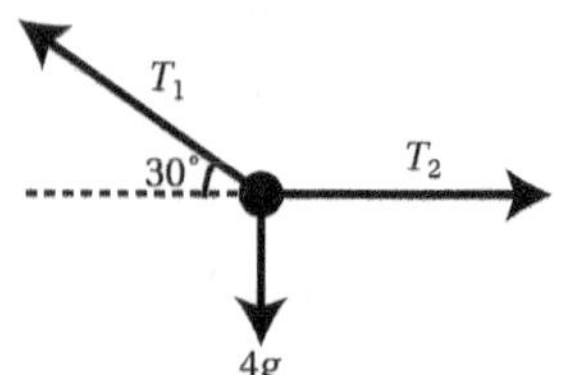

b Proof. c 67.9 N

6 a

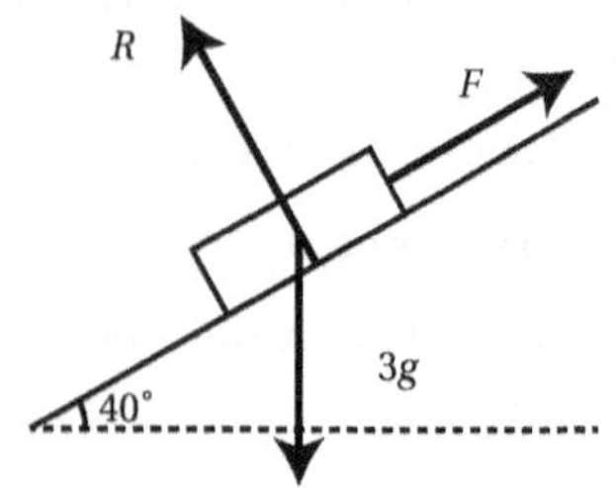

b 23 N (2 s.f.)

c 4.5 N (2 s.f.)

d 4.8 m s^{-2} (2 s.f.)

e For example: No air resistance force acting. No other forces acting on the box. They are the only forces that act. No turning effect (due to forces). Forces are concurrent.

7 a 72.1 N b 33.7°

8 a

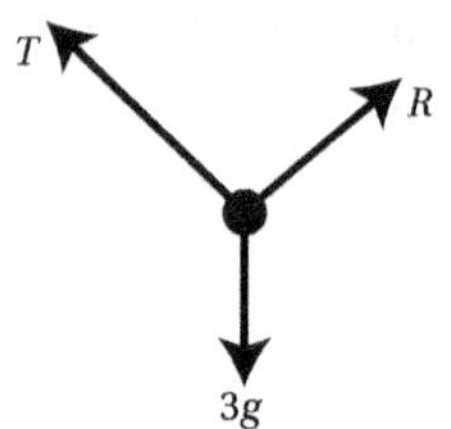

b Proof. c 25.5 N

9 a Proof.

b 5 m s^{-2}

c The acceleration is reduced because of air resistance or the fact that there is friction.

10 a

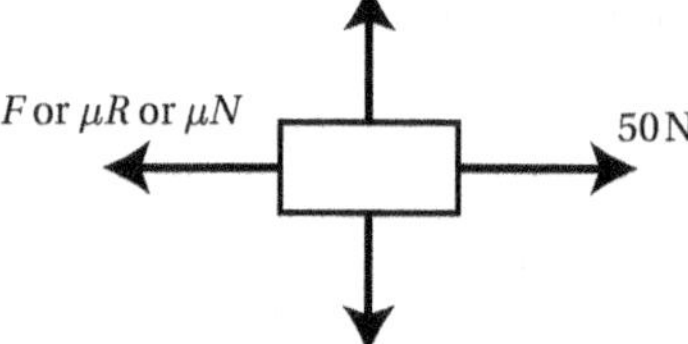

b 39 N (2 s.f.)

c 38 N

d 0.97 (2 s.f.)

e Less friction, so a smaller coefficient of friction.

11 C

12 a 24 N (2 s.f.)

b $a = 2.0\text{ m s}^{-2}$; $T = 39$ N (2 s.f.)

13 a 7 N (1 s.f.) b 100 N (1 s.f.)

14 a $F = 3.1$ N; $\mu = 0.37$ (2 s.f.)

b 1.8 m s^{-2}

15 a 29 m s^{-1} (2 s.f.) down the slope.

b 0.70 m (2 s.f.)

16 Proof.

17 a 34 N (2 s.f.)

b Proof.

c 51.5 N

d 1.4 m s^{-1}

e 3.4 m s^{-1} (2 s.f.)

18 a i 0.84 m s^{-2} ii 7.14 s

b 58.8 N

c i 4.9° (2 s.f.) ii 7.4° (2 s.f.)

d The air resistance force will increase (vary or change) with speed.

19 a

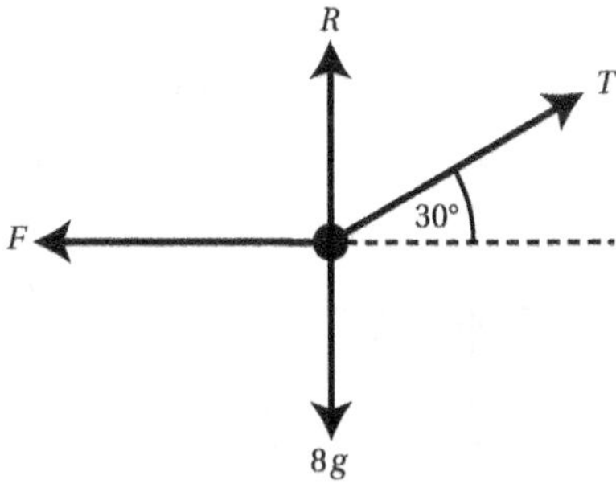

b $78.4 - 0.5T$

c 23.5 N

20 a $a = 1.96\left(\mu\left(\sqrt{3}+3\sqrt{2}\right)-1+1.5\sqrt{2}\right)$ (in direction of A moving up the slope)

b Yes; 7.51 seconds after the start of movement.

21 a i

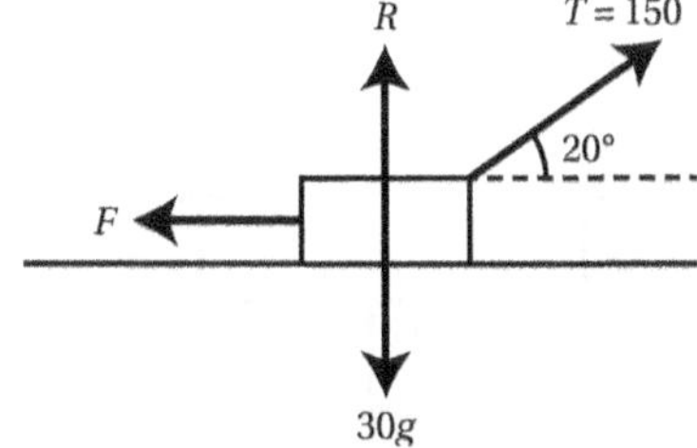

ii Proof. iii 97.1 N

iv 1.46 m s^{-2}

b 109 N c The same.

22 a 0.58 (2 s.f.)

b i $0.5X + 339.5$

ii 690 N (2 s.f.)

23 a

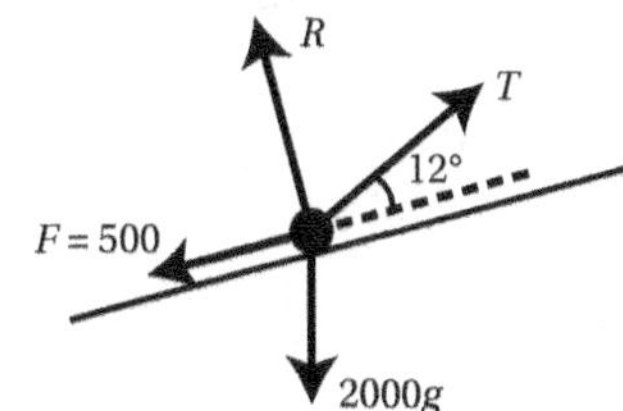

b Proof.

19 Moments

BEFORE YOU START

1

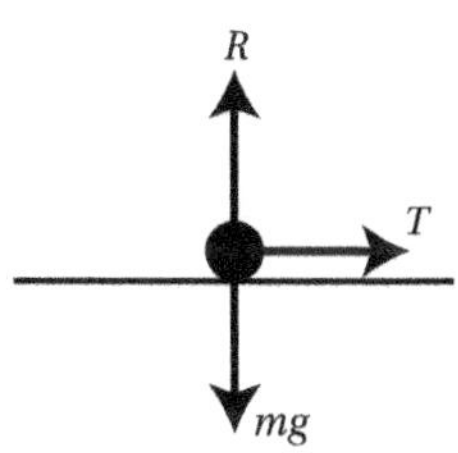

2 7 N

EXERCISE 19A

1 a i 1.5 N m anticlockwise.

ii 1 N m clockwise.

b i 3 N m clockwise.

ii 4 N m anticlockwise.

c i 0 N m ii 0 N m

2 a i 7.5 N m clockwise.

ii 14 N m anticlockwise.

b i 15 N m anticlockwise.

ii 3 N m clockwise.

c i 0 N m ii 0 N m

3 a i 100 N m clockwise.

ii 50 N m clockwise.

b i 0 N m

ii 8 N m clockwise.

c i 50 N m anticlockwise.

ii 100 N m anticlockwise.

4 a i 450 N m clockwise.

ii 450 N m anticlockwise.

b i 350 N m clockwise.

ii 750 N m clockwise.

c i 850 N m clockwise.

ii 650 N m clockwise.

d i 350 N m anticlockwise.

ii 150 N m anticlockwise.

e i 450 N m anticlockwise.

ii 350 N m clockwise.

5 0.9 N m clockwise.

6 250 N m clockwise.

7 a 4000 N m (1 s.f.)

b Since there is no bending, the diver is 3 m from the other end.

8 3 N m clockwise.

9 Proof.

10 0.18 N m (2 s.f.)

WORK IT OUT 19.1

Solution C is correct.

EXERCISE 19B

1 a i $x = 9$ m; $y = 10$ N

ii $x = 5.5$ m; $y = 40$ N

b i $x = 21.4$ N; $y = 28.6$ N

ii $x = 20$ N; $y = 30$ N

2 a i $T_1 = 100$ N; $T_2 = T_3 = 83.3$ N

ii $T_1 = 100$ N; $T_2 = T_3 = 100$ N

b i $T_1 = 100$ N; $T_2 = T_3 = 66.7$ N

ii $T_1 = 100$ N; $T_2 = T_3 = 10$ N

c i $x = 2$ m ii $x = 8$ m

3 0.5 m

4 60 N

5 a 0.333 m

b The crane attachment would be able to handle some net moment.

6 3.2 m

7 a Proof.

b 7 N

8 $3.75g$ N, $1.25g$ N

9 300 N; 700 N

10 a $\frac{80}{x}$ and $\frac{80}{x} - 40$

b The working in a assumes that the pole is rigid and horizontal.

11 20 N; 3 N (1 s.f.)

12 31 kg (2 s.f.)

13 a i $1.5(25 - x)$

ii 60 N

iii $1.5(25 - x)$

b x could be bigger than 25 cm.

14 At X: 90 N; At Y: 175 N

15 74.3 cm (3 s.f.)

16 4.8 m (2 s.f.)

MIXED PRACTICE 19

1 A

2 B

3 0.625

4 a 15 b 15 N upwards.

5 a

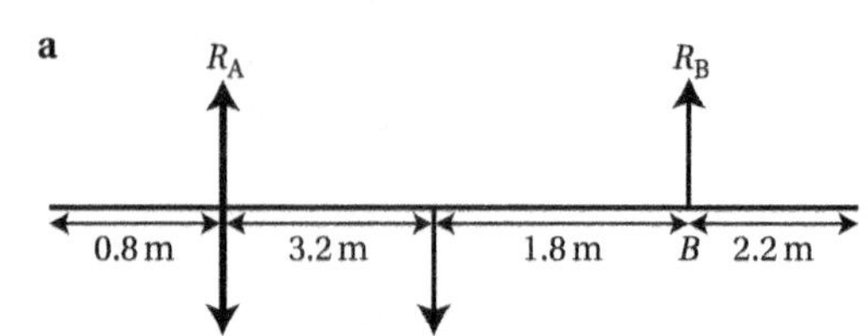

b Proof.

c 300 N (1 s.f.)

d The weight of the plank acts through its centre.

6 a i $T_P = 130$ N (2 s.f.); $T_Q = 120$ N (2 s.f.)

ii The weight of the plank acts through its centre.

b $\frac{25}{29}$

7 a

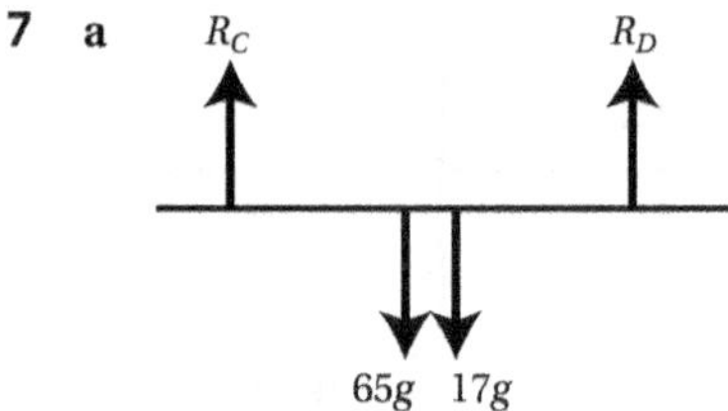

b 0.4 m

c The weight of the plank acts through its centre.

8 1.75 m

9 2 m (1 s.f.)

10 a 315 N b 1.4 m (2 s.f.)

11 a 60 N b 2.375 m

12 a Proof.

b $W = 60$ N; $x = 1.6$ m

13 a Proof.

b $0 \leq x < \frac{11}{4}$

14 $\frac{M - m}{M + m}$

15 $20x$

16 a Proof. b $\frac{g}{3}$ m s^{-2}

Focus on ... Proof 3

1 μW

2 Proof.

3 a i $\sqrt{1+\mu^2}$ ii $F_{\min} = \frac{\mu}{\sqrt{1+\mu^2}} W$

iii Proof.

b Proof.

4 a Proof. b $F_{\min} = \mu W$

Focus on ... Problem solving 3

The answers that are clearly wrong are:
1, 2, 4, 5, 7, 9, 10, 11.

Focus on ... Modelling 3

1 10^{23} m

2 The gravitational acceleration on Archimedes's planet is also 9.8 m s^{-2}.

3 The support needs to be further away from the point where the Earth is resting on the lever. If this distance is increased to d m, the length of the other end would need to be $d \times 10^{23}$ m.

4 5×10^{-16} m

Cross-topic review exercise 3

1 a $-3\mathbf{i}+3\mathbf{j}-6\mathbf{k}$ b $3\sqrt{6}$

2 a $(-5, 1, 0)$ b Proof.

c $(8, -7, 9)$

3 a $\left(e^{\frac{1}{2}t}-8\right)\mathbf{i}+(2t-6)\mathbf{j} \text{ m s}^{-1}$

b i 3.52 m s^{-1} ii West.

c $(0.5e^{\frac{3}{2}}\mathbf{i}+2\mathbf{j}) \text{ m s}^{-2}$

d 21.0 N

4 $7.54g$; $11.5g$

5 a 0.225 m to the right of the support.

b $11g$ N

6 a 37° (2 s.f.) b 7.6 (2 s.f.)

7 $a = \frac{11}{3}; b = \frac{4}{3}$

8 Proof.

9 a $12\mathbf{i}+8\mathbf{j}$ N b **Proof.**

c i Proof.

ii $-20\mathbf{i}-8\mathbf{j}+(3\mathbf{i}+2\mathbf{j})t$

iii 4 or 7.69 s

10 a $\left(2\cos\left(\frac{\pi}{2}t\right)\mathbf{i}+3t^2\mathbf{j}\right) \text{ m s}^{-2}$

b $\left(\frac{4}{\pi}\sin\left(\frac{\pi}{2}t\right)-3\right)\mathbf{i}+(t^3-8)\mathbf{j} \text{ m s}^{-1}$

c 2

d 3 m s^{-1}

11 a 3.04 N (3 s.f.) b 3.04 N upwards (3 s.f.)

12 a

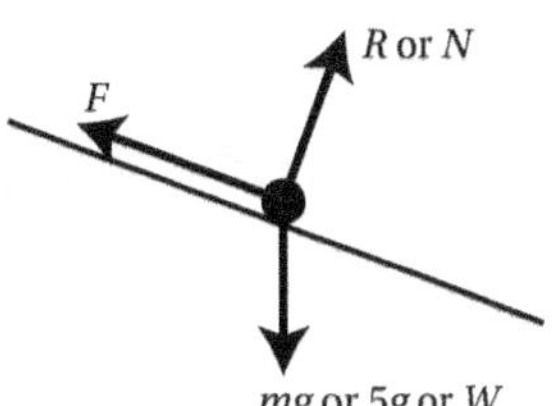

b Proof.

c 0.733

d There is less friction so the coefficient of friction must be less.

13 a $0.3g$

b 0.9 s (1 s.f.)

c 30 N (1 s.f.), 45° below the horizontal (to the left).

d i The tension is the same on both sides.

ii The two blocks have the same speed and acceleration.

14 a 19.6 N

b

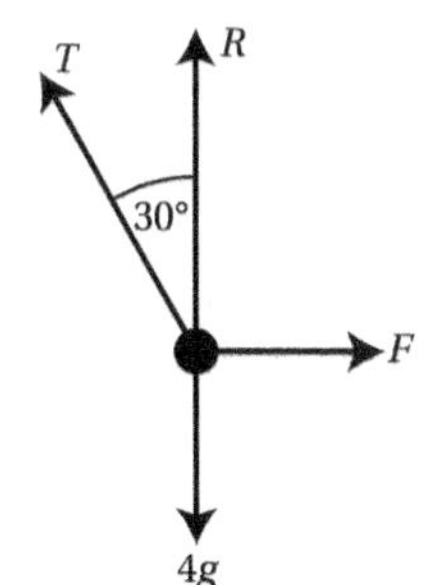

c Proof.

d 0.441 (3 s.f.)

15 a Proof.

b 14 m (2 s.f.)

c 12 m s^{-1} (2 s.f.)

d 33° (2 s.f.)

e The weight is the only force acting OR no air resistance.

16 a Proof.

b Proof.

17 Proof.

18 a It is a particle/no air resistance/lift forces act on the ball.

b Proof.

c $28\text{ ms}^{-1} < V < 29\text{ ms}^{-1}$ (2 s.f.)

19 a Proof.

b i $\sqrt{109}$ ii 6 m s^{-1} (1 s.f.)

20 Conditional probability

BEFORE YOU START

1 $\frac{1}{2}$

2 2

3 $\frac{1}{6}$

4 $\frac{15}{64}$

EXERCISE 20A

1 a i 0.5 ii 1

b i $\frac{1}{6}$ ii 0.05

c i $\frac{4}{15}$ ii 0.2

d i 0.7 ii 0.11

2 a i $\frac{7}{20}$ ii all (100%)

b i 27.5% ii $\frac{7}{30}$

3 a 0.2 b 0.4

4 55%

5 0.3

6 10%

7 0.2

8 a 0.6 or 0.25 b 0.6 or 0.25

EXERCISE 20B

1 a P(prime $\cap$ odd)

b P(Senegal $\cup$ Taiwan)

c P(French | A Levels)

d P(heart | red)

e P(lives in Munich | German)

f P(not black $\cap$ not white)

g P(potato | not cabbage)

h P(red | red $\cup$ blue)

2 a i $\frac{2}{3}$ ii $\frac{15}{28}$

b i $\frac{5}{14}$ ii $\frac{5}{6}$

3 a 0.5 b $\frac{1}{3}$

4 a $P(x > 4)$

b $P(y \leqslant 3)$

c 0

d $P(a \in R)$

e P(fruit)

f P(apple)

g P(multiple of 4)

h P(rectangle)

i P(cat)

j 0

k P(blue)

l P(blue $\cap$ red)

m 1

n 0

5 a

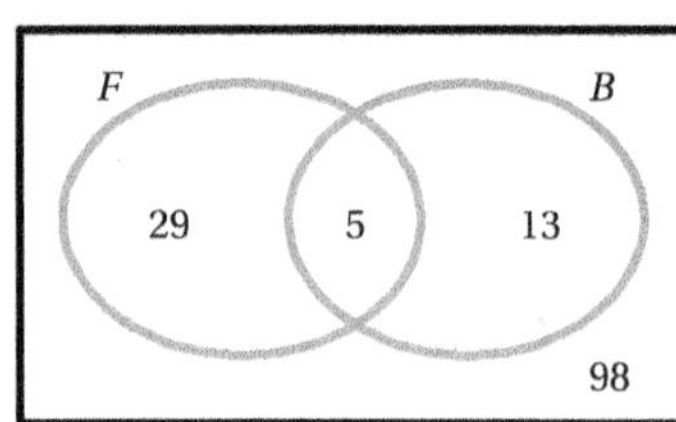

b 98

c $\frac{18}{145}$

d $\frac{5}{34}$

6 a

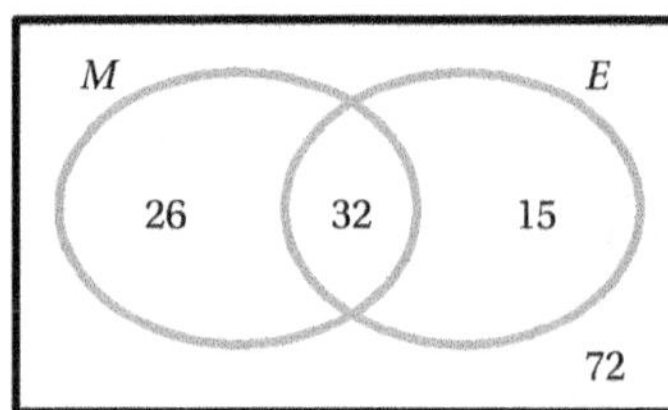

b 32

c $\frac{16}{29}$

7 a 5% b $\frac{5}{6}$

8 a 0.6 b $\frac{1}{6}$

9 a 0.166 b 0.041

c 0.247

10 a 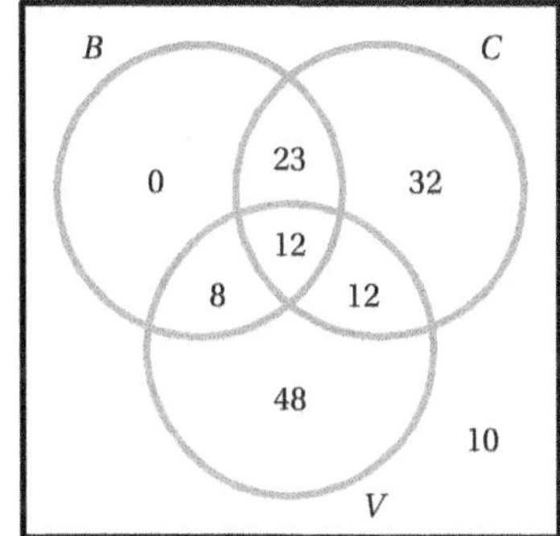

b 0 c 79

d $\frac{16}{29}$ e $\frac{3}{5}$

f $\frac{11}{29}$

11 a 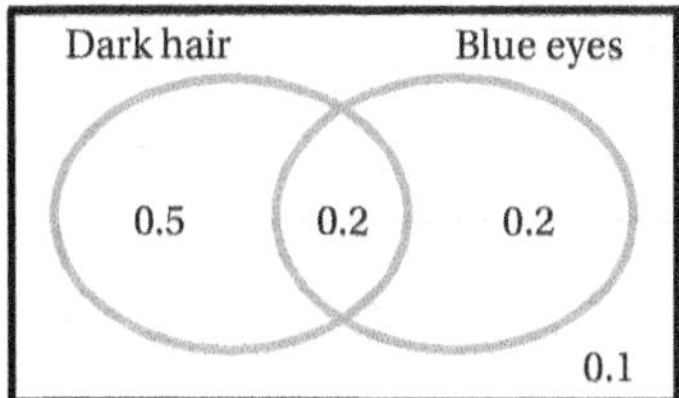

b 0.1

c $\frac{2}{7}$

d $\frac{2}{3}$

e No: $P(B \cap D) = 0.2 \neq P(B) \times P(D) = 0.28$

12 a 0.3

b 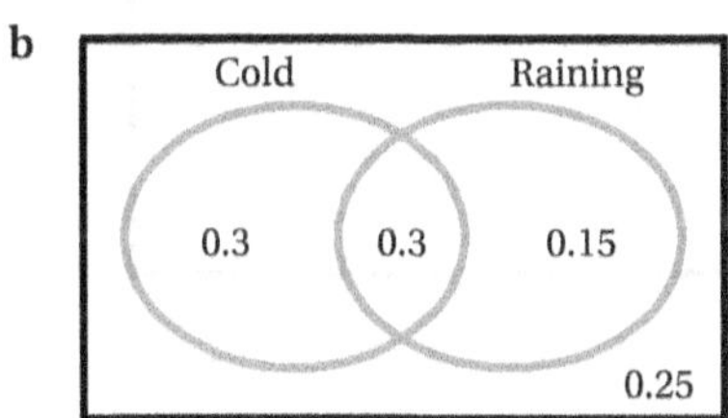

c $\frac{1}{3}$

d $\frac{3}{8}$

e No: $P(C \cap R) = 0.3 \neq P(C) \times P(R) = 0.27$

13 4

14 0.75

15 $0.5 \leqslant P(A|B) \leqslant 1$

16 a $0 \leqslant P(X) \leqslant 1$

b $P(A)(1 - P(B|A))$

c Proof.

EXERCISE 20C

1 a i 0.4 ii 0.3

b i $\frac{1}{3}$ ii $\frac{5}{13} = 0.38$

c i 0.25 ii 0.15

2 a

	Year 9	Year 10	Year 11	Total
Girls	98	85	86	269
Boys	88	75	77	240
Total	186	160	163	509

b $\frac{86}{509}$

c $\frac{86}{269}$

3 a $\frac{7}{15}$ b $\frac{7}{15}$

c Proof.

4 a $\frac{1}{3}$

b $\frac{7}{25}$

c $\frac{5}{14}$

5 a $\frac{a}{a+b+c+d}$

b $\frac{a+b+c}{a+b+c+d}$

c $\frac{a+c}{a+b+c+d}$

d $\frac{a}{a+c}$

e $\frac{b}{b+d}$

6 a $\frac{27}{67}$ b $\frac{99}{258}$

c $\frac{27}{258}$ d $\frac{139}{258}$

7 a $\frac{14}{223}$ b $\frac{101}{223}$

c $\frac{61}{223}$ d $\frac{7}{27}$

e $\frac{122}{169}$

8 a i 0.49 ii 0.10 iii 0.513

9 For example, numbers on a dice with A = 'even' and B = 'prime' shows **1** is wrong.

WORK IT OUT 20.1

Solution A is correct.

EXERCISE 20D

1 a i 0.12 ii 0

b i 0.24 ii 0.24

c i $\frac{13}{20}$ ii $\frac{3}{4}$

2 $\frac{15}{91}$

3 $\frac{1}{6}$

4 a

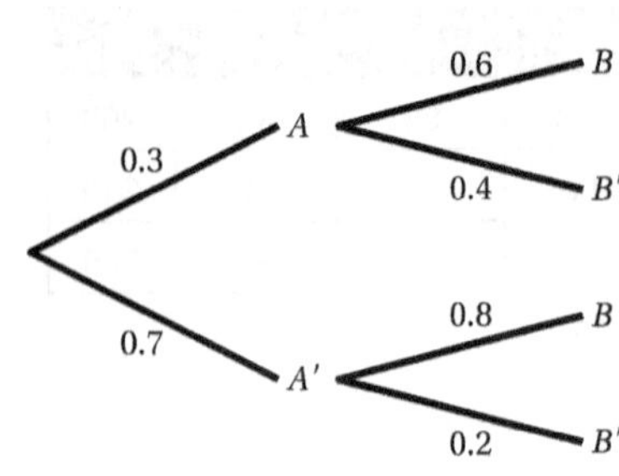

b 0.18 **c** 0.86

d 0.74

5 a $\frac{19}{27}$ **b** $\frac{25}{27}$

6 4.8%

7 $\frac{69}{110}$

8 a $\frac{149}{240}$ **b** $\frac{125}{149}$

9 a $\frac{64}{75}$ **b** $\frac{5}{32}$

10 a $\frac{9}{20}$ **b** $\frac{9}{44}$

11 a $\frac{2}{3}$ **b** $\frac{4}{5}$

12 0.0277

13 $\frac{1}{3}$

14 15 or 21

15 Proof.

MIXED PRACTICE 20

1 C

2 B

3 $\frac{49}{153}$

4 a

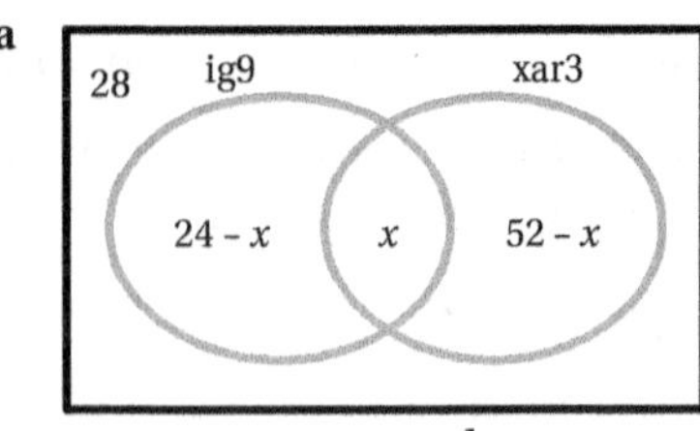

b 4% **c** $\frac{1}{13}$

5 a i 0.293 **ii** 0.032

iii 0.369 **iv** 0.291 **v** 0.328

6 a

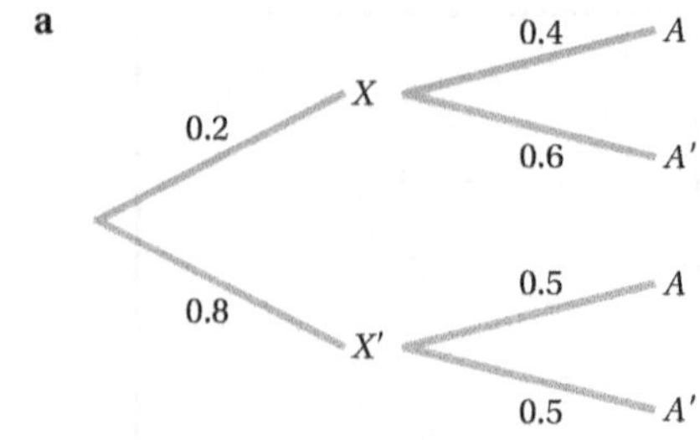

b 0.08

c 0.6

d 0.48

7 a $\frac{1}{2}$ **b** $\frac{171}{780}$

8 a i 0.855 **ii** 0.14

b i 0.72 **ii** 0.684

iii 0 **iv** 0.00425

9 a

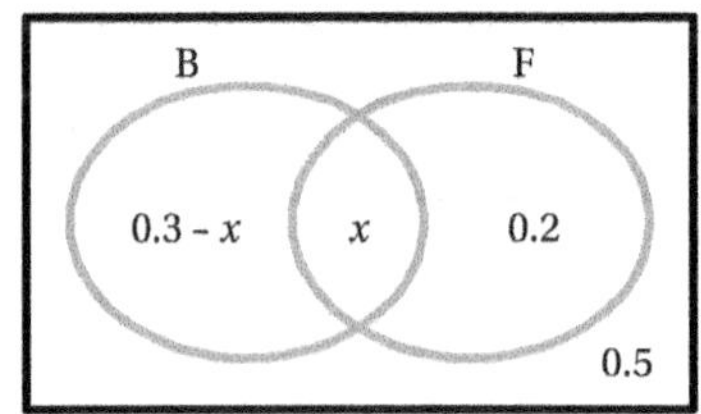

b $0.3 - x$ **c** 0.2

d

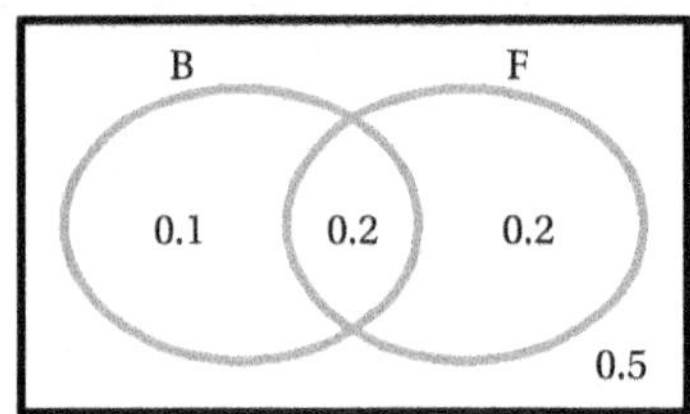

0.1

e $\frac{1}{3}$

10 $\frac{16}{41}$

11 $\frac{10}{19}$

12 $\frac{4}{9}$

13 a $\frac{14}{95}$ **b** Proof.

c $\frac{118}{855}$ **d** $\frac{21}{59}$

14 $\frac{1}{9}$

15 $\frac{1}{3}$

16 a $0.5 \leqslant k \leqslant 1$

b $2 - \frac{1}{k}$

21 Normal distribution

BEFORE YOU START

1 $x = 1.90$ (3 s.f.); $y = 1.15$ (3 s.f.)

2 6 people/kg

3 $\frac{256}{625}$

4 32%

5 $\frac{1}{4}$

EXERCISE 21A

1 a i 0.885 ii 0.212
 b i 0.401 ii 0.878
 c i 0.743 ii 0.191
 d i 0.807 ii 0.748
 e i 0.452 ii 0.689

2 a i 0.5 ii 1
 b i −1.67 ii −0.4

3 a i $P(Z < 1.6)$ ii $P(Z < 1.28)$
 b i $P(Z \geqslant -0.68)$ ii $P(Z \geqslant -2.96)$
 c i $P(-1.4 < Z < 0.2)$
 ii $P(-2.36 \leqslant Z \leqslant -0.2)$

4 a 1.16 b 0.123

5 a 0.278
 b i 0.127 ii 0.334
 c Assumes that they are independent, but Ali might be getting tired, OR he might get more determined.

6 1547

7 Mean 8; standard deviation 3.

8 Mean 12; standard deviation 4.

9 Mean 10; standard deviation 2.

10 a 0.309
 b Proof.
 c Best athletes likely to be chosen for the team; not independent as will depend on the weather/track/how well other athletes are doing; all four going under 59 seconds is a sufficient but not necessary condition.

11 a 0.160 b 0.171 c 0.727

12 a 0.707 b 0.663

13 a 67.3% b 0.314

14 a 0.952 b 0.838

15 a 0.841 b 0.811

16 0.640

17 0.0288

18 $1 - k$

WORK IT OUT 21.1

Solution 1 is correct.

EXERCISE 21B

1 a i 19.9 ii 13.3
 b i 32.4 ii 37.3
 c i 2.34 ii 4.44

2 141

3 3.61 kg

4 a 0.0548 b 154.2

5 a 0.691 b 39.7 c 0.240

6 99.7%

7 90

8 3.29

9 a 1 b 0.741

10 0

11 $2\Phi(k) - 1$

12 Take $X = \mu + \sigma\ \Phi^{-1}(U)$, where U is a uniform continuous distribution across [0, 1].

EXERCISE 21C

1 a i 5.68 ii 7.32
 b i 43.4 ii 15.3

2 a i $\mu = 8.91; \sigma = 2.27$
 ii $\mu = 130; \sigma = 38.6$
 b i $\mu = -0.201; \sigma = 1.19$
 ii $\mu = 870; \sigma = 202$

3 1.97 cm

4 3.56

5 $\mu = 70.6$ cm; $\sigma = 25.8$ cm

6 $\mu = 25.9$ s; $\sigma = 16.2$ s

7 3.23 hours

8 11.2°C

9 9.22×10^{-4}

10 70.6%

EXERCISE 21D

1 a Yes.

b No - not symmetrical.

c No - two modes.

d No - not a characteristic bell curve.

2 Three standard deviations below the mean would be −0.9 children, so the normal model would predict impossible results.

3 a $\bar{x} = 562$; $s = 89.7$

b $\bar{x} + 3s \approx 830$

$\bar{x} - 3s \approx 294$

All data lie within this range, so the normal distribution appears to be suitable, assuming the data are symmetrically distributed.

4 a 15, 17

b Mean 50.1; standard deviation 3.87.

c Symmetric, bell-shaped curve.

d About 10 students.

5 a i 0.0176 ii 0.0228

b 29.3%

6 a 0.344 b 3.12×10^{-5}

7 a $\mu \approx 180$; $\sigma \approx 12$

b For example: $p \approx 0.2$, $n \approx 900$.

8 $X < 645$

9 a Proof; For example $n = 20$, $p = 0.1$.

b 120

10 Proof.

MIXED PRACTICE 21

1 A

2 a 6.68% b 62

3 a 5.04 b 1.36

4 3.85

5 2.32

6 a i 0.788 ii 0.969

b 0.0294

7 B

8 a 0.412 b 0.149

9 a 0.249 b 0.935

10 a Proof. b $\sigma = 16.5$; $\mu = 68.1$

11 $\mu = 31.4°$; $\sigma = 4.52°$

12 a i 0.970

ii 0.939 iii 1

b 514.5 c 0.953

13 a For $n > 10$, $np > 5$ and $n(1 - p) > 5$.

b 0.234%

c Probably not independent.

14 a Within 3 standard deviations of the mean are scores from 18% to 90%, which are all possible. The mean is about the same as the mean, suggesting a symmetrical distribution.

b Distinction 64.1%; Merit 54%; Pass 38.6%.

15 a 0.0853 b 1.46 kN

16 a $\mu = 1.29$ mm; $\sigma = 0.554$ mm

b $\mu = 1.36$ mm; $\sigma = 0.764$ mm

22 Further hypothesis testing

BEFORE YOU START

1 A3, B2, C1

2 $p = 0.0543 < 0.10$; reject H_0

3 a 0.9332 b 0.8351

c 198.3

EXERCISE 22A

1 a i N(4, 25) ii N(20, 25)

b i N(0, 0.1) ii N(0, 2.5)

2 a i 0.655 ii 0.788

b i 0.942 ii 0.471

3 a N(40, 0.25); assuming independence of emissions.

b 0.159 c 0.954

4 0.788

5 0.0228

6 $\mu \geqslant 78.7$

7 608

8 6.62

9 a 0.138 b 15 c 4.73

10 a 0.9996 b 0.981

WORK IT OUT 22.1

Solution B is correct.

EXERCISE 22B

1 a i $H_0: \mu = 102; H_1: \mu \neq 102$

ii $H_0: \mu = 1.2; H_1: \mu \neq 1.2$

b i $H_0: \mu = 250; H_1: \mu < 250$

ii $H_0: \mu = 150\,000; H_1: \mu > 150\,000$

c i $H_0: \mu_T = 3000; H_1: \mu_T > 3000$

ii $H_0: \mu_t = 28; H_1: \mu_t < 28$

2 a i $\bar{X} < 55.1$ or $\bar{X} > 64.9$

ii $\bar{X} < 117$ or $\bar{X} > 123$

b i $\bar{X} > 85.5$ ii $\bar{X} > 753$

c i $\bar{X} < 79.2$ ii $\bar{X} < 92.2$

3 a i 0.0455; reject H_0.

ii 0.0578; do not reject H_0.

b i 0.0228; reject H_0.

ii 0.0288; reject H_0.

c i 0.0625; do not reject H_0.

ii 0.147; do not reject H_0.

d i 0.596; do not reject H_0.

ii 0.611; do not reject H_0

4 a $H_0: \mu = 168.8; H_1: \mu > 168.8$

b $p = 0.193$; do not reject H_0.

5 $p = 0.0918$; reject H_0.

6 a The scores of students in the school follow a normal distribution. The standard deviation is still 1.21. The 50 students were randomly selected from the cohort in the school.

b $p = 0.320$; do not reject H_0.

7 a $H_0: \mu = 2.7; H_1: \mu \neq 2.7$

b $\bar{X} < 2.53$ or $\bar{X} > 2.87$

c Reject H_0.

8 a $\bar{X} < 80.3$

b The post-diet masses follow a normal distribution.

c Reject H_0.

9 a $H_0: \mu = 20; H_1: \mu < 20$

b 2.28% c $n \geqslant 12$

10 a No ($p = 0.146$) b 25

11 Higher p-value if $\mu \neq 30$ as you have the probability from both tails.

EXERCISE 22C

1 a i Significant evidence.

ii Significant evidence.

b i No significant evidence.

ii No significant evidence.

2 a i No significant evidence.

ii No significant evidence.

b i Significant evidence.

ii Significant evidence.

3 a $H_0: \rho = 0; H_1: \rho \neq 0$

b Reject H_0.

4 a $H_0: \rho = 0; H_1: \rho > 0$

b Do not reject H_0.

5 a Countries in 2013.

b $H_0: \rho = 0; H_1: \rho < 0$

c Reject H_0.

6 a $H_0: \rho = 0; H_1: \rho \neq 0$

b Reject H_0.

c No – correlation does not imply causation and the test was not for positive correlation.

7 a $H_0: \rho = 0; H_1: \rho < 0$

b Reject H_0.

c Less likely to get a false positive, which would result in a useless drug being used.

8 a Significant evidence.

b $p = 0.064$; no significant evidence.

c Decrease, since increased value of r means it is less likely to be a chance result from a population with zero correlation.

9 25

10 With two data points there is always a perfectly fitted straight line.

MIXED PRACTICE 22

1 C

Many people think that p-values are a direct measure of how likely H_0 is to be true. Make sure you do not fall into this trap.

2 0.285

3 a Countries in 2015.

b $H_0: p = 0; H_1: p \neq 0$

c Do not reject H_0.

4 a $H_0: \mu = 300; H_1: \mu \neq 300$

b i 0.244 ii Do not reject H_0.

5 $p = 0.03 < 0.05$; reject H_0.

6 D

7 a Both variables must be normally distributed.

b Reject H_0.

8 a Still normally distributed. Still with a standard deviation of 5 seconds; $\bar{X} < 37.4$.

b Reject H_0.

c They would not be independent - the water needs to cool down to the original temperature.

9 a $H_0: \mu = 45; H_1: \mu \neq 45$

b $Z > 1.96$ or $Z < -1.96$

c $Z = -1.79$; do not reject H_0.

d Critical region now $Z < -1.64$ so now reject H_0.

10 a Random sample.

b $p = 0.046 > 0.025$; do not reject H_0.

c i H_1 becomes $\mu > 9.0$.

ii Becomes 1.6449. iii Reject H_0.

11 D

12 a $H_0: \mu = 5.84; H_1: \mu \neq 5.84$

b $5.84 - \frac{0.608}{\sqrt{n}} < \bar{X} < 5.84 + \frac{0.608}{\sqrt{n}}$

c 7

13 a $H_0: \mu = 6.4; H_1: \mu \neq 6.4$

b 10%

Focus on ... Proof 4

1 Very small (< 1 in 1 000 000)

2 a

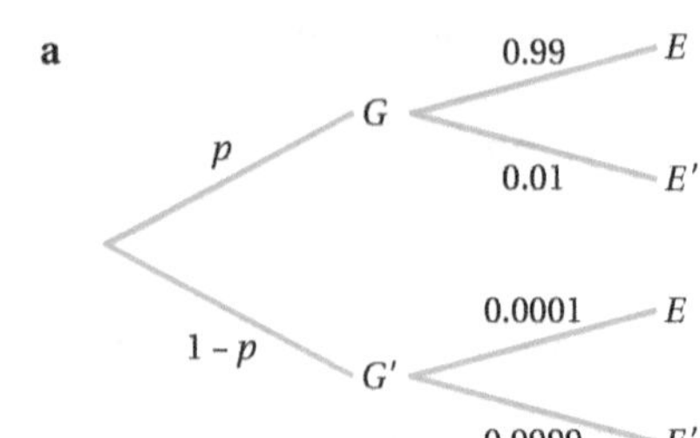

b i $0.0001 + 0.9899p$

ii $P(G \mid E) = \frac{9900p}{1+9899p}$, $P(G' \mid E) = \frac{1-p}{1+9899p}$

c Proof; the argument is flawed.

3 There are not a large number of diamond thieves in the population.

4 If $p \geq 0.1$, then $P(G' \mid E) < P(G \mid E)$.

5 Yes.

6 $P(E \mid G')$; $P(E \mid G) = \frac{999}{10000}$

7 Discussion.

Focus on ... Problem solving 4

1 Discussion.

2

		You choose			
		1	2	3	
Host opens	1	0	0	0	0
	2	50	0	100	150
	3	50	100	0	150
		100	100	100	300

$P(\text{win if switch}) = \frac{200}{300} = \frac{2}{3}$

3 a Discussion.

b Larger, because you already know that there are 9 girls.

c

		First child	
		girl	boy
Second child	girl	$\frac{1}{4}$	$\frac{1}{4}$
	boy	$\frac{1}{4}$	$\frac{1}{4}$

$P(\text{two girls} \mid \text{at least one girl}) = \frac{1}{3}$

Focus on ... Modelling 4

1 a 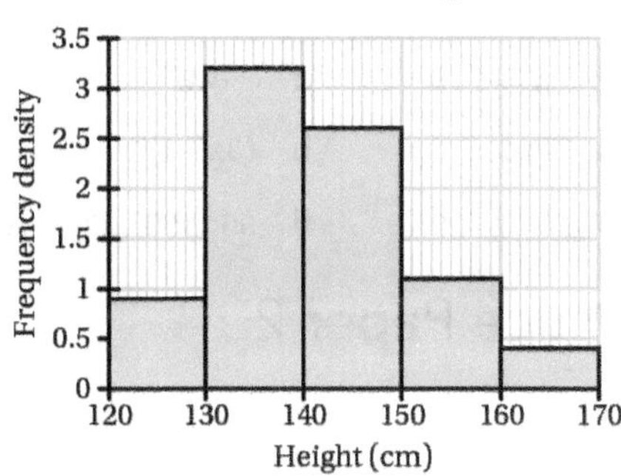

b No, not symmetrical.

2 a 99.7%

b Mean ≈ 44; standard deviation ≈ 2.7; interquartile range ≈ 3.64

c It could be a suitable model; the interquartile range is 4 (which is close to 3.64) and the box plot looks symmetrical.

3 a Yes

b No

4 It is discrete, whereas normal is continuous.

5 a It is likely that the distribution would not be symmetrical, as the minimum possible length of a phone call is 0, but the maximum can be more than 11.6 minutes.

b i Yes

ii No; bimodal.

iii No; discrete.

6 a Curves 1 and 2.

b Curve 1.

Cross-topic review exercise 4

1 $\frac{5}{6}$

2 a $H_0: \rho = 0; H_1: \rho \neq 0$

b Do not reject H_0; there is insufficient evidence of correlation between average daily temperature and rainfall.

3 a $H_0: \mu = 86; H_1: \mu > 86.$

b Reject H_0; there is evidence that the rose bushes are taller ($p = 0.0294 < 0.05$).

4 a 48.2

b 0.0672

c Not reasonable; predicts nearly all bulbs will last more than 570 hours.

5 a i 0.788

ii 0.758

iii 0.546

b 0.0494

c Other fuels; other vehicles; other types of customer; minimum purchase (policy); purchases in integer/fixed £s.

6 $\frac{10}{3}$

7 a $\frac{3}{20}$ b $\frac{9}{35}$

c $\frac{7}{12}$

8 a Proof.

b b $\frac{1}{6}\left(\frac{5}{6}\right)^{r-1}$

c Proof.

9 a i 0.3 ii 0.45

iii $\frac{6}{11}$ iv $\frac{5}{6}$

v $\frac{5}{12}$

b 0.1525

10 a 0.202 b 0.934

c 0.103

11 a 3.3

b 391 < 405 so mean unlikely to be correct.

95.5 > 3.3 so standard deviation unlikely to be correct.

c $\bar{y} = \frac{8210}{10} = 821 \approx 820$

$s = \sqrt{\frac{110}{9}} = 3.50 \approx 3.33$

12 a 0.341 34 b 0.340 78

c 0.164%

13 a $\mu \approx 83.5; \sigma \approx 9$

b $p \approx 0.03; n \approx 2780$

14 a $H_0: p = 0; H_1: p < 0$

b Reject H_0. Significant evidence of negative correlation; as the amount of time increases the test marks decrease.

c The two sections of the scatter graph represent two different students; for each student there is positive correlation, but one student got better marks despite spending less time on revision.

15 a Do not reject H_0.

b Reject H_0.

16 a $\frac{1}{5}$ b $\frac{4}{25}$

c $\frac{16}{125}$ d Proof.

e $\frac{4}{9}$ f $\frac{1}{2}$

17 a Reject H_0; sufficient evidence that the mean time has decreased ($z = -2.72, p = 0.0033$).

b 90

18 a $x = 5$ and $x = 9$

b 0.0968

c No, because some of the trapezia are above and some are below the curve.

Practice Paper 1

1 C

2 C

3 A

4 D

5 $\frac{e^3+2}{e^3-1}$

6 $1-4\theta-2\theta^2$

7 16.0 cm (3 s.f.)

8 $9x-6y = 9\ln 3-4$

9 a $x < -1.46$ or $x > 0.457$

b Proof.

10 $\frac{P^2}{16}$

11 a $A = 1, B = 2$ b $2-\frac{3}{2}x+\frac{29}{4}x^2$

c $|x| < \frac{2}{5}$

12 a $\ln 5$ or $-\ln 3$

b Proof.

13 $p = \frac{24}{5}$; $q = \frac{19}{40}$

14 a Proof.

b $L = 1.405, U = 2.48$

c 4.021; underestimate.

15 a $\frac{2x+2y}{y-2x}$

b Proof.

16 a, b Proof.

c $a = \frac{1}{2}, b = 2$ d 3.048

17 a Proof. b $(3x+1)(x-1)(3x+2)$

c i Proof. ii $90°, 199°, 222°, 318°, 341°$

Practice Paper 2

1 C

2 C

3 a $B(2, 4)$; $A(-3, 9)$

b

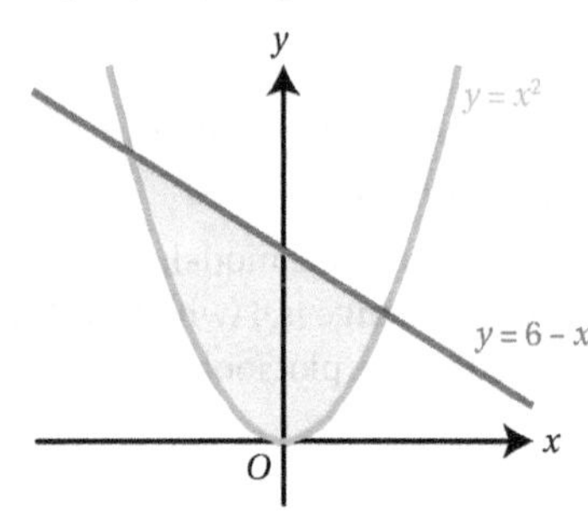

c −3

4 a $2\mathbf{i}-4\mathbf{j}+4\mathbf{k}$ b −7

5 $\left(\frac{2}{3}, \frac{2}{3}e^{-2}\right)$

6 Proof.

7 a $a = 3, b = 5, c = -2$

b 72°, 153°, 252°, 333°

8 $\ln 16-\frac{15}{16}$

9 a Proof. b $150-25\pi$

10 a $(0, 8)$

b ii $(\ln 5, 12)$ c $16\ln 5-16$

11 D

12 B

13 a 1.25 m s^{-2} b Proof. c 0 m s^{-1}

14 a 62 N (2 s.f.) b 2.1 m (2 s.f.)

15 a $(\mathbf{i}-5\mathbf{j})$ N

b i 8.50 m s^{-2}

ii 101°

16 a i $a = \frac{V-U}{T}$

ii Proof.

b 17.3 m s^{-1}

17 a Proof.

b $10.5\ \mathrm{m\,s^{-1}}$

c V would need to be larger.

18 a Proof.

b $m \leqslant \dfrac{5.2}{\sin 35° - \mu \cos 35°}$

c $5.4 \leqslant m \leqslant 29$

19 $5.6\ \mathrm{m\,s^{-1}}$

Practice Paper 3

1 B

2 C

3 202.5

4 $a = -5, b = -12$

5 a $6(x-1)^2 + 5$

b i Proof.

ii Yes; it is increasing and therefore one-to-one.

iii $x > 1$

6 a $\sqrt{12}\sin(x + 0.869)$

b $-12 + 12\sqrt{3}$

7 a i 2.46

ii

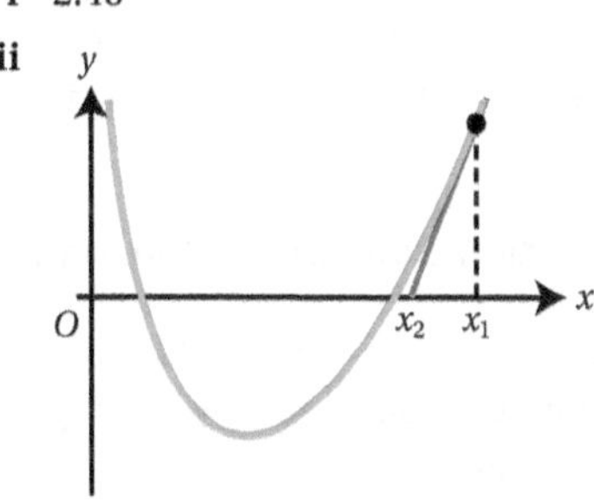

b i $\sqrt{\dfrac{3}{2}}$

ii The tangent does not cross the x-axis.

8 a Model 2; taking logs gives $\log P = \log a - k \log n$ which is of the form $y = 1.2 - 2.6x$.

b $a = 16$ (2 s.f); $k = 2.6$

9 a $\left(\dfrac{\sqrt{2}}{2}, 1\right)$

b i 0 at A; $\dfrac{\pi}{2}$ at O

ii $\dfrac{4}{3}$

10 a A b B

11 a $|r| < 0.3120$

b i No – there may be a non-linear relationship.

ii No – this sample does not provide significant evidence of correlation.

12 a $\dfrac{6}{217}$ b $\dfrac{107}{217}$

c $\dfrac{6}{14}$ d $\dfrac{15}{91}$

e

e.g. P(male | vegan) + P(male | vegan′) = 0.887

13 a 4.53%

b $H_0 : p = \dfrac{312}{655}$; $H_1 : p > \dfrac{312}{655}$, p-value = 6.42%
Do not reject H_0; this is not significantly above the expected number.

c Might not be independent; club not representative of whole school.

14 a 0.1 b 0.6 c 0.375

15 a $\dfrac{17}{35}$

b Proof.

16 a Mean is around middle of range. Range is about 6 standard deviations.

b 0.319

c 0.0375%, assuming independence.

d $H_0: \mu = 2502$; $H_1: \mu \neq 2502$; p-value = 22.7%;
Do not reject H_0. No significant change in energy intake.

e $\mu = 2440$ kilocalories; $\sigma^2 = 117\,000$ kilocalories^2

Glossary of terms

Absolute value: *See modulus.*

Acceptance region: The values of the observed data that lead to the null hypothesis not being rejected.

Arc: The part of the circumference between two points.

Arithmetic sequence/progression: A sequence that has a common difference between each term. To get from one term to the next you add the common difference (which could be negative).

Arithmetic series: The sum of the terms of an arithmetic sequence.

Cartesian equation: An equation involving just x and y.

Centre of mass: The point at which the object's weight acts.

Chain rule: A rule for differentiating composite functions. If $y = f(u)$ where $u = g(x)$, so that $y = f(g(x))$ then $\frac{dy}{dx} = \frac{dy}{du} \times \frac{du}{dx}$.

Chord: A straight line segment whose endpoints both lie on the circumference.

Cobweb diagram: A diagram that shows successive terms to an iterative sequence oscillating either side of a value to which the sequence is converging.

Coefficient of friction: A constant that measures the roughness of the surface a particle is moving over. Denoted by μ.

Complement: For a set A, the complement is everything that could happen other than A, written in set notation as A'.

Composite function: A function that results from applying a function to the output of another function.

Compound angle identities: Identities that express the sine, cosine or tangent of the sum and difference of angles in terms of trigonometric functions of the individual angles.

Concave: A curve that curves downwards. It has $\frac{d^2y}{dx^2} < 0$.

Conditional probability: A probability that takes into account information about events that have previously occurred.

Continuous function: A function whose graph can be drawn without taking the pen off the paper.

Converge: A sequence converges when the terms approach a limiting value.

Convergent: A sequence that converges to a limit.

Convex: A curve that curves upwards. It has $\frac{d^2y}{dx^2} > 0$.

Critical region: The values of the observed data that lead to the null hypothesis being rejected.

Decreasing sequence: A sequence where each term is smaller than the previous one.

Differential equation: An equation involving a derivative, for example $\frac{dy}{dx} = 3x^2$.

Distribution of the sample mean: If a random variable, X, is normally distributed, $X \sim N(\mu, \sigma^2)$, then the sample mean, $\bar{X}$ is also normal, and $\bar{X} \sim N\left(\mu, \frac{\sigma^2}{n}\right)$.

Diverge: A sequence diverges when the terms increase without limit.

Divergent: A sequence that diverges.

Domain: The set of allowed input values of a function.

Double angle identities: Identities that express the sine, cosine or tangent of twice the value of an angle in terms of trigonometric functions of the angle.

Explicit: A function expressed in the form $y = f(x)$.

Family of solutions: The set of solutions to a given differential equation obtained by varying the constant of integration.

Fixed point iteration: A method of creating a sequence that gets closer to a root of an equation of the form $x = g(x)$. A starting guess, x_1, generates a sequence $x_{n+1} = g(x_n)$. If this sequence converges to a limit, then this limit is a solution of the equation.

Function: A mapping where there is only one y value for each x value.

General solution: The solution to a differential equation that includes the constant of integration c.

Geometric sequence/progression: A sequence that has a common ratio between each term. To get from one term to the next you multiply by the common ratio.

Geometric series: The sum of the terms of a geometric sequence.

Horizontal line test: A function is one-to-one if any horizontal line will cross its graph at most once.

Image: The image of an x-value is the y-value it is mapped to.

Implicit: A function not expressed in the form $y = \mathrm{f}(x)$.

Increasing sequence: A sequence where each term is larger than the previous one.

Initial condition: The value of a function, y, at given input value, x.

Integration by parts: A method of integrating the product of two functions; $\int u \frac{\mathrm{d}v}{\mathrm{d}x}\,\mathrm{d}x = uv - \int v \frac{\mathrm{d}u}{\mathrm{d}x}\,\mathrm{d}x$.

Integration by substitution: A method of integration whereby the variable of integration is changed.

Intersection: The intersection of A and B means when both A and B happen, written in set notation as $A \cap B$.

Inverse function: A function that reverses the effect of another function. Denoted by f^{-1}.

Inverse normal distribution: For a given value of probability p, the inverse normal distribution gives the value of x such that $\mathrm{P}(X \leqslant x) = p$.

Inverse trigonometric function: The inverse function of the sine, cosine or tangent functions. Denoted by, for example, $\sin^{-1}$ or arcsin.

Limit: The value to which a sequence converges.

Limiting equilibrium: When friction is at its limiting value and an object subjected to a driving force is on the point of moving.

Limiting friction: The maximum value of friction before an object starts to move under the action of a force.

Lower bound: The lower bound on a definite integral is a value that the integral is definitely larger than. Found by underestimating the area under the curve.

Major arc: The longer arc between two points on the circumference of a circle.

Major sector: The larger part of the circle bounded by two radii and an arc.

Many-to-one: A function is many-to-one if there are some y-values that come from more than one x-value.

Mapping: Any rule that assigns to each input value (x) one or more output values (y).

Minor arc: The shorter arc between two points on the circumference of a circle.

Minor sector: The smaller part of the circle bounded by two radii and an arc.

Modulus: An operation that leaves positive numbers alone but makes negative numbers positive. $|x|$ denotes the modulus of number x. Also known as absolute value.

Moment: The turning effect of a force. The moment of a force F about a point P is moment $= Fd$, where d is the perpendicular distance of the line of action of the force from d.

Mutually exclusive: If there is no possibility of A and B occurring at the same time, then the events are mutually exclusive, written in set notation as $\mathrm{P}(A \cap B) = 0$.

Newton–Raphson method: An iterative method for finding the numerical solution to an equation which uses the tangent to the graph of $\mathrm{f}(x)$ to suggest where to look for the root. Given an approximate root x_0 of the equation $\mathrm{f}(x) = 0$, a better approximation is $x_1 = x_0 - \frac{\mathrm{f}(x)}{\mathrm{f}'(x)}$.

Normal distribution: A symmetrical distribution of values where a variable is very likely to be close to its average value, with values further away from the average becoming increasingly unlikely.

One-to-many: A mapping where a single input corresponds to more than one output (so is not a function).

One-to-one: A function is one-to-one if every y-value corresponds to only one x-value.

Ordinates: The y-values to be substituted into the trapezium rule formula for numerical integration.

Parameter: The third variable in parametric equations, which both x and y depend on, usually denoted by t.

Parametric equation: An equation in which both x and y are expressed in terms of a third variable, usually denoted by t.

Partial fraction: Two or more simpler algebraic fractions which add together to give a more complicated fraction.

Particular solution: A specific solution to a differential equation that does not depend on any unknown constants.

Periodic sequence: A sequence where the terms start repeating after a while; $u_{n+k} = u_n$ for some number k (the period of the sequence).

Point of inflection: A point at which a curve changes from convex to concave (or vice versa). At a point of inflection $\frac{d^2y}{dx^2} = 0$ and the second derivative changes sign.

Position-to-term rule: A rule that generates any term of the sequence from a formula (the nth term formula).

Probability density function: A curve such that the area under the curve represents probability.

Product rule: A rule for differentiating the products of two functions; if $y = f(x)g(x)$ then $\frac{dy}{dx} = f'(x)g(x) + f(x)g'(x)$.

Proof by contradiction: A method of proof that starts with the opposite of the statement you are trying to prove, and shows that this results in an impossible conclusion.

p-value: The probability of getting the observed data or more extreme if the null hypothesis is true.

Quotient rule: A rule for differentiating the quotient of two functions; if $y = \frac{f(x)}{g(x)}$ then $\frac{dy}{dx} = \frac{f'(x)g(x) - f(x)g'(x)}{[g(x)]^2}$.

Radian: The angle subtended at the centre of a circle by an arc equal in length to the radius. There are 2π radians in a complete rotation.

Range: The set of all possible outputs of a function.

Rational function: A fraction where both the denominator and numerator are polynomials.

Reciprocal trigonometric functions: The cosecant, secant and cotangent functions; $\operatorname{cosec} x = \frac{1}{\sin x}$, $\sec x = \frac{1}{\cos x}$ and $\cot x = \frac{1}{\tan x}$.

Rejection region: *See* critical region.

Resolve (forces): A force is resolved when it is split into two (often perpendicular) components.

Sector: A part of a circle bounded by two radii and an arc.

Segment: A segment of a circle is the region bounded by a chord and the arc subtended by the chord.

Separation of variables: A method used to solve a differential equation that is in the form $\frac{dy}{dx} = f(x)g(y)$.

Series: The sum of the terms of a sequence.

Sigma notation: A shorthand way to describe a series $\sum_{r=1}^{r=n} f(r)$, where r is a placeholder that increases by 1 with each new term.

Small angle approximation: For small θ, measured in radians, $\sin\theta \approx \theta$, $\cos\theta \approx 1 - \frac{1}{2}\theta^2$, and $\tan\theta \approx \theta$.

Smooth: A surface with no friction; $\mu = 0$.

Staircase diagram: A diagram that shows successive terms of an iterative sequence all increasing (or all decreasing).

Standard normal distribution: A random variable Z that has normal distribution with mean 0 and variance 1; $Z \sim N(0, 1)$.

Subtended: The angle at the centre of a circle between radii to the end of an arc is the angle subtended by the arc.

Sum to infinity: The limiting value of a geometric series. Exists only when $|r| < 1$, denoted by S_∞.

Term-to-term rule: A rule that generates the next term of a sequence from the previous term(s).

Trajectory: The path of a projectile.

Trapezium rule: A method of approximating a definite integral using n equal intervals with end-points $x_0, x_1, \ldots x_n$;

$$\int_a^b f(x)\,dx \approx \frac{h}{2}[y_0 + y_n + 2(y_1 + y_2 + \ldots + y_{n-1})]$$

where $y_i = f(x_i)$ and $h = \frac{b-a}{n}$.

Uniform lamina: A two-dimensional object, with constant mass per unit area.

Uniform rod: A one-dimensional shape with constant mass per unit length.

Union: The union of A and B means when either A happens, or B happens, or both happen, written in set notation as $A \cup B$.

Upper bound: The upper bound on a definite integral is a value that the integral is definitely smaller than. Found by overestimating the area under the curve.

Vertical line test: A mapping is a function if any vertical line will cross its graph at most once.

z-score: The number of standard deviations a value is away from the mean; $z = \frac{x-\mu}{\sigma}$.

Index

Acknowledgements

The authors and publishers acknowledge the following sources of copyright material and are grateful for the permissions granted. While every effort has been made, it has not always been possible to identify the sources of all the material used, or to trace all copyright holders. If any omissions are brought to our notice, we will be happy to include the appropriate acknowledgements on reprinting.

Thanks to the following for permission to reproduce images:

Cover image: Peter Medlicott Sola/Getty Images
Back cover: Fabian Oefner www.fabianoefner.com

Dulin/ Getty Images; Photo by Oxford Science Archive/Print Collector/ Getty Images; jallfree/istock/Getty Images; Photograph by Patrick Murphy / Getty Images; David Soanes Photography / Getty Images; Laura Lezza/Getty Images; Esben_H / Getty Images; Stefan Swalander/ EyeEm/Getty Images; Bettmann/Getty Images; Thai Yuan Lim / EyeEm / Getty Images; Dave Porter Peterborough Uk / Getty Images; Elli Thor Magnusson/Getty Images; Kalawin/Getty Images; Mitch Diamond/Getty Images; Mark Tipple/Getty Images; Graiki/Getty Images; Kizer13/Getty Images; Bernie Photo/Getty Images; robertcicchetti/ Getty Images; Doug Menuez/Getty Images; Mark Garlick/Getty Images; Panoramic Images/Getty Images; Jill Lehmann Photography/Getty Images; Titien Wattimena/EyeEm/Getty Images; Peter Beavis/Getty Images; Heritage Images/Getty Images; Manuela Schewe-Behnisch/EyeEm/Getty Images; Hans Neleman/Getty Images; Helmut Van Der Auweraer/EyeEm/Getty Images; Bettmann/Getty Images.

AQA material is reproduced by permission of AQA.